Student Solutions Manual

Applied Mathematics for the Managerial, Life, and Social Sciences

SIXTH EDITION

Soo T. Tan
Stonehill College

Prepared by

Soo T. Tan
Stonehill College

Australia • Brazil • Japan • Korea • Mexico • Singapore • Spain • United Kingdom • United States

For product information and technology assistance, contact us at **Cengage Learning Customer & Sales Support, 1-800-354-9706**

For permission to use material from this text or product, submit all requests online at **www.cengage.com/permissions**
Further permissions questions can be emailed to **permissionrequest@cengage.com**

ISBN-13: 978-1-133-10932-7
ISBN-10: 1-133-10932-2

Brooks/Cole
20 Davis Drive
Belmont, CA 94002-3098
USA

Cengage Learning is a leading provider of customized learning solutions with office locations around the globe, including Singapore, the United Kingdom, Australia, Mexico, Brazil, and Japan. Locate your local office at: **www.cengage.com/global**

Cengage Learning products are represented in Canada by Nelson Education, Ltd.

To learn more about Brooks/Cole, visit **www.cengage.com/brookscole**

Purchase any of our products at your local college store or at our preferred online store **www.cengagebrain.com**

Printed in the United States of America
1 2 3 4 5 16 15 14 13 12

ED185

CONTENTS

CHAPTER 1 Fundamentals of Algebra 1

1.1 Real Numbers 1
1.2 Polynomials 2
1.3 Factoring Polynomials 4
1.4 Rational Expressions 5
1.5 Integral Exponents 7
1.6 Solving Equations 9
1.7 Rational Exponents and Radicals 12
1.8 Quadratic Equations 14
1.9 Inequalities and Absolute Value 19
Chapter 1 Review 22
Chapter 1 Before Moving On 25

CHAPTER 2 Functions and Their Graphs 27

2.1 The Cartesian Coordinate System and Straight Lines 27
2.2 Equations of Lines 29
2.3 Functions and Their Graphs 34
2.4 The Algebra of Functions 40
2.5 Linear Functions 43
2.6 Quadratic Functions 47
2.7 Functions and Mathematical Models 52
Chapter 2 Review 57
Chapter 2 Before Moving On 60

CHAPTER 3 Exponential and Logarithmic Functions 61

3.1 Exponential Functions 61
3.2 Logarithmic Functions 64
3.3 Exponential Functions as Mathematical Models 67
Chapter 3 Review 70
Chapter 3 Before Moving On 72

CHAPTER 4 Mathematics of Finance 73

4.1 Compound Interest 73
4.2 Annuities 77
4.3 Amortization and Sinking Funds 80
4.4 Arithmetic and Geometric Progressions 86
Chapter 4 Review 88
Chapter 4 Before Moving On 90

CHAPTER 5 Systems of Linear Equations and Matrices 91

5.1 Systems of Linear Equations: An Introduction 91
5.2 Systems of Linear Equations: Unique Solutions 93
5.3 Systems of Linear Equations: Underdetermined and Overdetermined Systems 101
5.4 Matrices 106
5.5 Multiplication of Matrices 109
5.6 The Inverse of a Square Matrix 115
Chapter 5 Review 124
Chapter 5 Before Moving On 128

CHAPTER 6 Linear Programming 131

6.1 Graphing Systems of Linear Inequalities in Two Variables 131
6.2 Linear Programming Problems 134
6.3 Graphical Solutions of Linear Programming Problems 139
6.4 The Simplex Method: Standard Maximization Problems 150
6.5 The Simplex Method: Standard Minimization Problems 165
Chapter 6 Review 175
Chapter 6 Before Moving On 182

CHAPTER 7 Sets and Probability 185

7.1 Sets and Set Operations 185
7.2 The Number of Elements in a Finite Set 188
7.3 The Multiplication Principle 191
7.4 Permutations and Combinations 192
7.5 Experiments, Sample Spaces, and Events 196

7.6 Definition of Probability 198
7.7 Rules of Probability 200
Chapter 7 Review 203
Chapter 7 Before Moving On 206

CHAPTER 8 Additional Topics in Probability 207
8.1 Use of Counting Techniques in Probability 207
8.2 Conditional Probability and Independent Events 209
8.3 Bayes' Theorem 213
8.4 Distributions of Random Variables 218
8.5 Expected Value 220
8.6 Variance and Standard Deviation 222
Chapter 8 Review 226
Chapter 8 Before Moving On 228

CHAPTER 9 The Derivative 229
9.1 Limits 229
9.2 One-Sided Limits and Continuity 234
9.3 The Derivative 239
9.4 Basic Rules of Differentiation 246
9.5 The Product and Quotient Rules; Higher-Order Derivatives 251
9.6 The Chain Rule 258
9.7 Differentiation of Exponential and Logarithmic Functions 264
9.8 Marginal Functions in Economics 268
Chapter 9 Review 271
Chapter 9 Before Moving On 277

CHAPTER 10 Applications of the Derivative 279
10.1 Applications of the First Derivative 279
10.2 Applications of the Second Derivative 286
10.3 Curve Sketching 293
10.4 Optimization I 304
10.5 Optimization II 312
Chapter 10 Review 316
Chapter 10 Before Moving On 324

CHAPTER 11 Integration 327

11.1 Antiderivatives and the Rules of Integration 327

11.2 Integration by Substitution 331

11.3 Area and the Definite Integral 335

11.4 The Fundamental Theorem of Calculus 337

11.5 Evaluating Definite Integrals 340

11.6 Area Between Two Curves 343

11.7 Applications of the Definite Integral to Business and Economics 349

Chapter 11 Review 351

Chapter 11 Before Moving On 355

CHAPTER 12 Calculus of Several Variables 357

12.1 Functions of Several Variables 357

12.2 Partial Derivatives 359

12.3 Maxima and Minima of Functions of Several Variables 363

Chapter 12 Review 368

Chapter 12 Before Moving On 370

1 FUNDAMENTALS OF ALGEBRA

1.1 Real Numbers

Problem-Solving Tips

Suppose you are asked to determine whether a given statement is true or false, and you are also asked to explain your answer. How would you answer the question?

If you think the statement is true, then prove it. On the other hand, if you think the statement is false, then give an example that disproves the statement. For example, the statement "If a and b are real numbers, then $a - b = b - a$" is false, and an example that disproves it may be constructed by taking $a = 3$ and $b = 5$. For these values of a and b, we find $a - b = 3 - 5 = -2$, but $b - a = 5 - 3 = 2$, and this shows that $a - b \neq b - a$. Such an example is called a **counterexample**.

Concept Questions page 6

1. The set of natural numbers is $N = \{1, 2, 3, \ldots\}$; the set of whole numbers is $W = \{0, 1, 2, 3, \ldots\}$; the set of integers is $I = \{\ldots, -3, -2, -1, 0, 1, 2, 3, \ldots\}$; the set of rational numbers is $Q = \{a/b \mid a \text{ and } b \text{ are integers and } b \neq 0\}$ (example: $1/2$), and the set of irrational numbers contains all real numbers that cannot be expressed in the form a/b, where a and b are integers and $b \neq 0$ (example: π). The set of real numbers contains all irrational and rational numbers.

3. If $ab \neq 0$, then neither a nor b is equal to zero. If $abc \neq 0$, then none of a, b, and c is equal to zero.

Exercises page 6

1. The number -3 is an integer, a rational number, and a real number.

3. The number $\frac{3}{8}$ is a rational real number.

5. The number $\sqrt{11}$ is an irrational real number.

7. The number $\frac{\pi}{2}$ is an irrational real number.

9. The number $2.\overline{421}$ is a rational real number.

11. False. -2 is not a whole number.

13. True.

15. False. No natural number is irrational.

17. $(2x + y) + z = z + (2x + y)$: The Commutative Law of Addition.

19. $u(3v + w) = (3v + w)u$: The Commutative Law of Multiplication.

21. $u(2v + w) = 2uv + uw$: The Distributive Law.

23. $(2x+3y)+(x+4y)=2x+[3y+(x+4y)]$: The Associative Law of Addition.

25. $a-[-(c+d)]=a+(c+d)$: Property 1 of negatives.

27. $0(2a+3b)=0$: Property 1 involving zero.

29. If $(x-2)(2x+5)=0$, then $x=2$, or $x=-\frac{5}{2}$. Property 2 involving zero.

31. $\dfrac{(x+1)(x-3)}{(2x+1)(x-3)}=\dfrac{x+1}{2x+1}$. Property 2 of quotients.

33. $\dfrac{a+b}{b}\div\dfrac{a-b}{ab}=\dfrac{a(a+b)}{a-b}$. Properties 2 and 5 of quotients.

35. $\dfrac{a}{b+c}+\dfrac{c}{b}=\dfrac{ab+bc+c^2}{b(b+c)}$. Property 6 of quotients and the Distributive Law.

37. False. Consider $a=2$ and $b=\frac{1}{2}$. Then $ab=1$, but $a\neq 1$ and $b\neq 1$.

39. False. Consider $a=3$ and $b=2$. Then $a-b=3-2\neq b-a=2-3=-1$.

41. False. Consider $a=1$, $b=2$, and $c=3$. Then $(a-b)-c=(1-2)-3=-4\neq a-(b-c)=1-(2-3)=2$.

1.2 Polynomials

Concept Questions page 13

1. A polynomial of degree n in x is an expression of the form $a_nx^n+a_{n-1}x^{n-1}+\cdots+a_1x+a_0$, where n is a nonnegative integer and $a_0, a_1, \ldots, a_n$ are real numbers with $a_n\neq 0$. One polynomial of degree 4 in x is $x^4+2x^3-2x^2-5x-7$.

Exercises page 13

1. $3^4=3\cdot3\cdot3\cdot3=81$.

3. $\left(\frac{2}{3}\right)^3=\left(\frac{2}{3}\right)\left(\frac{2}{3}\right)\left(\frac{2}{3}\right)=\frac{8}{27}$

5. $-3^4=-3\cdot3\cdot3\cdot3=-81$.

7. $-3\left(\frac{3}{5}\right)^3=(-3)\left(\frac{3}{5}\right)\left(\frac{3}{5}\right)\left(\frac{3}{5}\right)=-\frac{81}{125}$.

9. $2^3\cdot2^5=2^8=256$.

11. $(3y)^2(3y)^3=(3y)^5=243y^5$.

13. $(2x+3)+(4x-6)=2x+3+4x-6=6x-3$.

15. $(7x^2-2x+5)+(2x^2+5x-4)=7x^2-2x+5+2x^2+5x-4=7x^2+2x^2-2x+5x+5-4=9x^2+3x+1$.

17. $(5y^2-2y+1)-(y^2-4y-8)=5y^2-2y+1-y^2+4y+8=5y^2-y^2-2y+4y+1+8=4y^2+2y+9$.

19.
$$\begin{aligned}(2.4x^3-3x^2+1.7x-6.2)-(1.2x^3+1.2x^2-0.8x+2)&=2.4x^3-3x^2+1.7x-6.2-1.2x^3-1.2x^2+0.8x-2\\&=1.2x^3-4.2x^2+2.5x-8.2.\end{aligned}$$

21. $(3x^2)(2x^3)=6x^5$.

23. $-2x\left(x^2-2\right)+4x^3=-2x^3+4x+4x^3=2x^3+4x.$

25. $2m\,(3m-4)+m\,(m-1)=6m^2-8m+m^2-m=7m^2-9m.$

27. $3\,(2a-b)-4\,(b-2a)=6a-3b-4b+8a=6a+8a-3b-4b=14a-7b.$

29. $(2x+3)\,(3x-2)=2x\,(3x-2)+3\,(3x-2)=6x^2-4x+9x-6=6x^2+5x-6.$

31. $(2x-3y)\,(3x+2y)=2x\,(3x+2y)-3y\,(3x+2y)=6x^2+4xy-9xy-6y^2=6x^2-5xy-6y^2.$

33. $(3r+2s)\,(4r-3s)=3r\,(4r-3s)+2s\,(4r-3s)=12r^2-9rs+8rs-6s^2=12r^2-rs-6s^2.$

35. $$\begin{aligned}(0.2x+1.2y)\,(0.3x-2.1y)&=0.2x\,(0.3x-2.1y)+1.2y\,(0.3x-2.1y)=0.06x^2-0.42xy+0.36xy-2.52y^2\\&=0.06x^2-0.06xy-2.52y^2.\end{aligned}$$

37. $(2x-y)\left(3x^2+2y\right)=2x\left(3x^2+2y\right)-y\left(3x^2+2y\right)=6x^3-3x^2y+4xy-2y^2.$

39. $(2x+3y)^2=(2x)^2+2\,(2x)\,(3y)+(3y)^2=4x^2+12xy+9y^2.$

41. $(2u-v)\,(2u+v)=(2u)^2-v^2=4u^2-v^2.$

43. $(2x-1)^2+3x-2\left(x^2+1\right)+3=4x^2-4x+1+3x-2x^2-2+3=2x^2-x+2.$

45. $$\begin{aligned}(2x+3y)^2-(2y+1)\,(3x-2)+2\,(x-y)&=4x^2+12xy+9y^2-6xy-3x+4y+2+2x-2y\\&=4x^2+6xy+9y^2-x+2y+2.\end{aligned}$$

47. $$\begin{aligned}\left(t^2-2t+4\right)\left(2t^2+1\right)&=\left(t^2-2t+4\right)\left(2t^2\right)+\left(t^2-2t+4\right)(1)=2t^4-4t^3+8t^2+t^2-2t+4\\&=2t^4-4t^3+9t^2-2t+4.\end{aligned}$$

49. $$\begin{aligned}2x-\{3x-[x-(2x-1)]\}&=2x-\{3x-[x-2x+1]\}=2x-[3x-(-x+1)]=2x-(3x+x-1)\\&=2x-(4x-1)=2x-4x+1=-2x+1.\end{aligned}$$

51. $$\begin{aligned}x-\{2x-[-x-(1+x)]\}&=x-[2x-(-x-1-x)]=x-[2x-(-2x-1)]=x-(2x+2x+1)\\&=x-4x-1=-3x-1.\end{aligned}$$

53. $$\begin{aligned}(2x-3)^2-3\,(x+4)\,(x-4)+2\,(x-4)+1&=(2x)^2-2\,(2x)\,(3)+3^2-3\left(x^2-16\right)+2x-8+1\\&=4x^2-12x+9-3x^2+48+2x-7=x^2-10x+50.\end{aligned}$$

55. $$\begin{aligned}&2x\,\{3x\,[2x-(3-x)]+(x+1)\,(2x-3)\}=2x\left[3x\,(2x-3+x)+2x^2-3x+2x-3\right]\\&=2x\left[3x\,(3x-3)+2x^2-x-3\right]=2x\left(9x^2-9x+2x^2-x-3\right)=2x\left(11x^2-10x-3\right)=22x^3-20x^2-6x.\end{aligned}$$

57. The total weekly profit is given by the revenue minus the cost:

$$\left(-0.04x^2 + 2000x\right) - \left(0.000002x^3 - 0.02x^2 + 1000x + 120{,}000\right)$$
$$= -0.04x^2 + 2000x - 0.000002x^3 + 0.02x^2 - 1000x - 120{,}000$$
$$= -0.000002x^3 - 0.02x^2 + 1000x - 120{,}000.$$

59. The total revenue is given by $\left(0.2t^2 + 150t\right) + \left(0.5t^2 + 200t\right) = 0.7t^2 + 350t$ thousand dollars t months from now, where $0 \le t \le 12$.

61. The gap is given by $\left(3.5t^2 + 26.7t + 436.2\right) - (24.3t + 365) = 3.5t^2 + 2.4t + 71.2$.

63. False. Let $a = 2$, $b = 3$, $m = 3$, and $n = 2$. Then $2^3 \cdot 3^2 = 8 \cdot 9 = 72 \neq (2 \cdot 3)^{3+2} = 6^5$.

65. False. For example, $x^2 + 1$ is a polynomial of degree 2 and x is a polynomial of degree 1, but $\left(x^2 + 1\right) x = x^3 + x$ is a polynomial of degree 3, not 2.

67. The degree of $p - q$ is m. To see this, suppose that $p = a_m x^m + \cdots + a_n x^n + \cdots + a_0$ and $q = b_n x^n + \cdots + b_0$. Because $m > n$, $p - q = a_m x^m + \cdots + (a_n - b_n) x^n + \cdots + (a_0 - b_0)$ has degree m.

1.3 Factoring Polynomials

Concept Questions page 19

1. A polynomial is completely factored over the set of integers if it is expressed as a product of prime polynomials with integral coefficients. An example is $4x^2 - 9y^2 = (2x - 3y)(2x + 3y)$.

Exercises page 19

1. $6m^2 - 4m = 2m(3m - 2)$.

3. $9ab^2 - 6a^2b = 3ab(3b - 2a)$.

5. $10m^2n - 15mn^2 + 20mn = 5mn(2m - 3n + 4)$.

7. $3x(2x + 1) - 5(2x + 1) = (2x + 1)(3x - 5)$.

9. $(3a + b)(2c - d) + 2a(2c - d)^2 = (2c - d)[3a + b + 2a(2c - d)] = (2c - d)(3a + b + 4ac - 2ad)$.

11. $2m^2 - 11m - 6 = (2m + 1)(m - 6)$.

13. $x^2 - xy - 6y^2 = (x - 3y)(x + 2y)$.

15. $x^2 - 3x - 1$ is prime.

17. $4a^2 - b^2 = (2a - b)(2a + b)$.

19. $u^2v^2 - w^2 = (uv)^2 - w^2 = (uv - w)(uv + w)$.

21. $z^2 + 4$ is prime.

23. $x^2 + 6xy + y^2$ is prime.

25. $x^2 + 3x - 4 = (x + 4)(x - 1)$.

27. $12x^2y - 10xy - 12y = 2y\left(6x^2 - 5x - 6\right) = 2y(3x + 2)(2x - 3)$.

29. $35r^2 + r - 12 = (7r - 4)(5r + 3)$.

31. $9x^3y - 4xy^3 = xy\left(9x^2 - 4y^2\right) = xy\left[(3x)^2 - (2y)^2\right] = xy(3x - 2y)(3x + 2y)$.

33. $x^4 - 16y^2 = (x^2)^2 - (4y)^2 = (x^2 - 4y)(x^2 + 4y)$.

35. $(a - 2b)^2 - (a + 2b)^2 = [(a - 2b) - (a + 2b)][(a - 2b) + (a + 2b)] = (-4b)(2a) = -8ab$.

37. $8m^3 + 1 = (2m)^3 + 1 = (2m + 1)(4m^2 - 2m + 1)$.

39. $8r^3 - 27s^3 = (2r)^3 - (3s)^3 = (2r - 3s)(4r^2 + 6rs + 9s^2)$.

41. $u^2v^6 - 8u^2 = u^2(v^6 - 8) = u^2(v^2 - 2)(v^4 + 2v^2 + 4)$.

43. $2x^3 + 6x + x^2 + 3 = 2x(x^2 + 3) + (x^2 + 3) = (x^2 + 3)(2x + 1)$.

45. $3ax + 6ay + bx + 2by = 3a(x + 2y) + b(x + 2y) = (x + 2y)(3a + b)$.

47. $u^4 - v^4 = (u^2)^2 - (v^2)^2 = (u^2 - v^2)(u^2 + v^2) = (u - v)(u + v)(u^2 + v^2)$.

49. $4x^3 - 9xy^2 + 4x^2y - 9y^3 = x(4x^2 - 9y^2) + y(4x^2 - 9y^2) = [(2x)^2 - (3y)^2](x + y)$

$$= (2x - 3y)(2x + 3y)(x + y).$$

51. $x^4 + 3x^3 - 2x - 6 = x^3(x + 3) - 2(x + 3) = (x + 3)(x^3 - 2)$.

53. $au^2 + (a + c)u + c = au^2 + au + cu + c = au(u + 1) + c(u + 1) = (u + 1)(au + c)$.

55. $P + Prt = P(1 + rt)$.

57. $8000x - 100x^2 = 100x(80 - x)$.

59. $kMx - kx^2 = kx(M - x)$.

61. $V = V_0 + \frac{V_0}{273}T = \frac{V_0}{273}(273 + T)$.

1.4 Rational Expressions

Concept Questions page 25

1. a. Quotients of polynomials are rational expressions; $\frac{2x^2 + 1}{3x^2 - 3x + 4}$.

b. Any polynomial P can be written in the form $\frac{P}{1}$, but not all rational expressions can be written as a polynomial.

Exercises page 25

1. $\frac{28x^2}{7x^3} = \frac{4}{x}$.

3. $\frac{4x + 12}{5x + 15} = \frac{4(x + 3)}{5(x + 3)} = \frac{4}{5}$.

5. $\frac{6x^2 - 3x}{6x^2} = \frac{3x(2x - 1)}{6x^2} = \frac{2x - 1}{2x}$.

7. $\frac{x^2 + x - 2}{x^2 + 3x + 2} = \frac{(x + 2)(x - 1)}{(x + 2)(x + 1)} = \frac{x - 1}{x + 1}$.

9. $\frac{x^2 - 9}{2x^2 - 5x - 3} = \frac{(x - 3)(x + 3)}{(2x + 1)(x - 3)} = \frac{x + 3}{2x + 1}$.

11. $\frac{x^3 + y^3}{x^2 - xy + y^2} = \frac{(x + y)(x^2 - xy + y^2)}{x^2 - xy + y^2} = x + y$.

13. $\frac{6x^3}{32} \cdot \frac{8}{3x^2} = \frac{1}{2}x$.

15. $\frac{3x^3}{8x^2} \div \frac{15x^4}{16x^5} = \frac{3x^3}{8x^2} \cdot \frac{16x^5}{15x^4} = \frac{2x^8}{5x^6} = \frac{2}{5}x^2$

17. $\frac{3x}{x+2y} \cdot \frac{5x+10y}{6} = \frac{(3x)\,5\,(x+2y)}{6\,(x+2y)} = \frac{5x}{2}.$

19. $\frac{2m+6}{3} \div \frac{3m+9}{6} = \frac{2\,(m+3)}{3} \cdot \frac{6}{3\,(m+3)} = \frac{4}{3}.$

21. $\frac{6r^2-r-2}{2r+4} \cdot \frac{6r+12}{4r+2} = \frac{(3r-2)\,(2r+1)\,6\,(r+2)}{2\,(r+2)\,2\,(2r+1)} = \frac{3\,(3r-2)}{2}.$

23. $\frac{k^2-2k-3}{k^2-k-6} \div \frac{k^2-6k+8}{k^2-2k-8} = \frac{(k-3)\,(k+1)}{(k-3)\,(k+2)} \cdot \frac{(k-4)\,(k+2)}{(k-4)\,(k-2)} = \frac{k+1}{k-2}.$

25. $\frac{2}{2x+3} + \frac{3}{2x-1} = \frac{2\,(2x-1)+3\,(2x+3)}{(2x+3)\,(2x-1)} = \frac{4x-2+6x+9}{(2x+3)\,(2x-1)} = \frac{10x+7}{(2x+3)\,(2x-1)}.$

27.
$$\frac{3}{x^2-x-6} + \frac{2}{x^2+x-2} = \frac{3}{(x-3)\,(x+2)} + \frac{2}{(x+2)\,(x-1)} = \frac{3\,(x-1)+2\,(x-3)}{(x-3)\,(x+2)\,(x-1)}$$
$$= \frac{3x-3+2x-6}{(x-3)\,(x+2)\,(x-1)} = \frac{5x-9}{(x-3)\,(x+2)\,(x-1)}.$$

29.
$$\frac{2m}{2m^2-2m-1} + \frac{3}{2m^2-3m+3} = \frac{2m\left(2m^2-3m+3\right)+3\left(2m^2-2m-1\right)}{\left(2m^2-2m-1\right)\left(2m^2-3m+3\right)}$$
$$= \frac{4m^3-6m^2+6m+6m^2-6m-3}{\left(2m^2-2m-1\right)\left(2m^2-3m+3\right)} = \frac{4m^3-3}{\left(2m^2-2m-1\right)\left(2m^2-3m+3\right)}.$$

31. $\frac{x}{1-x} + \frac{2x+3}{x^2-1} = -\frac{x}{x-1} + \frac{2x+3}{(x+1)\,(x-1)} = \frac{-x\,(x+1)+2x+3}{(x+1)\,(x-1)} = \frac{-x^2-x+2x+3}{(x+1)\,(x-1)} = -\frac{x^2-x-3}{(x+1)\,(x-1)}.$

33.
$$x - \frac{x^2}{x+2} + \frac{2}{x-2} = \frac{x\,(x+2)\,(x-2)-x^2\,(x-2)+2\,(x+2)}{(x+2)\,(x-2)} = \frac{x^3-4x-x^3+2x^2+2x+4}{(x+2)\,(x-2)}$$
$$= \frac{2x^2-2x+4}{(x+2)\,(x-2)} = \frac{2\left(x^2-x+2\right)}{(x+2)\,(x-2)}.$$

35.
$$\frac{x}{x^2+5x+6} + \frac{2}{x^2-4} - \frac{3}{x^2+3x+2} = \frac{x}{(x+3)\,(x+2)} + \frac{2}{(x-2)\,(x+2)} - \frac{3}{(x+1)\,(x+2)}$$
$$= \frac{x\,(x-2)\,(x+1)+2\,(x+3)\,(x+1)-3\,(x+3)\,(x-2)}{(x+3)\,(x+2)\,(x-2)\,(x+1)} = \frac{x^3-x^2-2x+2x^2+8x+6-3x^2-3x+18}{(x+3)\,(x+2)\,(x-2)\,(x+1)}$$
$$= \frac{x^3-2x^2+3x+24}{(x+3)\,(x+2)\,(x-2)\,(x+1)}.$$

37. $\frac{x}{ax-ay} + \frac{y}{by-bx} = \frac{x}{a\,(x-y)} - \frac{y}{b\,(x-y)} = \frac{bx-ay}{ab\,(x-y)}.$

39. $\dfrac{1+\frac{1}{x}}{1-\frac{1}{x}} = \dfrac{\frac{x+1}{x}}{\frac{x-1}{x}} = \frac{x+1}{x} \cdot \frac{x}{x-1} = \frac{x+1}{x-1}.$

41. $\dfrac{\frac{1}{x}+\frac{1}{y}}{1-\frac{1}{xy}} = \dfrac{\frac{y+x}{xy}}{\frac{xy-1}{xy}} = \frac{y+x}{xy} \cdot \frac{xy}{xy-1} = \frac{y+x}{xy-1}.$

43. $\dfrac{\dfrac{1}{x^2}-\dfrac{1}{y^2}}{x+y}=\dfrac{\dfrac{y^2-x^2}{x^2y^2}}{x+y}=\dfrac{(y+x)(y-x)}{x^2y^2}\cdot\dfrac{1}{x+y}=\dfrac{y-x}{x^2y^2}.$

45. $\dfrac{\dfrac{1}{2(x+h)}-\dfrac{1}{2x}}{h}=\dfrac{\dfrac{x-(x+h)}{2x(x+h)}}{h}=-\dfrac{h}{2x(x+h)}\cdot\dfrac{1}{h}=-\dfrac{1}{2x(x+h)}.$

47. a. $2.2+\dfrac{2500}{x}=\dfrac{2.2x+2500}{x}.$ **b.** The total cost is $x\left(\dfrac{2.2x+2500}{x}\right)=2.2x+2500.$

49. $P=\dfrac{R}{i}-\dfrac{R}{i(1+i)^n}=\dfrac{R(1+i)^n-R}{i(1+i)^n}=\dfrac{R\left[(1+i)^n-1\right]}{i(1+i)^n}.$

51. $A=\dfrac{136}{1+0.25(t-4.5)^2}+28=\dfrac{136+28\left[1+0.25(t-4.5)^2\right]}{1+0.25(t-4.5)^2}=\dfrac{164+7(t-4.5)^2}{1+0.25(t-4.5)^2}.$

1.5 Integral Exponents

Concept Questions page 30

1. If a is any real number and n is a natural number, then the expression a^n is defined as the number $a^n=\underbrace{a\cdot a\cdot a\cdot\cdots\cdot a}_{n\text{ factors}}$, where the number a is the base and the superscript n is the exponent, or power, to which the base is raised. For any real number a, $a^0=1$. If n is a negative number and $a\neq 0$, then $a^n=\dfrac{1}{a^{-n}}$.

Exercises page 30

1. $(-2)^3=-8.$

3. $7^{-2}=\dfrac{1}{7^2}=\dfrac{1}{49}.$

5. $-\left(-\dfrac{1}{4}\right)^{-2}=-\dfrac{1}{\left(-\frac{1}{4}\right)^2}=-\dfrac{1}{\frac{1}{16}}=-16.$

7. $2^{-2}+3^{-1}=\dfrac{1}{2^2}+\dfrac{1}{3}=\dfrac{1}{4}+\dfrac{1}{3}=\dfrac{7}{12}.$

9. $(0.03)^2=0.0009.$

11. $1996^0=1.$

13. $\left(ab^2\right)^0=1.$

15. $\dfrac{2^3\cdot 2^5}{2^4\cdot 2^9}=2^{3+5-4-9}=2^{-5}=\dfrac{1}{2^5}=\dfrac{1}{32}.$

17. $\dfrac{2^{-3}\cdot 2^{-4}}{2^{-5}\cdot 2^{-2}}=2^{-3-4+5+2}=2^0=1.$

19. $\left(\dfrac{3^4\cdot 3^{-3}}{3^{-2}}\right)^{-1}=\left(3^{4-3+2}\right)^{-1}=\left(3^3\right)^{-1}=\dfrac{1}{3^3}=\dfrac{1}{27}.$

21. $\left(2x^3\right)\left(\frac{1}{8}x^2\right)=\frac{1}{4}x^5.$

23. $\dfrac{3x^3}{2x^4}=\dfrac{3}{2x}.$

25. $\left(a^{-2}\right)^3=a^{-6}=\dfrac{1}{a^6}.$

27. $\left(2x^{-2}y^2\right)^3=8x^{-6}y^6=\dfrac{8y^6}{x^6}.$

29. $\left(4x^2y^{-3}\right)\left(2x^{-3}y^2\right) = 8x^{-1}y^{-1} = \dfrac{8}{xy}$.

31. $\left(-x^2y\right)^3\left(\dfrac{2y^2}{x^4}\right) = -\dfrac{2x^6y^3y^2}{x^4} = -2x^2y^5$.

33. $\left(\dfrac{2u^2v^3}{3uv}\right)^{-1} = \left(\dfrac{2uv^2}{3}\right)^{-1} = \dfrac{3}{2uv^2}$.

35. $\left(3x^{-2}\right)^3\left(2x^2\right)^5 = \left(27x^{-6}\right)\left(32x^{10}\right) = 864x^4$.

37. $\dfrac{3^0 \cdot 4x^{-2}}{16 \cdot \left(x^2\right)^3} = \dfrac{4x^{-2}}{16x^6} = \dfrac{1}{4x^8}$.

39. $\dfrac{2^2u^{-2}\left(v^{-1}\right)^3}{3^2\left(u^{-3}v\right)^2} = \dfrac{4u^{-2}v^{-3}}{9u^{-6}v^2} = \dfrac{4u^4}{9v^5}$.

41. $(-2x)^{-2}(3y)^{-3}(4z)^{-2} = (-2)^{-2}x^{-2}3^{-3}y^{-3}4^{-2}z^{-2} = \dfrac{1}{4 \cdot 27 \cdot 16x^2y^3z^2} = \dfrac{1}{1728x^2y^3z^2}$.

43. $\left(a^2b^{-3}\right)^2\left(a^{-2}b^2\right)^{-3} = a^4b^{-6}a^6b^{-6} = \dfrac{a^{10}}{b^{12}}$.

45. $\left[\left(\dfrac{a^{-2}b^{-2}}{3a^{-1}b^2}\right)^2\right]^{-1} = \left[\left(\dfrac{1}{3ab^4}\right)^2\right]^{-1} = \left(\dfrac{1}{9a^2b^8}\right)^{-1} = 9a^2b^8$.

47. $\left(\dfrac{3^2u^{-2}v^2}{2^2u^3v^{-3}}\right)^{-2}\left(\dfrac{3^2v^5}{4^2u}\right)^2 = \left(3^22^{-2}u^{-5}v^5\right)^{-2}\left(3^24^{-2}v^5u^{-1}\right)^2 = 3^{-4}2^4u^{10}v^{-10}3^44^{-4}v^{10}u^{-2} = 2^{-4}u^8v^0 = \dfrac{u^8}{16}$.

49. $\dfrac{x^{-1}-1}{x^{-1}+1} = \dfrac{\dfrac{1}{x}-1}{\dfrac{1}{x}+1} = \dfrac{\dfrac{1-x}{x}}{\dfrac{1+x}{x}} = \dfrac{1-x}{1+x}$.

51. $\dfrac{u^{-1}-v^{-1}}{v-u} = \dfrac{\dfrac{1}{u}-\dfrac{1}{v}}{v-u} = \dfrac{\dfrac{v-u}{uv}}{v-u} = \dfrac{v-u}{uv} \cdot \dfrac{1}{v-u} = \dfrac{1}{uv}$.

53. $\left(\dfrac{a^{-1}-b^{-1}}{a^{-1}+b^{-1}}\right)^{-1} = \dfrac{a^{-1}+b^{-1}}{a^{-1}-b^{-1}} = \dfrac{\dfrac{1}{a}+\dfrac{1}{b}}{\dfrac{1}{a}-\dfrac{1}{b}} = \dfrac{\dfrac{b+a}{ab}}{\dfrac{b-a}{ab}} = \dfrac{b+a}{b-a}$.

55. False. For example, if $a = 2$, $b = 3$, $m = 2$, and $n = 3$, then $a^mb^n = 2^2 \cdot 3^3 = 108$, and this is not equal to $(ab)^{mn} = 6^6 = 46{,}656$.

57. False. For example, if $a = 1$, $b = 2$, and $n = 2$, then $(a+b)^n = (1+2)^2 = 3^2 = 9$, whereas $a^n + b^n = 1^2 + 2^2 = 5$.

1.6 Solving Equations

Concept Questions page 35

1. An equation is a statement that two mathematical expressions are equal. A solution of an equation involving one variable is a number that renders the equation a true statement when it is substituted for the variable. The solution set of an equation is the set of all solutions to the equation.

One example: $2x = 3$ is an equation. Its solution is $x = \frac{3}{2}$ because $2\left(\frac{3}{2}\right) = 3$.

Another example: $\frac{5x}{2} = 10$ is an equation. Its solution is $x = 4$ because $\frac{5 \cdot 4}{2} = \frac{20}{2} = 10$.

3. A linear equation in the variable x is an equation that can be written in the form $ax + b = 0$, where a and b are constants with $a \neq 0$. Example: $3x + 4 = 5$. Solving for x, we have $3x = 1$, so $x = \frac{1}{3}$.

Exercises page 35

1.
$$3x = 12$$
$$\tfrac{1}{3}(3x) = \tfrac{1}{3}(12)$$
$$x = 4.$$

3.
$$0.3y = 2$$
$$\tfrac{1}{0.3}(0.3y) = \tfrac{1}{0.3}(2)$$
$$y = \tfrac{2}{0.3}$$
$$= \tfrac{20}{3}.$$

5.
$$3x + 4 = 2$$
$$3x + 4 - 4 = 2 - 4$$
$$3x = -2$$
$$\tfrac{1}{3}(3x) = \tfrac{1}{3}(-2)$$
$$x = -\tfrac{2}{3}.$$

7.
$$-2y + 3 = -7$$
$$-2y + 3 - 3 = -7 - 3$$
$$-2y = -10$$
$$-\tfrac{1}{2}(-2y) = -\tfrac{1}{2}(-10)$$
$$y = 5.$$

9.
$$\tfrac{1}{5}p - 3 = -\tfrac{1}{3}p + 5$$
$$15\left(\tfrac{1}{5}p - 3\right) = 15\left(-\tfrac{1}{3}p + 5\right)$$
$$3p - 45 = -5p + 75$$
$$3p - 45 + 45 = -5p + 75 + 45$$
$$3p = -5p + 120$$
$$8p = 120$$
$$p = 15.$$

11.
$$0.4 - 0.3p = 0.1(p + 4)$$
$$0.4 - 0.3p = 0.1p + 0.4$$
$$0.4 - 0.3p - 0.4 = 0.1p + 0.4 - 0.4$$
$$-0.3p = 0.1p$$
$$-0.3p - 0.1p = 0.1p - 0.1p$$
$$-0.4p = 0$$
$$p = 0.$$

13.
$$\tfrac{3}{5}(k+1)=\tfrac{1}{4}(2k+4)$$
$$12(k+1)=5(2k+4)$$
$$12k+12=10k+20$$
$$2k=8$$
$$k=4.$$

15.
$$\tfrac{2x-1}{3}+\tfrac{3x+4}{4}=\tfrac{7(x+3)}{10}$$
$$60\left(\tfrac{2x-1}{3}+\tfrac{3x+4}{4}\right)=60\left[\tfrac{7(x+3)}{10}\right]$$
$$20(2x-1)+15(3x+4)=42(x+3)$$
$$40x-20+45x+60=42x+126$$
$$85x+40=42x+126$$
$$85x=42x+86$$
$$43x=86$$
$$x=2.$$

17.
$$\tfrac{1}{2}[2x-3(x-4)]=\tfrac{2}{3}(x-5)$$
$$6\left\{\tfrac{1}{2}[2x-3(x-4)]\right\}=6\left[\tfrac{2}{3}(x-5)\right]$$
$$3(2x-3x+12)=4(x-5)$$
$$3(-x+12)=4x-20$$
$$-3x+36=4x-20$$
$$-7x+36=-20$$
$$-7x=-56$$
$$x=8.$$

19.
$$(2x+1)^2-(3x-2)^2=5x(2-x)$$
$$\left(4x^2+4x+1\right)-\left(9x^2-12x+4\right)=10x-5x^2$$
$$4x^2+4x+1-9x^2+12x-4=10x-5x^2$$
$$-5x^2+16x-3=10x-5x^2$$
$$16x-3=10x$$
$$6x-3=0$$
$$6x=3$$
$$x=\frac{1}{2}.$$

21.
$$\frac{8}{x}=24$$
$$8=24x$$
$$\frac{1}{3}=x.$$

23.
$$\frac{2}{y-1}=4$$
$$2=4(y-1)$$
$$2=4y-4$$
$$6=4y$$
$$\frac{3}{2}=y.$$

25.
$$\tfrac{2x-3}{x+1}=\tfrac{2}{5}$$
$$5(x+1)\left(\tfrac{2x-3}{x+1}\right)=5(x+1)\left(\tfrac{2}{5}\right)$$
$$5(2x-3)=2(x+1)$$
$$10x-15=2x+2$$
$$10x=2x+17$$
$$8x=17$$
$$x=\tfrac{17}{8}.$$

27.
$$\frac{2}{q-1}=\frac{3}{q-2}$$
$$(q-1)(q-2)\left(\frac{2}{q-1}\right)=(q-1)(q-2)\left(\frac{3}{q-2}\right)$$
$$(q-2)2=(q-1)3$$
$$2q-4=3q-3$$
$$-4=q-3$$
$$-1=q.$$

29. $$\begin{aligned} \frac{3k-2}{4}-\frac{3k}{4} &= \frac{k+3}{k} \\ -\frac{1}{2} &= \frac{k+3}{k} \\ -k &= 2k+6 \\ -3k &= 6 \\ k &= -2 \end{aligned}$$

31. $$\begin{aligned} \frac{m-2}{m}+\frac{2}{m} &= \frac{m+3}{m-3} \\ 1-\frac{2}{m}+\frac{2}{m} &= \frac{m+3}{m-3} \\ 1 &= \frac{m+3}{m-3} \\ m-3 &= m+3 \\ -3 &= 3 \end{aligned}$$

which is impossible. Thus, there is no solution.

33. $I = Prt$, so $r = \dfrac{I}{Pt}$.

35. $p = -3q + 1$, so $-3q = p - 1$. Thus, $q = \dfrac{p-1}{-3} = -\frac{1}{3}p + \frac{1}{3}$.

37. $iS = R\left[(1+i)^n - 1\right]$, so $R = \dfrac{iS}{(1+i)^n - 1}$.

39. $$\begin{aligned} V &= \frac{ax}{x+b} \\ V(x+b) &= ax \\ Vx + Vb &= ax \\ Vx - ax &= -Vb \\ x(V-a) &= -Vb \\ x &= -\frac{Vb}{V-a} \\ &= \frac{Vb}{a-V}. \end{aligned}$$

41. $$\begin{aligned} r &= \frac{2mI}{B(n+1)} \\ rB(n+1) &= 2mI \\ m &= \frac{rB(n+1)}{2I}. \end{aligned}$$

43. $$\begin{aligned} r &= \frac{2mI}{B(n+1)} \\ rBn + rB &= 2mI \\ rBn &= 2mI - rB \\ n &= \frac{2mI - rB}{rB}. \end{aligned}$$

45. $$\begin{aligned} \frac{1}{f} &= \frac{1}{p}+\frac{1}{q} \\ \frac{1}{p} &= \frac{1}{f}-\frac{1}{q} \\ &= \frac{q-f}{fq} \\ p &= \frac{fq}{q-f}. \end{aligned}$$

47. $I = Prt$, so $t = \dfrac{I}{Pr}$. If $I = 90$, $P = 1000$, and $r = 6\% = 0.06$, then $t = \dfrac{90}{(0.06)(1000)} = 1.5$, or 1.5 years.

49. $S = \dfrac{a}{t} + b = \dfrac{a+bt}{t}$, so $tS = a + bt$, $tS - bt = a$, $(S-b)t = a$, and $t = \dfrac{a}{S-b}$.

51. **a.** $V = C - \left(\dfrac{C-S}{N}\right)t$, so $V - \dfrac{St}{N} = C\left(1 - \dfrac{t}{N}\right)$ and $C = \dfrac{\dfrac{NV - St}{N}}{1 - \dfrac{t}{N}} = \dfrac{NV - St}{N - t}$.

b. If $N = 5$, $t = 3$, $S = 40{,}000$, and $V = 70{,}000$, we have $C = \dfrac{70{,}000\,(5) - 40{,}000\,(3)}{5-3} = \dfrac{230{,}000}{2} = 115{,}000$, or $\$115,000$.

53. a. $c = \left(\dfrac{t+1}{24}\right)a$, so $\dfrac{t+1}{24} = \dfrac{c}{a}$, $t+1 = \dfrac{24c}{a}$, and $t = \dfrac{24c}{a} - 1 = \dfrac{24c-a}{a}$.

b. Here $a = 500$ and $c = 125$, so the child's age is $t = \frac{24(125)-500}{500} = 5$, or 5 years.

1.7 Rational Exponents and Radicals

Concept Questions page 44

1. If n is a natural number and a and b are real numbers such that $a^n = b$, then a is the nth root of b. For example, 3 is the 4th root of 81; that is $\sqrt[4]{81} = 3$.

3. The process of eliminating a radical from the denominator of an algebraic expression is referred to as rationalizing the denominator. For example, $\dfrac{1}{1-\sqrt{6}} = \dfrac{1}{1-\sqrt{6}} \cdot \dfrac{1+\sqrt{6}}{1+\sqrt{6}} = \dfrac{1+\sqrt{6}}{1-6} = -\frac{1}{5}\left(1+\sqrt{6}\right)$.

Exercises page 44

1. $\sqrt{81} = 9$.

3. $\sqrt[4]{256} = 4$.

5. $16^{1/2} = 4$.

7. $8^{2/3} = 2^2 = 4$.

9. $-25^{1/2} = -5$.

11. $(-8)^{2/3} = (-2)^2 = 4$.

13. $\left(\dfrac{4}{9}\right)^{1/2} = \dfrac{2}{3}$.

15. $\left(\dfrac{27}{8}\right)^{2/3} = \left(\dfrac{3}{2}\right)^2 = \dfrac{9}{4}$.

17. $8^{-2/3} = \dfrac{1}{8^{2/3}} = \dfrac{1}{2^2} = \dfrac{1}{4}$.

19. $-\left(\dfrac{27}{8}\right)^{-1/3} = -\left(\dfrac{8}{27}\right)^{1/3} = -\dfrac{2}{3}$.

21. $3^{1/3} \cdot 3^{5/3} = 3^{(1/3)+(5/3)} = 3^2 = 9$.

23. $\dfrac{3^{1/2}}{3^{5/2}} = \dfrac{1}{3^2} = \dfrac{1}{9}$.

25. $\dfrac{2^{-1/2} \cdot 3^{2/3}}{2^{3/2} \cdot 3^{-1/3}} = \dfrac{3^{(2/3)+(1/3)}}{2^{(3/2)+(1/2)}} = \dfrac{3^1}{2^2} = \dfrac{3}{4}$.

27. $\left(2^{3/2}\right)^4 = 2^{(3/2)4} = 2^6 = 64$.

29. $x^{2/5} \cdot x^{-1/5} = x^{1/5}$.

31. $\dfrac{x^{3/4}}{x^{-1/4}} = x^{(3/4)+(1/4)} = x$.

33. $\left(\dfrac{x^3}{-27x^{-6}}\right)^{-2/3} = \left(\dfrac{x^9}{-27}\right)^{-2/3} = \dfrac{x^{-18/3}}{\frac{1}{9}} = 9x^{-6} = \dfrac{9}{x^6}$.

35. $\left(\dfrac{x^{-3}}{y^{-2}}\right)^{1/2}\left(\dfrac{y}{x}\right)^{3/2} = \dfrac{x^{-3/2}y^{3/2}}{y^{-1}x^{3/2}} = \dfrac{y^{5/2}}{x^3}$.

37. $x^{2/5}\left(x^2 - 2x^3\right) = x^{12/5} - 2x^{17/5}$.

39. $2p^{3/2}\left(2p^{1/2} - p^{-1/2}\right) = 4p^2 - 2p$.

41. $\sqrt{32} = \sqrt{4^2 \cdot 2} = 4\sqrt{2}$.

43. $\sqrt[3]{-54} = \sqrt[3]{(-1)(3^3)(2)} = -3\sqrt[3]{2}$.

45. $\sqrt{16x^2y^3} = \sqrt{4^2x^2y^2y} = 4xy\sqrt{y}$.

47. $\sqrt[3]{m^6n^3p^{12}} = \sqrt[3]{(m^2)^3\,n^3\,(p^4)^3} = m^2np^4$.

49. $\sqrt[3]{\sqrt{9}} = \sqrt[3]{3}$.

51. $\sqrt[3]{\sqrt{x}} = \sqrt[6]{x}$.

53. $\dfrac{2}{\sqrt{3}} \cdot \dfrac{\sqrt{3}}{\sqrt{3}} = \dfrac{2\sqrt{3}}{3}$.

55. $\dfrac{3}{2\sqrt{x}} \cdot \dfrac{\sqrt{x}}{\sqrt{x}} = \dfrac{3\sqrt{x}}{2x}$.

57. $\dfrac{2y}{\sqrt{3y}} \cdot \dfrac{\sqrt{3y}}{\sqrt{3y}} = \dfrac{2y\sqrt{3y}}{3y} = \frac{2}{3}\sqrt{3y}$.

59. $\dfrac{1}{\sqrt[3]{x}} \cdot \dfrac{\sqrt[3]{x^2}}{\sqrt[3]{x^2}} = \dfrac{\sqrt[3]{x^2}}{\sqrt[3]{x^3}} = \dfrac{\sqrt[3]{x^2}}{x}$.

61. $\dfrac{2}{1+\sqrt{3}} \cdot \dfrac{1-\sqrt{3}}{1-\sqrt{3}} = \dfrac{2\left(1-\sqrt{3}\right)}{1-3} = \dfrac{2\left(1-\sqrt{3}\right)}{-2} = \sqrt{3} - 1$.

63. $\dfrac{1+\sqrt{2}}{1-\sqrt{2}} \cdot \dfrac{1+\sqrt{2}}{1+\sqrt{2}} = \dfrac{\left(1+\sqrt{2}\right)^2}{1-2} = -\left(1+\sqrt{2}\right)^2$.

65. $\dfrac{q}{\sqrt{q}-1} \cdot \dfrac{\sqrt{q}+1}{\sqrt{q}+1} = \dfrac{q\left(\sqrt{q}+1\right)}{q-1}$.

67. $\dfrac{y}{\sqrt[3]{x^2z}} \cdot \dfrac{\sqrt[3]{xz^2}}{\sqrt[3]{xz^2}} = \dfrac{y\sqrt[3]{xz^2}}{\sqrt[3]{x^3z^3}} = \dfrac{y\sqrt[3]{xz^2}}{xz}$.

69. $\sqrt{\dfrac{16}{3}} = \dfrac{4}{\sqrt{3}} \cdot \dfrac{\sqrt{3}}{\sqrt{3}} = \dfrac{4\sqrt{3}}{3}$.

71. $\sqrt[3]{\dfrac{2}{3}} = \dfrac{\sqrt[3]{2}}{\sqrt[3]{3}} \cdot \dfrac{\sqrt[3]{3^2}}{\sqrt[3]{3^2}} = \dfrac{\sqrt[3]{18}}{3}$.

73. $\sqrt{\dfrac{3}{2x^2}} = \dfrac{\sqrt{3}}{x\sqrt{2}} \cdot \dfrac{\sqrt{2}}{\sqrt{2}} = \dfrac{\sqrt{6}}{2x}$.

75. $\sqrt[3]{\dfrac{2y^2}{3}} = \dfrac{\sqrt[3]{2y^2}}{\sqrt[3]{3}} \cdot \dfrac{\sqrt[3]{3^2}}{\sqrt[3]{3^2}} = \dfrac{\sqrt[3]{18y^2}}{3}$.

77. $\dfrac{1}{\sqrt{a}} + \sqrt{a} = \dfrac{1+a}{\sqrt{a}} = \dfrac{1+a}{\sqrt{a}} \cdot \dfrac{\sqrt{a}}{\sqrt{a}} = \dfrac{\sqrt{a}\,(1+a)}{a}$.

79. $\dfrac{\sqrt{x}}{\sqrt{x}+\sqrt{y}} + \dfrac{\sqrt{y}}{\sqrt{x}-\sqrt{y}} = \dfrac{\sqrt{x}\left(\sqrt{x}-\sqrt{y}\right) + \sqrt{y}\left(\sqrt{x}+\sqrt{y}\right)}{\left(\sqrt{x}+\sqrt{y}\right)\left(\sqrt{x}-\sqrt{y}\right)} = \dfrac{x - \sqrt{xy} + \sqrt{xy} + y}{x-y} = \dfrac{x+y}{x-y}$.

81. $(x+1)^{1/2} + \frac{1}{2}x\,(x+1)^{-1/2} = \frac{1}{2}(x+1)^{-1/2}\,[2\,(x+1) + x] = \frac{1}{2}(x+1)^{-1/2}\,(3x+2) = \dfrac{\sqrt{x+1}\,(3x+2)}{2\,(x+1)}$.

83. $\dfrac{\frac{1}{2}\left(1+x^{1/3}\right)x^{-1/2} - \frac{1}{3}x^{1/2}\cdot x^{-2/3}}{\left(1+x^{1/3}\right)^2} = \dfrac{\frac{1}{2}x^{-1/2} + \frac{1}{2}x^{-1/6} - \frac{1}{3}x^{-1/6}}{\left(1+x^{1/3}\right)^2} = \dfrac{\frac{1}{2}x^{-1/2} + \frac{1}{6}x^{-1/6}}{\left(1+x^{1/3}\right)^2} = \dfrac{\frac{1}{6}x^{-1/2}\left(3+x^{1/3}\right)}{\left(1+x^{1/3}\right)^2}$

$= \dfrac{3+x^{1/3}}{6x^{1/2}\left(1+x^{1/3}\right)^2}$.

85. $\sqrt{3x+1} = 2$

$$3x + 1 = 4$$
$$3x = 3$$
$$x = 1.$$

Check: $\sqrt{3(1)+1} \stackrel{?}{=} 2.$
Yes, $x = 1$ is a solution.

87. $\sqrt{k^2 - 4} = 4 - k$

$$k^2 - 4 = 16 - 8k + k^2$$
$$-4 = 16 - 8k$$
$$8k = 20$$
$$k = \tfrac{20}{8} = \tfrac{5}{2}.$$

Check: $\sqrt{\left(\tfrac{5}{2}\right)^2 - 4} \stackrel{?}{=} 4 - \tfrac{5}{2}$

$$\tfrac{3}{2} \stackrel{?}{=} \tfrac{3}{2}.$$

Yes, $k = \frac{5}{2}$ is a solution.

89. $\sqrt{k+1} + \sqrt{k} = 3\sqrt{k}$

$$\sqrt{k+1} = 2\sqrt{k}$$
$$k + 1 = 4k$$
$$3k = 1$$
$$k = \tfrac{1}{3}.$$

Check: $\sqrt{\tfrac{1}{3} + 1} + \sqrt{\tfrac{1}{3}} \stackrel{?}{=} 3\sqrt{\tfrac{1}{3}}$

$$\sqrt{\tfrac{4}{3}} + \sqrt{\tfrac{1}{3}} \stackrel{?}{=} 2\sqrt{\tfrac{1}{3}} + \sqrt{\tfrac{1}{3}}$$
$$3\sqrt{\tfrac{1}{3}} \stackrel{?}{=} 3\sqrt{\tfrac{1}{3}}.$$

Yes, $k = \frac{1}{3}$ is a solution.

91. $x = \sqrt{144 - p}$, so $x^2 = 144 - p$ and $p = 144 - x^2$.

93. True

95. True

1.8 Quadratic Equations

Concept Questions page 51

1. A quadratic equation in the variable x is any equation that can be written in the form $ax^2 + bx + c = 0$. For example, $4x^2 + 3x - 4 = 0$ is a quadratic equation.

3. The quadratic formula is $x = \dfrac{-b \pm \sqrt{b^2 - 4ac}}{2a}$. Using it to solve $2x^2 - 3x - 5 = 0$ for x, we substitute $a = 2$, $b = -3$, and $c = -5$, obtaining $x = \dfrac{-(-3) \pm \sqrt{(-3)^2 - 4(2)(-5)}}{2(2)} = \dfrac{3 \pm \sqrt{49}}{4}$. Simplifying, the solutions are $x = \frac{5}{2}$ and $x = -1$.

Exercises page 51

1. $(x + 2)(x - 3) = 0$. So $x + 2 = 0$ or $x - 3 = 0$; that is, $x = -2$ or $x = 3$.

3. $x^2 - 4 = (x - 2)(x + 2) = 0$, so $x = 2$ or $x = -2$.

5. $x^2 + x - 12 = (x+4)(x-3) = 0$, so $x = -4$ or $x = 3$.

7. $4t^2 + 2t - 2 = 2(t+1)(2t-1) = 0$, so $t = -1$ or $t = \frac{1}{2}$.

9. $\frac{1}{4}x^2 - x + 1 = 0$ is equivalent to $x^2 - 4x + 4 = 0$, or $(x-2)^2 = 0$. So $x = 2$ is a double root.

11. Rewrite the given equation in the form $2m^2 - 7m + 6 = 0$. Then $(2m-3)(m-2) = 0$ and $m = \frac{3}{2}$ or $m = 2$.

13. $4x^2 - 9 = (2x)^2 - 3^2 = (2x+3)(2x-3) = 0$, and so $x = -\frac{3}{2}$ or $x = \frac{3}{2}$.

15. $z(2z+1) = 6$ is equivalent to $2z^2 + z - 6 = 0$, so $(2z-3)(z+2) = 0$. Thus, $z = -2$ or $z = \frac{3}{2}$.

17. $x^2 + 2x + (1)^2 = 8 + 1$, so $(x+1)^2 = 9$, $x + 1 = \pm 3$, and the solutions are $x = -4$ and $x = 2$.

19. Rewrite the given equation in the form $6\left[x^2 - 2x + (-1)^2\right] = 3 + 6(-1)^2$. Then $6(x-1)^2 = 9$, $(x-1)^2 = \frac{3}{2}$, and $x - 1 = \pm\sqrt{\frac{3}{2}} = \pm\frac{1}{2}\sqrt{6}$. Therefore, $x = 1 - \frac{\sqrt{6}}{2}$ or $x = 1 + \frac{\sqrt{6}}{2}$.

21. $m^2 + m = 3$, so $m^2 + m + \left(\frac{1}{2}\right)^2 = 3 + \left(\frac{1}{2}\right)^2$, $\left(m + \frac{1}{2}\right)^2 = \frac{13}{4}$, and $m + \frac{1}{2} = \pm\frac{1}{2}\sqrt{13}$. Therefore, $m = -\frac{1}{2} - \frac{1}{2}\sqrt{13}$ or $m = -\frac{1}{2} + \frac{1}{2}\sqrt{13}$.

23. $2x^2 + 3x = 4$, so $2\left[x^2 + \frac{3}{2}x + \left(\frac{3}{4}\right)^2\right] = 4 + 2\left(\frac{3}{4}\right)^2$, $2\left(x + \frac{3}{4}\right)^2 = 4 + \frac{9}{8} = \frac{41}{8}$, $\left(x + \frac{3}{4}\right)^2 = \frac{41}{16}$, and $x + \frac{3}{4} = \pm\frac{\sqrt{41}}{4}$. Therefore, $x = -\frac{3}{4} - \frac{\sqrt{41}}{4}$ or $x = -\frac{3}{4} + \frac{\sqrt{41}}{4}$.

25. $4x^2 = 13$, so $x^2 = \frac{13}{4}$ and $x = \pm\frac{\sqrt{13}}{2}$.

27. Using the quadratic formula with $a = 2$, $b = -1$, and $c = -6$, we obtain

$$x = \frac{-(-1) \pm \sqrt{(-1)^2 - 4(2)(-6)}}{2(2)} = \frac{1 \pm \sqrt{1+48}}{4} = \frac{1 \pm 7}{4} = -\frac{3}{2} \text{ or } 2.$$

29. Rewrite the given equation in the form $m^2 - 4m + 1 = 0$. Then using the quadratic formula with $a = 1$, $b = -4$, and $c = 1$, we obtain $m = \dfrac{-(-4) \pm \sqrt{(-4)^2 - 4(1)(1)}}{2(1)} = \dfrac{4 \pm \sqrt{16-4}}{2} = \dfrac{4 \pm \sqrt{12}}{2} = \dfrac{4 \pm 2\sqrt{3}}{2} = 2 \pm \sqrt{3}$.

31. Rewrite the given equation in the form $8x^2 - 8x - 3 = 0$. Then using the quadratic formula with $a = 8$, $b = -8$, and $c = -3$, we obtain

$$x = \frac{-(-8) \pm \sqrt{(-8)^2 - 4(8)(-3)}}{2(8)} = \frac{8 \pm \sqrt{64+96}}{16} = \frac{8 \pm \sqrt{160}}{16} = \frac{8 \pm 4\sqrt{10}}{16} = \frac{1}{2} \pm \frac{1}{4}\sqrt{10}.$$

33. Rewrite the given equation in the form $2x^2 + 4x - 3 = 0$. Then using the quadratic formula with $a = 2$, $b = 4$, and $c = -3$, we obtain $x = \dfrac{-4 \pm \sqrt{4^2 - 4(2)(-3)}}{2(2)} = \dfrac{-4 \pm \sqrt{16+24}}{4} = \dfrac{-4 \pm \sqrt{40}}{4} = \dfrac{-4 \pm 2\sqrt{10}}{4} = -1 \pm \frac{1}{2}\sqrt{10}$.

35. Using the quadratic formula with $a = 2.1$, $b = -4.7$, and $c = -6.2$, we obtain

$$x = \frac{4.7 \pm \sqrt{(-4.7)^2 - 4(2.1)(-6.2)}}{2(2.1)} = \frac{4.7 \pm \sqrt{74.17}}{4.2} \approx \frac{4.7 \pm 8.6122}{4.2} \approx -0.93 \text{ or } 3.17.$$

37. $x^4 - 5x^2 + 6 = 0$. Let $m = x^2$. Then the equation reads $m^2 - 5m + 6 = 0$. Now, factoring, we obtain $(m-3)(m-2) = 0$, and so $m = 2$ or $m = 3$. Therefore, $x = \pm\sqrt{2}$ or $\pm\sqrt{3}$.

39. $y^4 - 7y^2 + 10 = 0$. Let $x = y^2$. Then we have $x^2 - 7x + 10 = 0$. Factoring, we obtain $(x-2)(x-5) = 0$, and so $x = 2$ or 5. Therefore, $y = \pm\sqrt{2}$ or $y = \pm\sqrt{5}$.

41. $6(x+2)^2 + 7(x+2) - 3 = 0$. Let $y = x + 2$. Then we have $6y^2 + 7y - 3 = 0$. Factoring, we obtain $(2y+3)(3y-1) = 0$, and so $y = -\frac{3}{2}$ or $\frac{1}{3}$. Therefore, $x + 2 = -\frac{3}{2}$ or $\frac{1}{3}$, and so $x = -\frac{7}{2}$ or $-\frac{5}{3}$.

43. $6w - 13\sqrt{w} + 6 = 0$. Let $x = \sqrt{w}$. Then $6x^2 - 13x + 6 = 0$, $(2x-3)(3x-2) = 0$, and so $x = \frac{3}{2}$ or $x = \frac{2}{3}$. Then the solutions are $w = x^2 = \frac{9}{4}$ or $\frac{4}{9}$.

Check $w = \frac{4}{9}$: $6\left(\frac{4}{9}\right) - 13\sqrt{\frac{4}{9}} + 6 = \frac{24}{9} - 13 \cdot \frac{2}{3} + 6 \stackrel{?}{=} 0$. Yes, $\frac{4}{9}$ is a solution.

Check $w = \frac{9}{4}$: $6\left(\frac{9}{4}\right) - 13\sqrt{\frac{9}{4}} + 6 = \frac{54}{4} - 13 \cdot \frac{3}{2} + 6 \stackrel{?}{=} 0$. Yes, $\frac{9}{4}$ is also a solution.

45.

$$\frac{2}{x+3} - \frac{4}{x} = 4$$
$$2(x) - 4(x+3) = 4(x)(x+3)$$
$$2x - 4x - 12 = 4x^2 + 12x$$
$$-2x - 12 = 4x^2 + 12x$$
$$4x^2 + 14x + 12 = 0$$
$$2x^2 + 7x + 6 = 0$$
$$(2x+3)(x+2) = 0.$$

Thus, the solutions are $x = -\frac{3}{2}$ and $x = -2$.

47.

$$x + 2 - \frac{3}{2x-1} = 0$$
$$x(2x-1) + 2(2x-1) - 3 = 0$$
$$2x^2 - x + 4x - 2 - 3 = 0$$
$$2x^2 + 3x - 5 = 0$$
$$(2x+5)(x-1) = 0.$$

Thus, the solutions are $x = -\frac{5}{2}$ and $x = 1$.

49.

$$2 - \frac{7}{2y} - \frac{15}{y^2} = 0$$
$$4y^2 - 7y - 30 = 0$$
$$(y+2)(4y-15) = 0.$$

Thus, $y = -2$ or $y = \frac{15}{4}$.

51.

$$\frac{3}{x^2-1} + \frac{2x}{x+1} = \frac{7}{3}$$
$$9 + 6x(x-1) = 7(x^2-1)$$
$$9 + 6x^2 - 6x = 7x^2 - 7$$
$$x^2 + 6x - 16 = 0$$
$$(x+8)(x-2) = 0.$$

Thus, $x = -8$ or $x = 2$.

53. $$\frac{3x}{x-2}+\frac{4}{x+2}=\frac{24}{x^2-4}$$
$$3x(x+2)+4(x-2)=24$$
$$3x^2+6x+4x-8=24$$
$$3x^2+10x-32=0$$
$$(3x+16)(x-2)=0.$$
Thus, $x=-\frac{16}{3}$ or $x=2$. But because $x=2$ results in division by zero in the original equation, we discard it. The only solution is $x=-\frac{16}{3}$.

55. $$\frac{2t+1}{t-2}-\frac{t}{t+1}=-1$$
$$(2t+1)(t+1)-t(t-2)=-1(t-2)(t+1)$$
$$2t^2+3t+1-t^2+2t=-t^2+t+2$$
$$2t^2+4t-1=0.$$
Using the quadratic formula with $a=2$, $b=4$, and $c=-1$, we obtain
$$t=\frac{-4\pm\sqrt{16-4(2)(-1)}}{4}=-1\pm\frac{\sqrt{24}}{4}$$
$$=-1\pm\tfrac{1}{2}\sqrt{6}.$$

57. $$\sqrt{u^2+u-5}=1$$
$$u^2+u-5=1$$
$$u^2+u-6=0$$
$$(u+3)(u-2)=0.$$
Thus, $u=-3$ or $u=2$.

Check $u=-3$: $\sqrt{(-3)^2-3-5}=\sqrt{1}\stackrel{?}{=}1$. Yes, so $u=-3$ is a solution.

Check $u=2$: $\sqrt{2^2+2-5}=\sqrt{1}\stackrel{?}{=}1$. Yes, so $u=2$ is also a solution.

59. $$\sqrt{2r+3}=r$$
$$2r+3=r^2$$
$$r^2-2r-3=0$$
$$(r-3)(r+1)=0.$$
Thus, $r=3$ or $r=-1$.

Check $r=3$: $\sqrt{2(3)+3}\stackrel{?}{=}3$. Yes, so $r=3$ is a solution.

Check $r=-1$: $\sqrt{2(-1)+3}\stackrel{?}{=}-1$. No, so $r=-1$ is not a solution.

61. $$\sqrt{s-2}-\sqrt{s+3}+1=0$$
$$\sqrt{s-2}=\sqrt{s+3}-1$$
$$s-2=s+3-2\sqrt{s+3}+1$$
$$2\sqrt{s+3}=6$$
$$s+3=3^2=9$$
$$s=6.$$
Check: $\sqrt{6-2}-\sqrt{6+3}+1\stackrel{?}{=}0$. Yes, so $s=6$ is the solution.

63. $$\frac{1}{(x-3)^2}-\frac{10}{x-3}+21=0$$
$$1-10(x-3)+21(x-3)^2=0$$
$$31-10x+21x^2-126x+189=0$$
$$21x^2-136x+220=0$$
$$(7x-22)(3x-10)=0.$$
Thus, $x=\frac{22}{7}$ or $x=\frac{10}{3}$.

65. $x^2-6x+5=0$. Here $a=1$, $b=-6$, and $c=5$. $b^2-4ac=(-6)^2-4(1)(5)=16>0$, and so the equation has two real solutions.

67. $3y^2-4y+5=0$. Here $a=3$, $b=-4$, and $c=5$. $b^2-4ac=(-4)^2-4(3)(5)=-44<0$, and so the equation has no real solution.

69. $4x^2+12x+9=0$. Here $a=4$, $b=12$, and $c=9$. $b^2-4ac=12^2-4(4)(9)=0$, and so the equation has one real solution.

71. $\frac{6}{k^2}+\frac{1}{k}-2=0$. Multiplying by k^2, we have $6+k-2k^2=0$ or $2k^2-k-6=0$. Here $a=2$, $b=-1$, and $c=-6$, so the discriminant is $b^2-4ac=(-1)^2-4(2)(-6)=49>0$, and the equation has two real solutions.

73. The ball reaches the ground when $h = 0$; that is, when $16t^2 - 64t - 768 = 0$, and $t^2 - 4t - 48 = 0$. Using the quadratic formula with $a = 1$, $b = -4$, and $c = -48$, we find $t = \frac{-(-4) \pm \sqrt{16 - 4(1)(-48)}}{2} = \frac{4 \pm \sqrt{208}}{2} \approx 9.21$, or approximately 9.2 seconds. (We discard the negative root.)

75. Substituting $u = 10$, $a = 4$, and $v = 22$ into the equation $v = ut + at^2$, we have $22 = 10t + 4t^2$. Then $4t^2 + 10t - 22 = 0$, or $2t^2 + 5t - 11 = 0$. Using the quadratic formula with $a = 2$, $b = 5$, and $c = -11$, we have $t = \frac{-5 \pm \sqrt{5^2 - 4(2)(-11)}}{2(2)} \approx \frac{-5 \pm \sqrt{113}}{4} \approx 1.41$ or -3.91. We reject the negative root, so the time taken is approximately 1.41 seconds after passing the tree.

77. Substituting $p = 10$ into $p = \frac{30}{0.02x^2 + 1}$, we have $10(0.02x^2 + 1) = 30$. Solving this equation for x, we have $0.2x^2 + 10 = 30$, $0.2x^2 = 20$, $x^2 = 100$, and $x = \pm 10$. Rejecting the negative root, we see that the quantity demanded is 10,000. (Remember that x is measured in units of one thousand.)

79. Substituting $p = 30$ into the equation $p = \frac{1}{10}\sqrt{x} + 10$, we have $300 = \sqrt{x} + 100$, so $\sqrt{x} = 200$ and $x = 200^2 = 40{,}000$. Thus, 40,000 satellite radios will be made available at the unit price of \$30.

81. We solve the equation $100\left(\frac{t^2 + 10t + 100}{t^2 + 20t + 100}\right) = 80$, obtaining $5(t^2 + 10t + 100) = 4(t^2 + 20t + 100)$, $5t^2 + 50t + 500 = 4t^2 + 80t + 400$, and $t^2 - 30t + 100 = 0$. Using the quadratic formula with $a = 1$, $b = -30$, and $c = 100$, we get $t = \frac{30 \pm \sqrt{30^2 - 4(1)(100)}}{2} = \frac{30 \pm \sqrt{500}}{2} \approx 3.82$ or 26.18. So the oxygen content first drops to 80% of its natural level approximately 4 days after the waste was dumped into the pond and is restored to that level approximately 26 days after the waste was dumped.

83. The total surface area is given by

$$S = (10 - 2x)(16 - 2x) + 2x(10 - 2x) + 2x(16 - 2x) = 160 - 20x - 32x + 4x^2 + 20x - 4x^2 + 32x - 4x^2$$
$$= -4x^2 + 160.$$

Since the total surface area is to be 144 square inches, we have $-4x^2 + 160 = 144$, $4x^2 = 16$, and $x^2 = 4$. Thus, $x = 2$ because x must be positive. The dimensions are therefore $12'' \times 6'' \times 2''$.

85. Let x denote the length of one piece of fencing so that the second piece has length $(120 - x)$ ft. The squares' side lengths are $\frac{x}{4}$ and $\frac{120 - x}{4}$, and so the sum of the areas is

$$A = \left(\frac{x}{4}\right)^2 + \left(\frac{120 - x}{4}\right)^2 = \frac{1}{16}\left[x^2 + (120 - x)^2\right] = \frac{1}{16}\left(x^2 + 14{,}400 - 240x + x^2\right)$$
$$= \frac{1}{16}\left(2x^2 - 240x + 14{,}400\right).$$

Since the sum of the areas of the two rectangles is to be 562.5 ft^2, we have $\frac{1}{16}(2x^2 - 240x + 14{,}400) = 562.5$, $2x^2 - 240x + 14{,}400 = 9000$, $2x^2 - 240x + 5400 = 0$, $x^2 - 120x + 2700 = 0$, and $(x - 30)(x - 90) = 0$. Therefore $x = 30$ or $x = 90$, and the lengths of the pieces of fencing are 30 ft and 90 ft.

87. Let x denote the width and y the length. Then $2x + y = 3000$. The area is given by $A = xy = x(3000 - 2x) = -2x^2 + 3000x$. The quadratic function $A = -2x^2 + 3000x$ has a maximum at $x = -\frac{b}{2a} = -\frac{3000}{2(-2)} = 750$. Therefore, $y = 3000 - 2(750) = 1500$. The dimensions are 750 yards by 1500 yards.

89. We solve the equation $2\pi r\ell + 4\pi r^2 = 28\pi$ with $\ell = 4$, obtaining $28\pi = 8\pi r + 4\pi r^2$, $4\pi r^2 + 8\pi r - 28\pi = 0$, and $r^2 + 2r - 7 = 0$. Using the quadratic formula with $a = 1$, $b = 2$, and $c = -7$, we get $r = \frac{-2 \pm \sqrt{4 + 4(1)(7)}}{2} = \frac{-2 \pm 4\sqrt{2}}{2} = -1 \pm 2\sqrt{2}$. Since r must be positive, we discard the negative root. Thus, $r = -1 + 2\sqrt{2} \approx 1.83$, and the radius of each hemisphere is approximately 1.83 ft.

91. False. In fact both a and b must be nonzero.

93. True.

1.9 Inequalities and Absolute Value

Concept Questions page 62

1. Let a, b, and c be any real numbers.

Property 1 If $a < b$ and $b < c$, then $a < c$. Example: $3 < 4$ and $5 < 9$, so $3 < 9$.

Property 2 If $a < b$, then $a + c < b + c$. Example: $-6 < -2$, so $-6 + 3 < -2 + 3$; that is, $-3 < 1$.

Property 3 If $a < b$ and $c > 0$, then $ac < bc$. Example: $-7 < -2$ and $3 > 0$, so $(-7)(3) < (-2)(3)$; that is, $-21 < -6$.

Property 4 If $a < b$ and $c < 0$, then $ac > bc$. Example: $-7 < -2$ and $-3 < 0$, so $(-7)(-3) > (-2)(-3)$; that is, $21 > 6$.

3. Let a, b, and c be any real numbers.

Property 1 $|-a| = |a|$. Example: $|-5| = |5| = 5$.

Property 2 $|ab| = |a|\,|b|$. Example: $|(3)(-4)| = |(3)|\,|(-4)| = 12$.

Property 3 $\left|\frac{a}{b}\right| = \frac{|a|}{|b|}$. Example: $\left|\frac{-4}{3}\right| = \frac{|-4|}{|3|} = \frac{4}{3}$.

Property 4 $|a + b| \le |a| + |b|$. Example: $|9 + (-4)| = |5| = 5 \le |9| + |-4| = 13$.

Exercises page 62

1. The statement is false because -3 is greater than -20. See the number line below.

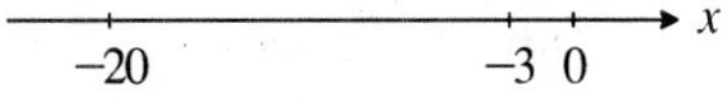

3. The statement is false because $\frac{2}{3} = \frac{4}{6}$ is less than $\frac{5}{6}$.

5. The interval $(3, 6)$ is shown on the number line below. Note that this is an open interval indicated by "(" and ")".

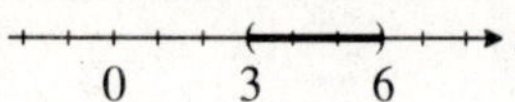

7. The interval $[-1, 4)$ is shown on the number line below. Note that this is a half-open interval indicated by "[" (closed) and ")"(open).

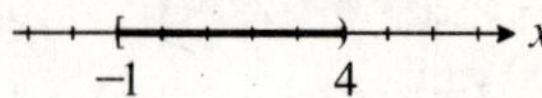

9. The infinite interval $(0, \infty)$ is shown on the number line below.

11. We are given $2x + 2 < 8$. Add -2 to each side of the inequality to obtain $2x < 6$, then multiply each side of the inequality by $\frac{1}{2}$ to obtain $x < 3$. We write this in interval notation as $(-\infty, 3)$.

13. We are given the inequality $-4x \geq 20$. Multiply both sides of the inequality by $-\frac{1}{4}$ and reverse the sign of the inequality to obtain $x \leq -5$. We write this in interval notation as $(-\infty, -5]$.

15. We are given the inequality $-6 < x - 2 < 4$. First add 2 to each member of the inequality to obtain $-6 + 2 < x < 4 + 2$ and $-4 < x < 6$, so the solution set is the open interval $(-4, 6)$.

17. We want to find the values of x that satisfy at least one of the inequalities $x + 1 > 4$ and $x + 2 < -1$. Adding -1 to both sides of the first inequality, we obtain $x + 1 - 1 > 4 - 1$, so $x > 3$. Similarly, adding -2 to both sides of the second inequality, we obtain $x + 2 - 2 < -1 - 2$, so $x < -3$. Therefore, the solution set is $(-\infty, -3) \cup (3, \infty)$.

19. We want to find the values of x that satisfy the inequalities $x + 3 > 1$ and $x - 2 < 1$. Adding -3 to both sides of the first inequality, we obtain $x + 3 - 3 > 1 - 3$, or $x > -2$. Similarly, adding 2 to each side of the second inequality, we obtain $x - 2 + 2 < 1 + 2$, so $x < 3$. Because both inequalities must be satisfied, the solution set is $(-2, 3)$.

21. We want to find the values of x that satisfy the inequality $(x + 3)(x - 5) \leq 0$. From the sign diagram, we see that the given inequality is satisfied when $-3 \leq x \leq 5$, that is, when the signs of the two factors are different or when one of the factors is equal to zero. The solution set is $[-3, 5]$.

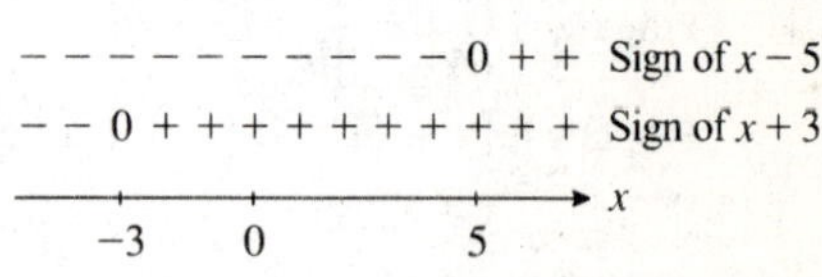

23. We want to find the values of x that satisfy the inequality $(2x - 3)(x - 1) \leq 0$. From the sign diagram, we see that the given inequality is satisfied when $x \geq 1$ and $x \leq \frac{3}{2}$; that is, when the signs of the two factors differ or one of the two factors is 0. The solution set is $\left[1, \frac{3}{2}\right]$.

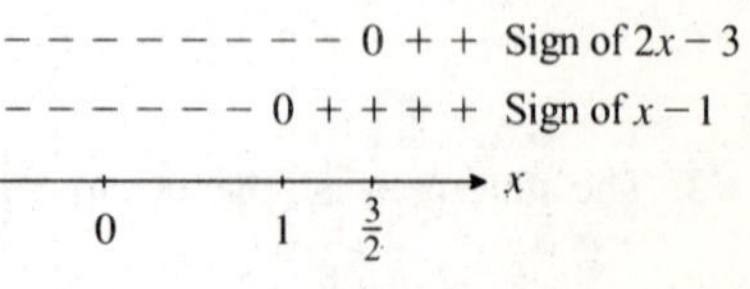

25. We want to find the values of x that satisfy the inequality $\frac{x+3}{x-2} \geq 0$. From the sign diagram, we see that the given inequality is satisfied when $x \leq -3$ or $x > 2$, that is, when the signs of the two factors are the same. The solution set is $(-\infty, -3] \cup (2, \infty)$. Notice that $x = 2$ is not included because the inequality is not defined at that value of x.

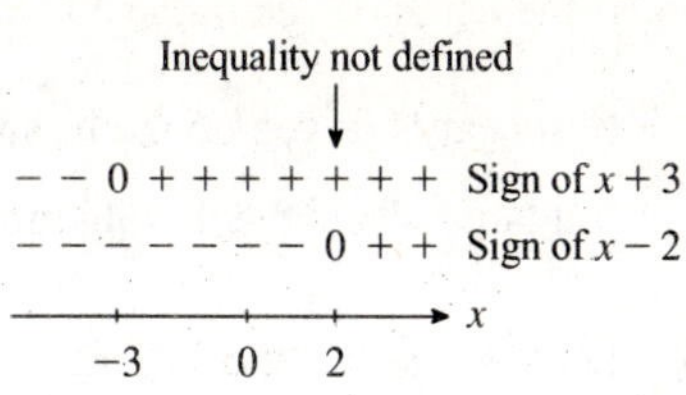

27. We want to find the values of x that satisfy the inequality $\frac{x-2}{x-1} \leq 2$. Subtracting 2 from each side of the given inequality and simplifying gives $\frac{x-2}{x-1} - 2 \leq 0$, $\frac{x-2-2(x-1)}{x-1} \leq 0$, and $-\frac{x}{x-1} \leq 0$. From the sign diagram, we see that the given inequality is satisfied when $x \leq 0$ or $x > 1$; that is, when the signs of the two factors differ. The solution set is $(-\infty, 0] \cup (1, \infty)$. Notice that $x = 1$ is not included because the inequality is undefined at that value of x.

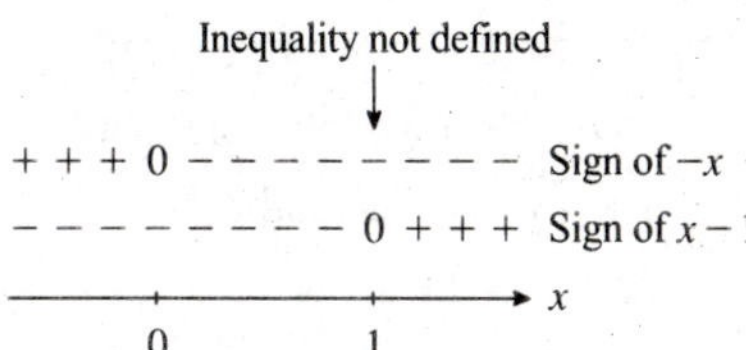

29. $|-6+2| = 4$.

31. $\frac{|-12+4|}{|16-12|} = \frac{|-8|}{|4|} = 2$.

33. $\sqrt{3}\,|-2| + 3\left|-\sqrt{3}\right| = \sqrt{3}\,(2) + 3\sqrt{3} = 5\sqrt{3}$.

35. $|\pi - 1| + 2 = \pi - 1 + 2 = \pi + 1$.

37. $\left|\sqrt{2} - 1\right| + \left|3 - \sqrt{2}\right| = \sqrt{2} - 1 + 3 - \sqrt{2} = 2$.

39. False. If $a > b$, then $-a < -b$, $-a + b < -b + b$, and $b - a < 0$.

41. False. Let $a = -2$ and $b = -3$. Then $a^2 = 4$ and $b^2 = 9$, and $4 < 9$. (Note that we need only provide a counterexample to show that the statement is not always true.)

43. True. There are three possible cases.
Case 1: If $a > 0$ and $b > 0$, then $a^3 > b^3$, since $a^3 - b^3 = (a-b)\left(a^2 + ab + b^2\right) > 0$.
Case 2: If $a > 0$ and $b < 0$, then $a^3 > 0$ and $b^3 < 0$, and it follows that $a^3 > b^3$.
Case 3: If $a < 0$ and $b < 0$, then $a^3 - b^3 = (a-b)\left(a^2 + ab + b^2\right) > 0$, and we see that $a^3 > b^3$. (Note that $a - b > 0$ and $ab > 0$.)

45. $|x - a| < b$ is equivalent to $-b < x - a < b$ or $a - b < x < a + b$.

47. False. If we take $a = -2$, then $|-a| = |-(-2)| = |2| = 2 \neq a$.

49. True. If $a - 4 < 0$, then $|a - 4| = 4 - a = |4 - a|$. If $a - 4 > 0$, then $|4 - a| = a - 4 = |a - 4|$.

51. False. If we take $a = 3$ and $b = -1$, then $|a + b| = |3 - 1| = 2 \neq |a| + |b| = 3 + 1 = 4$.

53. Simplifying $5(C - 25) \geq 1.75 + 2.5C$, we obtain $5C - 125 \geq 1.75 + 2.5C$, $5C - 2.5C \geq 1.75 + 125$, $2.5C \geq 126.75$, and finally $C \geq 50.7$. Therefore, the minimum cost is \$50.70.

55. If the car is driven in the city, then it can be expected to cover (18.1 gallons) $\left(20\ \frac{\text{miles}}{\text{gallon}}\right) = 362$ miles on a full tank. If the car is driven on the highway, then it can be expected to cover (18.1 gallons) $\left(27\ \frac{\text{miles}}{\text{gallon}}\right) = 488.7$ miles on a full tank. Thus, the driving range of the car may be described by the interval $[362, 488.7]$.

57. Let x represent the salesman's monthly sales in dollars. Then $0.15(x - 12{,}000) \geq 6000$, $15(x - 12{,}000) \geq 600{,}000$, $15x - 180{,}000 \geq 600{,}000$, $15x \geq 780{,}000$, and $x \geq 52{,}000$. We conclude that the salesman must have sales of at least \$52,000 to reach his goal.

59. We want to solve the inequality $-6x^2 + 30x - 10 \geq 14$. (Remember that x is expressed in thousands.) Adding -14 to both sides of this inequality, we have $-6x^2 + 30x - 10 - 14 \geq 14 - 14$, or $-6x^2 + 30x - 24 \geq 0$. Dividing both sides of the inequality by -6 (which reverses the sign of the inequality), we have $x^2 - 5x + 4 \leq 0$. Factoring this last expression, we have $(x - 4)(x - 1) \leq 0$.

From the sign diagram, we see that x must lie between 1 and 4. (The inequality is satisfied only when the two factors have different signs.) Because x is expressed in thousands of units, we see that the manufacturer must produce between 1000 and 4000 units of the commodity.

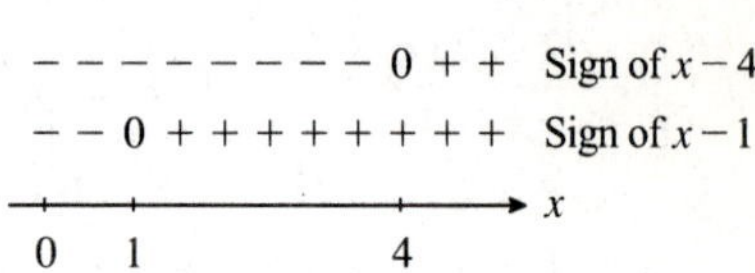

61. We solve the inequalities $25 \leq \frac{0.5x}{100 - x} \leq 30$, obtaining $2500 - 25x \leq 0.5x \leq 3000 - 30x$, which is equivalent to $2500 - 25x \leq 0.5x$ and $0.5x \leq 3000 - 30x$. Simplifying further, $25.5x \geq 2500$ and $30.5x \leq 3000$, so $x \geq \frac{2500}{25.5} \approx 98.04$ and $x \leq \frac{3000}{30.5} \approx 98.36$. Thus, the city could expect to remove between 98.04% and 98.36% of the toxic pollutant.

63. We solve $\frac{10{,}000}{t^2 + 1} + 2000 < 4000$, obtaining $\frac{10{,}000}{t^2 + 1} < 2000$, $10{,}000 < 2000\left(t^2 + 1\right)$, and $t^2 + 1 > 5$. Rewriting, we have $t^2 - 4 > 0$, or $(t - 2)(t + 2) > 0$. The solution of this inequality is $t < -2$ or $t > 2$. Because t must be positive, we conclude that the number of bacteria will have dropped below 4000 after 2 minutes.

65. The ball's height is 196 ft or greater when $128t - 16t^2 + 4 \geq 196$, that is, $16t^2 - 128t + 192 \leq 0$. Simplifying and factoring, this is equivalent to the inequality $t^2 - 8t + 12 = (t - 6)(t - 2) \leq 0$. The solution of this inequality is $2 \leq t \leq 6$. We conclude that the ball's height is greater than or equal to 196 ft for 4 seconds.

67. The rod is acceptable if $0.49 \leq x \leq 0.51$ or $-0.01 \leq x - 0.5 \leq 0.01$. This gives the required inequality, $|x - 0.5| \leq 0.01$.

CHAPTER 1 Concept Review Questions page 65

1. a. rational; repeating; terminating

b. irrational, terminates, repeats

3. a. a; $-(ab) = a(-b)$; ab

b. 0; 0

5. product; prime; $x(x + 2)(x - 1)$

7. complex; $\dfrac{1 + \frac{1}{x}}{1 - \frac{1}{y}}$

9. **a.** equation

b. number

c. $ax + b = 0$; 1

11. **a.** radical; $b^{1/n}$

b. radical

CHAPTER 1 Review Exercises page 65

1. The number $\frac{7}{8}$ is a rational number and a real number.

3. The number -2π is an irrational number and a real number.

5. The number $2.\overline{71}$ is a rational number and a real number.

7. $\left(\frac{9}{4}\right)^{3/2} = \frac{9^{3/2}}{4^{3/2}} = \frac{27}{8}$.

9. $(3 \cdot 4)^{-2} = 12^{-2} = \frac{1}{12^2} = \frac{1}{144}$.

11. $\left(\frac{16}{9}\right)^{3/2} = \left(\frac{4}{3}\right)^3 = \frac{64}{27}$.

13. $\sqrt[3]{\frac{27}{125}} = \frac{3}{5}$.

15. $\frac{4\left(x^2+y\right)^3}{x^2+y} = 4\left(x^2+y\right)^2$.

17. $\frac{\sqrt[4]{16x^5yz}}{\sqrt[4]{81xyz^5}} = \frac{\left(2^4x^5yz\right)^{1/4}}{\left(3^4xyz^5\right)^{1/4}} = \frac{2x^{5/4}y^{1/4}z^{1/4}}{3x^{1/4}y^{1/4}z^{5/4}} = \frac{2x}{3z}$.

19. $\left(\frac{3xy^2}{4x^3y}\right)^{-2}\left(\frac{3xy^3}{2x^2}\right)^3 = \left(\frac{3y}{4x^2}\right)^{-2}\left(\frac{3y^3}{2x}\right)^3 = \left(\frac{4x^2}{3y}\right)^2\left(\frac{3y^3}{2x}\right)^3 = \frac{\left(16x^4\right)\left(27y^9\right)}{\left(9y^2\right)\left(8x^3\right)} = 6xy^7$.

21. $\sqrt[3]{81x^5y^{10}}\sqrt[3]{9xy^2} = 3^{4/3}x^{5/3}y^{10/3} \cdot 3^{2/3}x^{1/3}y^{2/3} = 3^2x^2y^4 = 9x^2y^4$.

23. $\left(3x^4+10x^3+6x^2+10x+3\right)+\left(2x^4+10x^3+6x^2+4x\right)$
$= 3x^4+2x^4+10x^3+10x^3+6x^2+6x^2+10x+4x+3 = 5x^4+20x^3+12x^2+14x+3$.

25. $(2x+3y)^2-(3x+1)(2x-3) = 4x^2+12xy+9y^2-6x^2+7x+3 = -2x^2+9y^2+12xy+7x+3$.

27. $\frac{(t+6)(60)-(60t+180)}{(t+6)^2} = \frac{60t+360-60t-180}{(t+6)^2} = \frac{180}{(t+6)^2}$.

29. $\frac{2}{3}\left(\frac{4x}{2x^2-1}\right)+3\left(\frac{3}{3x-1}\right) = \frac{8x}{3\left(2x^2-1\right)}+\frac{9}{3x-1} = \frac{8x(3x-1)+27\left(2x^2-1\right)}{3\left(2x^2-1\right)(3x-1)} = \frac{78x^2-8x-27}{3\left(2x^2-1\right)(3x-1)}$.

31. $-2\pi^2r^3+100\pi r^2 = -2\pi r^2(\pi r-50)$.

33. $16-x^2 = 4^2-x^2 = (4-x)(4+x)$.

35. $-2x^2-4x+6 = -2\left(x^2+2x-3\right) = -2(x+3)(x-1)$.

37. $9a^2-25b^2 = (3a)^2-(5b)^2 = (3a-5b)(3a+5b)$.

39. $6a^4b^4c-3a^3b^2c-9a^2b^2 = 3a^2b^2\left(2a^2b^2c-ac-3\right)$.

41. $\dfrac{2x^2+3x-2}{2x^2+5x-3}=\dfrac{(2x-1)(x+2)}{(2x-1)(x+3)}=\dfrac{x+2}{x+3}$.

43. $\dfrac{2x-6}{x+3}\cdot\dfrac{x^2+6x+9}{x^2-9}=\dfrac{2(x-3)}{x+3}\cdot\dfrac{(x+3)(x+3)}{(x+3)(x-3)}=\dfrac{2(x+3)}{x+3}=2$.

45. $\dfrac{1+\dfrac{1}{x+2}}{x-\dfrac{9}{x}}=\dfrac{x+2+1}{x+2}\cdot\dfrac{x}{x^2-9}=\dfrac{x+3}{x+2}\cdot\dfrac{x}{(x+3)(x-3)}=\dfrac{x}{(x+2)(x-3)}$.

47. $8x^2+2x-3=(4x+3)(2x-1)=0$, so the solutions are $x=-\frac{3}{4}$ and $x=\frac{1}{2}$.

49. $2x^2-3x-4=0$. Using the quadratic formula with $a=2$, $b=-3$, and $c=-4$, we have $x=\dfrac{-(-3)\pm\sqrt{(-3)^2-4(2)(-4)}}{2(2)}=\dfrac{3\pm\sqrt{9+32}}{4}=\dfrac{3\pm\sqrt{41}}{4}$.

51. $2y^2-3y+1=(2y-1)(y-1)=0$, and so $y=\frac{1}{2}$ or 1.

53. $-x^3-2x^2+3x=-x\left(x^2+2x-3\right)=-x(x+3)(x-1)=0$, and so the roots of the equation are $x=0$, $x=-3$, and $x=1$.

55. $\frac{1}{4}x+2=\frac{3}{4}x-5$, so $-\frac{1}{2}x=-7$ and $x=14$.

57. $(x+2)^2-3x(1-x)=(x-2)^2$. Thus, $x^2+4x+4-3x+3x^2=x^2-4x+4$, $3x^2+5x=0$, and $x(3x+5)=0$, and so $x=0$ or $x=-\frac{5}{3}$.

59. $\sqrt{k-1}=\sqrt{2k-3}$, so $k-1=2k-3$ and $2=k$. Check: $\sqrt{2-1}=1$ and $\sqrt{2(2)-3}=1$, so $k=2$ is the solution.

61. Solve $C=\dfrac{20x}{100-x}$. $C(100-x)=20x$, $100C-Cx=20x$, $-Cx-20x=-100C$, $x(20+C)=100C$, and so $x=\dfrac{100C}{20+C}$.

63. $-x+3\le 2x+9$. Adding x to both sides yields $3\le 3x+9$, so $3x\ge -6$ and thus $x\ge -2$. We conclude that the solution set is $[-2,\infty)$.

65. The inequalities $x-3>2$ and $x+1<-1$ imply $x>5$ or $x<-4$, so the solution set is $(-\infty,-4)\cup(5,\infty)$.

67. $|-5+7|+|-2|=|2|+|-2|=2+2=4$.

69. $|2\pi-6|-\pi=2\pi-6-\pi=\pi-6$.

71. Factoring the left-hand side of $2x^2+3x-2\le 0$, we have $(2x-1)(x+2)\le 0$. From the sign diagram, we conclude that the given inequality is satisfied when $-2\le x\le\frac{1}{2}$. The solution set is $\left[-2,\frac{1}{2}\right]$.

73. $\dfrac{1}{x+2} > 2$ gives $\dfrac{1}{x+2} - 2 > 0$, $\dfrac{1-2x-4}{x+2} > 0$, and finally $\dfrac{-2x-3}{x+2} > 0$. From the sign diagram, we see that the given inequality is satisfied when $-2 < x < -\frac{3}{2}$. The solution set is $\left(-2, -\frac{3}{2}\right)$.

+ + + + 0 − − − − − −	Sign of $-2x-3$
− − 0 + + + + + + + +	Sign of $x+2$
-2 $\ -\frac{3}{2}$ $\quad 0$	x

75. The given inequality $|3x-4| \le 2$ is equivalent to $3x-4 \le 2$ or $3x-4 \ge -2$. Solving the first inequality, we have $3x \le 6$, so $x \le 2$. Similarly, we solve the second inequality and obtain $3x \ge 2$, so $x \ge \frac{2}{3}$. We conclude that $\frac{2}{3} \le x \le 2$. The solution set is $\left[\frac{2}{3}, 2\right]$.

77. $\dfrac{\sqrt{x}-1}{x-1} = \dfrac{\sqrt{x}-1}{x-1} \cdot \dfrac{\sqrt{x}+1}{\sqrt{x}+1} = \dfrac{(\sqrt{x})^2-1}{(x-1)(\sqrt{x}+1)} = \dfrac{x-1}{(x-1)(\sqrt{x}+1)} = \dfrac{1}{\sqrt{x}+1}$.

79. $\dfrac{\sqrt{x}-1}{2\sqrt{x}} = \dfrac{\sqrt{x}-1}{2\sqrt{x}} \cdot \dfrac{\sqrt{x}}{\sqrt{x}} = \dfrac{x-\sqrt{x}}{2x}$.

81. $x^2 - 2x - 5 = 0$. Using the quadratic formula with $a = 1$, $b = -2$, and $c = -5$, we have
$x = \dfrac{-b \pm \sqrt{b^2-4ac}}{2a} = \dfrac{-(-2) \pm \sqrt{(-2)^2 - 4(1)(-5)}}{2(1)} = \dfrac{2 \pm \sqrt{24}}{2} = 1 \pm \sqrt{6}$.

83. $2(1.5C + 80) \le 2(2.5C - 20)$. Simplifying, we obtain $1.5C + 80 \le 2.5C - 20$, so $C \ge 100$ and the minimum cost is \$100.

CHAPTER 1 Before Moving On... page 67

1. $2(3x-2)^2 - 3x(x+1) + 4 = 2(9x^2 - 12x + 4) - 3x^2 - 3x + 4 = 18x^2 - 24x + 8 - 3x^2 - 3x + 4$
$= 15x^2 - 27x + 12 = 3(5x^2 - 9x + 4)$.

2. a. $x^4 - x^3 - 6x^2 = x^2(x^2 - x - 6) = x^2(x-3)(x+2)$.

b. $(a-b)^2 - (a^2+b)^2 = [(a-b) - (a^2+b)][(a-b) + (a^2+b)] = (a - b - a^2 - b)(a + a^2 - b + b)$
$= (-a^2 - 2b + a)(a)(a+1)$.

3. $\dfrac{2x}{3x^2-5x-2} + \dfrac{x-1}{x^2-x-2} = \dfrac{2x}{(3x+1)(x-2)} + \dfrac{x-1}{(x-2)(x+1)} = \dfrac{2x(x+1) + (x-1)(3x+1)}{(3x+1)(x-2)(x+1)}$
$= \dfrac{2x^2 + 2x + 3x^2 - 2x - 1}{(3x+1)(x-2)(x+1)} = \dfrac{5x^2-1}{(3x+1)(x-2)(x+1)}$.

4. $\left(\dfrac{8x^2y^{-3}}{9x^{-3}y^2}\right)^{-1}\left(\dfrac{2x^2}{2y^3}\right)^2 = \dfrac{8^{-1}x^{-2}y^3}{9^{-1}x^3y^{-2}} \cdot \dfrac{2^2x^4}{3^2y^6} = \dfrac{1}{2} \cdot \dfrac{1}{xy} = \dfrac{1}{2xy}$.

5. $2s = \dfrac{r}{s+r}$, so $2s(s+r) = r$, $2s^2 + 2sr = r$, $r(1-2s) = 2s^2$, and $r = \dfrac{2s^2}{1-2s}$.

6. $\dfrac{2-\sqrt{3}}{2+\sqrt{3}} \cdot \dfrac{2-\sqrt{3}}{2-\sqrt{3}} = \dfrac{4-4\sqrt{3}+3}{2^2-(\sqrt{3})^2} = \dfrac{7-4\sqrt{3}}{4-3} = 7 - 4\sqrt{3}$.

7. **a.** $2x^2 + 5x - 12 = 0$, so $(2x - 3)(x + 4) = 0$. Thus, $x = \frac{3}{2}$ or $x = -4$.

b. $m^2 - 3m - 2 = 0$. Using the quadratic formula with $a = 1$, $b = -3$, and $c = -2$, we obtain

$$x = \frac{-(-3) \pm \sqrt{(-3)^2 - 4(1)(-2)}}{2} = \frac{3 \pm \sqrt{9+8}}{2} = \frac{3 \pm \sqrt{17}}{2}.$$

8. $\sqrt{x+4} - \sqrt{x-5} - 1 = 0$, so $\sqrt{x+4} - \sqrt{x-5} = 1$, $\sqrt{x+4} = 1 + \sqrt{x-5}$, $x + 4 = 1 + 2\sqrt{x-5} + x - 5$, $8 = 2\sqrt{x-5}$, $4 = \sqrt{x-5}$, $x - 5 = 16$, and $x = 21$.

9. We want to find the values of x for which $(3x + 2)(2x - 3) \le 0$. From the sign diagram, we conclude that the given inequality is satisfied when $-\frac{2}{3} \le x \le \frac{3}{2}$. The solution set is $\left[-\frac{2}{3}, \frac{3}{2}\right]$.

- - 0 + + + + + + + Sign of $3x + 2$
- - - - - - - - 0 + + Sign of $2x - 3$
x
$-\frac{2}{3}$ 0 $\frac{3}{2}$

10. $|2x + 3| \le 1$ is equivalent to $-1 \le 2x + 3 \le 1$. Thus, $-1 - 3 \le 2x \le 1 - 3$, or $-4 \le 2x \le -2$. We conclude that $-2 \le x \le -1$. The solution set is $[-2, -1]$.

2 FUNCTIONS AND THEIR GRAPHS

2.1 The Cartesian Coordinate System and Straight Lines

Problem-Solving Tips

When you solve a problem in the exercises that follow each section, first read the problem. Before you start computing or writing out a solution, try to formulate a strategy for solving the problem. Then proceed by using your strategy to solve the problem.

Here we summarize some general problem-solving techniques that are covered in Sections 2.1 and 2.2.

1. **To show that two lines are parallel**, you need to show that the slopes of the two lines are equal or that their slopes are both undefined.

2. **To show that two lines L_1 and L_2 are perpendicular** (Section 2.2) you need to show that the slope m_1 of L_1 is the negative reciprocal of the slope m_2 of L_2; that is, $m_1 = -1/m_2$.

3. **To find the equation of a line** (Section 2.2), you need the slope of the line and a point lying on the line. You can then find the equation of the line using the point-slope form of the equation of a line: $(y - y_1) = m(x - x_1)$.

Concept Questions page 74

1. a. $a < 0$ and $b > 0$. **b.** $a < 0$ and $b < 0$. **c.** $a > 0$ and $b < 0$.

Exercises page 74

1. The coordinates of A are $(3, 3)$ and it is located in Quadrant I.

3. The coordinates of C are $(2, -2)$ and it is located in Quadrant IV.

5. The coordinates of E are $(-4, -6)$ and it is located in Quadrant III.

7. A **9.** E, F, and G **11.** F

For Exercises 13–19, refer to the following figure.

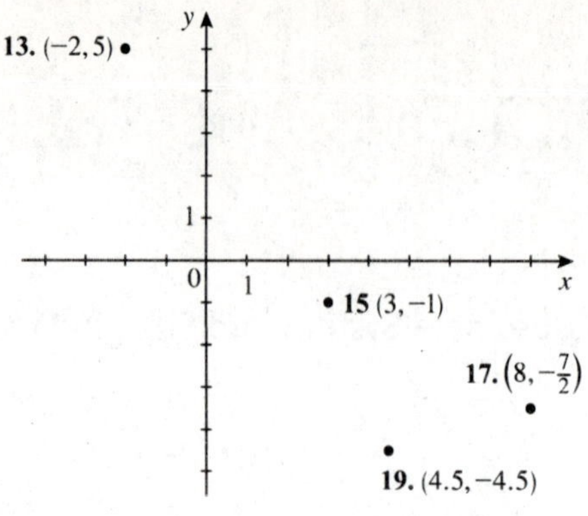

21. Referring to the figure shown in the text, we see that $m = \dfrac{2-0}{0-(-4)} = \dfrac{1}{2}$.

23. This is a vertical line, and hence its slope is undefined.

25. $m = \dfrac{y_2 - y_1}{x_2 - x_1} = \dfrac{8-3}{5-4} = 5$.

27. $m = \dfrac{y_2 - y_1}{x_2 - x_1} = \dfrac{8-3}{4-(-2)} = \dfrac{5}{6}$.

29. $m = \dfrac{y_2 - y_1}{x_2 - x_1} = \dfrac{d-b}{c-a}$, provided $a \neq c$.

31. Because the equation is already in slope-intercept form, we read off the slope $m = 4$.

a. If x increases by 1 unit, then y increases by 4 units.

b. If x decreases by 2 units, then y decreases by $4(-2) = -8$ units.

33. The slope of the line through A and B is $\dfrac{-10-(-2)}{-3-1} = \dfrac{-8}{-4} = 2$. The slope of the line through C and D is $\dfrac{1-5}{-1-1} = \dfrac{-4}{-2} = 2$. Because the slopes of these two lines are equal, the lines are parallel.

35. The slope of the line through the point $(1, a)$ and $(4, -2)$ is $m_1 = \dfrac{-2-a}{4-1}$ and the slope of the line through $(2, 8)$ and $(-7, a+4)$ is $m_2 = \dfrac{a+4-8}{-7-2}$. Because these two lines are parallel, m_1 is equal to m_2. Therefore, $\dfrac{-2-a}{3} = \dfrac{a-4}{-9}$, $-9(-2-a) = 3(a-4)$, $18 + 9a = 3a - 12$, and $6a = -30$, so $a = -5$.

37. Yes. A straight line with slope zero ($m = 0$) is a horizontal line, whereas a straight line whose slope does not exist (m cannot be computed) is a vertical line.

2.2 Equations of Lines

Problem-Solving Tips

See Section 2.1.

Concept Questions page 82

1. a. $y - y_1 = m(x - x_1)$ **b.** $y = mx + b$

c. $ax + by + c = 0$, where a and b are not both zero.

Exercises page 82

1. (e) **3.** (a) **5.** (f)

7. The slope of the line through A and B is $\dfrac{2-5}{4-(-2)} = -\dfrac{3}{6} = -\dfrac{1}{2}$. The slope of the line through C and D is $\dfrac{6-(-2)}{3-(-1)} = \dfrac{8}{4} = 2$. Because the slopes of these two lines are the negative reciprocals of each other, the lines are perpendicular.

9. An equation of a horizontal line is of the form $y = b$. In this case $b = -5$, so $y = -5$ is an equation of the line.

11. We use the point-slope form of an equation of a line with the point $(3, -4)$ and slope $m = 2$. Thus $y - y_1 = m(x - x_1)$ becomes $y - (-4) = 2(x - 3)$. Simplifying, we have $y + 4 = 2x - 6$, or $y = 2x - 10$.

13. Because the slope $m = 0$, we know that the line is a horizontal line of the form $y = b$. Because the line passes through $(-3, 2)$, we see that $b = 2$, and an equation of the line is $y = 2$.

15. We first compute the slope of the line joining the points $(2, 4)$ and $(3, 7)$, obtaining $m = \dfrac{7-4}{3-2} = 3$. Using the point-slope form of an equation of a line with the point $(2, 4)$ and slope $m = 3$, we find $y - 4 = 3(x - 2)$, or $y = 3x - 2$.

17. We first compute the slope of the line joining the points $(1, 2)$ and $(-3, -2)$, obtaining $m = \dfrac{-2-2}{-3-1} = \dfrac{-4}{-4} = 1$. Using the point-slope form of an equation of a line with the point $(1, 2)$ and slope $m = 1$, we find $y - 2 = x - 1$, or $y = x + 1$.

19. We use the slope-intercept form of an equation of a line: $y = mx + b$. Because $m = 3$ and $b = 5$, the equation is $y = 3x + 5$.

21. We use the slope-intercept form of an equation of a line: $y = mx + b$. Because $m = 0$ and $b = 5$, the equation is $y = 5$.

23. We first write the given equation in the slope-intercept form: $x - 2y = 0$, so $-2y = -x$, or $y = \frac{1}{2}x$. From this equation, we see that $m = \frac{1}{2}$ and $b = 0$.

25. We write the equation in slope-intercept form: $2x - 3y - 9 = 0$, $-3y = -2x + 9$, and $y = \frac{2}{3}x - 3$. From this equation, we see that $m = \frac{2}{3}$ and $b = -3$.

27. We write the equation in slope-intercept form: $2x + 4y = 14$, $4y = -2x + 14$, and $y = -\frac{2}{4}x + \frac{14}{4} = -\frac{1}{2}x + \frac{7}{2}$. From this equation, we see that $m = -\frac{1}{2}$ and $b = \frac{7}{2}$.

29. We first write the equation $2x - 4y - 8 = 0$ in slope-intercept form: $2x - 4y - 8 = 0$, $4y = 2x - 8$, $y = \frac{1}{2}x - 2$. Now the required line is parallel to this line, and hence has the same slope. Using the point-slope form of an equation of a line with $m = \frac{1}{2}$ and the point $(-2, 2)$, we have $y - 2 = \frac{1}{2}[x - (-2)]$ or $y = \frac{1}{2}x + 3$.

31. We first write the equation $3x + 4y - 22 = 0$ in slope-intercept form: $3x + 4y - 22 = 0$, so $4y = -3x + 22$ and $y = -\frac{3}{4}x + \frac{11}{2}$ Now the required line is perpendicular to this line, and hence has slope $\frac{4}{3}$ (the negative reciprocal of $-\frac{3}{4}$). Using the point-slope form of an equation of a line with $m = \frac{4}{3}$ and the point $(2, 4)$, we have $y - 4 = \frac{4}{3}(x - 2)$, or $y = \frac{4}{3}x + \frac{4}{3}$.

33. A line parallel to the x-axis has slope 0 and is of the form $y = b$. Because the line is 6 units below the axis, it passes through $(0, -6)$ and its equation is $y = -6$.

35. We use the point-slope form of an equation of a line to obtain $y - b = 0(x - a)$, or $y = b$.

37. Because the required line is parallel to the line joining $(-3, 2)$ and $(6, 8)$, it has slope $m = \dfrac{8 - 2}{6 - (-3)} = \dfrac{6}{9} = \dfrac{2}{3}$. We also know that the required line passes through $(-5, -4)$. Using the point-slope form of an equation of a line, we find $y - (-4) = \frac{2}{3}[x - (-5)]$, $y = \frac{2}{3}x + \frac{10}{3} - 4$, and finally $y = \frac{2}{3}x - \frac{2}{3}$.

39. Because the point $(-3, 5)$ lies on the line $kx + 3y + 9 = 0$, it satisfies the equation. Substituting $x = -3$ and $y = 5$ into the equation gives $-3k + 15 + 9 = 0$, or $k = 8$.

41. $3x - 2y + 6 = 0$. Setting $y = 0$, we have $3x + 6 = 0$ or $x = -2$, so the x-intercept is -2. Setting $x = 0$, we have $-2y + 6 = 0$ or $y = 3$, so the y-intercept is 3.

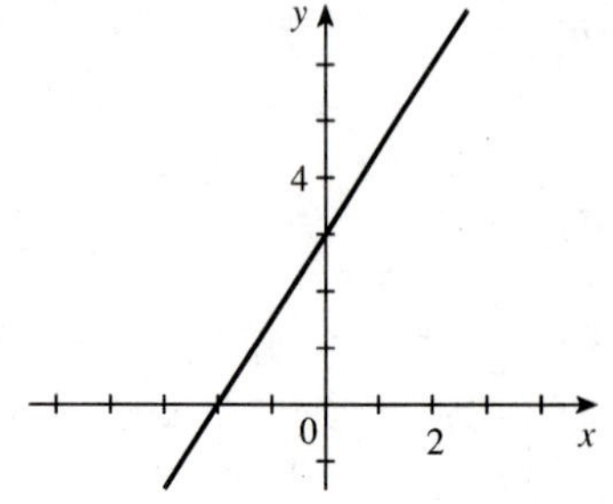

43. $x + 2y - 4 = 0$. Setting $y = 0$, we have $x - 4 = 0$ or $x = 4$, so the x-intercept is 4. Setting $x = 0$, we have $2y - 4 = 0$ or $y = 2$, so the y-intercept is 2.

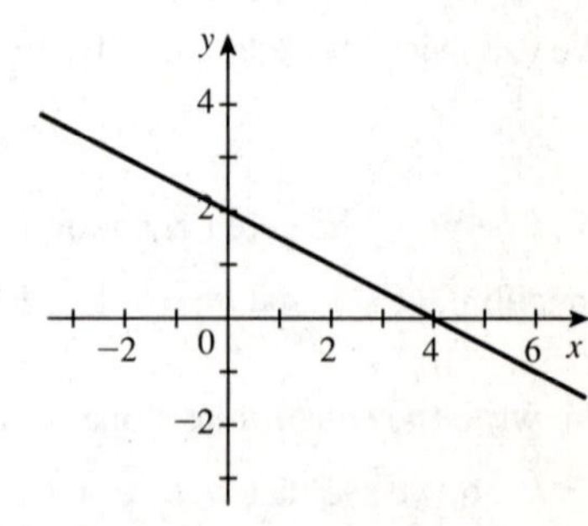

45. $y+5=0$. Setting $y=0$, we have $0+5=0$, which has no solution, so there is no x-intercept. Setting $x=0$, we have $y+5=0$ or $y=-5$, so the y-intercept is -5.

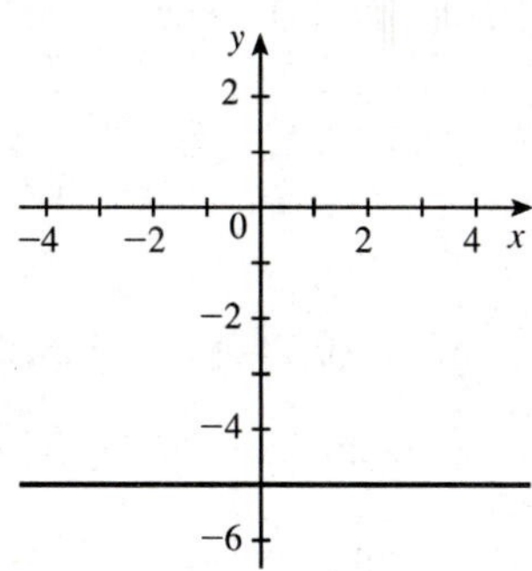

47. Because the line passes through the points $(a,0)$ and $(0,b)$, its slope is $m=\dfrac{b-0}{0-a}=-\dfrac{b}{a}$. Then, using the point-slope form of an equation of a line with the point $(a,0)$, we have $y-0=-\dfrac{b}{a}(x-a)$ or $y=-\dfrac{b}{a}x+b$, which may be written in the form $\dfrac{b}{a}x+y=b$. Multiplying this last equation by $\dfrac{1}{b}$, we have $\dfrac{x}{a}+\dfrac{y}{b}=1$.

49. Using the equation $\dfrac{x}{a}+\dfrac{y}{b}=1$ with $a=-2$ and $b=-4$, we have $-\dfrac{x}{2}-\dfrac{y}{4}=1$. Then $-4x-2y=8$, $2y=-8-4x$, and finally $y=-2x-4$.

51. Using the equation $\dfrac{x}{a}+\dfrac{y}{b}=1$ with $a=4$ and $b=-\frac{1}{2}$, we have $\dfrac{x}{4}+\dfrac{y}{-1/2}=1$, $-\frac{1}{4}x+2y=-1$, $2y=\frac{1}{4}x-1$, and so $y=\frac{1}{8}x-\frac{1}{2}$.

53. The slope of the line passing through A and B is $m=\dfrac{7-1}{1-(-2)}=\dfrac{6}{3}=2$, and the slope of the line passing through B and C is $m=\dfrac{13-7}{4-1}=\dfrac{6}{3}=2$. Because the slopes are equal, the points lie on the same line.

55. a.

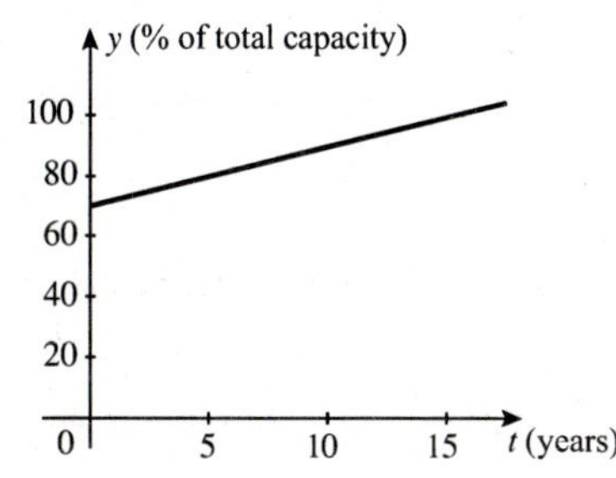

b. The slope is 1.9467 and the y-intercept is 70.082.

c. The output is increasing at the rate of 1.9467% per year. The output at the beginning of 1990 was 70.082%.

d. We solve the equation $1.9467t+70.082=100$, obtaining $1.9467t=29.918$ and $t\approx 15.37$. We conclude that the plants were generating at maximum capacity during April 2005.

57. a. $y=0.55x$

b. Solving the equation $1100=0.55x$ for x, we have $x=\dfrac{1100}{0.55}=2000$.

59. Using the points $(0,0.68)$ and $(10,0.80)$, we see that the slope of the required line is $m=\dfrac{0.80-0.68}{10-0}=\dfrac{0.12}{10}=0.012$. Next, using the point-slope form of the equation of a line, we have $y-0.68=0.012(t-0)$ or $y=0.012t+0.68$. Therefore, when $t=14$, we have $y=0.012(14)+0.68=0.848$, or 84.8%. That is, in 2004 women's wages were 84.8% of men's wages.

61. a, b.

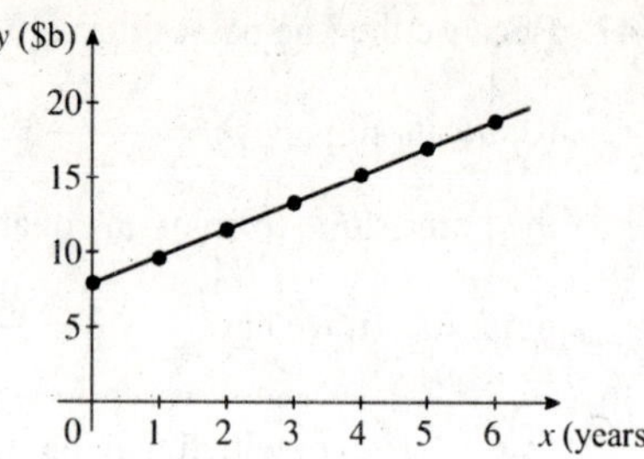

c. $m = \dfrac{18.8 - 7.9}{6 - 0} \approx 1.82$, so $y - 7.9 = 1.82(x - 0)$, or $y = 1.82x + 7.9$.

d. $y = 1.82(5) + 7.9 \approx 17$ or \$17 billion. This agrees with the actual data for that year.

63. a, b.

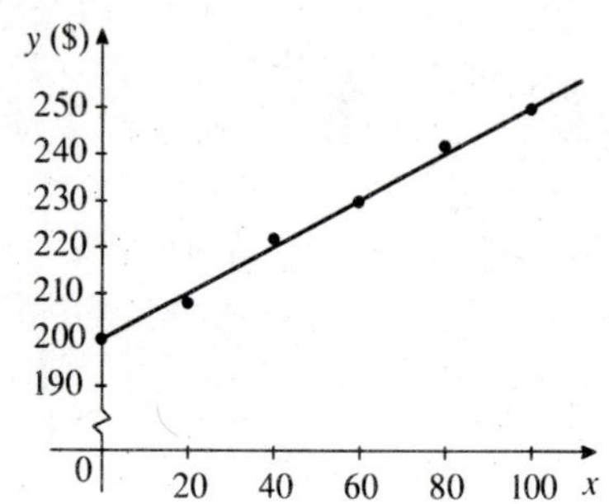

c. Using the points (0, 200) and (100, 250), we see that the slope of the required line is $m = \dfrac{250 - 200}{100} = \dfrac{1}{2}$. Therefore, an equation is $y - 200 = \frac{1}{2}x$ or $y = \frac{1}{2}x + 200$.

d. The approximate cost for producing 54 units of the commodity is $\frac{1}{2}(54) + 200$, or \$227.

65. a, b.

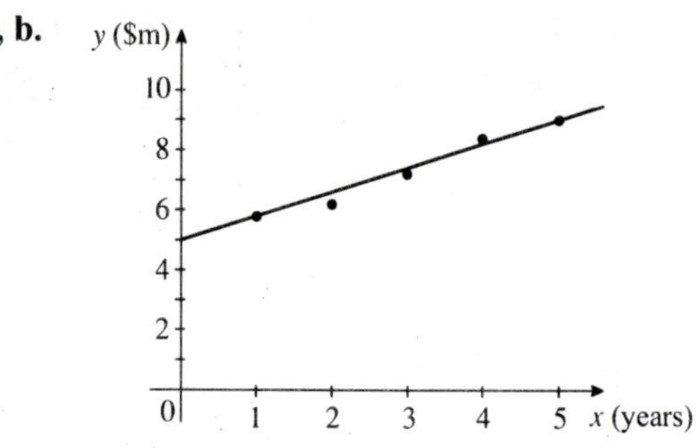

c. The slope of L is $m = \dfrac{9.0 - 5.8}{5 - 1} = \dfrac{3.2}{4} = 0.8$. Using the point-slope form of an equation of a line, we have $y - 5.8 = 0.8(x - 1) = 0.8x - 0.8$, or $y = 0.8x + 5$.

d. Using the equation from part c with $x = 9$, we have $y = 0.8(9) + 5 = 12.2$, or \$12.2 million.

67. False. Substituting $x = -1$ and $y = 1$ into the equation gives $3(-1) + 7(1) = 4$, and this is not equal to the right-hand side of the equation. Therefore, the equation is not satisfied and so the given point does not lie on the line.

69. True. The slope of the line $Ax + By + C = 0$ is $-\dfrac{A}{B}$. (Write it in slope-intercept form.) Similarly, the slope of the line $ax + by + c = 0$ is $-\dfrac{a}{b}$. They are parallel if and only if $-\dfrac{A}{B} = -\dfrac{a}{b}$, that is, if $Ab = aB$, or $Ab - aB = 0$.

71. True. The slope of the line $ax + by + c_1 = 0$ is $m_1 = -\dfrac{a}{b}$. The slope of the line $bx - ay + c_2 = 0$ is $m_2 = \dfrac{b}{a}$. Because $m_1 m_2 = -1$, the straight lines are indeed perpendicular.

73. Writing each equation in the slope-intercept form, we have $y = -\dfrac{a_1}{b_1}x - \dfrac{c_1}{b_1}$ $(b_1 \neq 0)$ and $y = -\dfrac{a_2}{b_2}x - \dfrac{c_2}{b_2}$ $(b_2 \neq 0)$. Because two lines are parallel if and only if their slopes are equal, we see that the lines are parallel if and only if $-\dfrac{a_1}{b_1} = -\dfrac{a_2}{b_2}$, or $a_1 b_2 - b_1 a_2 = 0$.

Technology Exercises

page 89

Graphing Utility

1.

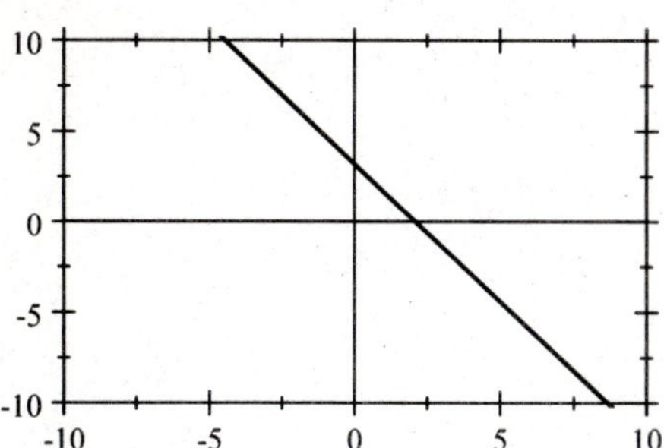

3.

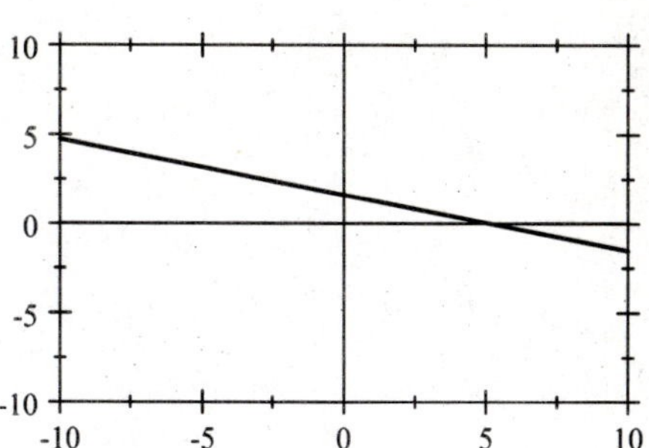

5. a.

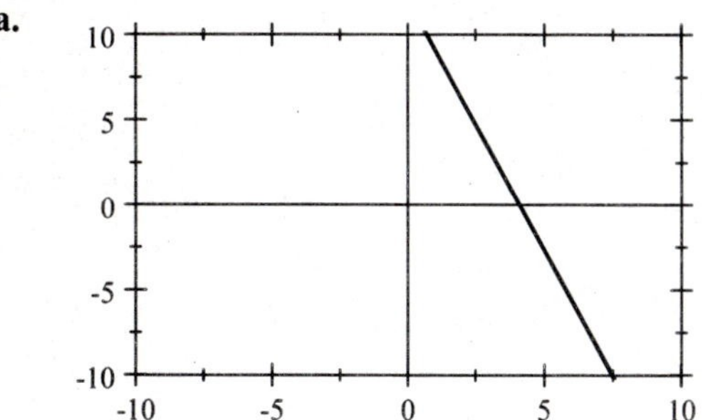

b.

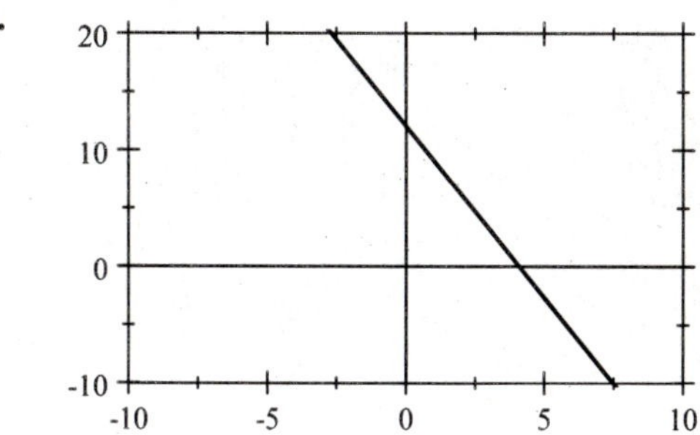

7. a.

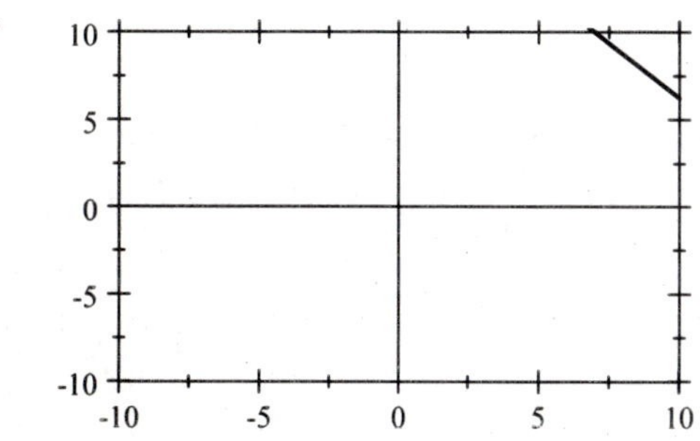

b.

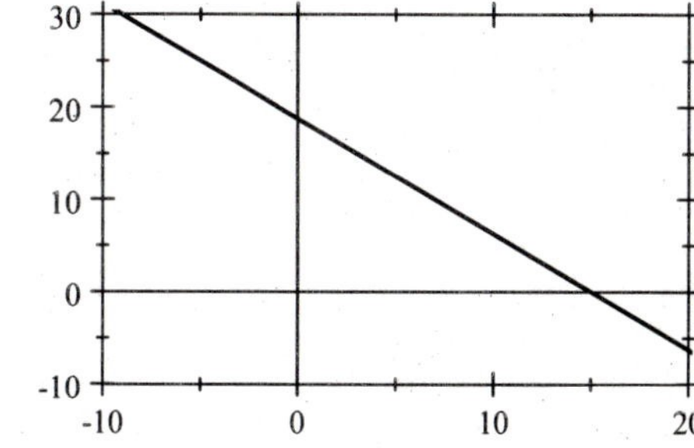

9.

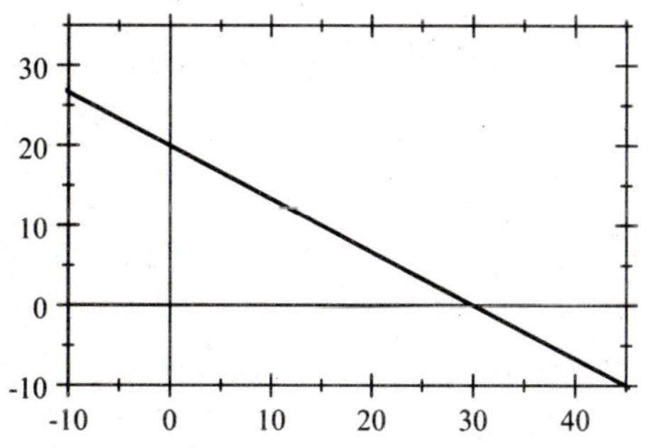

11.

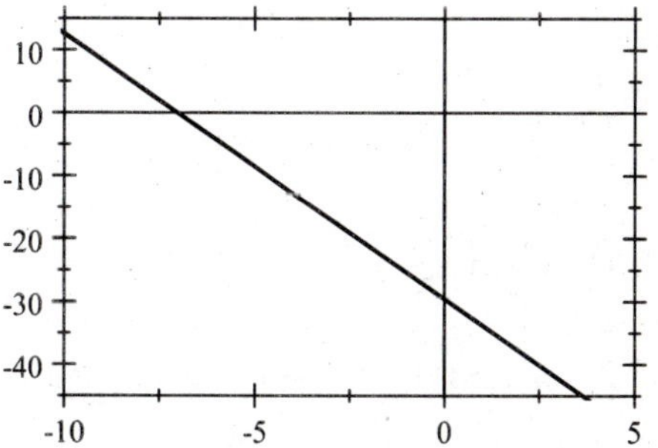

Excel

1.

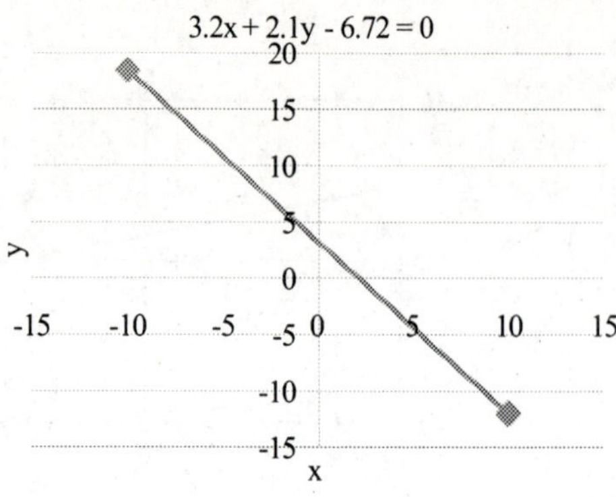

3.

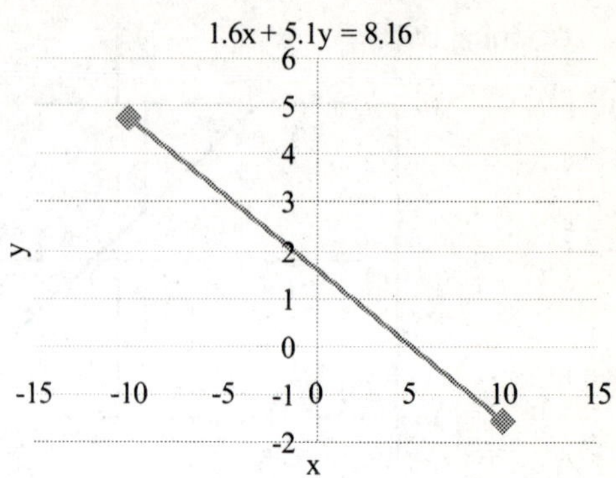

5.

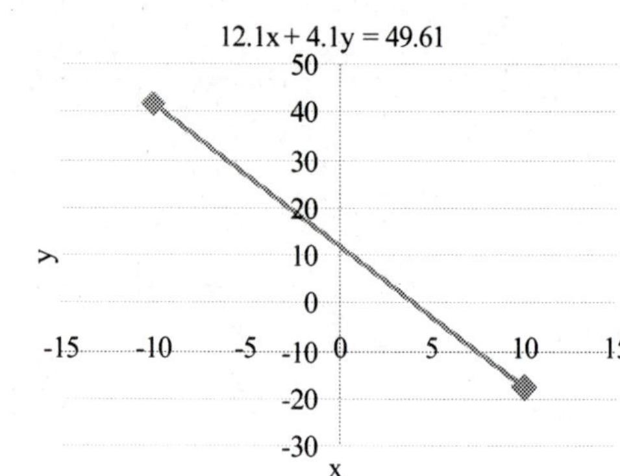

7.

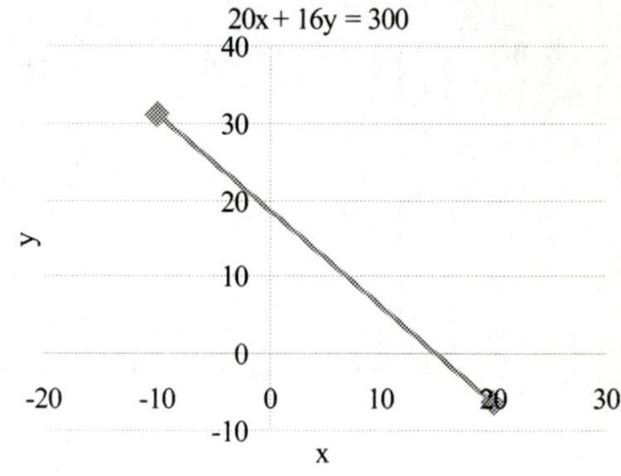

9.

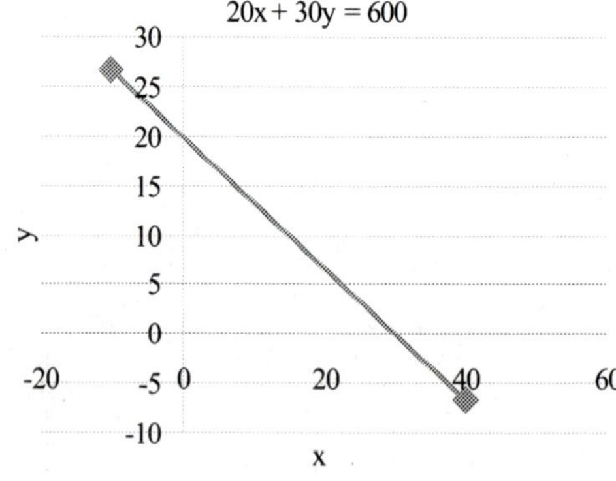

11.

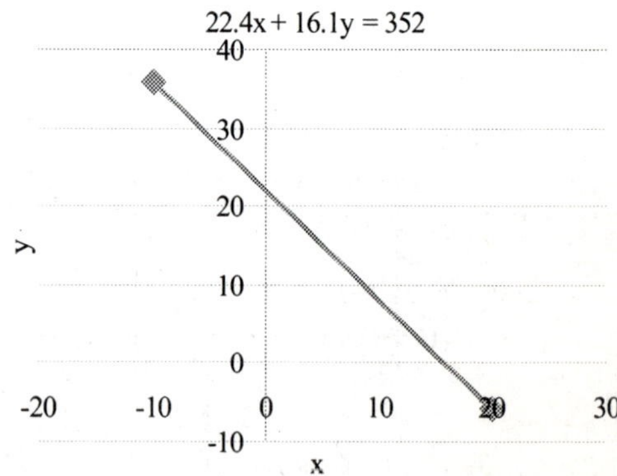

2.3 Functions and Their Graphs

Problem-Solving Tips

1. **To find the domain of a function** $f(x)$, find all values of x for which $f(x)$ is a real number.

   a. **If the function involves a quotient**, check to see if there are any values of x at which the denominator is equal to zero. (Remember, division by zero is not allowed.) Then exclude those points from the domain.

b. If the function involves the root of a real number, check to see if the root is an even or an odd root. If n is even, the nth root of a negative number is not defined, so values of x yielding the nth root of a negative number must be excluded from the domain of f. For example, $\sqrt{x-1}$ is defined only for $x \geq 1$, so the domain of $f(x) = \sqrt{x-1}$ is $[1, \infty)$.

2. To evaluate a piecewise-defined function $f(x)$ at a specific value of x, check to see which subdomain x lies in. Then evaluate the function using the rule for that subdomain.

3. To determine whether a curve is the graph of a function, use the vertical line test. If you can draw a vertical line through the curve that intersects the curve at more than one point, then the curve is not the graph of a function.

Concept Questions

page 96

1. a. A function is a rule that associates with each element in a set A exactly one element in a set B.

b. The domain of a function f is the set of all elements x in a set A such that $f(x)$ is an element in B. The range of f is the set of all elements $f(x)$ such that x is an element in its domain.

c. An independent variable is a variable in the domain of a function f. The dependent variable is $y = f(x)$.

3. a. Yes, every vertical line intersects the curve in at most one point.

b. No, a vertical line intersects the curve at more than one point.

c. No, a vertical line intersects the curve at more than one point.

d. Yes, every vertical line intersects the curve in at most one point.

Exercises

page 97

1. $f(x) = 5x + 6$. Therefore $f(3) = 5(3) + 6 = 21$, $f(-3) = 5(-3) + 6 = -9$, $f(a) = 5(a) + 6 = 5a + 6$, $f(-a) = 5(-a) + 6 = -5a + 6$, and $f(a+3) = 5(a+3) + 6 = 5a + 15 + 6 = 5a + 21$.

3. $g(x) = 3x^2 - 6x + 3$, so $g(0) = 3(0) - 6(0) + 3 = 3$, $g(-1) = 3(-1)^2 - 6(-1) + 3 = 3 + 6 + 3 = 12$, $g(a) = 3(a)^2 - 6(a) + 3 = 3a^2 - 6a + 3$, $g(-a) = 3(-a)^2 - 6(-a) + 3 = 3a^2 + 6a + 3$, and $g(x+1) = 3(x+1)^2 - 6(x+1) + 3 = 3\left(x^2 + 2x + 1\right) - 6x - 6 + 3 = 3x^2 + 6x + 3 - 6x - 3 = 3x^2$.

5. $f(x) = 2x + 5$, so $f(a+h) = 2(a+h) + 5 = 2a + 2h + 5$, $f(-a) = 2(-a) + 5 = -2a + 5$, $f\left(a^2\right) = 2\left(a^2\right) + 5 = 2a^2 + 5$, $f(a - 2h) = 2(a - 2h) + 5 = 2a - 4h + 5$, and $f(2a - h) = 2(2a - h) + 5 = 4a - 2h + 5$

7. $s(t) = \dfrac{2t}{t^2 - 1}$. Therefore, $s(4) = \dfrac{2(4)}{(4)^2 - 1} = \dfrac{8}{15}$, $s(0) = \dfrac{2(0)}{0^2 - 1} = 0$,

$s(a) = \dfrac{2(a)}{a^2 - 1} = \dfrac{2a}{a^2 - 1}$, $s(2+a) = \dfrac{2(2+a)}{(2+a)^2 - 1} = \dfrac{2(2+a)}{a^2 + 4a + 4 - 1} = \dfrac{2(2+a)}{a^2 + 4a + 3}$, and

$s(t+1) = \dfrac{2(t+1)}{(t+1)^2 - 1} = \dfrac{2(t+1)}{t^2 + 2t + 1 - 1} = \dfrac{2(t+1)}{t(t+2)}$.

9. $f(t) = \dfrac{2t^2}{\sqrt{t-1}}$. Therefore, $f(2) = \dfrac{2\left(2^2\right)}{\sqrt{2-1}} = 8$, $f(a) = \dfrac{2a^2}{\sqrt{a-1}}$, $f(x+1) = \dfrac{2(x+1)^2}{\sqrt{(x+1) - 1}} = \dfrac{2(x+1)^2}{\sqrt{x}}$,

and $f(x-1) = \dfrac{2(x-1)^2}{\sqrt{(x-1) - 1}} = \dfrac{2(x-1)^2}{\sqrt{x-2}}$.

11. For $x = -3 \le 0$, we calculate $f(-3) = (-3)^2 + 1 = 9 + 1 = 10$. For $x = 0 \le 0$, we calculate $f(0) = (0)^2 + 1 = 1$. For $x = 1 > 0$, we calculate $f(1) = \sqrt{1} = 1$.

13. For $x = -1 < 1$, $f(-1) = -\frac{1}{2}(-1)^2 + 3 = \frac{5}{2}$. For $x = 0 < 1$, $f(0) = -\frac{1}{2}(0)^2 + 3 = 3$. For $x = 1 \ge 1$, $f(1) = 2(1^2) + 1 = 3$. For $x = 2 \ge 1$, $f(2) = 2(2^2) + 1 = 9$.

15. **a.** $f(0) = -2$.

b. (i) $f(x) = 3$ when $x \approx 2$.
(ii) $f(x) = 0$ when $x = 1$.

c. $[0, 6]$

d. $[-2, 6]$

17. $g(2) = \sqrt{2^2 - 1} = \sqrt{3}$, so the point $\left(2, \sqrt{3}\right)$ lies on the graph of g.

19. $f(-2) = \dfrac{|-2-1|}{-2+1} = \dfrac{|-3|}{-1} = -3$, so the point $(-2, -3)$ does lie on the graph of f.

21. Because the point $(1, 5)$ lies on the graph of f it satisfies the equation defining f. Thus, $f(1) = 2(1)^2 - 4(1) + c = 5$, or $c = 7$.

23. Because $f(x)$ is a real number for any value of x, the domain of f is $(-\infty, \infty)$.

25. $f(x)$ is not defined at $x = 0$ and so the domain of f is $(-\infty, 0) \cup (0, \infty)$.

27. $f(x)$ is a real number for all values of x. Note that $x^2 + 1 \ge 1$ for all x. Therefore, the domain of f is $(-\infty, \infty)$.

29. Because the square root of a number is defined for all real numbers greater than or equal to zero, we have $5 - x \ge 0$, or $-x \ge -5$ and so $x \le 5$. (Recall that multiplying by -1 reverses the sign of an inequality.) Therefore, the domain of f is $(-\infty, 5]$.

31. The denominator of f is zero when $x^2 - 4 = 0$, or $x = \pm 2$. Therefore, the domain of f is $(-\infty, -2) \cup (-2, 2) \cup (2, \infty)$.

33. f is defined when $x + 3 \ge 0$, that is, when $x \ge -3$. Therefore, the domain of f is $[-3, \infty)$.

35. The numerator is defined when $1 - x \ge 0$, $-x \ge -1$ or $x \le 1$. Furthermore, the denominator is zero when $x = \pm 2$. Therefore, the domain is the set of all real numbers in $(-\infty, -2) \cup (-2, 1]$.

37. a. The domain of f is the set of all real numbers.

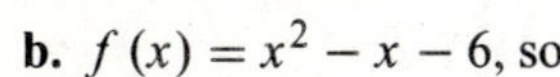

b. $f(x) = x^2 - x - 6$, so

$f(-3) = (-3)^2 - (-3) - 6 = 9 + 3 - 6 = 6$,

$f(-2) = (-2)^2 - (-2) - 6 = 4 + 2 - 6 = 0$,

$f(-1) = (-1)^2 - (-1) - 6 = 1 + 1 - 6 = -4$,

$f(0) = (0)^2 - (0) - 6 = -6$,

$f\left(\frac{1}{2}\right) = \left(\frac{1}{2}\right)^2 - \left(\frac{1}{2}\right) - 6 = \frac{1}{4} - \frac{2}{4} - \frac{24}{4} = -\frac{25}{4}$, $f(1) = (1)^2 - 1 - 6 = -6$,

$f(2) = (2)^2 - 2 - 6 = 4 - 2 - 6 = -4$, and $f(3) = (3)^2 - 3 - 6 = 9 - 3 - 6 = 0$.

c.

39. $f(x) = 2x^2 + 1$ has domain $(-\infty, \infty)$ and range $[1, \infty)$.

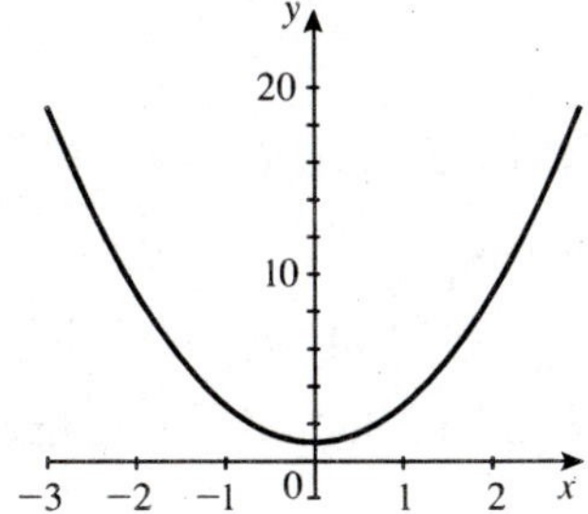

41. $f(x) = 2 + \sqrt{x}$ has domain $[0, \infty)$ and range $[2, \infty)$.

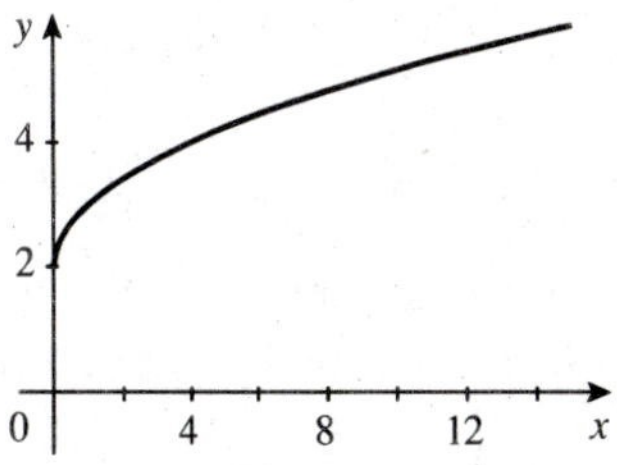

43. $f(x) = \sqrt{1 - x}$ has domain $(-\infty, 1]$ and range $[0, \infty)$

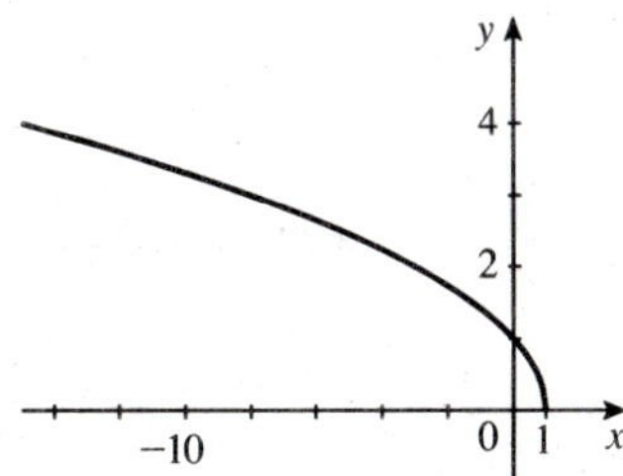

45. $f(x) = |x| - 1$ has domain $(-\infty, \infty)$ and range $[-1, \infty)$.

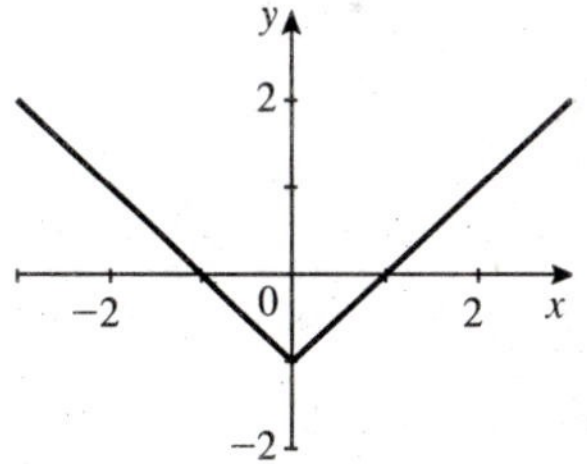

47. $f(x) = \begin{cases} x & \text{if } x < 0 \\ 2x + 1 & \text{if } x \ge 0 \end{cases}$ has domain $(-\infty, \infty)$ and range $(-\infty, 0) \cup [1, \infty)$.

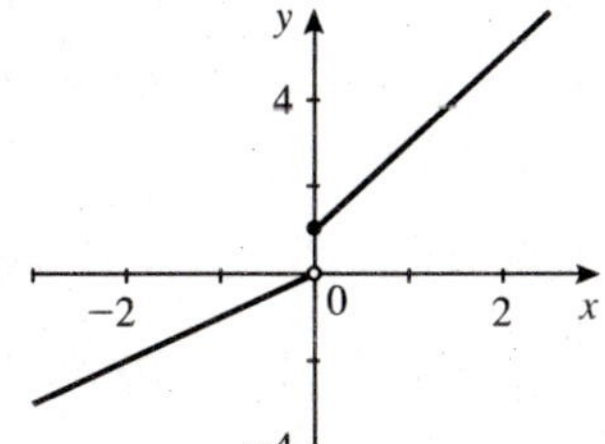

49. If $x \le 1$, the graph of f is the half-line $y = -x + 1$. For $x > 1$, we calculate a few points: $f(2) = 3$, $f(3) = 8$, and $f(4) = 15$. f has domain $(-\infty, \infty)$ and range $[0, \infty)$.

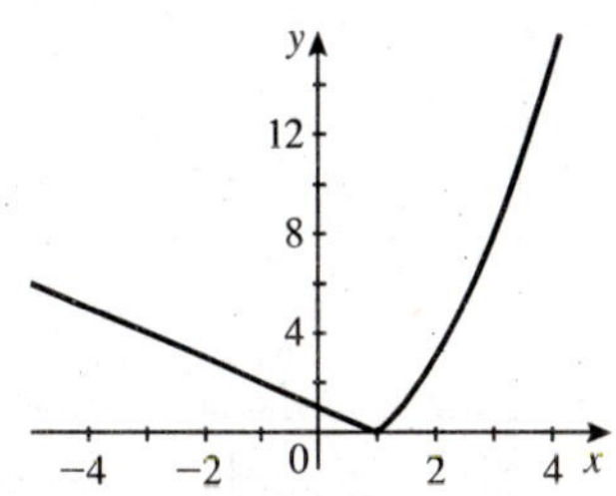

51. Each vertical line cuts the given graph at exactly one point, and so the graph represents y as a function of x.

53. Because there is a vertical line that intersects the graph at three points, the graph does not represent y as a function of x.

55. Each vertical line intersects the graph of f at exactly one point, and so the graph represents y as a function of x.

57. Each vertical line intersects the graph of f at exactly one point, and so the graph represents y as a function of x.

59. The circumference of a circle with a 5-inch radius is given by $C(5) = 2\pi(5) = 10\pi$, or 10π inches.

61. $\frac{4}{3}(\pi)(2r)^3 = \frac{4}{3}\pi 8r^3 = 8\left(\frac{4}{3}\pi r^3\right)$. Therefore, the volume of the tumor is increased by a factor of 8.

63. a. The slope of the straight line passing through the points (0, 0.58) and (20, 0.95) is $m_1 = \dfrac{0.95 - 0.58}{20 - 0} = 0.0185$, so an equation of the straight line passing through these two points is $y - 0.58 = 0.0185(t - 0)$ or $y = 0.0185t + 0.58$. Next, the slope of the straight line passing through the points (20, 0.95) and (30, 1.1) is $m_2 = \dfrac{1.1 - 0.95}{30 - 20} = 0.015$, so an equation of the straight line passing through the two points is $y - 0.95 = 0.015(t - 20)$ or $y = 0.015t + 0.65$. Therefore, a rule for f is

$$f(t) = \begin{cases} 0.0185t + 0.58 & \text{if } 0 \le t \le 20 \\ 0.015t + 0.65 & \text{if } 20 < t \le 30 \end{cases}$$

b. The ratios were changing at the rates of 0.0185/yr from 1960 through 1980 and 0.015/yr from 1980 through 1990.

c. The ratio was 1 when $t \approx 20.3$. This shows that the number of bachelor's degrees earned by women equaled the number earned by men for the first time around 1983.

65. $N(t) = -t^3 + 6t^2 + 15t$. Between 8 a.m. and 9 a.m., the average worker can be expected to assemble $N(1) - N(0) = (-1 + 6 + 15) - 0 = 20$, or 20 walkie-talkies. Between 9 a.m. and 10 a.m., we expect that $N(2) - N(1) = \left[-2^3 + 6(2^2) + 15(2)\right] - (-1 + 6 + 15) = 46 - 20 = 26$, or 26 walkie-talkies can be assembled by the average worker.

67. The amount spent in 2004 was $S(0) = 5.6$, or \$5.6 billion. The amount spent in 2008 was $S(4) = -0.03(4)^3 + 0.2(4)^2 + 0.23(4) + 5.6 = 7.8$, or \$7.8 billion.

69. a. The assets at the beginning of 2002 were \$0.6 trillion. At the beginning of 2003, they were $f(1) = 0.6$, or \$0.6 trillion.

b. The assets at the beginning of 2005 were $f(3) \approx 0.96$, or \$0.96 trillion. At the beginning of 2007, they were $f(5) \approx 1.20$, or \$1.2 trillion.

71. **a.** The amount of solids discharged in 1989 ($t = 0$) was 130 tons/day; in 1992 ($t = 3$), it was 100 tons/day; and in 1996 ($t = 7$), it was $f(7) = 1.25(7)^2 - 26.25(7) + 162.5 = 40$, or 40 tons/day.

b.

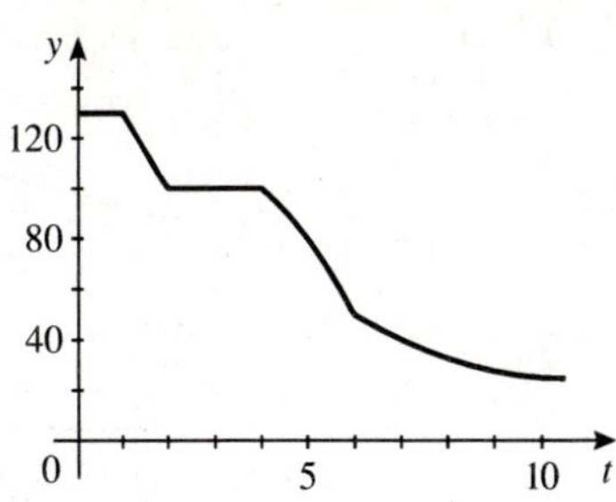

73. False. Take $f(x) = x^2$, $a = 1$, and $b = -1$. Then $f(1) = 1 = f(-1)$, but $a \neq b$.

75. False. It intersects the graph of a function in at most one point.

Technology Exercises

page 104

1. **a.**

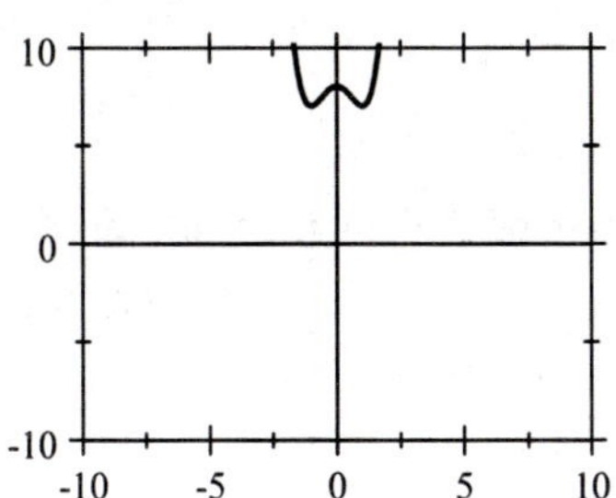

b.

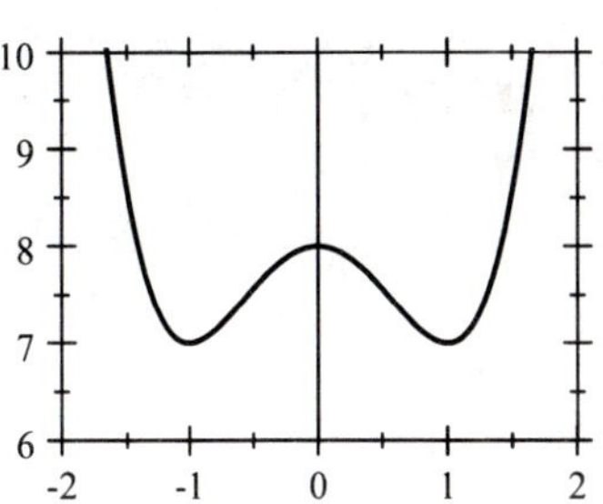

3. **a.**

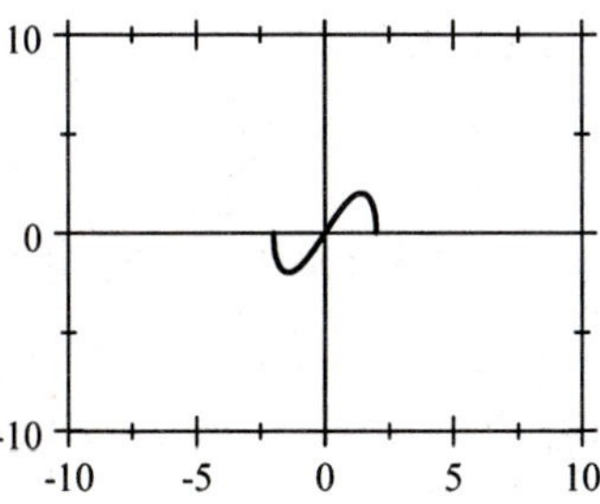

b.

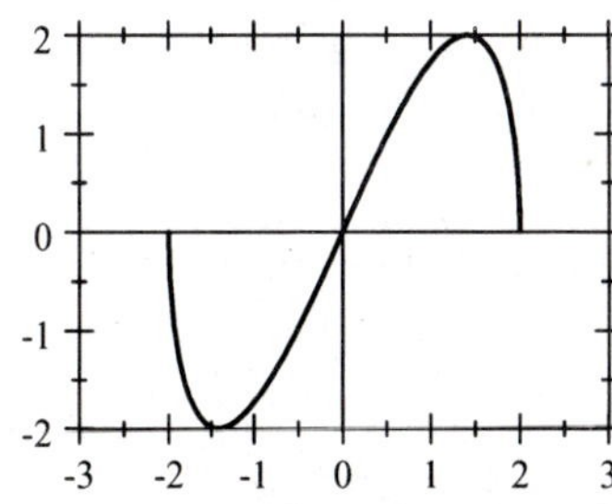

5.

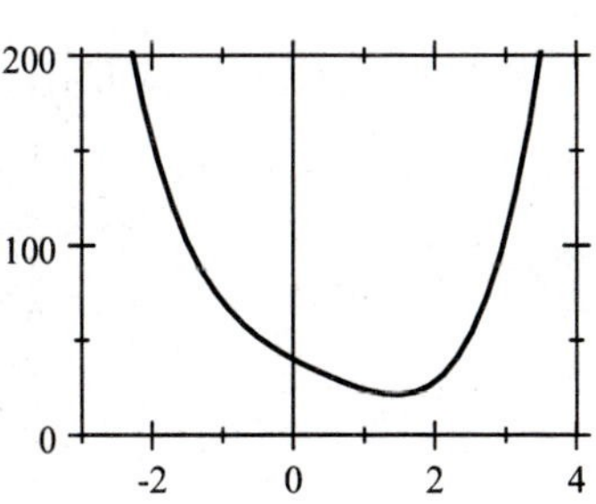

7.

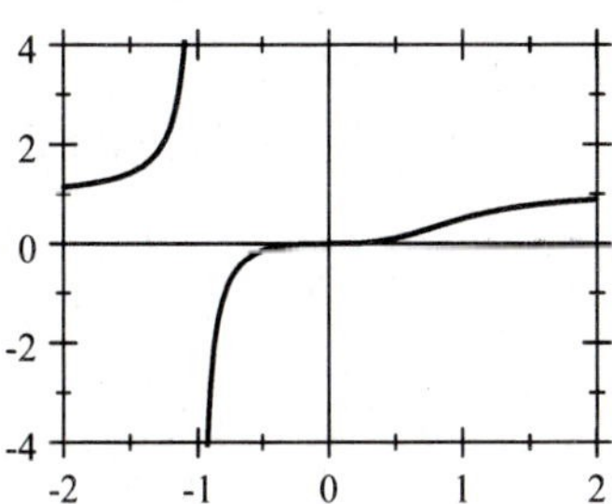

9. $f(2.145) \approx 18.5505$.

11. $f(2.41) \approx 4.1616$.

13. a.

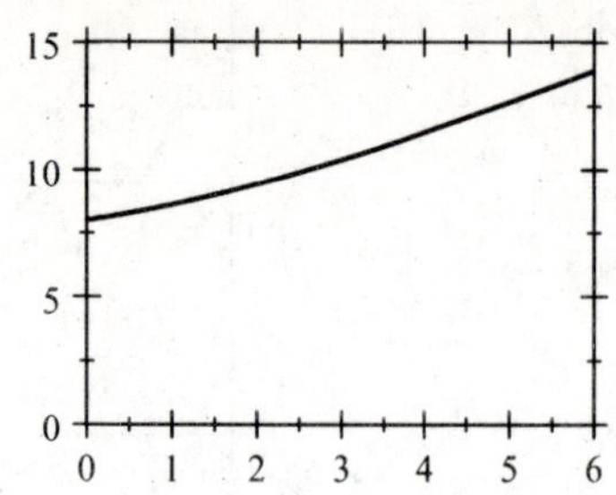

b. The amount spent in the year 2005 was $f(2) \approx 9.42$, or approximately \$9.4 billion. In 2009, it was $f(6) \approx 13.88$, or approximately \$13.9 billion.

15. a.

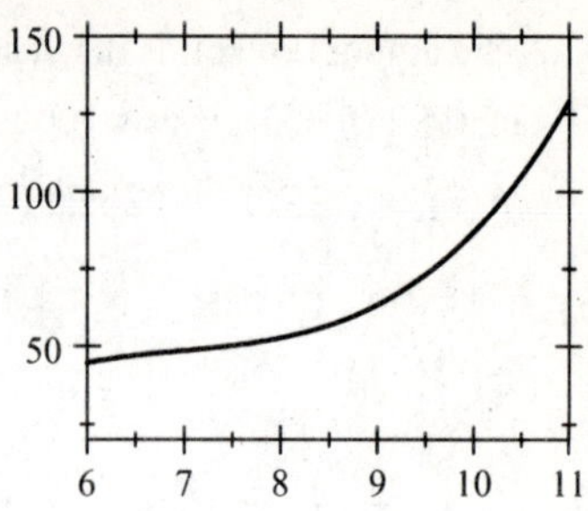

b. $f(6) = 44.7$, $f(8) = 52.7$, and $f(11) = 129.2$.

2.4 The Algebra of Functions

Problem-Solving Tips

When you come across new notation, make sure that you understand that notation. If you can't express the notation verbally, you haven't yet grasped its meaning. For example, in this section we introduced the notation $g \circ f$, read "g circle f."We use this notation to describe the composition of the functions g and f. Note that $g \circ f$ is not the same as $f \circ g$.

Here are some tips for solving the problems in the exercises that follow:

1. If f and g are functions with domains A and B, respectively, then the domain of $f + g$, $f - g$, and fg is $A \cap B$. The domain of the quotient f/g is $A \cap B$ excluding all numbers x such that $g(x) = 0$.

2. **To find the rule for the composite function** $g \circ f$, evaluate the function g at $f(x)$. Similarly, to find $f \circ g$, evaluate the function f at $g(x)$.

Concept Questions page 109

1. a. $P(x_1) = R(x_1) - C(x_1)$ gives the profit if x_1 units are sold.

b. $P(x_2) = R(x_2) - C(x_2)$. Because $P(x_2) < 0$, $|R(x_2) - C(x_2)| = -[R(x_2) - C(x_2)]$ gives the loss sustained if x_2 units are sold.

3. a. $y = (f + g)(x) = f(x) + g(x)$

b. $y = (f - g)(x) = f(x) - g(x)$

c. $y = (fg)(x) = f(x)g(x)$

d. $y = \left(\dfrac{f}{g}\right)(x) = \dfrac{f(x)}{g(x)}$

5. No. Let $A = (-\infty, \infty)$, $f(x) = x$, and $g(x) = \sqrt{x}$. Then $a = -1$ is in A, but $(g \circ f)(-1) = g(f(-1)) = g(-1) = \sqrt{-1}$ is not defined.

Exercises page 110

1. $(f+g)(x) = f(x) + g(x) = (x^3+5) + (x^2-2) = x^3 + x^2 + 3.$

3. $fg(x) = f(x)g(x) = (x^3+5)(x^2-2) = x^5 - 2x^3 + 5x^2 - 10.$

5. $\dfrac{f}{g}(x) = \dfrac{f(x)}{g(x)} = \dfrac{x^3+5}{x^2-2}.$

7. $\dfrac{fg}{h}(x) = \dfrac{f(x)g(x)}{h(x)} = \dfrac{(x^3+5)(x^2-2)}{2x+4} = \dfrac{x^5-2x^3+5x^2-10}{2x+4}.$

9. $(f+g)(x) = f(x) + g(x) = x - 1 + \sqrt{x+1}.$

11. $(fg)(x) = f(x)g(x) = (x-1)\sqrt{x+1}.$

13. $\dfrac{g}{h}(x) = \dfrac{g(x)}{h(x)} = \dfrac{\sqrt{x+1}}{2x^3-1}.$

15. $\dfrac{fg}{h}(x) = \dfrac{(x-1)(\sqrt{x+1})}{2x^3-1}.$

17. $\dfrac{f-h}{g}(x) = \dfrac{x-1-(2x^3-1)}{\sqrt{x+1}} = \dfrac{x-2x^3}{\sqrt{x+1}}.$

19. $(f+g)(x) = x^2+5+\sqrt{x}-2 = x^2+\sqrt{x}+3$, $(f-g)(x) = x^2+5-(\sqrt{x}-2) = x^2-\sqrt{x}+7$,
$(fg)(x) = (x^2+5)(\sqrt{x}-2)$, and $\left(\dfrac{f}{g}\right)(x) = \dfrac{x^2+5}{\sqrt{x}-2}.$

21. $(f+g)(x) = \sqrt{x+3} + \dfrac{1}{x-1} = \dfrac{(x-1)\sqrt{x+3}+1}{x-1}$, $(f-g)(x) = \sqrt{x+3} - \dfrac{1}{x-1} = \dfrac{(x-1)\sqrt{x+3}-1}{x-1}$,
$(fg)(x) = \sqrt{x+3}\left(\dfrac{1}{x-1}\right) = \dfrac{\sqrt{x+3}}{x-1}$, and $\left(\dfrac{f}{g}\right) = \sqrt{x+3}\,(x-1).$

23.
$$(f+g)(x) = \frac{x+1}{x-1} + \frac{x+2}{x-2} = \frac{(x+1)(x-2)+(x+2)(x-1)}{(x-1)(x-2)} = \frac{x^2-x-2+x^2+x-2}{(x-1)(x-2)}$$
$$= \frac{2x^2-4}{(x-1)(x-2)} = \frac{2(x^2-2)}{(x-1)(x-2)},$$
$$(f-g)(x) = \frac{x+1}{x-1} - \frac{x+2}{x-2} = \frac{(x+1)(x-2)-(x+2)(x-1)}{(x-1)(x-2)} = \frac{x^2-x-2-x^2-x+2}{(x-1)(x-2)}$$
$$= \frac{-2x}{(x-1)(x-2)},$$
$$(fg)(x) = \frac{(x+1)(x+2)}{(x-1)(x-2)}, \text{ and } \left(\frac{f}{g}\right)(x) = \frac{(x+1)(x-2)}{(x-1)(x+2)}.$$

25. $(f\circ g)(x) = f(g(x)) = f(x^2+1) = (x^2+1)^2 + (x^2+1) + 1 = (x^4+2x^2+1) + x^2 + 2 = x^4 + 3x^2 + 3$
and $(g\circ f)(x) = g(f(x)) = g(x^2+x+1) = (x^2+x+1)^2 + 1.$

27. $(f\circ g)(x) = f(g(x)) = f(x^2-1) = \sqrt{x^2-1} + 1$ and
$(g\circ f)(x) = g(f(x)) = g(\sqrt{x}+1) = (\sqrt{x}+1)^2 - 1 = x + 2\sqrt{x} + 1 - 1 = x + 2\sqrt{x}.$

29. $(f \circ g)(x) = f(g(x)) = f\left(\frac{1}{x}\right) = \frac{1}{x} \div \left(\frac{1}{x^2}+1\right) = \frac{1}{x} \cdot \frac{x^2}{x^2+1} = \frac{x}{x^2+1}$ and

$(g \circ f)(x) = g(f(x)) = g\left(\frac{x}{x^2+1}\right) = \frac{x^2+1}{x}$.

31. $h(2) = g(f(2))$. But $f(2) = 2^2+2+1 = 7$, so $h(2) = g(7) = 49$.

33. $h(2) = g(f(2))$. But $f(2) = \frac{1}{2(2)+1} = \frac{1}{5}$, so $h(2) = g\left(\frac{1}{5}\right) = \frac{1}{\sqrt{5}} = \frac{\sqrt{5}}{5}$.

35. $f(x) = 2x^3+x^2+1$, $g(x) = x^5$.

37. $f(x) = x^2-1$, $g(x) = \sqrt{x}$.

39. $f(x) = x^2-1$, $g(x) = \frac{1}{x}$.

41. $f(x) = 3x^2+2$, $g(x) = \frac{1}{x^{3/2}}$.

43. $f(a+h) - f(a) = [3(a+h)+4] - (3a+4) = 3a+3h+4-3a-4 = 3h$.

45. $f(a+h) - f(a) = 4-(a+h)^2 - (4-a^2) = 4-a^2-2ah-h^2-4+a^2 = -2ah-h^2 = -h(2a+h)$.

47. $$\frac{f(a+h)-f(a)}{h} = \frac{\left[2(a+h)^2+1\right]-\left(2a^2+1\right)}{h} = \frac{2a^2+4ah+2h^2+1-2a^2-1}{h} = \frac{4ah+2h^2}{h} = \frac{2h(2a+h)}{h} = 2(2a+h).$$

49. $$\frac{f(a+h)-f(a)}{h} = \frac{\left[(a+h)^3-(a+h)\right]-\left(a^3-a\right)}{h} = \frac{a^3+3a^2h+3ah^2+h^3-a-h-a^3+a}{h} = \frac{3a^2h+3ah^2+h^3-h}{h} = 3a^2+3ah+h^2-1.$$

51. $$\frac{f(a+h)-f(a)}{h} = \frac{\frac{1}{a+h}-\frac{1}{a}}{h} = \frac{\frac{a-(a+h)}{a(a+h)}}{h} = -\frac{1}{a(a+h)}.$$

53. $F(t)$ represents the total revenue for the two restaurants at time t.

55. $f(t)g(t)$ represents the dollar value of Nancy's holdings at time t.

57. $g \circ f$ is the function giving the amount of carbon monoxide pollution from cars in parts per million at time t.

59. $C(x) = 0.6x + 12{,}100$.

61. $D(t) = (D_2 - D_1)(t) = D_2(t) - D_1(t) = \left(0.035t^2+0.21t+0.24\right) - \left(0.0275t^2+0.081t+0.07\right)$

$= 0.0075t^2 + 0.129t + 0.17$.

The function D gives the difference in year t between the deficit without the \$160 million rescue package and the deficit with the rescue package.

63. a. $(g \circ f)(1) = g(f(1)) = g(406) = 23$. So in 2002, the percentage of reported serious crimes that end in arrests or in the identification of suspects was 23.

b. $(g \circ f)(6) = g(f(6)) = g(326) = 18$. In 2007, 18% of reported serious crimes ended in arrests or in the identification of suspects.

c. Between 2002 and 2007, the total number of detectives had dropped from 406 to 326 and as a result, the percentage of reported serious crimes that ended in arrests or in the identification of suspects dropped from 23 to 18.

65. a. $P(x) = R(x) - C(x) = -0.1x^2 + 500x - \left(0.000003x^3 - 0.03x^2 + 200x + 100{,}000\right)$

$= -0.000003x^3 - 0.07x^2 + 300x - 100{,}000.$

b. $P(1500) = -0.000003(1500)^3 - 0.07(1500)^2 + 300(1500) - 100{,}000 = 182{,}375$, or \$182,375.

67. a. The gap is $N(t) - C(t) = \left(3.5t^2 + 26.7t + 436.2\right) - (24.3t + 365) = 3.5t^2 + 2.4t + 71.2$.

b. At the beginning of 1983, the gap was $G(0) = 3.5(0)^2 + 2.4(0) + 71.2 = 71.2$, or 71,200.
At the beginning of 1986, the gap was $G(3) = 3.5(3)^2 + 2.4(3) + 71.2 = 109.9$, or 109,900.

69. a. The occupancy rate at the beginning of January is $r(0) = \frac{10}{81}(0)^3 - \frac{10}{3}(0)^2 + \frac{200}{9}(0) + 55 = 55$, or 55%.
$r(5) = \frac{10}{81}(5)^3 - \frac{10}{3}(5)^2 + \frac{200}{9}(5) + 55 \approx 98.2$, or approximately 98.2%.

b. The monthly revenue at the beginning of January is $R(r(0)) = R(55) = -\frac{3}{5000}(55)^3 + \frac{9}{50}(55)^2 \approx 444.68$, or approximately \$444,700.
The monthly revenue at the beginning of June is $R(r(5)) \approx R(98.2) = -\frac{3}{5000}(98.2)^3 + \frac{9}{50}(98.2)^2 \approx 1167.6$, or approximately \$1,167,600.

71. a. $s = f + g + h = (f + g) + h = f + (g + h)$. This suggests we define the sum s by $s(x) = (f + g + h)(x) = f(x) + g(x) + h(x)$.

b. Let f, g, and h define the revenue (in dollars) in week t of three branches of a store. Then its total revenue (in dollars) in week t is $s(t) = (f + g + h)(t) = f(t) + g(t) + h(t)$.

73. True. $(f + g)(x) = f(x) + g(x) = g(x) + f(x) = (g + f)(x)$.

75. False. Take $f(x) = \sqrt{x}$ and $g(x) = x + 1$. Then $(g \circ f)(x) = \sqrt{x} + 1$, but $(f \circ g)(x) = \sqrt{x + 1}$.

2.5 Linear Functions

Problem-Solving Tips

New mathematical terms in each section are defined either in blue boldface type or in green boxes. Each time you encounter a new term, read through the definition and then try to express the definition in your own words without looking at the book. Once you understand these definitions, it will be easier for you to work the exercise sets that follow each section.

Here are some hints for solving the problems in the exercises that follow:

1. **To determine whether a given equation defines y as a linear function of x**, check to see that the given equation has the form $Ax + By + C = 0$, where A, B, and C are constants and A and B are not both zero.

2. Because the demand for a commodity decreases as its unit price increases, **a demand function is generally a decreasing function**. Thus, a linear demand function has a negative slope and its graph slants downwards as we move from left to right along the x-axis. Similarly, because the supply of a commodity increases as the unit price increases, a supply function is generally an increasing function, and so a linear supply function has positive slope and its graph slants upwards as we move from left to right along the x-axis.

Concept Questions page 119

1. **a.** A linear function is a function of the form $f(x) = mx + b$, where m and b are constants. For example, $f(x) = 2x + 3$ is a linear function.

   b. The domain and range of a linear function are both $(-\infty, \infty)$.

   c. The graph of a linear function is a straight line.

3. **a.** The break-even point $P_0(x_0, p_0)$ is the solution of the simultaneous equations $p = R(x)$ and $p = C(x)$.

   b. The number x_0 is called the break-even quantity.

   c. The number p_0 is called the break-even revenue.

Exercises page 119

1. Yes. Solving for y in terms of x, we find $3y = -2x + 6$, or $y = -\frac{2}{3}x + 2$.

3. Yes. Solving for y in terms of x, we find $2y = x + 4$, or $y = \frac{1}{2}x + 2$.

5. Yes. Solving for y in terms of x, we have $4y = 2x + 9$, or $y = \frac{1}{2}x + \frac{9}{4}$.

7. y is not a linear function of x because of the quadratic term $2x^2$.

9. y is not a linear function of x because of the nonlinear term $-3y^2$.

11. **a.** $C(x) = 8x + 40{,}000$, where x is the number of units produced.

    b. $R(x) = 12x$, where x is the number of units sold.

    c. $P(x) = R(x) - C(x) = 12x - (8x + 40{,}000) = 4x - 40{,}000$.

    d. $P(8000) = 4(8000) - 40{,}000 = -8000$, or a loss of \$8,000. $P(12{,}000) = 4(12{,}000) - 40{,}000 = 8000$, or a profit of \$8000.

13. $f(0) = 4$ gives $m(0) + b = 4$, or $b = 4$. Thus, $f(x) = mx + 4$. Next, $f(3) = -2$ gives $m(3) + 4 = -2$, or $m = -2$.

15. We solve the system $y = 3x + 4$, $y = -2x + 19$. Substituting the first equation into the second yields $3x + 4 = -2x + 19$, $5x = 15$, and $x = 3$. Substituting this value of x into the first equation yields $y = 3(3) + 4$, so $y = 13$. Thus, the point of intersection is $(3, 13)$.

17. We solve the system $2x - 3y = 6$, $3x + 6y = 16$. Solving the first equation for y, we obtain $3y = 2x - 6$, so $y = \frac{2}{3}x - 2$ .Substituting this value of y into the second equation, we obtain $3x + 6\left(\frac{2}{3}x - 2\right) = 16$, $3x + 4x - 12 = 16$, $7x = 28$, and $x = 4$. Then $y = \frac{2}{3}(4) - 2 = \frac{2}{3}$, so the point of intersection is $\left(4, \frac{2}{3}\right)$.

19. We solve the system $y = \frac{1}{4}x - 5$, $2x - \frac{3}{2}y = 1$. Substituting the value of y given in the first equation into the second equation, we obtain $2x - \frac{3}{2}\left(\frac{1}{4}x - 5\right) = 1$, so $2x - \frac{3}{8}x + \frac{15}{2} = 1$, $16x - 3x + 60 = 8$, $13x = -52$, and $x = -4$. Substituting this value of x into the first equation, we have $y = \frac{1}{4}(-4) - 5 = -1 - 5$, so $y = -6$. Therefore, the point of intersection is $(-4, -6)$.

21. We solve the equation $R(x) = C(x)$, or $15x = 5x + 10{,}000$, obtaining $10x = 10{,}000$, or $x = 1000$. Substituting this value of x into the equation $R(x) = 15x$, we find $R(1000) = 15{,}000$. Therefore, the break-even point is $(1000, 15000)$.

23. We solve the equation $R(x) = C(x)$, or $0.4x = 0.2x + 120$, obtaining $0.2x = 120$, or $x = 600$. Substituting this value of x into the equation $R(x) = 0.4x$, we find $R(600) = 240$. Therefore, the break-even point is $(600, 240)$.

25. Let V be the book value of the office building after 2005. Since $V = 1{,}000{,}000$ when $t = 0$, the line passes through $(0, 1000000)$. Similarly, when $t = 50$, $V = 0$, so the line passes through $(50, 0)$. Then the slope of the line is given by $m = \dfrac{0 - 1{,}000{,}000}{50 - 0} = -20{,}000$. Using the point-slope form of the equation of a line with the point $(0, 1000000)$, we have $V - 1{,}000{,}000 = -20{,}000(t - 0)$, or $V = -20{,}000t + 1{,}000{,}000$.
In 2010, $t = 5$ and $V = -20{,}000(5) + 1{,}000{,}000 = 900{,}000$, or \$900,000.
In 2015, $t = 10$ and $V = -20{,}000(10) + 1{,}000{,}000 = 800{,}000$, or \$800,000.

27. **a.** $y = I(x) = 1.033x$, where x is the monthly benefit before adjustment and y is the adjusted monthly benefit.

b. His adjusted monthly benefit is $I(1520) = 1.033(1520) = 1570.16$, or \$1570.16.

29. Let the number of tapes produced and sold be x. Then $C(x) = 12{,}100 + 0.60x$, $R(x) = 1.15x$, and $P(x) = R(x) - C(x) = 1.15x - (12{,}100 + 0.60x) = 0.55x - 12{,}100$.

31. Let the value of the workcenter system after t years be V. When $t = 0$, $V = 60{,}000$ and when $t = 4$, $V = 12{,}000$.

a. Since $m = \dfrac{12{,}000 - 60{,}000}{4} = -\dfrac{48{,}000}{4} = -12{,}000$, the rate of depreciation $(-m)$ is \$12,000/yr.

b. Using the point-slope form of the equation of a line with the point $(4, 12000)$, we have $V - 12{,}000 = -12{,}000(t - 4)$, or $V = -12{,}000t + 60{,}000$.

c.

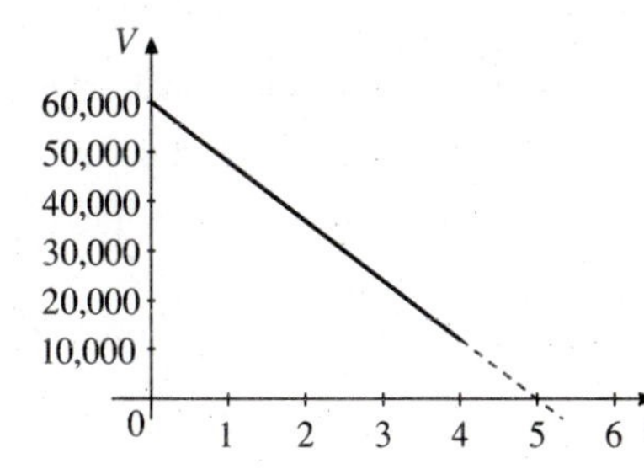

d. When $t = 3$, $V = -12{,}000(3) + 60{,}000 = 24{,}000$, or \$24,000.

33. The formula given in Exercise 32 is $V = C - \dfrac{C - S}{N}t$. When $C = 1{,}000{,}000$, $N = 50$, and $S = 0$, we have $V = 1{,}000{,}000 - \dfrac{1{,}000{,}000 - 0}{50}t$, or $V = 1{,}000{,}000 - 20{,}000t$. In 2010, $t = 5$ and $V = 1{,}000{,}000 - 20{,}000(5) = 900{,}000$, or \$900,000. In 2015, $t = 10$ and $V = 1{,}000{,}000 - 20{,}000(10) = 800{,}000$, or \$800,000.

35. **a.** $D(S) = \frac{Sa}{1.7}$. If we think of D as having the form $D(S) = mS + b$, then $m = \frac{a}{1.7}$, $b = 0$, and D is a linear function of S.

b. $D(0.4) = \frac{500(0.4)}{1.7} \approx 117.647$, or approximately 117.65 mg.

37. **a.** $f(t) = 6.5t + 20$, where $0 \le t \le 8$.

b. $f(8) = 6.5(8) + 20 = 72$, or 72 million.

39. **a.** Since the relationship is linear, we can write $F = mC + b$, where m and b are constants. Using the condition $C = 0$ when $F = 32$, we have $32 = b$, and so $F = mC + 32$. Next, using the condition $C = 100$ when $F = 212$, we have $212 = 100m + 32$, or $m = \frac{9}{5}$. Therefore, $F = \frac{9}{5}C + 32$.

b. From part a, we have $F = \frac{9}{5}C + 32$. When $C = 20$, $F = \frac{9}{5}(20) + 32 = 68$, and so the temperature equivalent to 20° C is 68° F.

c. Solving for C in terms of F, we find $\frac{9}{5}C = F - 32$, or $C = \frac{5}{9}F - \frac{160}{9}$. When $F = 70$, $C = \frac{5}{9}(70) - \frac{160}{9} = \frac{190}{9}$, or approximately 21.1° C.

41. **a.** $C(x) = 8x + 48{,}000$ and $R(x) = 14x$.

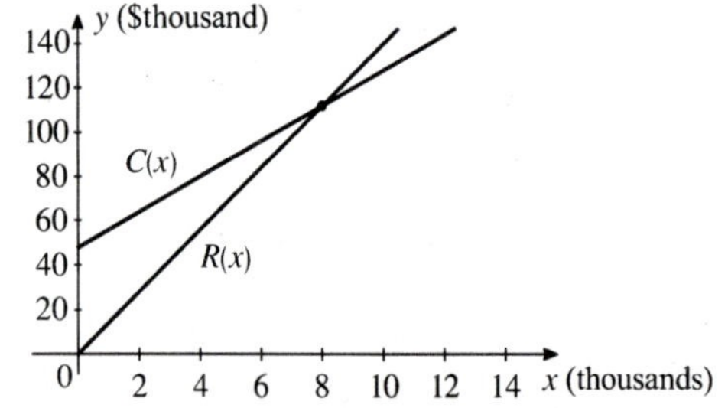

c. $P(x) = R(x) - C(x) = 14x - 8x - 48{,}000$

$= 6x - 48{,}000$.

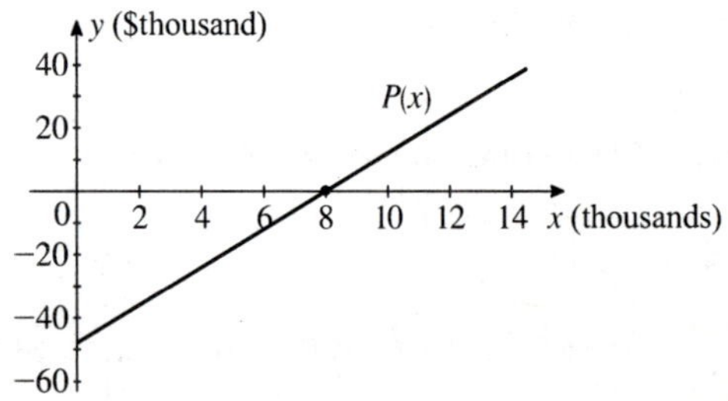

b. We solve the equation $R(x) = C(x)$ or $14x = 8x + 48{,}000$, obtaining $6x = 48{,}000$, so $x = 8000$. Substituting this value of x into the equation $R(x) = 14x$, we find $R(8000) = 14(8000) = 112{,}000$. Therefore, the break-even point is (8000, 112000).

d. The graph of the profit function crosses the x-axis when $P(x) = 0$, or $6x = 48{,}000$ and $x = 8000$. This means that the revenue is equal to the cost when 8000 units are produced and consequently the company breaks even at this point.

43. Let x denote the number of units sold. Then, the revenue function R is given by $R(x) = 9x$. Since the variable cost is 40% of the selling price and the monthly fixed costs are \$50,000, the cost function C is given by $C(x) = 0.4(9x) + 50{,}000 = 3.6x + 50{,}000$. To find the break-even point, we set $R(x) = C(x)$, obtaining $9x = 3.6x + 50{,}000$, $5.4x = 50{,}000$, and $x \approx 9259$, or 9259 units. Substituting this value of x into the equation $R(x) = 9x$ gives $R(9259) = 9(9259) = 83{,}331$. Thus, for a break-even operation, the firm should manufacture 9259 bicycle pumps, resulting in a break-even revenue of \$83,331.

45. a. The cost function associated with using machine I is $C_1(x) = 18{,}000 + 15x$. The cost function associated with using machine II is $C_2(x) = 15{,}000 + 20x$.

b.

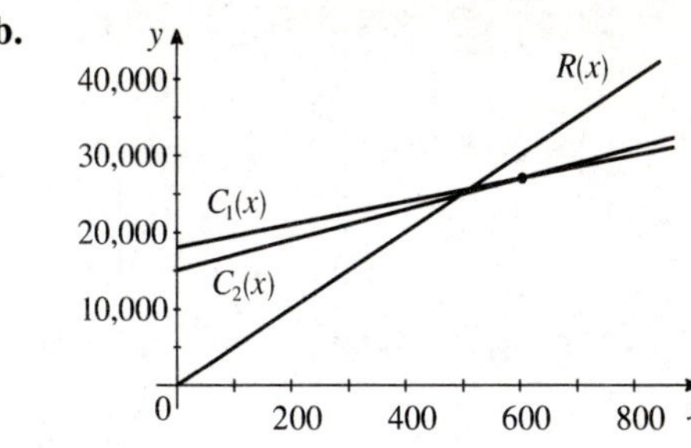

c. Comparing the cost of producing 450 units on each machine, we find $C_1(450) = 18{,}000 + 15(450) = 24{,}750$ or \$24,750 on machine I, and $C_2(450) = 15{,}000 + 20(450) = 24{,}000$ or \$24,000 on machine II. Therefore, machine II should be used in this case. Next, comparing the costs of producing 550 units on each machine, we find $C_1(550) = 18{,}000 + 15(550) = 26{,}250$ or \$26,250 on machine I, and $C_2(550) = 15{,}000 + 20(550) = 26{,}000$, or \$26,000 on machine II. Therefore, machine II should be used in this instance. Once again, we compare the cost of producing 650 units on each machine and find that $C_1(650) = 18{,}000 + 15(650) = 27{,}750$, or \$27,750 on machine I and $C_2(650) = 15{,}000 + 20(650) = 28{,}000$, or \$28,000 on machine II. Therefore, machine I should be used in this case.

d. We use the equation $P(x) = R(x) - C(x)$ and find $P(450) = 50(450) - 24{,}000 = -1500$, indicating a loss of \$1500 when machine II is used to produce 450 units. Similarly, $P(550) = 50(550) - 26{,}000 = 1500$, indicating a profit of \$1500 when machine II is used to produce 550 units. Finally, $P(650) = 50(650) - 27{,}750 = 4750$, for a profit of \$4750 when machine I is used to produce 650 units.

47. We solve the two equations simultaneously, obtaining $18t + 13.4 = -12t + 88$, $30t = 74.6$, and $t \approx 2.487$, or approximately 2.5 years. So shipments of LCDs will first overtake shipments of CRTs just before mid-2003.

49. a.

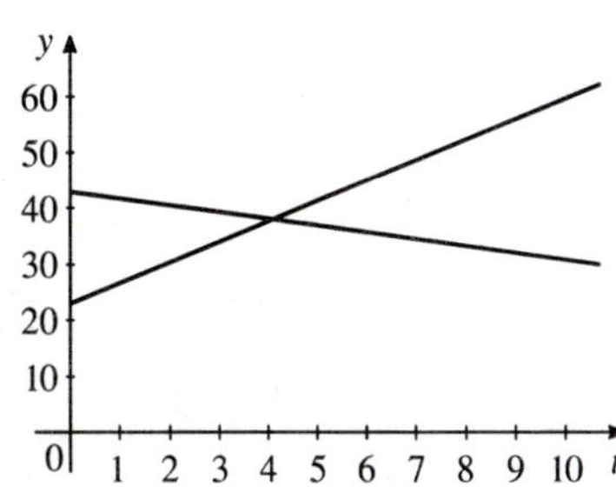

b. We solve the two equations simultaneously, obtaining $\frac{11}{3}t + 23 = -\frac{11}{9}t + 43$, $\frac{44}{9}t = 20$, and $t \approx 4.09$. Thus, electronic transactions first exceeded check transactions in early 2005.

51. True. The slope of the line is $-a$.

Technology Exercises page 126

1. 2.2875 **3.** 2.880952381 **5.** 7.2851648352 **7.** 2.4680851064

2.6 Quadratic Functions

Problem-Solving Tips

The graph of a quadratic function $f(x) = ax^2+bx+c = 0$ (where $a \neq 0$) is a parabola with vertex $\left(-\frac{b}{2a}, f\left(-\frac{b}{2a}\right)\right)$.

If $a > 0$, the parabola opens upward and its vertex is the lowest point (minimum) on the parabola.

If $a < 0$, then the parabola opens downward and its vertex is the highest point (maximum) on the parabola.

Concept Questions page 132

1. a. $(-\infty, \infty)$. **b.** It opens upward. **c.** $\left(-\frac{b}{2a}, f\left(-\frac{b}{2a}\right)\right)$. **d.** $-\frac{b}{2a}$.

Exercises page 132

1. $f(x) = x^2 + x - 6$; $a = 1$, $b = 1$, and $c = -6$. The x-coordinate of the vertex is $\frac{-b}{2a} = -\frac{1}{2(1)} = -\frac{1}{2}$ and the y-coordinate is $f\left(-\frac{1}{2}\right) = \left(-\frac{1}{2}\right)^2 + \left(-\frac{1}{2}\right) - 6 = -\frac{25}{4}$. Therefore, the vertex is $\left(-\frac{1}{2}, -\frac{25}{4}\right)$. Setting $x^2 + x - 6 = (x+3)(x-2) = 0$ gives -3 and 2 as the x-intercepts.

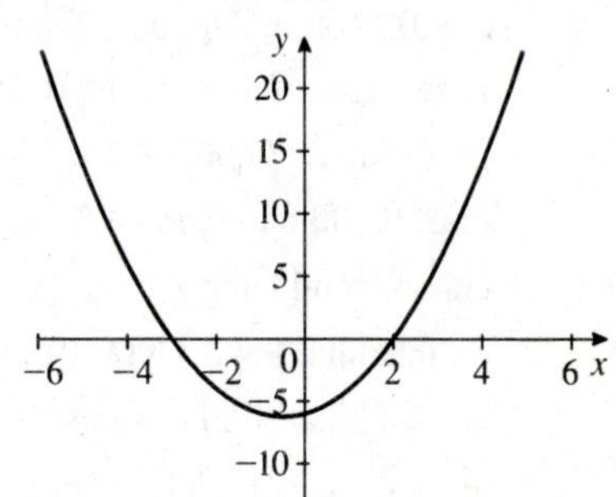

3. $f(x) = x^2 - 4x + 4$; $a = 1$, $b = -4$, and $c = 4$. The x-coordinate of the vertex is $\frac{-b}{2a} = -\frac{(-4)}{2} = 2$ and the y-coordinate is $f(2) = 2^2 - 4(2) + 4 = 0$.Therefore, the vertex is $(2, 0)$. Setting $x^2 - 4x + 4 = (x-2)^2 = 0$ gives 2 as the x-intercept.

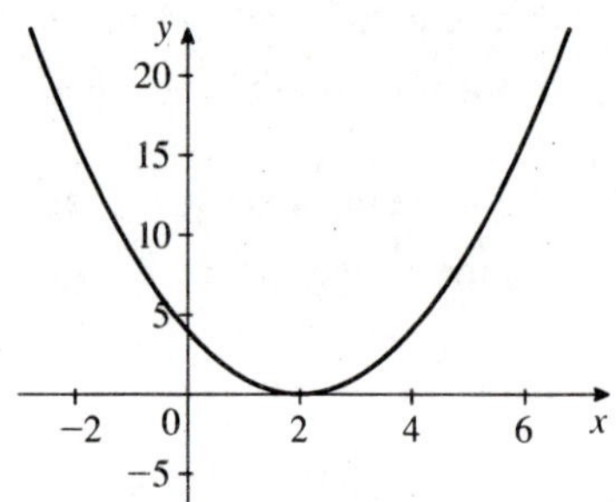

5. $f(x) = -x^2 + 5x - 6$; $a = -1$, $b = 5$, and $c = -6$. The x-coordinate of the vertex is $\frac{-b}{2a} = -\frac{5}{2(-1)} = \frac{5}{2}$ and the y-coordinate is $f\left(\frac{5}{2}\right) = -\left(\frac{5}{2}\right)^2 + 5\left(\frac{5}{2}\right) - 6 = \frac{1}{4}$.Therefore, the vertex is $\left(\frac{5}{2}, \frac{1}{4}\right)$. Setting $-x^2 + 5x - 6 = 0$ or $x^2 - 5x + 6 = (x-3)(x-2) = 0$ gives 2 and 3 as the x-intercepts.

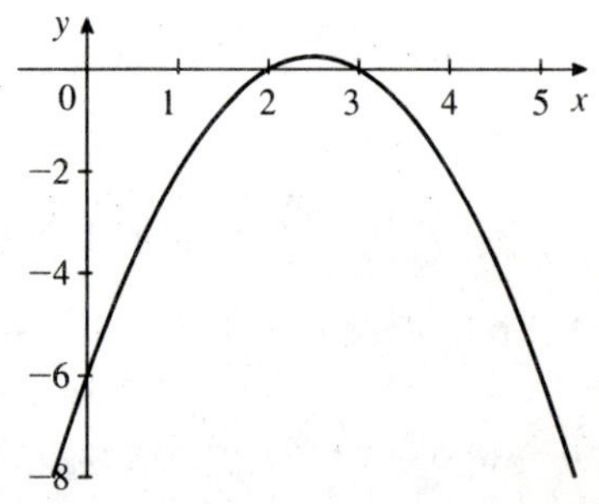

7. $f(x) = 3x^2 - 5x + 1$; $a = 3$, $b = -5$, and $c = 1$; The x-coordinate of the vertex is $\frac{-b}{2a} = -\frac{(-5)}{2(3)} = \frac{5}{6}$ and the y-coordinate is $f\left(\frac{5}{6}\right) = 3\left(\frac{5}{6}\right)^2 - 5\left(\frac{5}{6}\right) + 1 = -\frac{13}{12}$. Therefore, the vertex is $\left(\frac{5}{6}, -\frac{13}{12}\right)$. Next, solving $3x^2 - 5x + 1 = 0$, we use the quadratic formula and obtain

$$x = \frac{-(-5) \pm \sqrt{(-5)^2 - 4(3)(1)}}{2(3)} = \frac{5 \pm \sqrt{13}}{6}$$ and so the x-intercepts are 0.23241 and 1.43426.

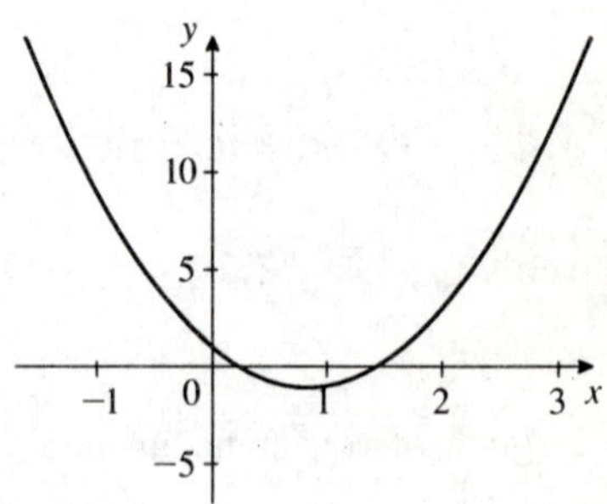

9. $f(x) = 2x^2 - 3x + 3$; $a = 2$, $b = -3$, and $c = 3$.The x-coordinate of the vertex is $\frac{-b}{2a} = -\frac{(-3)}{2(2)} = \frac{3}{4}$ and the y-coordinate is $f\left(\frac{3}{4}\right) = 2\left(\frac{3}{4}\right)^2 - 3\left(\frac{3}{4}\right) + 3 = \frac{15}{8}$. Therefore, the vertex is $\left(\frac{3}{4}, \frac{15}{8}\right)$. Next, observe that the discriminant of the quadratic equation $2x^2 - 3x + 3 = 0$ is $(-3)^2 - 4(2)(3) = 9 - 24 = -15 < 0$ and so it has no real roots. In other words, there are no x-intercepts.

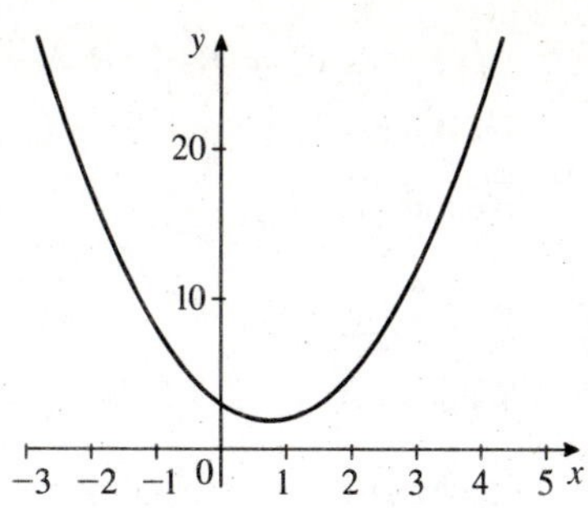

11. $f(x) = x^2 - 4$; $a = 1$, $b = 0$, and $c = -4$. The x-coordinate of the vertex is $\frac{-b}{2a} = -\frac{0}{2(1)} = 0$ and the y-coordinate is $f(0) = -4$. Therefore, the vertex is $(0, -4)$. The x-intercepts are found by solving $x^2 - 4 = (x + 2)(x - 2) = 0$ giving $x = -2$ or $x = 2$.

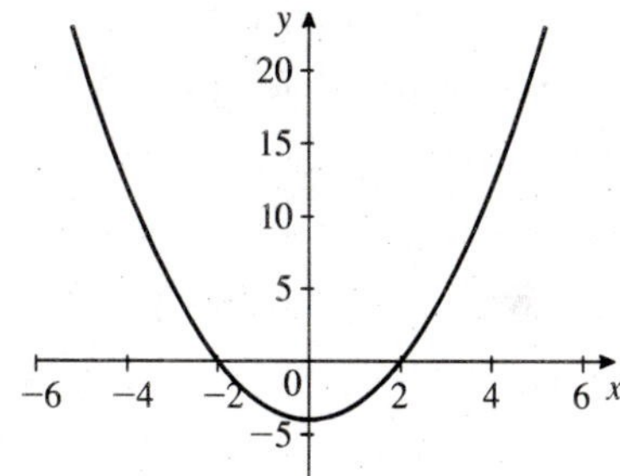

13. $f(x) = 16 - x^2$; $a = -1$, $b = 0$, and $c = 16$. The x-coordinate of the vertex is $\frac{-b}{2a} = -\frac{0}{2(-1)} = 0$ and the y-coordinate is $f(0) = 16$. Therefore, the vertex is $(0, 16)$. The x-intercepts are found by solving $16 - x^2 = 0$, giving $x = -4$ or $x = 4$.

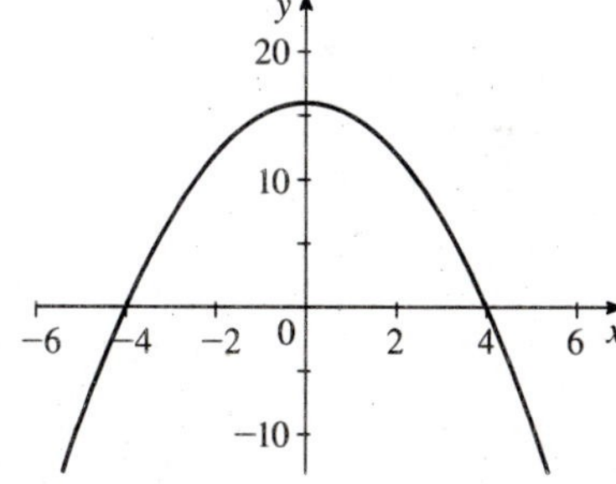

15. $f(x) = \frac{3}{8}x^2 - 2x + 2$; $a = \frac{3}{8}$, $b = -2$, and $c = 2$. The x-coordinate of the vertex is $\frac{-b}{2a} = -\frac{(-2)}{2\left(\frac{3}{8}\right)} = \frac{8}{3}$ and the y-coordinate is $f\left(\frac{8}{3}\right) = \frac{3}{8}\left(\frac{8}{3}\right)^2 - 2\left(\frac{8}{3}\right) + 2 = -\frac{2}{3}$. Therefore, the vertex is $\left(\frac{8}{3}, -\frac{2}{3}\right)$. The equation $f(x) = 0$ can be written $3x^2 - 16x + 16 = (3x - 4)(x - 4) = 0$ giving $x = \frac{4}{3}$ or $x = 4$ and so the x-intercepts are $\frac{4}{3}$ and 4.

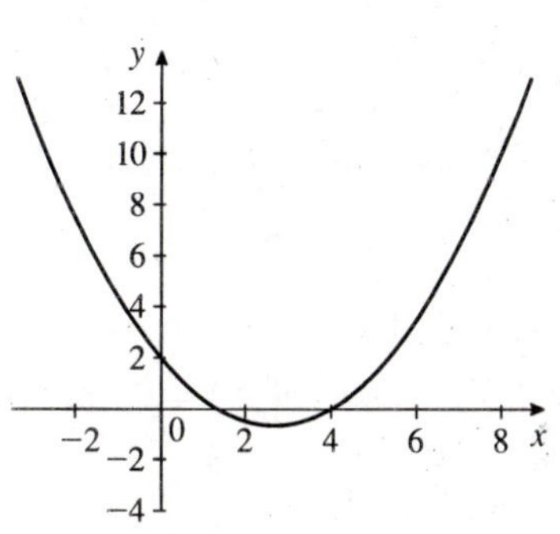

17. $f(x) = 1.2x^2 + 3.2x - 1.2$, so $a = 1.2$, $b = 3.2$, and $c = -1.2$. The x-coordinate of the vertex is $\frac{-b}{2a} = -\frac{3.2}{2(1.2)} = -\frac{4}{3}$ and the y-coordinate is $f\left(-\frac{4}{3}\right) = 1.2\left(-\frac{4}{3}\right)^2 + 3.2\left(-\frac{4}{3}\right)(1) - 1.2 = -\frac{10}{3}$. Therefore, the vertex is $\left(-\frac{4}{3}, -\frac{10}{3}\right)$. Next, we solve $f(x) = 0$ using the quadratic formula, obtaining

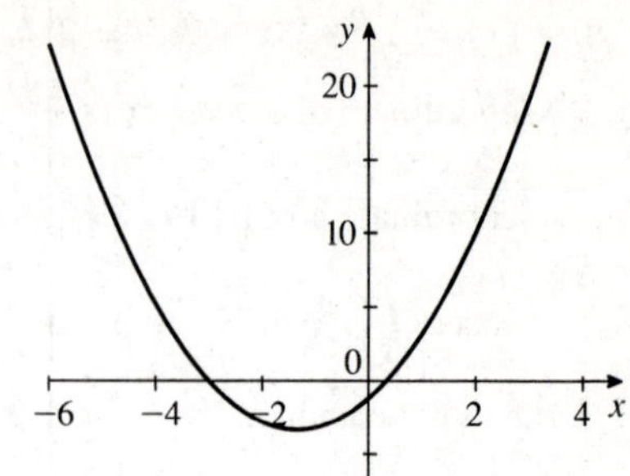

$$x = \frac{-3.2 \pm \sqrt{(3.2)^2 - 4(1.2)(-1.2)}}{2(1.2)} = \frac{-3.2 \pm \sqrt{16}}{2(1.2)} = \frac{-3.2 \pm 4}{2(1.2)} = -3$$

or $\frac{1}{3}$. Therefore, the x-intercepts are -3 and $\frac{1}{3}$.

19. We solve the equation $-x^2 + 4 = x - 2$. Rewriting, we have $x^2 + x - 6 = (x+3)(x-2) = 0$, giving $x = -3$ or $x = 2$. Therefore, the points of intersection are $(-3, -5)$ and $(2, 0)$.

21. We solve $-x^2 + 2x + 6 = x^2 - 6$, or $2x^2 - 2x - 12 = 0$. Rewriting, we have $x^2 - x - 6 = (x-3)(x+2) = 0$, giving $x = -2$ or 3. Therefore, the points of intersection are $(-2, -2)$ and $(3, 3)$.

23. We solve $2x^2 - 5x - 8 = -3x^2 + x + 5$, or $5x^2 - 6x - 13 = 0$. Using the quadratic formula, we obtain $x = \frac{-(-6) \pm \sqrt{(-6)^2 - 4(5)(-13)}}{2(5)} = \frac{6 \pm \sqrt{296}}{10} \approx -1.12047$ or 2.32047. Next, we find $f(-1.12047) = 2(-1.12047)^2 - 5(-1.12047) - 8 \approx 0.11326$ and $f(2.32047) = 2(2.32047)^2 - 5(2.32047) - 8 \approx -8.8332$. Therefore, the points of intersection are $(-1.1205, 0.1133)$ and $(2.3205, -8.8332)$.

25. a.

b. If $p = 11$, we have $11 = -x^2 + 36$, or $x^2 = 25$, so that $x = \pm 5$. Therefore, the quantity demanded when the unit price is \$11 is 5000 units.

27. a.

b. If $x = 2$, then $p = 2(2)^2 + 18 = 26$, or \$26.

29. We solve the equation $-2x^2 + 80 = 15x + 30$, or $-2x^2 + 80 = 15x + 30$, or $2x^2 + 15x - 50 = 0$, for x. Thus, $(2x-5)(x+10) = 0$, so $x = \frac{5}{2}$ or $x = -10$. Rejecting the negative root, we have $x = \frac{5}{2}$. The corresponding value of p is $p = -2\left(\frac{5}{2}\right)^2 + 80 = 67.5$. We conclude that the equilibrium quantity is 2500 and the equilibrium price is \$67.50.

31. Solving both equations for x, we have $x = -\frac{11}{3}p + 22$ and $x = 2p^2 + p - 10$. Equating the right-hand sides of these two equations, we have $-\frac{11}{3}p + 22 = 2p^2 + p - 10$, $-11p + 66 = 6p^2 + 3p - 30$, and $6p^2 + 14p - 96 = 0$. Dividing this last equation by 2 and then factoring, we have $(3p + 16)(p - 3) = 0$, so discarding the negative root $p = -\frac{16}{3}$, we conclude that $p = 3$. The corresponding value of x is $2(3)^2 + 3 - 10 = 11$. Thus, the equilibrium quantity is 11,000 and the equilibrium price is \$3.

33. **a.** $N(0) = 3.6$, or 3.6 million people; $N(25) = 0.0031(25)^2 + 0.16(25) + 3.6 = 9.5375$, or approximately 9.5 million people.

b. $N(30) = 0.0031(30)^2 + 0.16(30) + 3.6 = 11.19$, or approximately 11.2 million people.

35. **a.**

b. The time at which the stone reaches the highest point is given by the t-coordinate of the vertex of the parabola. This is $\frac{-b}{2a} = -\frac{64}{2(-16)} = 2$, so the stone it reaches its maximum height 2 seconds after it was thrown. Its maximum height is given by $h(2) = -16(2)^2 + 64(2) + 80 = 144$, or 144 ft.

37. $P(x) = -0.04x^2 + 240x - 10{,}000$. The optimal production level is given by the x-coordinate of the vertex of parabola; that is, by $\frac{-b}{2a} = -\frac{240}{2(-0.04)} = 3000$, or 3000 cameras.

39. **a.** $R(p) = -\frac{1}{2}p^2 + 30p$.

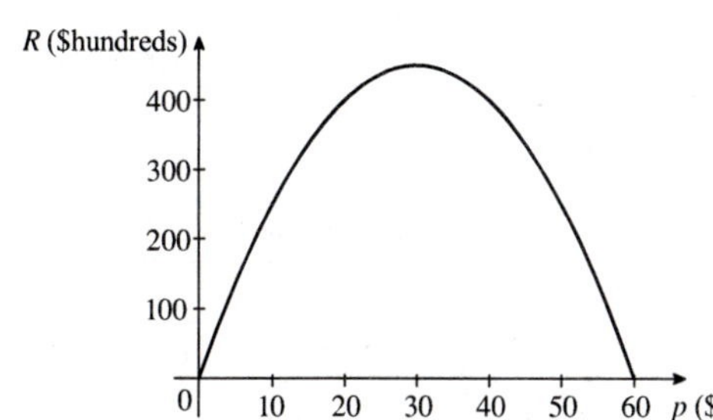

b. The monthly revenue is maximized when $p = -\frac{30}{2\left(-\frac{1}{2}\right)} = 30$; that is, when the unit price is \$30.

41. Equating the right-hand sides of the two equations, we have $0.1x^2 + 2x + 20 = -0.1x^2 - x + 40$, so $0.2x^2 + 3x - 20 = 0$, $2x^2 + 30x - 200 = 0$, $x^2 + 15x - 100 = 0$, and $(x + 20)(x - 5) = 0$. Thus, $x = -20$ or $x = 5$. Discarding the negative root and substituting $x = 5$ into the first equation, we obtain $p = -0.1(25) - 5 + 40 = 32.5$. Therefore, the equilibrium quantity is 500 tents and the equilibrium price is \$32.50.

43. **a.**

b. $v(r) = -1000r^2 + 40$. Its graph is a parabola, as shown in part a. $v(r)$ has a maximum value at $r = -\frac{0}{2(-1000)} = 0$ and a minimum value at $r = 0.2$ (r must be nonnegative). Thus the velocity of blood is greatest along the central artery (where $r = 0$) and smallest along the wall of the artery (where $r = 0.2$). The maximum velocity is $v(0) = 40$ cm/sec and the minimum velocity is $v(0.2) = 0$ cm/sec.

45. We want the window to have the largest possible area given the constraints. The area of the window is $A = 2xy + \frac{1}{2}\pi x^2$. The constraint on the perimeter dictates that $2x + 2y + \pi x = 28$. Solving for y gives $y = \frac{28 - 2x - \pi x}{2}$. Therefore, $A = 2x\left(\frac{28 - 2x - \pi x}{2}\right) + \frac{1}{2}\pi x^2 = \frac{56x - 4x^2 - 2\pi x^2 + \pi x^2}{2} = \frac{-(\pi + 4)x^2 + 56x}{2}$. A is maximized at $x = -\frac{b}{2a} = -\frac{56}{-2(\pi + 4)} = \frac{28}{\pi + 4}$ and $y = \frac{28 - \frac{56}{\pi+4} - \frac{28\pi}{\pi+4}}{2} = \frac{28\pi + 112 - 56 - 28\pi}{2(\pi + 4)} = \frac{28}{\pi + 4}$, or $\frac{28}{\pi + 4}$ ft.

47. True. $\frac{-b \pm \sqrt{b^2 - 4ac}}{2a}$ is a root of the equation $ax^2 + bx + c = 0$, and therefore $f\left(\frac{-b + \sqrt{b^2 - 4ac}}{2a}\right) = 0$.

49. True. If a and c have opposite signs then $b^2 - 4ac > 0$ and the equation has 2 roots.

51. True. The maximum occurs at the vertex of the parabola.

Technology Exercises

page 136

1. $(-3.0414, 0.1503)$, $(3.0414, 7.4497)$.

3. $(-2.3371, 2.4117)$, $(6.0514, -2.5015)$.

5. $(-1.1055, -6.5216)$ and $(1.1055, -1.8784)$

7. a.

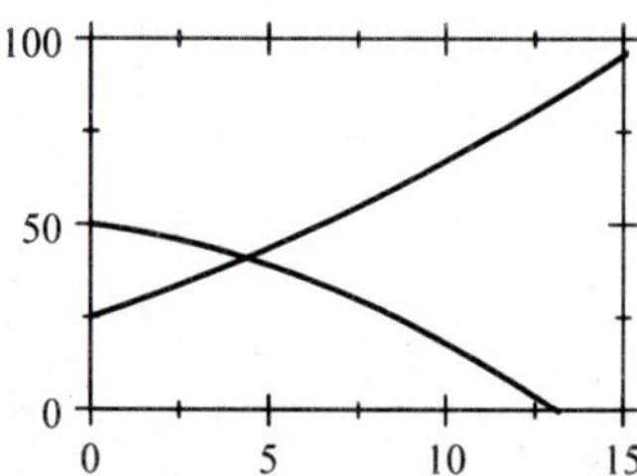

b. 438 wall clocks; \$40.92.

2.7 Functions and Mathematical Models

Problem-Solving Tips

When you solve a problem involving a function, it is helpful to identify the type of function you are working with. For example, if you wish to find the domain of a **polynomial function**, you know that there are no restrictions on the domain, becausae a polynomial is defined for all real numbers. If you want to find the domain of a **rational function**, you know that you have to check to see if there are any values for which the denominator is equal to 0.

Concept Questions

page 144

1. See page 137 of the text. Answers will vary.

Exercises page 144

1. f is a polynomial function in x of degree 6.

3. Expanding $G(x) = 2(x^2 - 3)^3$, we have $G(x) = 2x^6 - 18x^4 + 54x^2 - 54$, and we see that G is a polynomial function in x of degree 6.

5. f is neither a polynomial nor a rational function.

7. a. The number of enterprise IM accounts in 2006 is given by $N(0) = 59.7$, or 59.7 million.

b. The number of enterprise IM accounts in 2010, assuming a continuing trend, is given by $N(4) = 2.96(4)^2 + 11.37(4) + 59.7 = 152.54$ million.

9. a. The property tax in 1997 is given by $N(0) = 2360$, or \$2360.

b. The property tax in 2010, assuming a continuing trend, is given by $N(13) = 7.26(13)^2 + 91.7(13) + 2360 = 4779.04$, or \$4779.04.

11. a. The amount of benefits paid out in 2010 is $S(0) = 0.72$, or \$720 million.

b. The amount of benefits projected to be paid out in 2040 is $S(3) = 0.1375(3)^2 + 0.5185(3) + 0.72 = 3.513$, or \$3.513 trillion.

13. a. The given data imply that $R(40) = 50$, that is, $\frac{100(40)}{b+40} = 50$, so $50(b+40) = 4000$, or $b = 40$. Therefore, the required response function is $R(x) = \frac{100x}{40+x}$.

b. The response will be $R(60) = \frac{100(60)}{40+60} = 60$, or approximately 60 percent.

15. a. Total global mobile data traffic in 2009 was $f(0) = 0.06$, or 60,000 terabytes.

b. The total in 2012 will be $f(3) = 0.021(3)^3 + 0.015(3)^2 + 0.12(3) + 0.06 = 1.122$, or 1,122,000 terabytes.

17. a. $N(0) = 0.32$ or 320,000.

b. $N(4) = -0.0675(4)^4 + 0.5083(4)^3 - 0.893(4)^2 + 0.66(4) + 0.32 = 3.9232$, or 3,923,200.

19. a. We first construct a table.

t	$N(t)$
1	52
2	75
3	93
4	109
5	122

t	$N(t)$
6	135
7	146
8	157
9	167
10	177

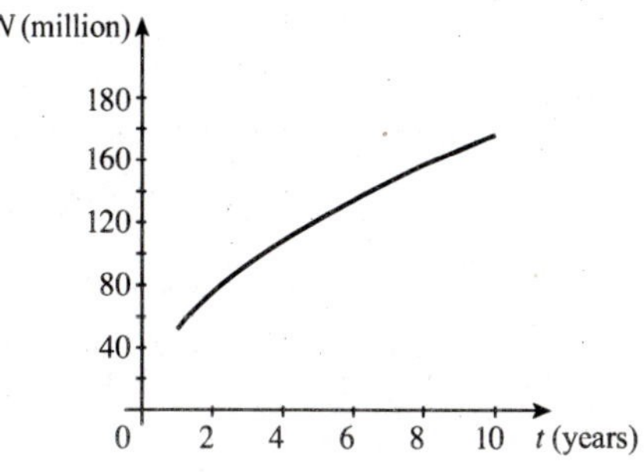

b. The number of viewers in 2012 is given by $N(10) = 52(10)^{0.531} \approx 176.61$, or approximately 177 million viewers.

21. $N(5) = 0.0018425(10)^{2.5} \approx 0.58265$, or approximately 582,650. $N(10) = 0.0018425(15)^{2.5} \approx 1.60559$, or approximately 1,605,590.

23. $A(0) = \frac{699}{1^{0.94}} = 699$ or \$699. $A(5) = \frac{699}{6^{0.94}} \approx 129.722$, or approximately \$130.

25. $h(t) = f(t) - g(t) = \frac{110}{\frac{1}{2}t+1} - 26\left(\frac{1}{4}t^2 - 1\right)^2 - 52.$

$h(0) = f(0) - g(0) = \frac{110}{\frac{1}{2}(0)+1} - 26\left[\frac{1}{4}(0)^2 - 1\right]^2 - 52 = 110 - 26 - 52 = 32$, or \$32.

$h(1) = f(1) - g(1) = \frac{110}{\frac{1}{2}(1)+1} - 26\left[\frac{1}{4}(1)^2 - 1\right]^2 - 52 \approx 6.71$, or approximately \$6.71.

$h(2) = f(2) - g(2) = \frac{110}{\frac{1}{2}(2)+1} - 26\left[\frac{1}{4}(2)^2 - 1\right]^2 - 52 = 3$, or \$3. We conclude that the price gap was narrowing.

27. The total cost by 2011 is given by $f(1) = 5$, or \$5 billion. The total cost by 2015 is given by $f(5) = -0.5278(5^3) + 3.012(5^2) + 49.23(5) - 103.29 = 152.185$, or approximately \$152 billion.

29. a.

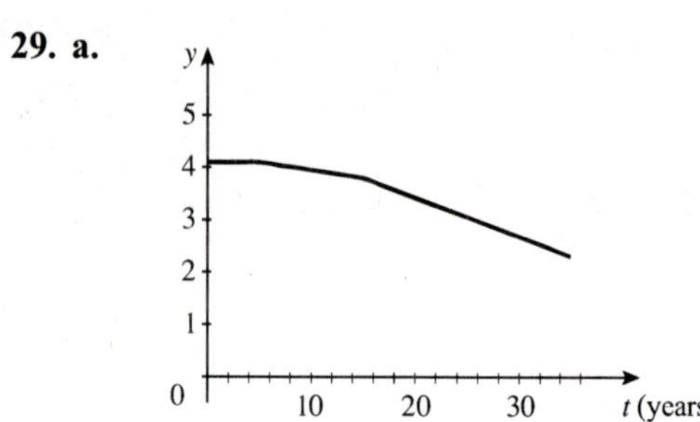

b. At the beginning of 2005, the ratio will be $f(10) = -0.03(10) + 4.25 = 3.95$. At the beginning of 2020, the ratio will be $f(25) = -0.075(25) + 4.925 = 3.05$.

c. The ratio is constant from 1995 to 2000.

d. The decline of the ratio is greatest from 2010 through 2030. It is $\frac{f(35) - f(15)}{35 - 15} = \frac{2.3 - 3.8}{20} = -0.075$.

31. Substituting $x = 6$ and $p = 8$ into the given equation gives $8 = \sqrt{-36a + b}$, or $-36a + b = 64$. Next, substituting $x = 8$ and $p = 6$ into the equation gives $6 = \sqrt{-64a + b}$, or $-64a + b = 36$. Solving the system $\begin{cases} -36a + b = 64 \\ -64a + b = 36 \end{cases}$ for a and b, we find $a = 1$ and $b = 100$. Therefore the demand equation is $p = \sqrt{-x^2 + 100}$.

When the unit price is set at \$7.50, we have $7.5 = \sqrt{-x^2 + 100}$, or $56.25 = -x^2 + 100$ from which we deduce that $x \approx \pm 6.614$. Thus, the quantity demanded is approximately 6614 units.

33. a. We solve the system of equations $p = cx + d$ and $p = ax + b$. Substituting the first equation into the second gives $cx + d = ax + b$, so $(c - a)x = b - d$ and $x = \frac{b - d}{c - a}$. Because $a < 0$ and $c > 0$, $c - a \neq 0$ and x is well-defined. Substituting this value of x into the second equation, we obtain $p = a\left(\frac{b-d}{c-a}\right) + b = \frac{ab - ad + bc - ab}{c - a} = \frac{bc - ad}{c - a}$. Therefore, the equilibrium quantity is $\frac{b-d}{c-a}$ and the equilibrium price is $\frac{bc - ad}{c - a}$.

b. If c is increased, the denominator in the expression for x increases and so x gets smaller. At the same time, the first term in the first equation for p decreases and so p gets larger. This analysis shows that if the unit price for producing the product is increased then the equilibrium quantity decreases while the equilibrium price increases.

c. If b is decreased, the numerator of the expression for x decreases while the denominator stays the same. Therefore, x decreases. The expression for p also shows that p decreases. This analysis shows that if the (theoretical) upper bound for the unit price of a commodity is lowered, then both the equilibrium quantity and the equilibrium price drop.

35. The area of Juanita's garden is 250 ft^2. Therefore $xy = 250$ and $y = \frac{250}{x}$. The amount of fencing needed is given by $2x + 2y$. Therefore, $f = 2x + 2\left(\frac{250}{x}\right) = 2x + \frac{500}{x}$. The domain of f is $x > 0$.

37. Because the volume of the box is the area of the base times the height of the box, we have $V = x^2y = 20$. Thus, we have $y = \frac{20}{x^2}$. Next, the amount of material used in constructing the box is given by the area of the base of the box, plus the area of the four sides, plus the area of the top of the box; that is, $A = x^2 + 4xy + x^2$. Then, the cost of constructing the box is given by $f(x) = 0.30x^2 + 0.40x \cdot \frac{20}{x^2} + 0.20x^2 = 0.5x^2 + \frac{8}{x}$, where $f(x)$ is measured in dollars and $x > 0$.

39. The average yield of the apple orchard is 36 bushels/tree when the density is 22 trees/acre. Let x be the unit increase in tree density beyond 22. Then the yield of the apple orchard in bushels/acre is given by $(22 + x)(36 - 2x)$.

41. a. Let x denote the number of bottles sold beyond 10,000 bottles. Then $P(x) = (10{,}000 + x)(5 - 0.0002x) = -0.0002x^2 + 3x + 50{,}000$.

b. He can expect a profit of $P(6000) = -0.0002\left(6000^2\right) + 3(6000) + 50{,}000 = 60{,}800$, or \$60,800.

43. False. $f(x) = 3x^{3/4} + x^{1/2} + 1$ is not a polynomial function. The powers of x must be nonnegative integers.

45. False. $f(x) = x^{1/2}$ is not defined for negative values of x.

Technology Exercises page 150

1. a. $f(t) = 1.85t + 16.9$.

b.

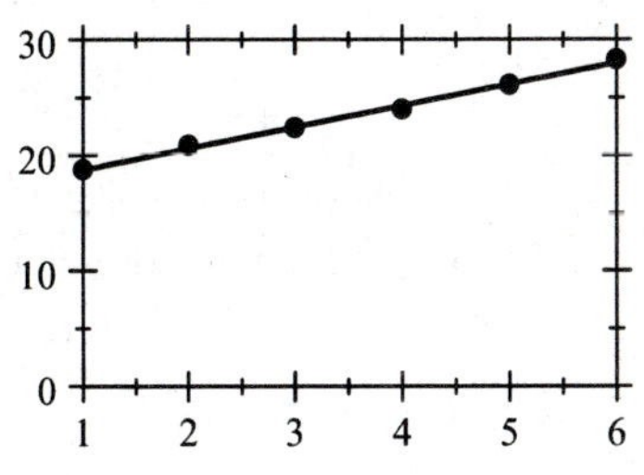

c.

t	y
1	18.8
2	20.6
3	22.5
4	24.3
5	26.2
6	28.0

These values are close to the given data.

d. $f(8) = 1.85(8) + 16.9 = 31.7$ gallons.

3. a. $f(t) = -0.221t^2 + 4.14t + 64.8.$

b.

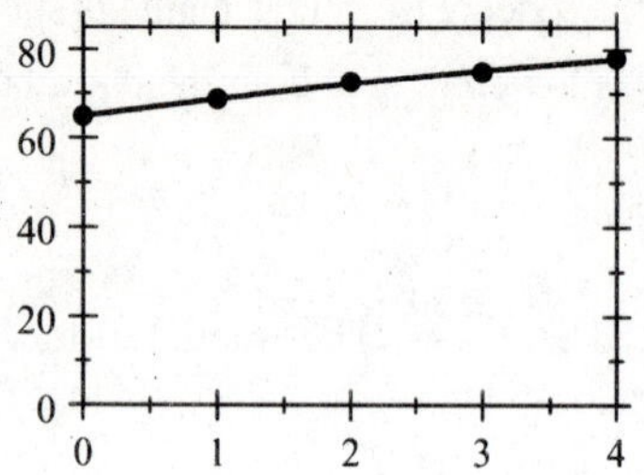

c. 77.8 million

5. a.

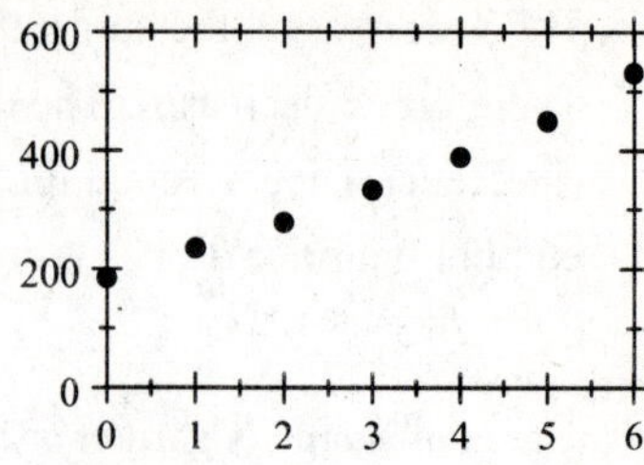

b. $f(t) = 2.94t^2 + 38.75t + 188.5$

c. The spending in 2010 was

$f(t) = 2.94(10^2) + 38.75(10) + 188.5 = 870,$

or approximately $870 billion.

7. a. $f(t) = -0.00081t^3 + 0.0206t^2 + 0.125t + 1.69.$

b.

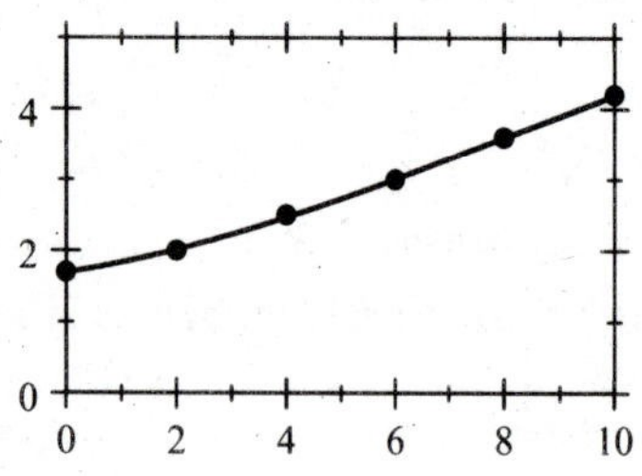

c.

t	y
1	1.8
5	2.7
10	4.2

The revenues were $1.8 trillion in 2001, $2.7 trillion in 2005, and $4.2 trillion in 2010.

9. a. $f(t) = -0.0056t^3 + 0.112t^2 + 0.51t + 8.$

b.

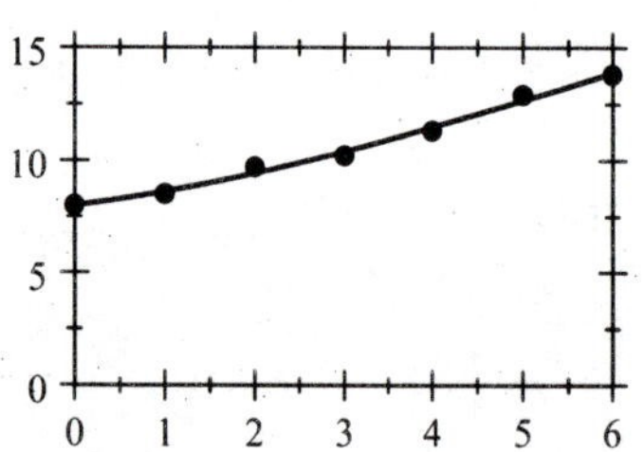

c.

t	0	3	6
$f(t)$	8	10.4	13.9

11. a. $f(t) = -2.4167t^3 + 24.5t^2 - 123.33t + 506.$

b.

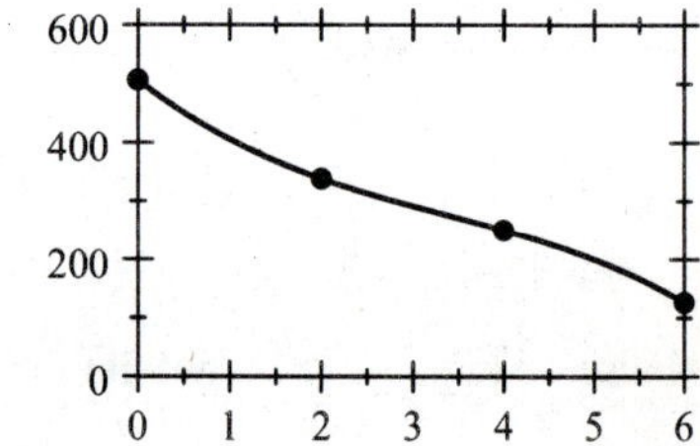

c. $f(0) = 506$, or 506,000; $f(2) = 338$, or 338,000; and $f(6) = 126$, or 126,000.

13. a. $f(t) = 0.00125t^4 - 0.0051t^3 - 0.0243t^2 + 0.129t + 1.71.$

b.

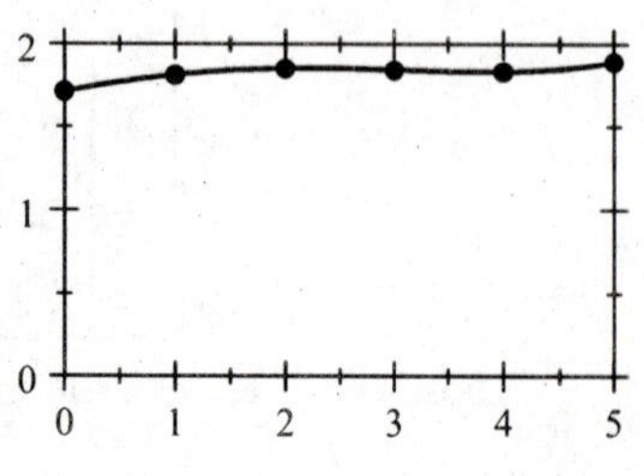

c.

t	0	1	2	3	4	5
$f(t)$	1.71	1.81	1.85	1.84	1.83	1.89

d. The average amount of nicotine in 2005 is $f(6) = 2.128$, or approximately 2.13 mg/cigarette.

CHAPTER 2 Concept Review Questions page 153

1. ordered, abscissa (x-coordinate), ordinate (y-coordinate)

3. a. $\dfrac{y_2 - y_1}{x_2 - x_1}$ b. undefined c. zero d. positive

5. a. $y - y_1 = m(x - x_1)$, point-slope form b. $y = mx + b$, slope-intercept

7. domain, range, B

9. $f(x) \pm g(x)$, $f(x)g(x)$, $\dfrac{f(x)}{g(x)}$, $A \cap B$, $A \cap B$, 0

11. $ax^2 + bx + c$, parabola, upward, downward, vertex, $-\frac{b}{2a}$, $x = -\frac{b}{2a}$.

CHAPTER 2 Review Exercises page 154

1. An equation is $x = -2$.

3. The slope of L is $m = \dfrac{\frac{7}{2} - 4}{3 - (-2)} = \dfrac{\frac{7-8}{2}}{5} = -\dfrac{1}{10}$ and an equation of L is $y - 4 = -\frac{1}{10}[x - (-2)] = -\frac{1}{10}x - \frac{1}{5}$, or $y = -\frac{1}{10}x + \frac{19}{5}$.

5. Writing the given equation in the form $y = \frac{5}{2}x - 3$, we see that the slope of the given line is $\frac{5}{2}$. Thus, an equation is $y - 4 = \frac{5}{2}(x + 2)$, or $y = \frac{5}{2}x + 9$.

7. Using the slope-intercept form of the equation of a line, we have $y = -\frac{1}{2}x - 3$.

9. Rewriting the given equation in slope-intercept form, we have $4y = -3x + 8$, or $y = -\frac{3}{4}x + 2$, and we conclude that the slope of the required line is $-\frac{3}{4}$. Using the point-slope form of the equation of a line with the point $(2, 3)$ and slope $-\frac{3}{4}$, we obtain $y - 3 = -\frac{3}{4}(x - 2)$, so $y = -\frac{3}{4}x + \frac{6}{4} + 3 = -\frac{3}{4}x + \frac{9}{2}$.

11. Rewriting the given equation in the slope-intercept form $y = \frac{2}{3}x - 8$, we see that the slope of the line with this equation is $\frac{2}{3}$. The slope of the required line is $-\frac{3}{2}$. Using the point-slope form of the equation of a line with the point $(-2, -4)$ and slope $-\frac{3}{2}$, we have $y - (-4) = -\frac{3}{2}[x - (-2)]$, or $y = -\frac{3}{2}x - 7$.

13. $-2x + 5y = 15$. Setting $x = 0$ gives $5y = 15$, or $y = 3$. Setting $y = 0$ gives $-2x = 15$, or $x = -\frac{15}{2}$.

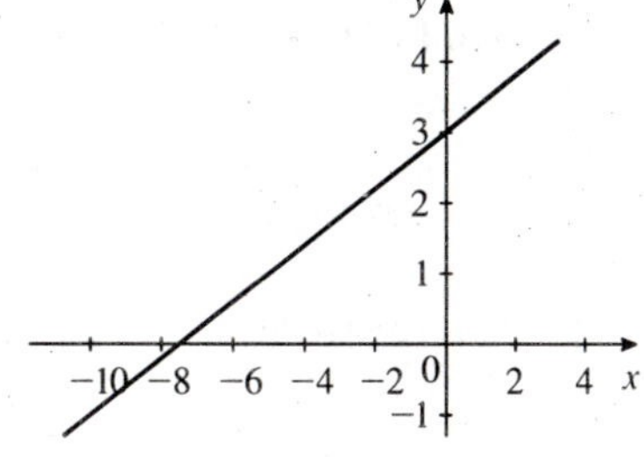

15. $2x^2 - x - 3 = (2x - 3)(x + 1)$, and $x = \frac{3}{2}$ or -1. Because the denominator of the given expression is zero at these points, we see that the domain of f cannot include these points and so the domain of f is $(-\infty, -1) \cup \left(-1, \frac{3}{2}\right) \cup \left(\frac{3}{2}, \infty\right)$.

17. a. From $t = 0$ to $t = 5$, the graph for cassettes lies above that for CDs, so from 1985 to 1990, the value of prerecorded cassettes sold was greater than that of CDs.

b. Sales of prerecorded CDs were greater than those of prerecorded cassettes from 1990 onward.

c. The graphs intersect at the point with coordinates $x = 5$ and $y \approx 3.5$, and this tells us that the sales of the two formats were the same in 1990 at the sales level of approximately \$3.5 billion.

19.

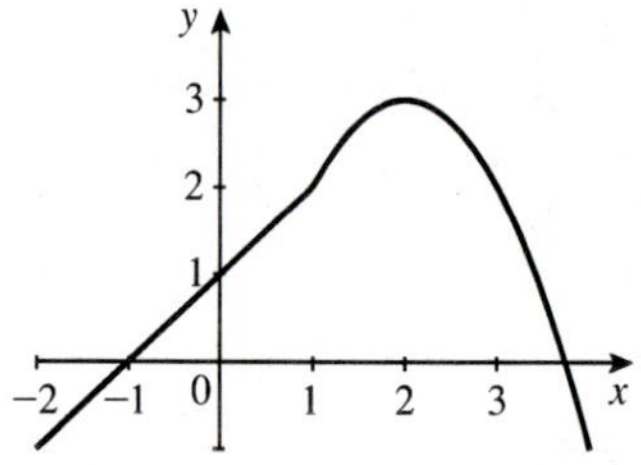

21. $y = 6x^2 - 11x - 10$. The x-coordinate of the vertex is $-\frac{-11}{2(6)} = \frac{11}{12}$ and the y-coordinate is $6\left(\frac{11}{12}\right)^2 - 11\left(\frac{11}{12}\right) - 10 = -\frac{361}{24}$. Therefore, the vertex is $\left(\frac{11}{12}, -\frac{361}{24}\right)$. Next, solving $6x^2 - 11x - 10 = (3x + 2)(2x - 5) = 0$ gives $-\frac{2}{3}$ and $\frac{5}{2}$ as the x-intercepts.

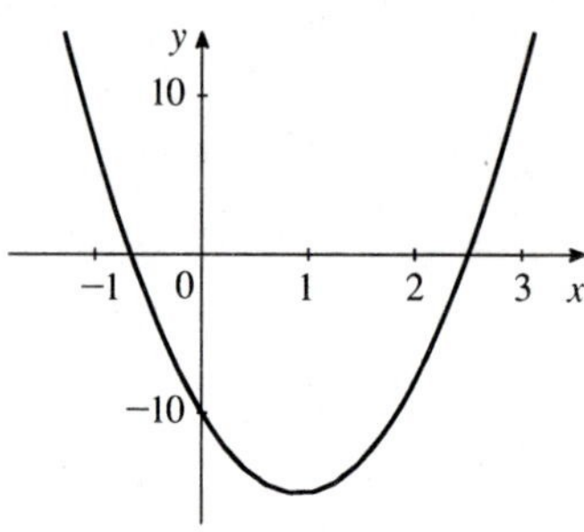

23. We solve the system $3x + 4y = -6$, $2x + 5y = -11$. Solving the first equation for x, we have $3x = -4y - 6$ and $x = -\frac{4}{3}y - 2$. Substituting this value of x into the second equation yields $2\left(-\frac{4}{3}y - 2\right) + 5y = -11$, so $-\frac{8}{3}y - 4 + 5y = -11$, $\frac{7}{3}y = -7$, and $y = -3$. Thus, $x = -\frac{4}{3}(-3) - 2 = 4 - 2 = 2$, so the point of intersection is $(2, -3)$.

25. We solve the system $7x + 9y = -11$, $3x = 6y - 8$. Multiplying the second equation by $\frac{1}{3}$, we have $x = 2y - \frac{8}{3}$. Substituting this value of x into the first equation, we have $7\left(2y - \frac{8}{3}\right) + 9y = -11$. Solving this equation for y, we have $14y - \frac{56}{3} + 9y = -11$, $69y = -33 + 56$, and $y = \frac{23}{69} = \frac{1}{3}$. Thus, $x = 2\left(\frac{1}{3}\right) - \frac{8}{3} = -2$. The lines intersect at $\left(-2, \frac{1}{3}\right)$.

27. The slope of L_2 is greater than that of L_1. This tells us that if the manufacturer lowers the unit price for each model clock radio by the same amount, the additional demand for model B radios will be greater than that for model A radios.

29. In 2012 (when $x = 5$), we have $S(5) = 6000(5) + 30{,}000 = 60{,}000$.

31. Let x denote the number of units produced and sold.

a. The cost function is $C(x) = 6x + 30{,}000$.

b. The revenue function is $R(x) = 10x$.

c. The profit function is $P(x) = R(x) - C(x) = 10x - (30{,}000 + 6x) = 4x - 30{,}000$.

d. $P(6000) = 4(6000) - 30{,}000 = -6{,}000$, a loss of \$6000; $P(8000) = 4(8000) - 30{,}000 = 2{,}000$, a profit of \$2000; and $P(12{,}000) = 4(12{,}000) - 30{,}000 = 18{,}000$, a profit of \$18,000.

33. The slope of the demand curve is $\dfrac{\Delta p}{\Delta x} = -\dfrac{10}{200} = -0.05$. Using the point-slope form of the equation of a line with the point $(0, 200)$, we have $p - 200 = -0.05(x)$, or $p = -0.05x + 200$.

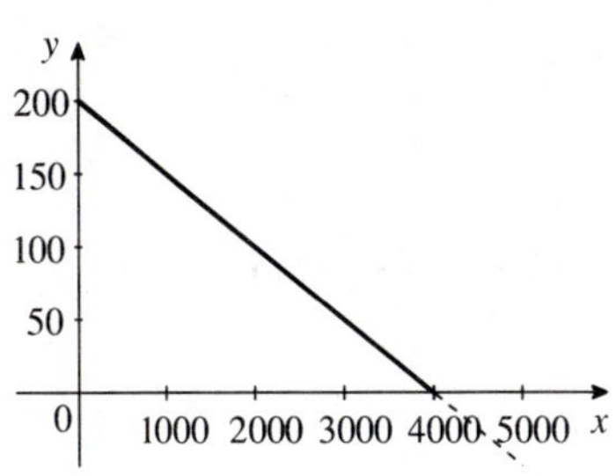

35. $D(w) = \dfrac{a}{150}w$. The given equation can be expressed in the form $y = mx + b$, where $m = \dfrac{a}{150}$ and $b = 0$. If $a = 500$ and $w = 35$, $D(35) = \frac{500}{150}(35) = 116\frac{2}{3}$, or approximately 117 mg.

37. $R(30) = -\frac{1}{2}(30)^2 + 30(30) = 450$, or \$45,000.

39. The population will increase by $P(9) - P(0) = \left[50{,}000 + 30(9)^{3/2} + 20(9)\right] - 50{,}000$, or 990, during the next 9 months. The population will increase by $P(16) - P(0) = \left[50{,}000 + 30(16)^{3/2} + 20(16)\right] - 50{,}000$, or 2240 during the next 16 months.

41. $N(0) = 648$, or 648,000, $N(1) = -35.8 + 202 + 87.7 + 648 = 901.9$, or 901,900, $N(2) = -35.8(2)^3 + 202(2)^2 + 87.8(2) + 648 = 1345.2$ or 1,345,200, and $N(3) = -35.8(3)^3 + 202(3)^2 + 87.8(3) + 648 = 1762.8$ or 1,762,800.

43. a. $f(t) = 267$ and $g(t) = 2t^2 + 46t + 733$.

b. $h(t) = (f + g)(t) = f(t) + g(t) = 267 + \left(2t^2 + 46t + 733\right) = 2t^2 + 46t + 1000$.

c. $h(13) = 2(13)^2 + 46(13) + 1000 = 1936$, or 1936 tons.

45. a. $V = \frac{4}{3}\pi r^3$, so $r^3 = \dfrac{3V}{4\pi}$ and $r = f(V) = \sqrt[3]{\dfrac{3V}{4\pi}}$.

b. $g(t) = \frac{9}{2}\pi t$.

c. $h(t) = (f \circ g)(t) = f(g(t)) = \left[\dfrac{3g(t)}{4\pi}\right]^{1/3} = \left[\dfrac{3(9)\pi t}{4\pi(2)}\right]^{1/3} = \frac{3}{2}\sqrt[3]{t}$.

d. $h(8) = \frac{3}{2}\sqrt[3]{8} = 3$, or 3 ft.

47. Measured in inches, the sides of the resulting box have length $20 - 2x$ and its height is x, so its volume is $V = x(20 - 2x)^2$ in^3.

CHAPTER 2 Before Moving On... page 156

1. $m = \dfrac{5-(-2)}{4-(-1)} = \dfrac{7}{5}$, so an equation is $y-(-2) = \frac{7}{5}[x-(-1)]$. Simplifying, $y = \frac{7}{5}x + \frac{7}{5} - 2$, or $y = \frac{7}{5}x - \frac{3}{5}$.

2. $m = -\frac{1}{3}$ and $b = \frac{4}{3}$, so an equation is $y = -\frac{1}{3}x + \frac{4}{3}$.

3. **a.** $f(-1) = -2(-1) + 1 = 3$. **b.** $f(0) = 2$. **c.** $f\left(\frac{3}{2}\right) = \left(\frac{3}{2}\right)^2 + 2 = \frac{17}{4}$.

4. **a.** $(f+g)(x) = f(x) + g(x) = \dfrac{1}{x+1} + x^2 + 1$. **b.** $(fg)(x) = f(x)g(x) = \dfrac{x^2+1}{x+1}$.

c. $(f \circ g)(x) = f(g(x)) = \dfrac{1}{g(x)+1} = \dfrac{1}{x^2+2}$. **d.** $(g \circ f)(x) = g(f(x)) = [f(x)]^2 + 1$
$= \dfrac{1}{(x+1)^2} + 1$.

5. $4x + h = 108$, so $h = 108 - 4x$. The volume is $V = x^2h = x^2(108-4x) = 108x^2 - 4x^3$.

3 EXPONENTIAL AND LOGARITHMIC FUNCTIONS

3.1 Exponential Functions

Problem-Solving Tips

1. Remember the order of operations when working with exponents. Note that $-5^2 \neq 25$, but rather $-5^2 = -(5)^2 = -25$. On the other hand, $(-5)^2 = 25$.

2. $b^{-x} = \dfrac{1}{b^x} = \left(\dfrac{1}{b}\right)^x$. If $b > 1$, then $0 < \dfrac{1}{b} < 1$, so the graph of b^{-x} for $b > 1$ is similar to the graph of $y = \left(\dfrac{1}{2}\right)^x$. (See Figure 3 in the text.)

Concept Questions page 162

1. $f(x) = b^x$ with $b > 0$ and $b \neq 1$.

Exercises page 162

1. a. $4^{-3} \cdot 4^5 = 4^{-3+5} = 4^2 = 16$.
   b. $3^{-3} \cdot 3^6 = 3^{6-3} = 3^3 = 27$.

3. a. $9(9)^{-1/2} = \dfrac{9}{9^{1/2}} = \dfrac{9}{3} = 3$.
   b. $5(5)^{-1/2} = 5^{1+(-1/2)} = 5^{1/2} = \sqrt{5}$.

5. a. $\dfrac{(-3)^4(-3)^5}{(-3)^8} = (-3)^{4+5-8} = (-3)^1 = -3$.
   b. $\dfrac{(2^{-4})(2^6)}{2^{-1}} = 2^{-4+6+1} = 2^3 = 8$.

7. a. $(64x^9)^{1/3} = 64^{1/3}(x^{9/3}) = 4x^3$.
   b. $(25x^3y^4)^{1/2} = (25^{1/2})(x^{3/2})(y^{4/2}) = 5x^{3/2}y^2$

9. a. $\dfrac{6a^{-4}}{3a^{-3}} = 2a^{-4+3} = 2a^{-1} = \dfrac{2}{a}$.
   b. $\dfrac{4b^{-4}}{12b^{-6}} = \frac{1}{3}b^{-4+6} = \frac{1}{3}b^2$.

11. a. $(2x^3y^2)^3 = 2^3 \cdot x^{3(3)} \cdot y^{2(3)} = 8x^9y^6$.
    b. $(4x^2y^2z^3)^2 = 4^2 \cdot x^{2(2)} \cdot y^{2(2)} \cdot z^{3(2)} = 16x^4y^4z^6$.

13. $6^{2x} = 6^6$ if and only if $2x = 6$; that is $x = 3$.

15. $3^{3x-4} = 3^5$ if and only if $3x - 4 = 5$; that is, $3x = 9$ and $x = 3$.

17. $(2.1)^{x+2} = (2.1)^5$ if and only if $x + 2 = 5$; that is, $x = 3$.

19. $8^x = \left(\frac{1}{32}\right)^{x-2}$, $(2^3)^x = (32)^{2-x} = (2^5)^{2-x}$, so $2^{3x} = 2^{5(2-x)}$. This is true if and only if $3x = 10 - 5x$, so $8x = 10$ and $x = \frac{5}{4}$.

21. Let $y = 3^x$. Then the given equation is equivalent to $y^2 - 12y + 27 = 0$, or $(y-9)(y-3) = 0$, giving $y = 3$ or 9. So $3^x = 3$ or $3^x = 9$, and therefore, $x = 1$ or $x = 2$.

23. $y = 2^x$, $y = 3^x$, and $y = 4^x$.

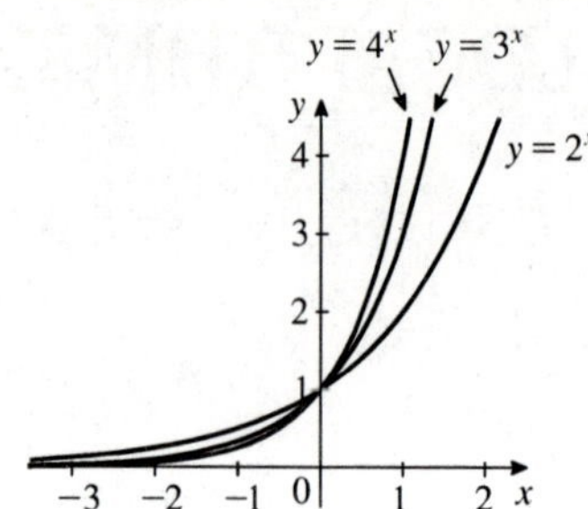

25. $y = 2^{-x}$, $y = 3^{-x}$, and $y = 4^{-x}$.

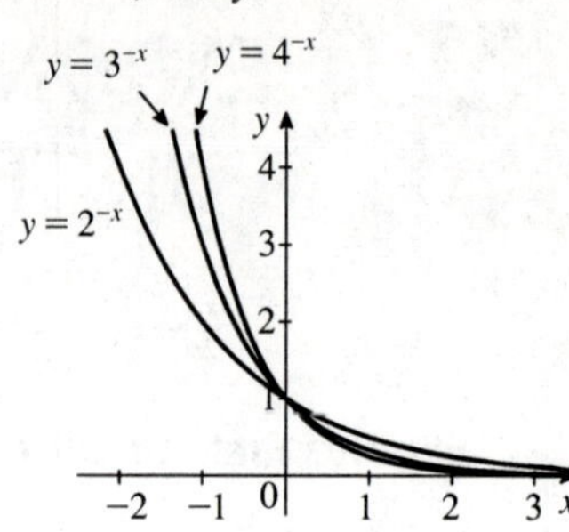

27. $y = 4^{0.5x}$, $y = 4^x$, and $y = 4^{2x}$.

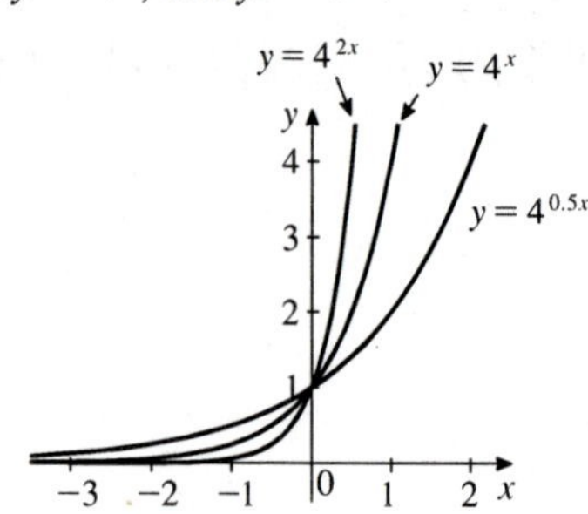

29. $y = e^{0.5x}$, $y = e^x$, $y = e^{1.5x}$.

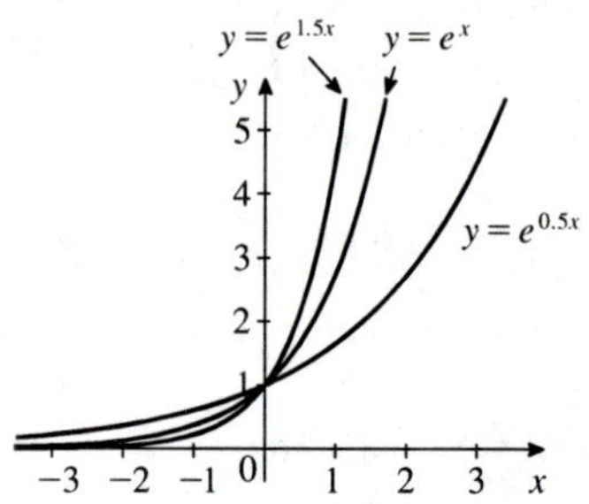

31. $y = 0.5e^{-x}$, $y = e^{-x}$, and $y = 2e^{-x}$.

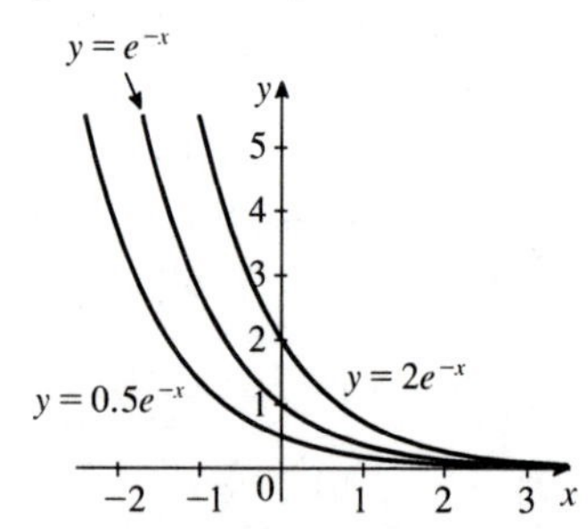

33. Because $f(0) = A = 100$ and $f(1) = 120$, we have $100e^k = 120$, and so $e^k = \frac{12}{10} = \frac{6}{5}$. Therefore, $f(x) = 100e^{kx} = 100\left(e^k\right) = 100\left(\frac{6}{5}\right)^x$.

35. $f(0) = 20$ implies that $\dfrac{1000}{1+B} = 20$, so $1000 = 20 + 20B$, or $B = \dfrac{980}{20} = 49$. Therefore, $f(t) = \dfrac{1000}{1+49e^{-kt}}$. Next, $f(2) = 30$, so $\dfrac{1000}{1+49e^{-2t}} = 30$. We have $1 + 49e^{-2k} = \dfrac{1000}{30} = \dfrac{100}{3}$, $49e^{-2k} = \dfrac{100}{3} - 1 = \dfrac{97}{3}$, $e^{-2k} = \dfrac{97}{147}$, and finally $e^{-k} = \left(\dfrac{97}{147}\right)^{1/2}$. Therefore, $f(t) = \dfrac{1000}{1+49\left(\frac{97}{147}\right)^{t/2}}$, so

$$f(5) = \frac{1000}{1+49\left(\frac{97}{147}\right)^{5/2}} \approx 54.6.$$

37. **a.** $R(t) = 26.3e^{-0.016t}$. In 1982 the rate was $R(0) = 26.3\%$, in 1986 it was $R(4) \approx 24.7\%$, in 1994 it was $R(12) \approx 21.7\%$, and in 2000 it was $R(18) \approx 19.7\%$.

b.

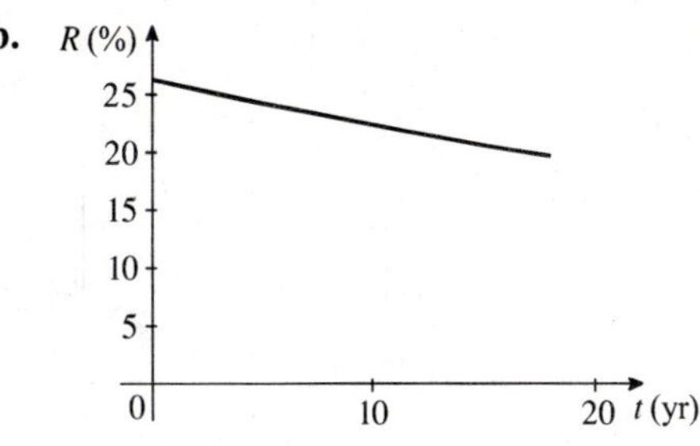

39. **a.**

Year	0	1	2	3	4	5
Number (billions)	0.45	0.80	1.41	2.49	4.39	7.76

b.

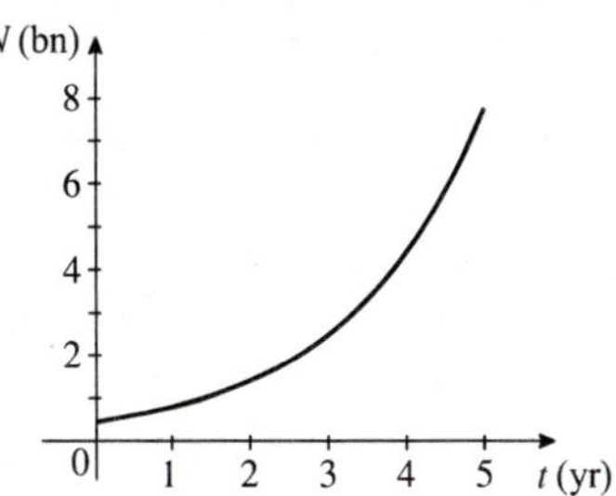

41. $N(t) = \dfrac{35.5}{1 + 6.89e^{-0.8674t}}$, so $N(6) = \dfrac{35.5}{1 + 6.89e^{-0.8674(6)}} \approx 34.2056$, or 34.21 million.

43. **a.** The initial concentration is given by $x(0) = 0.08 + 0.12\left(1 - e^{-0.02 \cdot 0}\right) = 0.08$, or 0.08 g/cm^3.

b. The concentration after 20 seconds is given by $x(20) = 0.08 + 0.12\left(1 - e^{-0.02 \cdot 20}\right) \approx 0.11956$, or 0.1196 g/cm^3.

45. False. $\left(x^2 + 1\right)^3 = x^6 + 3x^4 + 3x^2 + 1$.

47. True. The values of $f(x) = e^x$ get larger and larger as x increases, so if $x < y$, then $f(x) < f(y)$; that is, $e^x < e^y$.

Technology Exercises

page 165

1.

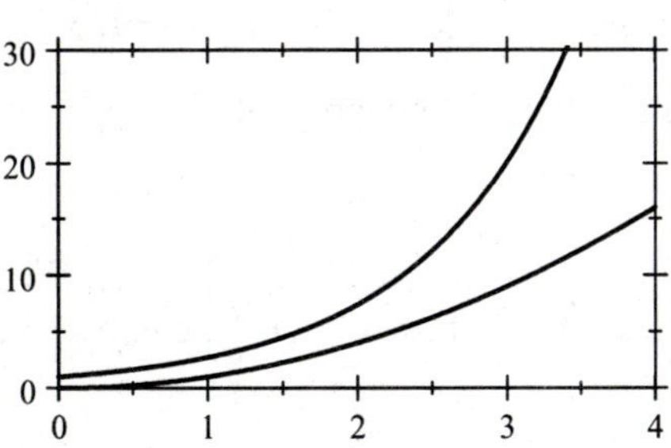

3.

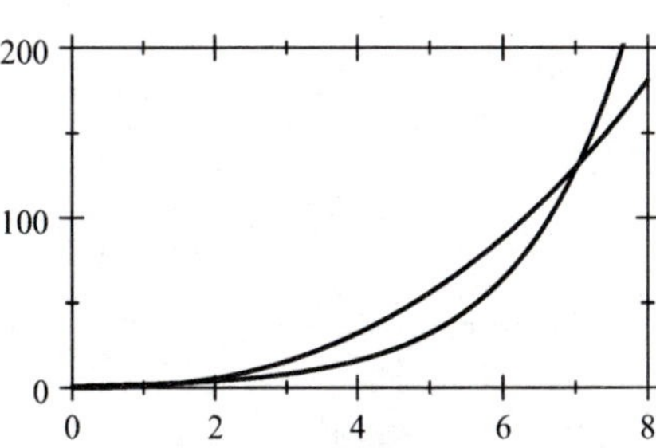

5.

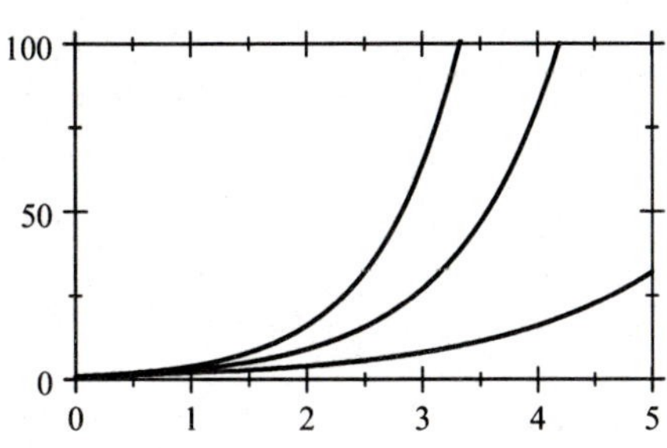

7.

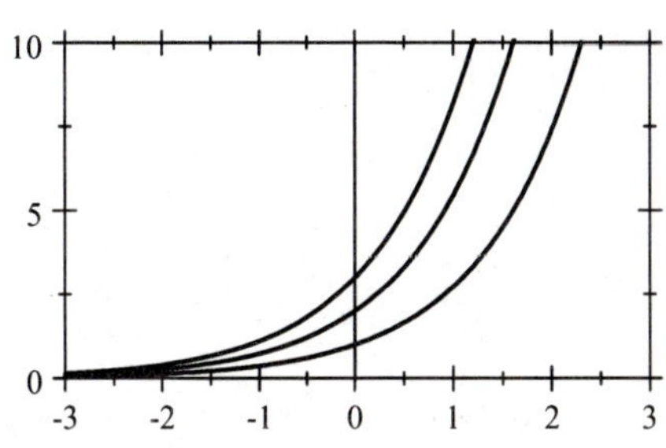

9.

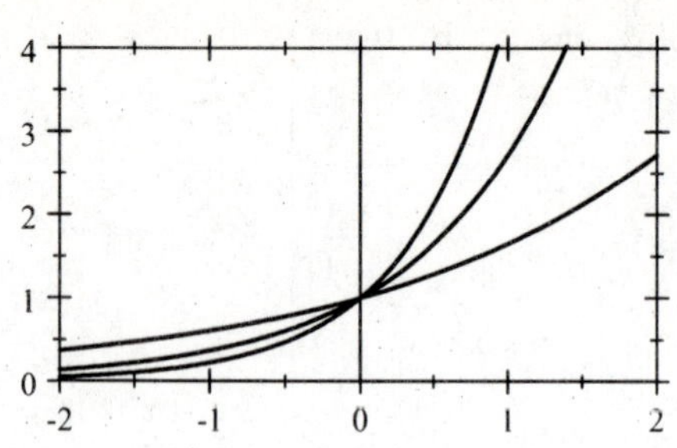

11. a.

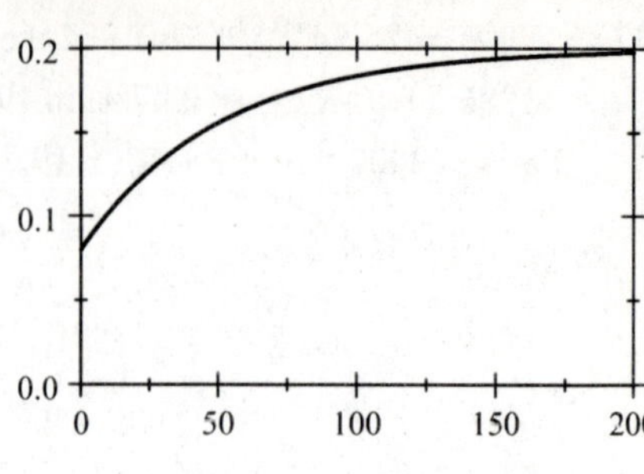

b. 0.08 g/cm^3. **c.** 0.12 g/cm^3.

13. a.

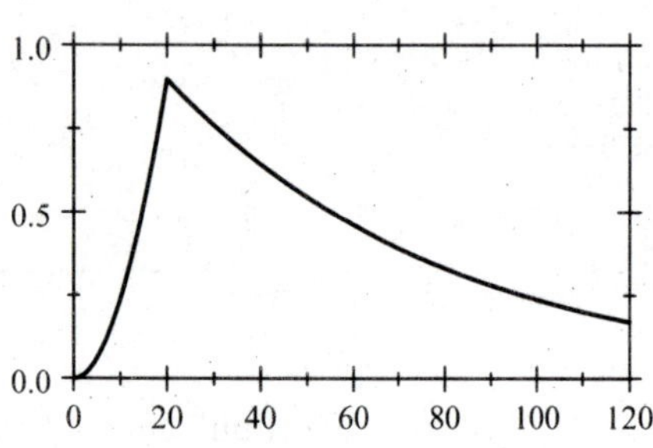

b. 20 seconds. **c.** 35.1 seconds.

3.2 Logarithmic Functions

Problem-Solving Tips

1. Property 1 of logarithms says that $\log_b mn = \log_b m + \log_b n$. However, $\log_b (m + n) \neq \log_b m + \log_b n$ and $\log_b \frac{m}{n} \neq \frac{\log_b m}{\log_b n}$.

2. When you work with logarithms be sure to distinguish between the following two operations: $\frac{\log 6}{\log 2} = \log 6 \div \log 2 \approx 2.585$ and $\log \frac{6}{2} = \log 6 - \log 2 \approx 0.477$. Property 2 of logarithms says that $\log_b \frac{m}{n} = \log_b m - \log_b n$.

3. The domain of the logarithmic function is $(0, \infty)$, so the logarithms of 0 and negative numbers are not defined.

Concept Questions page 172

1. a. $y = \log_b x$ if and only if $x = b^y$.

b. $f(x) = \log_b x$, $b > 0$, $b \neq 1$. Its domain is $(0, \infty)$.

3. a. $e^{\ln x} = x$. **b.** $\ln e^x = x$.

Exercises page 172

1. $\log_2 64 = 6$. **3.** $\log_4 \frac{1}{16} = -2$. **5.** $\log_{1/3} \frac{1}{3} = 1$. **7.** $\log_{32} 16 = \frac{4}{5}$.

9. $\log_{10} 0.001 = -3$.

11. $\log 12 = \log (4 \cdot 3) = \log 4 + \log 3 \approx 0.6021 + 0.4771 = 1.0792$.

13. $\log 16 = \log 4^2 = 2 \log 4 \approx 2\,(0.6021) = 1.2042$.

15. $\log 48 = \log \left(3 \cdot 4^2\right) = \log 3 + 2 \log 4 \approx 0.4771 + 2\,(0.6021) = 1.6813$.

17. $2 \ln a + 3 \ln b = \ln a^2 b^3$.

19. $\ln 3 + \frac{1}{2} \ln x + \ln y - \frac{1}{3} \ln z = \ln \dfrac{3\sqrt{x}y}{\sqrt[3]{z}}$.

21. $\log x\,(x+1)^4 = \log x + \log (x+1)^4 = \log x + 4 \log (x+1)$.

23. $\log \dfrac{\sqrt{x+1}}{x^2+1} = \log (x+1)^{1/2} - \log \left(x^2+1\right) = \frac{1}{2} \log (x+1) - \log \left(x^2+1\right)$.

25. $\ln x e^{-x^2} = \ln x + \ln e^{-x^2} = \ln x - x^2$.

27. $\ln \left(\dfrac{x^{1/2}}{x^2\sqrt{1+x^2}}\right) = \ln x^{1/2} - \ln x^2 - \ln \left(1+x^2\right)^{1/2} = \frac{1}{2} \ln x - 2 \ln x - \frac{1}{2} \ln \left(1+x^2\right) = -\frac{3}{2} \ln x - \frac{1}{2} \ln \left(1+x^2\right)$.

29. If $\log_2 x = 3$, then $2^3 = x$, and so $x = 8$.

31. If $\log_2 8 = x$, then $2^x = 8$, and so $x = 3$.

33. If $\log_x 10^3 = 3$, then $x^3 = 10^3$, and so $x = 10$.

35. If $\log_2 (2x+5) = 4$, then $2^4 = 2x+5$, $2x+5 = 16$, $2x = 11$, and so $x = \frac{11}{2}$.

37. If $\log_2 x - \log_2 (x-2) = 3$, then $\log_2 \dfrac{x}{x-2} = 3$, $\dfrac{x}{x-2} = 2^3 = 8$, $x = 8\,(x-2) = 8x - 16$, $-7x = -16$, and $x = \frac{16}{7}$.

39. If $\log_5 (2x+1) - \log_5 (x-2) = 1$, then $\log_5 \dfrac{2x+1}{x-2} = 1$, $\dfrac{2x+1}{x-2} = 5^1 = 5$, $2x+1 = 5\,(x-2) = 5x - 10$, $-3x = -11$, and so $x = \frac{11}{3}$.

41. If $\log x + \log (2x-5) = \log 3$, then $\log (x\,(2x-5)) = \log 3$, $x\,(2x-5) = 3$, $2x^2 - 5x - 3 = 0$, and $(2x+1)(x-3) = 0$. Thus, $x = -\frac{1}{2}$ or $x = 3$. Because x must be positive, the only solution is $x = 3$.

43. $y = \log_3 x$.

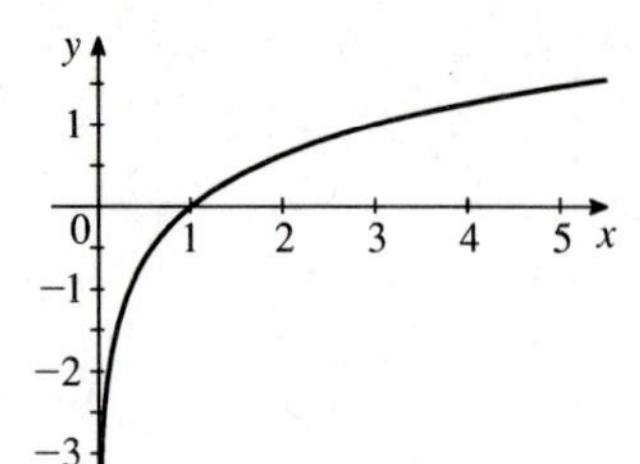

45. $y = \ln 2x$.

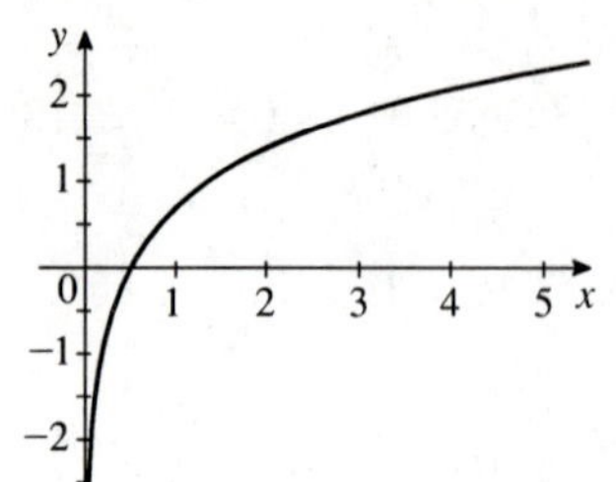

47. $y = 2^x$ and $y = \log_2 x$.

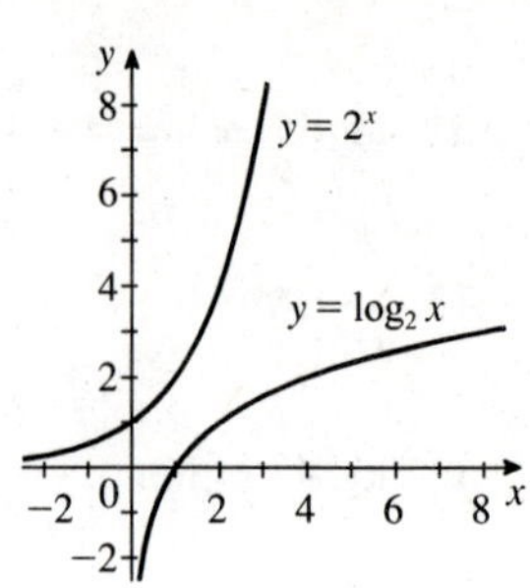

49. $e^{0.4t} = 8$, so $0.4t \ln e = \ln 8$ and thus $0.4t = \ln 8$ because $\ln e = 1$. Therefore, $t = \dfrac{\ln 8}{0.4} \approx 5.1986$.

51. $5e^{-2t} = 6$, so $e^{-2t} = \frac{6}{5} = 1.2$. Taking logarithms, we have $-2t \ln e = \ln 1.2$, so $t = -\dfrac{\ln 1.2}{2} \approx -0.0912$.

53. $2e^{-0.2t} - 4 = 6$, so $2e^{-0.2t} = 10$. Taking logarithms, we have $\ln e^{-0.2t} = \ln 5$, $-0.2t \ln e = \ln 5$, $-0.2t = \ln 5$, and $t = -\dfrac{\ln 5}{0.2} \approx -8.0472$.

55. $\dfrac{50}{1 + 4e^{0.2t}} = 20$, so $1 + 4e^{0.2t} = \dfrac{50}{20} = 2.5$, $4e^{0.2t} = 1.5$, $e^{0.2t} = \dfrac{1.5}{4} = 0.375$, $\ln e^{0.2t} = \ln 0.375$, and $0.2t = \ln 0.375$. Thus, $t = \dfrac{\ln 0.375}{0.2} \approx -4.9041$.

57. Taking logarithms of both sides, we obtain $\ln A = \ln Be^{-t/2}$, $\ln A = \ln B + \ln e^{-t/2}$, and $\ln A - \ln B = -\dfrac{t}{2} \ln e$, so $\ln \dfrac{A}{B} = -\dfrac{t}{2}$ and $t = -2 \ln \dfrac{A}{B} = 2 \ln \dfrac{B}{A}$.

59. $f(1) = 2$, so $a + b(0) = 2$. Thus, $a = 2$. Therefore, $f(x) = 2 + b \ln x$. Next, $f(2) = 4$, so $2 + b \ln 2 = 4$. Solving for b, we obtain $b = \dfrac{2}{\ln 2} \approx 2.8854$, so $f(x) = 2 + 2.8854 \ln x$.

61. $p(x) = 19.4 \ln x + 18$. For a child weighing 92 lb, we find $p(92) = 19.4 \ln 92 + 18 \approx 105.7$, or approximately 106 millimeters of mercury.

63. a. $30 = 10 \log \dfrac{I}{I_0}$, so $3 = \log \dfrac{I}{I_0}$, and $\dfrac{I}{I_0} = 10^3 = 1000$. Thus, $I = 1000 I_0$.

b. When $D = 80$, $I = 10^8 I_0$ and when $D = 30$, $I = 10^3 I_0$. Therefore, an 80-decibel sound is $10^8/10^3 = 10^5 = 100{,}000$ times louder than a 30-decibel sound.

c. If $D = 150$, then $I = 10^{15} I_0$, so this sound is $10^{15}/10^8 = 10^7 = 10{,}000{,}000$ times louder.

65. We solve the equation $\dfrac{160}{1 + 240e^{-0.2t}} = 80$ for t, obtaining $1 + 240e^{-0.2t} = \dfrac{160}{80}$, $240e^{-0.2t} = 2 - 1 = 1$, $e^{-0.2t} = \dfrac{1}{240}$, $-0.2t = \ln \dfrac{1}{240}$, and $t = -\dfrac{1}{0.2} \ln \dfrac{1}{240} \approx 27.40$, or approximately 27.4 years old.

67. We solve the equation $200\left(1-0.956e^{-0.18t}\right)=140$ for t, obtaining $1-0.956e^{-0.18t}=\frac{140}{200}=0.7$, $-0.956e^{-0.18t}=0.7-1=-0.3$, $e^{-0.18t}=\frac{0.3}{0.956}$, $-0.18t=\ln\left(\frac{0.3}{0.956}\right)$, and finally $t=-\dfrac{\ln\left(\frac{0.3}{0.956}\right)}{0.18}\approx 6.43875$. Thus, it is approximately 6.4 years old.

69. a. We solve the equation $0.08+0.12e^{-0.02t}=0.18$, obtaining $0.12e^{-0.02t}=0.1$, $e^{-0.02t}=\frac{0.1}{0.12}=\frac{1}{1.2}$, $\ln e^{-0.02t}=\ln\frac{1}{1.2}=\ln 1-\ln 1.2=-\ln 1.2$, $-0.02t=-\ln 1.2$, and $t=\frac{\ln 1.2}{0.02}\approx 9.116$, or approximately 9.1 seconds.

b. We solve the equation $0.08+0.12e^{-0.02t}=0.16$, obtaining $0.12e^{-0.02t}=0.08$, $e^{-0.02t}=\frac{0.08}{0.12}=\frac{2}{3}$, $-0.02t=\ln\frac{2}{3}$, and $t=-\frac{1}{0.02}\ln\frac{2}{3}\approx 20.2733$, or approximately 20.3 seconds.

71. False. Take $x=e$. Then $(\ln e)^3=1^3=1\neq 3\ln e=3$.

73. a. Taking logarithms of both sides gives $\ln 2^x=\ln e^{kx}$, so $x\ln 2=kx(\ln e)=kx$. Thus, $x(\ln 2-k)=0$ for all x, and this implies that $k=\ln 2$.

b. Proceeding as in part (a), we find that $k=\ln b$.

75. Let $\log_b m=p$. Then $m=b^p$. Therefore, $m^n=(b^p)^n=b^{np}$, and so $\log_b m^n=\log_b b^{np}=np\log_b b=np$ (since $\log_b b=1$) $=n\log_b m$.

3.3 Exponential Functions as Mathematical Models

Problem-Solving Tips

Four mathematical models were introduced in this section:

1. **Exponential growth:** $Q(t)=Q_0e^{kt}$ describes a quantity $Q(t)$ that is initially present in the amount $Q(0)=Q_0$ and whose rate of growth at any time t is directly proportional to the amount of the quantity present at time t.

2. **Exponential decay:** $Q(t)=Q_0e^{-kt}$ describes a quantity $Q(t)$ that is initially present in the amount $Q(0)=Q_0$ and decreases at a rate that is directly proportional to its size.

3. **Learning curves:** $Q(t)=C-Ae^{-kt}$ describes a quantity $Q(t)$, where $Q(0)=C-A$, and $Q(t)$ increases and approaches the number C as t increases without bound.

4. **Logistic growth functions:** $Q(t)=\dfrac{A}{1+Be^{-kt}}$ describes a quantity $Q(t)$, where $Q(0)=\dfrac{A}{1+B}$. Note that $Q(t)$ increases rapidly for small values of t but the rate of growth of $Q(t)$ decreases quickly as t increases. $Q(t)$ approaches the number A as t increases without bound.

Try to familiarize yourself with the examples and graphs for each of these models before you work through the applied problems in this section.

Concept Questions page 181

1. $Q(t)=Q_0e^{kt}$ where $k>0$ represents exponential growth and $k<0$ represents exponential decay. The larger the magnitude of k, the more quickly the former grows and the more quickly the latter decays.

3. $Q(t) = \dfrac{A}{1 + Be^{-kt}}$, where A, B, and k are positive constants. Q increases rapidly for small values of t but the rate of increase slows down as Q (always increasing) approaches the number A.

Exercises page 181

1. **a.** The growth constant is $k = 0.02$.

   b. Initially, there are 300 units present.

   c.

t	0	10	20	100	1000
Q	300	366	448	2217	1.46×10^{11}

3. **a.** $Q(t) = Q_0 e^{kt}$. Here $Q_0 = 100$ and so $Q(t) = 100e^{kt}$. Because the number of cells doubles in 20 minutes, we have $Q(20) = 100e^{20k} = 200$, $e^{20k} = 2$, $20k = \ln 2$, and so $k = \frac{1}{20}\ln 2 \approx 0.03466$. Thus, $Q(t) = 100e^{0.03466t}$.

   b. We solve the equation $100e^{0.03466t} = 1{,}000{,}000$, obtaining $e^{0.03466t} = 10{,}000$, $0.03466t = \ln 10{,}000$, and so $t = \dfrac{\ln 10{,}000}{0.03466} \approx 266$, or 266 minutes.

   c. $Q(t) = 1000e^{0.03466t}$.

5. **a.** We solve the equation $5.3e^{0.02t} = 3(5.3)$, obtaining $e^{0.02t} = 3$, $0.02t = \ln 3$, and so $t = \dfrac{\ln 3}{0.02} \approx 54.93$. Thus, the world population will triple in approximately 54.93 years.

   b. If the growth rate is 1.8%, then proceeding as before, we find $N(t) = 5.3e^{0.018t}$. If $t = 54.93$, the population would be $N(54.93) = 5.3e^{0.018(54.93)} \approx 14.25$, or approximately 14.25 billion.

7. $P(h) = p_0 e^{-kh}$, so $P(0) = p_0 = 15$. Thus, $P(4000) = 15e^{-4000k} = 12.5$, $e^{-4000k} = \frac{12.5}{15}$, $-4000k = \ln\left(\frac{12.5}{15}\right)$, and so $k \approx 0.00004558$. Therefore, $P(12{,}000) \approx 15e^{-0.00004558(12{,}000)} \approx 8.68$, or 8.7 lb/in^2.

9. Suppose the amount of P-32 at time t is given by $Q(t) = Q_0 e^{-kt}$, where Q_0 is the amount present initially and k is the decay constant. Because this element has a half-life of 14.2 days, we have $\frac{1}{2}Q_0 = Q_0 e^{-14.2k}$, so $e^{-14.2k} = \frac{1}{2}$, $-14.2k = \ln\frac{1}{2}$, and $k = -\frac{\ln(1/2)}{14.2} \approx 0.0488$. Therefore, the amount of P-32 present at any time t is given by $Q(t) = 100e^{-0.0488t}$. In particular, the amount left after 7.1 days is given by $Q(7.1) = 100e^{-0.0488(7.1)} = 100e^{-0.34648} \approx 70.717$, or 70.717 grams.

11. From Example 3, we have $k \approx 0.00012$. Next, we solve the equation $0.2Q_0 = Q_0 e^{-0.00012t}$, obtaining $\ln 0.2 = -0.00012t$ and $t = \dfrac{\ln 0.2}{-0.00012} \approx 13{,}412$, or approximately 13,412 years.

13.

   a. $Q(0) = 120(1 - e^0) + 60 = 60$, or 60 wpm.

   b. $Q(10) = 120(1 - e^{-0.5}) + 60 \approx 107.22$, or approximately 107 wpm.

   c. $Q(20) = 120(1 - e^{-1}) + 60 \approx 135.9$, or approximately 136 wpm.

15. The federal debt in 2001 was $f(1) = 5.37e^{0.078} \approx \$5.806$ trillion. In 2006, it was $f(6) = 5.37e^{0.078(6)} \approx \$8.575$ trillion.

17.

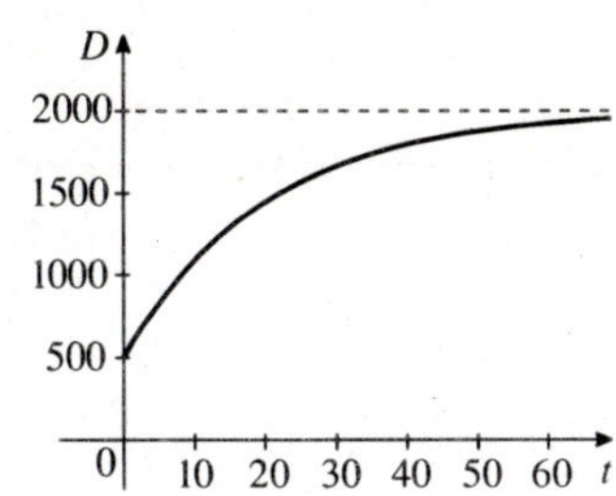

a. After 1 month, the demand is $D(1) = 2000 - 1500e^{-0.05} \approx 573$, after 12 months it is $D(12) = 2000 - 1500e^{-0.6} \approx 1177$, after 24 months it is $D(24) = 2000 - 1500e^{-1.2} \approx 1548$, and after 60 months, it is $D(60) = 2000 - 1500e^{-3} \approx 1925$.

b. As t gets larger, the graph of D approaches 2000. Also, $\lim_{t\to\infty} D(t) = \lim_{t\to\infty} \left(2000 - 1500e^{-0.05t}\right) = 2000$. We conclude that the demand is expected to stabilize at 2000 computers per month.

19. The length is given by $f(6) = 200\left(1 - 0.956e^{-0.18\cdot 6}\right) \approx 135.07$, or approximately 135.1 cm.

21. The proportion of lay teachers is $f(3) = \dfrac{98}{1 + 2.77e^{-3}} \approx 86.1228$, or 86.12%.

23. The projected population of citizens aged 45–64 in 2010 is $P(20) = \dfrac{197.9}{1 + 3.274e^{-0.0361(20)}} \approx 76.3962$, or 76.4 million. In 2015, it is $P(25) = \dfrac{197.9}{1 + 3.274e^{-0.0361(25)}} \approx 85.0164$, or 85.0 million.

25. The first of the given conditions implies that $f(0) = 300$, that is, $300 = \dfrac{3000}{1 + Be^0} = \dfrac{3000}{1 + B}$. Thus, $1 + B = 10$, and $B = 9$. Therefore, $f(t) = \dfrac{3000}{1 + 9e^{-kt}}$. Next, the condition $f(2) = 600$ gives the equation $600 = \dfrac{3000}{1 + 9e^{-2k}}$, so $1 + 9e^{-2k} = 5$, $e^{-2k} = \frac{4}{9}$, and $k = -\frac{1}{2}\ln\frac{4}{9}$. Therefore, $f(t) = \dfrac{3000}{1 + 9e^{(1/2)t\cdot\ln(4/9)}} = \dfrac{3000}{1 + 9\left(\frac{4}{9}\right)^{t/2}}$. The number of students who had heard about the policy four hours later is given by $f(4) = \dfrac{3000}{1 + 9\left(\frac{4}{9}\right)^2} = 1080$, or 1080 students.

27. $Q(t) = \dfrac{A}{1 + Be^{-kt}}$, so $Q(t_1) = \dfrac{A}{1 + Be^{-kt_1}} = Q_1$ implies that $A = Q_1 + Q_1e^{-kt_1}B$, so $e^{-kt_1} = \dfrac{A - Q_1}{BQ_1}$ (1). Next, we have $Q(t_2) = \dfrac{A}{1 + Be^{-kt_2}} = Q_2$, and this leads to $e^{-kt_2} = \dfrac{A - Q_2}{BQ_2}$ (2). Dividing equation (1) by equation (2) gives $\dfrac{e^{-kt_1}}{e^{-kt_2}} = \dfrac{A - Q_1}{BQ_1} \cdot \dfrac{BQ_2}{A - Q_2}$, so $e^{k(t_2 - t_1)} = \dfrac{Q_2(A - Q_1)}{Q_1(A - Q_2)}$, $k(t_2 - t_1) = \ln\dfrac{Q_2(A - Q_1)}{Q_1(A - Q_2)}$, and $k = \dfrac{1}{t_2 - t_1}\ln\dfrac{Q_2(A - Q_1)}{Q_1(A - Q_2)}$.

Technology Exercises page 185

1.

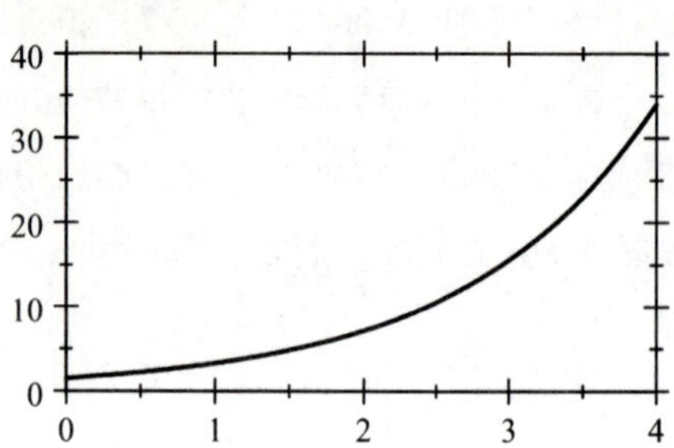

3. a.

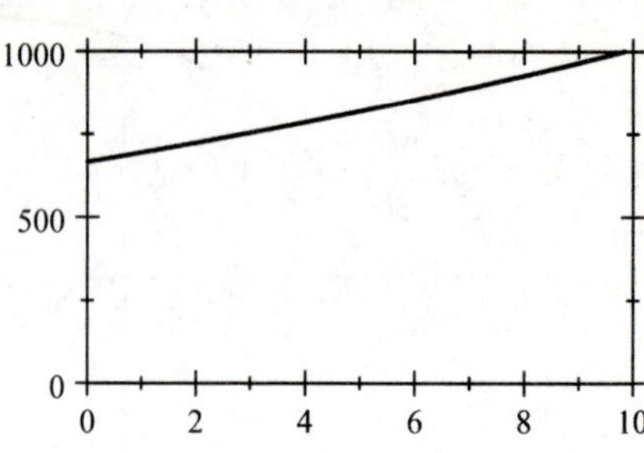

b. $T(0) = 666$ million; $T(10) \approx 1006.6$ million.

5. a.

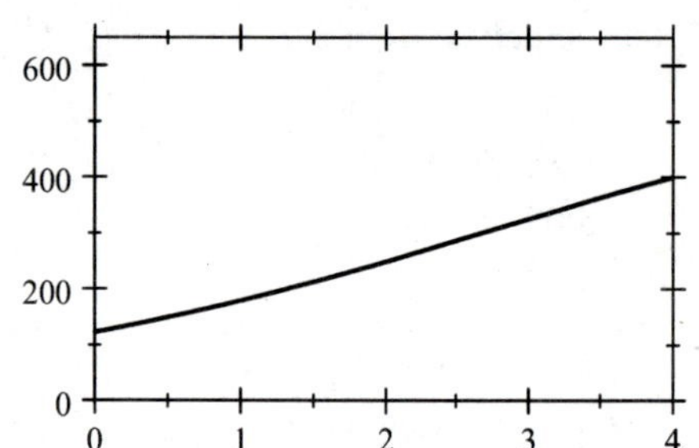

b. $P(3) \approx 325$ million.

7. a.

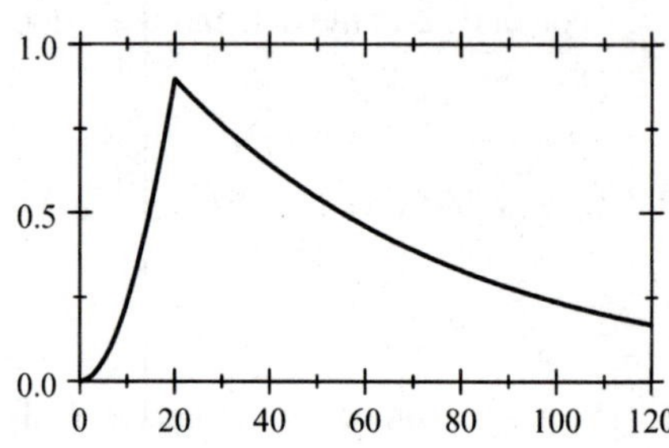

b. The initial concentration is 0.

c. $C(10) \approx 0.237\ \text{g/cm}^3$.

d. $C(30) \approx 0.760\ \text{g/cm}^3$.

9. a. $f(t) = \dfrac{544.61}{1 + 1.65e^{-0.1846t}}$.

b.

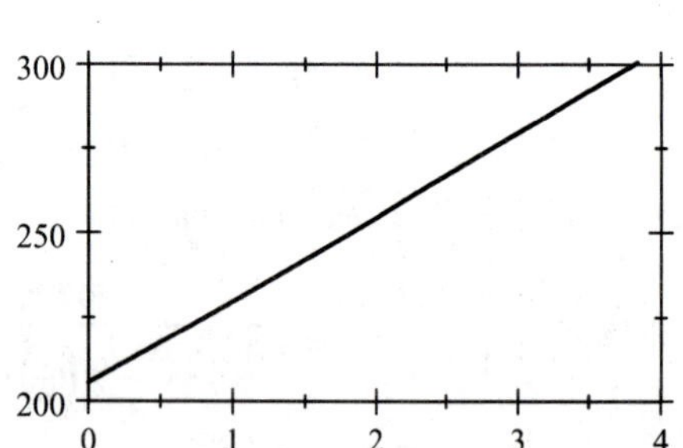

CHAPTER 3 Concept Review Questions page 186

1. power, 0, 1, exponential

3. a. $(0, \infty)$, $(-\infty, \infty)$, $(1, 0)$

b. falls, rises

5. a. initially, growth

b. decay

c. time, one-half

CHAPTER 3 Review Exercises page 187

1. $f(x) = 5^x$.

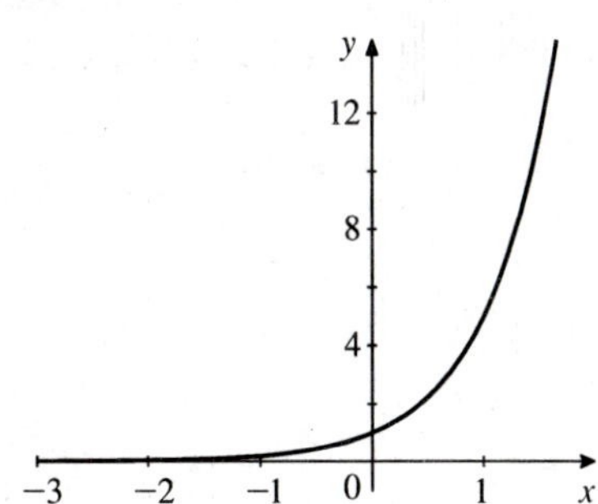

3. $f(x) = \log_4 x$.

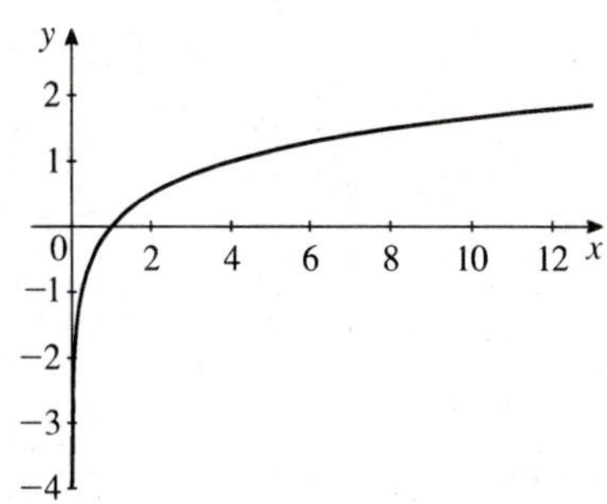

5. If $3^4 = 81$, then $\log_3 81 = 4$.

7. If $\left(\frac{2}{3}\right)^{-3} = \frac{27}{8}$, then $\log_{2/3}\left(\frac{27}{8}\right) = -3$.

9. $\ln 30 = \ln(2 \cdot 3 \cdot 5) = \ln 2 + \ln 3 + \ln 5 \approx 0.6931 + 1.0986 + 1.6094 = 3.4011$.

11. $\ln 3 \cdot 6 = \ln \frac{36}{10} = \ln \frac{18}{5} = \ln 18 - \ln 5 = \ln 3^2 \cdot 2 - \ln 5 = 2\ln 3 + \ln 2 - \ln 5 \approx 2(1.0986) + 0.6931 - 1.6094 = 1.2809$.

13. $\ln 30 = \ln(2 \cdot 3 \cdot 5) = \ln 2 + \ln 3 + \ln 5 = x + y + z$.

15. $\ln 75 = \ln\left(3 \cdot 5^2\right) = \ln 3 + 2\ln 5 = y + 2z$.

17. $e^{x^2+x} = e^2$, so $x^2 + x - 2 = (x+2)(x-1) = 0$ and $x = -2$ or 1.

19. $2^{x^2+x} = 4^{x^2-3} = \left(2^2\right)^{x^2-3} = 2^{2x^2-6}$, so $x^2 + x = 2x^2 - 6$, $x^2 - x - 6 = (x-3)(x+2) = 0$, and $x = -2$ or 3.

21. $\ln(x-1) + \ln 4 = \ln(2x+4) - \ln 2$, so $\ln(x-1) - \ln(2x+4) + \ln 4 + \ln 2 = 0$, $\ln \dfrac{(x-1)(4)(2)}{2x+4} = 0$, $\dfrac{8(x-1)}{2x+4} = 1$, $8(x-1) = 2x+4$, $8x - 8 = 2x + 4$, $6x = 12$, and $x = 2$.

23. $\ln 3^{-2x} = \ln 8$, $-2x \ln 3 = \ln 8$, so $x = -\dfrac{\ln 8}{2\ln 3} \approx -0.9464$.

25. $\ln 2e^{-x} = \ln 7$, so $\ln 2 + \ln e^{-x} = \ln 7$, $-x = \ln 7 - \ln 2$, and $x = \ln 2 - \ln 7 \approx -1.2528$.

27. $\ln e^{2x-1} = \ln 14$, so $2x - 1 = \ln 14$, $2x = 1 + \ln 14$, and $x = \frac{1}{2}(1 + \ln 14) \approx 1.8195$.

29. $\ln 2^{3x+1} = \ln 3^{2x-3}$, so $(3x+1)\ln 2 = (2x-3)\ln 3$, $3x\ln 2 + \ln 2 = 2x\ln 3 - 3\ln 3$, and $x(3\ln 2 - 2\ln 3) = -\ln 2 - 3\ln 3$. Thus, $x = \dfrac{\ln 2 + 3\ln 3}{2\ln 3 - 3\ln 2} \approx 33.8672$.

31. $e^{\sqrt{x}} = \frac{15}{3} = 5$, $\ln e^{\sqrt{x}} = \ln 5$, $\sqrt{x} = \ln 5$, and $x = (\ln 5)^2 \approx 2.5903$.

33. $e^{0.2x} = 8 - 2 = 6$, so $\ln e^{0.2x} = \ln 6$, $0.2x = \ln 6$, and $x = \dfrac{\ln 6}{0.2} \approx 8.9588$.

35. $1 + 2e^{-0.1x} = \frac{30}{5} = 6$, so $2e^{-0.1x} = 5$, $e^{-0.1x} = 2.5$, $-0.1x = \ln 2.5$, $0.1x = -\ln 2.5$, and $x = -\dfrac{\ln 2.5}{0.1} \approx -9.1629$.

37. We first sketch the graph of $y = 3^x - 1$, then reflect this graph with respect to the line $y = x$.

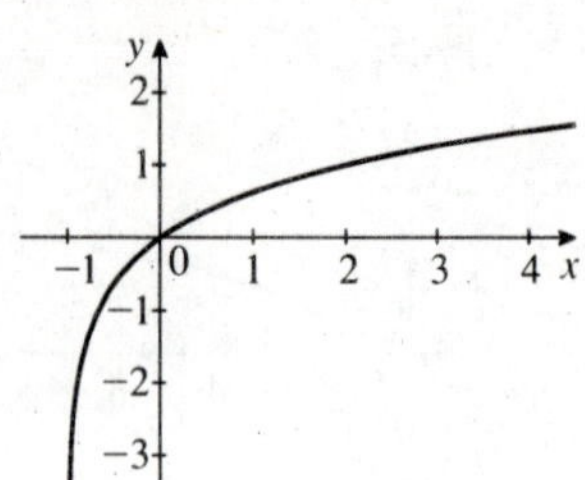

39. We have $Q(t) = Q_0 e^{-kt}$, where Q_0 is the amount of radium present initially. Because the half-life of radium is 1600 years, we have $\frac{1}{2}Q_0 = Q_0 e^{-1600k}$, $e^{-1600k} = \frac{1}{2}$, $-1600k = \ln\frac{1}{2} = -\ln 2$, and $k = \dfrac{\ln 2}{1600} \approx 0.0004332$.

41. We have $Q(10) = 90$, and so $\dfrac{3000}{1 + 499e^{-10k}} = 90$, $1 + 499e^{-10k} = \dfrac{3000}{90}$, $499e^{-10k} = \dfrac{2910}{90}$, $e^{-10k} = \dfrac{2910}{90(499)}$, $-10k = \ln\dfrac{2910}{90(499)}$, and $k = -\dfrac{1}{10}\ln\dfrac{2910}{90(499)} \approx 0.2737$. Thus, $N(t) = \dfrac{3000}{1 + 499e^{-0.2737t}}$. The number of students who have contracted the flu by the 20th day is $N(20) = \dfrac{3000}{1 + 499e^{-0.2737(20)}} \approx 969.93$, or approximately 970 students.

43. a. The initial concentration is $x(0) = 0.08\left(1 - e^{-0.02 \cdot 0}\right) = 0$, or $0\ \text{g/cm}^3$.

b. The concentration after 30 seconds is $x(30) = 0.08\left(1 - e^{-0.02 \cdot 30}\right) = 0.03609$, or $0.0361\ \text{g/cm}^3$.

CHAPTER 3 Before Moving On... page 188

1. $(2x^{-2})^2(9x^{-4})^{1/2} = (2^2x^{-4})(9x^{-4})^{1/2} = 4 \cdot 3x^{-4}x^{-4(1/2)} = 12x^{-6} = \dfrac{12}{x^6}$.

2. Let $u = e^x$. Substituting this value into $e^{2x} - e^x - 6 = 0$, we have $u^2 - u - 6 = 0$, or $(u-3)(u+2) = 0$. Thus, $u = 3$, or $u = -2$. Because e^x is never negative, we discard the negative root. Solving $e^x = 3$ for x, we find that $x = \ln 3$.

3. If $\log_2(x^2 - 8x + 1) = 0$, then $2^0 = x^2 - 8x + 1$, so $1 = x^2 - 8x + 1$. Thus, $x(x-8) = 0$, so $x = 0$ or $x = 8$.

Check $x = 0$: $\log_2 1 \stackrel{?}{=} 0$. Yes, so $x = 0$ is a solution.

Check $x = 8$: $\log_2(8^2 - 8(8) + 1) \stackrel{?}{=} \log_2(1) = 0$. Yes, so $x = 8$ is also a solution.

4. $\dfrac{100}{1 + 2e^{0.3t}} = 40$, so $1 + 2e^{0.3t} = \frac{100}{40} = 2.5$, $2e^{0.3t} = 1.5$, $e^{0.3t} = \frac{1.5}{2} = 0.75$, $0.3t = \ln 0.75$, and $t = \frac{\ln 0.75}{0.3} \approx -0.959$.

5. $T(0) = 200$ gives $70 + ce^0 = 70 + C = 200$, so $C = 130$. Thus, $T(t) = 70 + 130e^{-kt}$. $T(3) = 180$ implies $70 + 130e^{-3k} = 180$, so $130e^{-3k} = 110$, $e^{-3k} = \dfrac{110}{130}$, $-3k = \ln\dfrac{11}{13}$, and $k = -\frac{1}{3}\ln\frac{11}{13} \approx 0.0557$. Therefore, $T(t) = 70 + 130e^{-0.0557t}$. So when $T(t) = 150$, we have $70 + 130e^{-0.0557t} = 150$, $130e^{-0.0557t} = 80$, $e^{-0.0557t} = \frac{80}{130} = \frac{8}{13}$, $-0.0557t = \ln\frac{8}{13}$, and finally $t = -\dfrac{\ln\frac{8}{13}}{0.0557} \approx 8.716$, or approximately 8.7 minutes.

4 MATHEMATICS OF FINANCE

4.1 Compound Interest

Problem-Solving Tips

In this section, you encountered several formulas for computing interest. As you work through the exercises that follow, first decide which formula you need to solve the problem. Then write out your solution. After doing this a few times, you should have the formulas memorized. The key here is to try not to look at the formula in the text, and to work the problem just as if you were taking a test. If you train yourself to work in this manner, test-taking will be a lot easier.

1. First decide if the problem involves *simple interest* or *compound interest.* This will be stated in the problem.

2. Determine whether the problem is asking for the *present value* or *future value* of an amount. For example, if you are asked to determine the value of an investment 5 years from now with interest compounded each year, then use a compound interest formula giving the accumulated amount. If you are asked to determine the current value of an investment that will have a value of \$50,000 five years from now with interest compounded each year, then use a present value formula for compound interest. (If interest is compounded continuously, it will be stated in the problem.)

3. The effective rate of interest is the same as the APR rate that you see in advertisements involving loans. Because the interest for different loans may be compounded over different periods (daily, monthly, biannually, annually, or otherwise) it provides the consumer with a way to compare rates. It is the simple interest rate that would produce the same accumulated amount in 1 year as the nominal rate compounded m times per year.

Concept Questions page 202

1. In simple interest, the interest is based on the original principal. In compound interest, interest earned is periodically added to the principal and thereafter earns interest at the same rate.

3. The effective rate of interest is the simple interest that would produce the same amount in 1 year as the nominal rate compounded m times per year.

Exercises page 202

1. The interest is given by $I = (500)(0.08)(2) = 80$, or \$80. The accumulated amount is $500 + 80$, or \$580.

3. The interest is given by $I = (800)(0.06)(0.75) = 36$, or \$36. The accumulated amount is $800 + 36$, or \$836.

5. We are given that $A = 1160$, $t = 2$, and $r = 0.08$, and we are asked to find P. Because $A = P(1 + rt)$, we see that $P = \dfrac{A}{1 + rt} = \dfrac{1160}{1 + (0.08)(2)} = 1000$, or \$1000.

7. We use the formula $I = Prt$ and solve for t when $I = 20$, $P = 1000$, and $r = 0.03$. Thus, $20 = 1000\,(0.03)\left(\frac{t}{365}\right)$, and $t = \frac{365(20)}{30} = \frac{730}{3}$, or approximately 243 days.

9. We use the formula $A = P\,(1 + rt)$ with $A = 1075$, $P = 1000$, and $t = 0.75$, and solve for r. Thus, $1075 = 1000\,(1 + 0.75r)$, $75 = 750r$, and so $r = 0.10$. Therefore, the annual interest rate is 10%.

11. $A = 1000\,(1 + 0.07)^8 \approx 1718.19$, or \$1718.19.

13. $A = 2500\left(1 + \frac{0.07}{2}\right)^{20} \approx 4974.47$, or \$4974.47.

15. $A = 12{,}000\left(1 + \frac{0.08}{4}\right)^{42} \approx 27{,}566.93$, or \$27,566.93.

17. $A = 150{,}000\left(1 + \frac{0.14}{12}\right)^{48} \approx 261{,}751.04$, or \$261,751.04.

19. $A = 150{,}000\left(1 + \frac{0.12}{365}\right)^{1095} \approx 214{,}986.69$, or \$214,986.69.

21. Using the formula $r_{\text{eff}} = \left(1 + \frac{r}{m}\right)^m - 1$ with $r = 0.10$ and $m = 2$, we have $r_{\text{eff}} = \left(1 + \frac{0.10}{2}\right)^2 - 1 = 0.1025$, or 10.25% annually.

23. Using the formula $r_{\text{eff}} = \left(1 + \frac{r}{m}\right)^m - 1$ with $r = 0.08$ and $m = 12$, we have $r_{\text{eff}} = \left(1 + \frac{0.08}{12}\right)^{12} - 1 \approx 0.08300$, or 8.3% annually.

25. The present value is given by $P = 40{,}000\left(1 + \frac{0.06}{2}\right)^{-8} \approx 31{,}576.37$, or \$31,576.37.

27. The present value is given by $P = 40{,}000\left(1 + \frac{0.07}{12}\right)^{-48} \approx 30{,}255.95$, or \$30,255.95.

29. $A = 5000e^{0.08(4)} \approx 6885.64$, or approximately \$6885.64.

31. We use Formula 3 with $A = 6500$, $P = 5000$, $m = 12$, and $r = 0.12$. Thus, $6500 = 5000\left(1 + \frac{0.12}{12}\right)^{12t}$, so $(1.01)^{12t} = \frac{6500}{5000} = 1.3$, $12t \ln 1.01 = \ln 1.3$, and $t = \frac{\ln 1.3}{12 \ln 1.01} \approx 2.197$. Therefore, it will take approximately 2.2 years.

33. We use Formula 3 with $A = 4000$, $P = 2000$, $m = 12$, and $r = 0.09$. Thus, $4000 = 2000\left(1 + \frac{0.09}{12}\right)^{12t}$, $\left(1 + \frac{0.09}{12}\right)^{12t} = 2$, $12t \ln\left(1 + \frac{0.09}{12}\right) = \ln 2$, and $t = \dfrac{\ln 2}{12 \ln\left(1 + \frac{0.09}{12}\right)} \approx 7.73$. Therefore, it will take approximately 7.7 years.

35. We use Formula 5 with $A = 6000$, $P = 5000$, and $t = 3$. Thus, $6000 = 5000e^{3r}$, and so $e^{3r} = \frac{6000}{5000} = 1.2$. Next, taking logarithms of both sides of the equation, we have $3r = \ln 1.2$ (the natural logarithm of e^{3r} is $3r$), and so $r = \frac{\ln 1.2}{3} \approx 0.06077$. Therefore, the annual interest rate is 6.08%.

37. We use Formula 5 with $A = 7000$, $P = 6000$, and $r = 0.075$. Thus, $7000 = 6000e^{0.075t}$, giving $e^{0.075t} = \frac{7000}{6000} = \frac{7}{6}$. Taking logarithms of both sides and using the fact that the natural logarithm of $e^{0.075t}$ is $0.075t$, we have $0.075t \ln e = \ln \frac{7}{6}$, so $t = \dfrac{\ln \frac{7}{6}}{0.075} \approx 2.055$. Therefore, it will take 2.06 years.

39. Think of \$300 as the principal and \$306 as the accumulated amount at the end of 30 days. If r denotes the simple interest rate per annum, then we have $P = 300$, $A = 306$, and $t = \frac{1}{12}$, and we are required to find r. Using Equation 1(b), we have $306 = 300\left(1 + \frac{r}{12}\right) = 300 + r\left(\frac{300}{12}\right)$ and $r = \left(\frac{12}{300}\right) 6 = 0.24$, or 24% annually.

41. The Abdullahs will owe $A = P(1 + rt) = 120{,}000\left[1 + (0.10)\left(\frac{3}{12}\right)\right] = 123{,}000$, or \$123,000.

43. Here $P = 10{,}000$, $I = 3500$, and $t = 7$, and so from Formula 1(a), we have $3500 = 10{,}000\,(r)\,7$, and so $r = \frac{3500}{70{,}000} = 0.05$. Thus, the bond pays annual simple interest of 5%.

45. Using Equation 1(b) with $A = 15{,}000$, $P = 14{,}650$, and $t = \frac{52}{52} = 1$, we have $15{,}000 = 14{,}650\,(1 + r)$, so $r = \frac{15{,}000}{14{,}650} - 1 \approx 0.0239$. Thus, Maxwell's investment will earn simple interest at a rate of approximately 2.39% annually.

47. The rate that you would expect to pay is $A = 580\,(1 + 0.08)^5 \approx 852.21$, or \$852.21 per day.

49. The amount that they can expect to pay is given by $A = 210{,}000\,(1 + 0.05)^4 \approx 255{,}256$, or approximately \$255,256.

51. The investment will be worth $A = 1.5\left(1 + \frac{0.055}{2}\right)^{20} \approx 2.58064$, or approximately \$2.58 million.

53. We use Formula 3 with $P = 15{,}000$, $r = 0.098$, $m = 12$, and $t = 4$, giving the worth of Jodie's account as $A = 15{,}000\left(1 + \frac{0.098}{12}\right)^{(12)(4)} \approx 22{,}163.753$, or approximately \$22,163.75.

55. Using the formula $P = A\left(1 + \dfrac{r}{m}\right)^{-mt}$, we have $P = 40{,}000\left(1 + \frac{0.085}{4}\right)^{-20} \approx 26{,}267.49$, or \$26,267.49.

57. **a.** They should set aside $P = 100{,}000\,(1 + 0.085)^{-13} \approx 34{,}626.88$, or \$34,626.88.

b. They should set aside $P = 100{,}000\left(1 + \frac{0.085}{2}\right)^{-26} \approx 33{,}886.16$, or \$33,886.16.

c. They should set aside $P = 100{,}000\left(1 + \frac{0.085}{4}\right)^{-52} \approx 33{,}506.76$, or \$33,506.76.

59. The effective annual rate of interest for the Bendix Mutual Fund is $r_{\text{eff}} = \left(1 + \frac{0.104}{4}\right)^4 - 1 \approx 0.1081$, or10.81%, whereas the rate for the Acme Mutual Fund is $r_{\text{eff}} = \left(1 + \frac{0.106}{2}\right)^2 - 1 \approx 0.1088$, or 10.88%.We conclude that the Acme Mutual Fund has a better rate of return.

61. The present value of the \$8000 loan due in 3 years is given by $P = 8000\left(1+\frac{0.10}{2}\right)^{-6} \approx 5969.72$, or \$5969.72. The present value of the \$15,000 loan due in 6 years is given by $P = 15{,}000\left(1+\frac{0.10}{2}\right)^{-12} \approx 8352.56$, or \$8352.56. Therefore, the amount the proprietors of the inn will be required to pay at the end of 5 years is given by $A = 14{,}322.28\left(1+\frac{0.10}{2}\right)^{10} = 23{,}329.48$, or \$23,329.48.

63. Let $A = 10{,}000$, $r = 0.0525$, and $t = 10$. Using Formula 7, we have $P = 10{,}000\,(1+0.0525)^{-10} \approx 5994.86$. Thus, Juan should pay \$5994.86 for the bond.

65. The projected online retail sales for 2009 are $1.243\,(1.14)\,(1.305)\,(1.176)\,(1.105)\,(141.4) \approx 339.79$, or approximately \$339.79 billion.

67. Suppose \$1 is invested in each investment. For Investment A, the accumulated amount is $\left(1+\frac{0.1}{2}\right)^{8} \approx 1.47746$. For Investment B, the accumulated amount is $e^{0.0975(4)} \approx 1.47698$. Thus, Investment A has a higher rate of return.

69. If they invest the money at 10.5% compounded quarterly, they should set aside $P = 70{,}000\left(1+\frac{0.105}{4}\right)^{-28} \approx 33{,}885.14$, or \$33,885.14. If they invest the money at 10.5% compounded continuously, they should set aside $P = 70{,}000e^{-7(0.105)} \approx 33{,}565.38$, or \$33,565.38.

71. $P(t) = V(t)\,e^{-rt} = 80{,}000e^{\sqrt{t}/2}e^{-rt} = 80{,}000e^{(\sqrt{t}/2-0.09t)}$, so $P(4) = 80{,}000e^{1-0.09(4)} \approx 151{,}718.47$, or approximately \$151,718.

73. By definition, $A = P(1+r_{\text{eff}})^t$, so $(1+r_{\text{eff}})^t = \frac{A}{P}$, $1+r_{\text{eff}} = \left(\frac{A}{P}\right)^{1/t}$, and $r_{\text{eff}} = \left(\frac{A}{P}\right)^{1/t} - 1$.

75. Using the formula $r_{\text{eff}} = \left(\frac{A}{P}\right)^{1/t} - 1$ with $A = 256{,}000$, $P = 200{,}000$, and $t = 6$, we have $r_{\text{eff}} = \left(\frac{256{,}000}{200{,}000}\right)^{1/6} - 1 \approx 0.042$, or 4.2%.

77. Using the formula $r_{\text{eff}} = \left(\frac{A}{P}\right)^{1/t} - 1$ with $A = 10{,}000$, $P = 6{,}724.53$, and $t = 7$, we have $r_{\text{eff}} = \left(\frac{10{,}000}{6724.53}\right)^{1/7} - 1 \approx 0.0583$, or 5.83%.

79. True. $A = P(1+rt) = Prt$ is a linear function of t.

81. True. With $m = 1$, the effective rate is $r_{\text{eff}} = \left(1+\frac{r}{1}\right)^1 - 1 = r$.

Technology Exercises

page 208

1. \$5872.78

3. \$475.49

5. 8.95%/yr

7. 10.20%/yr

9. \$29,743.30

11. \$53,303.25

4.2 Annuities

Problem-Solving Tips

1. Note the difference between annuities and the compound interest problems solved in Section 4.1. An annuity is a *sequence of payments* made at regular intervals. If we are asked to find the value of a sequence of payments at some future time, then we use the formula for the future value of an annuity: $S = R\left[\frac{(1+i)^n - 1}{i}\right]$. If we are asked to find the current value of a sequence of payments that will be made over a certain period of time, then we use the formula for the present value of an annuity: $P = R\left[\frac{1-(1+i)^{-n}}{i}\right]$.

2. Note that the problems in this section deal with *ordinary annuities*—annuities in which the payments are made at the end of each payment period.

Concept Questions page 215

1. In an ordinary annuity, the term is fixed, the periodic payments are of the same size, the payments are made at the end of the payment period, and the payments coincide with the interest conversion periods.

3. The future value S of an annuity of n payments of R dollars each, paid at the end of each investment period into an account that earns interest at the rate of i per period, is $S = R\left[\frac{(1+i)^n - 1}{i}\right]$. One example is a retirement fund into which an employee makes a monthly deposit of a fixed amount for a certain period of time.

Exercises page 215

1. $S = 1000\left[\frac{(1+0.1)^{10} - 1}{0.1}\right] \approx 15{,}937.42$, or \$15,937.42.

3. $S = 1800\left[\frac{\left(1+\frac{0.08}{4}\right)^{24} - 1}{\frac{0.08}{4}}\right] \approx 54{,}759.35$, or \$54,759.35.

5. $S = 600\left[\frac{\left(1+\frac{0.12}{4}\right)^{36} - 1}{\frac{0.12}{4}}\right] \approx 37{,}965.57$, or \$37,965.57.

7. $S = 200\left[\frac{\left(1+\frac{0.09}{12}\right)^{243} - 1}{\frac{0.09}{12}}\right] \approx 137{,}209.97$, or \$137,209.97.

9. $P = 5000\left[\frac{1-(1+0.06)^{-8}}{0.06}\right] \approx 31{,}048.97$, or \$31,048.97.

11. $P = 4000\left[\frac{1-(1+0.09)^{-5}}{0.09}\right] \approx 15{,}558.61$, or \$15,558.61.

13. $P = 800\left[\dfrac{1-\left(1+\frac{0.12}{4}\right)^{-28}}{\frac{0.12}{4}}\right] \approx 15{,}011.29$, or \$15,011.29.

15. She will have $S = 1500\left[\dfrac{(1+0.08)^{20}-1}{0.08}\right] \approx 68{,}642.95$, or \$68,642.95.

17. On October 31, Linda's account will be worth $S = 40\left[\dfrac{\left(1+\frac{0.07}{12}\right)^{11}-1}{\frac{0.07}{12}}\right] \approx 453.06$, or \$453.06. One month later, this account will be worth $A = (453.06)\left(1+\frac{0.07}{12}\right) \approx 455.70$, or \$455.70.

19. The amount in Colin's employee retirement account is given by $S = 100\left[\dfrac{\left(1+\frac{0.07}{12}\right)^{144}-1}{\frac{0.07}{12}}\right] \approx 22{,}469.50$, or \$22,469.50. The amount in Colin's IRA is given by $S = 2000\left[\dfrac{(1+0.09)^{8}-1}{0.09}\right] \approx 22{,}056.95$, or \$22,056.95. Therefore, the total amount in his retirement fund is given by $22{,}469.50 + 22{,}056.95 = 44{,}526.45$, or \$44,526.45.

21. To find how much Karen has at age 65, we use Formula 9 with $R = 150$, $i = \frac{r}{m} = \frac{0.05}{12}$, and $n = mt = (12)(40) = 480$, giving $S = 150\left[\dfrac{\left(1+\frac{0.05}{12}\right)^{480}-1}{\frac{0.05}{12}}\right] \approx 228{,}903.02$, or \$228,903.02.

To find how much Matt will have upon attaining the age of 65, we use Formula 9 with $R = 250$, $i = \frac{r}{m} = \frac{0.05}{12}$, and $n = mt = (12)(30) = 360$, giving $S = 250\left[\dfrac{\left(1+\frac{0.05}{12}\right)^{360}-1}{\frac{0.05}{12}}\right] \approx 208{,}064.66$, or \$208,064.66. Therefore, Karen will have the bigger nest egg.

23. The equivalent cash payment is given by $P = 450\left[\dfrac{1-\left(1+\frac{0.09}{12}\right)^{-24}}{\frac{0.09}{12}}\right] \approx 9850.12$, or \$9850.12.

25. We use the formula for the present value of an annuity, obtaining $P = 22\left[\dfrac{1-\left(1+\frac{0.18}{12}\right)^{-36}}{\frac{0.18}{12}}\right] \approx 608.54$, or \$608.54.

27. With an \$2400 monthly payment, the present value of their loan would be

$$P = 2400\left[\frac{1-\left(1+\frac{0.075}{12}\right)^{-360}}{\frac{0.075}{12}}\right] \approx 343{,}242.31$$

or \$343,242.31. With a \$3000 monthly payment, the present value of their loan would be

$$P = 3000\left[\frac{1-\left(1+\frac{0.075}{12}\right)^{-360}}{\frac{0.075}{12}}\right] \approx 429{,}052.88$$

or \$429,052.88. Because they intend to make a \$40,000 down payment, they should consider homes priced from \$383,242 to \$469,053.

29. The lower limit of their house price range is

$$A = 2400\left[\frac{1-\left(1+\frac{0.07}{12}\right)^{-180}}{\frac{0.07}{12}}\right] + 40{,}000 \approx 307{,}014.30,$$

or approximately \$307,104. The upper limit of their house price range is

$$A = 3000\left[\frac{1-\left(1+\frac{0.07}{12}\right)^{-180}}{\frac{0.07}{12}}\right] + 40{,}000 \approx 373{,}767.87,$$

or approximately \$373,768. Therefore, they should consider homes priced from \$307,014 to \$373,768.

31. The deposits of \$200/month into the bank account for a period of 2 years will grow to a sum of

$$A_1 = 200\left[\frac{\left(1+\frac{0.06}{12}\right)^{24}-1}{\frac{0.06}{12}}\right] \approx 5086.391,$$

or \$5086.39. For the next 3 years, this amount will grow into a sum of

$$A_2 = A_1\left(1+\frac{0.06}{12}\right)^{36} = 5086.39\left(1+\frac{0.06}{12}\right)^{36} \approx 6086.784,$$

or \$6086.78. The deposits of \$300/month for a period of three years will grow to a sum of

$$A_3 = 300\left[\frac{\left(1+\frac{0.06}{12}\right)^{36}-1}{\frac{0.06}{12}}\right] \approx 11{,}800.831,$$

or \$11,800.83. Therefore, at the end of 5 years, he will have $A_2 + A_3 = 6086.78 + 11{,}800.83 \approx 17{,}887.61$, or approximately \$17,887.61.

33. Using Equation 9 with $R = 3000$, $r = 0.06$, $m = 1$, and $n = 10$, we see that the amount accumulated in Jacob's account at the end of 10 years is

$$S = 3000\left[\frac{(1+0.06)^{10}-1}{0.06}\right] \approx 39{,}542.38,$$

or \$39,542.38. Then, using Equation 3 with $P = 39{,}542.38$, $r = 0.06$, $m = 1$, and $n = 10$, we see that the amount accumulated in Jacob's account at the end of 20 years is $S = 39{,}542.38\,(1+0.06)^{10} \approx 70{,}814.38$, or \$70,814.38.

35. False. This statement is true only if the interest rate is zero.

Technology Exercises

page 219

1. \$59,622.15 **3.** \$8453.59 **5.** \$35,607.23 **7.** \$13,828.60

9. a.

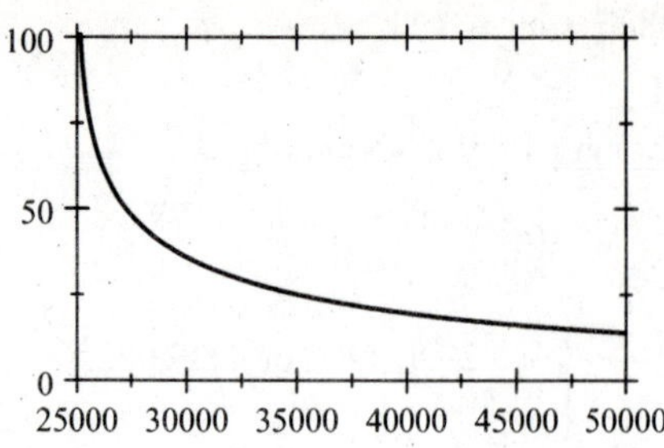

b. $f(x) = 25$ when $x \approx \$35{,}038.78/\text{yr}$.

4.3 Amortization and Sinking Funds

Problem-Solving Tips

1. If a problem asks for the periodic payment that will amortize a loan over n periods, then use the amortization formula $R = \dfrac{Pi}{1-(1+i)^{-n}}$. For example, if you want to calculate the payment for a home mortgage, use this formula to find the payment.

2. If a problem asks for the periodic payment required to accumulate a certain sum of money over n periods, then use the formula $R = \dfrac{iS}{(1+i)^n - 1}$. For example, if a businessman wants to set aside a certain sum of money through periodic payments for the purchase of new equipment, then use this formula to find the payment.

Concept Questions page 228

1. $R = \dfrac{Pi}{1-(1+i)^{-n}}$.

   a. We rewrite $R = \dfrac{Pi}{1-\frac{1}{(1+i)^n}}$. If n increases, then $(1+i)^n$ increases and $\dfrac{1}{(1+i)^n}$ decreases. Therefore $1 - \dfrac{1}{(1+i)^n}$ increases, and so R decreases.

   b. If the principal and interest rate are fixed, and the number of payments is allowed to increase, then the size of each monthly payment decreases.

Exercises page 228

1. The size of each installment is given by $R = \dfrac{100{,}000\,(0.08)}{1-(1+0.08)^{-10}} \approx 14{,}902.95$, or \$14,902.95.

3. The size of each installment is given by $R = \dfrac{5000\,(0.01)}{1-(1+0.01)^{-12}} \approx 444.24$, or \$444.24.

5. The size of each installment is given by $R = \dfrac{25{,}000\,(0.0075)}{1-(1+0.0075)^{-48}} \approx 622.13$, or \$622.13.

7. The size of each installment is $R = \dfrac{80{,}000\,(0.00875)}{1-(1+0.00875)^{-240}} \approx 798.70$, or \$798.70.

9. The required periodic payment is $R = \dfrac{20{,}000\,(0.02)}{(1+0.02)^{12}-1} \approx 1491.19$, or \$1491.19.

11. The required periodic payment is $R = \dfrac{100{,}000\,(0.0075)}{(1+0.0075)^{120}-1} \approx 516.76$, or \$516.76.

13. The required periodic payment is $R = \dfrac{250{,}000\,(0.00875)}{(1+0.00875)^{300}-1} \approx 172.95$, or \$172.95.

15. The required periodic payment is $R = \dfrac{60{,}000\left(\frac{0.10}{4}\right)}{\left(1+\frac{0.10}{4}\right)^{20}-1} \approx 2348.83$, or \$2348.83.

17. The required periodic payment is $R = \dfrac{35{,}000\left(\frac{0.075}{2}\right)}{1-\left(1+\frac{0.075}{2}\right)^{-13}} \approx 3450.87$, or \$3450.87.

19. The size of each installment is given by $R = \dfrac{100{,}000\,(0.10)}{1-(1+0.10)^{-10}} \approx 16{,}274.54$, or \$16,274.54.

21. The monthly payment in each case is given by $R = \dfrac{100{,}000\left(\frac{r}{12}\right)}{1-\left(1+\frac{r}{12}\right)^{-360}}$. Thus, if $r = 0.06$, then $R = \dfrac{100{,}000\left(\frac{0.06}{12}\right)}{1-\left(1+\frac{0.06}{12}\right)^{-360}} \approx 599.55$, or \$599.55. If $r = 0.07$, then $R = \dfrac{100{,}000\left(\frac{0.07}{12}\right)}{1-\left(1+\frac{0.07}{12}\right)^{-360}} \approx 665.30$, or \$665.30. If $r = 0.08$, then $R = \dfrac{100{,}000\left(\frac{0.08}{12}\right)}{1-\left(1+\frac{0.08}{12}\right)^{-360}} \approx 733.76$, or \$733.76. If $r = 0.09$, then $R = \dfrac{100{,}000\left(\frac{0.09}{12}\right)}{1-\left(1+\frac{0.09}{12}\right)^{-360}} \approx 804.62$, or \$804.62.

a. The difference in monthly payments in the two loans is \$877.57 − \$665.30 = \$212.27.

b. The monthly mortgage payment on a \$150,000 mortgage at 10% per year over 30 years would be 1.5 (877.57), or \$1316.36. The monthly mortgage payment on a \$50,000 mortgage at 10% per year over 30 years would be 0.5 (877.57), or \$438.79.

23. a. The amount of the loan required is 20,000 − (0.25) (20,000) or \$15,000. If the car is financed over 36 months, the monthly payments will be $R = \dfrac{15{,}000\left(\frac{0.06}{12}\right)}{1-\left(1+\frac{0.06}{12}\right)^{-36}} \approx 456.33$, or \$456.33. If the car is financed over 48 months, the monthly payments will be $R = \dfrac{15{,}000\left(\frac{0.06}{12}\right)}{1-\left(1+\frac{0.06}{12}\right)^{-48}} \approx 352.28$, or \$352.28.

b. The interest charges for the 36-month plan are 36 (456.33) − 15,000 = 1427.88, or \$1427.88. The interest charges for the 48-month plan are 48 (352.28) − 15,000 = 1909.44, or \$1909.44.

25. The amount borrowed is 270,000 − 30,000 = 240,000, or \$240,000. The size of the monthly installment is $R = \dfrac{240{,}000\left(\frac{0.06}{12}\right)}{1-\left(1+\frac{0.06}{12}\right)^{-360}} \approx 1438.92$, or \$1438.92. To find their equity after five years, we compute the present value of their remaining installments: $P = 1438.92\left[\dfrac{1-\left(1+\frac{0.06}{12}\right)^{-300}}{\frac{0.06}{12}}\right] \approx 223{,}330$, or \$223,330, and so their equity is 270,000 − 223,330 = 46,670, or \$46,670. To find their equity after ten years, we compute $P = 1438.92\left[\dfrac{1-\left(1+\frac{0.06}{12}\right)^{-240}}{\frac{0.06}{12}}\right] \approx 200{,}846$, or \$200,846, so their equity is 270,000 − 200, 846 = 69,154, or \$69,154. To find their equity after twenty years, we compute $P = 1438.92\left[\dfrac{1-\left(1+\frac{0.06}{12}\right)^{-120}}{\frac{0.06}{12}}\right] \approx 129{,}608$, or \$129,608, and their equity is 270,000 − 129,608 = 140,392, or \$140,392.

27. The amount that must be deposited annually into this fund is given by $R = \dfrac{(0.07)(2.5)}{(1+0.07)^{20}-1} = 0.06098231$ million, or approximately \$60,982.31.

29. The amount that must be deposited quarterly into this fund is $R = \dfrac{\left(\frac{0.09}{4}\right)200{,}000}{\left(1+\frac{0.09}{4}\right)^{40}-1} \approx 3{,}135.48$, or \$3,135.48.

31. The size of each monthly installment is given by $R = \dfrac{\left(\frac{0.066}{12}\right)250{,}000}{\left(1+\frac{0.066}{12}\right)^{300}-1} \approx 328.67$, or \$328.67.

33. Here $S = 450{,}000$, $i = \frac{0.06}{12}$, and $n = mt = (12)(30) = 360$. Thus, Formula 9 gives $450{,}000 = R\left[\dfrac{\left(1+\frac{0.06}{12}\right)^{360}-1}{\left(\frac{0.06}{12}\right)}\right]$, and so $R = \dfrac{450{,}000\left(\frac{0.06}{12}\right)}{\left(1+\frac{0.06}{12}\right)^{360}-1} \approx 447.98$. Her monthly payment is \$447.98.

35. The value of the IRA account after 20 years is $S = 375\left[\dfrac{\left(1+\frac{0.06}{4}\right)^{80}-1}{\frac{0.06}{4}}\right] \approx 57{,}266.57$, or $57,266.57. The payment he would receive at the end of each quarter for the next 15 years is given by $R = \dfrac{\left(\frac{0.06}{4}\right)57{,}266.57}{1-\left(1+\frac{0.06}{4}\right)^{-60}} \approx 1454.19$, or $1454.19. If he continues working and makes quarterly payments until age 65, the value of the IRA account would be $S = 375\left[\dfrac{\left(1+\frac{0.06}{4}\right)^{100}-1}{\frac{0.06}{4}}\right] \approx 85{,}801.14$, or $85,801.14. The payment he would receive at the end of each quarter for the next 10 years is given by $R = \dfrac{\left(\frac{0.06}{4}\right)85{,}801.14}{1-\left(1+\frac{0.06}{4}\right)^{-40}} \approx 2868.08$, or $2868.08.

37. The amount of the sinking fund Jason needs to accumulate is found using Equation 11 with $R = 8000$, $i = \dfrac{r}{m} = \dfrac{0.06}{12}$, and $n = 12 \cdot 20 = 240$. Thus, $S = \dfrac{8000\left[1-\left(1+\frac{0.06}{12}\right)^{-240}\right]}{\frac{0.06}{12}} \approx 1{,}116{,}646.173$, or $1,116,646.17. Next, using Equation 15 with $S = 1{,}116{,}646.17$, $i = \frac{0.06}{12}$, and $n = 360$, we obtain $R = \dfrac{1,116,646.17\left(\frac{0.06}{12}\right)}{\left(1+\frac{0.06}{12}\right)^{360}-1} \approx 1111.63$. Thus, he must contribute $1111.63 per month.

39. Using Equation 14 with $R = 400$, $i = \frac{0.072}{12}$, and $n = 48$, we have $P = 400\left[\dfrac{1-\left(1+\frac{0.072}{12}\right)^{-48}}{\frac{0.072}{12}}\right] \approx 16{,}639.53$. Because he can get $8000 for the trade-in, Dan can afford a car that costs no more than $24,639.53.

41. The monthly payment the Sandersons are required to make under the terms of their original loan is given by $R = \dfrac{100{,}000\left(\frac{0.10}{12}\right)}{1-\left(1+\frac{0.10}{12}\right)^{-240}} \approx 965.02$, or $965.02. Their monthly payment under the terms of their new loan is given by $R = \dfrac{100{,}000\left(\frac{0.078}{12}\right)}{1-\left(1+\frac{0.078}{12}\right)^{-240}} \approx 824.04$, or $824.04. The amount of money that the Sandersons can expect to save over the life of the loan by refinancing is given by $240\,(965.02 - 824.04) = 33{,}835.20$, or $33,835.20.

43. As of now, Paul owes his sister $A = Pe^{rt} = 10{,}000e^{(0.06)(2)} \approx 11{,}274.9685$, or $11,274.97. To repay the loan, Paul's monthly payment will be $R = \dfrac{Pi}{1-(1+i)^{-n}} = \dfrac{11{,}274.97\left(\frac{0.05}{12}\right)}{1-\left(1+\frac{0.05}{12}\right)^{-60}} \approx 212.773$, or approximately $212.77 per month.

45. Kim's monthly payment is found using Formula 13 with $P = 180{,}000$, $i = \dfrac{r}{m} = \dfrac{0.095}{12}$, and $n = (12)(30) = 360$.

Thus, $R = 180{,}000\left[\dfrac{\frac{0.095}{12}}{1-\left(1+\frac{0.095}{12}\right)^{-360}}\right] \approx 1513.5376$, or \$1513.54. After 8 years, he has made $8 \cdot 12 = 96$ payments. His outstanding principal is given by the sum of the remaining $360 - 96 = 264$ installments. Using Formula 11, we find $P = 1513.54\left[\dfrac{1-\left(1+\frac{0.095}{12}\right)^{-264}}{\frac{0.095}{12}}\right] \approx 167{,}341.592$, so his outstanding principal is \$167,341.59.

47. To find Emilio's monthly payment, we use Equation 13 with $P = 280{,}000$, $r = 0.075$, $m = 12$, and $t = 30$, obtaining $R = \dfrac{280{,}000\left(\frac{0.075}{12}\right)}{1-\left(1+\frac{0.075}{12}\right)^{-360}} \approx 1957.80$, or \$1957.80 per month. After $7 \cdot 12 = 84$ payments have been made, there are 276 payments remaining. The present value of an annuity with $n = 276$, $R = 1957.80$, and $i = \frac{0.075}{12}$ is $P = 1957.80\left[\dfrac{1-\left(1+\frac{0.075}{12}\right)^{-276}}{\frac{0.075}{12}}\right] \approx 257{,}135.23$, so Emilio's balloon payment is \$257,135.23.

49. **a.** Here $P = 200{,}000$, $i = \dfrac{r}{m} = \dfrac{0.095}{12}$, and $n = mt = 12 \cdot 30 = 360$. Therefore,

$$R = \frac{200{,}000\left(\frac{0.095}{12}\right)}{1-\left(1+\frac{0.095}{12}\right)^{-360}} \approx 1681.7084,$$ and so her monthly payment is \$1681.71.

b. After $4 \cdot 12 = 48$ monthly payments have been made, her outstanding principal is given by the sum of the present values of the remaining $360 - 48 = 312$ installments. Using Formula 11, we find it to be

$$P = 1681.71\left[\frac{1-\left(1+\frac{0.095}{12}\right)^{-312}}{\frac{0.095}{12}}\right] \approx 194{,}282.8524,$$ or approximately \$194,282.85.

c. Here $P = 194{,}282.85$, $i = \dfrac{r}{m} = \dfrac{0.0675}{12}$, and $n = 12 \cdot 30 = 360$. Thus,

$$R = \frac{194{,}282.85\left(\frac{0.0675}{12}\right)}{1-\left(1+\frac{0.0675}{12}\right)^{-360}} \approx 1260.1149,$$ and so her new monthly payment is \$1260.11.

d. The difference in the monthly payments is given by $1681.71 - 1260.11$, or \$421.60 per month. (Note that the term of her mortgage is also extended by 4 years.)

51. First we find Samantha's monthly payment on the original loan amount. Here $P = 150{,}000$, $i = \frac{r}{m} = \frac{0.075}{12}$, and $n = mt = 12 \cdot 30 = 360$. Therefore, $R = \frac{150{,}000\left(\frac{0.075}{12}\right)}{1 - \left(1 + \frac{0.075}{12}\right)^{-360}} \approx 1048.8218$, or \$1048.82. Next, to find her current outstanding principal, observe that this is just the sum of the present values of the $360 - 36 = 324$ remaining payments. Using Formula 11, we have $P = 1048.82\left[\frac{1 - \left(1 + \frac{0.075}{12}\right)^{-324}}{\frac{0.075}{12}}\right] \approx 145{,}521.1271$, or \$145,521.13.

Finally, using Formula 13 with $P = 145{,}521.13$, $i = \frac{r}{m} = \frac{0.07}{12}$, and $n = mt = 12 \cdot 27 = 324$, we find $R = \frac{145{,}521.13\left(\frac{0.07}{12}\right)}{1 - \left(1 + \frac{0.07}{12}\right)^{-324}} \approx 1000.9161$, and so Samantha's new monthly payment will be \$1000.92 per month.

53. The amount of the loan the Meyers need to secure is \$280,000. Using the bank's financing, the monthly payment would be $R = \frac{280{,}000\left(\frac{0.055}{12}\right)}{1 - \left(1 + \frac{0.055}{12}\right)^{-300}} \approx 1719.44$, or \$1719.44. Using the seller's financing, the monthly payment would be $R = \frac{280{,}000\left(\frac{0.049}{12}\right)}{1 - \left(1 + \frac{0.049}{12}\right)^{-300}} \approx 1620.58$, or \$1620.58. By choosing the seller's financing rather than the bank's, the Meyers would save $(1719.44 - 1620.58)(300) = 29{,}658$, or \$29,658 in interest.

55. a. Here $i = \frac{r}{m} = \frac{0.054}{12} = 0.0045$, $n = 12 \cdot 5 = 60$, and $R = 3000$. Using Equation 12, we find $P = 3000\left[\frac{1 - (1 + 0.0045)^{-360}}{0.0045}\right] \approx 534{,}253.87$, or \$534,253.87. Thus, if the Carlsons choose the 5/1 ARM, they can borrow at most \$534,253.87.

b. If P denotes the maximum amount that the Carlsons can borrow, then we have $12 \cdot 3000 = 0.0562P$, so $P = \frac{12(3000)}{0.0562} \approx 640{,}569.40$. Thus, if the Carlsons choose the interest-only loan, they can borrow at most \$640,569.40.

Technology Exercises page 234

1. \$628.02 **3.** \$1685.47 **5.** \$1960.96 **7.** \$894.12

9. The annual payment is \$18,288.92. The amortization schedule follows.

End of Period	Interest Charged	Repayment Made	Payment toward Principal	Outstanding Principal
0				\$120,000.00
1	\$10,200.00	\$18,288.92	\$ 8088.92	111,911.08
2	9512.44	18,288.92	8776.48	103,134.60
3	8766.44	18,288.92	9522.48	93,612.12
4	7957.03	18,288.92	10,331.89	83,280.23
5	7078.82	18,288.92	11,210.10	72,070.13
6	6125.96	18,288.92	12,162.96	59,907.17
7	5092.11	18,288.92	13,196.81	46,710.36
8	3970.38	18,288.92	14,318.54	32,391.82
9	2753.30	18,288.92	15,535.62	16,856.20
10	1432.78	18,288.98	16,856.14	0

4.4 Arithmetic and Geometric Progressions

Problem-Solving Tips

Note the difference between an arithmetic progression and a geometric progression:

An *arithmetic progression* is a sequence of numbers in which each term after the first is obtained by *adding a constant d to the preceding term.*

A *geometric progression* is a sequence of numbers in which each term after the first is obtained by *multiplying the preceding term by a constant r.*

Concept Questions page 240

1. a. $a_n = a + (n-1)\,d$

b. $S_n = \dfrac{n}{2}\left[2a + (n-1)\,d\right]$

Exercises page 241

1. $a_9 = 6 + (9-1)\,3 = 30$

3. $a_8 = -15 + (8-1)\left(\frac{3}{2}\right) = -\frac{9}{2} = -4.5$

5. $a_{11} - a_4 = (a_1 + 10d) - (a_1 + 3d) = 7d$. Also, $a_{11} - a_4 = 107 - 30 = 77$. Therefore, $7d = 77$ and $d = 11$. Next, $a_4 = a + 3d = a + 3\,(11) = a + 33 = 30$ and $a = -3$. Therefore, the first five terms are -3, 8, 19, 30, 41.

7. Here $a = x$, $n = 7$, and $d = y$. Therefore, the required term is $a_7 = x + (7-1)\,y = x + 6y$.

9. Using the formula for the sum of the terms of an arithmetic progression with $a = 4$, $d = 7$ and $n = 15$, we have $S_n = \frac{n}{2}\left[2a + (n-1)\,d\right]$, so $S_{15} = \frac{15}{2}\left[2\,(4) + (15-1)\,7\right] = \frac{15}{2}\,(106) = 795$.

11. The common difference is $d = 2$ and the first term is $a = 15$. Using the formula for the nth term $a_n = a + (n-1)\,d$, we have $57 = 15 + (n-1)\,(2) = 13 + 2n$, so $2n = 44$ and $n = 22$. Using the formula for the sum of the terms of an arithmetic progression with $a = 15$, $d = 2$ and $n = 22$, we have $S_n = \frac{n}{2}\left[2a + (n-1)\,d\right]$, so $S_{22} = \frac{22}{2}\left[2\,(15) + (22-1)\,2\right] = 11\,(72) = 792$.

13. $f(1)+f(2)+f(3)+\cdots+f(22)=[3(1)-4]+[3(2)-4]+[3(3)-4]+\cdots+[3(22)-4]$
$=3(1+2+3+\cdots+22)+22(-4)=3\left(\frac{22}{2}\right)[2(1)+(22-1)1]-88=671.$

15. $S_n=\frac{n}{2}[2a_1+(n-1)d]=\frac{n}{2}[a_1+a_1+(n-1)d]=\frac{n}{2}(a_1+a_n).$

a. $S_{11}=\frac{11}{2}(3+47)=275.$ **b.** $S_{20}=\frac{20}{2}[5+(-33)]=-280.$

17. Let n be the number of weeks until she reaches 10 miles. Then $a_n=1+(n-1)\frac{1}{4}=1+\frac{1}{4}n-\frac{1}{4}=\frac{1}{4}n+\frac{3}{4}=10.$ Therefore, $n+3=40$, so $n=37$; that is, at the beginning of the 37th week.

19. To compute Kunwoo's fare by taxi, take $a=2$, $d=1.20$, and $n=25$. Then the required fare is given by $a_{25}=2+(25-1)1.20=30.8$, or \$30.80. Therefore, by taking the airport limousine, Kunwoo will save $30.80-15.00=15.80$, or \$15.80.

21. **a.** Using the formula for the sum of an arithmetic progression, we have
$S_N=\frac{N}{2}[2a+(N-1)d]=\frac{N}{2}[2(1)+(N-1)(1)]=\frac{N}{2}(N+1).$

b. $S_{10}=\frac{10}{2}(10+1)=5(11)=55$, so
$$D_3=(C-S)\frac{N-(n-1)}{S_N}=(6000-500)\frac{10-(3-1)}{55}=5500\left(\frac{8}{55}\right)=800, \text{ or } \$800.$$

23. This is a geometric progression with $a=4$ and $r=2$. Next, $a_7=4(2)^6=256$ and $S_7=\dfrac{4(1-2^7)}{1-2}=508.$

25. If we compute the ratios $\dfrac{a_2}{a_1}=\dfrac{-\frac{3}{8}}{\frac{1}{2}}=-\dfrac{3}{4}$ and $\dfrac{a_3}{a_2}=\dfrac{\frac{1}{4}}{-\frac{3}{8}}=-\dfrac{2}{3}$, we see that the given sequence is not geometric because the ratios are not equal.

27. This is a geometric progression with $a=243$, and $r=\frac{1}{3}$. Thus, $a_7=243\left(\frac{1}{3}\right)^6=\frac{1}{3}$ and
$$S_7=\frac{243\left[1-\left(\frac{1}{3}\right)^7\right]}{1-\frac{1}{3}}=\frac{1093}{3}.$$

29. First, we compute $r=\dfrac{a_2}{a_1}=\dfrac{3}{-3}=-1$. Next, $a_{20}=-3(-1)^{19}=3$, and so $S_{20}=\dfrac{-3\left[1-(-1)^{20}\right]}{1-(-1)}=0.$

31. The population in five years is expected to be $200{,}000(1.08)^{6-1}=200{,}000(1.08)^5\approx 293{,}866.$

33. The salary of a union member whose salary was \$42,000 six years ago is given by the 7th term of a geometric progression whose first term is 42,000 and whose common ratio is 1.05. Thus, $a_7=(42{,}000)(1.05)^6\approx 56{,}284.02$, or \$56,284.

35. With 8% raises per year, the employee would make $S_4=48{,}000\left[\dfrac{1-(1.08)^4}{1-1.08}\right]\approx 216{,}293.38$, or \$216,293.38 over the next four years. With \$4000 raises per year, the employee would make $S_4=\frac{4}{2}[2(48{,}000)+(4-1)4000]=216{,}000$ or \$216,000 over the next four years. We conclude that the employee should choose annual raises of 8% per year.

37. a. During the sixth year, she will receive $a_6 = 10{,}000\,(1.15)^5 \approx 20{,}113.57$, or \$20,113.57.

b. The total amount of the six payments will be given by $S_6 = \dfrac{10{,}000[1-(1.15)^6]}{1-1.15} \approx 87{,}537.38$, or \$87,537.38.

39. The book value of the office equipment at the end of the eighth year is given by $V(8) = 150{,}000\left(1-\frac{2}{10}\right)^8 \approx 25{,}165.82$, or \$25,165.82.

41. The book value of the restaurant equipment at the end of six years is given by $V(6) = 150{,}000\,(0.8)^6 = 39{,}321.60$, or \$39,321.60. By the end of the sixth year, the equipment will have depreciated by $D(6) = 150{,}000 - 39{,}321.60 = 110{,}678.40$, or \$110,678.40.

43. True. Suppose d is the common difference of $a_1, a_2, \ldots, a_n$ and e is the common difference of $b_1, b_2, \ldots, b_n$. Then $d+e$ is the common difference of $a_1+b_1, a_2+b_2, \ldots, a_n+b_n$, and we see that the latter is indeed an arithmetic progression.

CHAPTER 4 Concept Review Questions page 243

1. a. original, $P(1+rt)$

b. interest, $P(1+i)^n$, $A(1+i)^{-n}$

3. annuity, ordinary annuity, simple annuity

5. $\dfrac{Pi}{1-(1+i)^{-n}}$

7. constant d, $a+(n-1)\,d$, $\frac{n}{2}[2a+(n-1)\,d]$

CHAPTER 4 Review Exercises page 244

1. a. Here $P = 5000$, $r = 0.1$, and $m = 1$. Thus, $i = r = 0.1$ and $n = 4$, so $A = 5000\,(1.1)^4 = 7320.5$, or \$7320.50.

b. Here $m = 2$, so $i = \frac{0.1}{2} = 0.05$ and $n = 4 \cdot 2 = 8$. Thus, $A = 5000\,(1.05)^8 \approx 7387.28$, or \$7387.28.

c. Here $m = 4$, so $i = \frac{0.1}{4} = 0.025$ and $n = 4 \cdot 4 = 16$. Thus, $A = 5000\,(1.025)^{16} \approx 7{,}422.53$, or \$7422.53.

d. Here $m = 12$, so that $i = 0.1/12$ and $n = 4 \cdot 12 = 48$. Thus, $A = 5000\left(1+\frac{0.10}{12}\right)^{48} \approx 7446.77$, or \$7446.77.

3. a. The effective rate of interest is given by $r_{\text{eff}} = \left(1+\frac{r}{m}\right)^m - 1 = (1+0.12) - 1 = 0.12$, or 12%.

b. The effective rate of interest is given by $r_{\text{eff}} = \left(1+\frac{r}{m}\right)^m - 1 = \left(1+\frac{0.12}{2}\right)^2 - 1 = 0.1236$, or 12.36%.

c. The effective rate of interest is given by $r_{\text{eff}} = \left(1+\frac{r}{m}\right)^m - 1 = \left(1+\frac{0.12}{4}\right)^4 - 1 \approx 0.125509$, or 12.5509%.

d. The effective rate of interest is given by $r_{\text{eff}} = \left(1+\frac{r}{m}\right)^m - 1 = \left(1+\frac{0.12}{12}\right)^{12} - 1 \approx 0.126825$, or 12.6825%.

5. The present value is given by $P = 41{,}413\left(1+\frac{0.065}{4}\right)^{-20} \approx 30{,}000.29$, or approximately \$30,000.

7. $S = 150\left[\dfrac{\left(1+\frac{0.08}{4}\right)^{28}-1}{\frac{0.08}{4}}\right] \approx 5557.68$, or \$5557.68.

9. Using the formula for the present value of an annuity with $R = 250$, $n = 36$, and $i = \frac{0.09}{12} = 0.0075$, we have

$$P = 250\left[\frac{1-(1.0075)^{-36}}{0.0075}\right] \approx 7861.70, \text{ or } \$7861.70.$$

11. Using the amortization formula with $P = 22{,}000$, $n = 36$, and $i = \frac{0.085}{12}$, we find

$$R = \frac{22{,}000\left(\frac{0.085}{12}\right)}{1-\left(1+\frac{0.085}{12}\right)^{-36}} \approx 694.49, \text{ or } \$694.49.$$

13. Using the sinking fund formula with $S = 18{,}000$, $n = 48$, and $i = \frac{0.06}{12}$, we have $R = \dfrac{\left(\frac{0.06}{12}\right)18{,}000}{\left(1+\frac{0.06}{12}\right)^{48}-1} \approx 332.73$, or \$332.73.

15. The effective rate of interest is given by $r_{\text{eff}} = \left(1+\frac{r}{m}\right)^m - 1 = \left(1+\frac{0.072}{12}\right)^{12} - 1 \approx 0.07442$, or 7.442%.

17. $P = 119{,}346e^{-(0.1)4} \approx 80{,}000$, or \$80,000.

19. At the end of five years, the investment will be worth $A = P\left(1+\frac{r}{m}\right)^{mt} = 4.2\left(1+\frac{0.054}{4}\right)^{4(5)} \approx 5.491922$ (million), or \$5,491,922.

21. Using the present value formula for compound interest, we have

$$P = A\left(1+\frac{r}{m}\right)^{-mt} = 19{,}440.31\left(1+\frac{0.065}{12}\right)^{-12(4)} \approx 15{,}000.00, \text{ or } \$15{,}000.$$

23. The future value of his investment is given by $A = P\left(1+r_{\text{eff}}\right)^t$. Therefore, $34{,}616 = 24{,}000\left(1+r_{\text{eff}}\right)^5$, $1+r_{\text{eff}} = \left(\frac{34{,}616}{24{,}000}\right)^{1/5}$, and $r_{\text{eff}} = \left(\frac{34{,}616}{24{,}000}\right)^{1/5} - 1 \approx 0.076002$, or approximately 7.6%.

25. Using the formula for the future value of an annuity with $R = 2(200) = 400$, $n = 120$, and $i = \frac{0.08}{12}$, we have

$$S = 400\left[\frac{\left(1+\frac{0.08}{12}\right)^{120}-1}{\frac{0.08}{12}}\right] \approx 73{,}178.41, \text{ or } \$73{,}178.41.$$

27. Using the formula for the present value of an annuity, we see that the purchase price of the furniture is

$$400 + P = 400 + 75.32\left[\frac{1-\left(1+\frac{0.12}{12}\right)^{-24}}{\frac{0.12}{12}}\right] \approx 400 + 1600.05, \text{ or approximately } \$2000.$$

29. **a.** The monthly payment is given by $P = \dfrac{(120{,}000)(0.0075)}{1-(1+0.0075)^{-180}} \approx 1217.1199$, or \$1217.12.

b. We can find the total interest payment by computing $180(1217.12) - 120{,}000 \approx 99{,}081.60$, or \$99,081.60.

c. We first compute the present value of their remaining payments:

$P = 1217.12\left[\dfrac{1-(1+0.0075)^{-60}}{0.0075}\right] \approx 58{,}632.78$ or \$58,632.78. Then their equity is $150{,}000 - 58{,}632.78$, or approximately \$91,367.

31. Using the sinking fund formula with $S = 120{,}000$, $n = 24$, and $i = \frac{0.058}{12}$, we find that the amount of each installment should be $R = \dfrac{\left(\frac{0.058}{12}\right)120{,}000}{\left(1+\frac{0.058}{12}\right)^{24}-1} \approx 4727.67$, or \$4727.67.

33. We use Formula 11 with $R = 250$, $r = 0.05$, $m = 12$, and $n = 9$, obtaining

$P = 250\left[\dfrac{1-\left(1+\frac{0.05}{12}\right)^{-9}}{\frac{0.05}{12}}\right] \approx 2203.83$, or \$2203.83. Thus, Matt's parents need to deposit \$2203.83.

CHAPTER 4 Before Moving On... page 245

1. Here $P = 2000$, $r = 0.08$, $t = 3$, and $m = 12$, so $i = \frac{0.08}{12}$. Therefore, $A = 2000\left(1+\frac{0.08}{12}\right)^{(12)(3)} \approx 2540.47$, or \$2540.47.

2. Here $r = 0.06$ and $m = 365$, so $r_{\text{eff}} = \left(1+\frac{0.06}{365}\right)^{365} - 1 \approx 0.0618$, or approximately 6.18% per year.

3. Here $R = 800$, $r = 0.06$, $m = 52$, and $n = 10 \cdot 52 = 520$, so $S = 800\left[\dfrac{\left(1+\frac{0.06}{52}\right)^{520}-1}{\frac{0.06}{52}}\right] \approx 569{,}565.47$, or \$569,565.47.

4. Here $P = 100{,}000$, $t = 10$, $r = 0.08$, $m = 12$, so $R = \dfrac{100{,}000\left(\frac{0.08}{12}\right)}{1-\left(1+\frac{0.08}{12}\right)^{-120}} \approx 1213.276$, or approximately \$1213.28.

5. Here $S = 15{,}000$, $t = 6$, $r = 0.1$, and $m = 52$, so $R = \dfrac{\left(\frac{0.1}{52}\right)15{,}000}{\left(1+\frac{0.1}{52}\right)^{(6)(52)}-1} \approx 35.132$, or \$35.13.

6. a. Here $a_1 = a = 3$ and $d = 4$, so with $n = 10$, $S_{10} = \frac{10}{2}\left[2(3)+9(4)\right] = 210$.

b. Here $a = \frac{1}{2}$ and $r = 2$, so with $n = 8$, $S_8 = \dfrac{\frac{1}{2}\left[1-(2)^8\right]}{1-2} = 127.5$

5 SYSTEMS OF LINEAR EQUATIONS AND MATRICES

5.1 Systems of Linear Equations: An Introduction

Concept Questions page 253

1. a. There may be no solution, a unique solution, or infinitely many solutions.

b. There is no solution if the two lines represented by the given system of linear equations are parallel and distinct; there is a unique solution if the two lines intersect at precisely one point; there are infinitely many solutions if the two lines are parallel and coincident.

Exercises page 253

1. Solving the first equation for x, we find $x = 3y - 1$. Substituting this value of x into the second equation yields $4(3y - 1) + 3y = 11$, so $12y - 4 + 3y = 11$, $15y = 15$, and $y = 1$. Substituting this value of y into the first equation gives $x = 3(1) - 1 = 2$. Therefore, the unique solution of the system is $(2, 1)$.

3. Solving the first equation for x, we have $x = 7 - 4y$. Substituting this value of x into the second equation, we have $\frac{1}{2}(7 - 4y) + 2y = 5$, so $7 - 4y + 4y = 10$, and $7 = 10$. Clearly, this is impossible and we conclude that the system of equations has no solution.

5. Solving the first equation for x, we obtain $x = 7 - 2y$. Substituting this value of x into the second equation, we have $2(7 - 2y) - y = 4$, so $14 - 4y - y = 4$, $-5y = -10$, and $y = 2$. Then $x = 7 - 2(2) = 7 - 4 = 3$. We conclude that the solution to the system is $(3, 2)$.

7. Solving the first equation for x, we have $2x = 5y + 10$, so $x = \frac{5}{2}y + 5$. Substituting this value of x into the second equation, we have $6\left(\frac{5}{2}y + 5\right) - 15y = 30$, $15y + 30 - 15y = 30$, and $0 = 0$. This result tells us that the second equation is equivalent to the first. Thus, any ordered pair of numbers (x, y) satisfying the equation $2x - 5y = 10$ (or $6x - 15y = 30$) is a solution to the system. In particular, by assigning the value t to x, where t is any real number, we find that $y = -2 + \frac{2}{5}t$ so the ordered pair, $\left(t, \frac{2}{5}t - 2\right)$ is a solution to the system, and we conclude that the system has infinitely many solutions.

9. Solving the first equation for x, we obtain $4x - 5y = 14$, so $4x = 14 + 5y$, and $x = \frac{14}{4} + \frac{5}{4}y = \frac{7}{2} + \frac{5}{4}y$. Substituting this value of x into the second equation gives $2\left(\frac{7}{2} + \frac{5}{4}y\right) + 3y = -4$, so $7 + \frac{5}{2}y + 3y = -4$, $\frac{11}{2}y = -11$, and $y = -2$. Thus, $x = \frac{7}{2} + \frac{5}{4}(-2) = 1$. We conclude that the ordered pair $(1, -2)$ satisfies the given system of equations.

11. Solving the first equation for x, we obtain $2x = 3y + 6$, so $x = \frac{3}{2}y + 3$. Substituting this value of x into the second equation gives $6\left(\frac{3}{2}y + 3\right) - 9y = 12$, so $9y + 18 - 9y = 12$ and $18 = 12$. which is impossible. We conclude that the system of equations has no solution.

13. Solving the first equation for x, we obtain $-3x = -5y + 1$, so $x = \frac{5}{3}y - \frac{1}{3}$. Substituting this value of y into the second equation yields $2\left(\frac{5}{3}y - \frac{1}{3}\right) - 4y = -1$, $\frac{10}{3}y - \frac{2}{3} - 4y = -1$, $-\frac{2}{3}y = -\frac{1}{3}$, and $y = \frac{1}{2}$. Thus, $x = \frac{5}{3}\left(\frac{1}{2}\right) - \frac{1}{3} = \frac{1}{2}$, and the system has the unique solution $\left(\frac{1}{2}, \frac{1}{2}\right)$.

15. Solving the first equation for x, we obtain $3x = 6y + 2$, so $x = 2y + \frac{2}{3}$. Substituting this value of y into the second equation yields $-\frac{3}{2}\left(2y + \frac{2}{3}\right) + 3y = -1$, $-3y - 1 + 3y = -1$, and $0 = 0$. We conclude that the system of equations has infinitely many solutions of the form $\left(2t + \frac{2}{3}, t\right)$, where t is a parameter.

17. Solving the first equation for y, we obtain $y = 2x - 3$. Substituting this value of y into the second equation yields $4x + k(2x - 3) = 4$, so $4x + 2xk - 3k = 4$, $2x(2 + k) = 4 + 3k$, and $x = \dfrac{4 + 3k}{2(2 + k)}$. Since x is not defined when the denominator of this last expression is zero, we conclude that the system has no solution when $k = -2$.

19. Let x and y denote the number of acres of corn and wheat planted, respectively. Then $x + y = 500$. Since the cost of cultivating corn is \$42/acre and that of wheat \$30/acre and Mr. Johnson has \$18,600 available for cultivation, we have $42x + 30y = 18{,}600$. Thus, the solution is found by solving the system of equations $\begin{cases} x + y = 500 \\ 42x + 30y = 18{,}600 \end{cases}$

21. Let x denote the number of pounds of the \$5.00/lb coffee and y denote the number of pounds of the \$6/lb coffee. Then $x + y = 100$. Since the blended coffee sells for \$5.60/lb, we know that the blended mixture is worth $(5.60)(100) = \$560$. Therefore, $5x + 6y = 560$. Thus, the solution is found by solving the system of equations

$$\begin{cases} x + y = 100 \\ 5x + 6y = 560 \end{cases}$$

23. Let x denote the number of children who ride the bus during the morning shift and y the number of adults who ride the bus during the morning shift. Then $x + y = 1000$. Since the total fare collected is \$1300, we have $0.5x + 1.5y = 1300$. Thus, the solution to the problem can be found by solving the system of equations

$$\begin{cases} x + y = 1000 \\ 0.5x + 1.5y = 1300 \end{cases}$$

25. Let x be the amount of money invested at 6% in a savings account, y the amount of money invested at 8% in mutual funds, and z the amount of money invested at 12% in bonds. Since the total interest was \$21,600, we have $0.06x + 0.08y + 0.12z = 21{,}600$. Also, since the amount of Sid's investment in bonds is twice the amount of the investment in the savings account, we have $z = 2x$. Finally, the interest earned from his investment in bonds was equal to the interest earned on his money mutual funds, so $0.08y = 0.12z$. Thus, the solution to the problem can be found by solving the system of equations $\begin{cases} 0.06x + 0.08y + 0.12z = 21{,}600 \\ 2x - z = 0 \\ 0.08y - 0.12z = 0 \end{cases}$

27. Let x, y, and z denote the number of 100-lb. bags of grade A, grade B, and grade C fertilizers to be produced. The amount of nitrogen required is $18x + 20y + 24z$, and this must be equal to 26,400, so we have $18x + 20y + 24z = 26{,}400$. Similarly, the constraints on the use of phosphate and potassium lead to the equations $4x + 4y + 3z = 4900$ and $5x + 4y + 6z = 6200$, respectively. Thus we have the problem of finding the solution to the system $\begin{cases} 18x + 20y + 24z = 26{,}400 & \text{(nitrogen)} \\ 4x + 4y + 3z = 4900 & \text{(phosphate)} \\ 5x + 4y + 6z = 6200 & \text{(potassium)} \end{cases}$

29. Let x, y, and z denote the number of compact, intermediate, and full-size cars to be purchased, respectively. The cost incurred in buying the specified number of cars is $18{,}000x + 27{,}000y + 36{,}000z$. Since the budget is \$2.25 million, we have the system $\begin{cases} 18{,}000x + 27{,}000y + 36{,}000z = 2{,}250{,}000 \\ x - 2y = 0 \\ x + y + z = 100 \end{cases}$

31. Let x be the number of ounces of Food I used in the meal, y the number of ounces of Food II used in the meal, and z the number of ounces of Food III used in the meal. Since 100% of the daily requirement of proteins, carbohydrates, and iron is to be met by this meal, we have the system of linear equations $\begin{cases} 10x + 6y + 8z = 100 \\ 10x + 12y + 6z = 100 \\ 5x + 4y + 12z = 100 \end{cases}$

33. True. If the three lines coincide, then the system has infinitely many solutions corresponding to all points on the (common) line. If at least one line is distinct from the others, then the system has no solution.

5.2 Systems of Linear Equations: Unique Solutions

Problem-Solving Tips

When you come across new notation, make sure that you understand that notation. If you can't express the notation verbally, you haven't yet grasped its use. For example, in this section we introduced the notation $R_i \leftrightarrow R_j$. This notation tells us to interchange row i with row j.

1. Make sure you are familiar with the three row operations $R_i \leftrightarrow R_j$, cR_i, and $R_i + aR_j$.

2. Before writing the augmented matrix, make sure that the variables in all of the equations are on the left and the constants are on the right side of the equality symbol.

3. The last step of the Gauss-Jordan elimination method states that the matrix must be in row-reduced form. This means that the following must hold:

a. Any row consisting entirely of zeros lies below any row with a nonzero entry.

b. The first nonzero entry in each row is a 1.

c. The leading 1 in any row lies to the right of any leading 1 in a row above that row.

d. All columns containing a leading 1 are unit columns.

Concept Questions page 266

1. a. The two systems are equivalent to each other if they have precisely the same solutions.

b. i. Interchange row i with row j.

ii. Replace row i with c times row i.

iii. Replace row i with the sum of row i and a times row j.

3. a. It lies below any other row having nonzero entries.

b. It is a 1.

c. The leading 1 in the lower row lies to the right of the leading 1 in the upper row.

d. They are all 0.

Exercises page 266

1. $\left[\begin{array}{cc|c} 2 & -3 & 7 \\ 3 & 1 & 4 \end{array}\right]$

3. $\left[\begin{array}{ccc|c} 0 & -1 & 2 & 5 \\ 2 & 2 & -8 & 4 \\ 0 & 3 & 4 & 0 \end{array}\right]$

5. $\begin{cases} 3x + 2y = -4 \\ x - y = 5 \end{cases}$

7. $\begin{cases} x + 3y + 2z = 4 \\ 2x = 5 \\ 3x - 3y + 2z = 6 \end{cases}$

9. Yes. Conditions 1–4 are satisfied (see page 259 of the text).

11. No. Condition 3 is violated. The first nonzero entry in the second row does not lie to the right of the first nonzero entry (1) in the first row.

13. Yes. Conditions 1–4 are satisfied.

15. No. Condition 2 and consequently condition 4 are not satisfied. The first nonzero entry in the last row is not a 1 and the column containing that entry does not have zeros elsewhere.

17. No. Condition 1 is violated. The first row consists entirely of zeros and it lies above row 2.

19. $\left[\begin{array}{cc|c} \textcircled{2} & 4 & 8 \\ 3 & 1 & 2 \end{array}\right] \xrightarrow{\frac{1}{2}R_1} \left[\begin{array}{cc|c} 1 & 2 & 4 \\ 3 & 1 & 2 \end{array}\right] \xrightarrow{R_2 - 3R_1} \left[\begin{array}{cc|c} 1 & 2 & 4 \\ 0 & -5 & -10 \end{array}\right]$

21. $\left[\begin{array}{cc|c} \textcircled{-1} & 2 & 3 \\ 6 & 8 & 2 \end{array}\right] \xrightarrow{-R_1} \left[\begin{array}{cc|c} 1 & -2 & -3 \\ 6 & 8 & 2 \end{array}\right] \xrightarrow{R_2 - 6R_1} \left[\begin{array}{cc|c} 1 & -2 & -3 \\ 0 & 20 & 20 \end{array}\right]$

23. $\left[\begin{array}{ccc|c} \textcircled{2} & 4 & 6 & 12 \\ 2 & 3 & 1 & 5 \\ 3 & -1 & 2 & 4 \end{array}\right] \xrightarrow{\frac{1}{2}R_1} \left[\begin{array}{ccc|c} 1 & 2 & 3 & 6 \\ 2 & 3 & 1 & 5 \\ 3 & -1 & 2 & 4 \end{array}\right] \xrightarrow[R_3 - 3R_1]{R_2 - 2R_1} \left[\begin{array}{ccc|c} 1 & 2 & 3 & 6 \\ 0 & -1 & -5 & -7 \\ 0 & -7 & -7 & -14 \end{array}\right]$

25. $\left[\begin{array}{ccc|c} 0 & 1 & 3 & 4 \\ 2 & 4 & \textcircled{1} & 3 \\ 5 & 6 & 2 & -4 \end{array}\right] \xrightarrow[R_3 - 2R_2]{R_1 - 3R_2} \left[\begin{array}{ccc|c} -6 & -11 & 0 & -5 \\ 2 & 4 & 1 & 3 \\ 1 & -2 & 0 & -10 \end{array}\right]$

27. $\left[\begin{array}{cc|c} \textcircled{3} & 9 & 6 \\ 2 & 1 & 4 \end{array}\right] \xrightarrow{\frac{1}{3}R_1} \left[\begin{array}{cc|c} 1 & 3 & 2 \\ 2 & 1 & 4 \end{array}\right] \xrightarrow{R_2 - 2R_1} \left[\begin{array}{cc|c} 1 & 3 & 2 \\ 0 & \textcircled{-5} & 0 \end{array}\right] \xrightarrow{-\frac{1}{5}R_2} \left[\begin{array}{cc|c} 1 & 3 & 2 \\ 0 & 1 & 0 \end{array}\right] \xrightarrow{R_1 - 3R_2} \left[\begin{array}{cc|c} 1 & 0 & 2 \\ 0 & 1 & 0 \end{array}\right]$

29. $\left[\begin{array}{ccc|c} 1 & 3 & 1 & 3 \\ 3 & 8 & 3 & 7 \\ 2 & -3 & 1 & -10 \end{array}\right] \xrightarrow[R_3-2R_1]{R_2-3R_1} \left[\begin{array}{ccc|c} 1 & 3 & 1 & 3 \\ 0 & -1 & 0 & -2 \\ 0 & -9 & -1 & -16 \end{array}\right] \xrightarrow{-R_2} \left[\begin{array}{ccc|c} 1 & 3 & 1 & 3 \\ 0 & 1 & 0 & 2 \\ 0 & -9 & -1 & -16 \end{array}\right] \xrightarrow[R_3+9R_2]{R_1-3R_2}$

$\left[\begin{array}{ccc|c} 1 & 0 & 1 & -3 \\ 0 & 1 & 0 & 2 \\ 0 & 0 & -1 & 2 \end{array}\right] \xrightarrow[-R_3]{R_1+R_3} \left[\begin{array}{ccc|c} 1 & 0 & 0 & -1 \\ 0 & 1 & 0 & 2 \\ 0 & 0 & 1 & -2 \end{array}\right]$

31. The augmented matrix is equivalent to the system of linear equations $\begin{cases} 3x+9y=6 \\ 2x+\ \ y=4 \end{cases}$ The solution to the system is $(2, 0)$.

33. The augmented matrix is equivalent to the system of linear equations $\begin{cases} x+3y+\ z=3 \\ 3x+8y+3z=7 \\ 2x-3y+\ z=-10 \end{cases}$ Reading off the solution from the last augmented matrix, $\left[\begin{array}{ccc|c} 1 & 0 & 0 & -1 \\ 0 & 1 & 0 & 2 \\ 0 & 0 & 1 & -2 \end{array}\right]$, which is in row-reduced form, we have $x=-1$, $y=2$, and $z=-2$.

35. Using the Gauss-Jordan method, we have

$\left[\begin{array}{cc|c} 1 & 1 & 3 \\ 2 & -1 & 3 \end{array}\right] \xrightarrow{R_2-2R_1} \left[\begin{array}{cc|c} 1 & 1 & 3 \\ 0 & -3 & -3 \end{array}\right] \xrightarrow{-\frac{1}{3}R_2} \left[\begin{array}{cc|c} 1 & 1 & 3 \\ 0 & 1 & 1 \end{array}\right] \xrightarrow{R_1-R_2} \left[\begin{array}{cc|c} 1 & 0 & 2 \\ 0 & 1 & 1 \end{array}\right]$. The solution is $(2, 1)$.

37. Using the Gauss-Jordan method, we have

$\left[\begin{array}{cc|c} 1 & -2 & 8 \\ 3 & 4 & 4 \end{array}\right] \xrightarrow{R_2-3R_1} \left[\begin{array}{cc|c} 1 & -2 & 8 \\ 0 & 10 & -20 \end{array}\right] \xrightarrow{\frac{1}{10}R_2} \left[\begin{array}{cc|c} 1 & -2 & 8 \\ 0 & 1 & -2 \end{array}\right] \xrightarrow{R_1+2R_2} \left[\begin{array}{cc|c} 1 & 0 & 4 \\ 0 & 1 & -2 \end{array}\right]$.

The solution is $(4, -2)$.

39. Using the Gauss-Jordan method, we have

$\left[\begin{array}{cc|c} 2 & -3 & -8 \\ 4 & 1 & -2 \end{array}\right] \xrightarrow{\frac{1}{2}R_1} \left[\begin{array}{cc|c} 1 & -\frac{3}{2} & -4 \\ 4 & 1 & -2 \end{array}\right] \xrightarrow{R_2-4R_1} \left[\begin{array}{cc|c} 1 & -\frac{3}{2} & -4 \\ 0 & 7 & 14 \end{array}\right] \xrightarrow{\frac{1}{7}R_2} \left[\begin{array}{cc|c} 1 & -\frac{3}{2} & -4 \\ 0 & 1 & 2 \end{array}\right] \xrightarrow{R_1+\frac{3}{2}R_2} \left[\begin{array}{cc|c} 1 & 0 & -1 \\ 0 & 1 & 2 \end{array}\right]$.

The solution is $(-1, 2)$.

41. Using the Gauss-Jordan method, we have

$\left[\begin{array}{cc|c} 6 & 8 & 15 \\ 2 & -4 & -5 \end{array}\right] \xrightarrow{\frac{1}{6}R_1} \left[\begin{array}{cc|c} 1 & \frac{4}{3} & \frac{5}{2} \\ 2 & -4 & -5 \end{array}\right] \xrightarrow{R_2-2R_1} \left[\begin{array}{cc|c} 1 & \frac{4}{3} & \frac{5}{2} \\ 0 & -\frac{20}{3} & -10 \end{array}\right] \xrightarrow{-\frac{3}{20}R_2} \left[\begin{array}{cc|c} 1 & \frac{4}{3} & \frac{5}{2} \\ 0 & 1 & \frac{3}{2} \end{array}\right] \xrightarrow{R_1-\frac{4}{3}R_2} \left[\begin{array}{cc|c} 1 & 0 & \frac{1}{2} \\ 0 & 1 & \frac{3}{2} \end{array}\right]$.

The solution is $\left(\frac{1}{2}, \frac{3}{2}\right)$.

43. Using the Gauss-Jordan method, we have

$$\left[\begin{array}{cc|c} \textcircled{3} & -2 & 1 \\ 2 & 4 & 2 \end{array}\right] \xrightarrow{\frac{1}{3}R_1} \left[\begin{array}{cc|c} 1 & -\frac{2}{3} & \frac{1}{3} \\ 2 & 4 & 2 \end{array}\right] \xrightarrow{R_2 - 2R_1} \left[\begin{array}{cc|c} 1 & -\frac{2}{3} & \frac{1}{3} \\ 0 & \textcircled{$\frac{16}{3}$} & \frac{4}{3} \end{array}\right] \xrightarrow{\frac{3}{16}R_2} \left[\begin{array}{cc|c} 1 & -\frac{2}{3} & \frac{1}{3} \\ 0 & 1 & \frac{1}{4} \end{array}\right] \xrightarrow{R_1 + \frac{2}{3}R_2} \left[\begin{array}{cc|c} 1 & 0 & \frac{1}{2} \\ 0 & 1 & \frac{1}{4} \end{array}\right].$$

The solution is $\left(\frac{1}{2}, \frac{1}{4}\right)$.

45. Using the Gauss-Jordan method, we have $\left[\begin{array}{ccc|c} \textcircled{1} & 1 & 1 & 0 \\ 2 & -1 & 1 & 1 \\ 1 & 1 & -2 & 2 \end{array}\right] \xrightarrow[R_3 - R_1]{R_2 - 2R_1} \left[\begin{array}{ccc|c} 1 & 1 & 1 & 0 \\ 0 & -3 & -1 & 1 \\ 0 & \textcircled{0} & -3 & 2 \end{array}\right] \xrightarrow{-\frac{1}{3}R_2}$

$$\left[\begin{array}{ccc|c} 1 & 1 & 1 & 0 \\ 0 & 1 & \frac{1}{3} & -\frac{1}{3} \\ 0 & 0 & -3 & 2 \end{array}\right] \xrightarrow{R_1 - R_2} \left[\begin{array}{ccc|c} 1 & 0 & \frac{2}{3} & \frac{1}{3} \\ 0 & 1 & \frac{1}{3} & -\frac{1}{3} \\ 0 & 0 & \textcircled{-3} & 2 \end{array}\right] \xrightarrow{-\frac{1}{3}R_3} \left[\begin{array}{ccc|c} 1 & 0 & \frac{2}{3} & \frac{1}{3} \\ 0 & 1 & \frac{1}{3} & -\frac{1}{3} \\ 0 & 0 & 1 & -\frac{2}{3} \end{array}\right] \xrightarrow[R_2 - \frac{1}{3}R_3]{R_1 - \frac{2}{3}R_3} \left[\begin{array}{ccc|c} 1 & 0 & 0 & \frac{7}{9} \\ 0 & 1 & 0 & -\frac{1}{9} \\ 0 & 0 & 1 & -\frac{2}{3} \end{array}\right].$$

The solution is $\left(\frac{7}{9}, -\frac{1}{9}, -\frac{2}{3}\right)$.

47. $\left[\begin{array}{ccc|c} 2 & 2 & 1 & 9 \\ 1 & 0 & 1 & 4 \\ 0 & 4 & -3 & 17 \end{array}\right] \xrightarrow{R_1 \leftrightarrow R_2} \left[\begin{array}{ccc|c} 1 & 0 & 1 & 4 \\ 2 & 2 & 1 & 9 \\ 0 & 4 & -3 & 17 \end{array}\right] \xrightarrow{R_2 - 2R_1} \left[\begin{array}{ccc|c} 1 & 0 & 1 & 4 \\ 0 & \textcircled{2} & -1 & 1 \\ 0 & 4 & -3 & 17 \end{array}\right] \xrightarrow{\frac{1}{2}R_2} \left[\begin{array}{ccc|c} 1 & 0 & 1 & 4 \\ 0 & 1 & -\frac{1}{2} & \frac{1}{2} \\ 0 & 4 & -3 & 17 \end{array}\right] \xrightarrow{R_3 - 4R_2}$

$\left[\begin{array}{ccc|c} 1 & 0 & 1 & 4 \\ 0 & 1 & -\frac{1}{2} & \frac{1}{2} \\ 0 & 0 & \textcircled{-1} & 15 \end{array}\right] \xrightarrow{-R_3} \left[\begin{array}{ccc|c} 1 & 0 & 1 & 4 \\ 0 & 1 & -\frac{1}{2} & \frac{1}{2} \\ 0 & 0 & 1 & -15 \end{array}\right] \xrightarrow[R_2 + \frac{1}{2}R_3]{R_1 - R_3} \left[\begin{array}{ccc|c} 1 & 0 & 0 & 19 \\ 0 & 1 & 0 & -7 \\ 0 & 0 & 1 & -15 \end{array}\right]$. The solution is $(19, -7, -15)$.

49. $\left[\begin{array}{ccc|c} 0 & -1 & 1 & 2 \\ 4 & -3 & 2 & 16 \\ 3 & 2 & 1 & 11 \end{array}\right] \xrightarrow{R_1 \leftrightarrow R_2} \left[\begin{array}{ccc|c} \textcircled{4} & -3 & 2 & 16 \\ 0 & -1 & 1 & 2 \\ 3 & 2 & 1 & 11 \end{array}\right] \xrightarrow{R_1 - R_3} \left[\begin{array}{ccc|c} 1 & -5 & 1 & 5 \\ 0 & \textcircled{-1} & 1 & 2 \\ 3 & 2 & 1 & 11 \end{array}\right] \xrightarrow[R_3 - 3R_1]{-R_2}$

$$\left[\begin{array}{ccc|c} 1 & -5 & 1 & 5 \\ 0 & 1 & -1 & -2 \\ 0 & 17 & -2 & -4 \end{array}\right] \xrightarrow[R_3 - 17R_2]{R_1 + 5R_2} \left[\begin{array}{ccc|c} 1 & 0 & -4 & -5 \\ 0 & 1 & -1 & -2 \\ 0 & 0 & \textcircled{15} & 30 \end{array}\right] \xrightarrow{\frac{1}{15}R_3} \left[\begin{array}{ccc|c} 1 & 0 & -4 & -5 \\ 0 & 1 & -1 & -2 \\ 0 & 0 & 1 & 2 \end{array}\right] \xrightarrow[R_2 + R_3]{R_1 + 4R_3} \left[\begin{array}{ccc|c} 1 & 0 & 0 & 3 \\ 0 & 1 & 0 & 0 \\ 0 & 0 & 1 & 2 \end{array}\right].$$

The solution is $(3, 0, 2)$.

51. Using the Gauss-Jordan method, we have $\left[\begin{array}{ccc|c} \textcircled{1} & -2 & 1 & 6 \\ 2 & 1 & -3 & -3 \\ 1 & -3 & 3 & 10 \end{array}\right] \xrightarrow[R_3 - R_1]{R_2 - 2R_1} \left[\begin{array}{ccc|c} 1 & -2 & 1 & 6 \\ 0 & \textcircled{5} & -5 & -15 \\ 0 & -1 & 2 & 4 \end{array}\right] \xrightarrow{\frac{1}{5}R_2}$

$\left[\begin{array}{ccc|c} 1 & -2 & 1 & 6 \\ 0 & 1 & -1 & -3 \\ 0 & -1 & 2 & 4 \end{array}\right] \xrightarrow[R_3 + R_2]{R_1 + 2R_2} \left[\begin{array}{ccc|c} 1 & 0 & -1 & 0 \\ 0 & 1 & -1 & -3 \\ 0 & 0 & \textcircled{1} & 1 \end{array}\right] \xrightarrow[R_2 + R_3]{R_1 + R_3} \left[\begin{array}{ccc|c} 1 & 0 & 0 & 1 \\ 0 & 1 & 0 & -2 \\ 0 & 0 & 1 & 1 \end{array}\right]$. The solution is $(1, -2, 1)$.

53. Using the Gauss-Jordan method, we have

$$\left[\begin{array}{ccc|c} 2 & 0 & 3 & -1 \\ 3 & -2 & 1 & 9 \\ 1 & 1 & 4 & 4 \end{array}\right] \xrightarrow{R_1 \leftrightarrow R_3} \left[\begin{array}{ccc|c} \textcircled{1} & 1 & 4 & 4 \\ 3 & -2 & 1 & 9 \\ 2 & 0 & 3 & -1 \end{array}\right] \xrightarrow[R_3 - 2R_1]{R_2 - 3R_1} \left[\begin{array}{ccc|c} 1 & 1 & 4 & 4 \\ 0 & \textcircled{-5} & -11 & -3 \\ 0 & -2 & -5 & -9 \end{array}\right] \xrightarrow{-\frac{1}{5}R_2}$$

$$\left[\begin{array}{ccc|c} 1 & 1 & 4 & 4 \\ 0 & 1 & \frac{11}{5} & \frac{3}{5} \\ 0 & -2 & -5 & -9 \end{array}\right] \xrightarrow[R_3 + 2R_2]{R_1 - R_2} \left[\begin{array}{ccc|c} 1 & 0 & \frac{9}{5} & \frac{17}{5} \\ 0 & 1 & \frac{11}{5} & \frac{3}{5} \\ 0 & 0 & \textcircled{-\frac{3}{5}} & -\frac{39}{5} \end{array}\right] \xrightarrow{-\frac{5}{3}R_3} \left[\begin{array}{ccc|c} 1 & 0 & \frac{9}{5} & \frac{17}{5} \\ 0 & 1 & \frac{11}{5} & \frac{3}{5} \\ 0 & 0 & 1 & 13 \end{array}\right] \xrightarrow[R_2 - \frac{11}{5}R_3]{R_1 - \frac{9}{5}R_3} \left[\begin{array}{ccc|c} 1 & 0 & 0 & -20 \\ 0 & 1 & 0 & -28 \\ 0 & 0 & 1 & 13 \end{array}\right].$$

The solution is $(-20, -28, 13)$.

55. Using the Gauss-Jordan method, we have

$$\left[\begin{array}{ccc|c} \textcircled{1} & -1 & 3 & 14 \\ 1 & 1 & 1 & 6 \\ -2 & -1 & 1 & -4 \end{array}\right] \xrightarrow[R_3 + 2R_1]{R_2 - R_1} \left[\begin{array}{ccc|c} 1 & -1 & 3 & 14 \\ 0 & \textcircled{2} & -2 & -8 \\ 0 & -3 & 7 & 24 \end{array}\right] \xrightarrow{\frac{1}{2}R_2} \left[\begin{array}{ccc|c} 1 & -1 & 3 & 14 \\ 0 & 1 & -1 & -4 \\ 0 & -3 & 7 & 24 \end{array}\right] \xrightarrow[R_3 + 3R_2]{R_1 + R_2} \left[\begin{array}{ccc|c} 1 & 0 & 2 & 10 \\ 0 & 1 & -1 & -4 \\ 0 & 0 & \textcircled{4} & 12 \end{array}\right] \xrightarrow{\frac{1}{4}R_3}$$

$$\left[\begin{array}{ccc|c} 1 & 0 & 2 & 10 \\ 0 & 1 & -1 & -4 \\ 0 & 0 & 1 & 3 \end{array}\right] \xrightarrow[R_2 + R_3]{R_1 - 2R_3} \left[\begin{array}{ccc|c} 1 & 0 & 0 & 4 \\ 0 & 1 & 0 & -1 \\ 0 & 0 & 1 & 3 \end{array}\right].$$ The solution is $(4, -1, 3)$.

57. We wish to solve the system of equations

$$\begin{aligned} x + \quad y &= 500 \\ 42x + 30y &= 18{,}600 \end{aligned}$$

where x is the number of acres of corn planted and y is the number of acres of wheat planted. Using the Gauss-Jordan method, we find

$$\left[\begin{array}{cc|c} \textcircled{1} & 1 & 500 \\ 42 & 30 & 18{,}600 \end{array}\right] \xrightarrow{R_2 - 42R_1} \left[\begin{array}{cc|c} 1 & 1 & 500 \\ 0 & \textcircled{-12} & -2400 \end{array}\right] \xrightarrow{-\frac{1}{12}R_2} \left[\begin{array}{cc|c} 1 & 1 & 500 \\ 0 & 1 & 200 \end{array}\right] \xrightarrow{R_1 - R_2} \left[\begin{array}{cc|c} 1 & 0 & 300 \\ 0 & 1 & 200 \end{array}\right].$$

The solution to this system of equations is $x = 300$, $y = 200$. We conclude that Jacob should plant 300 acres of corn and 200 acres of wheat.

59. Let x denote the number of pounds of \$5/lb coffee and y the number of pounds of \$6/lb coffee. Then we wish to solve the system

$$\begin{aligned} x + \quad y &= 100 \\ 5x + 6y &= 560 \end{aligned}$$

Using the Gauss-Jordan method, we have $\left[\begin{array}{cc|c} \textcircled{1} & 1 & 100 \\ 5 & 6 & 560 \end{array}\right] \xrightarrow{R_2 - 5R_1} \left[\begin{array}{cc|c} 1 & 1 & 100 \\ 0 & \textcircled{1} & 60 \end{array}\right] \xrightarrow{R_1 - R_2} \left[\begin{array}{cc|c} 1 & 0 & 40 \\ 0 & 1 & 60 \end{array}\right].$

Therefore, 40 pounds of \$5/lb coffee and 60 pounds of \$6/lb coffee should be used in the 100-lb. mixture.

61. Let x and y denote the numbers of children and adults respectively who rode the bus during the morning shift. Then the solution to the problem can be found by solving the system of equations

$$\begin{aligned} x + \quad y &= 1000 \\ 0.5x + 1.5y &= 1300 \end{aligned}$$

Using the Gauss-Jordan method, we have $\left[\begin{array}{cc|c} ① & 1 & 1000 \\ 0.5 & 1.5 & 1300 \end{array}\right] \xrightarrow{R_2 - 0.5R_1} \left[\begin{array}{cc|c} 1 & 1 & 1000 \\ 0 & ① & 800 \end{array}\right] \xrightarrow{R_1 - R_2} \left[\begin{array}{cc|c} 1 & 0 & 200 \\ 0 & 1 & 800 \end{array}\right]$.

We conclude that 800 adults and 200 children rode the bus during the morning shift.

63. Let x, y, and z, denote the amounts of money he should invest in a savings account, mutual funds, and bonds, respectively. Then we are required to solve the system

$$\begin{aligned} 0.06x + 0.08y + 0.12z &= 21{,}600 \\ 2x \qquad\qquad - \quad z &= 0 \\ 0.08y - 0.12z &= 0 \end{aligned}$$

Using the Gauss-Jordan method, we find

$$\left[\begin{array}{ccc|c} \boxed{0.06} & 0.08 & 0.12 & 21{,}600 \\ 2 & 0 & -1 & 0 \\ 0 & 0.08 & -0.12 & 0 \end{array}\right] \xrightarrow[\frac{1}{0.08}R_3]{\frac{1}{0.06}R_1} \left[\begin{array}{ccc|c} 1 & \frac{4}{3} & 2 & 360{,}000 \\ 2 & 0 & -1 & 0 \\ 0 & 1 & -\frac{3}{2} & 0 \end{array}\right] \xrightarrow{R_2 - 2R_1} \left[\begin{array}{ccc|c} 1 & \frac{4}{3} & 2 & 360{,}000 \\ 0 & \boxed{-\frac{8}{3}} & -5 & -720{,}000 \\ 0 & 1 & -\frac{3}{2} & 0 \end{array}\right] \xrightarrow{-\frac{3}{8}R_2}$$

$$\left[\begin{array}{ccc|c} 1 & \frac{4}{3} & 2 & 360{,}000 \\ 0 & 1 & \frac{15}{8} & 270{,}000 \\ 0 & 1 & -\frac{3}{2} & 0 \end{array}\right] \xrightarrow[R_3 - R_2]{R_1 - \frac{4}{3}R_2} \left[\begin{array}{ccc|c} 1 & 0 & -\frac{1}{2} & 0 \\ 0 & 1 & \frac{15}{8} & 270{,}000 \\ 0 & 0 & \boxed{-\frac{27}{8}} & -270{,}000 \end{array}\right] \xrightarrow{-\frac{8}{27}R_3} \left[\begin{array}{ccc|c} 1 & 0 & -\frac{1}{2} & 0 \\ 0 & 1 & \frac{15}{8} & 270{,}000 \\ 0 & 0 & 1 & 80{,}000 \end{array}\right] \xrightarrow[R_2 - \frac{15}{8}R_3]{R_1 + \frac{1}{2}R_3}$$

$$\left[\begin{array}{ccc|c} 1 & 0 & 0 & 40{,}000 \\ 0 & 1 & 0 & 120{,}000 \\ 0 & 0 & 1 & 80{,}000 \end{array}\right].$$

Therefore, Sid should invest \$40,000 in a savings account, \$120,000 in mutual funds, and \$80,000 in bonds.

65. Let x, y, and z denote the number of 100-lb bags of grade A, B, and C fertilizer produced. Then we are required to solve the system

$$\begin{aligned} 18x + 20y + 24z &= 26{,}400 \\ 4x + \ 4y + \ 3z &= 4900 \\ 5x + \ 4y + \ 6z &= 6200 \end{aligned}$$

We obtain the following augmented matrices: $\left[\begin{array}{ccc|c} 18 & 20 & 24 & 26{,}400 \\ 4 & 4 & 3 & 4900 \\ 5 & 4 & 6 & 6200 \end{array}\right] \xrightarrow{R_1 \leftrightarrow R_3} \left[\begin{array}{ccc|c} ⑤ & 4 & 6 & 6200 \\ 4 & 4 & 3 & 4900 \\ 18 & 20 & 24 & 26{,}400 \end{array}\right] \xrightarrow{R_1 - R_2}$

$$\left[\begin{array}{ccc|c} 1 & 0 & 3 & 1300 \\ 4 & 4 & 3 & 4900 \\ 18 & 20 & 24 & 26{,}400 \end{array}\right] \xrightarrow[R_3 - 18R_1]{R_2 - 4R_1} \left[\begin{array}{ccc|c} 1 & 0 & 3 & 1300 \\ 0 & ④ & -9 & -300 \\ 0 & 20 & -30 & 3000 \end{array}\right] \xrightarrow{\frac{1}{4}R_2} \left[\begin{array}{ccc|c} 1 & 0 & 3 & 1300 \\ 0 & 1 & -\frac{9}{4} & -75 \\ 0 & 20 & -30 & 3000 \end{array}\right] \xrightarrow{R_3 - 20R_2}$$

$$\left[\begin{array}{ccc|c} 1&0&3&1300\\ 0&1&-\frac{9}{4}&-75\\ 0&0&\textcircled{15}&4500\end{array}\right] \xrightarrow{\frac{1}{15}R_3} \left[\begin{array}{ccc|c}1&0&3&1300\\ 0&1&-\frac{9}{4}&-75\\ 0&0&1&300\end{array}\right] \xrightarrow[R_2+\frac{9}{4}R_3]{R_1-3R_3} \left[\begin{array}{ccc|c}1&0&0&400\\ 0&1&0&600\\ 0&0&1&300\end{array}\right].$$

We see that $x = 400$, $y = 600$, and $z = 300$. Therefore, Lawnco should produce 400, 600, and 300 bags of grades A, B, and C fertilizer, respectively.

67. Let x, y, and z denote the numbers of compact, intermediate, and full-size cars, respectively, to be purchased. Then the problem can be solved by solving the system

$$\begin{aligned} 18{,}000x + 27{,}000y + 36{,}000z &= 2{,}250{,}000 \\ x - \quad 2y \qquad\qquad &= 0 \\ x + \quad y + \quad z &= 100 \end{aligned}$$

Using the Gauss-Jordan method, we have

$$\left[\begin{array}{ccc|c} 18{,}000&27{,}000&36{,}000&2{,}250{,}000\\ 1&-2&0&0\\ 1&1&1&100\end{array}\right] \xrightarrow{R_1\leftrightarrow R_3} \left[\begin{array}{ccc|c} \textcircled{1}&1&1&100\\ 1&-2&0&0\\ 18{,}000&27{,}000&36{,}000&2{,}250{,}000\end{array}\right] \xrightarrow[R_3-18{,}000R_1]{R_2-R_1}$$

$$\left[\begin{array}{ccc|c} 1&1&1&100\\ 0&\textcircled{-3}&-1&-100\\ 0&9000&18{,}000&450{,}000\end{array}\right] \xrightarrow{-\frac{1}{3}R_2} \left[\begin{array}{ccc|c} 1&1&1&100\\ 0&1&\frac{1}{3}&\frac{100}{3}\\ 0&9000&18{,}000&450{,}000\end{array}\right] \xrightarrow[R_3-9000R_2]{R_1-R_2}$$

$$\left[\begin{array}{ccc|c} 1&0&\frac{2}{3}&\frac{200}{3}\\ 0&1&\frac{1}{3}&\frac{100}{3}\\ 0&0&\boxed{15{,}000}&150{,}000\end{array}\right] \xrightarrow{\frac{1}{15{,}000}R_3} \left[\begin{array}{ccc|c} 1&0&\frac{2}{3}&\frac{200}{3}\\ 0&1&\frac{1}{3}&\frac{100}{3}\\ 0&0&1&10\end{array}\right] \xrightarrow[R_2-\frac{1}{3}R_3]{R_1-\frac{2}{3}R_3} \left[\begin{array}{ccc|c} 1&0&0&60\\ 0&1&0&30\\ 0&0&1&10\end{array}\right].$$

We conclude that 60 compact cars, 30 intermediate cars, and 10 full-size cars will be purchased.

69. Let x, y, and z, represent the numbers of ounces of Foods I, II, III used in the meal, respectively. Then the problem reduces to solving the following system of linear equations:

$$\begin{aligned} 10x + 6y + 8z &= 100 \\ 10x + 12y + 6z &= 100 \\ 5x + 4y + 12z &= 100 \end{aligned}$$

Using the Gauss-Jordan method, we obtain

$$\left[\begin{array}{ccc|c} \textcircled{10}&6&8&100\\ 10&12&6&100\\ 5&4&12&100\end{array}\right] \xrightarrow{\frac{1}{10}R_1} \left[\begin{array}{ccc|c} 1&\frac{3}{5}&\frac{4}{5}&10\\ 10&12&6&100\\ 5&4&12&100\end{array}\right] \xrightarrow[R_3-5R_1]{R_2-10R_1} \left[\begin{array}{ccc|c} 1&\frac{3}{5}&\frac{4}{5}&10\\ 0&\textcircled{6}&-2&0\\ 0&1&8&50\end{array}\right] \xrightarrow{\frac{1}{6}R_2}$$

$$\left[\begin{array}{ccc|c} 1&\frac{3}{5}&\frac{4}{5}&10\\ 0&1&-\frac{1}{3}&0\\ 0&1&8&50\end{array}\right] \xrightarrow[R_3-R_2]{R_1-\frac{3}{5}R_2} \left[\begin{array}{ccc|c} 1&0&1&10\\ 0&1&-\frac{1}{3}&0\\ 0&0&\textcircled{\frac{25}{3}}&50\end{array}\right] \xrightarrow{\frac{3}{25}R_3} \left[\begin{array}{ccc|c} 1&0&1&10\\ 0&1&-\frac{1}{3}&0\\ 0&0&1&6\end{array}\right] \xrightarrow[R_2+\frac{1}{3}R_3]{R_1-R_3} \left[\begin{array}{ccc|c} 1&0&0&4\\ 0&1&0&2\\ 0&0&1&6\end{array}\right].$$

We conclude that 4 ounces of Food I, 2 ounces of Food II, and 6 ounces of Food III should be used to prepare the meal.

71. Let x, y, and z denote the numbers of front orchestra, rear orchestra, and front balcony seats sold for this performance. Then we are required to solve the system

$$\begin{aligned} x + y + z &= 1000 \\ 80x + 60y + 50z &= 62{,}800 \\ x + y - 2z &= 400 \end{aligned}$$

Using the Gauss-Jordan method, we find

$$\left[\begin{array}{ccc|c} \textcircled{1} & 1 & 1 & 1000 \\ 80 & 60 & 50 & 62{,}800 \\ 1 & 1 & -2 & 400 \end{array}\right] \xrightarrow[R_3 - R_1]{R_2 - 80R_1} \left[\begin{array}{ccc|c} 1 & 1 & 1 & 1000 \\ 0 & \textcircled{-20} & -30 & -17{,}200 \\ 0 & 0 & -3 & -600 \end{array}\right] \xrightarrow[-\frac{1}{3}R_3]{-\frac{1}{20}R_2} \left[\begin{array}{ccc|c} 1 & 1 & 1 & 1000 \\ 0 & 1 & \frac{3}{2} & 860 \\ 0 & 0 & 1 & 200 \end{array}\right] \xrightarrow{R_1 - R_2}$$

$$\left[\begin{array}{ccc|c} 1 & 0 & -\frac{1}{2} & 140 \\ 0 & 1 & \frac{3}{2} & 860 \\ 0 & 0 & \textcircled{1} & 200 \end{array}\right] \xrightarrow[R_2 - \frac{3}{2}R_3]{R_1 + \frac{1}{2}R_3} \left[\begin{array}{ccc|c} 1 & 0 & 0 & 240 \\ 0 & 1 & 0 & 560 \\ 0 & 0 & 1 & 200 \end{array}\right].$$

We conclude that tickets for 240 front orchestra seats, 560 rear orchestra seats, and 200 front balcony seats were sold.

73. Let x, y, and z denote the number of days spent in London, Paris, and Rome, respectively. We have

$$\begin{aligned} 180x + 230y + 160z &= 2660 \\ 110x + 120y + 90z &= 1520 \\ x - y - z &= 0 \end{aligned}$$

(since $x = y + z$). Using the Gauss-Jordan method to solve the system, we have

$$\left[\begin{array}{ccc|c} 180 & 230 & 160 & 2660 \\ 110 & 120 & 90 & 1520 \\ 1 & -1 & -1 & 0 \end{array}\right] \xrightarrow{R_1 \leftrightarrow R_3} \left[\begin{array}{ccc|c} \textcircled{1} & -1 & -1 & 0 \\ 110 & 120 & 90 & 1520 \\ 180 & 230 & 160 & 2660 \end{array}\right] \xrightarrow[R_3 - 180R_1]{R_2 - 110R_1} \left[\begin{array}{ccc|c} 1 & -1 & -1 & 0 \\ 0 & \textcircled{230} & 200 & 1520 \\ 0 & 410 & 340 & 2660 \end{array}\right] \xrightarrow{\frac{1}{230}R_2}$$

$$\left[\begin{array}{ccc|c} 1 & -1 & -1 & 0 \\ 0 & 1 & \frac{20}{23} & \frac{152}{23} \\ 0 & 410 & 340 & 2660 \end{array}\right] \xrightarrow[R_3 - 410R_2]{R_1 + R_2} \left[\begin{array}{ccc|c} 1 & 0 & -\frac{3}{23} & \frac{152}{23} \\ 0 & 1 & \frac{20}{23} & \frac{152}{23} \\ 0 & 0 & \textcircled{-\frac{380}{23}} & -\frac{1140}{23} \end{array}\right] \xrightarrow{-\frac{23}{380}R_3} \left[\begin{array}{ccc|c} 1 & 0 & -\frac{3}{23} & \frac{152}{23} \\ 0 & 1 & \frac{20}{23} & \frac{152}{23} \\ 0 & 0 & 1 & 3 \end{array}\right] \xrightarrow[R_2 - \frac{20}{23}R_3]{R_1 + \frac{3}{23}R_3}$$

$$\left[\begin{array}{ccc|c} 1 & 0 & 0 & 7 \\ 0 & 1 & 0 & 4 \\ 0 & 0 & 1 & 3 \end{array}\right].$$

The solution is $x = 7$, $y = 4$, and $z = 3$. Therefore, he spent 7 days in London, 4 days in Paris, and 3 days in Rome.

75. False. The constant cannot be zero. The system $\begin{cases} 2x + y = 1 \\ 3x - y = 2 \end{cases}$ is not equivalent to $\begin{cases} 2x + y = 1 \\ 0(3x - y) = 0(2) \end{cases}$ or

$$\begin{cases} 2x + y = 1 \\ 0 = 0 \end{cases}$$

Technology Exercises

page 271

1. $(3, 1, -1, 2)$ **3.** $(5, 4, -3, -4)$ **5.** $(1, -1, 2, 0, 3)$

5.3 Systems of Linear Equations: Underdetermined and Overdetermined Systems

Problem-Solving Tips

After reading Theorem 1, try to express it in your own words. While you will not usually be required to prove these theorems in this course, you should understand their results. For example, Theorem 1 helps us decide before we solve a problem what the nature of the solution may be. Two cases are described: **a.** A system has an equal or greater number of equations as variables, and **b.** A system has fewer equations than variables. In case (a), the system may have no solution, one solution, or infinitely many solutions. In case (b) the system may have no solution or infinitely many solutions.

1. A system does not have a solution if any row of the augmented matrix representing the system has all zeros to the left of the vertical line and a nonzero entry to the right of the line.

2. If $(t, 1-t, t)$, where t is a parameter, is a solution of a linear system of equations, then the system has infinitely many solutions, because t can be any real number. For example, if $t = 2$ then the solution is $(2, -1, 2)$, and if $t = -2$, then the solution is $(-2, 3, -2)$.

Concept Questions page 278

1. a. There may be no solution, a unique solution, or infinitely many solutions.

b. There may be no solution or infinitely many solutions.

Exercises page 278

1. a. The system has one solution. **b.** The solution is $(3, -1, 2)$.

3. a. The system has one solution. **b.** The solution is $(2, 5)$.

5. a. The system has infinitely many solutions.

b. Letting $x_3 = t$, we see that the solutions are given by $(4-t, -2, t)$, where t is a parameter.

7. a. The system has no solution.

b. The last row contains all zeros to the left of the vertical line and a nonzero number (1) to its right.

9. a. The system has infinitely many solutions.

b. Letting $x_4 = t$, we see that the solutions are given by $(4, -1, 3-t, t)$, where t is a parameter.

11. a. The system has infinitely many solutions.

b. Letting $x_3 = s$ and $x_4 = t$, the solutions are given by $(2-3s, 1+s, s, t)$, where s and t are parameters.

13. Using the Gauss-Jordan method, we have

$$\left[\begin{array}{cc|c} 2 & -1 & 3 \\ 1 & 2 & 4 \\ 2 & 3 & 7 \end{array}\right] \xrightarrow{R_1 \leftrightarrow R_2} \left[\begin{array}{cc|c} \textcircled{1} & 2 & 4 \\ 2 & -1 & 3 \\ 2 & 3 & 7 \end{array}\right] \xrightarrow[R_3 - 2R_1]{R_2 - 2R_1} \left[\begin{array}{cc|c} 1 & 2 & 4 \\ 0 & \textcircled{-5} & -5 \\ 0 & -1 & -1 \end{array}\right] \xrightarrow{-\frac{1}{5}R_2} \left[\begin{array}{cc|c} 1 & 2 & 4 \\ 0 & 1 & 1 \\ 0 & -1 & -1 \end{array}\right] \xrightarrow[R_3 + R_2]{R_1 - 2R_2} \left[\begin{array}{cc|c} 1 & 0 & 2 \\ 0 & 1 & 1 \\ 0 & 0 & 0 \end{array}\right].$$

The solution is $(2, 1)$.

15. Using the Gauss-Jordan method, we have

$$\left[\begin{array}{rr|r} 3 & -2 & -3 \\ 2 & 1 & 3 \\ 1 & -2 & -5 \end{array}\right] \xrightarrow{R_1 \leftrightarrow R_3} \left[\begin{array}{rr|r} \textcircled{1} & -2 & -5 \\ 2 & 1 & 3 \\ 3 & -2 & -3 \end{array}\right] \xrightarrow[R_3 - 3R_1]{R_2 - 2R_1} \left[\begin{array}{rr|r} 1 & -2 & -5 \\ 0 & \textcircled{5} & 13 \\ 0 & 4 & 12 \end{array}\right] \xrightarrow{\frac{1}{5}R_2} \left[\begin{array}{rr|r} 1 & -2 & -5 \\ 0 & 1 & \frac{13}{5} \\ 0 & 4 & 12 \end{array}\right] \xrightarrow[R_3 - 4R_2]{R_1 + 2R_2} \left[\begin{array}{rr|r} 1 & 0 & \frac{1}{5} \\ 0 & 1 & \frac{13}{5} \\ 0 & 0 & \frac{8}{5} \end{array}\right].$$

Since the last row implies the $0 = \frac{8}{5}$, we conclude that the system of equations is inconsistent and has no solution.

17. $$\left[\begin{array}{rr|r} 3 & -2 & 5 \\ -1 & 3 & -4 \\ 2 & -4 & 6 \end{array}\right] \xrightarrow{R_1 \leftrightarrow R_2} \left[\begin{array}{rr|r} \textcircled{-1} & 3 & -4 \\ 3 & -2 & 5 \\ 2 & -4 & 6 \end{array}\right] \xrightarrow{-R_1} \left[\begin{array}{rr|r} 1 & -3 & 4 \\ 3 & -2 & 5 \\ 2 & -4 & 6 \end{array}\right] \xrightarrow[R_3 - 2R_1]{R_2 - 3R_1} \left[\begin{array}{rr|r} 1 & -3 & 4 \\ 0 & \textcircled{7} & -7 \\ 0 & 2 & -2 \end{array}\right] \xrightarrow{\frac{1}{7}R_2}$$

$$\left[\begin{array}{rr|r} 1 & -3 & 4 \\ 0 & 1 & -1 \\ 0 & 2 & -2 \end{array}\right] \xrightarrow[R_3 - 2R_2]{R_1 + 3R_2} \left[\begin{array}{rr|r} 1 & 0 & 1 \\ 0 & 1 & -1 \\ 0 & 0 & 0 \end{array}\right].$$ We conclude that the solution is $(1, -1)$.

19. $$\left[\begin{array}{rr|r} \textcircled{1} & -2 & 2 \\ 7 & -14 & 14 \\ 3 & -6 & 6 \end{array}\right] \xrightarrow[R_3 - 3R_1]{R_2 - 7R_1} \left[\begin{array}{rr|r} 1 & -2 & 2 \\ 0 & 0 & 0 \\ 0 & 0 & 0 \end{array}\right].$$

We conclude that the infinitely many solutions are given by $(2t + 2, t)$, where t is a parameter.

21. $$\left[\begin{array}{rrr|r} \textcircled{1} & 2 & 1 & -2 \\ -2 & -3 & -1 & 1 \\ 2 & 4 & 2 & -4 \end{array}\right] \xrightarrow[R_3 - 2R_1]{R_2 + 2R_1} \left[\begin{array}{rrr|r} 1 & 2 & 1 & -2 \\ 0 & \textcircled{1} & 1 & -3 \\ 0 & 0 & 0 & 0 \end{array}\right] \xrightarrow{R_1 - 2R_2} \left[\begin{array}{rrr|r} 1 & 0 & -1 & 4 \\ 0 & 1 & 1 & -3 \\ 0 & 0 & 0 & 0 \end{array}\right].$$ Let $x_3 = t$ and we find that $x_1 = 4 + t$ and $x_2 = -3 - t$. The infinitely many solutions are given by $(4 + t, -3 - t, t)$.

23. $$\left[\begin{array}{rr|r} \textcircled{3} & 2 & 4 \\ -\frac{3}{2} & -1 & -2 \\ 6 & 4 & 8 \end{array}\right] \xrightarrow{\frac{1}{3}R_1} \left[\begin{array}{rr|r} 1 & \frac{2}{3} & \frac{4}{3} \\ -\frac{3}{2} & -1 & -2 \\ 6 & 4 & 8 \end{array}\right] \xrightarrow[R_3 - 6R_1]{R_2 + \frac{3}{2}R_1} \left[\begin{array}{rr|r} 1 & \frac{2}{3} & \frac{4}{3} \\ 0 & 0 & 0 \\ 0 & 0 & 0 \end{array}\right].$$

We conclude that the infinitely many solutions are given by $\left(\frac{4}{3} - \frac{2}{3}t, t\right)$, where t is a parameter.

25. $$\left[\begin{array}{rrr|r} \textcircled{1} & 1 & -2 & -3 \\ 2 & -1 & 3 & 7 \\ 1 & -2 & 5 & 0 \end{array}\right] \xrightarrow[R_3 - R_1]{R_2 - 2R_1} \left[\begin{array}{rrr|r} 1 & 1 & -2 & -3 \\ 0 & \textcircled{-3} & 7 & 13 \\ 0 & -3 & 7 & 3 \end{array}\right] \xrightarrow{-\frac{1}{3}R_2} \left[\begin{array}{rrr|r} 1 & 1 & -2 & -3 \\ 0 & 1 & -\frac{7}{3} & -\frac{13}{3} \\ 0 & -3 & 7 & 3 \end{array}\right] \xrightarrow[R_3 + 3R_2]{R_1 - R_2}$$

$$\left[\begin{array}{rrr|r} 1 & 0 & \frac{1}{3} & \frac{4}{3} \\ 0 & 1 & -\frac{7}{3} & -\frac{13}{3} \\ 0 & 0 & 0 & -10 \end{array}\right].$$

The last row implies that $0 = -10$, which is impossible. We conclude that the system of equations is inconsistent and has no solution.

27. $\left[\begin{array}{ccc|c} \textcircled{1} & -2 & 3 & 4 \\ 2 & 3 & -1 & 2 \\ 1 & 2 & -3 & -6 \end{array}\right] \xrightarrow[R_3-R_1]{R_2-2R_1} \left[\begin{array}{ccc|c} 1 & -2 & 3 & 4 \\ 0 & \textcircled{7} & -7 & -6 \\ 0 & 4 & -6 & -10 \end{array}\right] \xrightarrow{\frac{1}{7}R_2} \left[\begin{array}{ccc|c} 1 & -2 & 3 & 4 \\ 0 & 1 & -1 & -\frac{6}{7} \\ 0 & 4 & -6 & -10 \end{array}\right] \xrightarrow[R_3-4R_2]{R_1+2R_2}$

$\left[\begin{array}{ccc|c} 1 & 0 & 1 & \frac{16}{7} \\ 0 & 1 & -1 & -\frac{6}{7} \\ 0 & 0 & \textcircled{-2} & -\frac{46}{7} \end{array}\right] \xrightarrow{-\frac{1}{2}R_3} \left[\begin{array}{ccc|c} 1 & 0 & 1 & \frac{16}{7} \\ 0 & 1 & -1 & -\frac{6}{7} \\ 0 & 0 & 1 & \frac{23}{7} \end{array}\right] \xrightarrow[R_2+R_3]{R_1-R_3} \left[\begin{array}{ccc|c} 1 & 0 & 0 & -1 \\ 0 & 1 & 0 & \frac{17}{7} \\ 0 & 0 & 1 & \frac{23}{7} \end{array}\right].$

We conclude that the solution is $\left(-1, \frac{17}{7}, \frac{23}{7}\right)$.

29. $\left[\begin{array}{ccc|c} \textcircled{4} & 1 & -1 & 4 \\ 8 & 2 & -2 & 8 \end{array}\right] \xrightarrow{\frac{1}{4}R_1} \left[\begin{array}{ccc|c} 1 & \frac{1}{4} & -\frac{1}{4} & 1 \\ 8 & 2 & -2 & 8 \end{array}\right] \xrightarrow{R_2-8R_1} \left[\begin{array}{ccc|c} 1 & \frac{1}{4} & -\frac{1}{4} & 1 \\ 0 & 0 & 0 & 0 \end{array}\right].$

We conclude that the infinitely many solutions are given by $\left(1 - \frac{1}{4}s + \frac{1}{4}t, s, t\right)$, where s and t are parameters.

31. $\left[\begin{array}{ccc|c} 2 & 1 & -3 & 1 \\ 1 & -1 & 2 & 1 \\ 5 & -2 & 3 & 6 \end{array}\right] \xrightarrow{R_1 \leftrightarrow R_2} \left[\begin{array}{ccc|c} \textcircled{1} & -1 & 2 & 1 \\ 2 & 1 & -3 & 1 \\ 5 & -2 & 3 & 6 \end{array}\right] \xrightarrow[R_3-5R_1]{R_2-2R_1} \left[\begin{array}{ccc|c} 1 & -1 & 2 & 1 \\ 0 & \textcircled{3} & -7 & -1 \\ 0 & 3 & -7 & 1 \end{array}\right] \xrightarrow{\frac{1}{3}R_2}$

$\left[\begin{array}{ccc|c} 1 & -1 & 2 & 1 \\ 0 & 1 & -\frac{7}{3} & -\frac{1}{3} \\ 0 & 3 & -7 & 1 \end{array}\right] \xrightarrow[R_3-3R_2]{R_1+R_2} \left[\begin{array}{ccc|c} 1 & 0 & -\frac{1}{3} & \frac{2}{3} \\ 0 & 1 & -\frac{7}{3} & -\frac{1}{3} \\ 0 & 0 & 0 & 2 \end{array}\right].$

The last row implies that $0 = 2$, which is impossible. We conclude that the system of equations is inconsistent and has no solution.

33. $\left[\begin{array}{ccc|c} \textcircled{1} & 2 & -1 & -4 \\ 2 & 1 & 1 & 7 \\ 1 & 3 & 2 & 7 \\ 1 & -3 & 1 & 9 \end{array}\right] \xrightarrow[R_4-R_1]{\substack{R_2-2R_1 \\ R_3-R_1}} \left[\begin{array}{ccc|c} 1 & 2 & -1 & -4 \\ 0 & \textcircled{-3} & 3 & 15 \\ 0 & 1 & 3 & 11 \\ 0 & -5 & 2 & 13 \end{array}\right] \xrightarrow{-\frac{1}{3}R_2} \left[\begin{array}{ccc|c} 1 & 2 & -1 & -4 \\ 0 & 1 & -1 & -5 \\ 0 & 1 & 3 & 11 \\ 0 & -5 & 2 & 13 \end{array}\right] \xrightarrow[R_4+5R_2]{\substack{R_1-2R_2 \\ R_3-R_2}}$

$\left[\begin{array}{ccc|c} 1 & 0 & 1 & 6 \\ 0 & 1 & -1 & -5 \\ 0 & 0 & \textcircled{4} & 16 \\ 0 & 0 & -3 & -12 \end{array}\right] \xrightarrow{\frac{1}{4}R_3} \left[\begin{array}{ccc|c} 1 & 0 & 1 & 6 \\ 0 & 1 & -1 & -5 \\ 0 & 0 & 1 & 4 \\ 0 & 0 & -3 & -12 \end{array}\right] \xrightarrow[R_4+3R_3]{\substack{R_1-R_3 \\ R_2+R_3}} \left[\begin{array}{ccc|c} 1 & 0 & 0 & 2 \\ 0 & 1 & 0 & -1 \\ 0 & 0 & 1 & 4 \\ 0 & 0 & 0 & 0 \end{array}\right].$

We conclude that the solution of the system is $(2, -1, 4)$.

35. Let x, y, and z represent the numbers of compact, mid-sized, and full-size cars, respectively, to be purchased. Then the problem can be solved by solving the system

$$\begin{aligned} x + y + z &= 60 \\ 18{,}000x + 28{,}800y + 39{,}600z &= 1{,}512{,}000 \end{aligned}$$

Using the Gauss-Jordan method, we have

$$\left[\begin{array}{ccc|c} \textcircled{1} & 1 & 1 & 60 \\ 18{,}000 & 28{,}800 & 39{,}600 & 1{,}512{,}000 \end{array}\right] \xrightarrow{R_2 - 18{,}000R_1} \left[\begin{array}{ccc|c} 1 & 1 & 1 & 60 \\ 0 & \boxed{10{,}800} & 21{,}600 & 432{,}000 \end{array}\right] \xrightarrow{\frac{1}{10{,}800}R_2}$$

$$\left[\begin{array}{ccc|c} 1 & 1 & 1 & 60 \\ 0 & 1 & 2 & 40 \end{array}\right] \xrightarrow{R_1 - R_2} \left[\begin{array}{ccc|c} 1 & 0 & -1 & 20 \\ 0 & 1 & 2 & 40 \end{array}\right].$$ We conclude that there are infinitely many solutions of the form $(20 + z, 40 - 2z, z)$. Letting $z = 5$, we see that one possible solution is (25, 30, 5); that is Hartman should buy 25 compact, 30 mid-size, and 5 full-size cars. Letting $z = 10$, we see that another possible solution is (30, 20, 10); that is, 30 compact, 20 mid-size, and 10 full-size cars.

37. Let x, y, and z denote the numbers of ounces of Foods I, II, and III, respectively, that the dietician includes in the meal. Then the problem can be solved by solving the system

$$\begin{aligned} 400x + 1200y + 800z &= 8800 \\ 110x + 570y + 340z &= 2160 \\ 90x + 30y + 60z &= 1020 \end{aligned}$$

Using the Gauss-Jordan method, we have

$$\left[\begin{array}{ccc|c} \textcircled{400} & 1200 & 800 & 8800 \\ 110 & 570 & 340 & 2160 \\ 90 & 30 & 60 & 1020 \end{array}\right] \xrightarrow{\frac{1}{400}R_1} \left[\begin{array}{ccc|c} 1 & 3 & 2 & 22 \\ 110 & 570 & 340 & 2160 \\ 90 & 30 & 60 & 1020 \end{array}\right] \xrightarrow[R_3 - 90R_1]{R_2 - 110R_1} \left[\begin{array}{ccc|c} 1 & 3 & 2 & 22 \\ 0 & \textcircled{240} & 120 & -260 \\ 0 & -240 & -120 & -960 \end{array}\right] \xrightarrow{\frac{1}{240}R_2}$$

$$\left[\begin{array}{ccc|c} 1 & 3 & 2 & 22 \\ 0 & 1 & \frac{1}{2} & -\frac{13}{12} \\ 0 & -240 & -120 & -960 \end{array}\right] \xrightarrow[R_3 + 240R_2]{R_1 - 3R_2} \left[\begin{array}{ccc|c} 1 & 0 & \frac{1}{2} & \frac{101}{4} \\ 0 & 1 & \frac{1}{2} & -\frac{13}{12} \\ 0 & 0 & 0 & -1220 \end{array}\right].$$

The last row implies that $0 = -1220$, which is impossible. We conclude that the system of equations is inconsistent and has no solution—that is, the dietician cannot prepare a meal from these foods and meet the given requirements.

39. Let x, y, and z denote the amounts of money invested in stocks, bonds, and a money-market account, respectively. Then the problem can be solved by solving the system

$$\begin{aligned} x + y + z &= 100{,}000 \\ 12x + 8y + 4z &= 1{,}000{,}000 \\ x - y - 3z &= 0 \end{aligned}$$

(Note that we have multiplied the second equation by 100 to clear the decimals.) Using the Gauss-Jordan method, we have

$$\left[\begin{array}{ccc|c} \textcircled{1} & 1 & 1 & 100{,}000 \\ 12 & 8 & 4 & 1{,}000{,}000 \\ 1 & -1 & -3 & 0 \end{array}\right] \xrightarrow[R_3 - R_1]{R_2 - 12R_1} \left[\begin{array}{ccc|c} 1 & 1 & 1 & 100{,}000 \\ 0 & \textcircled{-4} & -8 & -200{,}000 \\ 0 & -2 & -4 & -100{,}000 \end{array}\right] \xrightarrow{-\frac{1}{4}R_2} \left[\begin{array}{ccc|c} 1 & 1 & 1 & 100{,}000 \\ 0 & 1 & 2 & 50{,}000 \\ 0 & -2 & -4 & -100{,}000 \end{array}\right] \xrightarrow[R_3 + 2R_2]{R_1 - R_2}$$

$$\left[\begin{array}{ccc|c} 1 & 0 & -1 & 50{,}000 \\ 0 & 1 & 2 & 50{,}000 \\ 0 & 0 & 0 & 0 \end{array}\right].$$

We conclude that there are infinitely many solutions of the form $(50000 + z, 50000 - 2z, z)$. Therefore, one possible solution for the Garcias is to invest \$10,000 in a money-market account, \$60,000 in stocks, and \$30,000 in bonds. Another possible solution is for the Garcias to invest \$20,000 in a money-market account, \$70,000 in stocks, and \$10,000 in bonds.

41. a. We write equations describing the fact that the number of vehicles entering each intersection is equal to the number of vehicles exiting that intersection.

$$\begin{aligned} x_2 + 600 &= 900 + x_3 \\ 1000 + x_3 &= 600 + x_4 \\ x_1 + x_7 &= 700 + x_2 \\ x_4 + 700 &= x_7 + x_5 \\ 800 + 900 &= x_1 + x_6 \\ x_6 + x_5 &= 1100 + 700 \end{aligned}$$

This is equivalent to the following system:

$$\begin{aligned} x_1 \qquad\qquad\qquad\qquad + x_6 \qquad &= 1700 \\ x_1 - x_2 \qquad\qquad\qquad\qquad + x_7 &= 700 \\ x_2 - x_3 \qquad\qquad\qquad\qquad &= 300 \\ -x_3 + x_4 \qquad\qquad\qquad &= 400 \\ -x_4 + x_5 \qquad + x_7 &= 700 \\ x_5 + x_6 \qquad &= 1800 \end{aligned}$$

b.

$$\left[\begin{array}{ccccccc|c} \textcircled{1} & 0 & 0 & 0 & 0 & 1 & 0 & 1700 \\ 1 & -1 & 0 & 0 & 0 & 0 & 1 & 700 \\ 0 & 1 & -1 & 0 & 0 & 0 & 0 & 300 \\ 0 & 0 & -1 & 1 & 0 & 0 & 0 & 400 \\ 0 & 0 & 0 & -1 & 1 & 0 & 1 & 700 \\ 0 & 0 & 0 & 0 & 1 & 1 & 0 & 1800 \end{array}\right] \xrightarrow{R_2 - R_1} \left[\begin{array}{ccccccc|c} 1 & 0 & 0 & 0 & 0 & 1 & 0 & 1700 \\ 0 & \textcircled{-1} & 0 & 0 & 0 & -1 & 1 & -1000 \\ 0 & 1 & -1 & 0 & 0 & 0 & 0 & 300 \\ 0 & 0 & -1 & 1 & 0 & 0 & 0 & 400 \\ 0 & 0 & 0 & -1 & 1 & 0 & 1 & 700 \\ 0 & 0 & 0 & 0 & 1 & 1 & 0 & 1800 \end{array}\right] \xrightarrow{-R_2}$$

$$\left[\begin{array}{ccccccc|c} 1 & 0 & 0 & 0 & 0 & 1 & 0 & 1700 \\ 0 & 1 & 0 & 0 & 0 & 1 & -1 & 1000 \\ 0 & 1 & -1 & 0 & 0 & 0 & 0 & 300 \\ 0 & 0 & -1 & 1 & 0 & 0 & 0 & 400 \\ 0 & 0 & 0 & -1 & 1 & 0 & 1 & 700 \\ 0 & 0 & 0 & 0 & 1 & 1 & 0 & 1800 \end{array}\right] \xrightarrow{R_3 - R_2} \left[\begin{array}{ccccccc|c} 1 & 0 & 0 & 0 & 0 & 1 & 0 & 1700 \\ 0 & 1 & 0 & 0 & 0 & 1 & -1 & 1000 \\ 0 & 0 & \textcircled{-1} & 0 & 0 & -1 & 1 & -700 \\ 0 & 0 & -1 & 1 & 0 & 0 & 0 & 400 \\ 0 & 0 & 0 & -1 & 1 & 0 & 1 & 700 \\ 0 & 0 & 0 & 0 & 1 & 1 & 0 & 1800 \end{array}\right] \xrightarrow{-R_3}$$

$$\left[\begin{array}{ccccccc|c} 1 & 0 & 0 & 0 & 0 & 1 & 0 & 1700 \\ 0 & 1 & 0 & 0 & 0 & 1 & -1 & 1000 \\ 0 & 0 & 1 & 0 & 0 & 1 & -1 & 700 \\ 0 & 0 & -1 & 1 & 0 & 0 & 0 & 400 \\ 0 & 0 & 0 & -1 & 1 & 0 & 1 & 700 \\ 0 & 0 & 0 & 0 & 1 & 1 & 0 & 1800 \end{array}\right] \xrightarrow{R_4 + R_3} \left[\begin{array}{ccccccc|c} 1 & 0 & 0 & 0 & 0 & 1 & 0 & 1700 \\ 0 & 1 & 0 & 0 & 0 & 1 & -1 & 1000 \\ 0 & 0 & 1 & 0 & 0 & 1 & -1 & 700 \\ 0 & 0 & 0 & 1 & 0 & 1 & -1 & 1100 \\ 0 & 0 & 0 & -1 & \textcircled{1} & 0 & 1 & 700 \\ 0 & 0 & 0 & 0 & 1 & 1 & 0 & 1800 \end{array}\right] \xrightarrow{R_5 + R_4}$$

$$\left[\begin{array}{ccccccc|c} 1 & 0 & 0 & 0 & 0 & 1 & 0 & 1700 \\ 0 & 1 & 0 & 0 & 0 & 1 & -1 & 1000 \\ 0 & 0 & 1 & 0 & 0 & 1 & -1 & 700 \\ 0 & 0 & 0 & 1 & 0 & 1 & -1 & 1100 \\ 0 & 0 & 0 & 0 & 1 & 1 & 0 & 1800 \\ 0 & 0 & 0 & 0 & 1 & 1 & 0 & 1800 \end{array}\right] \xrightarrow{R_6 - R_5} \left[\begin{array}{ccccccc|c} 1 & 0 & 0 & 0 & 0 & 1 & 0 & 1700 \\ 0 & 1 & 0 & 0 & 0 & 1 & -1 & 1000 \\ 0 & 0 & 1 & 0 & 0 & 1 & -1 & 700 \\ 0 & 0 & 0 & 1 & 0 & 1 & -1 & 1100 \\ 0 & 0 & 0 & 0 & 1 & 1 & 0 & 1800 \\ 0 & 0 & 0 & 0 & 0 & 0 & 0 & 0 \end{array}\right].$$

We conclude that the system has infinitely many solutions of the form $(1700 - s, 1000 - s + t, 700 - s + t, 1100 - s + t, 1800 - s, s, t)$. Two possible traffic patterns are $(900, 1000, 700, 1100, 1000, 800, 800)$ and $(1000, 1100, 800, 1200, 1100, 700, 800)$.

c. x_6 must have at least 300 cars/hour.

43. We solve the given system using the Gauss-Jordan method. We have

$$\left[\begin{array}{ccc|c} \textcircled{3} & -2 & 4 & 12 \\ -9 & 6 & -12 & k \end{array}\right] \xrightarrow{\frac{1}{3}R_1} \left[\begin{array}{ccc|c} 1 & -\frac{2}{3} & \frac{4}{3} & 4 \\ -9 & 6 & -12 & k \end{array}\right] \xrightarrow{R_2 + 9R_1} \left[\begin{array}{ccc|c} 1 & -\frac{2}{3} & \frac{4}{3} & 4 \\ 0 & 0 & 0 & k + 36 \end{array}\right].$$

Since this system has a solution only if the last row has all zero entries, we see that $k = -36$. We conclude that the system has infinitely many solutions of the form $\left(4 + \frac{2}{3}y - \frac{4}{3}z, y, z\right)$ and $k = -36$.

45. False.

Technology Exercises page 282

1. $(1 + t, 2 + t, t)$, where t is a parameter

3. $\left(-\frac{17}{7} + \frac{6}{7}t, 3 - t, -\frac{18}{7} + \frac{1}{7}t, t\right)$, where t is a parameter

5. No solution

5.4 Matrices

Problem-Solving Tips

1. If a matrix has size $m \times n$, then it has m rows and n columns. For example, a 4×3 matrix has 4 rows and 3 columns.

2. The sum and difference of two matrices A and B are defined only if A and B have the same size. To find the sum (difference) of A and B, we add (subtract) the corresponding entries in the two matrices.

3. To find the scalar product of a real number c and a matrix A, multiply each entry in A by c.

Concept Questions page 289

1. a. A matrix is an ordered rectangular array of real numbers.

b. A matrix has size (or dimension) $m \times n$ if it has m rows and n columns.

c. A row matrix has size $1 \times n$. **d.** A column matrix has size $m \times 1$. **e.** A square matrix has size $n \times n$.

3. $A = \begin{bmatrix} 1 & 2 & 4 \\ 2 & -2 & 1 \\ 4 & 1 & 3 \end{bmatrix}$. The entries satisfy $a_{ij} = a_{ji}$, that is A is symmetric with respect to the main diagonal.

Exercises page 289

1. The size of A is 4×4; the size of B is 4×3; the size of C is 1×5, and the size of D is 4×1.

3. These are entries of the matrix B. The entry b_{13} refers to the entry in the first row and third column and is equal to 2. Similarly, $b_{31} = 3$, and $b_{43} = 8$.

5. The column matrix is the matrix D. The transpose of the matrix D is $D^T = \begin{bmatrix} 1 & 3 & -2 & 0 \end{bmatrix}$.

7. A has size 3×2; B has size 3×2; C and D have size 3×3.

9. $A + B = \begin{bmatrix} -1 & 2 \\ 3 & -2 \\ 4 & 0 \end{bmatrix} + \begin{bmatrix} 2 & 4 \\ 3 & 1 \\ -2 & 2 \end{bmatrix} = \begin{bmatrix} 1 & 6 \\ 6 & -1 \\ 2 & 2 \end{bmatrix}$.

11. $C - D = \begin{bmatrix} 3 & -1 & 0 \\ 2 & -2 & 3 \\ 4 & 6 & 2 \end{bmatrix} - \begin{bmatrix} 2 & -2 & 4 \\ 3 & 6 & 2 \\ -2 & 3 & 1 \end{bmatrix} = \begin{bmatrix} 1 & 1 & -4 \\ -1 & -8 & 1 \\ 6 & 3 & 1 \end{bmatrix}$.

13. $\begin{bmatrix} 6 & 3 & 8 \\ 4 & 5 & 6 \end{bmatrix} - \begin{bmatrix} 1 & -2 & -1 \\ 2 & -5 & -7 \end{bmatrix} = \begin{bmatrix} 5 & 5 & 9 \\ 2 & 10 & 13 \end{bmatrix}$.

15. $\begin{bmatrix} 1 & 4 & -5 \\ 3 & -8 & 6 \end{bmatrix} + \begin{bmatrix} 4 & 0 & -2 \\ 3 & 6 & 5 \end{bmatrix} - \begin{bmatrix} 2 & 8 & 9 \\ -11 & 2 & -5 \end{bmatrix} = \begin{bmatrix} 3 & -4 & -16 \\ 17 & -4 & 16 \end{bmatrix}$.

17. $\begin{bmatrix} 1.2 & 4.5 & -4.2 \\ 8.2 & 6.3 & -3.2 \end{bmatrix} - \begin{bmatrix} 3.1 & 1.5 & -3.6 \\ 2.2 & -3.3 & -4.4 \end{bmatrix} = \begin{bmatrix} -1.9 & 3.0 & -0.6 \\ 6.0 & 9.6 & 1.2 \end{bmatrix}$.

19. $\frac{1}{2}\begin{bmatrix} 1 & 0 & 0 & -4 \\ 3 & 0 & -1 & 6 \\ -2 & 1 & -4 & 2 \end{bmatrix} + \frac{4}{3}\begin{bmatrix} 3 & 0 & -1 & 4 \\ -2 & 1 & -6 & 2 \\ 8 & 2 & 0 & -2 \end{bmatrix} - \frac{1}{3}\begin{bmatrix} 3 & -9 & -1 & 0 \\ 6 & 2 & 0 & -6 \\ 0 & 1 & -3 & 1 \end{bmatrix} = \begin{bmatrix} \frac{7}{2} & 3 & -1 & \frac{10}{3} \\ -\frac{19}{6} & \frac{2}{3} & -\frac{17}{2} & \frac{23}{3} \\ \frac{29}{3} & \frac{17}{6} & -1 & -2 \end{bmatrix}$.

21. $\begin{bmatrix} 2x-2 & 3 & 2 \\ 2 & 4 & y-2 \\ 2z & -3 & 2 \end{bmatrix} = \begin{bmatrix} 3 & u & 2 \\ 2 & 4 & 5 \\ 4 & -3 & 2 \end{bmatrix}$.

Now by the definition of equality of matrices, $u = 3$, $2x - 2 = 3$, and $2x = 5$ (so $x = \frac{5}{2}$), $y - 2 = 5$ (so $y - 7$), and $2z = 4$ (so $z = 2$).

23. $\begin{bmatrix} 1 & x \\ 2y & -3 \end{bmatrix} - 4\begin{bmatrix} 2 & -2 \\ 0 & 3 \end{bmatrix} = \begin{bmatrix} 3z & 10 \\ 4 & -u \end{bmatrix}$; $\begin{bmatrix} -7 & x+8 \\ 2y & -15 \end{bmatrix} = \begin{bmatrix} 3z & 10 \\ 4 & -u \end{bmatrix}$.

Now by the definition of equality of matrices, $-u = -15$ (so $u = 15$), $x + 8 = 10$ (so $x = 2$), $2y = 4$ (so $y = 2$), and $3z = -7$ (so $z = -\frac{7}{3}$).

25. To verify the Commutative Law for matrix addition, let us show that $A+B=B+A$.

Now $A+B=\begin{bmatrix}2&-4&3\\4&2&1\end{bmatrix}+\begin{bmatrix}4&3&2\\1&0&4\end{bmatrix}=\begin{bmatrix}6&-7&5\\5&2&5\end{bmatrix}=\begin{bmatrix}4&3&2\\1&0&4\end{bmatrix}+\begin{bmatrix}2&-4&3\\4&2&1\end{bmatrix}=B+A.$

27. $(3+5)A=8A=8\begin{bmatrix}3&1\\2&4\\-4&0\end{bmatrix}=\begin{bmatrix}24&8\\16&32\\-32&0\end{bmatrix}=3\begin{bmatrix}3&1\\2&4\\-4&0\end{bmatrix}+5\begin{bmatrix}3&1\\2&4\\-4&0\end{bmatrix}=3A+5A.$

29. $4(A+B)=4\left(\begin{bmatrix}3&1\\2&4\\-4&0\end{bmatrix}+\begin{bmatrix}1&2\\-1&0\\3&2\end{bmatrix}\right)=4\begin{bmatrix}4&3\\1&4\\-1&2\end{bmatrix}=\begin{bmatrix}16&12\\4&16\\-4&8\end{bmatrix}$ and

$$4A+4B=4\begin{bmatrix}3&1\\2&4\\-4&0\end{bmatrix}+4\begin{bmatrix}1&2\\-1&0\\3&2\end{bmatrix}=\begin{bmatrix}12&4\\8&16\\-16&0\end{bmatrix}+\begin{bmatrix}4&8\\-4&0\\12&8\end{bmatrix}=\begin{bmatrix}16&12\\4&16\\-4&8\end{bmatrix}.$$

31. $\begin{bmatrix}3&2&-1&5\end{bmatrix}^T=\begin{bmatrix}3\\2\\-1\\5\end{bmatrix}.$

33. $\begin{bmatrix}1&-1&2\\3&4&2\\0&1&0\end{bmatrix}^T=\begin{bmatrix}1&3&0\\-1&4&1\\2&2&0\end{bmatrix}.$

35.

	1	2	3	4
Mr. Cross	220	215	210	205
Mr. Jones	220	210	200	195
Mr. Smith	215	205	195	190

37. $B=(1.03)A=1.03\begin{bmatrix}340&360&380\\410&430&440\\620&660&700\end{bmatrix}=$

	M_1	M_2	M_3
I	350.2	370.8	391.4
II	422.3	442.9	453.2
III	638.6	679.8	721

39. a. $D=A+B-C=\begin{bmatrix}2820&1470&1120\\1030&520&480\\1170&540&460\end{bmatrix}+\begin{bmatrix}260&120&110\\140&60&50\\120&70&50\end{bmatrix}-\begin{bmatrix}120&80&80\\70&30&40\\60&20&40\end{bmatrix}=\begin{bmatrix}2960&1510&1150\\1100&550&490\\1230&590&470\end{bmatrix}.$

b. $E=1.1D=1.1\begin{bmatrix}2960&1510&1150\\1100&550&490\\1230&590&470\end{bmatrix}=\begin{bmatrix}3256&1661&1265\\1210&605&539\\1353&649&517\end{bmatrix}.$

41. $A=$

	2000	2001	2002
MA	6.88	7.05	7.18
U.S.	4.13	4.09	4.06

43. $A = \begin{array}{c} \\ \text{Women} \\ \text{Men} \end{array} \begin{array}{c} \begin{array}{ccc} \text{White} & \text{Black} & \text{Hispanic} \end{array} \\ \begin{bmatrix} 82.6 & 80.5 & 91.2 \\ 78 & 73.9 & 84.8 \end{bmatrix} \end{array}$ and $B = \begin{array}{c} \\ \text{White} \\ \text{Black} \\ \text{Hispanic} \end{array} \begin{array}{c} \begin{array}{cc} \text{Women} & \text{Men} \end{array} \\ \begin{bmatrix} 82.6 & 78 \\ 80.5 & 73.9 \\ 91.2 & 84.8 \end{bmatrix} \end{array}$.

45. True. Each element in $A + B$ is obtained by adding together the corresponding elements in A and B. Therefore, the matrix $c(A + B)$ is obtained by multiplying each element in $A + B$ by c. On the other hand, cA is obtained by multiplying each element in A by c and cB is obtained by multiplying each element in B by c, and $cA + cB$ is obtained by adding the corresponding elements in cA and cB. Thus $c(A + B) = cA + cB$.

47. False. Take $\begin{bmatrix} 1 & 2 \\ 3 & 4 \end{bmatrix}$ and $c = 2$. Then $cA = 2\begin{bmatrix} 1 & 2 \\ 3 & 4 \end{bmatrix} = \begin{bmatrix} 2 & 4 \\ 6 & 8 \end{bmatrix}$ and $(cA)^T = \begin{bmatrix} 2 & 6 \\ 4 & 8 \end{bmatrix}$. On the other hand, $\frac{1}{c}A^T = \frac{1}{2}\begin{bmatrix} 1 & 3 \\ 2 & 4 \end{bmatrix} = \begin{bmatrix} \frac{1}{2} & \frac{3}{2} \\ 1 & 2 \end{bmatrix} \neq (cA)^T$.

Technology Exercises

page 295

1. $\begin{bmatrix} 15 & 38.75 & -67.5 & 33.75 \\ 51.25 & 40 & 52.5 & -38.75 \\ 21.25 & 35 & -65 & 105 \end{bmatrix}$

3. $\begin{bmatrix} -5 & 6.3 & -6.8 & 3.9 \\ 1 & 0.5 & 5.4 & -4.8 \\ 0.5 & 4.2 & -3.5 & 5.6 \end{bmatrix}$

5. $\begin{bmatrix} 16.44 & -3.65 & -3.66 & 0.63 \\ 12.77 & 10.64 & 2.58 & 0.05 \\ 5.09 & 0.28 & -10.84 & 17.64 \end{bmatrix}$

7. $\begin{bmatrix} 22.2 & -0.3 & -12 & 4.5 \\ 21.6 & 17.7 & 9 & -4.2 \\ 8.7 & 4.2 & -20.7 & 33.6 \end{bmatrix}$

5.5 Multiplication of Matrices

Problem-Solving Tips

1. The **matrix product** of two matrices A and B is defined only if the number of columns in A is equal to the number of rows in B. The **scalar product** of a real number c and a matrix A is always defined.

2. We can write a system of equations in the matrix form $AX = B$, where A is the coefficient matrix, X is the column matrix of unknowns, and B is the column matrix of constants.

Concept Questions

page 302

1. Scalar multiplication involves multiplying a matrix A by a scalar c (result: cA); whereas matrix multiplication involves the product of two matrices. Example: $3\begin{bmatrix} 1 & 2 \\ 2 & 3 \end{bmatrix} = \begin{bmatrix} 3 & 6 \\ 6 & 9 \end{bmatrix}$ and $\begin{bmatrix} 2 & 1 \\ 3 & 0 \end{bmatrix}\begin{bmatrix} 1 & 3 & 2 \\ 1 & 4 & 3 \end{bmatrix} = \begin{bmatrix} 3 & 10 & 7 \\ 3 & 9 & 6 \end{bmatrix}$.

Exercises page 302

1. Size of A 2×3, Size of B 3×5: Same; (2×5) Size of AB. Size of B 3×5, Size of A 2×3: Not the same; BA is undefined

3. Size of A 1×7, Size of B 7×1: Same; (1×1) Size of AB. Size of B 7×1, Size of A 1×7: Same; (7×7) Size of BA

5. If AB and BA are defined, then $n = s$ and $m = t$.

7. $\begin{bmatrix} 1 & 2 \\ 3 & 0 \end{bmatrix} \begin{bmatrix} 1 \\ -1 \end{bmatrix} = \begin{bmatrix} -1 \\ 3 \end{bmatrix}$.

9. $\begin{bmatrix} 4 & 1 & 2 \\ -1 & 2 & 4 \end{bmatrix} \begin{bmatrix} 4 \\ 1 \\ -2 \end{bmatrix} = \begin{bmatrix} 13 \\ -10 \end{bmatrix}$.

11. $\begin{bmatrix} -1 & 2 \\ 3 & 1 \end{bmatrix} \begin{bmatrix} 2 & 4 \\ 3 & 1 \end{bmatrix} = \begin{bmatrix} 4 & -2 \\ 9 & 13 \end{bmatrix}$.

13. $\begin{bmatrix} 2 & 1 & 2 \\ 3 & 2 & 4 \end{bmatrix} \begin{bmatrix} -1 & 2 \\ 4 & 3 \\ 0 & 1 \end{bmatrix} = \begin{bmatrix} 2 & 9 \\ 5 & 16 \end{bmatrix}$.

15. $\begin{bmatrix} 0.1 & 0.9 \\ 0.2 & 0.8 \end{bmatrix} \begin{bmatrix} 1.2 & 0.4 \\ 0.5 & 2.1 \end{bmatrix} = \begin{bmatrix} 0.1(1.2) + 0.9(0.5) & 0.1(0.4) + 0.9(2.1) \\ 0.2(1.2) + 0.8(0.5) & 0.2(0.4) + 0.8(2.1) \end{bmatrix} = \begin{bmatrix} 0.57 & 1.93 \\ 0.64 & 1.76 \end{bmatrix}$.

17. $\begin{bmatrix} 6 & -3 & 0 \\ -2 & 1 & -8 \\ 4 & -4 & 9 \end{bmatrix} \begin{bmatrix} 1 & 0 & 0 \\ 0 & 1 & 0 \\ 0 & 0 & 1 \end{bmatrix} = \begin{bmatrix} 6 & -3 & 0 \\ -2 & 1 & -8 \\ 4 & -4 & 9 \end{bmatrix}$.

19. $\begin{bmatrix} 3 & 0 & -2 & 1 \\ 1 & 2 & 0 & -1 \end{bmatrix} \begin{bmatrix} 2 & 1 & -2 \\ -1 & 2 & 0 \\ 0 & 0 & 1 \\ -1 & -2 & 2 \end{bmatrix} = \begin{bmatrix} 5 & 1 & -6 \\ 1 & 7 & -4 \end{bmatrix}$.

21. $4 \begin{bmatrix} 1 & -2 & 0 \\ 2 & -1 & 1 \\ 3 & 0 & -1 \end{bmatrix} \begin{bmatrix} 1 & 3 & 1 \\ 1 & 4 & 0 \\ 0 & 1 & -2 \end{bmatrix} = \begin{bmatrix} -4 & -20 & 4 \\ 4 & 12 & 0 \\ 12 & 32 & 20 \end{bmatrix}$.

23. $\begin{bmatrix} 1 & 0 \\ 0 & 1 \end{bmatrix} \begin{bmatrix} 4 & -3 & 2 \\ 7 & 1 & -5 \end{bmatrix} \begin{bmatrix} 1 & 0 & 0 \\ 0 & 1 & 0 \\ 0 & 0 & 1 \end{bmatrix} = \begin{bmatrix} 1 & 0 \\ 0 & 1 \end{bmatrix} \begin{bmatrix} 4 & -3 & 2 \\ 7 & 1 & -5 \end{bmatrix} = \begin{bmatrix} 4 & -3 & 2 \\ 7 & 1 & -5 \end{bmatrix}$.

25. To verify the associative law for matrix multiplication, we will show that $(AB)C = A(BC)$:

$$AB = \begin{bmatrix} 1 & 0 & -2 \\ 1 & -3 & 2 \\ -2 & 1 & 1 \end{bmatrix}\begin{bmatrix} 3 & 1 & 0 \\ 2 & 2 & 0 \\ 1 & -3 & -1 \end{bmatrix} = \begin{bmatrix} 1 & 7 & 2 \\ -1 & -11 & -2 \\ -3 & -3 & -1 \end{bmatrix}, \text{ so}$$

$$(AB)C = \begin{bmatrix} 1 & 7 & 2 \\ -1 & -11 & -2 \\ -3 & -3 & -1 \end{bmatrix}\begin{bmatrix} 2 & -1 & 0 \\ 1 & -1 & 2 \\ 3 & -2 & 1 \end{bmatrix} = \begin{bmatrix} 15 & -12 & 16 \\ -19 & 16 & -24 \\ -12 & 8 & -7 \end{bmatrix}.$$

On the other hand, $BC = \begin{bmatrix} 3 & 1 & 0 \\ 2 & 2 & 0 \\ 1 & -3 & -1 \end{bmatrix}\begin{bmatrix} 2 & -1 & 0 \\ 1 & -1 & 2 \\ 3 & -2 & 1 \end{bmatrix} = \begin{bmatrix} 7 & -4 & 2 \\ 6 & -4 & 4 \\ -4 & 4 & -7 \end{bmatrix}$, so

$$A(BC) = \begin{bmatrix} 1 & 0 & -2 \\ 1 & -3 & 2 \\ -2 & 1 & 1 \end{bmatrix}\begin{bmatrix} 7 & -4 & 2 \\ 6 & -4 & 4 \\ -4 & 4 & -7 \end{bmatrix} = \begin{bmatrix} 15 & -12 & 16 \\ -19 & 16 & -24 \\ -12 & 8 & -7 \end{bmatrix}.$$

27. $AB = \begin{bmatrix} 1 & 2 \\ 3 & 4 \end{bmatrix}\begin{bmatrix} 2 & 1 \\ 4 & 3 \end{bmatrix} = \begin{bmatrix} 10 & 7 \\ 22 & 15 \end{bmatrix}$ and $BA = \begin{bmatrix} 2 & 1 \\ 4 & 3 \end{bmatrix}\begin{bmatrix} 1 & 2 \\ 3 & 4 \end{bmatrix} = \begin{bmatrix} 5 & 8 \\ 13 & 20 \end{bmatrix}$.

Therefore, $AB \neq BA$ and matrix multiplication is not commutative.

29. $AB = \begin{bmatrix} 3 & 0 \\ 8 & 0 \end{bmatrix}\begin{bmatrix} 0 & 0 \\ 4 & 5 \end{bmatrix} = \begin{bmatrix} 0 & 0 \\ 0 & 0 \end{bmatrix}$. Thus, $AB = 0$, but neither A nor B is the zero matrix. Therefore, $AB = 0$, does not imply that A or B is the zero matrix.

31. Let $A = \begin{bmatrix} a & b \\ c & d \end{bmatrix}$. Then $\begin{bmatrix} a & b \\ c & d \end{bmatrix}\begin{bmatrix} 1 & 0 \\ -1 & 3 \end{bmatrix} = \begin{bmatrix} a-b & 3b \\ c-d & 3d \end{bmatrix} = \begin{bmatrix} -1 & -3 \\ 3 & 6 \end{bmatrix}$. Thus, $3b = -3$ (so $b = -1$), $3d = 6$ (so $d = 2$), $a - b = -1$ (so $a = b - 1 = -2$), and $c - d = 3$ (so $c = d + 3 = 5$). Therefore, $A = \begin{bmatrix} -2 & -1 \\ 5 & 2 \end{bmatrix}$.

33. Let $A = \begin{bmatrix} a_1 & b_1 \\ 0 & d_1 \end{bmatrix}$ and $B = \begin{bmatrix} a_2 & b_2 \\ 0 & d_2 \end{bmatrix}$ be two upper triangular matrices.

a. $A + B = \begin{bmatrix} a_1 & b_1 \\ 0 & d_1 \end{bmatrix} + \begin{bmatrix} a_2 & b_2 \\ 0 & d_2 \end{bmatrix} = \begin{bmatrix} a_1 + a_2 & b_1 + b_2 \\ 0 & d_1 + d_2 \end{bmatrix}$, which is a 2×2 upper triangular matrix. Also,

$AB = \begin{bmatrix} a_1 & b_1 \\ 0 & d_1 \end{bmatrix}\begin{bmatrix} a_2 & b_2 \\ 0 & d_2 \end{bmatrix} = \begin{bmatrix} a_1a_2 & a_1b_2 + b_1d_2 \\ 0 & d_1d_2 \end{bmatrix}$ is a 2×2 upper triangular matrix.

b. We compute $BA = \begin{bmatrix} a_2 & b_2 \\ 0 & d_2 \end{bmatrix}\begin{bmatrix} a_1 & b_1 \\ 0 & d_1 \end{bmatrix} = \begin{bmatrix} a_1a_2 & a_2b_1 + b_2d_1 \\ 0 & d_1d_2 \end{bmatrix} \neq AB$. From the result of part (a), we see that $AB \neq BA$ in general.

35. **a.** $A^T = \begin{bmatrix} 2 & 5 \\ 4 & -6 \end{bmatrix}$ and $\left(A^T\right)^T = \begin{bmatrix} 2 & 4 \\ 5 & -6 \end{bmatrix} = A$.

b. $(A+B)^T = \begin{bmatrix} 6 & 12 \\ -2 & -3 \end{bmatrix}^T = \begin{bmatrix} 6 & -2 \\ 12 & -3 \end{bmatrix}$ and $A^T + B^T = \begin{bmatrix} 2 & 5 \\ 4 & -6 \end{bmatrix} + \begin{bmatrix} 4 & -7 \\ 8 & 3 \end{bmatrix} = \begin{bmatrix} 6 & -2 \\ 12 & -3 \end{bmatrix}$.

c. $AB = \begin{bmatrix} 2 & 4 \\ 5 & -6 \end{bmatrix}\begin{bmatrix} 4 & 8 \\ -7 & 3 \end{bmatrix} = \begin{bmatrix} -20 & 28 \\ 62 & 22 \end{bmatrix}$, so $(AB)^T = \begin{bmatrix} -20 & 62 \\ 28 & 22 \end{bmatrix}$, and

$$B^T A^T = \begin{bmatrix} 4 & -7 \\ 8 & 3 \end{bmatrix}\begin{bmatrix} 2 & 5 \\ 4 & -6 \end{bmatrix} = \begin{bmatrix} -20 & 62 \\ 28 & 22 \end{bmatrix} = (AB)^T.$$

37. The given system of linear equations can be represented by the matrix equation $AX = B$, where

$$A = \begin{bmatrix} 2 & -3 \\ 3 & -4 \end{bmatrix},\ X = \begin{bmatrix} x \\ y \end{bmatrix},\ \text{and } B = \begin{bmatrix} 7 \\ 8 \end{bmatrix}.$$

39. The given system of linear equations can be represented by the matrix equation $AX = B$, where

$$A = \begin{bmatrix} 2 & -3 & 4 \\ 0 & 2 & -3 \\ 1 & -1 & 2 \end{bmatrix},\ X = \begin{bmatrix} x \\ y \\ z \end{bmatrix},\ \text{and } B = \begin{bmatrix} 6 \\ 7 \\ 4 \end{bmatrix}.$$

41. The given system of linear equations can be represented by the matrix equation $AX = B$, where

$$A = \begin{bmatrix} -1 & 1 & 1 \\ 2 & -1 & -1 \\ -3 & 2 & 4 \end{bmatrix},\ X = \begin{bmatrix} x_1 \\ x_2 \\ x_3 \end{bmatrix},\ \text{and } B = \begin{bmatrix} 0 \\ 2 \\ 4 \end{bmatrix}.$$

43. a. $AB = \begin{bmatrix} 200 & 300 & 100 & 200 \\ 100 & 200 & 400 & 0 \end{bmatrix}\begin{bmatrix} 54 \\ 48 \\ 98 \\ 82 \end{bmatrix} = \begin{bmatrix} 51{,}400 \\ 54{,}200 \end{bmatrix}$.

b. The first entry shows that William's total stockholdings are \$51,400, while the second entry shows that Michael's stockholdings are \$54,200.

45. a. $A =$

	N. Kroner	S. Kronor	D. Kroner	Rubles
Kaitlin	82	68	62	1200
Emma	64	74	44	1600

b. $B =$

N. Kroner	0.1805
S. Kronor	0.1582
D. Kroner	0.1901
Rubles	0.0356

c. $AB = \begin{bmatrix} 82 & 68 & 62 & 1200 \\ 64 & 74 & 44 & 1600 \end{bmatrix}\begin{bmatrix} 0.1805 \\ 0.1582 \\ 0.1901 \\ 0.0356 \end{bmatrix} = \begin{bmatrix} 80.0648 \\ 88.5832 \end{bmatrix}$, so Kaitlin will have \$80.06 and Emma \$88.58.

47. a. $BA = \begin{bmatrix} 1 & 1 & 1 \end{bmatrix} \begin{bmatrix} 60 & 80 & 120 & 40 \\ 20 & 30 & 60 & 10 \\ 10 & 15 & 30 & 5 \end{bmatrix} = \begin{bmatrix} 90 & 125 & 210 & 55 \end{bmatrix}$. The entries give the total numbers of model I, II, III, and IV houses built in the three states.

b. $AC^T = \begin{bmatrix} 60 & 80 & 120 & 40 \\ 20 & 30 & 60 & 10 \\ 10 & 15 & 30 & 5 \end{bmatrix} \begin{bmatrix} 1 \\ 1 \\ 1 \\ 1 \end{bmatrix} = \begin{bmatrix} 300 \\ 120 \\ 60 \end{bmatrix}$. The entries give the total numbers of houses built in each of the three states.

49. The column vector that represents the admission prices is $B = \begin{bmatrix} 4 \\ 6 \\ 8 \end{bmatrix}$. The column vector that gives the gross receipts for each theater is $AB = \begin{bmatrix} 225 & 110 & 50 \\ 75 & 180 & 225 \\ 280 & 85 & 110 \\ 0 & 250 & 225 \end{bmatrix} \begin{bmatrix} 4 \\ 6 \\ 8 \end{bmatrix} = \begin{bmatrix} 1960 \\ 3180 \\ 2510 \\ 3300 \end{bmatrix}$.

The total revenue collected is given by $1960 + 3180 + 2510 + 3300$, or \$10,950.

51. $BA = \begin{bmatrix} 30{,}000 & 40{,}000 & 20{,}000 \end{bmatrix} \overset{\text{Dem. Rep. Ind.}}{\begin{bmatrix} 0.50 & 0.30 & 0.20 \\ 0.45 & 0.40 & 0.15 \\ 0.40 & 0.50 & 0.10 \end{bmatrix}} = \overset{\text{Dem. Rep. Ind.}}{\begin{bmatrix} 41{,}000 & 35{,}000 & 14{,}000 \end{bmatrix}}$.

53. $AB = \begin{bmatrix} 2700 & 3000 \\ 800 & 700 \\ 500 & 300 \end{bmatrix} \begin{bmatrix} 0.25 & 0.20 & 0.30 & 0.25 \\ 0.30 & 0.35 & 0.25 & 0.10 \end{bmatrix} = \begin{bmatrix} 1575 & 1590 & 1560 & 975 \\ 410 & 405 & 415 & 270 \\ 215 & 205 & 225 & 155 \end{bmatrix}$.

55. $AC = \begin{bmatrix} 80 & 60 & 40 \end{bmatrix} \begin{bmatrix} 0.34 \\ 0.42 \\ 0.48 \end{bmatrix} = \begin{bmatrix} 71.6 \end{bmatrix}$ and $BD = \begin{bmatrix} 300 & 150 & 250 \end{bmatrix} \begin{bmatrix} 0.24 \\ 0.31 \\ 0.35 \end{bmatrix} = \begin{bmatrix} 206 \end{bmatrix}$, so

$AC + BD = \begin{bmatrix} 277.60 \end{bmatrix}$, or \$277.60. This represents Cindy's long distance bill for phone calls to those three cities.

57. a. $MA^T = \begin{bmatrix} 400 & 1200 & 800 \\ 110 & 570 & 340 \\ 90 & 30 & 60 \end{bmatrix} \begin{bmatrix} 7 \\ 1 \\ 6 \end{bmatrix} = \begin{bmatrix} 8800 \\ 3380 \\ 1020 \end{bmatrix}$. The amounts of vitamin A, vitamin C, and calcium taken by a girl in the first meal are 8800, 3380, and 1020 units, respectively.

b. $MB^T = \begin{bmatrix} 400 & 1200 & 800 \\ 110 & 570 & 340 \\ 90 & 30 & 60 \end{bmatrix} \begin{bmatrix} 9 \\ 3 \\ 2 \end{bmatrix} = \begin{bmatrix} 8800 \\ 3380 \\ 1020 \end{bmatrix}$. The amounts of vitamin A, vitamin C, and calcium taken by a girl in the second meal are 8800, 3380, and 1020 units, respectively.

c. $M(A+B)^T = \begin{bmatrix} 400 & 1200 & 800 \\ 110 & 570 & 340 \\ 90 & 30 & 60 \end{bmatrix} \begin{bmatrix} 16 \\ 4 \\ 8 \end{bmatrix} = \begin{bmatrix} 17{,}600 \\ 6760 \\ 2040 \end{bmatrix}$. The amounts of vitamin A, vitamin C, and calcium taken by a girl in the two meals are 17,600, 6760, and 2040 units, respectively.

59. False. Let A be a matrix of order 2×3 and let B be a matrix of order 3×2. Then AB and BA are both defined, although neither A nor B is a square matrix.

61. True. In order for the sum $B + C$ to be defined, B and C must have the same size, and in order for the product of A and $B + C$ to be defined, the number of columns of A must be equal to the number of rows of $B + C$.

Technology Exercises

page 309

1. $\begin{bmatrix} 18.66 & 15.2 & -12 \\ 24.48 & 41.88 & 89.82 \\ 15.39 & 7.16 & -1.25 \end{bmatrix}$

3. $\begin{bmatrix} 20.09 & 20.61 & -1.3 \\ 44.42 & 71.6 & 64.89 \\ 20.97 & 7.17 & -60.65 \end{bmatrix}$

5. $\begin{bmatrix} 32.89 & 13.63 & -57.17 \\ -12.85 & -8.37 & 256.92 \\ 13.48 & 14.29 & 181.64 \end{bmatrix}$

7. $\begin{bmatrix} 128.59 & 123.08 & -32.50 \\ 246.73 & 403.12 & 481.52 \\ 125.06 & 47.01 & -264.81 \end{bmatrix}$

9. $AB = \begin{bmatrix} 87 & 68 & 110 & 82 \\ 119 & 176 & 221 & 143 \\ 51 & 128 & 142 & 94 \\ 28 & 174 & 174 & 112 \end{bmatrix}$ and $BA = \begin{bmatrix} 113 & 117 & 72 & 101 & 90 \\ 72 & 85 & 36 & 72 & 76 \\ 81 & 69 & 76 & 87 & 30 \\ 133 & 157 & 56 & 121 & 146 \\ 154 & 157 & 94 & 127 & 122 \end{bmatrix}$.

11. $AC + AD = \begin{bmatrix} 170 & 18.1 & 133.1 & -106.3 & 341.3 \\ 349 & 226.5 & 324.1 & 164 & 506.4 \\ 245.2 & 157.7 & 231.5 & 125.5 & 312.9 \\ 310 & 245.2 & 291 & 274.3 & 354.2 \end{bmatrix}$.

5.6 The Inverse of a Square Matrix

Problem-Solving Tips

The problem-solving skills that you learned in earlier sections are building blocks for the rest of the course. You can't skip a section or a concept and hope to understand the material in a new section—it just won't work. For example, in this section we discussed the process for finding the inverse of a matrix. You need to use the Gauss-Jordan method of elimination to find the inverse of a matrix, so if you don't know how to use that method to solve a system of equations, you won't be able to find the inverse of a matrix. If you are having difficulty, you may need to go back and review the earlier section before you go on.

1. Not every square matrix has an inverse. If there is a row to the left of the vertical line in the augmented matrix containing all zeros, then the matrix does not have an inverse.

2. You can use the formula $A^{-1} = \frac{1}{D}\begin{bmatrix} a & b \\ c & d \end{bmatrix}$, where $D = ad - bc \neq 0$, to find the inverse of a 2×2 matrix. Note that if $D = 0$, the matrix does not have an inverse.

3. The inverse of a matrix can be used to find the solution of a system of n equations in n unknowns.

Concept Questions page 317

1. The inverse of a square matrix A is the matrix A^{-1} satisfying the conditions

$$AA^{-1} = A^{-1}A = I$$

3. The formula for finding the inverse of a 2×2 matrix are given on page 314 of the text.

Exercises page 317

1. $\begin{bmatrix} 1 & -3 \\ 1 & -2 \end{bmatrix}\begin{bmatrix} -2 & 3 \\ -1 & 1 \end{bmatrix} = \begin{bmatrix} 1 & 0 \\ 0 & 1 \end{bmatrix}$ and $\begin{bmatrix} -2 & 3 \\ -1 & 1 \end{bmatrix}\begin{bmatrix} 1 & -3 \\ 1 & -2 \end{bmatrix} = \begin{bmatrix} 1 & 0 \\ 0 & 1 \end{bmatrix}$

3. $\begin{bmatrix} 3 & 2 & 3 \\ 2 & 2 & 1 \\ 2 & 1 & 1 \end{bmatrix}\begin{bmatrix} -\frac{1}{3} & -\frac{1}{3} & \frac{4}{3} \\ 0 & 1 & -1 \\ \frac{2}{3} & -\frac{1}{3} & -\frac{2}{3} \end{bmatrix} = \begin{bmatrix} 1 & 0 & 0 \\ 0 & 1 & 0 \\ 0 & 0 & 1 \end{bmatrix}$ and $\begin{bmatrix} -\frac{1}{3} & -\frac{1}{3} & \frac{4}{3} \\ 0 & 1 & -1 \\ \frac{2}{3} & -\frac{1}{3} & -\frac{2}{3} \end{bmatrix}\begin{bmatrix} 3 & 2 & 3 \\ 2 & 2 & 1 \\ 2 & 1 & 1 \end{bmatrix} = \begin{bmatrix} 1 & 0 & 0 \\ 0 & 1 & 0 \\ 0 & 0 & 1 \end{bmatrix}$.

5. Using Formula (13), we find $A^{-1} = \frac{1}{(2)(3) - (1)(5)}\begin{bmatrix} 3 & -5 \\ -1 & 2 \end{bmatrix} = \begin{bmatrix} 3 & -5 \\ -1 & 2 \end{bmatrix}$.

7. Since $ad - bc = (3)(2) - (-2)(-3) = 6 - 6 = 0$, the inverse does not exist.

9. $\left[\begin{array}{ccc|ccc} 2 & -3 & -4 & 1 & 0 & 0 \\ 0 & 0 & -1 & 0 & 1 & 0 \\ 1 & -2 & 1 & 0 & 0 & 1 \end{array}\right] \xrightarrow{R_1 \leftrightarrow R_3} \left[\begin{array}{ccc|ccc} 1 & -2 & 1 & 0 & 0 & 1 \\ 0 & 0 & -1 & 0 & 1 & 0 \\ 2 & -3 & -4 & 1 & 0 & 0 \end{array}\right] \xrightarrow{R_3 - 2R_1} \left[\begin{array}{ccc|ccc} 1 & -2 & 1 & 0 & 0 & 1 \\ 0 & 0 & -1 & 0 & 1 & 0 \\ 0 & 1 & -6 & 1 & 0 & -2 \end{array}\right] \xrightarrow{R_2 \leftrightarrow R_3}$

$\left[\begin{array}{ccc|ccc} 1 & -2 & 1 & 0 & 0 & 1 \\ 0 & 1 & -6 & 1 & 0 & -2 \\ 0 & 0 & -1 & 0 & 1 & 0 \end{array}\right] \xrightarrow[-R_3]{R_1 + 2R_2} \left[\begin{array}{ccc|ccc} 1 & 0 & -11 & 2 & 0 & -3 \\ 0 & 1 & -6 & 1 & 0 & -2 \\ 0 & 0 & 1 & 0 & -1 & 0 \end{array}\right] \xrightarrow[R_2 + 6R_3]{R_1 + 11R_3} \left[\begin{array}{ccc|ccc} 1 & 0 & 0 & 2 & -11 & -3 \\ 0 & 1 & 0 & 1 & -6 & -2 \\ 0 & 0 & 1 & 0 & -1 & 0 \end{array}\right].$

Therefore, the required inverse is $\begin{bmatrix} 2 & -11 & -3 \\ 1 & -6 & -2 \\ 0 & -1 & 0 \end{bmatrix}$.

11. $\left[\begin{array}{ccc|ccc} 4 & 2 & 2 & 1 & 0 & 0 \\ -1 & -3 & 4 & 0 & 1 & 0 \\ 3 & -1 & 6 & 0 & 0 & 1 \end{array}\right] \xrightarrow{R_1 - R_3} \left[\begin{array}{ccc|ccc} 1 & 3 & -4 & 1 & 0 & -1 \\ -1 & -3 & 4 & 0 & 1 & 0 \\ 3 & -1 & 6 & 0 & 0 & 1 \end{array}\right] \xrightarrow{R_2 + R_1} \left[\begin{array}{ccc|ccc} 1 & 3 & -4 & 1 & 0 & -1 \\ 0 & 0 & 0 & 1 & 1 & -1 \\ 3 & -1 & 6 & 0 & 0 & 1 \end{array}\right].$

Because there is a row of zeros to the left of the vertical line, we see that the inverse does not exist.

13. $\left[\begin{array}{ccc|ccc} 1 & 4 & -1 & 1 & 0 & 0 \\ 2 & 3 & -2 & 0 & 1 & 0 \\ -1 & 2 & 3 & 0 & 0 & 1 \end{array}\right] \xrightarrow[R_3 + R_1]{R_2 - 2R_1} \left[\begin{array}{ccc|ccc} 1 & 4 & -1 & 1 & 0 & 0 \\ 0 & -5 & 0 & -2 & 1 & 0 \\ 0 & 6 & 2 & 1 & 0 & 1 \end{array}\right] \xrightarrow{R_2 + R_3} \left[\begin{array}{ccc|ccc} 1 & 4 & -1 & 1 & 0 & 0 \\ 0 & 1 & 2 & -1 & 1 & 1 \\ 0 & 6 & 2 & 1 & 0 & 1 \end{array}\right] \xrightarrow[R_3 - 6R_2]{R_1 - 4R_2}$

$\left[\begin{array}{ccc|ccc} 1 & 0 & -9 & 5 & -4 & -4 \\ 0 & 1 & 2 & -1 & 1 & 1 \\ 0 & 0 & -10 & 7 & -6 & -5 \end{array}\right] \xrightarrow{-\frac{1}{10}R_3} \left[\begin{array}{ccc|ccc} 1 & 0 & -9 & 5 & -4 & -4 \\ 0 & 1 & 2 & -1 & 1 & 1 \\ 0 & 0 & 1 & -\frac{7}{10} & \frac{3}{5} & \frac{1}{2} \end{array}\right] \xrightarrow[R_2 - 2R_3]{R_1 + 9R_3} \left[\begin{array}{ccc|ccc} 1 & 0 & 0 & -\frac{13}{10} & \frac{7}{5} & \frac{1}{2} \\ 0 & 1 & 0 & \frac{2}{5} & -\frac{1}{5} & 0 \\ 0 & 0 & 1 & -\frac{7}{10} & \frac{3}{5} & \frac{1}{2} \end{array}\right]$, so

$A^{-1} = \begin{bmatrix} -\frac{13}{10} & \frac{7}{5} & \frac{1}{2} \\ \frac{2}{5} & -\frac{1}{5} & 0 \\ -\frac{7}{10} & \frac{3}{5} & \frac{1}{2} \end{bmatrix}.$

15. $\left[\begin{array}{cccc|cccc} 1 & 1 & -1 & 1 & 1 & 0 & 0 & 0 \\ 2 & 1 & 1 & 0 & 0 & 1 & 0 & 0 \\ 2 & 1 & 0 & 1 & 0 & 0 & 1 & 0 \\ 2 & -1 & -1 & 3 & 0 & 0 & 0 & 1 \end{array}\right] \xrightarrow[\substack{R_3 - 2R_1 \\ R_4 - 2R_1}]{R_2 - 2R_1} \left[\begin{array}{cccc|cccc} 1 & 1 & -1 & 1 & 1 & 0 & 0 & 0 \\ 0 & -1 & 3 & -2 & -2 & 1 & 0 & 0 \\ 0 & -1 & 2 & -1 & -2 & 0 & 1 & 0 \\ 0 & -3 & 1 & 1 & -2 & 0 & 0 & 1 \end{array}\right] \xrightarrow{-R_2}$

$\left[\begin{array}{cccc|cccc} 1 & 1 & -1 & 1 & 1 & 0 & 0 & 0 \\ 0 & 1 & -3 & 2 & 2 & -1 & 0 & 0 \\ 0 & -1 & 2 & -1 & -2 & 0 & 1 & 0 \\ 0 & -3 & 1 & 1 & -2 & 0 & 0 & 1 \end{array}\right] \xrightarrow[\substack{R_3 + R_2 \\ R_4 + 3R_2}]{R_1 - R_2} \left[\begin{array}{cccc|cccc} 1 & 0 & 2 & -1 & -1 & 1 & 0 & 0 \\ 0 & 1 & -3 & 2 & 2 & -1 & 0 & 0 \\ 0 & 0 & -1 & 1 & 0 & -1 & 1 & 0 \\ 0 & 0 & -8 & 7 & 4 & -3 & 0 & 1 \end{array}\right] \xrightarrow{-R_3}$

$$\left[\begin{array}{cccc|cccc} 1 & 0 & 2 & -1 & -1 & 1 & 0 & 0 \\ 0 & 1 & -3 & 2 & 2 & -1 & 0 & 0 \\ 0 & 0 & 1 & -1 & 0 & 1 & -1 & 0 \\ 0 & 0 & -8 & 7 & 4 & -3 & 0 & 1 \end{array}\right] \xrightarrow[R_4+8R_3]{\substack{R_1-2R_3 \\ R_2+3R_3}} \left[\begin{array}{cccc|cccc} 1 & 0 & 0 & 1 & -1 & -1 & 2 & 0 \\ 0 & 1 & 0 & -1 & 2 & 2 & -3 & 0 \\ 0 & 0 & 1 & -1 & 0 & 1 & -1 & 0 \\ 0 & 0 & 0 & \textcircled{-1} & 4 & 5 & -8 & 1 \end{array}\right] \xrightarrow[\substack{R_3-R_4 \\ -R_4}]{\substack{R_1+R_4 \\ R_2-R_4}}$$

$$\left[\begin{array}{cccc|cccc} 1 & 0 & 0 & 0 & 3 & 4 & -6 & 1 \\ 0 & 1 & 0 & 0 & -2 & -3 & 5 & -1 \\ 0 & 0 & 1 & 0 & -4 & -4 & 7 & -1 \\ 0 & 0 & 0 & 1 & -4 & -5 & 8 & -1 \end{array}\right].$$ Thus, the required inverse is $A^{-1} = \begin{bmatrix} 3 & 4 & -6 & 1 \\ -2 & -3 & 5 & -1 \\ -4 & -4 & 7 & -1 \\ -4 & -5 & 8 & -1 \end{bmatrix}$.

We can verify our result by showing that $A^{-1}A = A$: $\begin{bmatrix} 3 & 4 & -6 & 1 \\ -2 & -3 & 5 & -1 \\ -4 & -4 & 7 & -1 \\ -4 & -5 & 8 & -1 \end{bmatrix} \begin{bmatrix} 1 & 1 & -1 & 1 \\ 2 & 1 & 1 & 0 \\ 2 & 1 & 0 & 1 \\ 2 & -1 & -1 & 3 \end{bmatrix} = \begin{bmatrix} 1 & 0 & 0 & 0 \\ 0 & 1 & 0 & 0 \\ 0 & 0 & 1 & 0 \\ 0 & 0 & 0 & 1 \end{bmatrix}$.

17. a. $AX = B$, where $A = \begin{bmatrix} 2 & 5 \\ 1 & 3 \end{bmatrix}$, $X = \begin{bmatrix} x \\ y \end{bmatrix}$, and $B = \begin{bmatrix} 3 \\ 2 \end{bmatrix}$.

b. $X = A^{-1}B = \begin{bmatrix} 3 & -5 \\ -1 & 2 \end{bmatrix} \begin{bmatrix} 3 \\ 2 \end{bmatrix} = \begin{bmatrix} -1 \\ 1 \end{bmatrix}$.

19. a. $AX = B$, where $A = \begin{bmatrix} 2 & -3 & -4 \\ 0 & 0 & -1 \\ 1 & -2 & 1 \end{bmatrix}$, $X = \begin{bmatrix} x \\ y \\ z \end{bmatrix}$, and $B = \begin{bmatrix} 4 \\ 3 \\ -8 \end{bmatrix}$.

b. $X = A^{-1}B = \begin{bmatrix} 2 & -11 & -3 \\ 1 & -6 & -2 \\ 0 & -1 & 0 \end{bmatrix} \begin{bmatrix} 4 \\ 3 \\ -8 \end{bmatrix} = \begin{bmatrix} -1 \\ 2 \\ -3 \end{bmatrix}$.

21. a. $AX = B$, where $A = \begin{bmatrix} 1 & 4 & -1 \\ 2 & 3 & -2 \\ -1 & 2 & 3 \end{bmatrix}$, $X = \begin{bmatrix} x \\ y \\ z \end{bmatrix}$, and $B = \begin{bmatrix} 3 \\ 1 \\ 7 \end{bmatrix}$.

b. $X = A^{-1}B = \begin{bmatrix} -\frac{13}{10} & \frac{7}{5} & \frac{1}{2} \\ \frac{2}{5} & -\frac{1}{5} & 0 \\ -\frac{7}{10} & \frac{3}{5} & \frac{1}{2} \end{bmatrix} \begin{bmatrix} 3 \\ 1 \\ 7 \end{bmatrix} = \begin{bmatrix} 1 \\ 1 \\ 2 \end{bmatrix}$.

23. a. $AX = B$, where $A = \begin{bmatrix} 1 & 1 & -1 & 1 \\ 2 & 1 & 1 & 0 \\ 2 & 1 & 0 & 1 \\ 2 & -1 & -1 & 3 \end{bmatrix}$, $X = \begin{bmatrix} x_1 \\ x_2 \\ x_3 \\ x_4 \end{bmatrix}$, and $B = \begin{bmatrix} 6 \\ 4 \\ 7 \\ 9 \end{bmatrix}$.

b. $X = A^{-1}B = \begin{bmatrix} 3 & 4 & -6 & 1 \\ -2 & -3 & 5 & -1 \\ -4 & -4 & 7 & -1 \\ -4 & -5 & 8 & -1 \end{bmatrix}\begin{bmatrix} 6 \\ 4 \\ 7 \\ 9 \end{bmatrix} = \begin{bmatrix} 1 \\ 2 \\ 0 \\ 3 \end{bmatrix}$.

25. a. $AX = B$, where $A = \begin{bmatrix} 1 & 2 \\ 2 & -1 \end{bmatrix}$, $X = \begin{bmatrix} x \\ y \end{bmatrix}$, and $B = \begin{bmatrix} b_1 \\ b_2 \end{bmatrix}$. Using the formula for the inverse of a 2×2 matrix, we find that $A^{-1} = \begin{bmatrix} 0.2 & 0.4 \\ 0.4 & -0.2 \end{bmatrix}$.

b. i. $X = A^{-1}B = \begin{bmatrix} 0.2 & 0.4 \\ 0.4 & -0.2 \end{bmatrix}\begin{bmatrix} 14 \\ 5 \end{bmatrix} = \begin{bmatrix} 4.8 \\ 4.6 \end{bmatrix}$, and we conclude that $x = 4.8$ and $y = 4.6$.

ii. $X = A^{-1}B = \begin{bmatrix} 0.2 & 0.4 \\ 0.4 & -0.2 \end{bmatrix}\begin{bmatrix} 4 \\ -1 \end{bmatrix} = \begin{bmatrix} 0.4 \\ 1.8 \end{bmatrix}$, and we conclude that $x = 0.4$ and $y = 1.8$.

27. a. The matrix equation is $\begin{bmatrix} 1 & 2 & 1 \\ 1 & 1 & 1 \\ 3 & 1 & 1 \end{bmatrix}\begin{bmatrix} x \\ y \\ z \end{bmatrix} = \begin{bmatrix} b_1 \\ b_2 \\ b_3 \end{bmatrix}$. To find A^{-1}, we compute $\left[\begin{array}{ccc|ccc} 1 & 2 & 1 & 1 & 0 & 0 \\ 1 & 1 & 1 & 0 & 1 & 0 \\ 3 & 1 & 1 & 0 & 0 & 1 \end{array}\right] \xrightarrow[R_3 - 3R_1]{R_2 - R_1}$

$\left[\begin{array}{ccc|ccc} 1 & 2 & 1 & 1 & 0 & 0 \\ 0 & \textcircled{-1} & 0 & -1 & 1 & 0 \\ 0 & -5 & -2 & -3 & 0 & 1 \end{array}\right] \xrightarrow{-R_2} \left[\begin{array}{ccc|ccc} 1 & 2 & 1 & 1 & 0 & 0 \\ 0 & 1 & 0 & 1 & -1 & 0 \\ 0 & -5 & -2 & -3 & 0 & 1 \end{array}\right] \xrightarrow[R_3 + 5R_2]{R_1 - 2R_2} \left[\begin{array}{ccc|ccc} 1 & 0 & 1 & -1 & 2 & 0 \\ 0 & 1 & 0 & 1 & -1 & 0 \\ 0 & 0 & \textcircled{-2} & 2 & -5 & 1 \end{array}\right] \xrightarrow{-\frac{1}{2}R_3}$

$\left[\begin{array}{ccc|ccc} 1 & 0 & 1 & -1 & 2 & 0 \\ 0 & 1 & 0 & 1 & -1 & 0 \\ 0 & 0 & 1 & -1 & \frac{5}{2} & -\frac{1}{2} \end{array}\right] \xrightarrow{R_1 - R_3} \left[\begin{array}{ccc|ccc} 1 & 0 & 0 & 0 & -\frac{1}{2} & \frac{1}{2} \\ 0 & 1 & 0 & 1 & -1 & 0 \\ 0 & 0 & 1 & -1 & \frac{5}{2} & -\frac{1}{2} \end{array}\right]$. Therefore, $A^{-1} = \begin{bmatrix} 0 & -\frac{1}{2} & \frac{1}{2} \\ 1 & -1 & 0 \\ -1 & \frac{5}{2} & -\frac{1}{2} \end{bmatrix}$.

b. i. $\begin{bmatrix} x \\ y \\ z \end{bmatrix} = \begin{bmatrix} 0 & -\frac{1}{2} & \frac{1}{2} \\ 1 & -1 & 0 \\ -1 & \frac{5}{2} & -\frac{1}{2} \end{bmatrix}\begin{bmatrix} 7 \\ 4 \\ 2 \end{bmatrix} = \begin{bmatrix} -1 \\ 3 \\ 2 \end{bmatrix}$, and we conclude that $x = -1$, $y = 3$, and $z = 2$.

ii. $\begin{bmatrix} x \\ y \\ z \end{bmatrix} = \begin{bmatrix} 0 & -\frac{1}{2} & \frac{1}{2} \\ 1 & -1 & 0 \\ -1 & \frac{5}{2} & -\frac{1}{2} \end{bmatrix}\begin{bmatrix} 5 \\ -3 \\ -1 \end{bmatrix} = \begin{bmatrix} 1 \\ 8 \\ -12 \end{bmatrix}$, and we conclude that $x = 1$, $y = 8$, and $z = -12$.

29. a. The matrix equation is $\begin{bmatrix} 3 & 2 & -1 \\ 2 & -3 & 1 \\ 1 & -1 & -1 \end{bmatrix}\begin{bmatrix} x \\ y \\ z \end{bmatrix} = \begin{bmatrix} b_1 \\ b_2 \\ b_3 \end{bmatrix}$. To find A^{-1}, we compute

$$\left[\begin{array}{ccc|ccc} 3 & 2 & -1 & 1 & 0 & 0 \\ 2 & -3 & 1 & 0 & 1 & 0 \\ 1 & -1 & -1 & 0 & 0 & 1 \end{array}\right] \xrightarrow{R_1 \leftrightarrow R_3} \left[\begin{array}{ccc|ccc} \textcircled{1} & -1 & -1 & 0 & 0 & 1 \\ 2 & -3 & 1 & 0 & 1 & 0 \\ 3 & 2 & -1 & 1 & 0 & 0 \end{array}\right] \xrightarrow[R_3 - 3R_1]{R_2 - 2R_1} \left[\begin{array}{ccc|ccc} 1 & -1 & -1 & 0 & 0 & 1 \\ 0 & \textcircled{-1} & 3 & 0 & 1 & -2 \\ 0 & 5 & 2 & 1 & 0 & -3 \end{array}\right] \xrightarrow{-R_2}$$

$$\left[\begin{array}{ccc|ccc} 1 & -1 & -1 & 0 & 0 & 1 \\ 0 & 1 & -3 & 0 & -1 & 2 \\ 0 & 5 & 2 & 1 & 0 & -3 \end{array}\right] \xrightarrow[R_3 - 5R_2]{R_1 + R_2} \left[\begin{array}{ccc|ccc} 1 & 0 & -4 & 0 & -1 & 3 \\ 0 & 1 & -3 & 0 & -1 & 2 \\ 0 & 0 & \textcircled{17} & 1 & 5 & -13 \end{array}\right] \xrightarrow{\frac{1}{17}R_3} \left[\begin{array}{ccc|ccc} 1 & 0 & -4 & 0 & -1 & 3 \\ 0 & 1 & -3 & 0 & -1 & 2 \\ 0 & 0 & 1 & \frac{1}{17} & \frac{5}{17} & -\frac{13}{17} \end{array}\right] \xrightarrow[R_2 + 3R_3]{R_1 + 4R_3}$$

$$\left[\begin{array}{ccc|ccc} 1 & 0 & 0 & \frac{4}{17} & \frac{3}{17} & -\frac{1}{17} \\ 0 & 1 & 0 & \frac{3}{17} & -\frac{2}{17} & -\frac{5}{17} \\ 0 & 0 & 1 & \frac{1}{17} & \frac{5}{17} & -\frac{13}{17} \end{array}\right]. \text{ Therefore, } A^{-1} = \begin{bmatrix} \frac{4}{17} & \frac{3}{17} & -\frac{1}{17} \\ \frac{3}{17} & -\frac{2}{17} & -\frac{5}{17} \\ \frac{1}{17} & \frac{5}{17} & -\frac{13}{17} \end{bmatrix}.$$

b. i. $\begin{bmatrix} x \\ y \\ z \end{bmatrix} = \begin{bmatrix} \frac{4}{17} & \frac{3}{17} & -\frac{1}{17} \\ \frac{3}{17} & -\frac{2}{17} & -\frac{5}{17} \\ \frac{1}{17} & \frac{5}{17} & -\frac{13}{17} \end{bmatrix}\begin{bmatrix} 2 \\ -2 \\ 4 \end{bmatrix} = \begin{bmatrix} -\frac{2}{17} \\ -\frac{10}{17} \\ -\frac{60}{17} \end{bmatrix}$. We conclude that $x = -\frac{2}{17}$, $y = -\frac{10}{17}$, and $z = -\frac{60}{17}$.

ii. $\begin{bmatrix} x \\ y \\ z \end{bmatrix} = \begin{bmatrix} \frac{4}{17} & \frac{3}{17} & -\frac{1}{17} \\ \frac{3}{17} & -\frac{2}{17} & -\frac{5}{17} \\ \frac{1}{17} & \frac{5}{17} & -\frac{13}{17} \end{bmatrix}\begin{bmatrix} 8 \\ -3 \\ 6 \end{bmatrix} = \begin{bmatrix} 1 \\ 0 \\ -5 \end{bmatrix}$. We conclude that $x = 1$, $y = 0$, and $z = -5$.

31. a. $AX = B_1$ and $AX = B_2$, where $A = \begin{bmatrix} 1 & 1 & 1 & 1 \\ 1 & -1 & -1 & 1 \\ 0 & 1 & 2 & 2 \\ 1 & 2 & 1 & -2 \end{bmatrix}$, $X = \begin{bmatrix} x_1 \\ x_2 \\ x_3 \\ x_4 \end{bmatrix}$, $B_1 = \begin{bmatrix} 1 \\ -1 \\ 4 \\ 0 \end{bmatrix}$, and $B_2 = \begin{bmatrix} 2 \\ 8 \\ 4 \\ -1 \end{bmatrix}$.

We first find A^{-1}: $\left[\begin{array}{cccc|cccc} 1 & 1 & 1 & 1 & 1 & 0 & 0 & 0 \\ 1 & -1 & -1 & 1 & 0 & 1 & 0 & 0 \\ 0 & 1 & 2 & 2 & 0 & 0 & 1 & 0 \\ 1 & 2 & 1 & -2 & 0 & 0 & 0 & 1 \end{array}\right] \xrightarrow[R_4 - R_1]{R_2 - R_1} \left[\begin{array}{cccc|cccc} 1 & 1 & 1 & 1 & 1 & 0 & 0 & 0 \\ 0 & -2 & -2 & 0 & -1 & 1 & 0 & 0 \\ 0 & 1 & 2 & 2 & 0 & 0 & 1 & 0 \\ 0 & 1 & 0 & -3 & -1 & 0 & 0 & 1 \end{array}\right] \xrightarrow{R_2 \leftrightarrow R_3}$

$$\left[\begin{array}{cccc|cccc} 1 & 1 & 1 & 1 & 1 & 0 & 0 & 0 \\ 0 & \textcircled{1} & 2 & 2 & 0 & 0 & 1 & 0 \\ 0 & -2 & -2 & 0 & -1 & 1 & 0 & 0 \\ 0 & 1 & 0 & -3 & -1 & 0 & 0 & 1 \end{array}\right] \xrightarrow[\substack{R_3 + 2R_2 \\ R_4 - R_2}]{R_1 - R_2} \left[\begin{array}{cccc|cccc} 1 & 0 & -1 & -1 & 1 & 0 & -1 & 0 \\ 0 & 1 & 2 & 2 & 0 & 0 & 1 & 0 \\ 0 & 0 & \textcircled{2} & 4 & -1 & 1 & 2 & 0 \\ 0 & 0 & -2 & -5 & -1 & 0 & -1 & 1 \end{array}\right] \xrightarrow{\frac{1}{2}R_3}$$

$$\left[\begin{array}{cccc|cccc} 1 & 0 & -1 & -1 & 1 & 0 & -1 & 0 \\ 0 & 1 & 2 & 2 & 0 & 0 & 1 & 0 \\ 0 & 0 & 1 & 2 & -\frac{1}{2} & \frac{1}{2} & 1 & 0 \\ 0 & 0 & -2 & -5 & -1 & 0 & -1 & 1 \end{array}\right] \xrightarrow[R_4+2R_3]{\substack{R_1+R_3 \\ R_2-2R_3}} \left[\begin{array}{cccc|cccc} 1 & 0 & 0 & 1 & \frac{1}{2} & \frac{1}{2} & 0 & 0 \\ 0 & 1 & 0 & -2 & 1 & -1 & -1 & 0 \\ 0 & 0 & 1 & 2 & -\frac{1}{2} & \frac{1}{2} & 1 & 0 \\ 0 & 0 & 0 & \textcircled{-1} & -2 & 1 & 1 & 1 \end{array}\right] \xrightarrow[\substack{R_3+2R_4 \\ -R_4}]{\substack{R_1+R_4 \\ R_2-2R_4}}$$

$$\left[\begin{array}{cccc|cccc} 1 & 0 & 0 & 0 & -\frac{3}{2} & \frac{3}{2} & 1 & 1 \\ 0 & 1 & 0 & 0 & 5 & -3 & -3 & -2 \\ 0 & 0 & 1 & 0 & -\frac{9}{2} & \frac{5}{2} & 3 & 2 \\ 0 & 0 & 0 & 1 & 2 & -1 & -1 & -1 \end{array}\right]. \text{ Therefore, } A^{-1} = \begin{bmatrix} -\frac{3}{2} & \frac{3}{2} & 1 & 1 \\ 5 & -3 & -3 & -2 \\ -\frac{9}{2} & \frac{5}{2} & 3 & 2 \\ 2 & -1 & -1 & -1 \end{bmatrix}.$$

b. i. $\begin{bmatrix} x_1 \\ x_2 \\ x_3 \\ x_4 \end{bmatrix} = \begin{bmatrix} -\frac{3}{2} & \frac{3}{2} & 1 & 1 \\ 5 & -3 & -3 & -2 \\ -\frac{9}{2} & \frac{5}{2} & 3 & 2 \\ 2 & -1 & -1 & -1 \end{bmatrix} \begin{bmatrix} 1 \\ -1 \\ 4 \\ 0 \end{bmatrix} = \begin{bmatrix} 1 \\ -4 \\ 5 \\ -1 \end{bmatrix}$, so $x_1 = 1$, $x_2 = -4$, $x_3 = 5$, and $x_4 = -1$.

ii. $\begin{bmatrix} x_1 \\ x_2 \\ x_3 \\ x_4 \end{bmatrix} = \begin{bmatrix} -\frac{3}{2} & \frac{3}{2} & 1 & 1 \\ 5 & -3 & -3 & -2 \\ -\frac{9}{2} & \frac{5}{2} & 3 & 2 \\ 2 & -1 & -1 & -1 \end{bmatrix} \begin{bmatrix} 2 \\ 8 \\ 4 \\ -1 \end{bmatrix} = \begin{bmatrix} 12 \\ -24 \\ 21 \\ -7 \end{bmatrix}$, so $x_1 = 12$, $x_2 = -24$, $x_3 = 21$, and $x_4 = -7$.

33. a. Using Formula (13), we find $A^{-1} = \dfrac{1}{(2)(-5) - (-4)(3)} \begin{bmatrix} -5 & -3 \\ 4 & 2 \end{bmatrix} = \begin{bmatrix} -\frac{5}{2} & -\frac{3}{2} \\ 2 & 1 \end{bmatrix}$.

b. Using Formula (13) once again, we find $\left(A^{-1}\right)^{-1} = \dfrac{1}{\left(-\frac{5}{2}\right)(1) - 2\left(-\frac{3}{2}\right)} \begin{bmatrix} 1 & \frac{3}{2} \\ -2 & -\frac{5}{2} \end{bmatrix} = \begin{bmatrix} 2 & 3 \\ -4 & -5 \end{bmatrix} = A$.

35. a. $ABC = \begin{bmatrix} 2 & -5 \\ 1 & -3 \end{bmatrix} \begin{bmatrix} 4 & 3 \\ 1 & 1 \end{bmatrix} \begin{bmatrix} 2 & 3 \\ -2 & 1 \end{bmatrix} = \begin{bmatrix} 2 & -5 \\ 1 & -3 \end{bmatrix} \begin{bmatrix} 2 & 15 \\ 0 & 4 \end{bmatrix} = \begin{bmatrix} 4 & 10 \\ 2 & 3 \end{bmatrix}$. Using the formula for the inverse of a 2×2 matrix, we find $A^{-1} = \begin{bmatrix} 3 & -5 \\ 1 & -2 \end{bmatrix}$, $B^{-1} = \begin{bmatrix} 1 & -3 \\ -1 & 4 \end{bmatrix}$, and $C^{-1} = \begin{bmatrix} \frac{1}{8} & -\frac{3}{8} \\ \frac{1}{4} & \frac{1}{4} \end{bmatrix}$.

b. Using the formula for the inverse of a 2×2 matrix, we find $(ABC)^{-1} = \begin{bmatrix} -\frac{3}{8} & \frac{5}{4} \\ \frac{1}{4} & -\frac{1}{2} \end{bmatrix}$, while

$C^{-1}B^{-1}A^{-1} = \begin{bmatrix} \frac{1}{8} & -\frac{3}{8} \\ \frac{1}{4} & \frac{1}{4} \end{bmatrix} \begin{bmatrix} 1 & -3 \\ -1 & 4 \end{bmatrix} \begin{bmatrix} 3 & -5 \\ 1 & -2 \end{bmatrix} = \begin{bmatrix} \frac{1}{8} & -\frac{3}{8} \\ \frac{1}{4} & \frac{1}{4} \end{bmatrix} \begin{bmatrix} 0 & 1 \\ 1 & -3 \end{bmatrix} = \begin{bmatrix} -\frac{3}{8} & \frac{5}{4} \\ \frac{1}{4} & -\frac{1}{2} \end{bmatrix}$.

Therefore, $(ABC)^{-1} = C^{-1}B^{-1}A^{-1}$.

37. Multiplying both sides of the equation on the right by $\begin{bmatrix} 1 & 2 \\ 3 & -1 \end{bmatrix}^{-1}$, we obtain

$A \begin{bmatrix} 1 & 2 \\ 3 & -1 \end{bmatrix} \begin{bmatrix} 1 & 2 \\ 3 & -1 \end{bmatrix}^{-1} = \begin{bmatrix} 2 & 1 \\ 3 & -2 \end{bmatrix} \begin{bmatrix} 1 & 2 \\ 3 & -1 \end{bmatrix}^{-1}$, so $A = \begin{bmatrix} 2 & 1 \\ 3 & -2 \end{bmatrix} \begin{bmatrix} \frac{1}{7} & \frac{2}{7} \\ \frac{3}{7} & -\frac{1}{7} \end{bmatrix} = \begin{bmatrix} \frac{5}{7} & \frac{3}{7} \\ -\frac{3}{7} & \frac{8}{7} \end{bmatrix}$.

39. Let x denote the number of copies of the deluxe edition and y the number of copies of the standard edition demanded per month when the unit prices are p and q dollars, respectively. Then the three systems of linear equations

$$\begin{aligned} 5x + y &= 20{,}000 \\ x + 3y &= 15{,}000 \end{aligned} \qquad \begin{aligned} 5x + y &= 25{,}000 \\ x + 3y &= 15{,}000 \end{aligned} \qquad \begin{aligned} 5x + y &= 25{,}000 \\ x + 3y &= 20{,}000 \end{aligned}$$

give the quantity demanded of each edition at the stated price. These systems may be written in the form $AX = B_1$, $AX = B_2$, and $AX = B_3$, where $A = \begin{bmatrix} 5 & 1 \\ 1 & 3 \end{bmatrix}$, $B_1 = \begin{bmatrix} 20{,}000 \\ 15{,}000 \end{bmatrix}$, $B_2 = \begin{bmatrix} 25{,}000 \\ 15{,}000 \end{bmatrix}$, and $B_3 = \begin{bmatrix} 25{,}000 \\ 20{,}000 \end{bmatrix}$. Using the formula for the inverse of a 2×2 matrix, with $a = 5$, $b = 1$, $c = 1$, $d = 3$, and $D = ad - bc = (5)(3) - (1)(1) = 14$, we find that $A^{-1} = \begin{bmatrix} \frac{3}{14} & -\frac{1}{14} \\ -\frac{1}{14} & \frac{5}{14} \end{bmatrix}$.

a. $\begin{bmatrix} x \\ y \end{bmatrix} = \begin{bmatrix} \frac{3}{14} & -\frac{1}{14} \\ -\frac{1}{14} & \frac{5}{14} \end{bmatrix} \begin{bmatrix} 20{,}000 \\ 15{,}000 \end{bmatrix} = \begin{bmatrix} 3214 \\ 3929 \end{bmatrix}$. **b.** $\begin{bmatrix} x \\ y \end{bmatrix} = \begin{bmatrix} \frac{3}{14} & -\frac{1}{14} \\ -\frac{1}{14} & \frac{5}{14} \end{bmatrix} \begin{bmatrix} 25{,}000 \\ 15{,}000 \end{bmatrix} = \begin{bmatrix} 4286 \\ 3571 \end{bmatrix}$.

c. $\begin{bmatrix} x \\ y \end{bmatrix} = \begin{bmatrix} \frac{3}{14} & -\frac{1}{14} \\ -\frac{1}{14} & \frac{5}{14} \end{bmatrix} \begin{bmatrix} 25{,}000 \\ 20{,}000 \end{bmatrix} = \begin{bmatrix} 3929 \\ 5357 \end{bmatrix}$.

41. Let x, y, and z denote the number of acres of soybeans, corn, and wheat to be cultivated, respectively. Furthermore, let a, b, and c denote the amount of land available, the amount of labor available, and the amount of money available for seeds, respectively. Then we have the system

$$\begin{aligned} x + y + z &= a && \text{(land)} \\ 2x + 6y + 6z &= b && \text{(labor)} \\ 12x + 20y + 8z &= c && \text{(seeds)} \end{aligned}$$

The system can be written in the form $AX = B$, where $A = \begin{bmatrix} 1 & 1 & 1 \\ 2 & 6 & 6 \\ 12 & 20 & 8 \end{bmatrix}$, $X = \begin{bmatrix} x \\ y \\ z \end{bmatrix}$, and $B = \begin{bmatrix} a \\ b \\ c \end{bmatrix}$. To find A^{-1}, we compute $\left[\begin{array}{ccc|ccc} \textcircled{1} & 1 & 1 & 1 & 0 & 0 \\ 2 & 6 & 6 & 0 & 1 & 0 \\ 12 & 20 & 8 & 0 & 0 & 1 \end{array}\right] \xrightarrow[R_3 - 12R_1]{R_2 - 2R_1} \left[\begin{array}{ccc|ccc} 1 & 1 & 1 & 1 & 0 & 0 \\ 0 & \textcircled{4} & 4 & -2 & 1 & 0 \\ 0 & 8 & -4 & -12 & 0 & 1 \end{array}\right] \xrightarrow{\frac{1}{4}R_2}$

$\left[\begin{array}{ccc|ccc} 1 & 1 & 1 & 1 & 0 & 0 \\ 0 & 1 & 1 & -\frac{1}{2} & \frac{1}{4} & 0 \\ 0 & 8 & -4 & -12 & 0 & 1 \end{array}\right] \xrightarrow[R_3 - 8R_2]{R_1 - R_2} \left[\begin{array}{ccc|ccc} 1 & 0 & 0 & \frac{3}{2} & -\frac{1}{4} & 0 \\ 0 & 1 & 1 & -\frac{1}{2} & \frac{1}{4} & 0 \\ 0 & 0 & \textcircled{-12} & -8 & -2 & 1 \end{array}\right] \xrightarrow{-\frac{1}{12}R_3} \left[\begin{array}{ccc|ccc} 1 & 0 & 0 & \frac{3}{2} & -\frac{1}{4} & 0 \\ 0 & 1 & 1 & -\frac{1}{2} & \frac{1}{4} & 0 \\ 0 & 0 & 1 & \frac{2}{3} & \frac{1}{6} & -\frac{1}{12} \end{array}\right] \xrightarrow{R_2 - R_3}$

$\left[\begin{array}{ccc|ccc} 1 & 0 & 0 & \frac{3}{2} & -\frac{1}{4} & 0 \\ 0 & 1 & 0 & -\frac{7}{6} & \frac{1}{12} & \frac{1}{12} \\ 0 & 0 & 1 & \frac{2}{3} & \frac{1}{6} & -\frac{1}{12} \end{array}\right]$. Therefore, $A^{-1} = \begin{bmatrix} \frac{3}{2} & -\frac{1}{4} & 0 \\ -\frac{7}{6} & \frac{1}{12} & \frac{1}{12} \\ \frac{2}{3} & \frac{1}{6} & -\frac{1}{12} \end{bmatrix}$.

a. Here $a = 1000$, $b = 4400$, and $c = 13{,}200$. Therefore

$$X = A^{-1}B = \begin{bmatrix} \frac{3}{2} & -\frac{1}{4} & 0 \\ -\frac{7}{6} & \frac{1}{12} & \frac{1}{12} \\ \frac{2}{3} & \frac{1}{6} & -\frac{1}{12} \end{bmatrix} \begin{bmatrix} 1000 \\ 4400 \\ 13{,}200 \end{bmatrix} = \begin{bmatrix} 400 \\ 300 \\ 300 \end{bmatrix}$$

so Jackson Farms should cultivate 400, 300, and 300 acres of soybeans, corn, and wheat, respectively.

b. Here $a = 1200$, $b = 5200$, and $c = 16{,}400$. Therefore

$$X = A^{-1}B = \begin{bmatrix} \frac{3}{2} & -\frac{1}{4} & 0 \\ -\frac{7}{6} & \frac{1}{12} & \frac{1}{12} \\ \frac{2}{3} & \frac{1}{6} & -\frac{1}{12} \end{bmatrix} \begin{bmatrix} 1200 \\ 5200 \\ 16{,}400 \end{bmatrix} = \begin{bmatrix} 500 \\ 400 \\ 300 \end{bmatrix}$$

so Jackson Farms should cultivate 500, 400, and 300 acres of soybeans, corn, and wheat, respectively.

43. Let x, y, and z denote the amount to be invested in high-, medium-, and low-risk stocks, respectively. Next, let a denote the amount to be invested and let c denote the return on the investments. Then we have the system

$$\begin{aligned} x + \quad y + \quad z &= a \\ x + \quad y - \quad z &= 0 \qquad \text{(because } z = x + y\text{)} \\ 0.15x + 0.1y + 0.06z &= c \end{aligned}$$

The system is equivalent to the matrix equation $AX = B$, where

$A = \begin{bmatrix} 1 & 1 & 1 \\ 1 & 1 & -1 \\ 0.15 & 0.10 & 0.06 \end{bmatrix}$, $X = \begin{bmatrix} x \\ y \\ z \end{bmatrix}$, and $B = \begin{bmatrix} a \\ 0 \\ c \end{bmatrix}$. To find A^{-1}, we compute

$$\left[\begin{array}{ccc|ccc} \textcircled{1} & 1 & 1 & 1 & 0 & 0 \\ 1 & 1 & -1 & 0 & 1 & 0 \\ 0.15 & 0.10 & 0.06 & 0 & 0 & 1 \end{array}\right] \xrightarrow[R_3 - 0.15R_1]{R_2 - R_1} \left[\begin{array}{ccc|ccc} 1 & 1 & 1 & 1 & 0 & 0 \\ 0 & 0 & -2 & -1 & 1 & 0 \\ 0 & -0.05 & -0.09 & -0.15 & 0 & 1 \end{array}\right] \xrightarrow{R_2 \leftrightarrow R_3}$$

$$\left[\begin{array}{ccc|ccc} 1 & 1 & 1 & 1 & 0 & 0 \\ 0 & \boxed{-0.05} & -0.09 & -0.15 & 0 & 1 \\ 0 & 0 & -2 & -1 & 1 & 0 \end{array}\right] \xrightarrow{-20R_2} \left[\begin{array}{ccc|ccc} 1 & 1 & 1 & 1 & 0 & 0 \\ 0 & 1 & 1.8 & 3 & 0 & -20 \\ 0 & 0 & -2 & -1 & 1 & 0 \end{array}\right] \xrightarrow[-0.5R_3]{R_1 - R_2}$$

$$\left[\begin{array}{ccc|ccc} 1 & 0 & 1 & -2 & 0 & 20 \\ 0 & 1 & 1.8 & 3 & 0 & -20 \\ 0 & 0 & \textcircled{1} & 0.5 & -0.5 & 0 \end{array}\right] \xrightarrow[R_2 - 1.8R_3]{R_1 + 0.8R_3} \left[\begin{array}{ccc|ccc} 1 & 0 & 0 & -1.6 & -0.4 & 20 \\ 0 & 1 & 0 & 2.1 & 0.9 & -20 \\ 0 & 0 & 1 & 0.5 & -0.5 & 0 \end{array}\right].$$

Therefore, $A^{-1} = \begin{bmatrix} -1.6 & -0.4 & 20 \\ 2.1 & 0.9 & -20 \\ 0.5 & -0.5 & 0 \end{bmatrix}$.

a. Here $a = 200{,}000$ and $c = 20{,}000$, so $X = A^{-1}B = \begin{bmatrix} -1.6 & -0.4 & 20 \\ 2.1 & 0.9 & -20 \\ 0.5 & -0.5 & 0 \end{bmatrix} \begin{bmatrix} 200{,}000 \\ 0 \\ 20{,}000 \end{bmatrix} = \begin{bmatrix} 80{,}000 \\ 20{,}000 \\ 100{,}000 \end{bmatrix}$.

Thus, the club should invest \$80,000 in high-risk, \$20,000 in medium-risk, and \$100,000 in low risk stocks.

b. Here $a = 220{,}000$ and $c = 22{,}000$, so $X = A^{-1}B = \begin{bmatrix} -1.6 & -0.4 & 20 \\ 2.1 & 0.9 & -20 \\ 0.5 & -0.5 & 0 \end{bmatrix} \begin{bmatrix} 220{,}000 \\ 0 \\ 22{,}000 \end{bmatrix} = \begin{bmatrix} 88{,}000 \\ 22{,}000 \\ 110{,}000 \end{bmatrix}$.

Thus, the club should invest \$88,000 in high-risk, \$22,000 in medium-risk, and \$110,000 in low-risk stocks.

c. Here $a = 240{,}000$ and $c = 22{,}000$, so $X = A^{-1}B = \begin{bmatrix} -1.6 & -0.4 & 20 \\ 2.1 & 0.9 & -20 \\ 0.5 & -0.5 & 0 \end{bmatrix} \begin{bmatrix} 240{,}000 \\ 0 \\ 22{,}000 \end{bmatrix} = \begin{bmatrix} 56{,}000 \\ 64{,}000 \\ 120{,}000 \end{bmatrix}$.

Thus, the club should invest \$56,000 in high-risk stocks, \$64,000 in medium-risk stocks, and \$120,000 in low-risk stocks.

45. In order for the inverse of A to exist, $D = ad - bc \neq 0$. Here $a = 1$, $b = 2$, $c = k$, and $d = 3$, so we must have $(1)(3) - (2)(k) \neq 0$, or $k \neq \frac{3}{2}$. Therefore, A^{-1} has an inverse provided $k \neq \frac{3}{2}$. Using Formula (13), we have

$$A^{-1} = \frac{1}{3 - 2k} \begin{bmatrix} 3 & -2 \\ -k & 1 \end{bmatrix}.$$

47. From the computation $\left[\begin{array}{cc|cc} a & 0 & 1 & 0 \\ 0 & d & 0 & 1 \end{array}\right] \xrightarrow[\frac{1}{d}R_2]{\frac{1}{a}R_1} \left[\begin{array}{cc|cc} 1 & 0 & \frac{1}{a} & 0 \\ 0 & 1 & 0 & \frac{1}{d} \end{array}\right]$, we see that A^{-1} exists provided $ad \neq 0$. Also, we can see that an $n \times n$ diagonal matrix A has an inverse provided all main diagonal entries are nonzero. In fact, the inverse is an $n \times n$ matrix whose main diagonal entries are the reciprocals of the corresponding entries of A.

49. True. Multiplying both sides of the equation by cA yields

$$I = (cA)(cA)^{-1} = (cA)\left[\frac{1}{c}\left(A^{-1}\right)\right] = c\left(\frac{1}{c}\right)AA^{-1} = I.$$

51. True. $AX = B$ can have a unique solution only if A^{-1} exists, in which case the solution is found as follows: $A^{-1}(AX) = A^{-1}B$, $\left(A^{-1}A\right)X = A^{-1}B$, $IX = A^{-1}B$, and $X = A^{-1}B$.

Technology Exercises page 323

1. $\begin{bmatrix} 0.36 & 0.04 & -0.36 \\ 0.06 & 0.05 & 0.20 \\ -0.19 & 0.10 & 0.09 \end{bmatrix}$

3. $\begin{bmatrix} 0.01 & -0.09 & 0.31 & -0.11 \\ -0.25 & 0.58 & -0.15 & -0.02 \\ 0.86 & -0.42 & 0.07 & -0.37 \\ -0.27 & 0.01 & -0.05 & 0.31 \end{bmatrix}$

5. $\begin{bmatrix} 0.30 & 0.85 & -0.10 & -0.77 & -0.11 \\ -0.21 & 0.10 & 0.01 & -0.26 & 0.21 \\ 0.03 & -0.16 & 0.12 & -0.01 & 0.03 \\ -0.14 & -0.46 & 0.13 & 0.71 & -0.05 \\ 0.10 & -0.05 & -0.10 & -0.03 & 0.11 \end{bmatrix}$

7. $x = 1.2$, $y = 3.6$, and $z = 2.7$.

9. $x_1 \approx 2.50$, $x_2 \approx -0.88$, $x_3 \approx 0.70$, and $x_4 \approx 0.51$.

CHAPTER 5 Concept Review Questions page 325

1. a. one, many, no **b.** one, many, no

3. $R_i \leftrightarrow R_j$, cR_i, $R_i + aR_j$, solution

5. size, entries

7. $m \times n$, $n \times m$, a_{ji}

9. a. columns, rows **b.** $m \times p$

11. $A^{-1}A$, AA^{-1}, singular

CHAPTER 5 Review Exercises page 325

1. $\begin{bmatrix} 1 & 2 \\ -1 & 3 \\ 2 & 1 \end{bmatrix} + \begin{bmatrix} 1 & 0 \\ 0 & 1 \\ 1 & 2 \end{bmatrix} = \begin{bmatrix} 2 & 2 \\ -1 & 4 \\ 3 & 3 \end{bmatrix}$.

3. $\begin{bmatrix} -3 & 2 & 1 \end{bmatrix} \begin{bmatrix} 2 & 1 \\ -1 & 0 \\ 2 & 1 \end{bmatrix} = \begin{bmatrix} -6 & -2 \end{bmatrix}$.

5. By the equality of matrices, $x = 2$, $z = 1$, $y = 3$, and $w = 3$.

7. By the equality of matrices, $a + 3 = 6$ (so $a = 3$), $-1 = e + 2$ (so $e = -3$), $b = 4$, $c + 1 = -1$ (so $c = -2$), and $d = 2$.

9. $2A + 3B = 2\begin{bmatrix} 1 & 3 & 1 \\ -2 & 1 & 3 \\ 4 & 0 & 2 \end{bmatrix} + 3\begin{bmatrix} 2 & 1 & 3 \\ -2 & -1 & -1 \\ 1 & 4 & 2 \end{bmatrix} = \begin{bmatrix} 2 & 6 & 2 \\ -4 & 2 & 6 \\ 8 & 0 & 4 \end{bmatrix} + \begin{bmatrix} 6 & 3 & 9 \\ -6 & -3 & -3 \\ 3 & 12 & 6 \end{bmatrix} = \begin{bmatrix} 8 & 9 & 11 \\ -10 & -1 & 3 \\ 11 & 12 & 10 \end{bmatrix}$.

11. $3A = 3\begin{bmatrix} 1 & 3 & 1 \\ -2 & 1 & 3 \\ 4 & 0 & 2 \end{bmatrix} = \begin{bmatrix} 3 & 9 & 3 \\ -6 & 3 & 9 \\ 12 & 0 & 6 \end{bmatrix}$ and $2(3A) = 2\begin{bmatrix} 3 & 9 & 3 \\ -6 & 3 & 9 \\ 12 & 0 & 6 \end{bmatrix} = \begin{bmatrix} 6 & 18 & 6 \\ -12 & 6 & 18 \\ 24 & 0 & 12 \end{bmatrix}$.

13. $B - C = \begin{bmatrix} 2 & 1 & 3 \\ -2 & -1 & -1 \\ 1 & 4 & 2 \end{bmatrix} - \begin{bmatrix} 3 & -1 & 2 \\ 1 & 6 & 4 \\ 2 & 1 & 3 \end{bmatrix} = \begin{bmatrix} -1 & 2 & 1 \\ -3 & -7 & -5 \\ -1 & 3 & -1 \end{bmatrix}$ and so

$$A(B - C) = \begin{bmatrix} 1 & 3 & 1 \\ -2 & 1 & 3 \\ 4 & 0 & 2 \end{bmatrix} \begin{bmatrix} -1 & 2 & 1 \\ -3 & -7 & -5 \\ -1 & 3 & -1 \end{bmatrix} = \begin{bmatrix} -11 & -16 & -15 \\ -4 & -2 & -10 \\ -6 & 14 & 2 \end{bmatrix}.$$

15. $BC = \begin{bmatrix} 2 & 1 & 3 \\ -2 & -1 & -1 \\ 1 & 4 & 2 \end{bmatrix} \begin{bmatrix} 3 & -1 & 2 \\ 1 & 6 & 4 \\ 2 & 1 & 3 \end{bmatrix} = \begin{bmatrix} 13 & 7 & 17 \\ -9 & -5 & -11 \\ 11 & 25 & 24 \end{bmatrix}$, so

$$ABC = \begin{bmatrix} 1 & 3 & 1 \\ -2 & 1 & 3 \\ 4 & 0 & 2 \end{bmatrix} \begin{bmatrix} 13 & 7 & 17 \\ -9 & -5 & -11 \\ 11 & 25 & 24 \end{bmatrix} = \begin{bmatrix} -3 & 17 & 8 \\ -2 & 56 & 27 \\ 74 & 78 & 116 \end{bmatrix}.$$

17. Using the Gauss-Jordan method, we find

$$\left[\begin{array}{cc|c} \textcircled{2} & -3 & 5 \\ 3 & 4 & -1 \end{array}\right] \xrightarrow{\frac{1}{2}R_1} \left[\begin{array}{cc|c} 1 & -\frac{3}{2} & \frac{5}{2} \\ 3 & 4 & -1 \end{array}\right] \xrightarrow{R_2-3R_1} \left[\begin{array}{cc|c} 1 & -\frac{3}{2} & \frac{5}{2} \\ 0 & \textcircled{\frac{17}{2}} & -\frac{17}{2} \end{array}\right] \xrightarrow{\frac{2}{17}R_2} \left[\begin{array}{cc|c} 1 & -\frac{3}{2} & \frac{5}{2} \\ 0 & 1 & -1 \end{array}\right] \xrightarrow{R_1+\frac{3}{2}R_2}$$

$$\left[\begin{array}{cc|c} 1 & 0 & 1 \\ 0 & 1 & -1 \end{array}\right].$$

We conclude that $x = 1$ and $y = -1$.

19.
$$\left[\begin{array}{ccc|c} \textcircled{1} & -1 & 2 & 5 \\ 3 & 2 & 1 & 10 \\ 2 & -3 & -2 & -10 \end{array}\right] \xrightarrow[R_3-2R_1]{R_2-3R_1} \left[\begin{array}{ccc|c} 1 & -1 & 2 & 5 \\ 0 & \textcircled{5} & -5 & -5 \\ 0 & -1 & -6 & -20 \end{array}\right] \xrightarrow{\frac{1}{5}R_2} \left[\begin{array}{ccc|c} 1 & -1 & 2 & 5 \\ 0 & 1 & -1 & -1 \\ 0 & -1 & -6 & -20 \end{array}\right] \xrightarrow[R_3+R_2]{R_1+R_2}$$

$$\left[\begin{array}{ccc|c} 1 & 0 & 1 & 4 \\ 0 & 1 & -1 & -1 \\ 0 & 0 & \textcircled{-7} & -21 \end{array}\right] \xrightarrow{-\frac{1}{7}R_3} \left[\begin{array}{ccc|c} 1 & 0 & 1 & 4 \\ 0 & 1 & -1 & -1 \\ 0 & 0 & 1 & 3 \end{array}\right] \xrightarrow[R_2+R_3]{R_1-R_3} \left[\begin{array}{ccc|c} 1 & 0 & 0 & 1 \\ 0 & 1 & 0 & 2 \\ 0 & 0 & 1 & 3 \end{array}\right].$$ Therefore, $x = 1$, $y = 2$, and $z = 3$.

21.
$$\left[\begin{array}{ccc|c} \textcircled{3} & -2 & 4 & 11 \\ 2 & -4 & 5 & 4 \\ 1 & 2 & -1 & 10 \end{array}\right] \xrightarrow{R_1-R_2} \left[\begin{array}{ccc|c} 1 & 2 & -1 & 7 \\ 2 & -4 & 5 & 4 \\ 1 & 2 & -1 & 10 \end{array}\right] \xrightarrow[R_3-R_1]{R_2-2R_1} \left[\begin{array}{ccc|c} 1 & 2 & -1 & 7 \\ 0 & -8 & 7 & -10 \\ 0 & 0 & 0 & 3 \end{array}\right].$$

Since this last row implies that $0 = 3$, we conclude that the system has no solution.

23.
$$\left[\begin{array}{ccc|c} \textcircled{3} & -2 & 1 & 4 \\ 1 & 3 & -4 & -3 \\ 2 & -3 & 5 & 7 \\ 1 & -8 & 9 & 10 \end{array}\right] \xrightarrow{R_1-R_3} \left[\begin{array}{ccc|c} 1 & 1 & -4 & -3 \\ 1 & 3 & -4 & -3 \\ 2 & -3 & 5 & 7 \\ 1 & -8 & 9 & 10 \end{array}\right] \xrightarrow[R_4-R_1]{\substack{R_2-R_1 \\ R_3-2R_1}} \left[\begin{array}{ccc|c} 1 & 1 & -4 & -3 \\ 0 & \textcircled{2} & 0 & 0 \\ 0 & -5 & 13 & 13 \\ 0 & -9 & 13 & 13 \end{array}\right] \xrightarrow{\frac{1}{2}R_2}$$

$$\left[\begin{array}{ccc|c} 1 & 1 & -4 & -3 \\ 0 & 1 & 0 & 0 \\ 0 & -5 & 13 & 13 \\ 0 & -9 & 13 & 13 \end{array}\right] \xrightarrow[R_4+9R_2]{\substack{R_1-R_2 \\ R_3+5R_2}} \left[\begin{array}{ccc|c} 1 & 0 & -4 & -3 \\ 0 & 1 & 0 & 0 \\ 0 & 0 & \textcircled{13} & 13 \\ 0 & 0 & 13 & 13 \end{array}\right] \xrightarrow{\frac{1}{13}R_3} \left[\begin{array}{ccc|c} 1 & 0 & -4 & -3 \\ 0 & 1 & 0 & 0 \\ 0 & 0 & 1 & 1 \\ 0 & 0 & 13 & 13 \end{array}\right] \xrightarrow[R_4-13R_3]{R_1+4R_3} \left[\begin{array}{ccc|c} 1 & 0 & 0 & 1 \\ 0 & 1 & 0 & 0 \\ 0 & 0 & 1 & 1 \\ 0 & 0 & 0 & 0 \end{array}\right].$$

Thus, $x = 1$, $y = 0$, and $z = 1$.

25. $A^{-1} = \dfrac{1}{(3)(2)-(1)(1)}\begin{bmatrix} 2 & -1 \\ -1 & 3 \end{bmatrix} = \begin{bmatrix} \frac{2}{5} & -\frac{1}{5} \\ -\frac{1}{5} & \frac{3}{5} \end{bmatrix}.$

27. $A^{-1} = \dfrac{1}{(3)(2)-(2)(4)}\begin{bmatrix} 2 & -4 \\ -2 & 3 \end{bmatrix} = \begin{bmatrix} -1 & 2 \\ 1 & -\frac{3}{2} \end{bmatrix}.$

29. $$\left[\begin{array}{ccc|ccc} \textcircled{2} & 3 & 1 & 1 & 0 & 0 \\ 1 & -1 & 2 & 0 & 1 & 0 \\ 1 & 2 & 1 & 0 & 0 & 1 \end{array}\right] \xrightarrow{R_1 - R_2} \left[\begin{array}{ccc|ccc} 1 & 4 & -1 & 1 & -1 & 0 \\ 1 & -1 & 2 & 0 & 1 & 0 \\ 1 & 2 & 1 & 0 & 0 & 1 \end{array}\right] \xrightarrow[R_3 - R_1]{R_2 - R_1} \left[\begin{array}{ccc|ccc} 1 & 4 & -1 & 1 & -1 & 0 \\ 0 & \textcircled{-5} & 3 & -1 & 2 & 0 \\ 0 & -2 & 2 & -1 & 1 & 1 \end{array}\right]$$

$$\xrightarrow{R_2 - 3R_3} \left[\begin{array}{ccc|ccc} 1 & 4 & -1 & 1 & -1 & 0 \\ 0 & 1 & -3 & 2 & -1 & -3 \\ 0 & -2 & 2 & -1 & 1 & 1 \end{array}\right] \xrightarrow[R_3 + 2R_2]{R_1 - 4R_2} \left[\begin{array}{ccc|ccc} 1 & 0 & 11 & -7 & 3 & 12 \\ 0 & 1 & -3 & 2 & -1 & -3 \\ 0 & 0 & \textcircled{-4} & 3 & -1 & -5 \end{array}\right] \xrightarrow{-\frac{1}{4}R_3}$$

$$\left[\begin{array}{ccc|ccc} 1 & 0 & 11 & -7 & 3 & 12 \\ 0 & 1 & -3 & 2 & -1 & -3 \\ 0 & 0 & 1 & -\frac{3}{4} & \frac{1}{4} & \frac{5}{4} \end{array}\right] \xrightarrow[R_2 + 3R_3]{R_1 - 11R_3} \left[\begin{array}{ccc|ccc} 1 & 0 & 0 & \frac{5}{4} & \frac{1}{4} & -\frac{7}{4} \\ 0 & 1 & 0 & -\frac{1}{4} & -\frac{1}{4} & \frac{3}{4} \\ 0 & 0 & 1 & -\frac{3}{4} & \frac{1}{4} & \frac{5}{4} \end{array}\right], \text{ so } A^{-1} = \begin{bmatrix} \frac{5}{4} & \frac{1}{4} & -\frac{7}{4} \\ -\frac{1}{4} & -\frac{1}{4} & \frac{3}{4} \\ -\frac{3}{4} & \frac{1}{4} & \frac{5}{4} \end{bmatrix}.$$

31. $$\left[\begin{array}{ccc|ccc} \textcircled{1} & 2 & 4 & 1 & 0 & 0 \\ 3 & 1 & 2 & 0 & 1 & 0 \\ 1 & 0 & -6 & 0 & 0 & 1 \end{array}\right] \xrightarrow[R_3 - R_1]{R_2 - 3R_1} \left[\begin{array}{ccc|ccc} 1 & 2 & 4 & 1 & 0 & 0 \\ 0 & \textcircled{-5} & -10 & -3 & 1 & 0 \\ 0 & -2 & -10 & -1 & 0 & 1 \end{array}\right] \xrightarrow{R_2 - 3R_3} \left[\begin{array}{ccc|ccc} 1 & 2 & 4 & 1 & 0 & 0 \\ 0 & 1 & 20 & 0 & 1 & -3 \\ 0 & -2 & -10 & -1 & 0 & 1 \end{array}\right] \xrightarrow[R_3 + 2R_2]{R_1 - 2R_2}$$

$$\left[\begin{array}{ccc|ccc} 1 & 0 & -36 & 1 & -2 & 6 \\ 0 & 1 & 20 & 0 & 1 & -3 \\ 0 & 0 & \textcircled{30} & -1 & 2 & -5 \end{array}\right] \xrightarrow{\frac{1}{30}R_3} \left[\begin{array}{ccc|ccc} 1 & 0 & -36 & 1 & -2 & 6 \\ 0 & 1 & 20 & 0 & 1 & -3 \\ 0 & 0 & 1 & -\frac{1}{30} & \frac{1}{15} & -\frac{1}{6} \end{array}\right] \xrightarrow[R_2 - 20R_3]{R_1 + 36R_3} \left[\begin{array}{ccc|ccc} 1 & 0 & 0 & -\frac{1}{5} & \frac{2}{5} & 0 \\ 0 & 1 & 0 & \frac{2}{3} & -\frac{1}{3} & \frac{1}{3} \\ 0 & 0 & 1 & -\frac{1}{30} & \frac{1}{15} & -\frac{1}{6} \end{array}\right], \text{ so}$$

$$A^{-1} = \begin{bmatrix} -\frac{1}{5} & \frac{2}{5} & 0 \\ \frac{2}{3} & -\frac{1}{3} & \frac{1}{3} \\ -\frac{1}{30} & \frac{1}{15} & -\frac{1}{6} \end{bmatrix}.$$

33. $\left(A^{-1}B\right)^{-1} = B^{-1}\left(A^{-1}\right)^{-1} = B^{-1}A$. Now

$$B^{-1} = \frac{1}{(3)(2) - 4(1)} \begin{bmatrix} 2 & -1 \\ -4 & 3 \end{bmatrix} = \begin{bmatrix} 1 & -\frac{1}{2} \\ -2 & \frac{3}{2} \end{bmatrix}, \text{ so } B^{-1}A = \begin{bmatrix} 1 & -\frac{1}{2} \\ -2 & \frac{3}{2} \end{bmatrix} \begin{bmatrix} 1 & 2 \\ -1 & 2 \end{bmatrix} = \begin{bmatrix} \frac{3}{2} & 1 \\ -\frac{7}{2} & -1 \end{bmatrix}.$$

35. $2A - C = \begin{bmatrix} 2 & 4 \\ -2 & 4 \end{bmatrix} - \begin{bmatrix} 1 & 1 \\ -1 & 2 \end{bmatrix} = \begin{bmatrix} 1 & 3 \\ -1 & 2 \end{bmatrix}$, so $(2A - C)^{-1} = \dfrac{1}{(1)(2) - (-1)(3)} \begin{bmatrix} 2 & -3 \\ 1 & 1 \end{bmatrix} = \begin{bmatrix} \frac{2}{5} & -\frac{3}{5} \\ \frac{1}{5} & \frac{1}{5} \end{bmatrix}$.

37. $AX = C$, where $A = \begin{bmatrix} 2 & 3 \\ 1 & -2 \end{bmatrix}$, $X = \begin{bmatrix} x \\ y \end{bmatrix}$, and $C = \begin{bmatrix} -8 \\ 3 \end{bmatrix}$, so

$$A^{-1} = \frac{1}{(-2)(2) - (1)(3)} \begin{bmatrix} -2 & -3 \\ -1 & 2 \end{bmatrix} = \begin{bmatrix} \frac{2}{7} & \frac{3}{7} \\ \frac{1}{7} & -\frac{2}{7} \end{bmatrix}.$$

Thus, $\begin{bmatrix} x \\ y \end{bmatrix} = A^{-1}B = \begin{bmatrix} \frac{2}{7} & \frac{3}{7} \\ \frac{1}{7} & -\frac{2}{7} \end{bmatrix} \begin{bmatrix} -8 \\ 3 \end{bmatrix} = \begin{bmatrix} -1 \\ -2 \end{bmatrix}$.

39. Put $X = \begin{bmatrix} x \\ y \\ z \end{bmatrix}$, $A = \begin{bmatrix} 1 & -2 & 4 \\ 2 & 3 & -2 \\ 1 & 4 & -6 \end{bmatrix}$, and $C = \begin{bmatrix} 13 \\ 0 \\ -15 \end{bmatrix}$. Then $AX = C$ and $X = A^{-1}C$.

To find A^{-1}, we calculate $\left[\begin{array}{ccc|ccc} \textcircled{1} & -2 & 4 & 1 & 0 & 0 \\ 2 & 3 & -2 & 0 & 1 & 0 \\ 1 & 4 & -6 & 0 & 0 & 1 \end{array}\right] \xrightarrow[R_3 - R_1]{R_2 - 2R_1} \left[\begin{array}{ccc|ccc} 1 & -2 & 4 & 1 & 0 & 0 \\ 0 & \textcircled{7} & -10 & -2 & 1 & 0 \\ 0 & 6 & -10 & -1 & 0 & 1 \end{array}\right] \xrightarrow{R_2 - R_3}$

$$\left[\begin{array}{ccc|ccc} 1 & -2 & 4 & 1 & 0 & 0 \\ 0 & 1 & 0 & -1 & 1 & -1 \\ 0 & 6 & -10 & -1 & 0 & 1 \end{array}\right] \xrightarrow[R_3 - 6R_2]{R_1 + 2R_2} \left[\begin{array}{ccc|ccc} 1 & 0 & 4 & -1 & 2 & -2 \\ 0 & 1 & 0 & -1 & 1 & -1 \\ 0 & 0 & \textcircled{-10} & 5 & -6 & 7 \end{array}\right] \xrightarrow{-\frac{1}{10}R_3} \left[\begin{array}{ccc|ccc} 1 & 0 & 4 & -1 & 2 & -2 \\ 0 & 1 & 0 & -1 & 1 & -1 \\ 0 & 0 & 1 & -\frac{1}{2} & \frac{3}{5} & -\frac{7}{10} \end{array}\right] \xrightarrow{R_1 - 4R_3}$$

$$\left[\begin{array}{ccc|ccc} 1 & 0 & 0 & 1 & -\frac{2}{5} & \frac{4}{5} \\ 0 & 1 & 0 & -1 & 1 & -1 \\ 0 & 0 & 1 & -\frac{1}{2} & \frac{3}{5} & -\frac{7}{10} \end{array}\right], \text{ so } A^{-1} = \begin{bmatrix} 1 & -\frac{2}{5} & \frac{4}{5} \\ -1 & 1 & -1 \\ -\frac{1}{2} & \frac{3}{5} & -\frac{7}{10} \end{bmatrix}.$$

Therefore, $X = A^{-1}C = \begin{bmatrix} 1 & -\frac{2}{5} & \frac{4}{5} \\ -1 & 1 & -1 \\ -\frac{1}{2} & \frac{3}{5} & -\frac{7}{10} \end{bmatrix} \begin{bmatrix} 13 \\ 0 \\ -15 \end{bmatrix} = \begin{bmatrix} 1 \\ 2 \\ 4 \end{bmatrix}$; that is, $x = 1$, $y = 2$, and $z = 4$.

41. Let $S = \begin{bmatrix} 600 & 800 & 1000 & 700 \\ 700 & 600 & 1200 & 400 \\ 900 & 700 & 1400 & 800 \end{bmatrix}$ (columns: Premium, Super, Regular, Diesel) and $T = \begin{bmatrix} 3.70 \\ 3.48 \\ 3.30 \\ 3.60 \end{bmatrix}$ (rows: Premium, Super, Regular, Diesel) be the matrices representing the sales in the three gasoline stations and the unit prices for the various fuels. Then the total revenue at each station is found by computing $ST = \begin{bmatrix} 600 & 800 & 1000 & 700 \\ 700 & 600 & 1200 & 400 \\ 900 & 700 & 1400 & 800 \end{bmatrix} \begin{bmatrix} 3.70 \\ 3.48 \\ 3.30 \\ 3.60 \end{bmatrix} = \begin{bmatrix} 10{,}824 \\ 10{,}078 \\ 13{,}266 \end{bmatrix}$.

We conclude that the total revenue of station A is \$10,824, that of station B is \$10,078, and that of station C is \$13,266.

43. a. $A = \begin{bmatrix} 800 & 1200 & 400 & 1500 \\ 600 & 1400 & 600 & 2000 \end{bmatrix}$ **b.** $B = \begin{bmatrix} 50.26 \\ 31.00 \\ 103.07 \\ 38.67 \end{bmatrix}$

c. $AB = \begin{bmatrix} 800 & 1200 & 400 & 1500 \\ 600 & 1400 & 600 & 2000 \end{bmatrix} \begin{bmatrix} 50.26 \\ 31.00 \\ 103.07 \\ 38.67 \end{bmatrix} = \begin{bmatrix} 176{,}641 \\ 212{,}738 \end{bmatrix}$

45. We wish to solve the system of equations

$$\begin{aligned} 2x + 2y + 3z &= 210 \\ 2x + 3y + 4z &= 270 \\ 3x + 4y + 3z &= 300 \end{aligned}$$

Using the Gauss–Jordan method, we find $\left[\begin{array}{ccc|c} \textcircled{2} & 2 & 3 & 210 \\ 2 & 3 & 4 & 270 \\ 3 & 4 & 3 & 300 \end{array}\right] \xrightarrow{\frac{1}{2}R_1} \left[\begin{array}{ccc|c} 1 & 1 & \frac{3}{2} & 105 \\ 2 & 3 & 4 & 270 \\ 3 & 4 & 3 & 300 \end{array}\right] \xrightarrow[R_3 - 3R_1]{R_2 - 2R_1}$

$$\left[\begin{array}{ccc|c} 1 & 1 & \frac{3}{2} & 105 \\ 0 & \textcircled{1} & 1 & 60 \\ 0 & 1 & -\frac{3}{2} & -15 \end{array}\right] \xrightarrow[R_3 - R_2]{R_1 - R_2} \left[\begin{array}{ccc|c} 1 & 0 & \frac{1}{2} & 45 \\ 0 & 1 & 1 & 60 \\ 0 & 0 & \textcircled{$-\frac{5}{2}$} & -75 \end{array}\right] \xrightarrow{-\frac{2}{5}R_3} \left[\begin{array}{ccc|c} 1 & 0 & \frac{1}{2} & 45 \\ 0 & 1 & 1 & 60 \\ 0 & 0 & 1 & 30 \end{array}\right] \xrightarrow[R_2 - R_3]{R_1 - \frac{1}{2}R_3} \left[\begin{array}{ccc|c} 1 & 0 & 0 & 30 \\ 0 & 1 & 0 & 30 \\ 0 & 0 & 1 & 30 \end{array}\right].$$

Thus, $x = y = z = 30$, and so Desmond should produce 30 of each type of pendant.

CHAPTER 5 Before Moving On... page 327

1. $$\left[\begin{array}{ccc|c} 2 & 1 & -1 & -1 \\ 1 & 3 & 2 & 2 \\ 3 & 3 & -3 & -5 \end{array}\right] \xrightarrow{R_1 \leftrightarrow R_2} \left[\begin{array}{ccc|c} \textcircled{1} & 3 & 2 & 2 \\ 2 & 1 & -1 & -1 \\ 3 & 3 & -3 & -5 \end{array}\right] \xrightarrow[R_3 - 3R_1]{R_2 - 2R_1} \left[\begin{array}{ccc|c} 1 & 3 & 2 & 2 \\ 0 & \textcircled{-5} & -5 & -5 \\ 0 & -6 & -9 & -11 \end{array}\right] \xrightarrow{-\frac{1}{5}R_2}$$

$$\left[\begin{array}{ccc|c} 1 & 3 & 2 & 2 \\ 0 & 1 & 1 & 1 \\ 0 & -6 & -9 & -11 \end{array}\right] \xrightarrow[R_3 + 6R_2]{R_1 - 3R_2} \left[\begin{array}{ccc|c} 1 & 0 & -1 & -1 \\ 0 & 1 & 1 & 1 \\ 0 & 0 & \textcircled{-3} & -5 \end{array}\right] \xrightarrow{-\frac{1}{3}R_3} \left[\begin{array}{ccc|c} 1 & 0 & -1 & -1 \\ 0 & 1 & 1 & 1 \\ 0 & 0 & 1 & \frac{5}{3} \end{array}\right] \xrightarrow[R_2 - R_3]{R_1 + R_3} \left[\begin{array}{ccc|c} 1 & 0 & 0 & \frac{2}{3} \\ 0 & 1 & 0 & -\frac{2}{3} \\ 0 & 0 & 1 & \frac{5}{3} \end{array}\right].$$

The solution is $x = \frac{2}{3}$, $y = -\frac{2}{3}$, $z = \frac{5}{3}$.

2. a. $x = 2$, $y = -3$, $z = 1$

b. No solution.

c. $x = 2$, $y = 1 - 3t$, $z = t$, where t is a parameter.

d. $x = y = z = w = 0$.

e. $x = 2 + t$, $y = 3 - 2t$, $z = t$, where t is a parameter

3. a. $$\left[\begin{array}{cc|c} \textcircled{1} & 2 & 3 \\ 3 & -1 & -5 \\ 4 & 1 & -2 \end{array}\right] \xrightarrow[R_3 - 4R_1]{R_2 - 3R_1} \left[\begin{array}{cc|c} 1 & 2 & 3 \\ 0 & \textcircled{-7} & -14 \\ 0 & -7 & -14 \end{array}\right] \xrightarrow{-\frac{1}{7}R_2} \left[\begin{array}{cc|c} 1 & 2 & 3 \\ 0 & 1 & 2 \\ 0 & -7 & -14 \end{array}\right] \xrightarrow[R_3 + 7R_2]{R_1 - 2R_2} \left[\begin{array}{cc|c} 1 & 0 & -1 \\ 0 & 1 & 2 \\ 0 & 0 & 0 \end{array}\right].$$

The solution is $x = -1$, $y = 2$.

b. $$\left[\begin{array}{ccc|c} \textcircled{1} & -2 & 4 & 2 \\ 3 & 1 & -2 & 1 \end{array}\right] \xrightarrow{R_2 - 3R_1} \left[\begin{array}{ccc|c} 1 & -2 & 4 & 2 \\ 0 & \textcircled{7} & -14 & -5 \end{array}\right] \xrightarrow{\frac{1}{7}R_2} \left[\begin{array}{ccc|c} 1 & -2 & 4 & 2 \\ 0 & 1 & -2 & -\frac{5}{7} \end{array}\right] \xrightarrow{R_1 + 2R_2} \left[\begin{array}{ccc|c} 1 & 0 & 0 & \frac{4}{7} \\ 0 & 1 & -2 & -\frac{5}{7} \end{array}\right].$$

The solution is $x = \frac{4}{7}$, $y = -\frac{5}{7} + 2t$, $z = t$, where t is a parameter.

4. a. $$AB = \begin{bmatrix} 1 & -2 & 4 \\ 3 & 0 & 1 \end{bmatrix} \begin{bmatrix} 1 & -1 & 2 \\ 3 & 1 & -1 \\ 2 & 1 & 0 \end{bmatrix} = \begin{bmatrix} 3 & 1 & 4 \\ 5 & -2 & 6 \end{bmatrix}.$$

b. $A + C^T = \begin{bmatrix} 1 & -2 & 4 \\ 3 & 0 & 1 \end{bmatrix} + \begin{bmatrix} 2 & 1 & 3 \\ -2 & 1 & 4 \end{bmatrix} = \begin{bmatrix} 3 & -1 & 7 \\ 1 & 1 & 5 \end{bmatrix}$ and

$$\left(A + C^T\right) B = \begin{bmatrix} 3 & -1 & 7 \\ 1 & 1 & 5 \end{bmatrix} \begin{bmatrix} 1 & -1 & 2 \\ 3 & 1 & -1 \\ 2 & 1 & 0 \end{bmatrix} = \begin{bmatrix} 14 & 3 & 7 \\ 14 & 5 & 1 \end{bmatrix}.$$

c. $C^T B - AB^T = \begin{bmatrix} 2 & 1 & 3 \\ -2 & 1 & 4 \end{bmatrix} \begin{bmatrix} 1 & -1 & 2 \\ 3 & 1 & -1 \\ 2 & 1 & 0 \end{bmatrix} \begin{bmatrix} 1 & -2 & 4 \\ 3 & 0 & 1 \end{bmatrix} \begin{bmatrix} 1 & 3 & 2 \\ -1 & 1 & 1 \\ 2 & -1 & 0 \end{bmatrix}$

$$= \begin{bmatrix} 11 & 2 & 3 \\ 9 & 7 & -5 \end{bmatrix} - \begin{bmatrix} 11 & -3 & 0 \\ 5 & 8 & 6 \end{bmatrix} = \begin{bmatrix} 0 & 5 & 3 \\ 4 & -1 & -11 \end{bmatrix}.$$

5. $\left[\begin{array}{ccc|ccc} 2 & 1 & 2 & 1 & 0 & 0 \\ 0 & -1 & 3 & 0 & 1 & 0 \\ 1 & 1 & 0 & 0 & 0 & 1 \end{array}\right] \xrightarrow{R_1 \leftrightarrow R_3} \left[\begin{array}{ccc|ccc} \textcircled{1} & 1 & 0 & 0 & 0 & 1 \\ 0 & -1 & 3 & 0 & 1 & 0 \\ 2 & 1 & 2 & 1 & 0 & 0 \end{array}\right] \xrightarrow[R_3 - 2R_1]{R_1 + R_2,\ -R_2} \left[\begin{array}{ccc|ccc} 1 & 0 & 3 & 0 & 1 & 1 \\ 0 & \textcircled{1} & -3 & 0 & -1 & 0 \\ 0 & -1 & 2 & 1 & 0 & -2 \end{array}\right] \xrightarrow{R_3 + R_2}$

$\left[\begin{array}{ccc|ccc} 1 & 0 & 3 & 0 & 1 & 1 \\ 0 & 1 & -3 & 0 & -1 & 0 \\ 0 & 0 & \textcircled{-1} & 1 & -1 & -2 \end{array}\right] \xrightarrow{-R_3} \left[\begin{array}{ccc|ccc} 1 & 0 & 3 & 0 & 1 & 1 \\ 0 & 1 & -3 & 0 & -1 & 0 \\ 0 & 0 & 1 & -1 & 1 & 2 \end{array}\right] \xrightarrow[R_2 + 3R_3]{R_1 - 3R_3} \left[\begin{array}{ccc|ccc} 1 & 0 & 0 & 3 & -2 & -5 \\ 0 & 1 & 0 & -3 & 2 & 6 \\ 0 & 0 & 1 & -1 & 1 & 2 \end{array}\right]$, so

$$A^{-1} = \begin{bmatrix} 3 & -2 & -5 \\ -3 & 2 & 6 \\ -1 & 1 & 2 \end{bmatrix}.$$

6. $A = \begin{bmatrix} 2 & 0 & 1 \\ 2 & 1 & -1 \\ 3 & 1 & -1 \end{bmatrix}$, $B = \begin{bmatrix} 4 \\ -1 \\ 0 \end{bmatrix}$, and $X = \begin{bmatrix} x \\ y \\ z \end{bmatrix}$. To find A^{-1}, we calculate

$\left[\begin{array}{ccc|ccc} 2 & 0 & 1 & 1 & 0 & 0 \\ 2 & 1 & -1 & 0 & 1 & 0 \\ 3 & 1 & -1 & 0 & 0 & 1 \end{array}\right] \xrightarrow{R_1 \leftrightarrow R_3} \left[\begin{array}{ccc|ccc} 3 & 1 & -1 & 0 & 0 & 1 \\ 2 & \textcircled{1} & -1 & 0 & 1 & 0 \\ 2 & 0 & 1 & 1 & 0 & 0 \end{array}\right] \xrightarrow{R_1 - R_2} \left[\begin{array}{ccc|ccc} \textcircled{1} & 0 & 0 & 0 & -1 & 1 \\ 2 & 1 & -1 & 0 & 1 & 0 \\ 2 & 0 & 1 & 1 & 0 & 0 \end{array}\right] \xrightarrow[R_3 - 2R_1]{R_2 - 2R_1}$

$\left[\begin{array}{ccc|ccc} 1 & 0 & 0 & 0 & -1 & 1 \\ 0 & 1 & -1 & 0 & 3 & -2 \\ 0 & 0 & \textcircled{1} & 1 & 2 & -2 \end{array}\right] \xrightarrow{R_2 + R_3} \left[\begin{array}{ccc|ccc} 1 & 0 & 0 & 0 & -1 & 1 \\ 0 & 1 & 0 & 1 & 5 & -4 \\ 0 & 0 & 1 & 1 & 2 & -2 \end{array}\right]$. Therefore, $A^{-1} = \begin{bmatrix} 0 & -1 & 1 \\ 1 & 5 & -4 \\ 1 & 2 & -2 \end{bmatrix}$, so

$X = \begin{bmatrix} x \\ y \\ z \end{bmatrix} = A^{-1}B = \begin{bmatrix} 0 & -1 & 1 \\ 1 & 5 & -4 \\ 1 & 2 & -2 \end{bmatrix} \begin{bmatrix} 4 \\ -1 \\ 0 \end{bmatrix} = \begin{bmatrix} 1 \\ -1 \\ 2 \end{bmatrix}$, and we conclude that $x = 1$, $y = -1$, and $z = 2$.

6 LINEAR PROGRAMMING

6.1 Graphing Systems of Linear Inequalities in Two Variables

Problem-Solving Tips

1. When you graph an inequality, use a solid line to show inclusion in the solution set ($\leq$ or $\geq$) and a dashed line to show exclusion from the solution set ($<$ or $>$).

2. Pick the point $(0, 0)$ as your test point if it does not lie on the line you are considering. Otherwise, pick a point at which it is easy to evaluate the quantities you are working with.

3. If the solution set of a system of linear inequalities can be enclosed by a circle, the solutions set is **bounded**. Otherwise, it is **unbounded**.

Concept Questions page 335

1. **a.** The solution set of $ax + by < c$ is a half-plane that does not include the line with equation $ax + by = c$. The solution set of $ax + by \leq c$, on the other hand, includes the line.

   b. It is the line with equation $ax + by = c$.

Exercises page 335

1. $4x - 8 < 0$ implies $x < 2$.

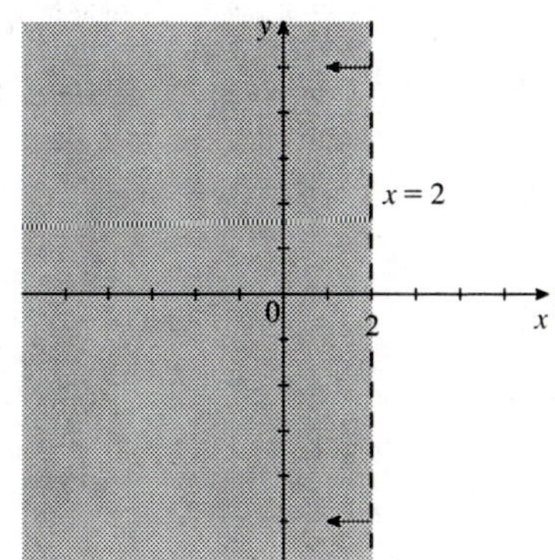

3. $x - y \leq 0$ implies $x \leq y$.

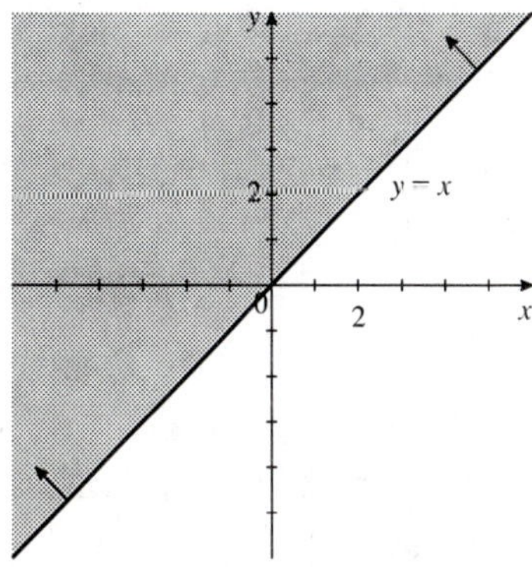

5. The graph of $x \le -3$ is a half-plane.

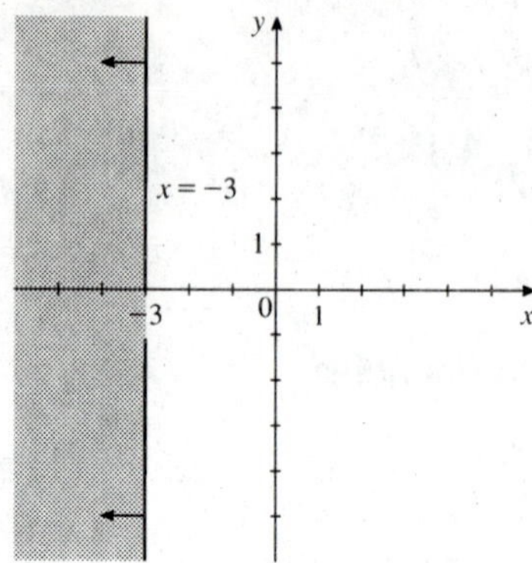

7. We first sketch the straight line with equation $2x + y = 4$. Next, picking the test point $(0, 0)$, we have $2(0) + (0) = 0 \le 4$. We conclude that the half-plane containing the origin is the required half-plane.

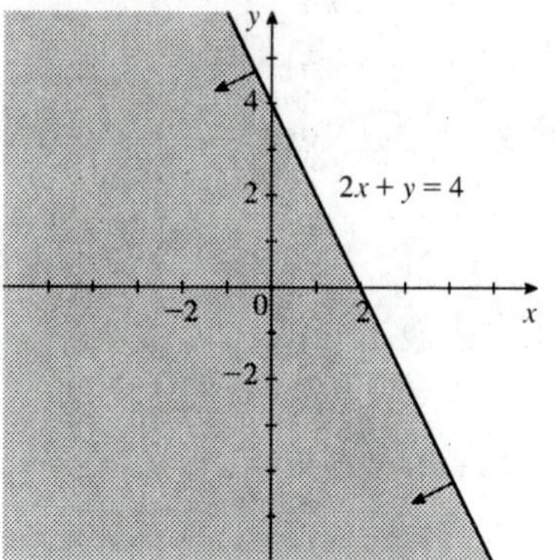

9. We first sketch the graph of the straight line $4x - 3y = -24$. Next, picking the test point $(0, 0)$, we see that $4(0) - 3(0) = 0 \not\le -24$. We conclude that the half-plane not containing the origin is the required half-plane.

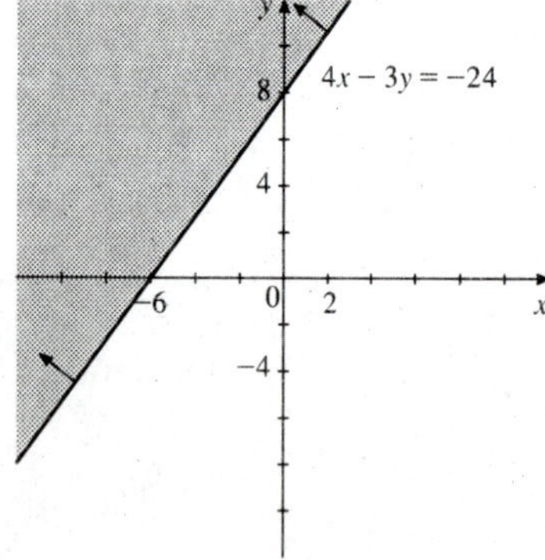

11. The system of linear inequalities that describes the shaded region is $x \ge 1$, $x \le 5$, $y \ge 2$, $y \le 4$. We may also combine the first and second inequalities and the third and fourth inequalities and write $1 \le x \le 5$ and $2 \le y \le 4$.

13. The system of linear inequalities that describes the shaded region is $2x - y \ge 2$, $5x + 7y \ge 35$, $x \le 4$.

15. The system of linear inequalities that describes the shaded region is $7x + 4y \le 140$, $x + 3y \ge 30$, $x - y \ge -10$.

17. The system of linear inequalities that describes the shaded region is $x + y \ge 7$, $x \ge 2$, $y \ge 3$, $y \le 7$. We may also combine the third and fourth inequalities and write $3 \le y \le 7$.

19. The vertex $A(2, 3)$ is found by solving the system $\begin{cases} 2x + 4y = 16 \\ -x + 3y = 7 \end{cases}$ Observe that a dashed line is used to show that no point on the line constitutes a solution to the given problem. Observe also that this is an unbounded solution set.

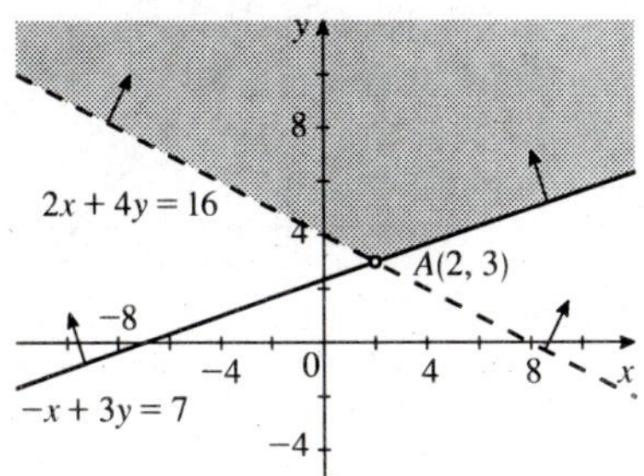

21. The vertex $A(2, 2)$ is found by solving the system $\begin{cases} x - y = 0 \\ 2x + 3y = 10 \end{cases}$ The solution set is unbounded.

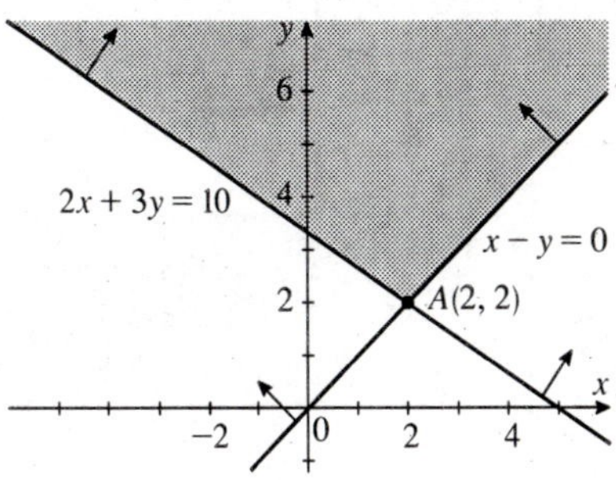

23. Because the two half-planes defined by the system of inequalities $\begin{cases} x + 2y \geq 3 \\ 2x + 4y \leq -2 \end{cases}$ have no point in common, we conclude that the system has no solution. The (empty) set is bounded.

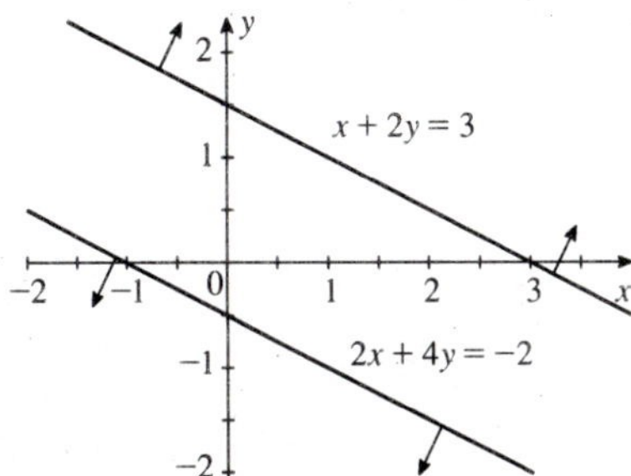

25. The vertex $A(3, 3)$ is found by solving the system $\begin{cases} x + y = 6 \\ x = 3 \end{cases}$ Observe that the solution set is bounded.

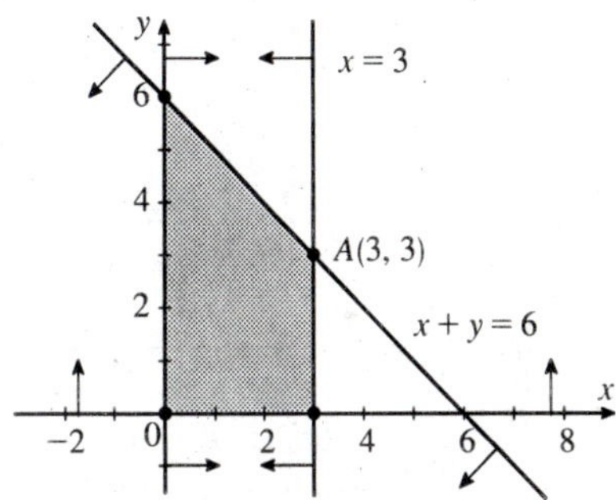

27. Observe that the two lines described by the equations $3x - 6y = 12$ and $-x + 2y = 4$ do not intersect because they are parallel. The solution set is unbounded.

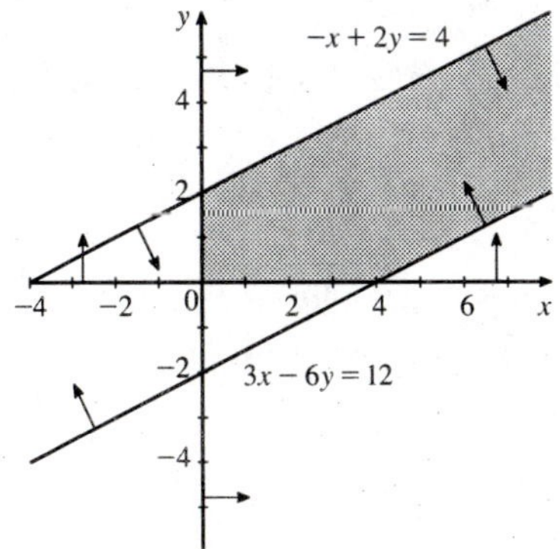

29. The vertices are $A\left(0, \frac{24}{7}\right)$, $B\left(0, \frac{8}{3}\right)$, and $C(8, 0)$. The solution set is unbounded.

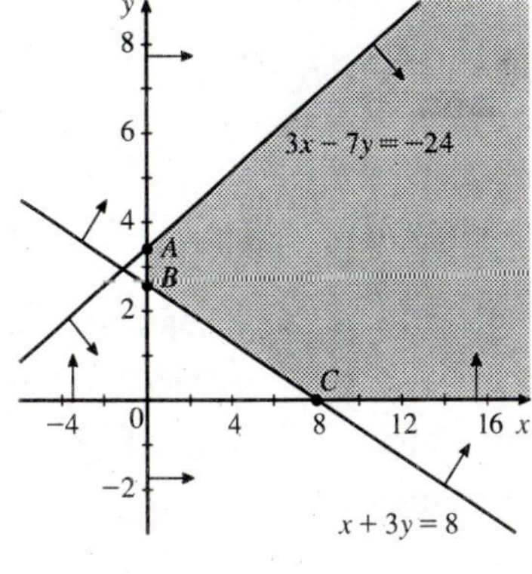

31. The corners of the bounded solution set are $\left(0, \frac{3}{2}\right)$, $(0, 2)$, $\left(\frac{24}{5}, 2\right)$, $\left(\frac{16}{5}, 0\right)$, and $(3, 0)$.

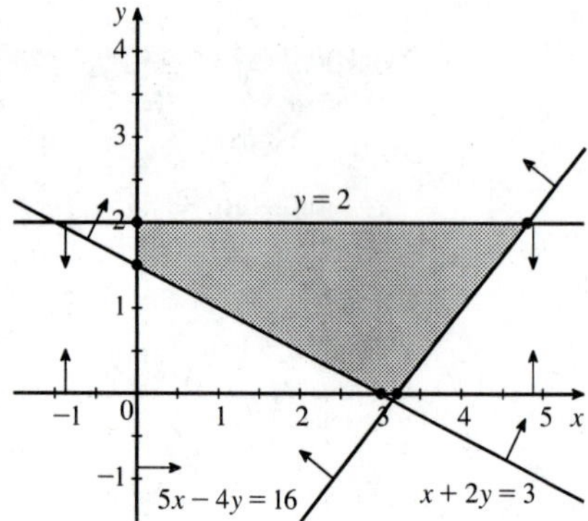

33. The bounded solution set has vertices $(0, 6)$, $(5, 0)$, $(4, 0)$, and $(1, 3)$.

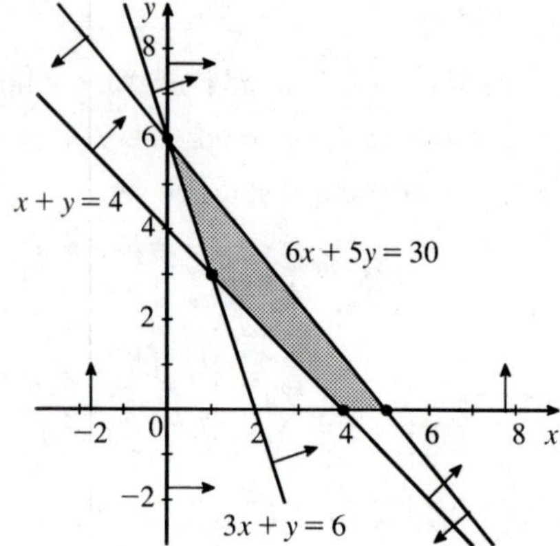

35. The unbounded solution set has vertices $(2, 8)$, $(0, 6)$, $(0, 3)$, and $(2, 2)$.

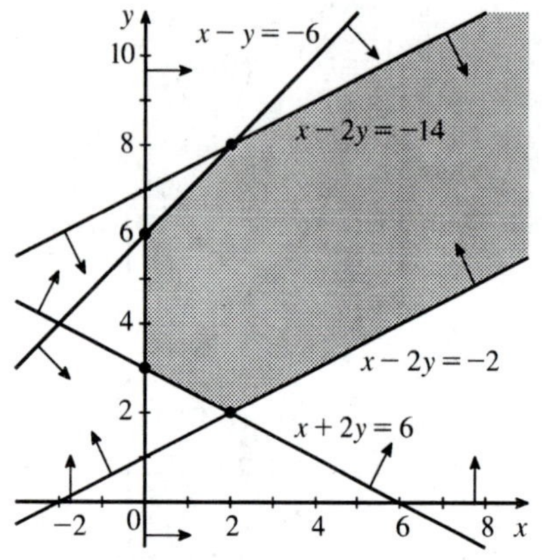

37. False. It is always a half-plane. A straight line is the graph of a linear equation and vice versa.

39. True. Since a circle can always be enclosed by a rectangle, the solution set of such a system is bounded if it can be enclosed by a rectangle.

6.2 Linear Programming Problems

Problem-Solving Tips

1. In a linear programming problem, a linear objective function is maximized or minimized subject to certain constraints. These constraints can be **linear equations** or **inequalities**.

2. When you solve an applied linear programming problem, it is important to understand the question in mathematical terms. If you are asked to maximize the profit, then the problem involves a linear objective function that should be maximized. If you are asked to minimize the cost, then the problem involves a linear objective function that should be minimized.

3. It is helpful to organize information in a table, as done in Examples 1–4, before you formulate a problem.

Concept Questions page 342

1. See the definition on page 338 of the text.

3. In a maximization linear programming problem, we find the greatest value of the objective function. In a minimization problem, we find its least value.

Exercises page 342

1. We tabulate the given information:

	Product		
Machine	**A**	**B**	**Time Available**
I	6 minutes	9 minutes	300 minutes
II	5 minutes	4 minutes	180 minutes
Profit per unit	\$3	\$4	

Let x and y denote the numbers of units of Products A and B to be produced. Then the linear programming problem is:

$$\begin{array}{ll} \text{Maximize} & P = 3x + 4y \\ \text{subject to} & 6x + 9y \leq 300 \\ & 5x + 4y \leq 180 \\ & x \geq 0, y \geq 0 \end{array}$$

3. Let x and y denote the numbers of model A and model B grates to be produced. Since only 1000 pounds of cast iron are available, we must have $3x + 4y \leq 1000$. The restriction that only 20 hours (or 1200 minutes) of labor are available per day implies that $6x + 3y \leq 1200$. Then the profit on the production of these grates is given by $P = 2x + 1.5y$.

Summarizing, we have the following linear programming problem:

$$\begin{array}{ll} \text{Maximize} & P = 2x + 1.5y \\ \text{subject to} & 3x + 4y \leq 1000 \\ & 6x + 3y \leq 1200 \\ & x \geq 0, y \geq 0 \end{array}$$

5. Let x denote the number of tables and y the number of chairs to be manufactured. Since 3200 board feet are available, we have $40x + 16y \leq 3200$. Next, since 520 hours of labor are available, we have $3x + 4y \leq 520$. Then the profit for the production of tables and chairs is given by $P = 45x + 20y$.

Summarizing, we have the following linear programming problem:

$$\begin{array}{ll} \text{Maximize} & P = 45x + 20y \\ \text{subject to} & 40x + 16y \leq 3200 \\ & 3x + 4y \leq 520 \\ & x \geq 0, y \geq 0 \end{array}$$

7. Suppose the company extends x million dollars in homeowner loans and y million dollars in automobile loans. Then, the returns on these loans are given by $P = 0.1x + 0.12y$ million dollars. Since the company has a total of \$20 million for these loans, we have $x + y \leq 20$. Furthermore, since the total amount of homeowner loans should be greater than or equal to four times the total amount of automobile loans, we have $x \geq 4y$. Therefore, the linear programming problem is:

$$\begin{aligned} \text{Maximize} \quad & P = 0.1x + 0.12y \\ \text{subject to} \quad & x + \ \ y \leq 20 \\ & x - 4y \geq \ \ 0 \\ & x \geq 0, y \geq 0 \end{aligned}$$

9. Let x denote the number of fully assembled units to be produced daily and y the number of kits to be produced. Then the fraction of the day the fabrication department works on the fully assembled cabinets is $\frac{1}{200}x$. Similarly the fraction of the day the fabrication department works on kits is $\frac{1}{200}y$. Since the fraction of the day during which the fabrication department is busy cannot exceed one, we must have $\frac{1}{200}x + \frac{1}{200}y \leq 1$. Similarly, the restrictions on the assembly department lead to the inequality $\frac{1}{100}x + \frac{1}{300}y \leq 1$. The profit (objective) function is $P = 50x + 40y$. Summarizing, the linear programming problem is:

$$\begin{aligned} \text{Maximize} \quad & P = 50x + 40y \\ \text{subject to} \quad & \tfrac{1}{200}x + \tfrac{1}{200}y \leq 1 \\ & \tfrac{1}{100}x + \tfrac{1}{300}y \leq 1 \\ & x \geq 0, y \geq 0 \end{aligned}$$

11. Let x and y denote the number of days the Saddle and Horseshoe mines are operated, respectively. Then the operating cost is $C = 14{,}000x + 16{,}000y$. The amount of gold produced in the two mines is $(50x + 75y)$ oz., and this amount must be at least 650 oz. Thus, we have $50x + 75y \geq 650$. Similarly, the requirement for silver production leads to the inequality $3000x + 1000y \geq 18{,}000$. So the linear programming problem is:

$$\begin{aligned} \text{Minimize} \quad & C = 14{,}000x + 16{,}000y \\ \text{subject to} \quad & \ \ 50x + \ \ 75y \geq \ \ 650 \\ & 3000x + 1000y \geq 18{,}000 \\ & x \geq 0, y \geq 0 \end{aligned}$$

13. Let x and y denote the numbers of gallons of water (in millions) obtained each day from the local reservoir and the pipeline, respectively. The requirement that at least 10 million gallons of water be supplied per day implies that $x + y \geq 10$. Next, because the maximum yield of the local reservoir is 5 million gallons per day, we have $x \leq 5$. The maximum yield of the pipeline is 10 million gallons per day and the pipeline has been contracted to supply at least 6 million gallons per day, so we have $6 \leq y \leq 10$. Then the cost function is given by $C = 300x + 500y$. Summarizing, we have the following linear programming problem:

$$\begin{aligned} \text{Minimize} \quad & C = 300x + 500y \\ \text{subject to} \quad & x + y \geq 10 \\ & x \leq 5 \\ & 6 \leq y \leq 10 \\ & x \geq 0 \end{aligned}$$

15. Let x and y denote the amount of food A and food B, respectively, used to prepare a meal. Then the requirement that the meal contain a minimum of 400 mg of calcium implies $30x + 25y \geq 400$. Similarly, the requirements that the meal contain at least 10 mg of iron and 40 mg of vitamin C imply that $x + 0.5y \geq 10$ and $2x + 5y \geq 40$. The cholesterol content is given by $C = 2x + 5y$. Therefore, the linear programming problem is:

$$\begin{aligned} \text{Minimize} \quad & C = 2x + 5y \\ \text{subject to} \quad & 30x + 25y \geq 400 \\ & x + 0.5y \geq 10 \\ & 2x + 5y \geq 40 \\ & x \geq 0, y \geq 0 \end{aligned}$$

17. Let x and y denote the numbers of advertisements to be placed in newspapers I and II, respectively. Then the linear programming problem is:

$$\begin{aligned} \text{Minimize} \quad & C = 1000x + 800y \\ \text{subject to} \quad & 70{,}000x + 10{,}000y \geq 2{,}000{,}000 \\ & 40{,}000x + 20{,}000y \geq 1{,}400{,}000 \\ & 20{,}000x + 40{,}000y \geq 1{,}000{,}000 \\ & x \geq 0, y \geq 0 \end{aligned}$$

19. Let x, y, and z denote the amounts of money invested in projects A, B, and C, respectively. Since she plans to invest up to \$2 million, we must have $x + y + z \leq 2{,}000{,}000$. Because she decides to put not more than 20% of her total investment in project C, we have $z \leq 0.2\,(x + y + z)$, or $-2x - 2y + 8z \leq 0$. Since her investments in projects B and C should not exceed 60% of her total investment, we have $y + z \leq 0.6\,(x + y + z)$, or $-6x + 4y + 4z \leq 0$. Also, because her investment in project A should be at least 60% of her investments in projects B and C, we have $x \geq 0.6\,(y + z)$, or $-10x + 6y + 6z \leq 0$. Finally, the returns on her investments are given by $P = 0.1x + 0.15y + 0.2z$. To summarize, the linear programming problem is:

$$\begin{aligned} \text{Maximize} \quad & P = 0.1x + 0.15y + 0.2z \\ \text{subject to} \quad & x + y + z \leq 2{,}000{,}000 \\ & -2x - 2y + 8z \leq 0 \\ & -6x + 4y + 4z \leq 0 \\ & -10x + 6y + 6z \leq 0 \\ & x \geq 0, y \geq 0, z \geq 0 \end{aligned}$$

21. Let x, y, and z denote the numbers of units produced of products A, B, and C, respectively. From the given information, we see that the time required by department I is given by $2x + y + 2z$, and this must not exceed 900 minutes. The time required by department II is given by $3x + y + 2z$, and this must not exceed 1080 minutes. The time required by department III is given by $2x + 2y + z$, and this must not exceed 840 minutes. The profit is given by $P = 18x + 12y + 15z$, and this is the quantity to be maximized. Thus, the linear programming problem is:

$$\begin{aligned} \text{Maximize} \quad & P = 18x + 12y + 15z \\ \text{subject to} \quad & 2x + y + 2z \leq 900 \\ & 3x + y + 2z \leq 1080 \\ & 2x + 2y + z \leq 840 \\ & x \geq 0, y \geq 0, z \geq 0 \end{aligned}$$

23. We first tabulate the given information:

	Model			
Department	**A**	**B**	**C**	**Time Available**
Fabrication	$\frac{5}{4}$	$\frac{3}{2}$	$\frac{3}{2}$	310
Assembly	1	1	$\frac{3}{4}$	205
Finishing	1	1	$\frac{1}{2}$	190

Let x, y, and z denote the numbers of units of models A, B, and C to be produced, respectively. Then the required linear programming problem is:

$$\begin{aligned} \text{Maximize} \quad & P = 26x + 28y + 24z \\ \text{subject to} \quad & \tfrac{5}{4}x + \tfrac{3}{2}y + \tfrac{3}{2}z \le 310 \\ & x + y + \tfrac{3}{4}z \le 205 \\ & x + y + \tfrac{1}{2}z \le 190 \\ & x \ge 0, y \ge 0, z \ge 0 \end{aligned}$$

25. The shipping costs are tabulated in the first table. Letting x_1 denote the number of pianos shipped from plant I to warehouse A, x_2 the number of pianos shipped from plant I to warehouse B, and so on, we have the second table.

	Warehouse		
Plant	**A**	**B**	**C**
I	60	60	80
II	80	70	50

	Warehouse			
Plant	**A**	**B**	**C**	**Max. Prod.**
I	x_1	x_2	x_3	300
II	x_4	x_5	x_6	250
Min. Req.	200	150	200	

From the two tables, we see that the total monthly shipping cost is given by $C = 60x_1 + 60x_2 + 80x_3 + 80x_4 + 70x_5 + 50x_6$. Next, the production constraints on plants I and II lead to the inequalities $x_1 + x_4 \ge 200$, $x_2 + x_5 \ge 150$, $x_3 + x_6 \ge 200$.

Summarizing, we have the following linear programming problem:

$$\begin{aligned} \text{Minimize} \quad & C = 60x_1 + 60x_2 + 80x_3 + 80x_4 + 70x_5 + 50x_6 \\ \text{subject to} \quad & x_1 + x_2 + x_3 \le 300 \\ & x_4 + x_5 + x_6 \le 250 \\ & x_1 + x_4 \ge 200 \\ & x_2 + x_5 \ge 150 \\ & x_3 + x_6 \ge 200 \\ & x_1 \ge 0, x_2 \ge 0, x_3 \ge 0, x_4 \ge 0, x_5 \ge 0, x_6 \ge 0 \end{aligned}$$

27. The given data can be summarized as follows:

	Concentrate			
Juice	**Pineapple**	**Orange**	**Banana**	**Profit ($)**
Pineapple-orange	8	8	0	1
Orange-banana	0	12	4	0.80
Pineapple-orange-banana	4	8	4	0.90
Maximum available (oz.)	16,000	24,000	5000	

Suppose x, y, and z cartons of pineapple-orange, orange-banana, and pineapple-orange-banana juice are to be produced, respectively. The linear programming problem is:

$$\begin{aligned} \text{Maximize } P &= x + 0.8y + 0.9z \\ \text{subject to } 8x + 4z &\leq 16{,}000 \\ 8x + 12y + 8z &\leq 24{,}000 \\ 4y + 4z &\leq 5000 \\ z &\leq 800 \\ x \geq 0, y \geq 0, z &\geq 0 \end{aligned}$$

29. False. The objective function $P = xy$ is not a linear function in x and y.

6.3 Graphical Solution of Linear Programming Problems

Problem-Solving Tips

1. To solve a linear programming problem using the method of corners, perform the following steps:

a. Graph the feasible set.

b. Find the coordinates of all corner points (vertices) of the feasible set.

c. Evaluate the objective function at each corner point.

Then check to see which vertex yields the maximum (or minimum). If only one vertex yields the maximum (or minimum) then the problem has a unique solution. If there are two adjacent vertices that yield the same maximum (or minimum), then any point lying on the line segment joining these two vertices is a solution.

2. It's helpful to set up a table, as done in Examples 1–3, to evaluate the objective function at each vertex.

Concept Questions page 353

1. a. The feasible set is the set of points satisfying the constraints associated with the linear programming problem.

b. A feasible solution of a linear programming problem is a point in the feasible set.

c. An optimal solution of a linear programming problem is a feasible solution that also optimizes (maximizes or minimizes) the objective function.

Exercises page 353

1. Evaluating the objective function at each of the corner points, we obtain the following table.

Vertex	$Z = 2x + 3y$
(1, 1)	$\boxed{5}$
(8, 5)	31
(4, 9)	$\boxed{35}$
(2, 8)	28

From the table, we conclude that the maximum value of Z is 35 and it occurs at the vertex (4, 9). The minimum value of Z is 5 and it occurs at the vertex (1, 1).

3. Evaluating the objective function at each of the corner points, we obtain the following table.

Vertex	$Z = 2x + 3y$
(0, 20)	60
(3, 10)	36
(4, 6)	26
(9, 0)	$\boxed{18}$

From the graph, we conclude that there is no maximum value since Z is unbounded. The minimum value of Z is 18 and it occurs at the vertex (9, 0).

5. Evaluating the objective function at each of the corner points, we obtain the following table.

Vertex	$Z = x + 4y$
(0, 6)	24
(4, 10)	$\boxed{44}$
(12, 8)	$\boxed{44}$
(15, 0)	$\boxed{15}$

From the table, we conclude that the maximum value of Z is 44 and it occurs at every point on the line segment joining the points (4, 10) and (12, 8). The minimum value of Z is 15 and it occurs at the vertex (15, 0).

7. The linear programming problem is:

$$\begin{aligned} \text{Maximize} \quad & P = 3x + 2y \\ \text{subject to} \quad & x + y \le 6 \\ & x \le 3 \\ & x \ge 0, y \ge 0 \end{aligned}$$

The feasible set S for the problem is shown in the figure, and the values of the function P at the vertices of S are summarized in the table.

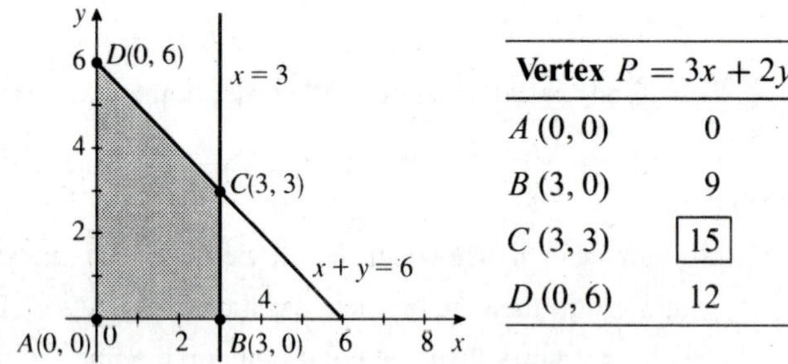

Vertex	$P = 3x + 2y$
A (0, 0)	0
B (3, 0)	9
C (3, 3)	$\boxed{15}$
D (0, 6)	12

We conclude that P attains a maximum value of 15 when $x = 3$ and $y = 3$.

9. The linear programming problem is:

$$\begin{aligned} \text{Maximize} \quad & P = 2x + y \\ \text{subject to} \quad & x + y \le 4 \\ & 2x + y \le 5 \\ & x \ge 0, y \ge 0 \end{aligned}$$

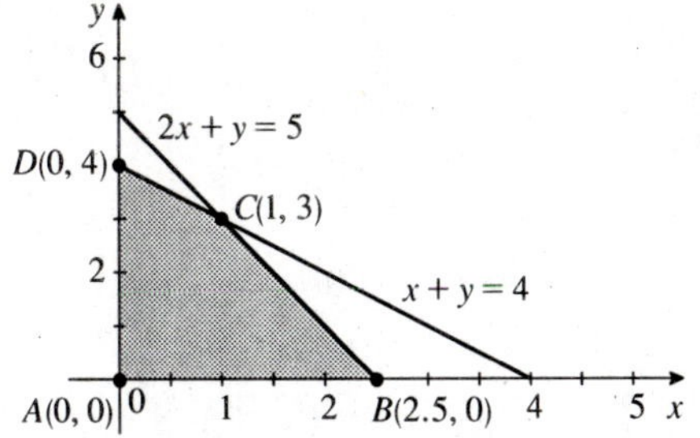

Vertex	$P = 2x + y$
$A\ (0, 0)$	0
$B\ (2.5, 0)$	$\boxed{5}$
$C\ (1, 3)$	$\boxed{5}$
$D\ (0, 4)$	4

From the figure and the table, we conclude that P attains a maximum value of 5 at any point (x, y) lying on the line segment joining $(1, 3)$ and $(2.5, 0)$.

11. The linear programming problem is:

$$\begin{aligned} \text{Maximize} \quad & P = x + 8y \\ \text{subject to} \quad & x + y \le 8 \\ & 2x + y \le 10 \\ & x \ge 0, y \ge 0 \end{aligned}$$

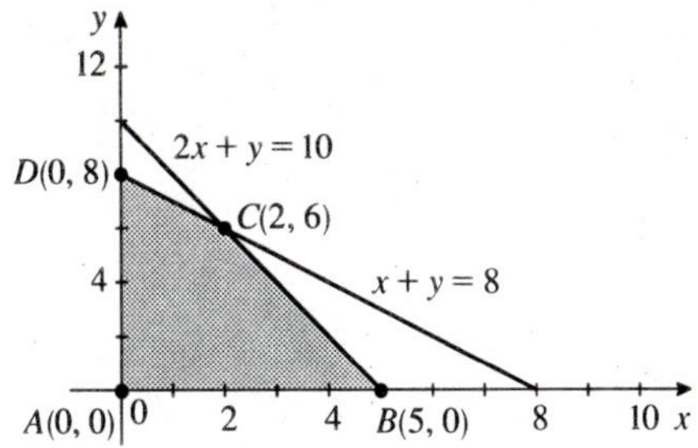

Vertex	$P = x + 8y$
$A\ (0, 0)$	0
$B\ (5, 0)$	5
$C\ (2, 6)$	50
$D\ (0, 8)$	$\boxed{64}$

From the figure and the table, we conclude that P attains a maximum value of 64 when $x = 0$ and $y = 8$.

13. The linear programming problem is:

$$\begin{aligned} \text{Maximize} \quad & P = x + 3y \\ \text{subject to} \quad & 2x + y \le 6 \\ & x + y \le 4 \\ & x \le 1 \\ & x \ge 0, y \ge 0 \end{aligned}$$

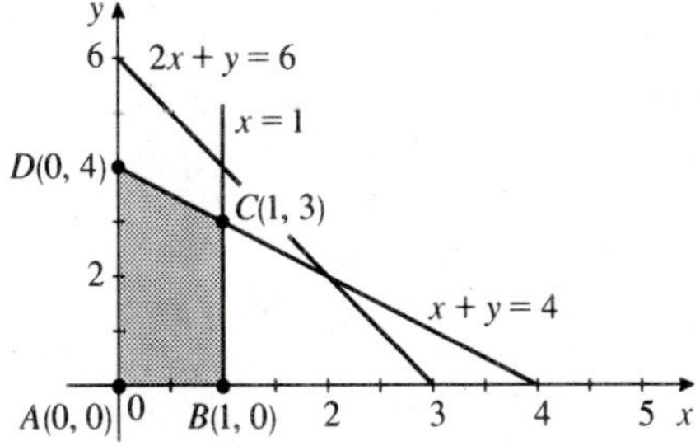

Vertex	$P = x + 3y$
$A\ (0, 0)$	0
$B\ (1, 0)$	1
$C\ (1, 3)$	10
$D\ (0, 4)$	$\boxed{12}$

From the figure and the table, we conclude that P attains a maximum value of 12 when $x = 0$ and $y = 4$.

15. The linear programming problem is:

$$\begin{aligned} \text{Minimize} \quad & C = 2x + 5y \\ \text{subject to} \quad & x + y \geq 3 \\ & x + 2y \geq 4 \\ & x \geq 0, y \geq 0 \end{aligned}$$

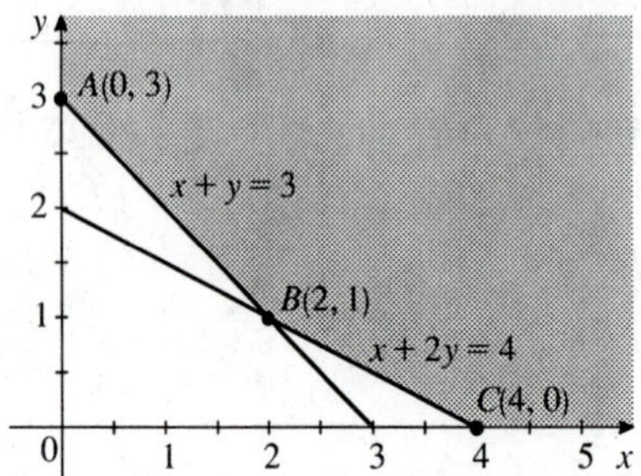

Vertex	$C = 2x + 5y$
A (0, 3)	15
B (2, 1)	9
C (4, 0)	$\boxed{8}$

From the figure and the table, we conclude that C attains a minimum value of 8 when $x = 4$ and $y = 0$.

17. The linear programming problem is:

$$\begin{aligned} \text{Minimize} \quad & C = 3x + 6y \\ \text{subject to} \quad & x + 2y \geq 40 \\ & x + y \geq 30 \\ & x \geq 0, y \geq 0 \end{aligned}$$

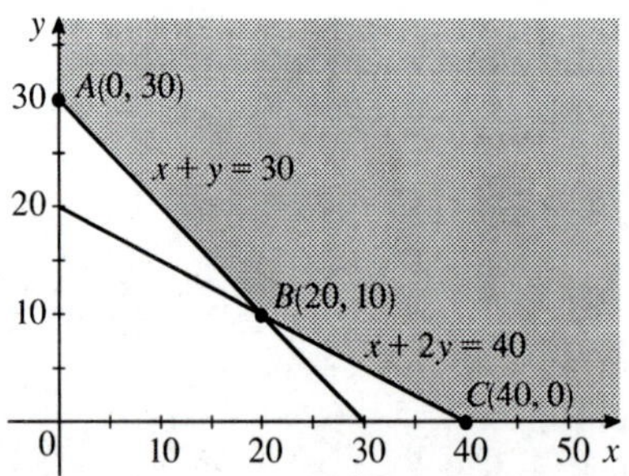

Vertex	$C = 3x + 6y$
A (0, 30)	180
B (20, 10)	$\boxed{120}$
C (40, 0)	$\boxed{120}$

From the figure and the table, we conclude that C attains a minimum value of 120 at any point on the line segment joining (20, 10) to (40, 0).

19. The linear programming problem is:

$$\begin{aligned} \text{Minimize} \quad & C = 2x + 10y \\ \text{subject to} \quad & 5x + 2y \geq 40 \\ & x + 2y \geq 20 \\ & y \geq 3, x \geq 0 \end{aligned}$$

The feasible set S for the problem is shown in the figure, and the values of the function C at the vertices of S are shown in the table.

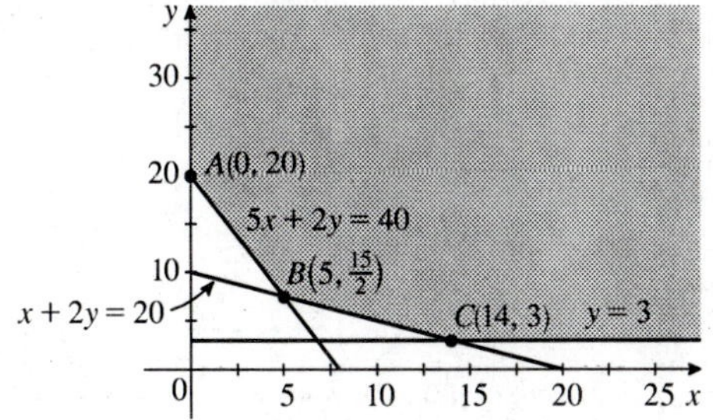

Vertex	$C = 2x + 10y$
$A\,(0, 20)$	200
$B\left(5, \frac{15}{2}\right)$	85
$C\,(14, 3)$	$\boxed{58}$

We conclude that C attains a minimum value of 58 when $x = 14$ and $y = 3$.

21. The linear programming problem is:

$$\begin{aligned} \text{Minimize} \quad & C = 10x + 15y \\ \text{subject to} \quad & x + y \le 10 \\ & 3x + y \ge 12 \\ & -2x + 3y \ge 3 \\ & x \ge 0,\ y \ge 0 \end{aligned}$$

The feasible set is shown in the figure, and the values of C at each of the vertices of S are shown in the table.

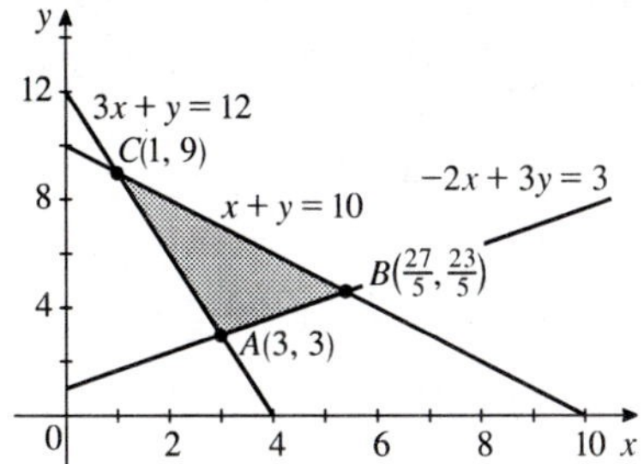

Vertex	$C = 10x + 15y$
$A\,(3, 3)$	$\boxed{75}$
$B\left(\frac{27}{5}, \frac{23}{5}\right)$	123
$C\,(1, 9)$	145

We conclude that C attains a minimum value of 75 when $x = 3$ and $y = 3$.

23. The linear programming problem is:

$$\begin{aligned} \text{Maximize} \quad & P = 3x + 4y \\ \text{subject to} \quad & x + 2y \le 50 \\ & 5x + 4y \le 145 \\ & 2x + y \ge 25 \\ & x \ge 0,\ y \ge 5 \end{aligned}$$

The feasible set S is shown in the figure, and the values of P at each of the vertices of S are shown in the table.

Vertex	$P = 3x + 4y$
$A\,(10, 5)$	50
$B\,(25, 5)$	95
$C\left(15, \frac{35}{2}\right)$	$\boxed{115}$
$D\,(0, 25)$	100

We conclude that P attains a maximum value of 115 when $x = 15$ and $y = \frac{35}{2}$.

25. The linear programming problem is:

$$\begin{aligned} \text{Maximize} \quad & P = 2x + 3 \\ \text{subject to} \quad & x + y \le 48 \\ & x + 3y \ge 60 \\ & 9x + 5y \le 320 \\ & x \ge 10,\ y \ge 0 \end{aligned}$$

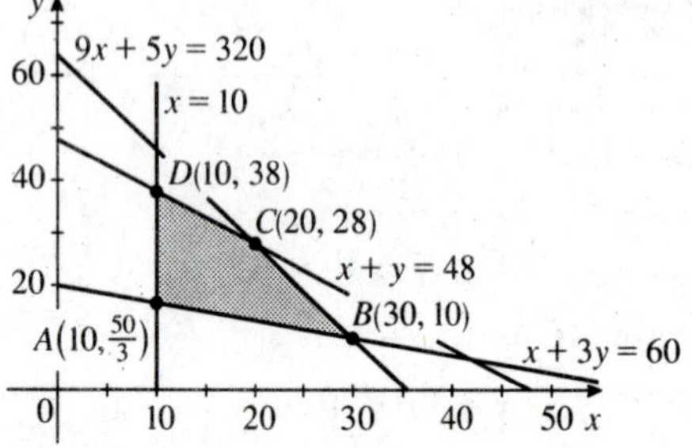

Vertex	$P = 2x + 3y$
$A\left(10, \frac{50}{3}\right)$	70
$B\ (30, 10)$	90
$C\ (20, 28)$	124
$D\ (10, 38)$	134

From the graph and the table, we conclude that P attains a maximum value of 134 when $x = 10$ and $y = 38$.

27. The linear programming problem is:

$$\begin{aligned} \text{Maximize (and minimize)} \quad & P = 8x + 5y \\ \text{subject to} \quad & 5x + 2y \ge 63 \\ & x + y \ge 18 \\ & 3x + 2y \le 51 \\ & x \ge 0,\ y \ge 0 \end{aligned}$$

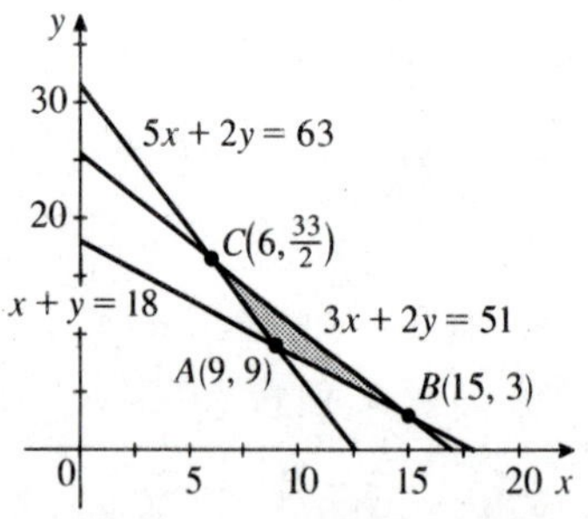

Vertex	$P = 8x + 5y$
$A\ (9, 9)$	117
$B\ (15, 3)$	135
$C\left(6, \frac{33}{2}\right)$	130.5

From the graph and the table, we see that P attains a maximum value of 135 when $x = 15$ and $y = 3$. The minimum value of P is 117, attained when $x = 9$ and $y = 9$.

29. Refer to the solution to Exercise 6.2.1 on page 135 of this manual. The linear programming problem is:

$$\begin{aligned} \text{Maximize} \quad & P = 3x + 4y \\ \text{subject to} \quad & 6x + 9y \le 300 \\ & 5x + 4y \le 180 \\ & x \ge 0,\ y \ge 0 \end{aligned}$$

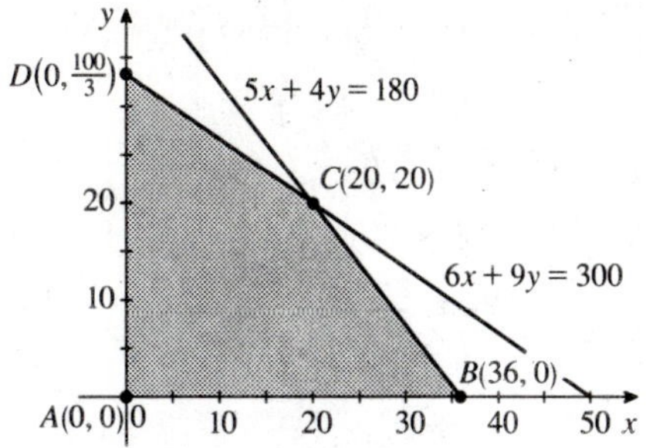

Vertex	$C = 3x + 4y$
$A\ (0, 0)$	0
$B\ (36, 0)$	108
$C\ (20, 20)$	$\boxed{140}$
$D\left(0, \frac{100}{3}\right)$	$\frac{400}{3}$

From the graph and the table, we see that P attains a maximum value of 140 when $x = y = 20$. Thus, by producing 20 units of each product in each shift, the company will realize an optimal profit of $140.

31. Refer to the solution to Exercise 6.2.3 on page 135 of this manual. The linear programming problem is:

$$\begin{aligned} \text{Maximize} \quad & P = 2x + 1.5y \\ \text{subject to} \quad & 3x + 4y \le 1000 \\ & 6x + 3y \le 1200 \\ & x \ge 0, y \ge 0 \end{aligned}$$

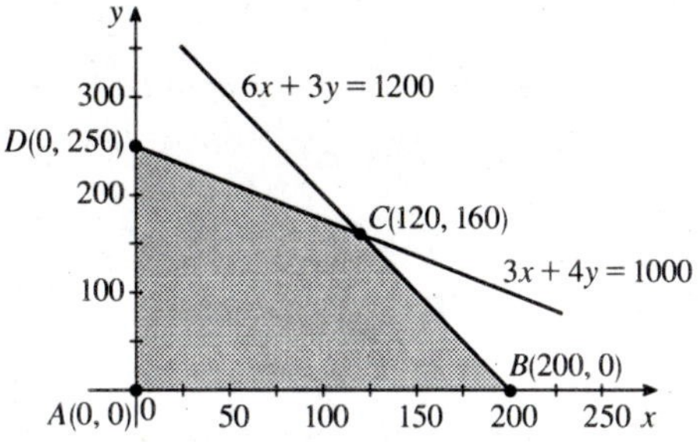

Vertex	$P = 2x + 1.5y$
$A\ (0, 0)$	0
$B\ (200, 0)$	400
$C\ (120, 160)$	$\boxed{480}$
$D\ (0, 250)$	375

From the graph and the table, we see that P attains a maximum value of 480 when $x = 120$ and $y = 160$. Thus, by producing 120 model A grates and 160 model B grates in each shift, the company will realize an optimal profit of $480.

33. Let x denote the number of tables and y denote the number of chairs to be manufactured. Then the linear programming problem is:

$$\begin{aligned} \text{Maximize} \quad & P = 45x + 20y \\ \text{subject to} \quad & 40x + 16y \le 3200 \\ & 3x + 4y \le 520 \\ & x \ge 0, y \ge 0 \end{aligned}$$

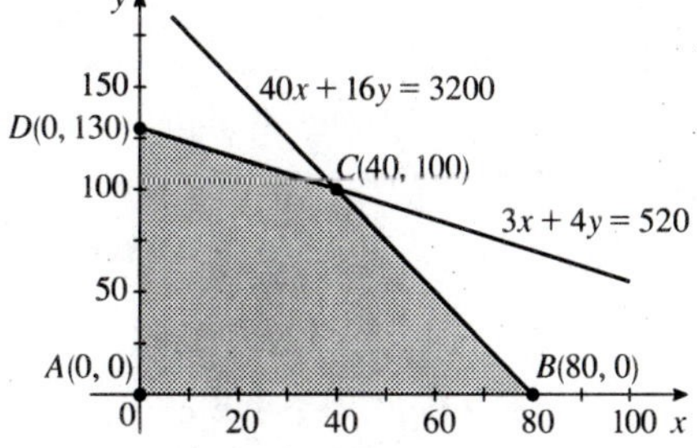

Vertex	$P = 45x + 20y$
$A\ (0, 0)$	0
$B\ (80, 0)$	3600
$C\ (40, 100)$	$\boxed{3800}$
$D\ (0, 130)$	2600

From the graph and the table, we see that Winston should manufacture 40 tables and 100 chairs for a maximum profit of $3800.

35. Refer to the solution to Exercise 6.2.7 on page 135 of this manual. The linear programming problem is:

$$\begin{aligned} \text{Maximize} \quad & P = 0.1x + 0.12y \\ \text{subject to} \quad & x + y \le 20 \\ & x - 4y \ge 0 \\ & x \ge 0, y \ge 0 \end{aligned}$$

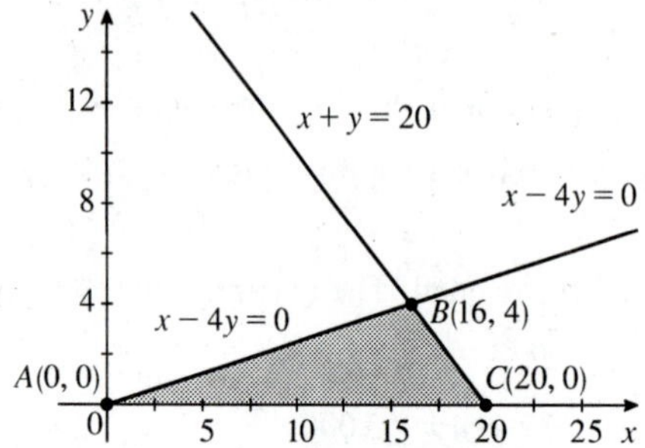

Vertex	$P = 0.1x + 0.12y$
A (0, 0)	0
B (16, 4)	2.08
C (20, 0)	2.00

The maximum value of P is attained when $x = 16$ and $y = 4$. Thus, by extending \$16 million in housing loans and \$4 million in automobile loans, the company will realize a return of \$2.08 million on its loans.

37. Refer to the solution to Exercise 6.2.9 on page 136 of this manual. The linear programming problem is:

$$\begin{aligned} \text{Maximize} \quad & P = 50x + 40y \\ \text{subject to} \quad & \tfrac{1}{200}x + \tfrac{1}{200}y \le 1 \\ & \tfrac{1}{100}x + \tfrac{1}{300}y \le 1 \\ & x \ge 0, y \ge 0 \end{aligned}$$

We can rewrite the constraints as $x + y \le 200$, $3x + y \le 300$, $x \le 0$, $y \le 0$.

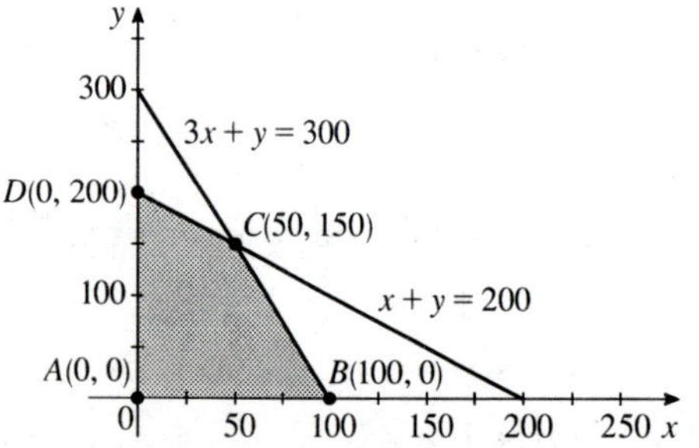

Vertex	$P = 50x + 40y$
A (0, 0)	0
B (100, 0)	5000
C (50, 150)	8500
D (0, 200)	8000

From the figure and the table, we conclude that the company should produce 50 fully assembled units and 150 kits daily in order to realize a profit of \$8500.

39. Let x and y denote the numbers of days the Saddle Mine and the Horseshoe Mine are operated, respectively. Then the operating cost is $C = 14{,}000x + 16{,}000y$. The amount of gold produced in the two mines is $(50x + 75y)$ oz., and this amount must be at least 650 oz., so we have $50x + 75y \ge 650$. Similarly, the requirement for silver production leads to the inequality $3000x + 1000y \ge 18{,}000$. So the linear programming problem is:

$$\begin{aligned} \text{Minimize} \quad & C = 14{,}000x + 16{,}000y \\ \text{subject to} \quad & 50x + 75y \ge 650 \\ & 3000x + 1000y \ge 18{,}000 \\ & x \ge 0, y \ge 0 \end{aligned}$$

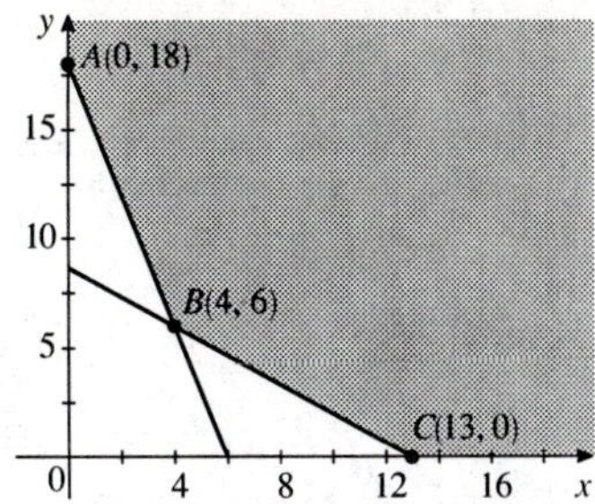

Vertex	$C = 14{,}000x + 16{,}000y$
A (0, 18)	288,000
B (4, 6)	152,000
C (13, 0)	182,000

From the figure and the table, we see that the minimum value of $C = 152{,}000$ is attained at $x = 4$ and $y = 6$. So, the Saddle Mine should be operated for 4 days and the Horseshoe Mine should be operated for 6 days at a minimum cost of \$152,000/day.

41. Let x and y denote the numbers of gallons of water in millions obtained from the local reservoir and the pipeline per day, respectively. Then, we have the following linear programming problem:

$$\begin{array}{ll} \text{Minimize} & C = 300x + 500y \\ \text{subject to} & x + y \geq 10 \\ & x \leq 5 \\ & 6 \leq y \leq 10 \\ & x \geq 0 \end{array}$$

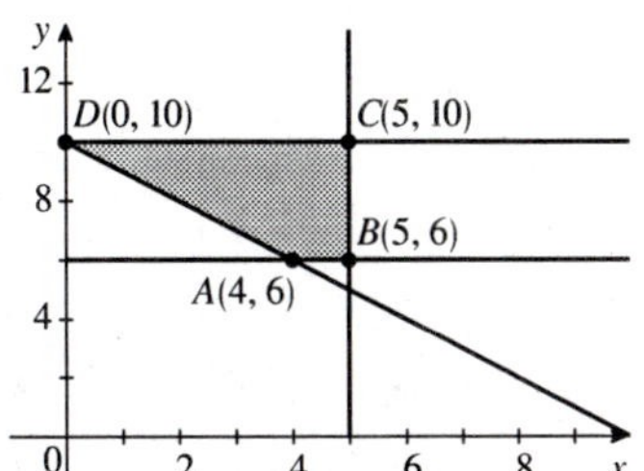

Vertex	$C = 300x + 500y$
A (4, 6)	4200
B (5, 6)	4500
C (5, 10)	6500
D (0, 10)	5000

From the figure and the table, we see that the minimum value of C is 4200 and it is attained at $x = 4$ and $y = 6$. Thus, 4 million gallons should be obtained from the reservoir and 6 million gallons from the pipeline at a minimum cost of \$4200.

43. Refer to the solution to Exercise 6.2.15 on page 136 of this manual. The linear programming problem is:

$$\begin{array}{ll} \text{Minimize} & C = 2x + 5y \\ \text{subject to} & 30x + 25y \geq 400 \\ & x + 0.5y \geq 10 \\ & 2x + 5y \geq 40 \\ & x \geq 0, y \geq 0 \end{array}$$

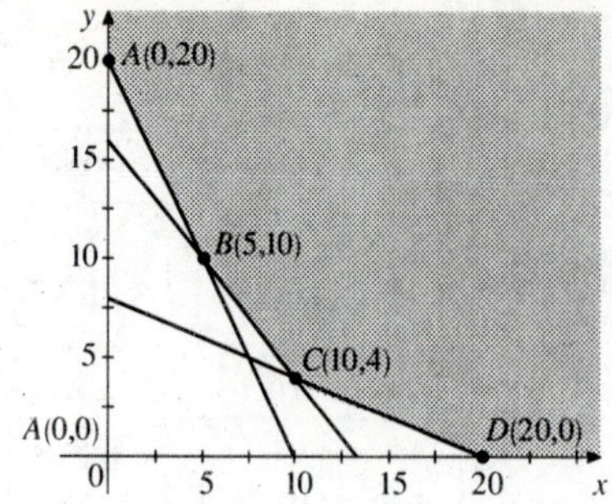

Vertex	$C = 2x + 5y$
A (0, 20)	100
B (5, 10)	60
C (10, 4)	40
D (20, 0)	40

From the figure and the table, we see that C attains a minimum value of 40 when $x = 10$ and $y = 4$ and when $x = 20$ and $y = 0$. This means that any point lying on the line joining the points (10, 4) and (20, 0) will satisfy these constraints. For example, we could use 10 ounces of food A and 4 ounces of food B, or we could use 20 ounces of food A and zero ounces of food B.

45. Let x and y denote the numbers of advertisements to be placed in newspapers I and II, respectively. Then the linear programming problem is:

$$\begin{aligned} \text{Minimize} \quad & C = 1000x + 800y \\ \text{subject to} \quad & 70{,}000x + 10{,}000y \geq 2{,}000{,}000 \\ & 40{,}000x + 20{,}000y \geq 1{,}400{,}000 \\ & 20{,}000x + 40{,}000y \geq 1{,}000{,}000 \\ & x \geq 0, y \geq 0 \end{aligned}$$

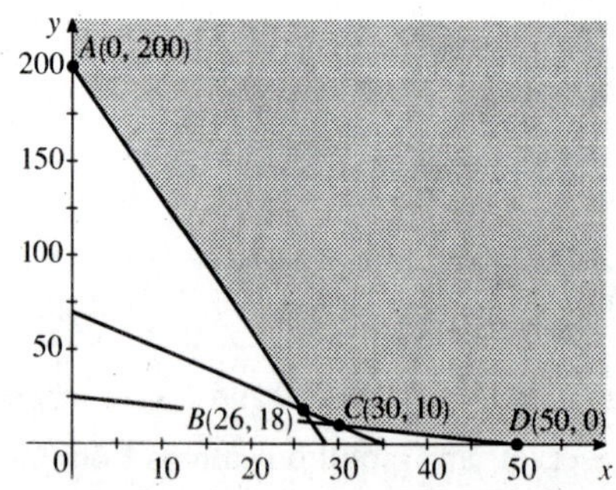

Vertex	$C = 1000x + 800y$
A (0, 200)	200,000
B (26, 18)	40,400
C (30, 10)	38,000
D (50, 0)	50,000

From the figure and the table, we see that the minimum value of C of 38,000 is attained at $x = 30$ and $y = 10$. Thus, Everest Deluxe World Travel should place 30 advertisements in newspaper I and 10 advertisements in newspaper II at a total (minimum) cost of \$38,000.

47. The linear programming problem is:

$$\begin{aligned} \text{Minimize} \quad & C = 14{,}500 - 20x - 10y \\ \text{subject to} \quad & x + y \geq 40 \\ & x + y \leq 100 \\ & 0 \leq x \leq 80, 0 \leq y \leq 70 \end{aligned}$$

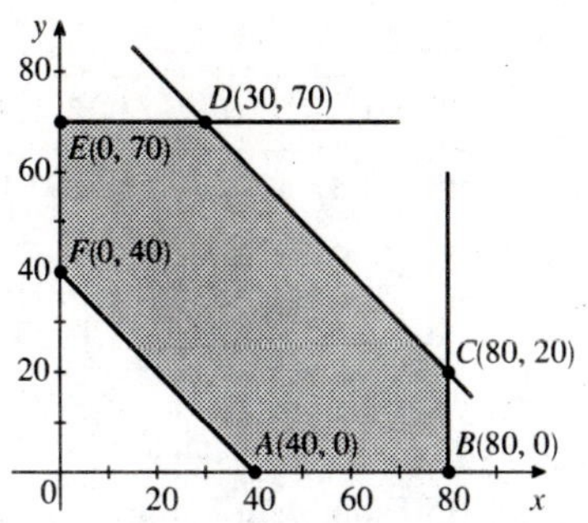

Vertex	$C = 14{,}500 - 20x - 10y$
A (40, 0)	13,700
B (80, 0)	12,900
C (80, 20)	12,700
D (30, 70)	13,200
E (0, 70)	13,800
F (0, 40)	14,100

From the figure and the table, we conclude that the minimum value of C occurs when $x = 80$ and $y = 20$. Thus, 80 engines should be shipped from plant I to assembly plant A, and 20 engines should be shipped from plant I to assembly plant B; whereas $80 - x = 80 - 80 = 0$ and $70 - y = 70 - 20 = 50$ engines should be shipped from plant II to assembly plants A and B, respectively, at a total cost of \$12,700.

49. Let x and y denote Patricia's investments in growth and speculative stocks, respectively, where both x and y are measured in thousands of dollars. Then the return on her investments is given by $P = 0.15x + 0.25y$. Since her investment may not exceed \$30,000, we have the constraint $x + y \leq 30$. The condition that her investment in growth stocks be at least 3 times as much as her investment in speculative stocks translates into the inequality $x \geq 3y$. Thus, we have the following linear programming problem:

$$\begin{aligned} \text{Maximize} \quad & P = 0.15x + 0.25y \\ \text{subject to} \quad & x + y \leq 30 \\ & x - 3y \geq 0 \\ & x \geq 0,\ y \geq 0 \end{aligned}$$

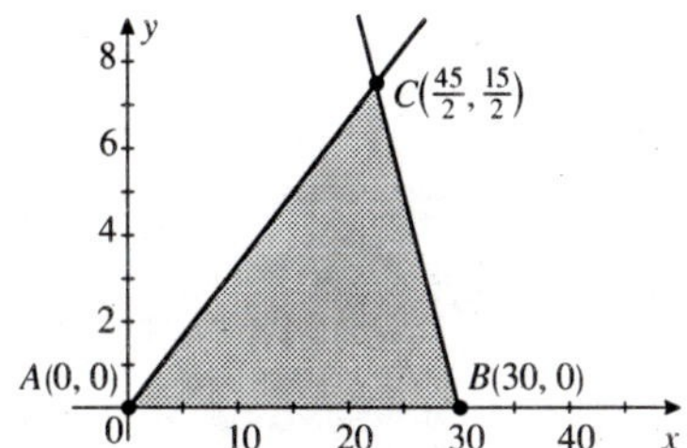

Vertex	$P = 0.15x + 0.25y$
A (0, 0)	0
B (30, 0)	4.5
$C\left(\frac{45}{2}, \frac{15}{2}\right)$	5.25

From the figure and the table, we conclude that the maximum value of P occurs when $x = \frac{45}{2}$ and $y = \frac{15}{2}$. Thus, by investing \$22,500 in growth stocks and \$7500 in speculative stocks. Patricia's maximum return is \$5250.

51. Let x denote the number of urban families and let y denote the number of suburban families interviewed by the company. Then, the amount of money paid to Trendex will be $P = 6000 + 8(x + y) - 4.4x - 5y = 6000 + 3.6x + 3y$. Since a maximum of 1500 families are to be interviewed, we have $x + y \leq 1500$. Next, the condition that at least 500 urban families are to be interviewed translates into the condition $x \geq 500$. Finally the condition that at least half of the families interviewed must be from the suburban area gives $y \geq \frac{1}{2}(x + y)$, or $y - x \geq 0$. Thus, we are led to the following programming problem:

$$\begin{aligned} \text{Maximize} \quad & P = 6000 + 3.6x + 3y \\ \text{subject to} \quad & x + y \leq 1500 \\ & y - x \geq 0 \\ & x \geq 500,\ y \geq 0 \end{aligned}$$

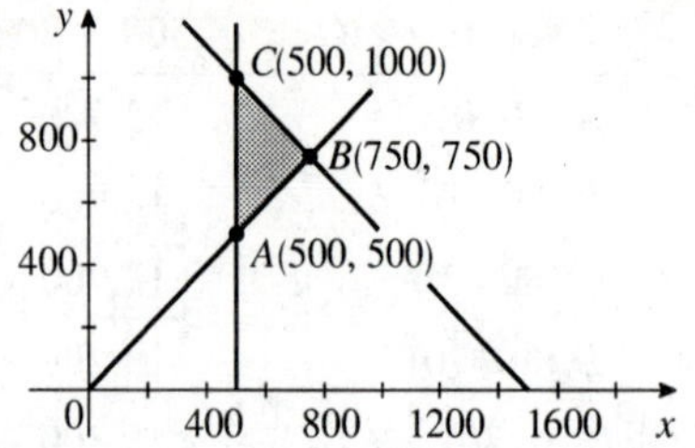

Vertex	$P = 6000 + 3.6x + 3y$
A (500, 500)	9300
B (750, 750)	10,950
C (500, 1000)	10,800

From the figure and the table, we conclude that the profit will be maximized when $x = 750$ and $y = 750$. Thus, a maximum profit of \$10,950 will be realized when 750 urban and 750 suburban families are interviewed.

53. False.

55. a. True. Because $a > 0$, the term ax can be made as large as we please by taking x sufficiently large (because S is unbounded) and therefore P is unbounded as well.

b. True. Maximizing $P = ax + by$ on S is the same as minimizing $Q = -P = -(ax + by) = -ax - by = Ax + By$, where $A \geq 0$ and $B \geq 0$. Since $x \geq 0$ and $y \geq 0$, the linear function Q (and therefore P) has at least one optimal solution.

57. Refer to the figure in the text. Let $A(x_1, y_1)$ and $B(x_2, y_2)$. Then you can verify that $Q(\overline{x}, \overline{y})$, where $\overline{x} = x_1 + t(x_2 - x_1)$, $\overline{y} = y_1 + t(y_2 - y_1)$, and t is a number satisfying $0 < t < 1$. Therefore, the value of P at Q is $P = a\overline{x} + b\overline{y} = a[x_1 + t(x_2 - x_1)] + b[y_1 + t(y_2 - y_1)] = ax_1 + by_1 + [a(x_2 - x_1) + b(y_2 - y_1)]t$. Now if $c = a(x_2 - x_1) + b(y_2 - y_1) = 0$, then P has the (maximum) value $ax_1 + by_1$ on the line segment joining A and B; that is, the infinitely many solutions lie on this line segment. If $c > 0$, then a point a little to the right of Q will give a larger value of P. In this case, P is not maximal at Q. (Such a point can be found because Q lies in the interior of the line segment.) A similar statement holds for the case $c < 0$. Thus, the maximum of P cannot occur at Q unless it occurs in every point on the line segment joining A and B.

59. a.

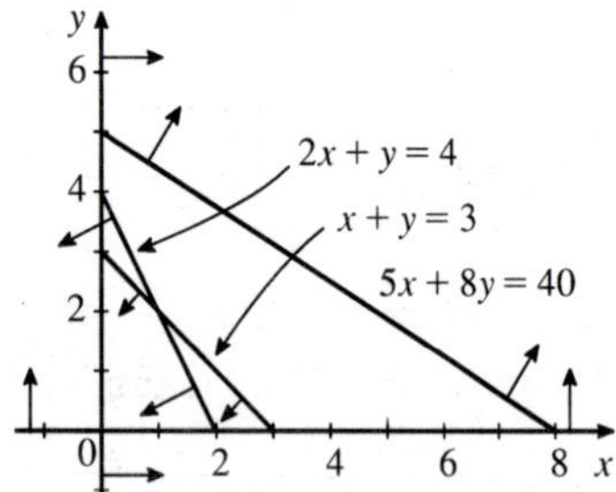

b. There is no point that satisfies all the given inequalities. Therefore, there is no solution.

6.4 The Simplex Method: Standard Maximization Problems

Problem-Solving Tips

1. Make sure that you set up the initial simplex tableau correctly:

a. First, rewrite the linear inequalities as equalities by introducing slack variables.

b. Next, rewrite the objective function so that all variables are on the left-hand side and the coefficient of P is 1. Then place this equation below the other equations.

2. In the simplex method, the optimal solution has been reached if *all the entries* in the last row to the left of the vertical line are *nonnegative.*

3. In the simplex method, you will know that the optimal solution has not been reached if there are *negative entries in the last row to the left of the vertical line.* If so, locate the most negative entry in that row. This gives you the pivot column. Proceed to find the pivot row by dividing each positive entry in the pivot column by its corresponding entry in the column of constants. Look for the smallest ratio. The corresponding entry will be in the pivot row. Proceed to make the pivot column a unit column by pivoting about the pivot element.

Concept Questions page 374

1. **a.** The objective function is to be maximized.
   b. All the variables involved in the problem are nonnegative.
   c. Each other linear constraint may be written so that the expression involving the variables is less than or equal to a nonnegative constant.

3. To find the *pivot column,* locate the most negative entry to the left of the vertical line in the last row. The column containing this entry is the pivot column. To find the *pivot row,* divide each positive entry in the pivot column into its corresponding entry in the column of constants. The pivot row is the row corresponding to the smallest ratio thus obtained. The *pivot element* is the element common to the pivot column and the pivot row.

Exercises page 374

1. **a.** It is already in standard form.

   b.

x	y	u	v	P	Constant
1	4	1	0	0	12
1	3	0	1	0	10
−2	−4	0	0	1	0

3. **a.** We multiply the second inequality by −1, obtaining the following standard maximization problem:

$$\begin{aligned} \text{Maximize} \quad & P = 2x + 3y \\ \text{subject to} \quad & x + y \le 10 \\ & x + 2y \le 12 \\ & 2x + y \le 12 \\ & x \ge 0, y \ge 0 \end{aligned}$$

   b.

x	y	u	v	w	P	Constant
1	1	1	0	0	0	10
1	2	0	1	0	0	12
2	1	0	0	1	0	12
−2	−3	0	0	0	1	0

5. **a.** We multiply the second inequality by −1, obtaining the following standard maximization problem:

$$\begin{aligned} \text{Maximize} \quad & P = x + 3y + 4z \\ \text{subject to} \quad & x + 2y + z \le 40 \\ & x + y + z \le 30 \\ & x \ge 0, y \ge 0, z \ge 0 \end{aligned}$$

   b.

x	y	z	u	v	P	Constant
1	2	1	1	0	0	40
1	1	1	0	1	0	30
−1	−3	−4	0	0	1	0

7. All entries in the last row of the simplex tableau are nonnegative and an optimal solution has been reached. We find $x = \frac{30}{7}$, $y = \frac{20}{7}$, $u = 0$, $v = 0$, and $P = \frac{220}{7}$.

9. The simplex tableau is not in final form because there is a negative entry in the last row. The entry in the first row, second column, is the next pivot element and has a value of $\frac{1}{2}$.

11. The simplex tableau is in final form. We find $x = \frac{1}{3}$, $y = 0$, $z = \frac{13}{3}$, $u = 0$, $v = 6$, $w = 0$ and $P = 17$.

13. The simplex tableau is not in final form because there are two negative entries in the last row. The entry in the third row, second column, is the pivot element and has a value of 1.

15. The simplex tableau is in final form. The solutions are $x = 30$, $y = 10$, $z = 0$, $u = 0$, $v = 0$, $P = 60$ and $x = 30$, $y = 0$, $z = 0$, $u = 10$, $v = 0$, $P = 60$. (There are infinitely many solutions.)

17. We calculate the following sequence of tableaus:

	x	y	u	v	P	Constant	Ratio
Pivot row →	1	①	1	0	0	4	4
	2	1	0	1	0	5	5
	−3	−4	0	0	1	0	
		↑ Pivot column					

$\xrightarrow[R_3+4R_1]{R_2-R_1}$

x	y	u	v	P	Constant
1	1	①	0	0	4
1	0	−1	1	0	1
1	0	4	0	1	16

The last tableau is in final form and we conclude that $x = 0$, $y = 4$, $u = 0$, $v = 1$, and $P = 16$.

19. We calculate the following sequence of tableaus:

	x	y	u	v	P	Constant	Ratio
Pivot row →	1	②	1	0	0	12	6
	3	2	0	1	0	24	12
	−10	−12	0	0	1	0	
		↑ Pivot column					

$\xrightarrow{\frac{1}{2}R_1}$

x	y	u	v	P	Constant
$\frac{1}{2}$	①	$\frac{1}{2}$	0	0	6
3	2	0	1	0	24
−10	−12	0	0	1	0

$\xrightarrow[R_3+12R_1]{R_2-2R_1}$

	x	y	u	v	P	Constant	Ratio
	$\frac{1}{2}$	1	$\frac{1}{2}$	0	0	6	12
Pivot row →	②	0	−1	1	0	12	6
	−4	0	6	0	1	72	
	↑ Pivot column						

$\xrightarrow{\frac{1}{2}R_2}$

x	y	u	v	P	Constant
$\frac{1}{2}$	1	$\frac{1}{2}$	0	0	6
①	0	$-\frac{1}{2}$	$\frac{1}{2}$	0	6
−4	0	6	0	1	72

$\xrightarrow[R_3+4R_2]{R_1-\frac{1}{2}R_2}$

x	y	u	v	P	Constant
0	1	$\frac{3}{4}$	$-\frac{1}{4}$	0	3
1	0	$-\frac{1}{2}$	$\frac{1}{2}$	0	6
0	0	4	2	1	96

The last tableau is in final form. We conclude that $x = 6$, $y = 3$, $u = 0$, $v = 0$, and $P = 96$.

21. We calculate the following sequence of tableaus:

	x	y	u	v	w	P	Constant	Ratio
	3	1	1	0	0	0	24	24
	2	1	0	1	0	0	18	18
Pivot row →	1	(3)	0	0	1	0	24	8
	−4	−6	0	0	0	1	0	
		↑ Pivot column						

$\xrightarrow{\frac{1}{3}R_3}$

x	y	u	v	w	P	Constant
3	1	1	0	0	0	24
2	1	0	1	0	0	18
$\frac{1}{3}$	(1)	0	0	$\frac{1}{3}$	0	8
−4	−6	0	0	0	1	0

$\xrightarrow[R_4+6R_3]{R_1-R_3,\ R_2-R_3}$

	x	y	u	v	w	P	Constant	Ratio
Pivot row →	($\frac{8}{3}$)	0	1	0	$-\frac{1}{3}$	0	16	6
	$\frac{5}{3}$	0	0	1	$-\frac{1}{3}$	0	10	6
	$\frac{1}{3}$	1	0	0	$\frac{1}{3}$	0	8	24
	−2	0	0	0	2	1	48	
	↑ Pivot column							

$\xrightarrow{\frac{3}{8}R_1}$

x	y	u	v	w	P	Constant
(1)	0	$\frac{3}{8}$	0	$-\frac{1}{8}$	0	6
$\frac{5}{3}$	0	0	1	$-\frac{1}{3}$	0	10
$\frac{1}{3}$	1	0	0	$\frac{1}{3}$	0	8
−2	0	0	0	2	1	48

$\xrightarrow[R_4+2R_1]{R_2-\frac{5}{3}R_1,\ R_3-\frac{1}{3}R_1}$

x	y	u	v	w	P	Constant
1	0	$\frac{3}{8}$	0	$-\frac{1}{8}$	0	6
0	0	$-\frac{5}{8}$	1	$-\frac{1}{8}$	0	0
0	1	$-\frac{1}{8}$	0	$\frac{3}{8}$	0	6
0	0	$\frac{3}{4}$	0	$\frac{7}{4}$	1	60

Observe that we have a choice after the third tableau, in which the ratios on the first and second lines are both 6. The last tableau is in final form, and we deduce that $x = 6$, $y = 6$, $u = 0$, $v = 0$, $w = 0$, and $P = 60$.

23. We calculate the following sequence of tableaus:

	x	y	z	u	v	P	Constant	Ratio
	1	1	1	1	0	0	8	8
Pivot row →	3	2	(4)	0	1	0	24	6
	−3	−4	−5	0	0	1	0	
			↑ Pivot column					

$\xrightarrow{\frac{1}{4}R_2}$

x	y	z	u	v	P	Constant
1	1	1	1	0	0	8
$\frac{3}{4}$	$\frac{1}{2}$	(1)	0	$\frac{1}{4}$	0	6
−3	−4	−5	0	0	1	0

$\xrightarrow{R_1-R_2,\ R_3+5R_2}$

	x	y	z	u	v	P	Constant	Ratio
Pivot row →	$\frac{1}{4}$	($\frac{1}{2}$)	0	1	$-\frac{1}{4}$	0	2	4
	$\frac{3}{4}$	$\frac{1}{2}$	1	0	$\frac{1}{4}$	0	6	12
	$\frac{3}{4}$	$-\frac{3}{2}$	0	0	$\frac{5}{4}$	1	30	
		↑ Pivot column						

$\xrightarrow{2R_1}$

x	y	z	u	v	P	Constant
$\frac{1}{2}$	(1)	0	2	$-\frac{1}{2}$	0	4
$\frac{3}{4}$	$\frac{1}{2}$	1	0	$\frac{1}{4}$	0	6
$\frac{3}{4}$	$-\frac{3}{2}$	0	0	$\frac{5}{4}$	1	30

$\xrightarrow{R_2-\frac{1}{2}R_1,\ R_3+\frac{3}{2}R_1}$

x	y	z	u	v	P	Constant
$\frac{1}{2}$	1	0	2	$-\frac{1}{2}$	0	4
$\frac{1}{2}$	0	1	-1	$\frac{1}{2}$	0	4
$\frac{3}{2}$	0	0	3	$\frac{1}{2}$	1	36

The last tableau is in final form, and we deduce that $x = 0$, $y = 4$, $z = 4$, $u = 0$, $v = 0$, and $P = 36$.

25. We calculate the following sequence of tableaus:

	x	y	z	u	v	w	P	Constant	Ratio
	3	10	5	1	0	0	0	120	12
Pivot row →	5	(2)	8	0	1	0	0	6	3
	8	10	3	0	0	1	0	105	$\frac{21}{2}$
	-3	-4	-1	0	0	0	1	0	
		↑ Pivot column							

$\xrightarrow{\frac{1}{2}R_2}$

x	y	z	u	v	w	P	Constant
3	10	5	1	0	0	0	120
$\frac{5}{2}$	(1)	4	0	$\frac{1}{2}$	0	0	3
8	10	3	0	0	1	0	105
-3	-4	-1	0	0	0	1	0

$\xrightarrow[R_4+4R_2]{R_1-10R_2,\ R_3-10R_2}$

x	y	z	u	v	w	P	Constant
-22	0	-35	1	-5	0	0	90
$\frac{5}{2}$	1	4	0	$\frac{1}{2}$	0	0	3
-17	0	-37	0	-5	1	0	75
7	0	15	0	2	0	1	12

The last tableau is in final form. We conclude that $x = 0$, $y = 3$, $z = 0$, $u = 90$, $v = 0$, $w = 75$, and $P = 12$.

27. We calculate the following sequence of tableaus:

	x	y	z	u	v	w	P	Constant	Ratio
	1	1	1	1	0	0	0	20	20
Pivot row →	2	(4)	3	0	1	0	0	42	$\frac{21}{2}$
	2	0	3	0	0	1	0	30	–
	-4	-6	-5	0	0	0	1	0	
		↑ Pivot column							

$\xrightarrow{\frac{1}{4}R_2}$

x	y	z	u	v	w	P	Constant
1	1	1	1	0	0	0	20
$\frac{1}{2}$	(1)	$\frac{3}{4}$	0	$\frac{1}{4}$	0	0	$\frac{21}{2}$
2	0	3	0	0	1	0	30
-4	-6	-5	0	0	0	1	0

$\xrightarrow[R_4+6R_2]{R_1-R_2}$

	x	y	z	u	v	w	P	Constant	Ratio
	$\frac{1}{2}$	0	$\frac{1}{4}$	1	$-\frac{1}{4}$	0	0	$\frac{19}{2}$	19
	$\frac{1}{2}$	1	$\frac{3}{4}$	0	$\frac{1}{4}$	0	0	$\frac{21}{2}$	21
Pivot row →	(2)	0	3	0	0	1	0	30	15
	-1	0	$-\frac{1}{2}$	0	$\frac{3}{2}$	0	1	63	
	↑ Pivot column								

$\xrightarrow{\frac{1}{2}R_3}$

x	y	z	u	v	w	P	Constant
$\frac{1}{2}$	0	$\frac{1}{4}$	1	$-\frac{1}{4}$	0	0	$\frac{19}{2}$
$\frac{1}{2}$	1	$\frac{3}{4}$	0	$\frac{1}{4}$	0	0	$\frac{21}{2}$
(1)	0	$\frac{3}{2}$	0	0	$\frac{1}{2}$	0	15
-1	0	$-\frac{1}{2}$	0	$\frac{3}{2}$	0	1	63

$\xrightarrow[R_4+R_3]{R_1-\frac{1}{2}R_3,\ R_2-\frac{1}{2}R_3}$

x	y	z	u	v	w	P	Constant
0	0	$-\frac{1}{2}$	1	$-\frac{1}{4}$	$-\frac{1}{4}$	0	2
0	1	0	0	$\frac{1}{4}$	$-\frac{1}{4}$	0	3
1	0	$\frac{3}{2}$	0	0	$\frac{1}{2}$	0	15
0	0	1	0	$\frac{3}{2}$	$\frac{1}{2}$	1	78

The last tableau is in final form. We conclude that $x = 15$, $y = 3$, $z = 0$, $u = 2$, $v = 0$, $w = 0$, and $P = 78$.

29. We calculate the following sequence of tableaus:

	x	y	z	u	v	w	P	Constant	Ratio
Pivot row →	(2)	1	1	1	0	0	0	10	5
	3	5	1	0	1	0	0	45	15
	2	5	1	0	0	1	0	40	20
	−12	−10	−5	0	0	0	1	0	
	↑ Pivot column								

$\xrightarrow{\frac{1}{2}R_1}$

x	y	z	u	v	w	P	Constant
(1)	$\frac{1}{2}$	$\frac{1}{2}$	$\frac{1}{2}$	0	0	0	5
3	5	1	0	1	0	0	45
2	5	1	0	0	1	0	40
−12	−10	−5	0	0	0	1	0

$\xrightarrow[R_4+12R_1]{R_2-3R_1,\ R_3-2R_1}$

	x	y	z	u	v	w	P	Constant	Ratio
	1	$\frac{1}{2}$	$\frac{1}{2}$	$\frac{1}{2}$	0	0	0	5	10
	0	$\frac{7}{2}$	$-\frac{1}{2}$	$-\frac{3}{2}$	1	0	0	30	$\frac{60}{7}$
Pivot row →	0	(4)	0	−1	0	1	0	30	$\frac{15}{2}$
	0	−4	1	6	0	0	1	60	
		↑ Pivot column							

$\xrightarrow{\frac{1}{4}R_3}$

x	y	z	u	v	w	P	Constant
1	$\frac{1}{2}$	$\frac{1}{2}$	$\frac{1}{2}$	0	0	0	5
0	$\frac{7}{2}$	$-\frac{1}{2}$	$-\frac{3}{2}$	1	0	0	30
0	(1)	0	$-\frac{1}{4}$	0	$\frac{1}{4}$	0	$\frac{15}{2}$
0	−4	1	6	0	0	1	60

$\xrightarrow[R_4+4R_3]{R_1-\frac{1}{2}R_3,\ R_2-\frac{7}{2}R_3}$

x	y	z	u	v	w	P	Constant
1	0	$\frac{1}{2}$	$\frac{5}{8}$	0	$-\frac{1}{8}$	0	$\frac{5}{4}$
0	0	$-\frac{1}{2}$	$-\frac{5}{8}$	1	$-\frac{7}{8}$	0	$\frac{15}{4}$
0	1	0	$-\frac{1}{4}$	0	$\frac{1}{4}$	0	$\frac{15}{2}$
0	0	1	5	0	1	1	90

The last tableau is in final form, and we conclude that $x = \frac{5}{4}$, $y = \frac{15}{2}$, $z = 0$, $u = 0$, $v = \frac{15}{4}$, $w = 0$, and $P = 90$.

31. We calculate the following sequence of tableaus:

	x	y	z	u	v	w	P	Constant	Ratio
Pivot row →	(2)	1	2	1	0	0	0	7	$\frac{7}{2}$
	2	3	1	0	1	0	0	8	4
	1	2	3	0	0	1	0	7	7
	−24	−16	−23	0	0	0	1	0	
	↑ Pivot column								

$\xrightarrow{\frac{1}{2}R_1}$

x	y	z	u	v	w	P	Constant
(1)	$\frac{1}{2}$	1	$\frac{1}{2}$	0	0	0	$\frac{7}{2}$
2	3	1	0	1	0	0	8
1	2	3	0	0	1	0	7
−24	−16	−23	0	0	0	1	0

$\xrightarrow[R_4+24R_1]{R_2-2R_1,\ R_3-R_1}$

	x	y	z	u	v	w	P	Constant	Ratio
	1	$\frac{1}{2}$	1	$\frac{1}{2}$	0	0	0	$\frac{7}{2}$	7
Pivot row →	0	②	-1	-1	1	0	0	1	$\frac{1}{2}$
	0	$\frac{3}{2}$	2	$-\frac{1}{2}$	0	1	0	$\frac{7}{2}$	$\frac{7}{3}$
	0	-4	1	12	0	0	1	84	
		↑ Pivot column							

$\xrightarrow{\frac{1}{2}R_2}$

x	y	z	u	v	w	P	Constant
1	$\frac{1}{2}$	1	$\frac{1}{2}$	0	0	0	$\frac{7}{2}$
0	①	$-\frac{1}{2}$	$-\frac{1}{2}$	$\frac{1}{2}$	0	0	$\frac{1}{2}$
0	$\frac{3}{2}$	2	$-\frac{1}{2}$	0	1	0	$\frac{7}{2}$
0	-4	1	12	0	0	1	84

$\xrightarrow[R_4+4R_2]{R_1-\frac{1}{2}R_2,\ R_3-\frac{3}{2}R_2}$

	x	y	z	u	v	w	P	Constant	Ratio
	1	0	$\frac{5}{4}$	$\frac{3}{4}$	$-\frac{1}{4}$	0	0	$\frac{13}{4}$	$\frac{13}{5}$
	0	1	$-\frac{1}{2}$	$-\frac{1}{2}$	$\frac{1}{2}$	0	0	$\frac{1}{2}$	–
Pivot row →	0	0	($\frac{11}{4}$)	$\frac{1}{4}$	$-\frac{3}{4}$	1	0	$\frac{11}{4}$	1
	0	0	-1	10	2	0	1	86	
			↑ Pivot column						

$\xrightarrow{\frac{4}{11}R_3}$

x	y	z	u	v	w	P	Constant
1	0	$\frac{5}{4}$	$\frac{3}{4}$	$-\frac{1}{4}$	0	0	$\frac{13}{4}$
0	1	$-\frac{1}{2}$	$-\frac{1}{2}$	$\frac{1}{2}$	0	0	$\frac{1}{2}$
0	0	①	$\frac{1}{11}$	$-\frac{3}{11}$	$\frac{4}{11}$	0	1
0	0	-1	10	2	0	1	86

$\xrightarrow[R_4+R_3]{R_1-\frac{5}{4}R_3,\ R_2+\frac{1}{2}R_3}$

x	y	z	u	v	w	P	Constant
1	0	0	$\frac{7}{11}$	$\frac{1}{11}$	$-\frac{5}{11}$	0	2
0	1	0	$-\frac{5}{11}$	$\frac{4}{11}$	$\frac{2}{11}$	0	1
0	0	1	$\frac{1}{11}$	$-\frac{3}{11}$	$\frac{4}{11}$	0	1
0	0	0	$\frac{111}{11}$	$\frac{19}{11}$	$\frac{4}{11}$	1	87

The last tableau is in final form, and we conclude that P attains a maximum value of 87 when $x = 2$, $y = 1$, $z = 1$, $u = 0$, $v = 0$, and $w = 0$.

33. Pivoting about the x-column in the initial simplex tableau, we have

	x	y	z	u	v	P	Constant	Ratio
	3	3	-2	1	0	0	100	$\frac{100}{3}$
Pivot row →	⑤	5	3	0	1	0	150	30
	-2	-2	4	0	0	1	0	
	↑ Pivot column							

$\xrightarrow{\frac{1}{5}R_2}$

x	y	z	u	v	P	Constant
3	3	-2	1	0	0	100
①	1	$\frac{3}{5}$	0	$\frac{1}{5}$	0	30
-2	-2	4	0	0	1	0

$\xrightarrow[R_3+2R_2]{R_1-3R_2}$

x	y	z	u	v	P	Constant
0	0	$-\frac{19}{5}$	1	$-\frac{3}{5}$	0	10
1	1	$\frac{3}{5}$	0	$\frac{1}{5}$	0	30
0	0	$\frac{26}{5}$	0	$\frac{2}{5}$	1	60

and we see that one optimal solution occurs when $x = 30$, $y = 0$, $z = 0$, and $P = 60$. Similarly, pivoting about the y-column, we obtain another optimal solution: $x = 0$, $y = 30$, $z = 0$, $P = 60$.

35. Let x and y denote the numbers of model A and model B fax machines produced each shift, respectively. Then we have the following linear programming problem:

$$\begin{aligned} \text{Maximize} \quad & P = 30x + 40y \\ \text{subject to} \quad & 100x + 150y \le 600{,}000 \\ & x + y \le 2500 \\ & x \ge 0,\ y \ge 0 \end{aligned}$$

Using the simplex method, we calculate the following sequence of tableaus:

	x	y	u	v	P	Constant	Ratio
	100	150	1	0	0	600,000	4000
Pivot row →	1	(1)	0	1	0	2500	2500
	−30	−40	0	0	1	0	
		↑ Pivot column					

$\xrightarrow[R_3 + 40R_2]{R_1 - 150R_2}$

x	y	u	v	P	Constant
−50	0	1	−150	0	225,000
1	(1)	0	1	0	2500
10	0	0	40	1	100,000

We conclude that the maximum monthly profit is \$100,000, and this occurs when no model A and 2500 model B fax machines are produced.

37. Suppose the farmer plants x acres of Crop A and y acres of Crop B. Then the linear programming problem is:

$$\begin{aligned} \text{Maximize} \quad & P = 150x + 200y \\ \text{subject to} \quad & x + y \le 150 \\ & 40x + 60y \le 7400 \\ & 20x + 25y \le 3300 \\ & x \ge 0,\ y \ge 0 \end{aligned}$$

Using the simplex method, we calculate the following sequence of tableaus:

	x	y	u	v	w	P	Constant	Ratio
	1	1	1	0	0	0	150	150
Pivot row →	40	(60)	0	1	0	0	7400	$\frac{370}{3}$
	20	25	0	0	1	0	3300	132
	−150	−200	0	0	0	1	0	
		↑ Pivot column						

$\xrightarrow{\frac{1}{60}R_2}$

x	y	u	v	w	P	Constant
1	1	1	0	0	0	150
$\frac{2}{3}$	(1)	0	$\frac{1}{60}$	0	0	$\frac{370}{3}$
20	25	0	0	1	0	3300
−150	−200	0	0	0	1	0

$\xrightarrow[R_4 + 200R_2]{R_1 - R_2,\ R_3 - 25R_2}$

	x	y	u	v	w	P	Constant	Ratio
	$\frac{1}{3}$	0	1	$-\frac{1}{60}$	0	0	$\frac{80}{3}$	80
	$\frac{2}{3}$	1	0	$\frac{1}{60}$	0	0	$\frac{370}{3}$	185
Pivot row →	($\frac{10}{3}$)	0	0	$-\frac{5}{12}$	1	0	$\frac{650}{3}$	65
	$-\frac{50}{3}$	0	0	$\frac{10}{3}$	0	1	$\frac{74{,}000}{3}$	
	↑ Pivot column							

$\xrightarrow[R_4 + \frac{50}{3}R_3]{R_1 - \frac{1}{3}R_3,\ R_2 - \frac{2}{3}R_3}$

x	y	u	v	w	P	Constant
0	0	1	$\frac{1}{40}$	$-\frac{1}{10}$	0	5
0	1	0	$\frac{1}{10}$	$-\frac{1}{5}$	0	80
1	0	0	$-\frac{1}{8}$	$\frac{3}{10}$	0	65
0	0	0	$\frac{5}{4}$	5	1	25,750

The last tableau is in final form. We conclude that $x = 65$, $y = 80$, and $P = 25{,}750$, so the maximum profit of \$25,750 is realized by planting 65 acres of Crop A and 80 acres of Crop B. Because $u = 5$, there are 5 acres of land left unused.

39. Suppose Ashley invests x, y, and z dollars in the money market fund, the international equity fund, and the growth-and-income fund, respectively. Then the objective function is $P = 0.06x + 0.1y + 0.15z$. The constraints are $x + y + z \le 250{,}000$, $z \le 0.25\,(x + y + z)$, and $y \le 0.5\,(x + y + z)$. The last two inequalities simplify to $-\frac{1}{4}x - \frac{1}{4}y + \frac{3}{4}z \le 0$, or $-x - y + 3z \le 0$, and $-\frac{1}{2}x + \frac{1}{2}y - \frac{1}{2}z \le 0$, or $-x + y - z \le 0$. Thus, the required linear programming problem is:

$$\begin{aligned} \text{Maximize} \quad & P = 0.06x + 0.1y + 0.15z = \tfrac{3}{50}x + \tfrac{1}{10}y + \tfrac{3}{20}z \\ \text{subject to} \quad & x + y + z \le 250{,}000 \\ & -x - y + 3z \le 0 \\ & -x + y - z \le 0 \\ & x \ge 0, y \ge 0, z \ge 0 \end{aligned}$$

Let u, v, and w be slack variables. We calculate the following tableaus:

	x	y	z	u	v	w	P	Constant	Ratio
	1	1	1	1	0	0	0	250,000	250,000
Pivot row →	−1	−1	(3)	0	1	0	0	0	0
	−1	1	−1	0	0	1	0	0	–
	$-\frac{3}{50}$	$-\frac{1}{10}$	$-\frac{3}{20}$ ↑ Pivot column	0	0	0	1	0	

$\xrightarrow{\frac{1}{3}R_2}$

x	y	z	u	v	w	P	Constant
1	1	1	1	0	0	0	250,000
$-\frac{1}{3}$	$-\frac{1}{3}$	(1)	0	$\frac{1}{3}$	0	0	0
−1	1	−1	0	0	1	0	0
$-\frac{3}{50}$	$-\frac{1}{10}$	$-\frac{3}{20}$	0	0	0	1	0

$\xrightarrow[R_4 + \frac{3}{20}R_2]{R_1 - R_2,\ R_3 + R_2}$

	x	y	z	u	v	w	P	Constant	Ratio
	$\frac{4}{3}$	$\frac{4}{3}$	0	1	$-\frac{1}{3}$	0	0	250,000	187,500
	$-\frac{1}{3}$	$-\frac{1}{3}$	1	0	$\frac{1}{3}$	0	0	0	–
Pivot row →	$-\frac{4}{3}$	$(\frac{2}{3})$	0	0	$\frac{1}{3}$	1	0	0	0
	$-\frac{11}{100}$	$-\frac{3}{20}$ ↑ Pivot column	0	0	$\frac{1}{20}$	0	1	0	

$\xrightarrow{\frac{3}{2}R_3}$

x	y	z	u	v	w	P	Constant
$\frac{4}{3}$	$\frac{4}{3}$	0	1	$-\frac{1}{3}$	0	0	250,000
$-\frac{1}{3}$	$-\frac{1}{3}$	1	0	$\frac{1}{3}$	0	0	0
−2	(1)	0	0	$\frac{1}{2}$	$\frac{3}{2}$	0	0
$-\frac{11}{100}$	$-\frac{3}{20}$	0	0	$\frac{1}{20}$	0	1	0

$\xrightarrow[R_4 + \frac{3}{20}R_3]{R_1 - \frac{4}{3}R_3,\ R_2 + \frac{1}{3}R_3}$

	x	y	z	u	v	w	P	Constant	Ratio
Pivot row →	(4)	0	0	1	−1	−2	0	250,000	62,500
	−1	0	1	0	$\frac{1}{2}$	$\frac{1}{2}$	0	0	–
	−2	1	0	0	$\frac{1}{2}$	$\frac{3}{2}$	0	0	–
	$-\frac{41}{100}$ ↑ Pivot column	0	0	0	$\frac{1}{8}$	$\frac{9}{40}$	1	0	

$\xrightarrow{\frac{1}{4}R_1}$

x	y	z	u	v	w	P	Constant
(1)	0	0	$\frac{1}{4}$	$-\frac{1}{4}$	$-\frac{1}{2}$	0	62,500
−1	0	1	0	$\frac{1}{2}$	$\frac{1}{2}$	0	0
−2	1	0	0	$\frac{1}{2}$	$\frac{3}{2}$	0	0
$-\frac{41}{100}$	0	0	0	$\frac{1}{8}$	$\frac{9}{40}$	1	0

$\xrightarrow[R_4 + \frac{41}{100}R_1]{R_2 + R_1,\ R_3 + 2R_1}$

x	y	z	u	v	w	P	Constant
1	0	0	$\frac{1}{4}$	$-\frac{1}{4}$	$-\frac{1}{2}$	0	62,500
0	0	1	$\frac{1}{4}$	$\frac{1}{4}$	0	0	62,500
0	1	0	$\frac{1}{2}$	0	$\frac{1}{2}$	0	125,000
0	0	0	$\frac{41}{100}$	$\frac{9}{400}$	$\frac{1}{50}$	1	25,625

The last tableau is in final form, and we see that $x = 62{,}500$, $y = 125{,}000$, $z = 62{,}500$, and $P = 25{,}625$. Thus, Ashley should invest \$62,500 in the money market fund, \$125,000 in the international equity fund, and \$62,500 in the growth-and-income fund. Her maximum return is \$25,625.

41. We have the following linear programming problem:

$$\begin{aligned} \text{Maximize} \quad & P = 18x + 12y + 15z \\ \text{subject to} \quad & 2x + y + 2z \le 900 \\ & 3x + y + 2z \le 1080 \\ & 2x + 2y + z \le 840 \\ & x \ge 0, y \ge 0, z \ge 0 \end{aligned}$$

Let u, v, and w be slack variables. We calculate the following tableaus:

	x	y	z	u	v	w	P	Constant	Ratio
	2	1	2	1	0	0	0	900	450
Pivot row →	(3)	1	2	0	1	0	0	1080	360
	2	2	1	0	0	1	0	840	420
	−18	−12	−15	0	0	0	1	0	
	↑ Pivot column								

$\xrightarrow{\frac{1}{3}R_2}$

x	y	z	u	v	w	P	Constant
2	1	2	1	0	0	0	900
(1)	$\frac{1}{3}$	$\frac{2}{3}$	0	$\frac{1}{3}$	0	0	360
2	2	1	0	0	1	0	840
−18	−12	−15	0	0	0	1	0

$\xrightarrow[R_4 + 18R_2]{R_1 - 2R_2,\ R_3 - 2R_2}$

	x	y	z	u	v	w	P	Constant	Ratio
	0	$\frac{1}{3}$	$\frac{2}{3}$	1	$-\frac{2}{3}$	0	0	180	540
	1	$\frac{1}{3}$	$\frac{2}{3}$	0	$\frac{1}{3}$	0	0	360	1080
Pivot row →	0	($\frac{4}{3}$)	$-\frac{1}{3}$	0	$-\frac{2}{3}$	1	0	120	90
	0	−6	−3	0	6	0	1	6480	
		↑ Pivot column							

$\xrightarrow{\frac{3}{4}R_3}$

x	y	z	u	v	w	P	Constant
0	$\frac{1}{3}$	$\frac{2}{3}$	1	$-\frac{2}{3}$	0	0	180
1	$\frac{1}{3}$	$\frac{2}{3}$	0	$\frac{1}{3}$	0	0	360
0	(1)	$-\frac{1}{4}$	0	$-\frac{1}{2}$	$\frac{3}{4}$	0	90
0	−6	−3	0	6	0	1	6480

$\xrightarrow[R_4 + 6R_3]{R_1 - \frac{1}{3}R_3,\ R_2 - \frac{1}{3}R_3}$

	x	y	z	u	v	w	P	Constant	Ratio
Pivot row →	0	0	($\frac{3}{4}$)	1	$-\frac{1}{2}$	$-\frac{1}{4}$	0	150	200
	1	0	$\frac{3}{4}$	0	$\frac{1}{2}$	$-\frac{1}{4}$	0	330	440
	0	1	$-\frac{1}{4}$	0	$-\frac{1}{2}$	$\frac{3}{4}$	0	90	–
	0	0	$-\frac{9}{2}$	0	3	$\frac{9}{2}$	1	7020	
			↑ Pivot column						

$\xrightarrow{\frac{4}{3}R_1}$

x	y	z	u	v	w	P	Constant
0	0	(1)	$\frac{4}{3}$	$-\frac{2}{3}$	$-\frac{1}{3}$	0	200
1	0	$\frac{3}{4}$	0	$\frac{1}{2}$	$-\frac{1}{4}$	0	330
0	1	$-\frac{1}{4}$	0	$-\frac{1}{2}$	$\frac{3}{4}$	0	90
0	0	$-\frac{9}{2}$	0	3	$\frac{9}{2}$	1	7020

$\xrightarrow[R_4 + \frac{9}{2}R_1]{R_2 - \frac{3}{4}R_1,\ R_3 + \frac{1}{4}R_1}$

x	y	z	u	v	w	P	Constant
0	0	1	$\frac{4}{3}$	$-\frac{2}{3}$	$-\frac{1}{3}$	0	200
1	0	0	−1	1	0	0	180
0	1	0	$\frac{1}{3}$	$-\frac{2}{3}$	$\frac{2}{3}$	0	140
0	0	0	6	0	3	1	7920

The last tableau is in final form, and we conclude that the company will realize a maximum profit of \$7920 by producing 180 units of product A, 140 units of product B, and 200 units of product C. Since $u = v = w = 0$, there are no resources left over.

43. Suppose the Excelsior Company buys x, y, and z minutes of morning, afternoon, and evening commercials, respectively. Then the linear programming problem is:

$$\begin{aligned} \text{Maximize} \quad & P = 200{,}000x + 100{,}000y + 600{,}000z \\ \text{subject to} \quad & 3000x + 1000y + 12{,}000z \le 102{,}000 \\ & z \le 6 \\ & x + y + z \le 25 \\ & x \ge 0, y \ge 0, z \ge 0 \end{aligned}$$

We calculate the following tableaus:

	x	y	z	u	v	w	P	Constant	Ratio
	3000	1000	12,000	1	0	0	0	102,000	$\frac{17}{2}$
Pivot row →	0	0	(1)	0	1	0	0	6	6
	1	1	1	0	0	1	0	25	25
	−200,000	−100,000	−600,000	0	0	0	1	0	
			↑ Pivot column						

$\xrightarrow[R_4 + 600{,}000R_2]{R_1 - 12{,}000R_2 \quad R_3 - R_2}$

	x	y	z	u	v	w	P	Constant	Ratio
Pivot row →	(3000)	1000	0	1	−12,000	0	0	30,000	10
	0	0	1	0	1	0	0	6	–
	1	1	0	0	−1	1	0	19	19
	−200,000	−100,000	0	0	600,000	0	1	3,600,000	
	↑ Pivot column								

$\xrightarrow{\frac{1}{3000}R_1}$

x	y	z	u	v	w	P	Constant
(1)	$\frac{1}{3}$	0	$\frac{1}{3000}$	−4	0	0	10
0	0	1	0	1	0	0	6
1	1	0	0	−1	1	0	19
−200,000	−100,000	0	0	600,000	0	1	3,600,000

$\xrightarrow[R_4 + 200{,}000R_1]{R_3 - R_1}$

x	y	z	u	v	w	P	Constant
1	$\frac{1}{3}$	0	$\frac{1}{3000}$	−4	0	0	10
0	0	1	0	1	0	0	6
0	$\frac{2}{3}$	0	$-\frac{1}{3000}$	(3)	1	0	9
0	$-\frac{100{,}000}{3}$	0	$\frac{200}{3}$	−200,000	0	1	5,600,000

$\xrightarrow{\frac{1}{3}R_3}$

x	y	z	u	v	w	P	Constant
1	$\frac{1}{3}$	0	$\frac{1}{3000}$	−4	0	0	10
0	0	1	0	1	0	0	6
0	$\frac{2}{9}$	0	$-\frac{1}{9000}$	(1)	$\frac{1}{3}$	0	3
0	$-\frac{100{,}000}{3}$	0	$\frac{200}{3}$	−200,000	0	1	5,600,000

$\xrightarrow[R_4 + 200{,}000R_3]{R_1 + 4R_3 \quad R_2 - R_3}$

x	y	z	u	v	w	P	Constant
1	$\frac{11}{9}$	0	$-\frac{1}{9000}$	0	$\frac{4}{3}$	0	22
0	$-\frac{2}{9}$	1	$\frac{1}{9000}$	0	$-\frac{1}{3}$	0	3
0	$\frac{2}{9}$	0	$-\frac{1}{9000}$	1	$\frac{1}{3}$	0	3
0	$\frac{100{,}000}{9}$	0	$\frac{400}{9}$	0	$\frac{200{,}000}{3}$	1	6,200,000

We conclude that $x = 22$, $y = 0$, $z = 3$, $u = 0$, $v = 3$, $w = 0$, and $P = 6{,}200{,}000$. Therefore, the company should buy 22 minutes of morning and 3 minutes of evening advertising time, thereby maximizing their exposure at 6,200,000 viewers.

45. We first tabulate the given information:

	Model			
Department	**A**	**B**	**C**	**Time Available**
Fabrication	$\frac{5}{4}$	$\frac{3}{2}$	$\frac{3}{2}$	310
Assembly	1	1	$\frac{3}{4}$	205
Finishing	1	1	$\frac{1}{2}$	190
Profit	26	28	24	–

Let x, y, and z denote the numbers of units of models A, B, and C to be produced, respectively. Then the required linear programming problem is:

$$\begin{aligned} \text{Maximize} \quad & P = 26x + 28y + 24z \\ \text{subject to} \quad & \tfrac{5}{4}x + \tfrac{3}{2}y + \tfrac{3}{2}z \le 310 \\ & x + y + \tfrac{3}{4}z \le 205 \\ & x + y + \tfrac{1}{2}z \le 190 \\ & x \ge 0, y \ge 0, z \ge 0 \end{aligned}$$

Using the simplex method, we obtain the following tableaus:

	x	y	z	u	v	w	P	Constant	Ratio
	$\frac{5}{4}$	$\frac{3}{2}$	$\frac{3}{2}$	1	0	0	0	310	$\frac{620}{3}$
	1	1	$\frac{3}{4}$	0	1	0	0	205	205
Pivot row →	1	(1)	$\frac{1}{2}$	0	0	1	0	190	190
	−26	−28	−24	0	0	0	1	0	
		↑ Pivot column							

$\xrightarrow[R_4 + 28R_3]{R_1 - \frac{3}{2}R_3,\ R_2 - R_3}$

	x	y	z	u	v	w	P	Constant	Ratio
Pivot row →	$-\frac{1}{4}$	0	($\frac{3}{4}$)	1	0	$-\frac{3}{2}$	0	25	$\frac{100}{3}$
	0	0	$\frac{1}{4}$	0	1	−1	0	15	60
	1	1	$\frac{1}{2}$	0	0	1	0	190	380
	2	0	−10	0	0	28	1	5320	
			↑ Pivot column						

$\xrightarrow{\frac{4}{3}R_1}$

x	y	z	u	v	w	P	Constant
$-\frac{1}{3}$	0	(1)	$\frac{4}{3}$	0	−2	0	$\frac{100}{3}$
0	0	$\frac{1}{4}$	0	1	−1	0	15
1	1	$\frac{1}{2}$	0	0	1	0	190
2	0	−10	0	0	28	1	5320

$\xrightarrow[R_4 + 10R_1]{R_2 - \frac{1}{4}R_1,\ R_3 - \frac{1}{2}R_1}$

	x	y	z	u	v	w	P	Constant	Ratio
	$-\frac{1}{3}$	0	1	$\frac{4}{3}$	0	−2	0	$\frac{100}{3}$	–
Pivot row →	($\frac{1}{12}$)	0	0	$-\frac{1}{3}$	1	$-\frac{1}{2}$	0	$\frac{20}{3}$	80
	$\frac{7}{6}$	1	0	$-\frac{2}{3}$	0	2	0	$\frac{520}{3}$	$\frac{1040}{7}$
	$-\frac{4}{3}$	0	0	$\frac{40}{3}$	0	8	1	$\frac{16{,}960}{3}$	
	↑ Pivot column								

$\xrightarrow{12R_2}$

x	y	z	u	v	w	P	Constant
$-\frac{1}{3}$	0	1	$\frac{4}{3}$	0	-2	0	$\frac{100}{3}$
(1)	0	0	-4	12	-6	0	80
$\frac{7}{6}$	1	0	$-\frac{2}{3}$	0	2	0	$\frac{520}{3}$
$-\frac{4}{3}$	0	0	$\frac{40}{3}$	0	8	1	$\frac{16,960}{3}$

$\xrightarrow{R_1+\frac{1}{3}R_2,\ R_3-\frac{7}{6}R_2,\ R_4+\frac{4}{3}R_2}$

x	y	z	u	v	w	P	Constant
0	0	1	0	4	-4	0	60
1	0	0	-4	12	-6	0	80
0	1	0	4	-14	9	0	80
0	0	0	8	16	0	1	5760

The last tableau is in final form. We see that $x = 80$, $y = 80$, $z = 60$, $u = 0$, $v = 0$, $w = 0$, and $P = 5760$. Thus, by producing 80 units each of Models A and B and 60 units of Model C, the company stands to make a maximum profit of \$5760. Because $u = v = w = 0$, there are no resources left over.

47. Let x, y, and z denote the numbers (in thousands) of bottles of formulas I, II, and III produced, respectively. The resulting linear programming problem is:

$$\begin{aligned} \text{Maximize} \quad & P = 180x + 200y + 300z \\ \text{subject to} \quad & \tfrac{5}{2}x + 3y + 4z \le 70 \\ & x \le 9,\ y \le 12,\ z \le 6 \\ & x \ge 0,\ y \ge 0,\ z \ge 0 \end{aligned}$$

Using the simplex method, we have

	x	y	z	s	t	u	v	P	Constant	Ratio
	$\frac{5}{2}$	3	4	1	0	0	0	0	70	$\frac{35}{2}$
	1	0	0	0	1	0	0	0	9	–
	0	1	0	0	0	1	0	0	12	–
Pivot row →	0	0	(1)	0	0	0	1	0	6	6
	-180	-200	-300 ↑ Pivot column	0	0	0	0	1	0	

$\xrightarrow{R_1-4R_4,\ R_5+300R_4}$

	x	y	z	s	t	u	v	P	Constant	Ratio
	$\frac{5}{2}$	3	0	1	0	0	-4	0	46	$\frac{46}{3}$
	1	0	0	0	1	0	0	0	9	–
Pivot row →	0	(1)	0	0	0	1	0	0	12	12
	0	0	1	0	0	0	1	0	6	–
	-180	-200 ↑ Pivot column	0	0	0	0	300	1	1800	

$\xrightarrow{R_1-3R_3,\ R_5+200R_3}$

	x	y	z	s	t	u	v	P	Constant	Ratio
Pivot row →	($\frac{5}{2}$)	0	0	1	0	-3	-4	0	10	4
	1	0	0	0	1	0	0	0	9	9
	0	1	0	0	0	1	0	0	12	–
	0	0	1	0	0	0	1	0	6	–
	-180 ↑ Pivot column	0	0	0	0	200	300	1	4200	

$\xrightarrow{\frac{2}{5}R_1}$

x	y	z	s	t	u	v	P	Constant
(1)	0	0	$\frac{2}{5}$	0	$-\frac{6}{5}$	$-\frac{8}{5}$	0	4
1	0	0	0	1	0	0	0	9
0	1	0	0	0	1	0	0	12
0	0	1	0	0	0	1	0	6
-180	0	0	0	0	200	300	1	4200

$\xrightarrow[R_5+180R_1]{R_2-R_1}$

	x	y	z	s	t	u	v	P	Constant	Ratio
	1	0	0	$\frac{2}{5}$	0	$-\frac{6}{5}$	$-\frac{8}{5}$	0	4	–
Pivot row →	0	0	0	$-\frac{2}{5}$	1	($\frac{6}{5}$)	$\frac{8}{5}$	0	5	$\frac{25}{6}$
	0	1	0	0	0	1	0	0	12	12
	0	0	1	0	0	0	1	0	6	–
	0	0	0	72	0	-16	12	1	4920	
						↑ Pivot column				

$\xrightarrow{\frac{5}{6}R_2}$

x	y	z	s	t	u	v	P	Constant
1	0	0	$\frac{2}{5}$	0	$-\frac{6}{5}$	$-\frac{8}{5}$	0	4
0	0	0	$-\frac{1}{3}$	$\frac{5}{6}$	(1)	$\frac{4}{3}$	0	$\frac{25}{6}$
0	1	0	0	0	1	0	0	12
0	0	1	0	0	0	1	0	6
0	0	0	72	0	-16	12	1	4920

$\xrightarrow[R_5+16R_2]{R_1+\frac{6}{5}R_2,\; R_3-R_2}$

x	y	z	s	t	u	v	P	Constant
1	0	0	0	1	0	0	0	9
0	0	0	$-\frac{1}{3}$	$\frac{5}{6}$	1	$\frac{4}{3}$	0	$\frac{25}{6}$
0	1	0	$\frac{1}{3}$	$-\frac{5}{6}$	0	$-\frac{4}{3}$	0	$\frac{47}{6}$
0	0	1	0	0	0	1	0	6
0	0	0	$\frac{200}{3}$	$\frac{40}{3}$	0	$\frac{100}{3}$	1	$\frac{14{,}960}{3}$

The last tableau is in final form. We conclude that $x = 9$, $y = \frac{47}{6}$, $z = 6$, $s = 0$, $t = 0$, $u = \frac{25}{6}$, and $P = \frac{14{,}960}{3} \approx 4986.67$; that is, the company should manufacture 9000 bottles of formula I, 7833 bottles of formula II, and 6000 bottles of formula III for a maximum profit of \$4986.60, leaving ingredients for 4167 bottles of formula II unused.

49. Refer to the solution to Exercise 6.2.19 on page 137 of this manual. The linear programming problem is:

$$\begin{aligned} \text{Maximize}\quad & P = 0.1x + 0.15y + 0.2z \\ \text{subject to}\quad & x + y + z \le 2{,}000{,}000 \\ & -2x - 2y + 8z \le 0 \\ & -6x + 4y + 4z \le 0 \\ & -10x + 6y + 6z \le 0 \\ & x \ge 0,\ y \ge 0,\ z \le 0 \end{aligned}$$

Letting u, v, w, and s be slack variables, we calculate the following sequence of tableaus:

	x	y	z	u	v	w	s	P	Constant	Ratio
	1	1	1	1	0	0	0	0	2,000,000	2,000,000
Pivot row →	-2	-2	(8)	0	1	0	0	0	0	–
	-6	4	4	0	0	1	0	0	0	0
	-10	6	6	0	0	0	1	0	0	–
	$-\frac{1}{10}$	$-\frac{3}{20}$	$-\frac{1}{5}$	0	0	0	0	1	0	
			↑ Pivot column							

$\xrightarrow{\frac{1}{8}R_2}$

x	y	z	u	v	w	s	P	Constant
1	1	1	1	0	0	0	0	2,000,000
$-\frac{1}{4}$	$-\frac{1}{4}$	①	0	$\frac{1}{8}$	0	0	0	0
-6	4	4	0	0	1	0	0	0
-10	6	6	0	0	0	1	0	0
$-\frac{1}{10}$	$-\frac{3}{20}$	$-\frac{1}{5}$	0	0	0	0	1	0

$\xrightarrow[R_4 - 6R_2,\; R_5 + \frac{1}{5}R_2]{R_1 - R_2,\; R_3 - 4R_2}$

	x	y	z	u	v	w	s	P	Constant	Ratio
	$\frac{5}{4}$	$\frac{5}{4}$	0	1	$-\frac{1}{8}$	0	0	0	2,000,000	1,600,000
	$-\frac{1}{4}$	$-\frac{1}{4}$	1	0	$\frac{1}{8}$	0	0	0	0	–
Pivot row →	-5	⑤	0	0	$-\frac{1}{2}$	1	0	0	0	0
	$-\frac{17}{2}$	$\frac{15}{2}$	0	0	$-\frac{3}{4}$	0	1	0	0	0
	$-\frac{3}{20}$	$-\frac{1}{5}$	0	0	$\frac{1}{40}$	0	0	1	0	
		↑ Pivot column								

$\xrightarrow{\frac{1}{5}R_3}$

x	y	z	u	v	w	s	P	Constant
$\frac{5}{4}$	$\frac{5}{4}$	0	1	$-\frac{1}{8}$	0	0	0	2,000,000
$-\frac{1}{4}$	$-\frac{1}{4}$	1	0	$\frac{1}{8}$	0	0	0	0
-1	1	0	0	$-\frac{1}{10}$	$\frac{1}{5}$	0	0	0
$-\frac{17}{2}$	$\frac{15}{2}$	0	0	$-\frac{3}{4}$	0	1	0	0
$-\frac{3}{20}$	$-\frac{1}{5}$	0	0	$\frac{1}{40}$	0	0	1	0

$\xrightarrow[R_4 - \frac{15}{2}R_3,\; R_5 + \frac{1}{5}R_3]{R_1 - \frac{5}{4}R_3,\; R_2 + \frac{1}{4}R_3}$

	x	y	z	u	v	w	s	P	Constant	Ratio
Pivot row →	$\left(\frac{5}{2}\right)$	0	0	1	0	$-\frac{1}{4}$	0	0	2,000,000	800,000
	$-\frac{1}{2}$	0	1	0	$\frac{1}{10}$	$\frac{1}{20}$	0	0	0	–
	-1	1	0	0	$-\frac{1}{10}$	$\frac{1}{5}$	0	0	0	0
	-1	0	0	0	0	$-\frac{3}{2}$	1	0	0	0-
	$-\frac{7}{20}$	0	0	0	$\frac{1}{200}$	$\frac{1}{25}$	0	1	0	
	↑ Pivot column									

$\xrightarrow{\frac{2}{5}R_1}$

x	y	z	u	v	w	s	P	Constant
①	0	0	$\frac{2}{5}$	0	$-\frac{1}{10}$	0	0	800,000
$-\frac{1}{2}$	0	1	0	$\frac{1}{10}$	$\frac{1}{20}$	0	0	0
-1	1	0	0	$-\frac{1}{10}$	$\frac{1}{5}$	0	0	0
-1	0	0	0	0	$-\frac{3}{2}$	1	0	0
$-\frac{7}{20}$	0	0	0	$\frac{1}{200}$	$\frac{1}{25}$	0	1	0

$\xrightarrow[R_4 + R_1,\; R_5 + \frac{7}{20}R_1]{R_2 + \frac{1}{2}R_1,\; R_3 + R_1}$

x	y	z	u	v	w	s	P	Constant
1	0	0	$\frac{2}{5}$	0	$-\frac{1}{10}$	0	0	800,000
0	0	1	$\frac{1}{5}$	$\frac{1}{10}$	0	0	0	400,000
0	1	0	$\frac{2}{5}$	$-\frac{1}{10}$	$\frac{1}{10}$	0	0	800,000
0	0	0	$\frac{2}{5}$	0	$-\frac{8}{5}$	1	0	800,000
0	0	0	$\frac{7}{50}$	$\frac{1}{200}$	$\frac{1}{200}$	0	1	280,000

The last tableau is in final form, and we see that $x = 800{,}000$, $y = 800{,}000$, $z = 400{,}000$, and $P = 280{,}000$. Thus, the financier should invest \$800,000 each in Projects A and B and \$400,000 in Project C for a maximum return of \$280,000.

51. False. Consider the following linear programming problem:

$$\begin{aligned} \text{Maximize} \quad & P = 2x + 3y \\ \text{subject to} \quad & -x + y \leq 0 \\ & x \geq 0, y \geq 0 \end{aligned}$$

53. True. Consider the objective function $P = c_1x_1 + c_2x_2 + \cdots + c_nx_n$, which may be written in the form $-c_1x_1 - c_2x_2 - \cdots - c_nx_n + P = 0$. Observe that the most negative of the numbers $-c_1, -c_2, \ldots, -c_n$ (which are the numbers comprising the last row of the simplex tableau) is just the largest coefficient of x_i in the expression for P. Thus, moving in the direction of the variable with this coefficient ensures the maximal increase in P.

Technology Exercises page 384

1. $x = 1.2, y = 0, z = 1.6, w = 0; P = 8.8$.

3. $x = 1.6, y = 0, z = 0, w = 3.6; P = 12.4$.

6.5 The Simplex Method: Standard Minimization Problems

Problem-Solving Tips

1. The given problem is called the **primal problem** and the problem related to it is called the **dual problem**. Dual problems are standard minimization problems.

2. To solve a dual problem, first *write down the tableau* for the primal problem. (Note that this is not a simplex tableau as there are no slack variables.) Then *interchange the columns and rows* of this tableau. Use this tableau to *write the dual problem* and then use the simplex method to complete the solution to the problem. The minimum value of C will appear in the lower right corner of the final simplex tableau.

Concept Questions page 394

1. Maximize $P = -C = 3x + 5y$ subject to

$$\begin{aligned} 5x + 2y &\leq 30 \\ x + 3y &\leq 21 \\ x \geq 0, y &\geq 0 \end{aligned}$$

3. The primal problem is the linear programming (maximization) problem associated with a minimization linear programming problem. The dual problem is the linear programming (minimization) problem associated with the maximization linear programming problem.

Exercises page 394

1. We solve the associated regular problem:

$$\begin{aligned} \text{Maximize} \quad & P = -C = 2x - y \\ \text{subject to} \quad & x + 2y \le 6 \\ & 3x + 2y \le 12 \\ & x \ge 0, y \ge 0 \end{aligned}$$

Using the simplex method with slack variables u and v, we have

	x	y	u	v	P	Constant	Ratio
	1	2	1	0	0	6	6
Pivot row →	(3)	2	0	1	0	12	4
	−2	1	0	0	1	0	
	↑ Pivot column						

$\xrightarrow{\frac{1}{3}R_2}$

x	y	u	v	P	Constant
1	2	1	0	0	6
(1)	$\frac{2}{3}$	0	$\frac{1}{3}$	0	4
−2	1	0	0	1	0

$\xrightarrow[R_3 + 2R_2]{R_1 - R_2}$

x	y	u	v	P	Constant
0	$\frac{4}{3}$	1	$-\frac{1}{3}$	0	2
1	$\frac{2}{3}$	0	$\frac{1}{3}$	0	4
0	$\frac{7}{3}$	0	$\frac{2}{3}$	1	8

Therefore, $x = 4$, $y = 0$, and $C = -P = -8$.

3. We maximize $P = -C = 3x + 2y$. Using the simplex method, we obtain

	x	y	u	v	P	Constant	Ratio
	3	4	1	0	0	24	8
Pivot row →	(7)	−4	0	1	0	16	$\frac{16}{7}$
	−3	−2	0	0	1	0	
	↑ Pivot column						

$\xrightarrow{\frac{1}{7}R_2}$

x	y	u	v	P	Constant
3	4	1	0	0	24
(1)	$-\frac{4}{7}$	0	$\frac{1}{7}$	0	$\frac{16}{7}$
−3	−2	0	0	1	0

$\xrightarrow[R_3 + 3R_2]{R_1 - 3R_2}$

	x	y	u	v	P	Constant	Ratio
Pivot row →	0	$\left(\frac{40}{7}\right)$	1	$-\frac{3}{7}$	0	$\frac{120}{7}$	3
	1	$-\frac{4}{7}$	0	$\frac{1}{7}$	0	$\frac{16}{7}$	–
	0	$-\frac{26}{7}$	0	$\frac{3}{7}$	1	$\frac{48}{7}$	
		↑ Pivot column					

$\xrightarrow{\frac{7}{40}R_1}$

x	y	u	v	P	Constant
0	(1)	$\frac{7}{40}$	$-\frac{3}{40}$	0	3
1	$-\frac{4}{7}$	0	$\frac{1}{7}$	0	$\frac{16}{7}$
0	$-\frac{26}{7}$	0	$\frac{3}{7}$	1	$\frac{48}{7}$

$\xrightarrow[R_3 + \frac{26}{7}R_1]{R_2 + \frac{4}{7}R_1}$

x	y	u	v	P	Constant
0	1	$\frac{7}{40}$	$-\frac{3}{40}$	0	3
1	0	$\frac{1}{10}$	$\frac{1}{10}$	0	4
0	0	$\frac{13}{20}$	$\frac{3}{20}$	1	18

The last tableau is in final form. We find $x = 4$, $y = 3$, and $C = -P = -18$.

5. We maximize $P = -C = -2x + 3y + 4z$ subject to the given constraints. Using the simplex method, we obtain

	x	y	z	u	v	w	P	Constant	Ratio
	-1	2	-1	1	0	0	0	8	–
Pivot row →	1	-2	$\textcircled{2}$	0	1	0	0	10	5
	2	4	-3	0	0	1	0	12	–
	2	-3	-4	0	0	0	1	0	

↑ Pivot column (z)

$\xrightarrow{\frac{1}{2}R_2}$

x	y	z	u	v	w	P	Constant
-1	2	-1	1	0	0	0	8
$\frac{1}{2}$	-1	$\textcircled{1}$	0	$\frac{1}{2}$	0	0	5
2	4	-3	0	0	1	0	12
2	-3	-4	0	0	0	1	0

$\xrightarrow[R_4+4R_2]{R_1+R_2,\ R_3+3R_2}$

	x	y	z	u	v	w	P	Constant	Ratio
Pivot row →	$-\frac{1}{2}$	$\textcircled{1}$	0	1	$\frac{1}{2}$	0	0	13	13
	$\frac{1}{2}$	-1	1	0	$\frac{1}{2}$	0	0	5	–
	$\frac{7}{2}$	1	0	0	$\frac{3}{2}$	1	0	27	27
	4	-7	0	0	2	0	1	20	

↑ Pivot column (y)

$\xrightarrow[R_4+7R_1]{R_2+R_1,\ R_3-R_1}$

x	y	z	u	v	w	P	Constant
$-\frac{1}{2}$	$\textcircled{1}$	0	1	$\frac{1}{2}$	0	0	13
0	0	1	1	1	0	0	18
4	0	0	-1	1	1	0	14
$\frac{1}{2}$	0	0	7	$\frac{11}{2}$	0	1	111

The last tableau is in final form. We see that $x = 0$, $y = 13$, $z = 18$, $u = 0$, $v = 0$, $w = 14$, and $C = -P = -111$.

7. $x = \frac{5}{4}$, $y = \frac{1}{4}$, $u = 2$, $v = 3$, and $C = P = 13$.

9. $x = 5$, $y = 10$, $z = 0$, $u = 1$, $v = 2$, and $C = P = 80$.

11. We first write out the primal tableau:

x	y	Constant
1	2	4
3	2	6
2	5	

Then we obtain a dual tableau by interchanging rows and columns:

u	v	Constant
1	3	2
2	2	5
4	6	

From this table we construct the dual problem:

$$\begin{aligned} \text{Maximize} \quad & P = 4u + 6v \\ \text{subject to} \quad & u + 3v \le 2 \\ & 2u + 2v \le 5 \\ & u \ge 0, v \ge 0 \end{aligned}$$

Solving the dual problem using the simplex method with x and y as the slack variables, we obtain

	u	v	x	y	P	Constant	Ratio
Pivot row →	1	(3)	1	0	0	2	$\frac{2}{3}$
	2	2	0	1	0	5	$\frac{5}{2}$
	−4	−6	0	0	1	0	
		↑ Pivot column					

$\xrightarrow{\frac{1}{3}R_1}$

u	v	x	y	P	Constant
$\frac{1}{3}$	(1)	$\frac{1}{3}$	0	0	$\frac{2}{3}$
2	2	0	1	0	5
−4	−6	0	0	1	0

$\xrightarrow[R_3+6R_1]{R_2-2R_1}$

	u	v	x	y	P	Constant	Ratio
Pivot row →	$(\frac{1}{3})$	1	$\frac{1}{3}$	0	0	$\frac{2}{3}$	2
	$\frac{4}{3}$	0	$-\frac{2}{3}$	1	0	$\frac{11}{3}$	$\frac{11}{4}$
	−2	0	2	0	1	4	
	↑ Pivot column						

$\xrightarrow{3R_1}$

u	v	x	y	P	Constant
(1)	3	1	0	0	2
$\frac{4}{3}$	0	$-\frac{2}{3}$	1	0	$\frac{11}{3}$
−2	0	2	0	1	4

$\xrightarrow[R_3+2R_1]{R_2-\frac{4}{3}R_1}$

u	v	x	y	P	Constant
1	3	1	0	0	2
0	−4	−2	1	0	1
0	6	4	0	1	8

Interpreting the final tableau, we see that $x = 4$, $y = 0$, $u = 2$, $v = 0$, and $P = C = 8$.

13. We write the primal tableau, then obtain a dual tableau by interchanging rows and columns.

x	y	Constant
6	1	60
2	1	40
1	1	30
6	4	

u	v	w	Constant
6	2	1	6
1	1	1	4
60	40	30	

From this table we construct the dual problem:

$$\begin{aligned} \text{Maximize} \quad & P = 60u + 40v + 30w \\ \text{subject to} \quad & 6u + 2v + w \le 6 \\ & u + v + w \le 4 \\ & u \ge 0, v \ge 0, w \ge 0 \end{aligned}$$

We solve the problem as follows:

	u	v	w	x	y	P	Constant	Ratio
Pivot row →	(6)	2	1	1	0	0	6	1
	1	1	1	0	1	0	4	4
	−60	−40	−30	0	0	1	0	–
	↑ Pivot column							

$\xrightarrow{\frac{1}{6}R_1}$

u	v	w	x	y	P	Constant
(1)	$\frac{1}{3}$	$\frac{1}{6}$	$\frac{1}{6}$	0	0	1
1	1	1	0	1	0	4
−60	−40	−30	0	0	1	0

$\xrightarrow[R_3+60R_1]{R_2-R_1}$

	u	v	w	x	y	P	Constant	Ratio
	1	$\frac{1}{3}$	$\frac{1}{6}$	$\frac{1}{6}$	0	0	1	6
Pivot row →	0	$\frac{2}{3}$	($\frac{5}{6}$)	$-\frac{1}{6}$	1	0	3	$\frac{18}{5}$
	0	−20	−20 ↑ Pivot column	10	0	1	60	–

$\xrightarrow{\frac{6}{5}R_2}$

u	v	w	x	y	P	Constant
1	$\frac{1}{3}$	$\frac{1}{6}$	$\frac{1}{6}$	0	0	1
0	$\frac{4}{5}$	(1)	$-\frac{1}{5}$	$\frac{6}{5}$	0	$\frac{18}{5}$
0	−20	−20	10	0	1	60

$\xrightarrow[R_3+20R_2]{R_1-\frac{1}{6}R_2}$

	u	v	w	x	y	P	Constant	Ratio
Pivot row →	1	($\frac{1}{5}$)	0	$\frac{1}{5}$	$-\frac{1}{5}$	0	$\frac{2}{5}$	2
	0	$\frac{4}{5}$	1	$-\frac{1}{5}$	$\frac{6}{5}$	0	$\frac{18}{5}$	$\frac{9}{2}$
	0	−4 ↑ Pivot column	0	6	24	1	132	–

$\xrightarrow{5R_1}$

u	v	w	x	y	P	Constant
5	(1)	0	1	−1	0	2
0	$\frac{4}{5}$	1	$-\frac{1}{5}$	$\frac{6}{5}$	0	$\frac{18}{5}$
0	−4	0	6	24	1	132

$\xrightarrow[R_3+4R_1]{R_2-\frac{4}{5}R_1}$

u	v	w	x	y	P	Constant
5	1	0	1	−1	0	2
−4	0	1	−1	2	0	2
20	0	0	10	20	1	140

The last tableau is in final form. We find that $x = 10$, $y = 20$, $u = 0$, $v = 2$, $w = 0$, and $C = 140$.

15. We write the primal tableau, then obtain a dual tableau by interchanging rows and columns.

x	y	z	Constant
20	10	1	10
1	1	2	20
200	150	120	

u	v	Constant
20	1	200
10	1	150
1	2	120
10	20	

From this table we construct the dual problem:

$$\begin{aligned} \text{Maximize} \quad & P = 10u + 20v \\ \text{subject to} \quad & 20u + v \le 200 \\ & 10u + v \le 150 \\ & u + 2v \le 120 \\ & u \ge 0, v \ge 0 \end{aligned}$$

Solving the dual problem using the simplex method with x, y, and z as slack variables, we obtain the following tableaus:

	u	v	x	y	z	P	Constant	Ratio
	20	1	1	0	0	0	200	200
	10	1	0	1	0	0	150	150
Pivot row →	1	(2)	0	0	1	0	120	60
	−10	−20 ↑ Pivot column	0	0	0	1	0	

$\xrightarrow{\frac{1}{2}R_3}$

u	v	x	y	z	P	Constant
20	1	1	0	0	0	200
10	1	0	1	0	0	150
$\frac{1}{2}$	(1)	0	0	$\frac{1}{2}$	0	60
−10	−20	0	0	0	1	0

$\xrightarrow[R_4+20R_3]{\substack{R_1-R_3 \\ R_2-R_3}}$

u	v	x	y	z	P	Constant
$\frac{39}{2}$	0	1	0	$-\frac{1}{2}$	0	140
$\frac{19}{2}$	0	0	1	$-\frac{1}{2}$	0	90
$\frac{1}{2}$	1	0	0	$\frac{1}{2}$	0	60
0	0	0	0	10	1	1200

This last tableau is in final form. We find that $x = 0$, $y = 0$, $z = 10$, and $C = 1200$.

17. We write the primal tableau, then obtain a dual tableau by interchanging rows and columns.

x	y	z	Constant
1	2	2	10
2	1	1	24
1	1	1	16
6	8	4	

u	v	w	Constant
1	2	1	6
2	1	1	8
2	1	1	4
10	24	16	

From this table we construct the dual problem:

$$\begin{aligned} \text{Maximize} \quad & P = 10u + 24v + 16w \\ \text{subject to} \quad & u + 2v + w \leq 6 \\ & 2u + v + w \leq 8 \\ & 2u + v + w \leq 4 \\ & u \geq 0, v \geq 0, w \geq 0 \end{aligned}$$

Solving the dual problem using the simplex method with x, y, and z as slack variables, we obtain the following tableaus:

	u	v	w	x	y	z	P	Constant	Ratio
Pivot row →	1	(2)	1	1	0	0	0	6	3
	2	1	1	0	1	0	0	8	8
	2	1	1	0	0	1	0	4	4
	−10	−24	−16	0	0	0	1	0	
		↑ Pivot column							

$\xrightarrow{\frac{1}{2}R_1}$

u	v	w	x	y	z	P	Constant
$\frac{1}{2}$	(1)	$\frac{1}{2}$	$\frac{1}{2}$	0	0	0	3
2	1	1	0	1	0	0	8
2	1	1	0	0	1	0	4
−10	−24	−16	0	0	0	1	0

$\xrightarrow[R_4+24R_1]{\substack{R_2-R_1 \\ R_3-R_1}}$

u	v	w	x	y	z	P	Constant	Ratio
$\frac{1}{2}$	1	$\frac{1}{2}$	$\frac{1}{2}$	0	0	0	3	6
$\frac{3}{2}$	0	$\frac{1}{2}$	$-\frac{1}{2}$	1	0	0	5	10
$\frac{3}{2}$	0	($\frac{1}{2}$)	$-\frac{1}{2}$	0	1	0	1	2
2	0	−4	12	0	0	1	72	

$\xrightarrow{2R_3}$

u	v	w	x	y	z	P	Constant
$\frac{1}{2}$	1	$\frac{1}{2}$	$\frac{1}{2}$	0	0	0	3
$\frac{3}{2}$	0	$\frac{1}{2}$	$-\frac{1}{2}$	1	0	0	5
3	0	1	−1	0	2	0	2
2	0	−4	12	0	0	1	72

$\xrightarrow[R_4+4R_3]{\substack{R_1-\frac{1}{2}R_3 \\ R_2-\frac{1}{2}R_3}}$

u	v	w	x	y	z	P	Constant
−1	1	0	1	0	−1	0	2
0	0	0	0	1	−1	0	4
3	0	1	−1	0	2	0	2
14	0	0	8	0	8	1	80

The solution to the primal problem is thus $x = 8$, $y = 0$, $z = 8$, $u = 0$, $v = 2$, $w = 2$, and $C = 80$.

19. We write the primal tableau, then obtain a dual tableau by interchanging rows and columns.

x	y	z	Constant
2	4	3	6
6	0	1	2
0	6	2	4
30	12	20	

u	v	w	Constant
2	6	0	30
4	0	6	12
3	1	2	20
6	2	4	

From this table we construct the dual problem:

$$\begin{aligned}\text{Maximize}\quad & P = 6u + 2v + 4w\\ \text{subject to}\quad & 2u + 6v \le 30\\ & 4u + 6w \le 12\\ & 3u + v + 2w \le 20\\ & u \ge 0,\ v \ge 0,\ w \ge 0\end{aligned}$$

Solving the dual problem using the simplex method with x, y, and z as slack variables, we obtain the following tableaus:

	u	v	w	x	y	z	P	Constant	Ratio
	2	6	0	1	0	0	0	30	15
Pivot row →	(4)	0	6	0	1	0	0	12	3
	3	1	2	0	0	1	0	20	$\frac{20}{3}$
	−6	−2	−4	0	0	0	1	0	
	↑ Pivot column								

$\xrightarrow{\frac{1}{4}R_2}$

u	v	w	x	y	z	P	Constant
2	6	0	1	0	0	0	30
(1)	0	$\frac{3}{2}$	0	$\frac{1}{4}$	0	0	3
3	1	2	0	0	1	0	20
−6	−2	−4	0	0	0	1	0

$\xrightarrow[R_4+6R_2]{R_1-2R_2,\ R_3-3R_2}$

	u	v	w	x	y	z	P	Constant	Ratio
Pivot row →	0	(6)	−3	1	$-\frac{1}{2}$	0	0	24	4
	1	0	$\frac{3}{2}$	0	$\frac{1}{4}$	0	0	3	–
	0	1	$-\frac{5}{2}$	0	$-\frac{3}{4}$	1	0	11	11
	0	−2	5	0	$\frac{3}{2}$	0	1	18	
		↑ Pivot column							

$\xrightarrow{\frac{1}{6}R_1}$

u	v	w	x	y	z	P	Constant
0	(1)	$-\frac{1}{2}$	$\frac{1}{6}$	$-\frac{1}{12}$	0	0	4
1	0	$\frac{3}{2}$	0	$\frac{1}{4}$	0	0	3
0	1	$-\frac{5}{2}$	0	$-\frac{3}{4}$	1	0	11
0	−2	5	0	$\frac{3}{2}$	0	1	18

$\xrightarrow[R_4+2R_1]{R_3-R_1}$

u	v	w	x	y	z	P	Constant
0	1	$-\frac{1}{2}$	$\frac{1}{6}$	$-\frac{1}{12}$	0	0	4
1	0	$\frac{3}{2}$	0	$\frac{1}{4}$	0	0	3
0	0	−2	$-\frac{1}{6}$	$-\frac{2}{3}$	1	0	7
0	0	4	$\frac{1}{3}$	$\frac{4}{3}$	0	1	26

The last tableau is in final form. We find $x = \frac{1}{3}$, $y = \frac{4}{3}$, $z = 0$, $u = 3$, $v = 4$, $w = 0$, and $C = 26$.

21. Let x denote the number of type A vessels and y the number of type B vessels to be operated. Then the linear programming problem is:

$$\begin{aligned} \text{Minimize} \quad & C = 44{,}000x + 54{,}000y \\ \text{subject to} \quad & 60x + 80y \geq 360 \\ & 160x + 120y \geq 680 \\ & x \geq 0, y \geq 0 \end{aligned}$$

We first write a tableau for the primal problem, then obtain a dual tableau by interchanging rows and columns.

x	y	Constant
60	80	360
160	120	680
44,000	54,000	

u	v	Constant
60	160	44,000
80	120	54,000
360	680	

Proceeding, we are led to the dual problem:

$$\begin{aligned} \text{Maximize} \quad & P = 360u + 680v \\ \text{subject to} \quad & 60u + 160v \leq 44{,}000 \\ & 80u + 120v \leq 54{,}000 \\ & u \geq 0, v \geq 0 \end{aligned}$$

Let x and y be slack variables. We obtain the following tableaus:

	u	v	x	y	P	Constant	Ratio
Pivot row →	60	(160)	1	0	0	44, 000	275
	80	120	0	1	0	54, 000	450
	−360	−680	0	0	1	0	
		↑ Pivot column					

$\xrightarrow{\frac{1}{160}R_1}$

u	v	x	y	P	Constant
$\frac{3}{8}$	(1)	$\frac{1}{160}$	0	0	275
80	120	0	1	0	54, 000
−360	−680	0	0	1	0

$\xrightarrow[R_3+680R_1]{R_2-120R_1}$

	u	v	x	y	P	Constant	Ratio
	$\frac{3}{8}$	1	$\frac{1}{160}$	0	0	275	$\frac{2200}{3}$
Pivot row →	(35)	0	$-\frac{3}{4}$	1	0	21,000	600
	−105	0	$\frac{17}{4}$	0	1	187,000	
	↑ Pivot column						

$\xrightarrow{\frac{1}{35}R_2}$

u	v	x	y	P	Constant
$\frac{3}{8}$	1	$\frac{1}{160}$	0	0	275
(1)	0	$-\frac{3}{140}$	$\frac{1}{35}$	0	600
−105	0	$\frac{17}{4}$	0	1	187,000

$\xrightarrow[R_3+105R_2]{R_1-\frac{3}{8}R_2}$

u	v	x	y	P	Constant
0	1	$\frac{1}{70}$	$-\frac{3}{280}$	0	50
1	0	$-\frac{3}{140}$	$\frac{1}{35}$	0	600
0	0	2	3	1	250,000

The last tableau is in final form. The fundamental theorem of duality tells us that the solution to the primal problem is $x = 2$, $y = 3$ with a minimum value for C of 250,000. Thus, Deluxe River Cruises should use two type A vessels and three type B vessels. The minimum operating cost is $250,000.

23. Let x and y denote the numbers of advertisements to be placed in newspapers I and II, respectively. Then the linear programming problem is:

$$\begin{aligned} \text{Minimize} \quad & C = 1000x + 800y \\ \text{subject to} \quad & 70{,}000x + 10{,}000y \ge 2{,}000{,}000 \\ & 40{,}000x + 20{,}000y \ge 1{,}400{,}000 \\ & 20{,}000x + 40{,}000y \ge 1{,}000{,}000 \\ & x \ge 0,\ y \ge 0 \end{aligned}$$

Or, upon simplification:

$$\begin{aligned} \text{Minimize} \quad & C = 1000x + 800y \\ \text{subject to} \quad & 7x + y \ge 200 \\ & 2 + y \ge 70 \\ & x + 2y \ge 50 \\ & x \ge 0,\ y \ge 0 \end{aligned}$$

We write down a tableau for the primal problem, then interchange columns and rows to obtain a duplex tableau.

x	y	Constant
7	1	200
2	1	70
1	2	50
1000	800	

u	v	w	Constant
7	2	1	1000
1	1	2	800
200	70	50	

Thus, the dual problem is:

$$\begin{aligned} \text{Maximize} \quad & P = 200u + 70v + 50w \\ \text{subject to} \quad & 7u + 2v + w \le 1000 \\ & u + v + 2w \le 800 \\ & u \ge 0,\ v \ge 0,\ w \ge 0 \end{aligned}$$

Let x and y denote the slack variables. We obtain the following sequence of tableaus:

	u	v	w	x	y	P	Constant	Ratio
Pivot row →	⑦	2	1	1	0	0	1000	$\frac{1000}{7}$
	1	1	2	0	1	0	800	800
	−200	−70	−50	0	0	1	0	
	↑ Pivot column							

$\xrightarrow{\frac{1}{7}R_1}$

u	v	w	x	y	P	Constant
①	$\frac{2}{7}$	$\frac{1}{7}$	$\frac{1}{7}$	0	0	$\frac{1000}{7}$
1	1	2	0	1	0	800
−200	−70	−50	0	0	1	0

$\xrightarrow[R_3 + 200R_1]{R_2 - R_1}$

	u	v	w	x	y	P	Constant	Ratio
	1	$\frac{2}{7}$	$\frac{1}{7}$	$\frac{1}{7}$	0	0	$\frac{1000}{7}$	1000
Pivot row →	0	$\frac{5}{7}$	($\frac{13}{7}$)	$-\frac{1}{7}$	1	0	$\frac{4600}{7}$	$\frac{4600}{13}$
	0	$-\frac{90}{7}$	$-\frac{150}{7}$	$\frac{200}{7}$	0	1	$\frac{200{,}000}{7}$	
			↑ Pivot column					

$\xrightarrow{\frac{7}{13}R_2}$

u	v	w	x	y	P	Constant
1	$\frac{2}{7}$	$\frac{1}{7}$	$\frac{1}{7}$	0	0	$\frac{1000}{7}$
0	$\frac{5}{13}$	①	$-\frac{1}{13}$	$\frac{7}{13}$	0	$\frac{4600}{13}$
0	$-\frac{90}{7}$	$-\frac{150}{7}$	$\frac{200}{7}$	0	1	$\frac{200{,}000}{7}$

$\xrightarrow[R_3 + \frac{150}{7}R_2]{R_1 - \frac{1}{7}R_2}$

	u	v	w	x	y	P	Constant	Ratio
Pivot row →	1	$\textcircled{\frac{3}{13}}$	0	$\frac{2}{13}$	$-\frac{1}{13}$	0	$\frac{1200}{13}$	400
	0	$\frac{5}{13}$	1	$-\frac{1}{13}$	$\frac{7}{13}$	0	$\frac{4600}{13}$	920
	0	$-\frac{60}{13}$	0	$\frac{350}{13}$	$\frac{150}{13}$	1	$\frac{470{,}000}{13}$	
		↑ Pivot column						

$\xrightarrow{\frac{13}{3}R_1}$

u	v	w	x	y	P	Constant
$\frac{13}{3}$	①	0	$\frac{2}{3}$	$-\frac{1}{3}$	0	400
0	$\frac{5}{13}$	1	$-\frac{1}{13}$	$\frac{7}{13}$	0	$\frac{4600}{13}$
0	$-\frac{60}{13}$	0	$\frac{350}{13}$	$\frac{150}{13}$	1	$\frac{470{,}000}{13}$

$\xrightarrow[R_3+\frac{60}{13}R_1]{R_2-\frac{5}{13}R_1}$

u	v	w	x	y	P	Constant
$\frac{13}{3}$	1	0	$\frac{2}{3}$	$-\frac{1}{3}$	0	400
$-\frac{5}{2}$	0	1	$-\frac{1}{3}$	$\frac{2}{3}$	0	200
20	0	0	30	10	1	38,000

The last tableau is in final form. The fundamental theorem of duality tells us that the solution to the primal problem is $x = 30$, $y = 10$, and $C = 38{,}000$. Therefore, Everest Deluxe World Travel should place 30 advertisements in newspaper I and 10 in newspaper II, for a minimum cost of \$38,000.

25. The given data may be summarized as follows:

	Orange Juice	**Grapefruit Juice**
Vitamin A	60 I.U.	120 I.U.
Vitamin C	16 I.U.	12 I.U.
Calories	14	11

Suppose x ounces of orange juice and y ounces of pink grapefruit juice are required for each glass of the blend. Then the linear programming problem is:

$$\begin{aligned} \text{Minimize} \quad & C = 14x + 11y \\ \text{subject to} \quad & 60x + 120y \geq 1200 \\ & 16x + 12y \geq 200 \\ & x \geq 0, y \geq 0 \end{aligned}$$

We write down a tableau for the primal problem, then interchange columns and rows to obtain a duplex tableau.

x	y	Constant
60	120	1200
16	12	200
14	11	

u	v	Constant
60	16	14
120	12	11
1200	200	

Thus, the dual problem is:

$$\begin{aligned} \text{Maximize} \quad & P = 1200u + 200v \\ \text{subject to} \quad & 60u + 16v \leq 14 \\ & 120u + 12v \leq 11 \\ & u \geq 0, v \geq 0 \end{aligned}$$

Using the slack variables x and y, we obtain the following sequence of tableaus:

	u	v	x	y	P	Constant	Ratio
	60	16	1	0	0	14	$\frac{7}{30}$
Pivot row →	(120)	12	0	1	0	11	$\frac{11}{120}$
	−1200	−200	0	0	1	0	
	↑ Pivot column						

$\xrightarrow{\frac{1}{120}R_2}$

u	v	x	y	P	Constant
60	16	1	0	0	14
(1)	$\frac{1}{10}$	0	$\frac{1}{120}$	0	$\frac{11}{120}$
−1200	−200	0	0	1	0

$\xrightarrow[R_3 + 1200R_2]{R_1 - 60R_2}$

	u	v	x	y	P	Constant	Ratio
Pivot row →	0	(10)	1	$-\frac{1}{2}$	0	$\frac{17}{2}$	$\frac{17}{20}$
	1	$\frac{1}{10}$	0	$\frac{1}{120}$	0	$\frac{11}{120}$	$\frac{11}{12}$
	0	−80	0	10	1	110	
		↑ Pivot column					

$\xrightarrow{\frac{1}{10}R_1}$

u	v	x	y	P	Constant
0	(1)	$\frac{1}{10}$	$-\frac{1}{20}$	0	$\frac{17}{20}$
1	$\frac{1}{10}$	0	$\frac{1}{120}$	0	$\frac{11}{120}$
0	−80	0	10	1	110

$\xrightarrow[R_3 + 80R_1]{R_2 - \frac{1}{10}R_1}$

u	v	x	y	P	Constant
0	1	$\frac{1}{10}$	$-\frac{1}{20}$	0	$\frac{17}{20}$
1	0	$-\frac{1}{100}$	$\frac{1}{75}$	0	$\frac{1}{150}$
0	0	8	6	1	178

We conclude that the owner should use 8 ounces of orange juice and 6 ounces of pink grapefruit juice per glass of the blend for a minimal calorie count of 178.

27. True. To maximize P, one maximizes $-C$. Because the minimization problem has a unique solution, the negative of that solution is the solution of the maximization problem.

Technology Exercises page 400

1. $x = \frac{4}{3}$, $y = \frac{10}{3}$, $z = 0$, and $C = \frac{14}{3}$

3. $x = 0.9524$, $y = 4.2857$, $z = 0$, and $C = 6.09524$

CHAPTER 6 Concept Review Questions page 400

1. a. half-plane, line

b. $ax + by \leq c$; $ax + by = c$

3. objective function, maximized, minimized, linear, inequalities

5. maximized, nonnegative, less than, equal to

7. minimized, nonnegative, greater than, equal to

CHAPTER 6 Review Exercises page 401

1. We evaluate Z at each of the corner points of the feasible set S.

Vertex	$Z = 2x + 3y$
$(0, 0)$	$\boxed{0}$
$(5, 0)$	10
$(3, 4)$	$\boxed{18}$
$(0, 6)$	$\boxed{18}$

From the table, we conclude that Z attains a minimum value of 0 when $x = 0$ and $y = 0$, and a maximum value of 18 when x and y lie on the line segment joining $(3, 4)$ and $(0, 6)$.

3.

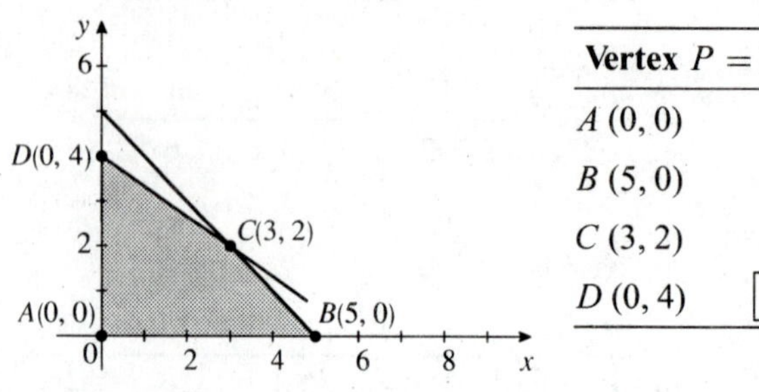

Vertex	$P = 3x + 5y$
$A\ (0, 0)$	0
$B\ (5, 0)$	15
$C\ (3, 2)$	19
$D\ (0, 4)$	$\boxed{20}$

From the graph and the table, we conclude that the maximum value of P is 20 when $x = 0$ and $y = 4$.

5.

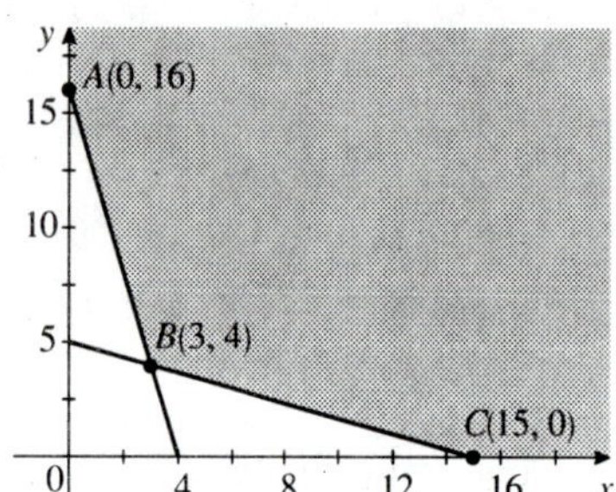

Vertex	$C = 2x + 5y$
$A\ (0, 16)$	80
$B(3, 4)$	$\boxed{26}$
$C\ (15, 0)$	30

From the graph and the table, we conclude that the minimum value of C is 26 when $x = 3$ and $y = 4$.

7.

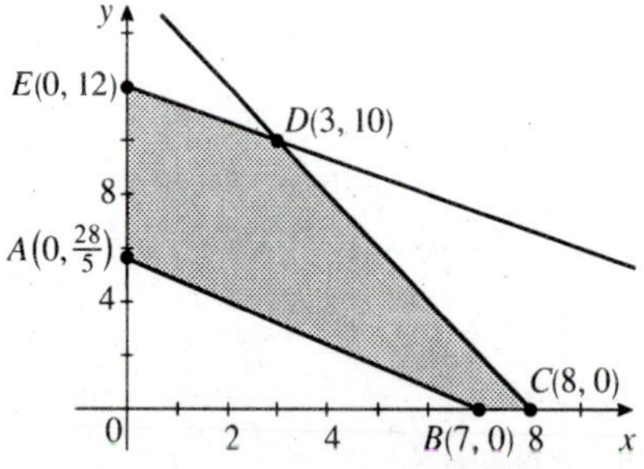

Vertex	$P = 3x + 2y$
$A\left(0, \frac{28}{5}\right)$	$\frac{56}{5}$
$B\ (7, 0)$	21
$C\ (8, 0)$	24
$D\ (3, 10)$	$\boxed{29}$
$E\ (0, 12)$	24

From the graph and the table, we conclude that P attains a maximum value of 29 when $x = 3$ and $y = 10$.

9.

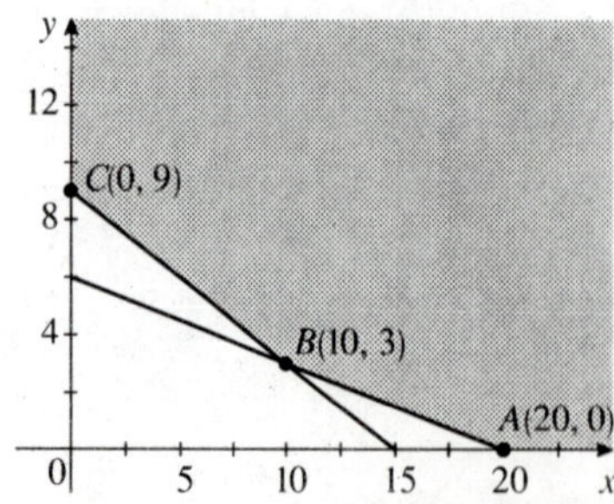

Vertex	$C = 2x + 7y$
$A\ (20, 0)$	$\boxed{40}$
$B\ (10, 3)$	41
$C\ (0, 9)$	63

From the graph and the table, we conclude that C attains a minimum value of 40 when $x = 20$ and $y = 0$.

11.

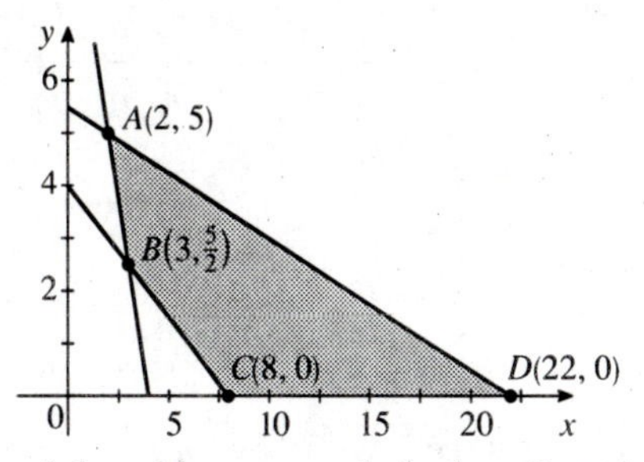

Vertex	$Q = x + y$
$A\ (2, 5)$	7
$B\left(3, \frac{5}{2}\right)$	$\boxed{\frac{11}{2}}$
$C\ (8, 0)$	8
$D\ (22, 0)$	$\boxed{22}$

From the graph and the table, we conclude that Q attains a maximum value of 22 when $x = 22$ and $y = 0$ and a minimum value of $\frac{11}{2}$ when $x = 3$ and $y = \frac{5}{2}$.

13. This is a regular linear programming problem. Using the simplex method with u and v as slack variables, we obtain the following sequence of tableaus:

	x	y	u	v	P	Constant	Ratio
Pivot row →	1	(3)	1	0	0	15	5
	4	1	0	1	0	16	16
	−3	−4	0	0	1	0	
		↑ Pivot column					

$\xrightarrow{\frac{1}{3}R_1}$

x	y	u	v	P	Constant
$\frac{1}{3}$	(1)	$\frac{1}{3}$	0	0	5
4	1	0	1	0	16
−3	−4	0	0	1	0

$\xrightarrow[R_3+4R_1]{R_2-R_1}$

	x	y	u	v	P	Constant	Ratio
	$\frac{1}{3}$	1	$\frac{1}{3}$	0	0	5	15
Pivot row →	$\left(\frac{11}{3}\right)$	0	$-\frac{1}{3}$	1	0	11	3
	$-\frac{5}{3}$	0	$\frac{4}{3}$	0	1	20	
	↑ Pivot column						

$\xrightarrow{\frac{3}{11}R_2}$

x	y	u	v	P	Constant
$\frac{1}{3}$	1	$\frac{1}{3}$	0	0	5
(1)	0	$-\frac{1}{11}$	$\frac{3}{11}$	0	3
$-\frac{5}{3}$	0	$\frac{4}{3}$	0	1	20

$\xrightarrow[R_3+\frac{5}{3}R_2]{R_1-\frac{1}{3}R_2}$

x	y	u	v	P	Constant
0	1	$\frac{4}{11}$	$-\frac{1}{11}$	0	4
1	0	$-\frac{1}{11}$	$\frac{3}{11}$	0	3
0	0	$\frac{13}{11}$	$\frac{5}{11}$	1	25

We conclude that $x = 3$, $y = 4$, $u = 0$, $v = 0$, and $P = 25$.

15. Using the simplex method to solve this regular linear programming problem, we calculate the following tableaus.

	x	y	z	u	v	P	Constant	Ratio
Pivot row →	1	2	(3)	1	0	0	12	4
	1	−3	2	0	1	0	10	5
	−2	−3	−5	0	0	1	0	
			↑ Pivot column					

$\xrightarrow{\frac{1}{3}R_1}$

x	y	z	u	v	P	Constant
$\frac{1}{3}$	$\frac{2}{3}$	(1)	$\frac{1}{3}$	0	0	4
1	−3	2	0	1	0	10
−2	−3	−5	0	0	1	0

$\xrightarrow[R_3+5R_1]{R_2-2R_1}$

	x	y	z	u	v	P	Constant	Ratio
	$\frac{1}{3}$	$\frac{2}{3}$	1	$\frac{1}{3}$	0	0	4	12
Pivot row →	($\frac{1}{3}$)	$-\frac{13}{3}$	0	$-\frac{2}{3}$	1	0	2	6
	$-\frac{1}{3}$	$\frac{1}{3}$	0	$\frac{5}{3}$	0	1	20	
	↑ Pivot column							

$\xrightarrow{3R_2}$

x	y	z	u	v	P	Constant
$\frac{1}{3}$	$\frac{2}{3}$	1	$\frac{1}{3}$	0	0	4
(1)	-13	0	-2	3	0	6
$-\frac{1}{3}$	$\frac{1}{3}$	0	$\frac{5}{3}$	0	1	20

$\xrightarrow[R_3+\frac{1}{3}R_2]{R_1-\frac{1}{3}R_2}$

	x	y	z	u	v	P	Constant	Ratio
Pivot row →	0	(5)	1	1	-1	0	2	$\frac{5}{2}$
	0	-13	0	-2	3	0	6	–
	0	-4	0	1	1	1	22	
		↑ Pivot column						

$\xrightarrow{\frac{1}{5}R_1}$

x	y	z	u	v	P	Constant
0	1	$\frac{1}{5}$	$\frac{1}{5}$	$-\frac{1}{5}$	0	$\frac{2}{5}$
1	-13	0	-2	3	0	6
0	-4	0	1	1	1	22

$\xrightarrow[R_3+4R_1]{R_2+13R_1}$

x	y	z	u	v	P	Constant
0	1	$\frac{1}{5}$	$\frac{1}{5}$	$-\frac{1}{5}$	0	$\frac{2}{5}$
1	0	$\frac{13}{5}$	$\frac{3}{5}$	$\frac{2}{5}$	0	$\frac{56}{5}$
0	0	$\frac{4}{5}$	$\frac{9}{5}$	$\frac{1}{5}$	1	$\frac{118}{5}$

We conclude that the P attains a maximum value of $\frac{118}{5}$ when $x = \frac{56}{5}$, $y = \frac{2}{5}$, $z = 0$, $u = 0$, and $v = 0$.

17. We write a tableau for the primal problem, then interchange rows and columns to obtain a tableau from which we construct the dual problem:

x	y	Constant
2	3	6
2	1	4
3	2	

u	v	Constant
2	2	3
3	1	2
6	4	

Maximize $P = 6u + 4v$
subject to $2u + 2v \le 3$
$3u + v \le 2$
$u \ge 0, v \ge 0$

Using the simplex method with x and y as slack variables, we obtain the following tableaus:

	u	v	x	y	P	Constant	Ratio
	2	2	1	0	0	3	$\frac{3}{2}$
Pivot row →	(3)	1	0	1	0	2	$\frac{2}{3}$
	-6	-4	0	0	1	0	
	↑ Pivot column						

$\xrightarrow{\frac{1}{3}R_2}$

u	v	x	y	P	Constant
2	2	1	0	0	3
(1)	$\frac{1}{3}$	0	$\frac{1}{3}$	0	$\frac{2}{3}$
-6	-4	0	0	1	0

$\xrightarrow[R_3+6R_2]{R_1-2R_2}$

	u	v	x	y	P	Constant	Ratio
Pivot row →	0	($\frac{4}{3}$)	1	$-\frac{2}{3}$	0	$\frac{5}{3}$	$\frac{5}{4}$
	1	$\frac{1}{3}$	0	$\frac{1}{3}$	0	$\frac{2}{3}$	2
	0	-2	0	2	1	4	
		↑ Pivot column					

$\xrightarrow{\frac{3}{4}R_1}$

u	v	x	y	P	Constant
0	(1)	$\frac{3}{4}$	$-\frac{1}{2}$	0	$\frac{5}{4}$
1	$\frac{1}{3}$	0	$\frac{1}{3}$	0	$\frac{2}{3}$
0	-2	0	2	1	4

$\xrightarrow[R_3+2R_1]{R_2-\frac{1}{3}R_1}$

u	v	x	y	P	Constant
0	1	$\frac{3}{4}$	$-\frac{1}{2}$	0	$\frac{5}{4}$
1	0	$-\frac{1}{4}$	$\frac{1}{2}$	0	$\frac{1}{4}$
0	0	$\frac{3}{2}$	1	1	$\frac{13}{2}$

Therefore, C attains a minimum value of $\frac{13}{2}$ when $x = \frac{3}{2}$, $y = 1$, $u = 0$, and $v = 0$.

19. We write a tableau for the primal problem, then interchange rows and columns to obtain a tableau from which we construct the dual problem:

x	y	z	Constant
3	2	1	4
1	1	3	6
24	18	24	

u	v	Constant
3	1	24
2	1	18
1	3	24
4	6	

Maximize $P = 4u + 6v$
subject to
$$3u + v \le 24$$
$$2u + v \le 18$$
$$u + 3v \le 24$$
$$u \ge 0, v \ge 0$$

Using the simplex method with x and y as slack variables, we obtain the following tableaus:

	u	v	x	y	z	P	Constant	Ratio
	3	1	1	0	0	0	24	36
	2	1	0	1	0	0	18	18
Pivot row →	1	(3)	0	0	1	0	24	8
	−4	−6	0	0	0	1	0	
		↑ Pivot column						

$\xrightarrow{\frac{1}{3}R_3}$

u	v	x	y	z	P	Constant
3	1	1	0	0	0	24
2	1	0	1	0	0	18
$\frac{1}{3}$	(1)	0	0	$\frac{1}{3}$	0	8
−4	−6	0	0	0	1	0

$\xrightarrow[R_4+6R_3]{R_1-R_3,\ R_2-R_3}$

	u	v	x	y	z	P	Constant	Ratio
Pivot row →	($\frac{8}{3}$)	0	1	0	$-\frac{1}{3}$	0	16	6
	$\frac{5}{3}$	0	0	1	$-\frac{1}{3}$	0	10	6
	$\frac{1}{3}$	1	0	0	$\frac{1}{3}$	0	8	24
	−2	0	0	0	2	1	48	
	↑ Pivot column							

$\xrightarrow{\frac{3}{8}R_1}$

u	v	x	y	z	P	Constant
(1)	0	$\frac{3}{8}$	0	$-\frac{1}{8}$	0	6
$\frac{5}{3}$	0	0	1	$-\frac{1}{3}$	0	10
$\frac{1}{3}$	1	0	0	$\frac{1}{3}$	0	8
−2	0	0	0	2	1	48

$\xrightarrow[R_4+2R_1]{R_2-\frac{5}{3}R_1,\ R_3-\frac{1}{3}R_1}$

u	v	x	y	z	P	Constant
1	0	$\frac{3}{8}$	0	$-\frac{1}{8}$	0	6
0	0	$-\frac{5}{8}$	1	$-\frac{1}{8}$	0	0
0	1	$-\frac{1}{8}$	0	$\frac{3}{8}$	0	6
0	0	$\frac{3}{4}$	0	$\frac{7}{4}$	1	60

We conclude that C attains a minimum value of 60 when $x = \frac{3}{4}$, $y = 0$, $z = \frac{7}{4}$, $u = 0$, and $v = 0$.

21. Suppose the investor puts x and y thousand dollars into the stocks of companies A and B, respectively. Then we have the following linear programming problem:

$$\begin{aligned} \text{Maximize} \quad & P = 0.14x + 0.20y \\ \text{subject to} \quad & x + y \le 80 \\ & 0.01x + 0.04y \le 2 \\ & x \ge 0, y \ge 0 \end{aligned}$$

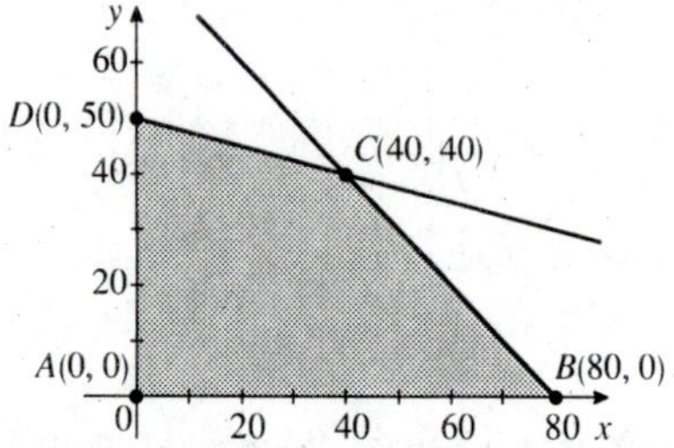

Vertex	$P = 0.14x + 0.20y$
$A\ (0, 0)$	0
$B\ (80, 0)$	11.2
$C\ (40, 40)$	13.6
$D\ (0, 50)$	10

From the graph and the table, we conclude that P attains a maximum value of 13.6 when $x = 40$ and $y = 40$. Thus, by investing \$40,000 in the stocks of each company, the investor will achieve a maximum return of \$13,600.

23. Let x and y denote the numbers of model A and model B grates to be produced. Then the constraint on the amount of cast iron available leads to $3x + 4y \le 1000$ and the constraint on the number of minutes of labor used each day leads to $6x + 3y \le 1200$. One additional constraint specifies that $y \ge 180$. The daily profit is $P = 2x + 1.5y$. Therefore, we have the following linear programming problem:

$$\begin{aligned} \text{Maximize} \quad & P = 2x + 1.5y \\ \text{subject to} \quad & 3x + 4y \le 1000 \\ & 6x + 3y \le 1200 \\ & x \ge 0, y \ge 180 \end{aligned}$$

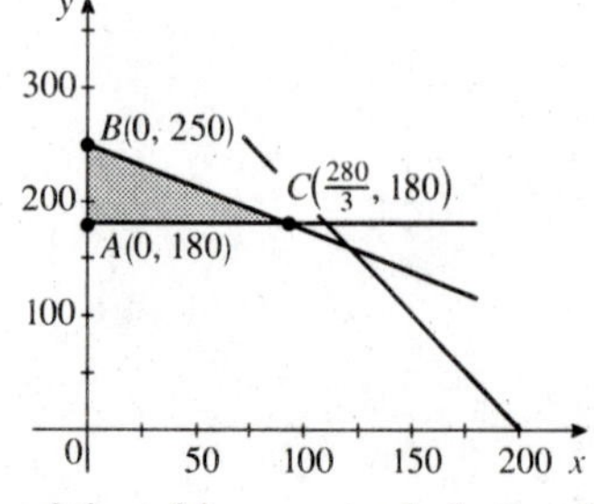

Vertex	$P = 2x + 1.5y$
$A\ (0, 180)$	270
$B\ (0, 250)$	375
$C\ \left(\frac{280}{3}, 180\right)$	$\frac{1370}{3}$

From the graph and the table, we conclude that the optimal profit of \$456 is realized when 93 model A grates and 180 model B grates are produced.

25. Let the amounts invested in blue chip, growth, and speculative stocks be x, y, and z, respectively. Then the linear programming problem is:

$$\begin{aligned} \text{Maximize} \quad & P = 0.10x + 0.15y + 0.20z \\ \text{subject to} \quad & x + y + z \le 100{,}000 \\ & -0.3x + 0.7y + 0.7z \le 0 \\ & -0.5x + 0.5y - 0.5z \le 0 \\ & x \ge 0, y \ge 0, z \ge 0 \end{aligned}$$

Using the simplex method, we obtain the following tableaus:

	x	y	z	u	v	w	P	Constant	Ratio
	1	1	1	1	0	0	0	100,000	100,000
Pivot row →	$-\frac{3}{10}$	$\frac{7}{10}$	($\frac{7}{10}$)	0	1	0	0	0	0
	$-\frac{1}{2}$	$\frac{1}{2}$	$-\frac{1}{2}$	0	0	1	0	0	–
	$-\frac{1}{10}$	$-\frac{3}{20}$	$-\frac{1}{5}$	0	0	0	1	0	
			↑ Pivot column						

$\xrightarrow{\frac{10}{7}R_2}$

x	y	z	u	v	w	P	Constant
1	1	1	1	0	0	0	100,000
$-\frac{3}{7}$	1	(1)	0	$\frac{10}{7}$	0	0	0
$-\frac{1}{2}$	$\frac{1}{2}$	$-\frac{1}{2}$	0	0	1	0	0
$-\frac{1}{10}$	$-\frac{3}{20}$	$-\frac{1}{5}$	0	0	0	1	0

$\xrightarrow[R_4 + \frac{1}{5}R_2]{R_1 - R_2,\ R_3 + \frac{1}{2}R_2}$

	x	y	z	u	v	w	P	Constant	Ratio
Pivot row →	($\frac{10}{7}$)	0	0	1	$-\frac{10}{7}$	0	0	100,000	70,000
	$-\frac{3}{7}$	1	1	0	$\frac{10}{7}$	0	0	0	–
	$-\frac{5}{7}$	1	0	0	$\frac{5}{7}$	1	0	0	–
	$-\frac{13}{70}$	$\frac{1}{20}$	0	0	$\frac{2}{7}$	0	1	0	
	↑ Pivot column								

$\xrightarrow{\frac{7}{10}R_1}$

x	y	z	u	v	w	P	Constant
(1)	0	0	$\frac{7}{10}$	-1	0	0	70,000
$-\frac{3}{7}$	1	1	0	$\frac{10}{7}$	0	0	0
$-\frac{5}{7}$	1	0	0	$\frac{5}{7}$	1	0	0
$-\frac{13}{70}$	$\frac{1}{20}$	0	0	$\frac{2}{7}$	0	1	0

$\xrightarrow[R_4 + \frac{13}{70}R_1]{R_2 + \frac{3}{7}R_1,\ R_3 + \frac{5}{7}R_1}$

x	y	z	u	v	w	P	Constant
1	0	0	$\frac{7}{10}$	-1	0	0	70,000
0	1	1	$\frac{3}{10}$	1	0	0	30,000
0	1	0	$\frac{1}{2}$	0	1	0	50,000
0	$\frac{1}{20}$	0	$\frac{13}{100}$	$\frac{1}{10}$	0	1	13,000

We conclude that Jorge should invest \$70,000 in blue-chip stocks, nothing in growth stocks, and \$30,000 in speculative stocks to realize a maximum return of \$13,000 on his investments.

CHAPTER 6 Before Moving On... page 403

1.

Vertex	$Z = 3x - y$
(8, 2)	22
(28, 8)	$\boxed{76}$
(16, 24)	24
(3, 16)	$\boxed{-7}$

From the table, we see that the maximum value is $Z = 76$ at (28, 8) and the minimum value is $Z = -7$ at (3, 16).

2. Maximize $P = x + 3y$ subject to $2x + 3y \leq 11$, $3x + 7y \leq 24$, $x \geq 0$, $y \geq 0$. To find the coordinates of C, we solve $2x + 3y = 11$ and $3x + 7y = 24$ simultaneously, obtaining $6x + 9y = 33$, $6x + 14y = 48$, $5y = 15$, and $y = 3$, so $x = 1$.

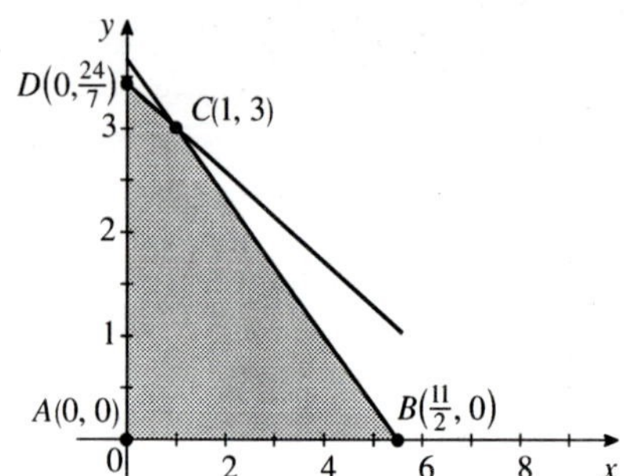

Vertex	$P = x + 3y$
$A\,(0, 0)$	0
$B\left(\frac{11}{2}, 0\right)$	$\frac{11}{2}$
$C\,(1, 3)$	10
$D\left(0, \frac{24}{7}\right)$	$\boxed{\frac{72}{7}}$

From the graph and the table, we see that the maximum value of P is $\frac{72}{7}$, attained at $x = 0$, $y = \frac{24}{7}$.

3. We introduce slack variables u, v, and w.

	x	y	z	u	v	w	P	Constant	Ratio
Pivot row →	2	⓵	−1	1	0	0	0	3	3
	1	−2	3	0	1	0	0	1	–
	3	2	4	0	0	1	0	17	$\frac{17}{2}$
	−1	−2	3	0	0	0	1	0	
		↑ Pivot column							

The pivot element is 1, as shown.

4. $x = 2$, $y = 0$, $z = 11$, $u = 2$, $v = w = 0$, and $P = 28$.

5. We introduce slack variables u and v.

	x	y	u	v	P	Constant	Ratio
	4	3	1	0	0	30	$\frac{15}{2}$
Pivot row →	(2)	−3	0	1	0	6	3
	−5	−2	0	0	1	0	
	↑ Pivot column						

$\xrightarrow{\frac{1}{2}R_2}$

x	y	u	v	P	Constant
4	3	1	0	0	30
(1)	$-\frac{3}{2}$	0	$\frac{1}{2}$	0	3
−5	−2	0	0	1	0

$\xrightarrow[R_3+5R_2]{R_1-4R_2}$

	x	y	u	v	P	Constant	Ratio
Pivot row →	0	(9)	1	−2	0	18	2
	1	$-\frac{3}{2}$	0	$\frac{1}{2}$	0	3	–
	0	$-\frac{19}{2}$	0	$\frac{5}{2}$	1	15	
		↑ Pivot column					

$\xrightarrow{\frac{1}{9}R_1}$

x	y	u	v	P	Constant
0	(1)	$\frac{1}{9}$	$-\frac{2}{9}$	0	2
1	$-\frac{3}{2}$	0	$\frac{1}{2}$	0	3
0	$-\frac{19}{2}$	0	$\frac{5}{2}$	1	15

$\xrightarrow[R_3+\frac{19}{2}R_1]{R_2+\frac{3}{2}R_1}$

x	y	u	v	P	Constant
0	1	$\frac{1}{9}$	$-\frac{2}{9}$	0	2
1	0	$\frac{1}{6}$	$\frac{1}{6}$	0	6
0	0	$\frac{19}{8}$	$\frac{7}{18}$	1	34

The optimal solution is $x = 6$, $y = 2$, $u = v = 0$, and $P = 34$.

7 SETS AND PROBABILITY

7.1 Sets and Set Operations

Problem-Solving Tips

It's often easier to remember a formula if you learn to express the formula in words. For example, DeMorgan's Laws can be expressed as follows:

$(A \cup B)^c = A^c \cap B^c$ says that the complement of the union of two sets is equal to the intersection of their complements.

$(A \cap B)^c = A^c \cup B^c$ says that the complement of the intersection of two sets is equal to the union of their complements.

Concept Questions page 412

1. a. A set is a well-defined collection of objects. As an example, consider the set of all freshmen students in a college.

b. Two sets A and B are equal if they have exactly the same elements.

c. The empty set is the set that contains no element.

3. a. If $A \subset B$, then $B^c \subset A^c$.

b. If $A^c = \varnothing$, then $A = U$, the universal set.

Exercises page 412

1. $\{x \mid x$ is a gold medalist in the 2010 Winter Olympic Games$\}$

3. $\{x \mid x$ is an integer greater than 2 and less than 8$\}$

5. $\{2, 3, 4, 5, 6\}$

7. $\{-2\}$

9. a. True. The order in which the elements are listed is not important.

b. False. No set contains itself.

11. a. False. The empty set has no element.

b. False. 0 is not a set.

13. True.

15. a. True. 2 belongs to A.

b. False. For example, 5 belongs to A but $5 \notin \{2, 4, 6\}$.

17. a and b.

19. a. $\varnothing, \{1\}, \{2\}, \{1, 2\}$

b. $\varnothing, \{1\}, \{2\}, \{3\}, \{1, 2\}, \{1, 3\}, \{2, 3\}, \{1, 2, 3\}$

c. $\varnothing, \{1\}, \{2\}, \{3\}, \{4\}, \{1, 2\}, \{1, 3\}, \{1, 4\}, \{2, 3\}, \{2, 4\}, \{3, 4\}, \{1, 2, 3\}, \{1, 2, 4\}, \{1, 3, 4\}, \{2, 3, 4\}, \{1, 2, 3, 4\}$

21. $\{1, 2, 3, 4, 6, 8, 10\}$.

23. {Jill, John, Jack, Susan, Sharon}.

25. a.

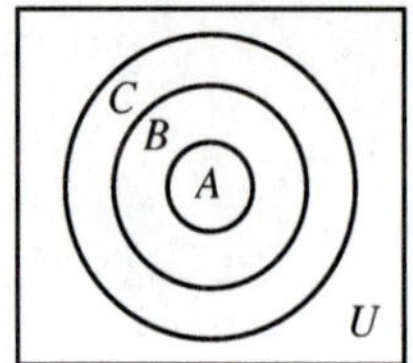

b.

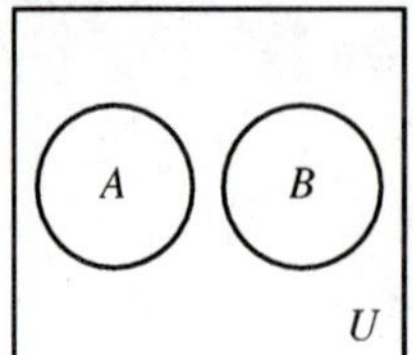

c.

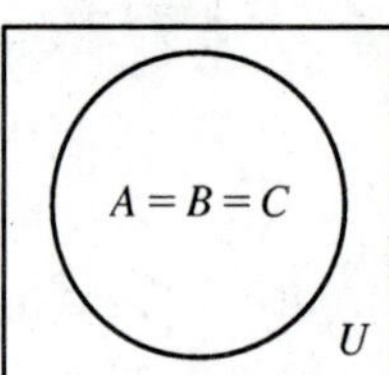

27. a. $A \cap B^c$

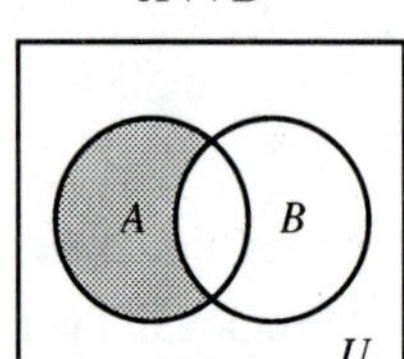

b. $A^c \cap B$

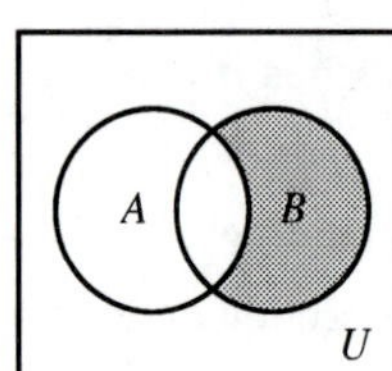

29. a. $A \cup B \cup C$

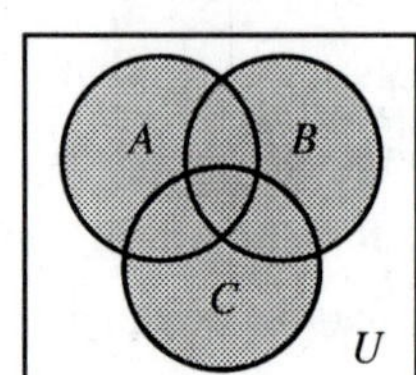

b. $A \cap B \cap C$

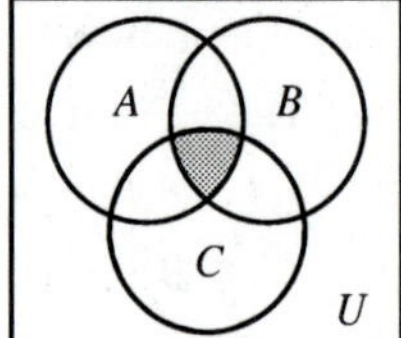

31. a. $A^c \cap B^c \cap C^c$

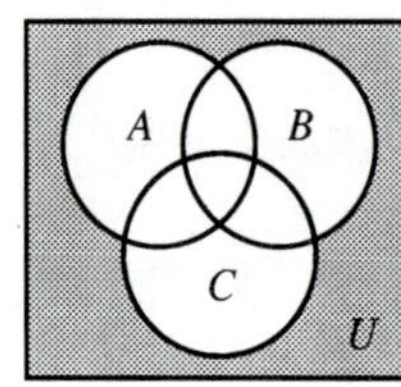

b. $(A \cup B)^c \cap C$

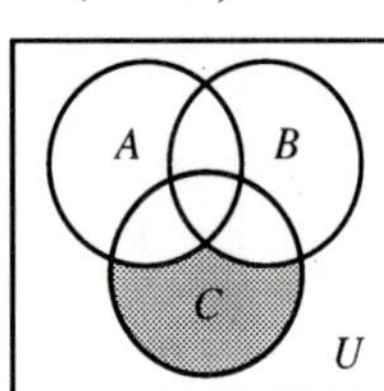

33. a. $A^c = \{2, 4, 6, 8, 10\}$

b. $B \cup C^c = \{2, 4, 6, 8, 10\} \cup \{3, 6, 7, 10\} = \{2, 3, 4, 6, 7, 8, 10\}$

c. $C \cup C^c = U = \{1, 2, 3, 4, 5, 6, 7, 8, 9, 10\}$

35. a. $(A \cap B) \cup C = C = \{1, 2, 4, 5, 8, 9\}$

b. $(A \cup B \cup C)^c = \{1, 2, 3, 4, 5, 6, 7, 8, 9, 10\}^c = \varnothing$

c. $(A \cap B \cap C)^c = U = \{1, 2, 3, 4, 5, 6, 7, 8, 9, 10\}$

37. a. The sets are not disjoint because 4 is an element of both sets.

b. The sets are disjoint as they have no common element.

39. a. The set of all employees at the Universal Life Insurance Company who do not drink tea.

b. The set of all employees at the Universal Life Insurance Company who do not drink coffee.

41. a. The set of all employees at the Universal Life Insurance Company who drink tea but not coffee.

b. The set of all employees at the Universal Life Insurance Company who drink coffee but not tea.

43. a. The set of all employees at the hospital who are not doctors.

b. The set of all employees at the hospital who are not nurses.

45. a. The set of all employees at the hospital who are female doctors.

b. The set of all employees at the hospital who are both doctors and administrators.

47. a. $D \cap F$ **b.** $R \cap F^c \cap L^c$ **49. a.** B^c **b.** $A \cap B$ **c.** $A \cap B \cap C^c$

51. a. Region 1: $A \cap B \cap C$ is the set of tourists who used all three modes of transportation over a one-week period in London.

b. Regions 1 and 4: $A \cap C$ is the set of tourists who have taken the underground and a bus over a one-week period in London.

c. Regions 4, 5, 7, and 8: B^c is the set of tourists who have not taken a cab over a one-week period in London.

53. 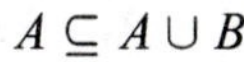 $A \subseteq A \cup B$ $B \subseteq A \cup B$

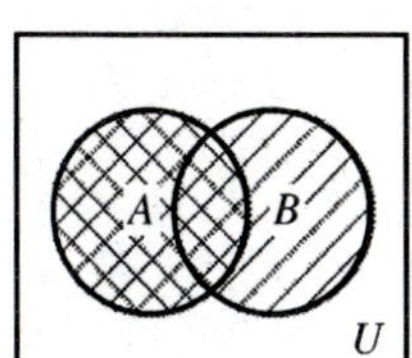

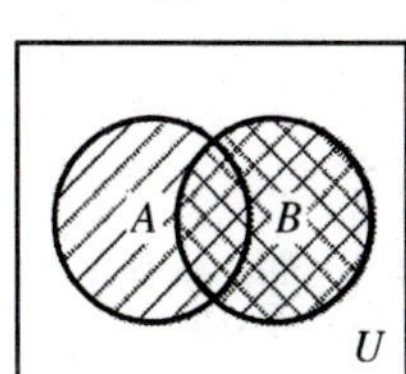

55. $A \cup (B \cup C) = (A \cup B) \cup C$

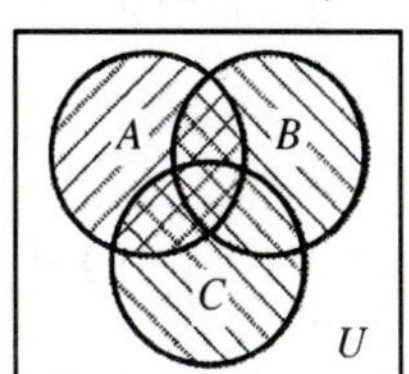

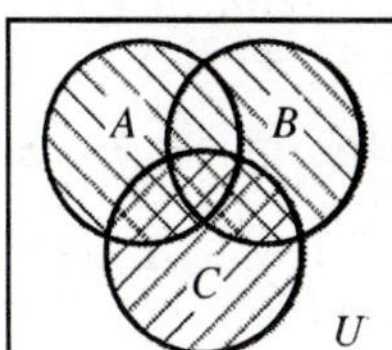

57. $A \cap (B \cup C) = (A \cap B) \cup (A \cap C)$

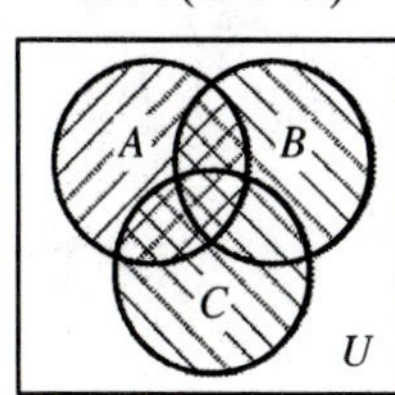

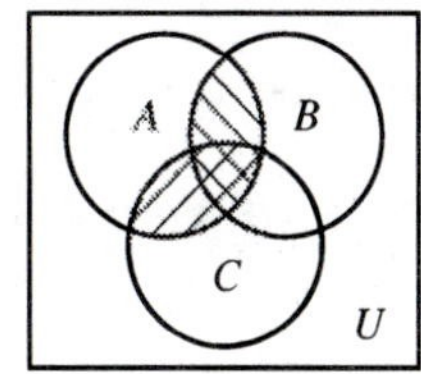

59. a. $A \cup (B \cup C) = \{1, 3, 5, 7, 9\} \cup (\{1, 2, 4, 7, 8\} \cup \{2, 4, 6, 8\}) = \{1, 3, 5, 7, 9\} \cup \{1, 2, 4, 6, 7, 8\}$

$= \{1, 2, 3, 4, 5, 6, 7, 8, 9\}$

$(A \cup B) \cup C = (\{1, 3, 5, 7, 9\} \cup (\{1, 2, 4, 7, 8\}) \cup \{2, 4, 6, 8\}) = \{1, 2, 3, 4, 5, 7, 8, 9\} \cup \{2, 4, 6, 8\}$

$= \{1, 2, 3, 4, 5, 6, 7, 8, 9\}$

b. $A \cap (B \cap C) = \{1, 3, 5, 7, 9\} \cap (\{1, 2, 4, 7, 8\} \cap \{2, 4, 6, 8\}) = \{1, 3, 5, 7, 9\} \cap \{2, 4, 8\} = \varnothing$

$(A \cap B) \cap C = (\{1, 3, 5, 7, 9\} \cap \{1, 2, 4, 7, 8\}) \cap \{2, 4, 6, 8\} = \{1, 7\} \cap \{2, 4, 6, 8\} = \varnothing$

61. a. $r, u, v, w, x\ y$ **b.** v, r **63. a.** t, y, s **b.** t, s, w, x, z

65. $A \subset C$

67. False. Because every element in a set A belongs to A, A is a subset of itself.

69. True. If at least one of the sets A or B is nonempty, then $A \cup B \neq \varnothing$.

71. True. $(A \cup A^c)^c = U^c = \varnothing$.

73. True. Because $A \subseteq B$, all of the elements in A are also in B, so $A \cup B = B$.

75. True. Because A is a proper subset of B, all of the elements in A are also in B and there is at least one element in B that is not in A. Therefore, there is at least one element in A^c that is not in B^c, and so $B^c \subset A^c$.

7.2 The Number of Elements in a Finite Set

Problem-Solving Tips

In the problems that follow, it is often helpful to draw a Venn diagram.

Concept Questions page 418

1. a. If A and B are sets with $A \cap B = \varnothing$, then $n(A) + n(B) = n(A \cup B)$.

b. If $n(A \cup B) \neq n(A) + n(B)$, then $A \cap B \neq \varnothing$.

Exercises page 418

1. $A \cup B = \{a, e, g, h, i, k, l, m, o, u\}$ and so $n(A \cup B) = 10$. Next, $n(A) + n(B) = 5 + 5 = 10$.

3. a. $A = \{2, 4, 6, 8\}$ and so $n(A) = 4$.

b. $B = \{6, 7, 8, 9, 10\}$ and so $n(B) = 5$.

c. $A \cup B = \{2, 4, 6, 7, 8, 9, 10\}$ and so $n(A \cup B) = 7$.

d. $A \cap B = \{6, 8\}$ and so $n(A \cap B) = 2$.

5. Using the results of Exercise 3, we see that $n(A \cup B) = 7$ and $n(A) + n(B) - n(A \cap B) = 4 + 5 - 2 = 7$.

7. Because $n(A \cup B) = n(A) + n(B) - n(A \cap B)$, $n(B) = n(A \cup B) + n(A \cap B) - n(A) = 30 + 4 - 15 = 19$.

9. a. $n(A) = 12 + 3 = 15$

b. $n(A \cup B) = 12 + 3 + 15 = 30$

c. $n(A^c \cap B) = 15$

d. $n(A \cap B^c) = 12$

e. $n(U) = 20 + 12 + 3 + 15 = 50$

f. $n\left[(A \cup B)^c\right] = 20$

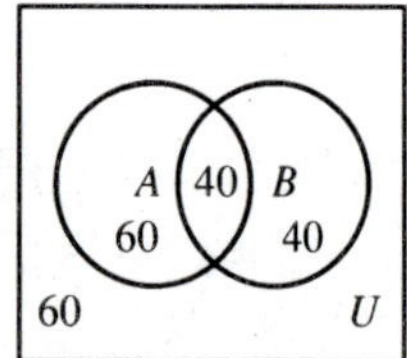

11. Refer to the Venn diagram at left.

a. $n(A \cup B) = 60 + 40 + 40 = 140$.

b. $n(A^c) = 40 + 60 = 100$.

c. $n(A \cap B^c) = 60$.

13. $n(A \cup B) = n(A) + n(B) - n(A \cap B) = 6 + 10 - 4 = 12$.

15. $n(A \cup B) = n(A) + n(B) - n(A \cap B)$, so $n(A \cap B) = n(A) + n(B) - n(A \cup B) = 4 + 5 - 9 = 0$.

17. $n(A \cap B \cap C) = n(A \cup B \cup C) - n(A) - n(B) - n(C) + n(A \cap B) + n(A \cap C) + n(B \cap C)$, so

$$n(C) = n(A \cup B \cup C) - n(A \cap B \cap C) - n(A) - n(B) + n(A \cap B) + n(A \cap C) + n(B \cap C)$$
$$= 24 - 2 - 12 - 12 + 5 + 5 + 4 = 12.$$

19. Let A denote the set of prisoners in the Wilton County Jail who were accused of a felony and B the set of prisoners in that jail who were accused of a misdemeanor. Then we are given that $n(A \cup B) = 190$. Referring to the Venn diagram, the number of prisoners who were accused of both a felony and a misdemeanor is given by $n(A \cap B) = n(A) + n(B) - n(A \cup B) = 130 + 121 - 190 = 61$.

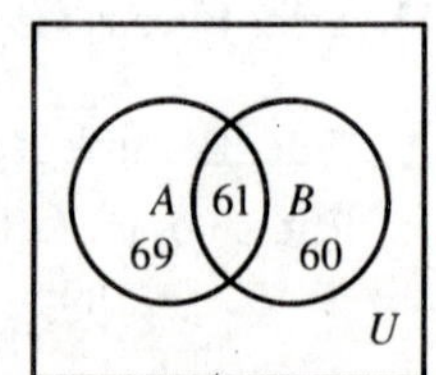

21. Let U denote the set of all customers surveyed, $A = \{x \in U \mid x \text{ buys brand } A\}$, and $B = \{x \in U \mid x \text{ buys brand } B\}$. Then $n(U) = 120$, $n(A) = 80$, $n(B) = 68$, and $n(A \cap B) = 42$.

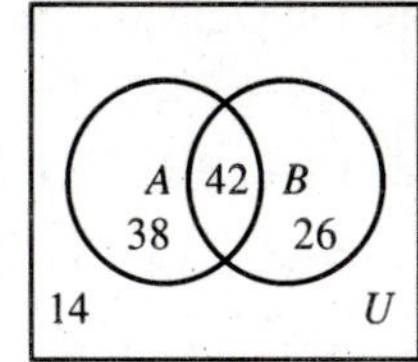

a. The number of customers who buy at least one of these brands is $n(A \cup B) = 80 + 68 - 42 = 106$.

b. The number who buy exactly one of these brands is $n(A \cap B^c) + n(A^c \cap B) = 38 + 26 = 64$.

c. The number who buy only brand A is $n(A \cap B^c) = 38$.

d. The number who buy none of these brands is $n\left[(A \cup B)^c\right] = 120 - 106 = 14$.

23. Let U denote the set of 200 investors, $A = \{x \in U \mid x \text{ uses a discount broker}\}$, and $B = \{x \in U \mid x \text{ uses a full-service broker}\}$.

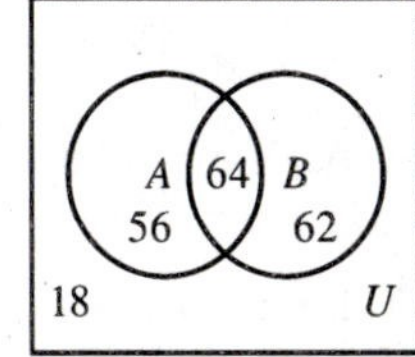

a. The number of investors who use at least one kind of broker is $n(A \cup B) = n(A) + n(B) - n(A \cap B) = 120 + 126 - 64 = 182$.

b. The number of investors who use exactly one kind of broker is $n(A \cap B^c) + n(A^c \cap B) = 56 + 62 = 118$.

c. The number of investors who use only discount brokers is $n(A \cap B^c) = 56$.

d. The number of investors who do not use a broker is $n(A \cup B)^c = n(U) - n(A \cup B) = 200 - 182 = 18$.

25. Let U denote the set of 200 households in the survey, $A = \{x \in U \mid x \text{ owns a desktop computer}\}$, and $B = \{x \in U \mid x \text{ owns a laptop computer}\}$. Referring to the figure, we see that the number of households that own both desktop and laptop computers is $n(A \cap B) = 200 - 120 - 10 - 40 = 30$.

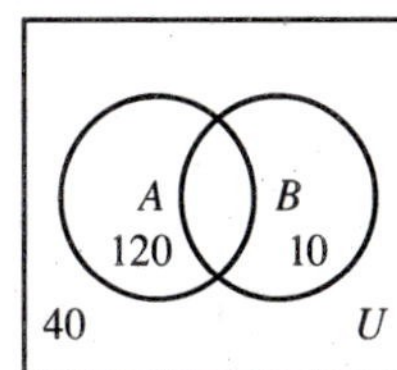

27. **a.** $n(A) = 7 + 2 + 3 + 4 = 16$

b. $n(A \cup B) = 7 + 2 + 3 + 5 + 10 + 4 = 31$

c. $n(A \cap B \cap C^c) = 4$

d. $n\left[(A \cup B) \cap C^c\right] = 7 + 4 + 10 = 21$

e. $n\left[(A \cup B \cup C)^c\right] = 11$

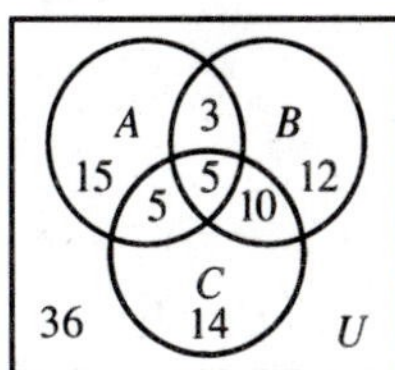

29. **a.** $n(A \cup B \cup C) = 64$

b. $n(A^c \cap B \cap C) = 10$

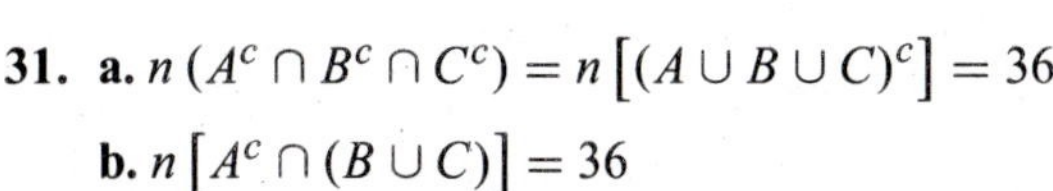

31. **a.** $n(A^c \cap B^c \cap C^c) = n\left[(A \cup B \cup C)^c\right] = 36$

b. $n\left[A^c \cap (B \cup C)\right] = 36$

33. Let U denote the set of all economists surveyed, $A = \{x \in U \mid x \text{ had lowered his or her estimate of the consumer inflation rate}\}$, and $B = \{x \in U \mid x \text{ had raised his or her estimate of the GNP growth rate}\}$. Then $n(U) = 10$, $n(A) = 7$, $n(B) = 8$, and $n(A \cap B^c) = 2$, so the number of economists who had both lowered their estimate of the consumer inflation rate and raised their estimate of the GNP growth rate is given by $n(A \cap B) = 5$.

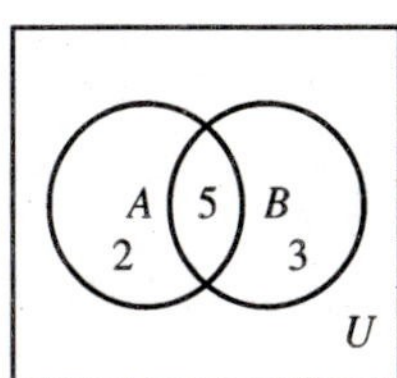

35. Let U denote the set of 100 college students who were surveyed, $A = \{x \in U \mid x \text{ reads } \textit{Time}\}$, $B = \{x \in U \mid x \text{ reads } \textit{Newsweek}\}$, and $C = \{x \in U \mid x \text{ reads } \textit{U.S. News and World Report}\}$. Then $n(A) = 40$, $n(B) = 30$, $n(C) = 25$, $n(A \cap B) = 15$, $n(A \cap C) = 12$, $n(B \cap C) = 10$, and $n(A \cap B \cap C) = 4$.

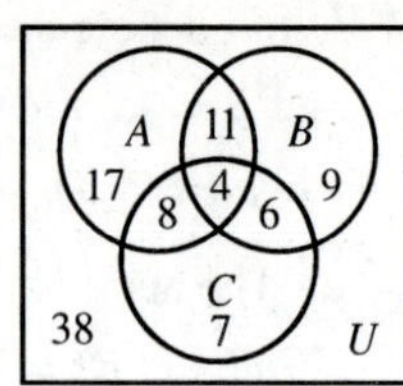

a. The number of students surveyed who read at least one magazine is $n(A \cup B \cup C) = 17 + 11 + 4 + 8 + 6 + 7 + 9 = 62$.

b. The number of students surveyed who read exactly one magazine is $n(A \cap B^c \cap C^c) + n(A^c \cap B \cap C^c) + n(A^c \cap B^c \cap C) = 17 + 9 + 7 = 33$.

c. The number of students surveyed who read exactly two magazines is $n(A \cap B \cap C^c) + n(A^c \cap B \cap C) + n(A \cap B^c \cap C) = 11 + 6 + 8 = 25$.

d. The number of students surveyed who did not read any of these magazines is $n(A \cup B \cup C)^c = 100 - 62 = 38$.

37. Let U denote the set of all customers surveyed, $A = \{x \in U \mid x \text{ buys brand } A\}$, $B = \{x \in U \mid x \text{ buys brand } B\}$, and $C = \{x \in U \mid x \text{ buys brand } C\}$. Then $n(U) = 120$, $n(A \cap B \cap C^c) = 15$, $n(A^c \cap B \cap C^c) = 25$, $n(A^c \cap B^c \cap C) = 26$, $n(A \cap B \cap C^c) = 15$, $n(A \cap B^c \cap C) = 10$, $n(A^c \cap B \cap C) = 12$, and $n(A \cap B \cap C) = 8$.

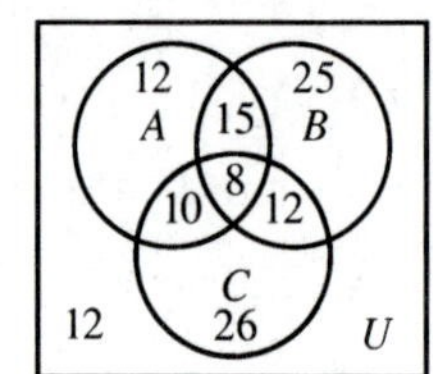

a. The number of customers who buy at least one of these brands is $n(A \cup B \cup C) = 12 + 15 + 25 + 12 + 8 + 10 + 26 = 108$.

b. The number who buy brands A and B but not C is $n(A \cap B \cap C^c) = 15$.

c. The number who buy brand A is $n(A) = 12 + 10 + 15 + 8 = 45$.

d. The number who buy none of these brands is $n\left[(A \cup B \cup C)^c\right] = 120 - 108 = 12$.

39. Let U denote the set of 200 employees surveyed, $A = \{x \in U \mid x \text{ had investments in stock funds}\}$, $B = \{x \in U \mid x \text{ had investments in bond funds}\}$, and $C = \{x \in U \mid x \text{ had investments in money market funds}\}$. Then $n(U) = 200$, $n(A) = 141$, $n(B) = 91$, $n(C) = 60$, $n(A \cap B) = 47$, $n(A \cap C) = 36$, $n(B \cap C) = 36$, and $n(A^c \cap B^c \cap C^c) = n\left[(A \cup B \cup C)^c\right] = 5$.

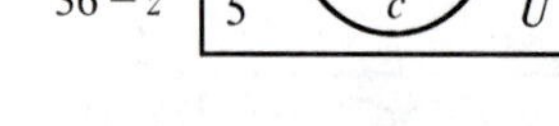

Letting $n(A \cap B \cap C) = z$, we arrive at the Venn diagram shown.

Next, using the facts that $n(A) = 141$, $n(B) = 91$, and $n(C) = 60$, we obtain $a + (36 - z) + (47 - z) + z = 141$, $b + (47 - z) + (36 - z) + z = 91$, $c + (36 - z) + (36 - z) + z = 60$, and $a + b + c + (36 - z) + (47 - z) + (36 - z) + z + 5 = 200$. These equations simplify to $a - z = 58$, $b - z = 8$, $c - z = -12$, and $a + b + c - 2z = 76$. Solving, we find $a = 80$, $b = 30$, $c = 10$, and $z = 22$.

a. The number of employees surveyed who had invested in all three investments is $n(A \cap B \cap C) = z = 22$.

b. The number who had invested in stock funds only is given by $n(A \cap B^c \cap C^c) = a = 80$.

41. True. $n(A \cup B) = n(A) + n(B) - n(A \cap B)$.

43. True. If $A \cap B \neq \varnothing$, then $n(A \cup B) = n(A) + n(B) - n(A \cap B)$.

45. Write Equation 4 as $n(D \cup E) = n(D) + n(E) - n(D \cap E)$ and let $D = A \cup B$ and $E = C$. Then

$$n(A \cup B \cup C) = n(A \cup B) + n(C) - n[(A \cup B) \cap C] = n(A) + n(B) - n(A \cap B) + n(C) - n[(A \cup B) \cap C]$$

$$= n(A) + n(B) - n(A \cap B) + n(C) - n[(A \cap C) \cup (B \cap C)]$$

$$= n(A) + n(B) - n(A \cap B) + n(C) - [n(A \cap C) + n(B \cap C) - n(A \cap C \cap B \cap C)]$$

$$= n(A) + n(B) + n(C) - n(A \cap B) - n(A \cap C) - n(B \cap C) + n(A \cap B \cap C).$$

7.3 The Multiplication Principle

Concept Questions page 425

1. If task T_1 can be performed in N_1 ways, task T_2 can be performed in N_2 ways, …, and task T_n can be performed in N_n ways, then the number of ways of performing the tasks $T_1, T_2, \ldots, T_n$ in succession is given by $N_1 N_2 \cdots N_n$.

Exercises page 425

1. By the multiplication principle, the number of rates is given by $4 \times 3 = 12$.

3. By the multiplication principle, the number of ways that a blackjack hand can be dealt is $4 \times 16 = 64$.

5. By the multiplication principle, she can create $3 \cdot 4 \cdot 3 = 36$ different ensembles.

7. The number of paths is $2 \times 4 \times 3 = 24$.

9. By the multiplication principle, we see that the number of ways a health-care plan can be selected is $(10)(3)(2) = 60$.

11. $10^9 = 1{,}000{,}000{,}000$

13. There are six optional toppings, each of which can be selected or not. Therefore, the number of different pizzas that can be made is $(2)(2)(2)(2)(2)(2) = 64$.

15. The number of different responses is $\underbrace{(5)(5)\cdots(5)}_{50 \text{ terms}} = 5^{50}$.

17. The number of selections is given by $(5)(2)(4)(5)(2) = 400$.

19. The number of different selections is $(10)(10)(10)(10) - 10 = 10{,}000 - 10 = 9990$.

21. **a.** The number of license plate numbers that may be formed is $(26)(26)(26)(10)(10)(10) = 17{,}576{,}000$.

b. The number of license plate numbers that may be formed is $(10)(10)(10)(26)(26)(26) = 17,576,000$.

23. If every question is answered, there are $2^{10} = 1024$ ways. In the second case, there are three ways to answer each question, and so we have $3^{10} = 59{,}049$ ways.

25. The number of ways the first, second and third prizes can be awarded is $(15)(14)(13) = 2730$.

27. The number of ways in which the nine symbols on the wheels can appear in the window slot is $(9)(9)(9) = 729$. The number of ways in which the eight symbols other than the "lucky dollar" can appear in the window slot is $(8)(8)(8) = 512$. Therefore, the number of ways in which the "lucky dollars" can appear in the window slot is $729 - 512 = 217$.

29. True. There are four choices for the digit in the hundreds position, four choices in the tens position, and two choices in the units position, for a total of $4 \cdot 4 \cdot 2 = 32$ such numbers.

7.4 Permutations and Combinations

Problem-Solving Tips

1. Note the difference between a permutation and a combination. A permutation is an arrangement of a set of distinct objects in a *definite order* whereas a combination is an arrangement of a set of distinct objects without regard to order. In a permutation of two distinct objects A and B, we distinguish between the selections AB and BA, whereas in a combination these selections would be considered the same.

2. Sometimes the solution of an applied problem involves the multiplication principle and a permutation and/or a combination. (See Example 11 on page 434 of the text.)

Concept Questions page 436

1. a. Given a set of distinct objects, a permutation of the set is an arrangement of these objects in a *definite order*.

b. $P(n,r) = \dfrac{n!}{(n-r)!}$, so $P(5,3) = \dfrac{5!}{(5-3)!} = \dfrac{5 \cdot 4}{2 \cdot 1} = 10$.

3. a. $C(n,r) = \dfrac{n!}{r!\,(n-r)!}$

b. $C(6,3) = \dfrac{6!}{3!\,(6-3)!} = \dfrac{6 \cdot 5 \cdot 4}{3 \cdot 2} = 20$

Exercises page 436

1. $3(5!) = 3(5)(4)(3)(2)(1) = 360$

3. $\dfrac{5!}{2!\,2!} = \dfrac{5 \cdot 4 \cdot 3 \cdot 2}{2 \cdot 2} = 30$

5. $P(5,5) = \dfrac{5!}{(5-5)!} = \dfrac{5!}{0!} = 120$

7. $P(5,3) = \dfrac{5!}{(5-3)!} = \dfrac{5!}{2!} = (5)(4)(3) = 60$

9. $P(n,1) = \dfrac{n!}{(n-1)!} = n$

11. $C(6,6) = \dfrac{6!}{6!\,0!} = 1$

13. $C(7,2) = \dfrac{7!}{2!\,5!} = \dfrac{7 \cdot 6}{2} = 21$

15. $C(5,0) = \dfrac{5!}{5!\,0!} = 1$

17. $C(9,6) = \dfrac{9!}{3!\,6!} = \dfrac{9 \cdot 8 \cdot 7}{3 \cdot 2} = 84$

19. $C(n,3) = \dfrac{n!}{(n-3)!3!} = \dfrac{n(n-1)(n-2)}{3 \cdot 2}$

21. $P(n,n-2) = \dfrac{n!}{[n-(n-2)]!} = \dfrac{n!}{(n-n+2)!} = \dfrac{n!}{2}$

23. Order is important here because the word *GLACIER* is different from the word *REICALG*. Therefore, this is a permutation.

25. Order is not important here. Therefore, we are dealing with a combination. If we consider a sample of three cellphones of which one is defective, it does not matter whether the first, second, or third member of our sample is defective. The net result is a sample of three cellphones, of which one is defective.

27. The order is important here. Therefore, we are dealing with a permutation. Consider, for example, nine books on a library shelf. Each of the nine books has a call number, and the books are filed in order of their call numbers; that is, a call number of 902 comes before a call number of 910.

29. The order is not important here, and consequently we are dealing with a combination. There is no difference between the hand Q Q Q 5 5 and the hand 5 5 Q Q Q. In each case the hand consists of three queens and a pair of fives.

31. The number of four-letter permutations is $P(4,4) = \frac{4!}{0!} = 4 \cdot 3 \cdot 2 \cdot 1 = 24$.

33. The number of seating arrangements is $P(4,4) = \frac{4!}{0!} = 24$.

35. The number of different batting orders is $P(9,9) = \frac{9!}{0!} = 362{,}880$.

37. The number of different ways the four candidates can be selected is $C(12,4) = \frac{12!}{8!\,4!} = \frac{12 \cdot 11 \cdot 10 \cdot 9}{4 \cdot 3 \cdot 2 \cdot 1} = 495$.

39. There are ten letters in the word *ANTARCTICA*, including 3 *A* s, 2 *C* s, 1 *I*, 1 *N*, 1 *R*, and 2 *T* s. Therefore, we use the formula for the permutation of n objects, not all distinct: $\frac{n!}{n_1!\,n_2!\cdots n_r!} = \frac{10!}{3!\,2!\,2!} = 151{,}200$.

41. The vowels cannot be permuted among themselves and may be considered as identical, so we can view the problem as that of finding the number of permutations of seven letters, taken all together, where two of the letters are identical. Thus, the result is $\frac{7!}{2!\,(1!)^5} = (7)(6)(5)(4)(3) = 2520$.

43. Here we use Formula 7. The number of distinct numbers is given by $\frac{5!}{3!\,1!\,1!} = 20$.

45. The number of ways the three sites can be selected is $C(12,3) = \frac{12!}{9!\,3!} = \frac{12 \cdot 11 \cdot 10}{3 \cdot 2 \cdot 1} = 220$.

47. The number of ways in which the sample of three microprocessors can be selected is $C(100,3) = \frac{100!}{97!\,3!} = \frac{100 \cdot 99 \cdot 98}{3 \cdot 2 \cdot 1} = 161{,}700$.

49. In this case order is important, as it makes a difference whether a commercial is shown first, last, or in between. The number of ways that the director can schedule the commercials is given by $P(6,6) = 6! = 720$.

51. The inquiries can be directed in $P(12,5) = \frac{12!}{7!} = 12 \cdot 11 \cdot 10 \cdot 9 \cdot 8 = 95{,}040$ ways.

53. **a.** The ten books can be arranged in $P(10,10) = 10! = 3{,}628{,}800$ ways.

b. The books on the same subject are grouped together, so we multiply the number of ways the mathematics books can be arranged, the number of ways the social science books can be arranged, the number of ways the biology books can be arranged, and the number of ways the three sets of books can be arranged. Thus, there are $P(3,3) \times P(4,4) \times P(3,3) \times P(3,3) = 5184$ ways.

55. Notice that order is certainly important here.

a. The number of ways that the 20 featured items can be arranged is given by $P(20,20) = 20! \approx 2.43 \times 10^{18}$.

b. If items from the same department must appear in the same row, then the number of ways they can be arranged on the page is

$$\begin{pmatrix}\text{number of ways}\\ \text{of arranging the rows}\end{pmatrix} \cdot \begin{pmatrix}\text{number of ways of arranging}\\ \text{the items in each of the five rows}\end{pmatrix}$$
$$= P(5,5) \cdot [P(4,4) \times P(4,4) \times P(4,4) \times P(4,4) \times P(4,4)] = 5! \cdot (4!)^5 = 955{,}514{,}880.$$

57. **a.** $P(12,9) = \dfrac{12!}{3!} = 79{,}833{,}600.$ **b.** $C(12,9) = \dfrac{12!}{3!\,9!} = 220$ **c.** $C(12,9) \cdot C(3,2) = 220 \cdot 3 = 660$

59. The number of ways is given by

$$(\text{number of players})\left[\begin{pmatrix}\text{number of ways to win}\\ \text{in exactly two sets}\end{pmatrix} + \begin{pmatrix}\text{number of ways to win}\\ \text{in exactly three sets}\end{pmatrix}\right]$$
$$= 2\{C(2,2) + [C(3,2) - C(2,2)]\} = 2[1 + (3-1)] = 2 \cdot 3 = 6.$$

61. The number of ways the measure can be passed is $C(3,3)[C(8,6) + C(8,7) + C(8,8)] = 37$. Here three of the three permanent members must vote for passage of the bill and this can be done in $C(3,3) = 1$ way. Of the eight nonpermanent members who are voting, six, seven, or eight can vote for passage of the bill. Therefore, there are $C(8,6) + C(8,7) + C(8,8) = 28 + 8 + 1 = 37$ ways that the nonpermanent members can vote to ensure passage of the measure. This gives $1 \times 37 = 37$ ways that the members can vote so that the bill is passed.

63. **a.** If no preference is given to any student, then the number of ways of awarding the three teaching assistantships is $C(12,3) = \dfrac{12!}{3!\,9!} = 220.$

b. If it is stipulated that one particular student receive one of the assistantships, then the remaining two assistantships must be awarded to two of the remaining 11 students. Thus, the number of ways is $C(11,2) = \dfrac{11!}{2!\,9!} = 55.$

c. If at least one woman is to be awarded one of the assistantships, and the group of students consists of seven men and five women, then the number of ways the assistantships can be awarded is

$$C(5,1) \times C(7,2) + C(5,2) \times C(7,1) + C(5,3) = \frac{5!}{4!\,1!} \cdot \frac{7!}{5!\,2!} + \frac{5!}{3!\,2!} \cdot \frac{7!}{6!\,1!} + \frac{5!}{3!\,2!} = 105 + 70 + 10 = 185.$$

65. The number of ways of awarding the three contracts to seven different firms is given by $P(7,3) = \dfrac{7!}{4!} = 210$. The number of ways of awarding the three contracts to two different firms (so one firm gets two contracts) from a choice of seven different firms is $C(7,2) \times P(3,2) = 126$ (first pick the two firms, and then award the 3 contracts). Therefore, the number of ways the contracts can be awarded if no firm is to receive more than two contracts is $210 + 126 = 336$.

67. The number of different curricula that are available for the student's consideration is given by

$$C(5,1)\times C(3,1)\times C(6,2)\times[C(4,1)+C(3,1)]=\frac{5!}{4!\,1!}\cdot\frac{3!}{2!\,1!}\cdot\frac{6!}{4!\,2!}\cdot\left(\frac{4!}{3!\,1!}+\frac{3!}{2!\,1!}\right)$$
$$=(5)(3)(15)(4)+(5)(3)(15)(3)=900+675=1575.$$

69. The number of ways of dealing a straight flush (five cards in sequence in the same suit) is given by

$$\begin{pmatrix}\text{number of ways of selecting five cards}\\ \text{in sequence in the same suit}\end{pmatrix}\cdot\begin{pmatrix}\text{number of ways}\\ \text{of selecting a suit}\end{pmatrix}=10\cdot 4=40.$$

71. The number of ways of dealing a flush (five cards in one suit that are not all in sequence) is given by

$$\begin{pmatrix}\text{number of ways of selecting}\\ \text{five cards in the same suit}\end{pmatrix}-\begin{pmatrix}\text{number of}\\ \text{straight flushes}\end{pmatrix}=4C(13,5)-40=5148-40=5108.$$

73. The number of ways of dealing a full house (three of a kind and a pair) is given by

$$\begin{pmatrix}\text{number of}\\ \text{different ranks}\end{pmatrix}\cdot\begin{pmatrix}\text{number of ways of picking}\\ \text{three of a kind of that rank}\end{pmatrix}\cdot\begin{pmatrix}\text{number of ways of picking a pair}\\ \text{from the 12 remaining ranks}\end{pmatrix}$$
$$=13\cdot C(4,3)\cdot 12C(4,2)=13\cdot 4\cdot 12(6)=3744.$$

75. The bus will travel a total of 6 blocks. Each route must include 2 blocks running north and south and 4 blocks running east and west. To compute the total number of possible routes, it suffices to compute the number of ways the 2 blocks running north and south can be selected from the 6 blocks. Thus, the number of possible routes is $C(6,2)=\frac{6!}{2!\,4!}=15$.

77. The number of ways that the quorum can be formed is given by

$$C(12,6)+C(12,7)+C(12,8)+C(12,9)+C(12,10)+C(12,11)+C(12,12)$$
$$=\frac{12!}{6!\,6!}+\frac{12!}{7!\,5!}+\frac{12!}{8!\,4!}+\frac{12!}{9!\,3!}+\frac{12!}{10!\,2!}+\frac{12!}{11!\,1!}+\frac{12!}{12!\,0!}=924+792+495+220+66+12+1=2510.$$

79. Using the formula from Exercise 78, we see that the number of ways of seating the five commentators at a round table is $(5-1)!=4!=24$.

81. The number of possible corner points is $C(8,3)=\frac{8!}{5!\,3!}=56$.

83. True.

85. True. $C(n,r)=\frac{n!}{(n-r)!\,r!}$ and $C(n,n-r)=\frac{n!}{[n-(n-r)]!\,(n-r)!}=\frac{n!}{r!\,(n-r)!}$, so $C(n,r)=C(n,n-r)$.

Technology Exercises page 441

1. $1.307674368 \times 10^{12}$ 3. $2.56094948229 \times 10^{16}$ 5. 674,274,182,400 7. 133,784,560

9. 4,656,960

11. Using the multiplication principle, the number of 10-question exams she can set is given by $C(25, 3) \times C(40, 5) \times C(30, 2) = 658{,}337{,}004{,}000$.

7.5 Experiments, Sample Spaces, and Events

Problem-Solving Tips

1. The **union of two events** A **and** B, written $A \cup B$, is the set of outcomes in A and/or B. The **intersection of two events** A **and** B, written $A \cap B$, is the set of outcomes in both A and B. **The complement of an event** A, written A^c, is the set of outcomes in the sample space S that are not in A.

2. Two events A and B are mutually exclusive if $A \cap B = \varnothing$. In other words, the two events cannot occur at the same time.

Concept Questions page 446

1. An experiment is an activity with observable results. Examples vary.

Exercises page 446

1. $E \cup F = \{a, b, d, f\}$, $E \cap F = \{a\}$.

3. $F^c = \{b, c, e\}$, $E \cap G^c = \{a, b\} \cap \{a, d, f\} = \{a\}$.

5. Because $E \cap F = \{a\}$ is not a null set, we conclude that E and F are not mutually exclusive.

7. $E \cup F \cup G = \{2, 4, 6\} \cup \{1, 3, 5\} \cup \{5, 6\} = \{1, 2, 3, 4, 5, 6\} = S$.

9. $(E \cup F \cup G)^c = \{1, 2, 3, 4, 5, 6\}^c = \varnothing$.

11. Yes, $E \cap F = \varnothing$, that is, E and F do not contain any common elements.

13. $E^c = \{2, 4, 6\}^c = \{1, 3, 5\} = F$ and so E and F are complementary.

15. $E \cup G$ 17. F^c 19. $(E \cup F \cup G)^c$

21. a. Refer to Example 4 on page 444 of the text.
$E = \{(2, 1), (3, 1), (4, 1), (5, 1), (6, 1), (3, 2), (4, 2), (5, 2), (6, 2), (4, 3), (5, 3), (6, 3), (5, 4), (6, 4), (6, 5)\}$.
b. $E = \{(1, 2), (2, 4), (3, 6)\}$.

23. $\varnothing$, $\{a\}$, $\{b\}$, $\{c\}$, $\{a, b\}$, $\{a, c\}$, $\{b, c\}$, $\{a, b, c\}$.

25. a. $S = \{R, B\}$ b. $\varnothing$, $\{B\}$, $\{R\}$, $\{B, R\}$

27. **a.** $S = \{(\text{H}, 1), (\text{H}, 2), (\text{H}, 3), (\text{H}, 4), (\text{H}, 5), (\text{H}, 6), (\text{T}, 1), (\text{T}, 2), (\text{T}, 3), (\text{T}, 4), (\text{T}, 5), (\text{T}, 6)\}$

b. $E = \{(\text{H}, 1), (\text{H}, 3), (\text{H}, 5)\}$

29. Here $S = \{1, 2, 3, 4, 5, 6\}$, $E = \{2\}$, and $F = \{2, 4, 6\}$.

a. Because $E \cap F = \{2\} \neq \varnothing$, we conclude that E and F are not mutually exclusive.

b. $E^c = \{1, 3, 4, 5, 6\} \neq F$, and so E and F are not complementary.

31. $S = \{ddd, ddn, dnd, ndd, dnn, ndn, nnd, nnn\}$

33. **a.**

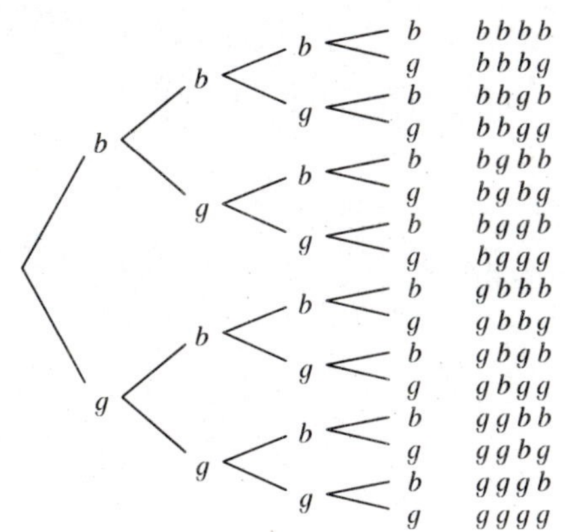

From the tree diagram, we see that the required sample space is

$S = \{bbbb, bbbg, bbgb, bbgg, bgbb, bgbg, bggb, bggg, gbbb, gbbg, gbgb, gbgg, ggbb, ggbg, gggb, gggg\}$.

b. $E = \{bbbg, bbgb, bgbb, gbbb\}$

c. $F = \{bbbg, bbgg, bgbg, bggg, gbbg, gbgg, ggbg, gggg\}$

d. $G = \{gbbg, gbgg, ggbg, gggg\}$

35. **a.** $\{ABC, ABD, ABE, ACD, ACE, ADE, BCD, BCE, BDE, CDE\}$. **b.** 6 **c.** 3 **d.** 6

37. **a.** E^c **b.** $E^c \cap F^c$ **c.** $E \cup F$ **d.** $(E \cap F^c) \cup (E^c \cap F)$

39. **a.** $S = \{t \mid t > 0\}$ **b.** $E = \{t \mid 0 < t \leq 2\}$ **c.** $F = \{t \mid t > 2\}$

41. **a.** $S = \{0, 1, 2, 3, \ldots, 10\}$ **b.** $E = \{0, 1, 2, 3\}$ **c.** $F = \{5, 6, 7, 8, 9, 10\}$

43. **a.** $S = \{0, 1, 2, \ldots, 20\}$ **b.** $E = \{0, 1, 2, \ldots, 9\}$ **c.** $F = \{20\}$

45. Let S denote the sample space of the experiment that is the set of 52 cards. Then $E = \{x \in S \mid x \text{ is an ace}\}$, $F = \{x \in S \mid x \text{ is a spade}\}$, and $E \cap F = \{x \in S \mid x \text{ is the ace of spades}\}$. Now $n(E) = 4$, $n(F) = 13$, and $n(E \cap F) = 1$. Also, $E \cup F = \{x \in S \mid x \text{ is an ace or a spade}\}$ and $n(E \cup F) = 16$, so $n(E) + n(F) - n(E \cap F) = 4 + 13 - 1 = 16 = n(E \cup F)$.

47. $E^c \cap F^c = (E \cup F)^c$ by DeMorgan's Law. Because $(E \cup F) \cap (E \cup F)^c = \varnothing$, they are mutually exclusive.

49. False. Let $E = \{1, 2, 3\}$, $F = \{4, 5, 6\}$, and $G = \{4, 5\}$. Then $E \cap F = \varnothing$ and $E \cap G = \varnothing$, but $F \cap G = \{4, 5\} \neq \varnothing$.

7.6 Definition of Probability

Problem-Solving Tips

1. **Uniform sample spaces** are sample spaces in which the outcomes are equally likely.

2. Events consisting of a single outcome are called **simple events**.

3. To find the probability of an event E, follow these steps:

   a. Determine an appropriate space S associated with the experiment.

   b. Assign probabilities to the simple events of the experiment. Then $P(E) = P(s_1) + P(s_2) + P(s_3) + \cdots + P(s_n)$, where $E = \{s_1, s_2, s_3, \ldots, s_n\}$ and $\{s_1\}, \{s_2\}, \{s_3\}, \ldots, \{s_n\}$ are the simple events of S.

Concept Questions page 454

1. a. By assigning probabilities to each simple event of an experiment, we obtain a *probability distribution* that gives the probability of each simple event. Examples vary.

b. The function P that assigns a probability to each of the simple events is called a *probability function.* Examples vary.

3. $P(E) = P(s_1) + P(s_2) + \cdots + P(s_n)$; $P(\varnothing) = 0$.

Exercises page 454

1. $\{(H, H)\}, \{(H, T)\}, \{(T, H)\}, \{(T, T)\}$.

3. $\{(D, m)\}, \{(D, f)\}, \{(R, m)\}, \{(R, f)\}, \{(I, m)\}, \{(I, f)\}$

5. $\{(1, i)\}, \{(1, d)\}, \{(1, s)\}, \{(2, i)\}, \{(2, d)\}, \{(2, s)\}, \ldots, \{(5, i)\}, \{(5, d)\}, \{(5, s)\}$

7. $\{(\text{A}, \text{Rh}^+)\}, \{(\text{A}, \text{Rh}^-)\}, \{(\text{B}, \text{Rh}^+)\}, \{(\text{B}, \text{Rh}^-)\}, \{(\text{AB}, \text{Rh}^+)\}, \{(\text{AB}, \text{Rh}^-)\}, \{(\text{O}, \text{Rh}^+)\}, \{(\text{O}, \text{Rh}^-)\}$

9.

Grade	A	B	C	D	F
Probability	.10	.25	.45	.15	.05

11.

Answer	Falling behind	Staying even	Increasing faster	Don't know
Probability	.40	.44	.12	.04

13. The probability is $\dfrac{360 + 192}{800} = .69$.

15.

Number of times	0	1	2	3	4	5	6	7
Probability	.05	.06	.09	.15	.11	.20	.17	.17

17.

Rating	A	B	C	D	E
Probability	.026	.199	.570	.193	.012

19. **a.** $S = \{(0 < x \leq 200), (200 < x \leq 400), (400 < x \leq 600), (600 < x \leq 800), (800 < x \leq 1000), (x > 1000)\}$

b.

Number of cars (x)	Probability
$0 < x \leq 200$	.075
$200 < x \leq 400$	.1
$400 < x \leq 600$	.175
$600 < x \leq 800$	.35
$800 < x \leq 1000$	.225
$x > 1000$	.075

21. The probability is $\frac{84{,}000{,}000}{179{,}000{,}000} \approx .469$.

23. **a.** The probability that a person killed by lightning is a male is $\frac{376}{439} \approx .856$.

b. The probability that a person killed by lightning is a female is $\frac{439-376}{439} = \frac{63}{439} \approx .144$.

25. The probability that the retailer uses electronic tags as antitheft devices is $\frac{81}{176} \approx .460$.

27. **a.** $P(D) = \frac{13}{52} = \frac{1}{4}$ **b.** $P(B) = \frac{26}{52} = \frac{1}{2}$ **c.** $P(A) = \frac{4}{52} = \frac{1}{13}$

29. The probability of arriving at the traffic light when it is red is $\frac{30}{30+5+45} = \frac{30}{80} = \frac{3}{8}$.

31. The required sample space is $S = \{ab, ac, ad, ae, ba, bc, bd, be, ca, cb, cd, ce, da, db, dc, de, ea, eb, ec, ed\}$. The set of interviewee choices including applicant a is $E = \{ab, ac, ad, ae, ba, ca, da, ea\}$, the set of choices including a and c is $F = \{ac, ca\}$, and the set including d and e is $G = \{de, ed\}$.

a. The required probability is $\dfrac{n(E)}{n(S)} = \dfrac{8}{20} = .4$. **b.** The required probability is $\dfrac{n(F)}{n(S)} = \dfrac{2}{20} = .1$.

c. The required probability is $\dfrac{n(G)}{n(S)} = \dfrac{2}{20} = .1$.

33. **a.** $P(E) = \frac{62}{9+62+27} = \frac{62}{98} \approx .633$ **b.** $P(E) = \frac{27}{98} \approx .276$

35. **a.** The probability that a registered voter favors the proposition is .35.

b. The probability that a registered voter is undecided about the proposition is $1 - .35 - .32 = .33$.

37. The required probability is $\dfrac{281+251}{382+281+251+90} \approx .530$.

39. **a.** The required probability is $\dfrac{25+15}{37+14+25+15+9} \approx .4$.

b. The required probability is $\dfrac{14+9}{37+14+25+15+9} \approx .23$.

41. **a.** The required probability is $\dfrac{448}{448+169+155+100+22+106} = .448$.

b. The required probability is $\dfrac{155+100}{448+169+155+100+22+106} = .255$.

43. The probability that the primary cause of the crash was due to pilot error or bad weather is given by

$$\frac{327+22}{327+49+14+22+19+15}=\frac{349}{446}\approx .783.$$

45. There are six ways of getting a 7: (2, 5), (5, 2), (3, 4), (4, 3), (1, 6), and (6, 1).

47. No, the outcomes are not equally likely. For example, there is only one way to obtain a sum of 2, but there are two ways of obtaining a sum of 3.

49. a. $P(A)=P(s_1)+P(s_3)=\frac{1}{12}+\frac{1}{12}=\frac{1}{6}$.

b. $P(B)=P(s_2)+P(s_4)+P(s_5)+P(s_6)=\frac{1}{4}+\frac{1}{6}+\frac{1}{3}+\frac{1}{12}=\frac{5}{6}$.

c. $P(C)=P(S)=1$.

51. True

7.7 Rules of Probability

Problem-Solving Tips

If S is a sample space of an experiment and E and F are events of the experiment, then the following are always true:

a. $P(E)\geq 0$ for any E.

b. $P(S)=1$.

c. If E and F are mutually exclusive, then $P(E\cup F)=P(E)+P(F)$. More generally, if E and F are any two events of an experiment, then $P(E\cup F)=P(E)+P(F)-P(E\cap F)$.

d. $P(E^c)=1-P(E)$.

Concept Questions page 463

1. a. The event E cannot occur.

b. There is a 50% chance that the event F will occur.

c. The event S will certainly occur.

d. The probability of the event $E\cup F$ occurring is given by the sum of the probabilities of E and F minus the probability of $E\cap F$.

Exercises page 463

1. Refer to Example 4 on page 444 of the text. Let E denote the event of interest. Then $P(E)=\frac{18}{36}=\frac{1}{2}$

3. Refer to Example 4 on page 444 of the text. The event of interest is $E=\{1,1\}$, and $P(E)=\frac{1}{36}$.

5. Let E denote the event of interest. Then $E=\{(6,3),(6,2),(6,1),(1,6),(2,6),(3,6)\}$ and $P(E)=\frac{6}{36}=\frac{1}{6}$.

7. Let E denote the event that the card drawn is a king and F the event that the card drawn is a diamond. Then the required probability is $P(E\cap F)=\frac{1}{52}$.

9. Let E denote the event that a face card is drawn. Then $P(E) = \frac{12}{52} = \frac{3}{13}$.

11. Let E denote the event that an ace is drawn. Then $P(E) = \frac{1}{13}$. Then E^c is the event that an ace is not drawn and $P(E^c) = 1 - P(E) = \frac{12}{13}$.

13. Let E denote the event that a ticketholder will win first prize. Then $P(E) = \frac{1}{500} = .002$, and the probability of the event that a ticketholder will not win first prize is $P(E^c) = 1 - .002 = .998$.

15. Property 2 of the laws of probability is violated. The sum of the probabilities must add up to 1. In this case $P(S) = 1.1$, which is not possible.

17. The five events are not mutually exclusive: the probability of winning at least one purse is
$1 - (\text{probability of losing all five times}) = 1 - \dfrac{9^5}{10^5} \approx 1 - .5905 = .4095$.

19. The two events are not mutually exclusive; hence, the probability of the given event is $\frac{1}{6} + \frac{1}{6} - \frac{1}{36} = \frac{11}{36}$.

21. $E^c \cap F^c = \{c, d, e\} \cap \{a, b, e\} = \{e\} \neq \varnothing$.

23. Let G denote the event that a customer purchases a pair of glasses and let C denote the event that the customer purchases a pair of contact lenses. Then $P\left[(G \cup C)^c\right] \neq 1 - P(G) - P(C)$. Mr. Owens has not considered the case in which the customer buys both glasses and contact lenses.

25. **a.** $P(E \cap F) = 0$ because E and F are mutually exclusive.

b. $P(E \cup F) = P(E) + P(F) - P(E \cap F) = .2 + .5 = .7$.

c. $P(F^c) = 1 - P(F) = 1 - .5 = .5$.

d. $P(E^c \cap F^c) = P\left[(E \cup F)^c\right] = 1 - P(E \cup F) = 1 - .7 = .3$.

27. **a.** $P(A) = P(s_1) + P(s_2) = \frac{1}{8} + \frac{3}{8} = \frac{1}{2}$; $P(B) = P(s_1) + P(s_3) = \frac{1}{8} + \frac{1}{4} = \frac{3}{8}$.

b. $P(A^c) = 1 - P(A) = 1 - \frac{1}{2} = \frac{1}{2}$ and $P(B^c) = 1 - P(B) = 1 - \frac{3}{8} = \frac{5}{8}$.

c. $P(A \cap B) = P(s_1) = \frac{1}{8}$.

d. $P(A \cup B) = P(A) + P(B) - P(A \cap B) = \frac{1}{2} + \frac{3}{8} - \frac{1}{8} = \frac{3}{4}$.

29. Let U denote the set of cars in the experiment,
$A = \{x \in U \mid x \text{ failed the tread test}\}$, and
$B = \{x \in U \mid x \text{ failed the pressure test}\}$.

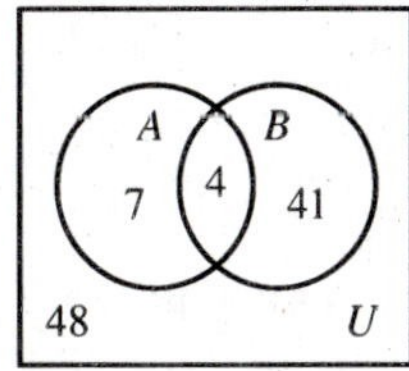

a. The probability is $P(B \cap A^c) = \frac{41}{100} = .41$.

b. The probability is $P\left[(A \cup B)^c\right] = \frac{48}{100} = .48$.

31.

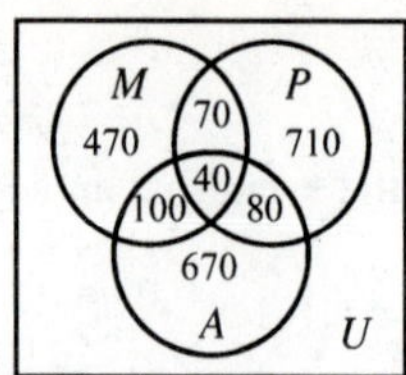

P: Lack of parental support

M: Malnutrition

A: Abused or neglected

From the diagram, we see that the probability that a teacher selected at random from this group said that lack of parental support is the only problem hampering a student's schooling is $\frac{710}{2140} \approx .332$.

33.

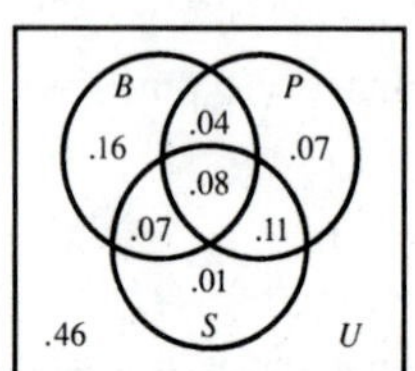

B: Probability of buying a blouse

P: Probability of buying pants

S: Probability of buying a skirt

From the given information and the diagram, we have $P(B) = .35$, $P(P) = .30$, $P(S) = .27$, $P(B \cap S) = .15$, $P(S \cap P) = .19$, $P(B \cap P) = .12$, and $P(B \cap P \cap S) = .08$.

a. The probability of exactly one item is

$$P(B \cap S^c \cap P^c) + P(P \cap B^c \cap S^c) + P(S \cap B^c \cap P^c) = .16 + .07 + .01 = .24.$$

b. The probability of buying no item is

$$1 - P(A \cup B \cup C) = 1 - (.16 + .07 + .01 + .08 + .04 + .11 + .07) = .46.$$

35.

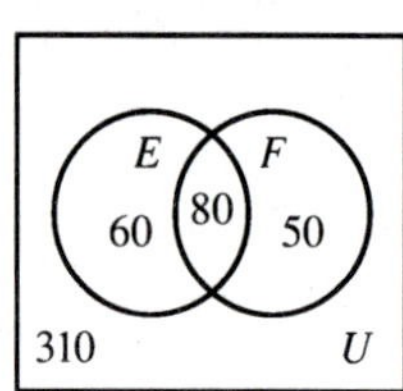

Let E and F denote the events that the person surveyed learned of the products from *Good Housekeeping* and *Ladies' Home Journal*, respectively. Then $P(E) = \frac{140}{500} = \frac{7}{25}$, $P(F) = \frac{130}{500} = \frac{13}{50}$, and $P(E \cap F) = \frac{80}{500} = .16$.

a. $P(E \cap F) = \frac{80}{500} = .16$.

b. $P(E \cup F) = \frac{14}{50} + \frac{13}{50} - \frac{8}{50} = \frac{19}{50} = .38$.

c. $P(E \cap F^c) + P(E^c \cap F) = \frac{60}{500} + \frac{50}{500} = \frac{110}{500} = .22$.

37. a. The required probability is $.236 + .174 = .41$.

b. The probability that they were planning to use computer software to prepare their taxes or to do their taxes by hand is $.339 + .143 = .482$. The probability that they were not planning to either use computer software to prepare their taxes or do their taxes by hand is $1 - .482 = .518$.

39.

Answer	Very likely	Somewhat likely	Somewhat unlikely	Very unlikely	Don't know
Probability	$\frac{40}{200}$	$\frac{28}{200}$	$\frac{26}{200}$	$\frac{104}{200}$	$\frac{2}{200}$

a. The required probability is $\frac{104}{200} = .52$.

b. The required probability is $\frac{28}{200} + \frac{40}{200} = \frac{68}{200} = .34$.

41. a. The required probability is $.16 + .279 = .439$.

b. The required probability is $.142 + .243 = .385$.

43. a. The sum of the numbers $45.1 + 16.5 + 6.9 + 6.1 + 4.2 + 3.8 + 2.5 + 14.9$ is 100, and so the table does give a probability distribution.

b. The probability is $45.1 + 6.9 = 52$ percent, or .52.

c. The probability is $1 - (.061 + .042 + .038) = .859$.

45. Let $A = \{t \mid t < 3\}$, $B = \{t \mid t \leq 4\}$, and $C = \{t \mid t > 5\}$.

a. $D = \{t \mid t \leq 5\}$ and $P(D) = 1 - P(C) = 1 - .1 = .9$.

b. $E = \{t \mid t > 4\}$ and $P(E) = 1 - P(B) = 1 - .6 = .4$.

c. $F = \{t \mid 3 \leq t \leq 4\}$ and $P(F) = P(A^c \cap B) = .4$.

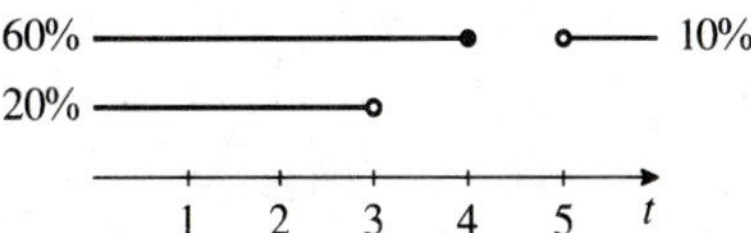

47. a. The probability that the participant favors tougher gun control laws is $\frac{150}{250} = .6$.

b. The probability that the participant owns a handgun is $\frac{58+25}{250} = .332$.

c. The probability that the participant owns a handgun but not a rifle is $\frac{58}{250} = .232$.

d. The probability that the participant favors tougher gun control laws and does not own a handgun is $\frac{12+138}{250} = .6$.

49. The probability that Bill will fail to solve the problem is $1 - p_1$ and the probability that Mike will fail to solve the problem is $1 - p_2$. Therefore, the probability that both Bill and Mike will fail to solve the problem is $(1 - p_1)(1 - p_2)$. So, the probability that at least one of them will solve the problem is $1 - (1 - p_1)(1 - p_2) = 1 - (1 - p_2 - p_1 + p_1 p_2) = p_1 + p_2 - p_1 p_2$.

51. True. Write $B = A \cup (B - A)$. Because A and $B - A$ are mutually exclusive, we have $P(B) = P(A) + P(B - A)$. Because $P(B) = 0$, we have $P(A) + P(B - A) = 0$. If $P(A) > 0$, then $P(B - A) < 0$ and this is not possible. Therefore, $P(A) = 0$.

53. False. Take $E_1 = \{1, 2\}$ and $E_2 = \{2, 3\}$, where $S = \{1, 2, 3\}$. Then $P(E_1) = \frac{2}{3}$ and $P(E_2) = \frac{2}{3}$, but $P(E_1 \cup E_2) = P(S) = 1$.

CHAPTER 7 Concept Review Questions page 469

1. set, elements, set

3. subset

5. union, intersection

7. $A^c \cap B^c \cap C^c$

9. experiment, sample, space, event

11. uniform, $1/n$

CHAPTER 7 Review Exercises page 469

1. $\{3\}$. The set consists of all solutions to the equation $3x - 2 = 7$.

3. $\{4, 6, 8, 10\}$

5. Yes

7. Yes

9. $A \cup (B \cap C)$

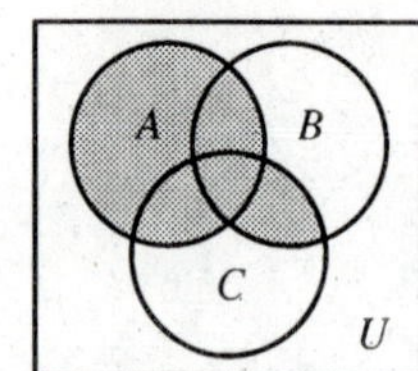

11. $A^c \cap B^c \cap C^c$

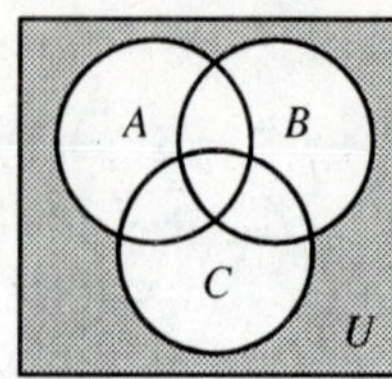

13. $A \cup (B \cup C) = \{a, b\} \cup [\{b, c, d\} \cup \{a, d, e\}] = \{a, b\} \cup \{a, b, c, d, e\} = \{a, b, c, d, e\}$, while $(A \cup B) \cup C = [\{a, b\} \cup \{b, c, d\}] \cup \{a, d, e\} = \{a, b, c, d\} \cup \{a, d, e\} = \{a, b, c, d, e\}$.

15. $A \cap (B \cup C) = \{a, b\} \cap [\{b, c, d\} \cup \{a, d, e\}] = \{a, b\} \cap \{a, b, c, d, e\} = \{a, b\}$, while $(A \cap B) \cup (A \cap C) = [\{a, b\} \cap \{b, c, d\}] \cup [\{a, b\} \cap \{a, d, e\}] = \{b\} \cup \{a\} = \{a, b\}$.

17. The set of all participants in a consumer behavior survey who both avoided buying a product because it is not recyclable and boycotted a company's products because of its record on the environment.

19. The set of all participants in a consumer behavior survey who both did not use cloth diapers rather than disposable diapers and voluntarily recycled their garbage.

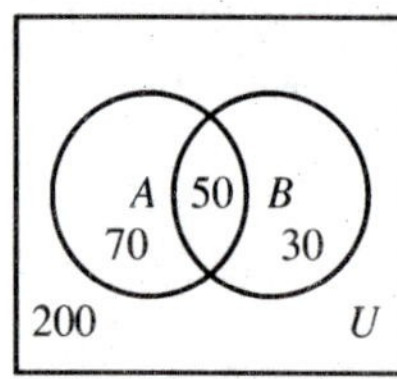

21. $n(A \cup B) = n(A) + n(B) - n(A \cap B) = 120 + 80 - 50 = 150$.

23. $n(B^c) = n(U) - n(B) = 350 - 80 = 270$.

25. $n(A \cap B^c) = n(A) - n(A \cap B) = 120 - 50 = 70$.

27. $C(20, 18) = \dfrac{20!}{18!\,2!} = 190$.

29. $C(5, 3) \cdot P(4, 2) = \dfrac{5!}{3!\,2!} \cdot \dfrac{4!}{2!} = 10 \cdot 12 = 120$.

31. a. $P(E \cap F) = 0$ because E and F are mutually exclusive.

b. $P(E \cup F) = P(E) + P(F) - P(E \cap F) = .4 + .2 = .6$.

c. $P(E^c) = 1 - P(E) = 1 - .4 = .6$.

d. $P(E^c \cap F^c) = P(E \cup F)^c = 1 - P(E \cup F) = 1 - .6 = .4$.

e. $P(E^c \cup F^c) = P(E \cap F)^c = 1 - P(E \cap F) = 1 - 0 = 1$.

33. a. The probability of the number being even is $P(2) + P(4) + P(6) = .12 + .18 + .19 = .49$.

b. The probability that the number is either a 1 or a 6 is $P(1) + P(6) = .20 + .19 = .39$.

c. The probability that the number is less than 4 is $P(1) + P(2) + P(3) = .20 + .12 + .16 = .48$.

35. Let U denote the set of 5 major cards, $A = \{x \in U \mid x \text{ offered cash advances}\}$, $B = \{x \in U \mid x \text{ offered extended payments for all goods and services purchased}\}$, and $C = \{x \in U \mid x \text{ required an annual fee that was less than \$35}\}$. Thus, $n(A) = 3$, $n(B) = 3$, $n(C) = 2$, $n(A \cap B) = 2$, $n(B \cap C) = 1$, and $n(A \cap B \cap C) = 0$. From the diagram, we have $x + y + 2 = 3$ and $y + 2 = 2$. Solving, we find $x = 1$, and $y = 0$. Therefore, the number of cards that offer cash advances and have an annual fee that is less than \$35 is given by $n(A \cap C) = y = 0$.

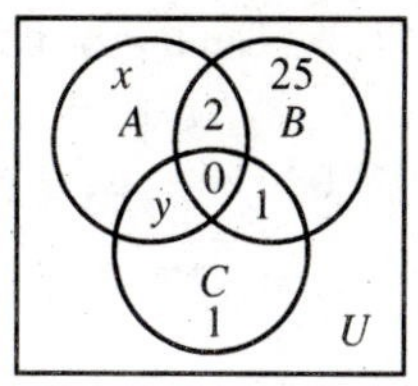

37. The number of ways the compact discs can be arranged on a shelf is $P(6, 6) = 6! = 720$.

39. **a.** Because there is repetition of the letters C, I, and N, we use the formula for the permutation of n objects (not all distinct) with $n = 10$, $n_1 = 2$, $n_2 = 3$, $n_3 = 3$, $n_4 = 1$, $n_5 = 1$, and $n_1 + n_2 + \cdots + n_k = 10$. Then the number of permutations that can be formed is given by $\dfrac{10!}{2!\,3!\,3!} = 50{,}400$.

b. Again we use the formula for the permutation of n objects (not all distinct), this time with $n = 8$, $n_1 = 2$, $n_2 = 2$, $n_3 = 2$, $n_4 = 1$, $n_5 = 1$, and $n_1 + n_2 + \cdots + n_k = 8$. Then the number of permutations is given by $\dfrac{8!}{2!\,2!\,2!} = 5040$.

41. Let U denote the set comprising Halina's clients, $A = \{x \in U \mid x \text{ owns stocks}\}$, $B = \{x \in U \mid x \text{ owns bonds}\}$, and $C = \{x \in U \mid x \text{ owns mutual funds}\}$. Then $n(A) = 300$, $n(B) = 180$, $n(C) = 160$, $n(A \cap B) = 110$, $n(A \cap C) = 120$, and $n(B \cap C) = 90$. With $n(A \cap B \cap C) = z$, we have the Venn diagram shown. Using the facts that $n(A) = 300$, $n(B) = 180$, and $n(C) = 160$, we have the system

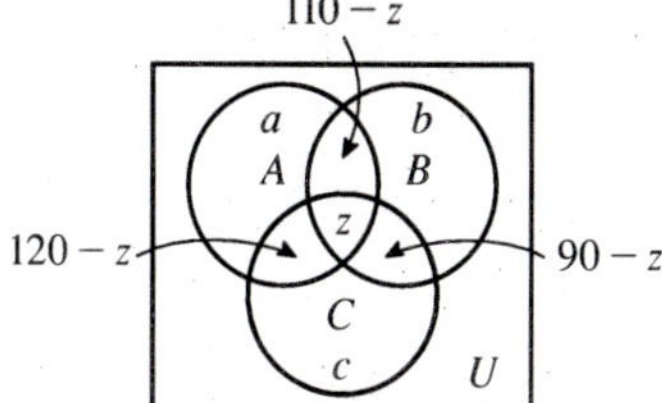

$$\begin{cases} a + (110 - z) + (120 - z) + z = 300 \\ b + (110 - z) + (90 - z) + z = 180 \\ c + (120 - z) + (90 - z) + z = 160 \\ a + b + c + (110 - z) + (120 - z) + (90 - z) + z = 400 \end{cases}$$

which simplifies to $a - z = 70$, $b - z = -20$, $c - z = -50$, and $a + b + c - 2z = 80$. Solving, we find $a = 150$, $b = 60$, $c = 30$, and $z = 80$. Therefore, the number who own stocks, bonds and mutual funds is $n(A \cap B \cap C) = z = 80$.

43. The number of possible outcomes is $(6)(4)(5)(6) = 720$.

45. **a.** The number of ways the seven students can be assigned to seats is $P(7, 7) = 7! = 5040$.

b. The number of ways two specified students can be seated next to each other is $2(6) = 12$. (Think of seven numbered seats. Then the students can be seated in seats 1-2, or 2-3, or 3-4, or 4-5, or 5-6, or 6-7. Because there are six such possibilities and the pair of students can be seated in two different orders, we conclude that there are $2 \cdot 6$ possible arrangements.) Then the remaining five students can be seated in $P(5, 5) = 5!$ ways. Therefore, the number of ways the seven students can be seated if two specified students sit next to each other is $2 \cdot 6 \cdot 5! = 1440$. Finally, the number of ways the students can be seated if the two students do not sit next to each other is $P(7, 7) - 2 \cdot 6 \cdot 5! = 5040 - 1440 = 3600$.

47. a. The number of samples that can be selected is $C(15, 4) = \dfrac{15!}{4!\,11!} = 1365$.

b. There are $C(5, 0) \cdot C(10, 4) = 1 \cdot \dfrac{10!}{4!\,6!} = 210$ ways of selecting four balls none of which is white. Therefore, there are $1365 - 210 = 1155$ ways of selecting four balls of which at least one is white.

49. Let A, B, C, and D denote the events that Olivia buys from Epson, Brother, Canon, and Hewlett-Packard, respectively. Then $P(A) + P(B) + P(C) + P(D) = .23 + .18 + .31 + .28 = 1$.

a. The probability is $P(A \cup C) = P(A) + P(C) = .23 + .31 = .54$.

b. The probability is $P(A \cup B \cup C) = P(A) + P(B) + P(C) = .23 + .18 + .31 = .72$. It can also be calculated as $1 - P(D) = 1 - .28 = .72$, as obtained earlier.

51. a. The required probability is $.44 + .12 = .56$.

b. The required probability is .4.

CHAPTER 7 Before Moving On... page 472

1. a. $B \cup C = \{b, c, d, e, f, g\}$, so $A \cap (B \cup C) = \{d, f, g\}$.

b. $A \cap C = \{f\}$, and so $(A \cap C) \cup (B \cup C) = \{b, c, d, e, f, g\}$.

c. $A^c = \{b, c, e\}$.

2. The number of possibilities is $C(6, 4) = \dfrac{6!}{4!\,2!} = 15$.

3. There are $C(6, 3) = \dfrac{6!}{3!\,3!} = 20$ ways of picking the three seniors and $C(5, 2) = \dfrac{5!}{2!\,3!} = 10$ ways of picking the two juniors. Therefore, there are $20 \cdot 10 = 200$ possible teams.

4. $P(s_1, s_3, s_6) = \frac{1}{12} + \frac{3}{12} + \frac{1}{12} = \frac{5}{12}$.

5. The number of ways of drawing a deuce or face card is 16. Therefore, the required probability is $\frac{16}{52} = \frac{4}{13}$.

6.

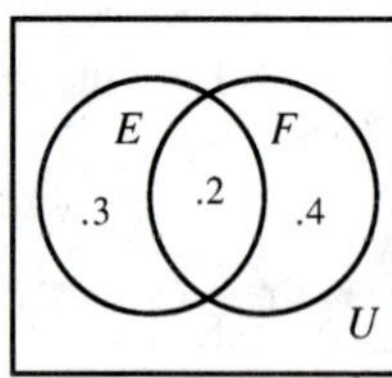

a. $P(E \cup F) = .3 + .2 + .4 = .9$.

b. $P(E \cap F^c) = .3$.

8 ADDITIONAL TOPICS IN PROBABILITY

8.1 Use of Counting Techniques in Probability

Concept Questions page 478

1. If E is an event in a uniform sample space S, then the probability of E occurring is
$P(E) = \frac{\text{Number of outcomes in } E}{\text{Number of outcomes in } S} = \frac{n(E)}{n(S)}$.

Exercises page 478

1. Let E denote the event that the coin lands heads all five times. Then $P(E) = \frac{1}{2^5} = \frac{1}{32}$.

3. Let E denote the event that the coin lands tails all five times. Then E^c is the event that the coin lands heads at least once, and $P(E^c) = 1 - P(E) = 1 - \frac{1}{32} = \frac{31}{32}$.

5. Let E denote the event that a pair is drawn. Then $P(E) = \frac{13 \cdot C(4,2)}{C(52,2)} = \frac{78}{1326} \approx .059$.

7. Let E denote the event that two black cards are drawn. Then $P(E) = \frac{C(26,2)}{C(52,2)} = \frac{325}{1326} = \frac{25}{102} \approx .245$.

9. The probability of the event that two of the balls will be white and two will be blue is
$P(E) = \frac{n(E)}{n(S)} = \frac{C(3,2) \cdot C(5,2)}{C(8,4)} = \frac{3 \cdot 10}{70} = \frac{3}{7}$.

11. The probability of the event that exactly one of the balls is blue is $P(E) = \frac{n(E)}{n(S)} = \frac{C(5,1)\,C(3,3)}{C(8,4)} = \frac{5}{70} = \frac{1}{14}$.

13. The probability that a family will have two boys and one girl is $P(E) = \frac{C(3,2)}{8} = \frac{3}{8}$.

15. The probability that a family will have no girl is $P(E) = \frac{C(3,3)}{8} = \frac{1}{8}$.

17. The number of elements in the sample space is 2^{10}. There are $C(10,6) = \frac{10!}{6!\,4!} = 210$ ways of answering exactly six questions correctly. Therefore, the required probability is $\frac{210}{2^{10}} = \frac{210}{1024} \approx .205$.

19. a. Let E denote the event that both of the bulbs are defective. Then
$P(E) = \frac{C(4,2)}{C(24,2)} = \frac{\frac{4!}{2!\,2!}}{\frac{24!}{22!\,2!}} = \frac{4 \cdot 3}{24 \cdot 23} = \frac{1}{46} \approx .022$.

b. Let F denote the event that none of the bulbs is defective. Then $P(F) = \frac{C(20,2)}{C(24,2)} = \frac{20!}{18!\,2!} \cdot \frac{22!\,2!}{24!} = \frac{20}{24} \cdot \frac{19}{23} \approx .6884$. Therefore, the probability that at least one of the light bulbs is defective is $1 - P(F) \approx 1 - .6884 = .3116$.

21. a. The probability that both of the cartridges are defective is $P(E) = \frac{C(6,2)}{C(80,2)} = \frac{15}{3160} \approx .005$.

b. Let F denote the event that none of the cartridges is defective. Then $P(F) = \frac{C(74,2)}{C(80,2)} = \frac{2701}{3160} \approx .855$, and so $P(F^c) = 1 - P(F) \approx 1 - .855 = .145$ is the probability that at least one of the cartridges is defective.

23. a. The probability that Mary will be selected is $P(E) = \frac{12}{100} = .12$. The probability that both Mary and John will be selected is $P(F) = \frac{C(98,10)}{C(100,12)} = \frac{\frac{98!}{88!\,10!}}{\frac{100!}{88!\,12!}} = \frac{12 \cdot 11}{100 \cdot 99} \approx .013$.

b. The probability that Mary will be selected is $P(M) = \frac{6}{40} = .15$. The probability that both Mary and John will be selected is $P(M) \cdot P(J) = \frac{6}{60} \cdot \frac{6}{40} = \frac{36}{2400} = .015$.

25. The probability is $\frac{C(12,8) \cdot C(8,2)}{C(20,10)} + \frac{C(12,9)\,C(8,1)}{C(20,10)} + \frac{C(12,10)}{C(20,10)} = \frac{(28)(495) + (220)(8) + 66}{184{,}756} \approx .085$.

27. a. The probability that he will select brand B is $\left(\frac{\text{the number of selections that include brand } B}{\text{the number of possible selections}}\right) = \frac{C(4,2)}{C(5,3)} = \frac{6}{10} = \frac{3}{5}$.

b. The probability that he will select brands B and C is $\frac{C(3,1)}{C(5,3)} = .3$.

c. The probability that he will select at least one of the two brands B and C is $(1 - \text{probability that he selects neither of brands } B \text{ and } C) = 1 - \frac{C(3,3)}{C(5,3)} = .9$.

29. The probability that the three "Lucky Dollar" symbols will appear in the window of the slot machine is $P(E) = \frac{n(E)}{n(S)} = \frac{(1)(1)(1)}{C(9,1)\,C(9,1)\,C(9,1)} = \frac{1}{729}$.

31. The probability of a ticket holder having all four digits in exact order is $\frac{1}{C(10,1) \cdot C(10,1) \cdot C(10,1) \cdot C(10,1)} = \frac{1}{10{,}000} = .0001$.

33. The probability of a ticket holder having one specified digit is $\frac{C(1,1) \cdot C(10,1) \cdot C(10,1) \cdot C(10,1)}{10^4} = .1$.

35. The number of ways of selecting a five-card hand from 52 cards is $C(52,5) = 2{,}598{,}960$. The number of straight flushes that can be dealt in each suit is 10, so there are $4 \cdot 10 = 40$ possible straight flushes. Therefore, the probability of being dealt a straight flush is $\frac{4(10)}{C(52,5)} = \frac{40}{2{,}598{,}960} \approx .0000154$.

37. The number of ways of being dealt a flush in one suit is $C(13,5)$. Because there are four suits, the number of ways of being dealt a flush is $4 \cdot C(13,5)$. Because we wish to exclude the hands that are straight flushes we subtract the number of possible straight flushes from $4 \cdot C(13,5)$. Therefore, the probability of being drawn a flush (but not a straight flush) is $\frac{4 \cdot C(13,5) - 40}{C(52,5)} = \frac{5108}{2{,}598{,}960} \approx .00197$.

39. The total number of ways to select three cards of one rank is $13 \cdot C(4,3)$. The remaining two cards must form a pair of another rank, and there are $12 \cdot C(4,2)$ ways of selecting the pair. Thus, the total number of ways to be dealt a full house is $13 \cdot C(4,3) \cdot 12 \cdot C(4,2) = 3744$. Hence, the probability of being dealt a full house is $\dfrac{13 \cdot C(4,3) \cdot 12 \cdot C(4,2)}{C(52,5)} = \dfrac{3{,}744}{2{,}598{,}960} \approx .00144$.

41. a. Let E denote the event that in a group of five, no two have the same sign. Then $P(E) = \dfrac{12 \cdot 11 \cdot 10 \cdot 9 \cdot 8}{12^5} \approx .3819$. Therefore, the probability that at least two will have the same sign is given by $1 - P(E) = 1 - .3819 \approx .618$.

b. $P(\text{no Aries}) = \dfrac{11 \cdot 11 \cdot 11 \cdot 11 \cdot 11}{12^5} \approx .647$ and $P(\text{one Aries}) = \dfrac{C(5,1) \cdot (1)(11)(11)(11)(11)}{12^5} \approx .294$. Therefore, the probability that at least two will have the sign Aries is given by $1 - [P(\text{no Aries}) + P(\text{one Aries})] = 1 - .941 \approx .059$.

43. Referring to the table on page 477 of the text, we see that in a group of 50 people, the probability that no two have the same birthday is approximately $1 - .970 = .030$.

8.2 Conditional Probability and Independent Events

Concept Questions page 490

1. The conditional probability of an event is the probability of an event occurring given that another event has already occurred. Examples will vary.

3. $P(A \cap B) = P(A)P(B \mid A)$.

Exercises page 490

1. a. $P(A \mid B) = \dfrac{P(A \cap B)}{P(B)} = \dfrac{.3}{.5} = \dfrac{3}{5}$.

b. $P(B \mid A) = \dfrac{P(A \cap B)}{P(A)} = \dfrac{.3}{.6} = \dfrac{1}{2}$.

3. $P(A \cap B) = P(A)P(B \mid A) = (.6)(.5) = .3$.

5. $P(A) \cdot P(B) = (.3)(.6) \neq .2 = P(A \cap B)$. Therefore, the events are not independent.

7. $P(A \cap B) = P(A) + P(B) - P(A \cup B) = .5 + .7 - .85 = .35 = P(A) \cdot P(B)$, so A and B are independent events.

9. a. $P(A \cap B) = P(A)P(B) = (.5)(.6) = .3$.

b. $P(A \cup B) = P(A) + P(B) - P(A \cap B) = .5 + .6 - .3 = .8$.

c. $P(A \mid B) = P(A) = .4$ because A and B are independent.

d. $P(A^c \cup B^c) = P\left[(A \cap B)^c\right] = 1 - P(A \cap B) = 1 - .3 = .7$.

11. **a.** $P(A) = .5.$

b. $P(E \mid A) = .4.$

c. $P(A \cap E) = P(A)\,P(E \mid A) = (.5)(.4) = .2.$

d. $P(E) = (.5)(.4) + (.5)(.3) = .35.$

e. No. $P(A \cap E) \neq P(A) \cdot P(E) = (.5)(.35).$

f. A and E are not independent events.

13.

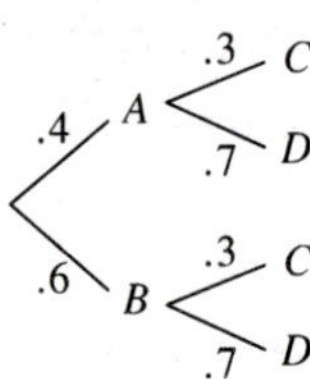

a. $P(A) = .4.$

b. $P(C \mid A) = .3.$

c. $P(A \cap C) = P(A)\,P(C \mid A) = (.4)(.3) = .12.$

d. $P(C) = (.4)(.3) + (.6)(.3) = .30.$

e. Yes. $P(A \cap C) = .12 = P(A)\,P(C).$

f. Yes.

15. **a.** Refer to Figure 3 on page 483 of the text. Here $E = \{(5,1), (5,2), (5,3), (5,4), (5,5), (5,6)\}$ and $F = \{(6,4), (5,5), (4,6)\}$, so $P(F) = \frac{3}{36} = \frac{1}{12}$.

b. $P(E \cap F) = \frac{1}{36}$ because $E \cap F = \{(5,5)\}$.

c. $P(F \mid E) = \frac{1}{6}$.

d. $P(E) = \frac{6}{36} = \frac{1}{6}$.

e. $P(F \mid E) = \dfrac{P(E \cap F)}{P(E)} = \dfrac{\frac{1}{36}}{\frac{1}{6}} = \dfrac{1}{6} \neq P(F) = \dfrac{1}{12}$, and so the events are not independent.

17. Let A denote the event that the sum of the numbers is less than 9 and B the event that at least one of the numbers is a 6. Then, $P(A \mid B) = \dfrac{P(A \cap B)}{P(B)} = \dfrac{\frac{4}{36}}{\frac{11}{36}} = \dfrac{4}{11}$.

19. Refer to Figure 3 on page 483 of the text. Here $E = \{(3,1), (3,2), (3,3), (3,4), (3,5), (3,6)\}$ and $F = \{(1,6), (6,1), (2,5), (5,2), (3,4), (4,3)\}$, so $E \cap F = \{(3,4)\}$. Now $P(E \cap F) = \frac{1}{36}$, and this is equal to $P(E) \cdot P(F) = \frac{6}{36} \cdot \frac{6}{36} = \frac{1}{36}$, so E and F are independent events.

21. $P(E \cap F) = \frac{13}{24} = \frac{1}{4}$, $P(E) = \frac{26}{52} = \frac{1}{2}$, and $P(F) = \frac{13}{52} = \frac{1}{4}$. Now $P(E) \cdot P(F) = \frac{1}{2} \cdot \frac{1}{4} = \frac{1}{8} \neq P(E \cap F) = \frac{1}{4}$, so E and F are not independent events. The fact that the card drawn is black increases the probability that it is a spade.

23. Let A denote the event that the battery lasts 10 or more hours and let B denote the event that the battery lasts 15 or more hours. Then $P(A) = .8$, $P(B) = .15$, and $P(A \cap B) = .15$. Therefore, the probability that the battery will last 15 hours or more is $P(B \mid A) = \dfrac{P(A \cap B)}{P(A)} = \dfrac{.15}{.8} = \dfrac{3}{16} = .1875$.

25.

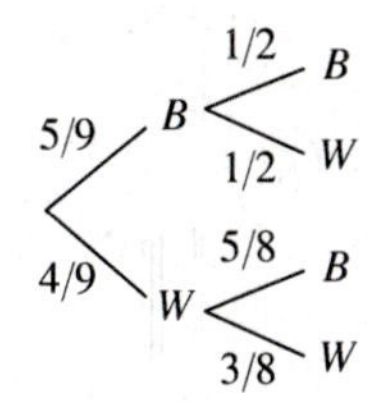

Without replacement

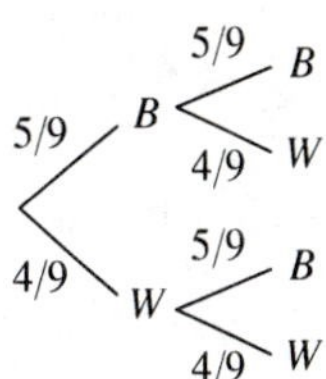

With replacement

a. The probability that the second ball drawn is a white ball if the second ball is drawn without replacing the first is $P(B)P(W \mid B) + P(W)P(W \mid W) = \frac{5}{9} \cdot \frac{1}{2} + \frac{4}{9} \cdot \frac{3}{8} = \frac{4}{9}$.

b. The probability that the second ball drawn is a white ball if the first ball is replaced before the second is drawn is $\frac{5}{9} \cdot \frac{4}{9} + \frac{4}{9} \cdot \frac{4}{9} = \frac{4}{9}$.

27.

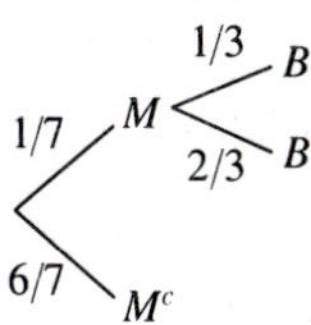

a. The probability that a student selected at random from this medical school is black is $\left(\frac{1}{7}\right)\left(\frac{1}{3}\right) = \frac{1}{21}$.

b. The probability that a student selected at random from this medical school is black if it is known that the student is a member of a minority group is $P(B \mid M) = \frac{1}{3}$.

29. Let A denote the event that Sandy takes Olivia to the supermarket on Friday and B the event that Sandy buys Olivia a popsicle. Then $P(A) = .6$ and $P(B \mid A) = .8$, so the probability that Sandy takes Olivia to the supermarket on Friday and buys her a popsicle is $P(A \cap B) = P(A)P(B \mid A) = (.6)(.8) = .48$.

31. Let A denote the event that a potential buyer will read the ad and B the event that a reader will buy Jack's car. Then $P(A) = .3$ and $P(B \mid A) = .2$, so the probability that the person who reads the ad will buy Jack's car is $P(A \cap B) = P(A)P(B \mid A) = (.3)(.2) = .06$.

33. The sample space for a three-child family is $S = \{GGG, GGB, GBG, GBB, BGG, BGB, BBG, BBB\}$. Because we know that there is at least one girl in the three-child family we are dealing with a reduced sample space $S_1 = \{GGG, GGB, GBG, GBB, BGG, BGB, BBG\}$ in which there are 7 outcomes. Then the probability that all three children are girls is $P(E) = \dfrac{n(E)}{n(S)} = \dfrac{1}{7}$.

35. a.

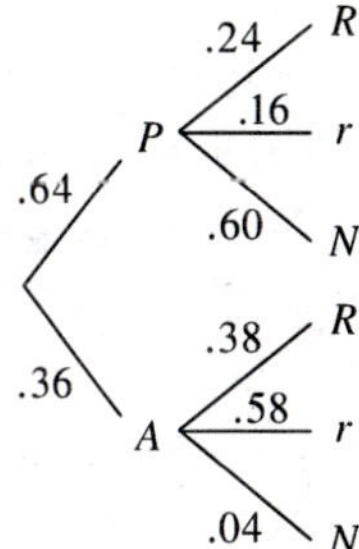

P Professional

A Amateur

R Recovered within 48 hours

r Recovered after more than 48 hours

N Never recovered

b. The required probability is .24.

c. The required probability is $(.64)(.60) + (.36)(.04) = .3984$.

37. M: 39/75; A: 4/10; E: 4/10; E^c; A^c; F: 36/75; A: 45/100; E: 4/10; E^c; A^c

a. $P(A \cap E \mid M) = (.4)(.4) = .16.$

b. $P(A) = P(M \cap A) + P(F \cap A) = \frac{39}{75} \cdot \frac{4}{10} + \frac{36}{75} \cdot \frac{45}{100} = .424.$

c. $P(M \cap A \cap E) + P(F \cap A \cap E) = \frac{39}{75} \cdot \frac{4}{10} \cdot \frac{4}{10} + \frac{36}{75} \cdot \frac{45}{100} \cdot \frac{4}{10}$

$= .0832 + .0864 = .1696.$

39. a. The probability that none of the dozen eggs is broken is $(.992)^{12} \approx .908$. Therefore, the probability that at least one egg is broken is $1 - .908 = .092$.

b. Using the results of part (a), we see that the required probability is $(.092)(.092)(.908) \approx .008$.

41. a. $P(A) = \dfrac{1120}{4000} = .28$, $P(B) = \dfrac{1560}{4000} = .39$, $P(A \cap B) = \dfrac{720}{4000} = .18$,

$$P(B \mid A) = \frac{P(A \cap B)}{P(A)} = \frac{n(A \cap B)}{n(A)} = \frac{720}{1120} \approx .643, \text{ and}$$

$$P(B \mid A^c) = \frac{P(A^c \cap B)}{P(A^c)} = \frac{n(A^c \cap B)}{n(A^c)} = \frac{840}{2880} \approx .292.$$

b. $P(B \mid A) \neq P(B)$, so A and B are not independent events.

43. Let C denote the event that a person in the survey was a heavy coffee drinker and Q the event that a person in the survey had cancer of the pancreas. Then $P(C) = \frac{3200}{10{,}000} = .32$, $P(Q) = \frac{160}{10{,}000} = .016$, $P(C \cap Q) = \frac{132}{160} = .825$, and $P(C) \cdot P(Q) = .00512 \neq P(C \cap Q)$. Therefore, the events are not independent.

45. The probability that the first test will fail is .03, the probability that the second test will fail is .015, and the probability that the third test will fail is .015. Because these are independent events, the probability that all three tests will fail is $(.03)(.015)(.015) = .0000068$.

47. a. Let $P(A)$, $P(B)$, and $P(C)$ denote the probabilities that the first, second, and third patient suffers a rejection, respectively. Then $P(A) = \frac{1}{2}$, $P(B) = \frac{1}{3}$, and $P(C) = \frac{1}{10}$, so the probabilities that each patient does not suffer a rejection are given by $P(A^c) = \frac{1}{2}$, $P(B^c) = \frac{2}{3}$, and $P(C^c) = \frac{9}{10}$. Thus, the probability that none of the three patients suffers a rejection is given by $P(A^c) \cdot P(B^c) \cdot P(C^c) = \frac{1}{2} \cdot \frac{2}{3} \cdot \frac{9}{10} = \frac{18}{60} = \frac{3}{10}$. Finally, the probability that at least one patient will suffer a rejection is $1 - P(A^c) \cdot P(B^c) \cdot P(C^c) = 1 - \frac{3}{10} = \frac{7}{10}$.

b. The probability that exactly two patients will suffer rejections is

$$P(A)P(B)P(C^c) + P(A)P(B^c)P(C) + P(A^c)P(B)P(C) = \tfrac{1}{2} \cdot \tfrac{1}{3} \cdot \tfrac{9}{10} + \tfrac{1}{2} \cdot \tfrac{2}{3} \cdot \tfrac{1}{10} + \tfrac{1}{2} \cdot \tfrac{1}{3} \cdot \tfrac{1}{10}$$

$$= \tfrac{9+2+1}{60} = \tfrac{12}{60} = \tfrac{1}{5}.$$

49. Let A denote the event that at least one of the floodlights remain functional over the one-year period. Then $P(A) = .99999$ and $P(A^c) = 1 - P(A) = .00001$. Letting n represent the minimum number of floodlights needed, we have $(.01)^n = .00001$, so $n \log(.01) = -5$, $n(-2) = -5$, and $n = \frac{5}{2} = 2.5$. Therefore, the minimum number of floodlights needed is 3.

51. The probability that the event will not occur in one trial is $1 - p$. Therefore, the probability that it will not occur in n independent trials is $(1 - p)^n$. Therefore, the probability that it will occur at least once in n independent trials is $1 - (1 - p)^n$.

53. $P(E \mid F) = \dfrac{P(E \cap F)}{P(F)} = \dfrac{P(F)}{P(F)} = 1$. (Note that $E \cap F = F$ because $F \subset E$.)

Interpretation: Because $F \subset E$, an occurrence of F implies an occurrence of E. In other words, given that F has occurred, it is a certainty that E will occur; that is, $P(E \mid F) = 1$.

55. $E = E \cap (F \cup F^c) = (E \cap F) \cup (E \cap F^c)$. Because $(E \cap F) \cap (E \cap F^c) = \varnothing$, we see that $(E \cap F)$ and $(E \cap F^c)$ are mutually exclusive, so $P(E) = P(E \cap F) + P(E \cap F^c)$. Thus, $P(E \cap F^c) = P(E) - P(E \cap F) = P(E) - P(E)P(F)$ (E and F are independent). Continuing, $P(E \cap F^c) = P(E)[1 - P(F)] = P(E)P(F^c)$, showing that E and F^c are independent.

57. True. Because A and B are mutually exclusive, $A \cap B = \varnothing$ and $P(A \mid B) = \dfrac{P(A \cap B)}{P(B)} = \dfrac{P(\varnothing)}{P(B)} = 0$.

59. True. This follows from Formula 3: $P(A \cap B) = P(A) \cdot P(B \mid A)$.

8.3 Bayes' Theorem

Problem-Solving Tips

If $A_1, A_2, \ldots A_n$ is a partition of a sample space S and E is an event of the experiment such that $P(E) \neq 0$, then

$$P(A_i \mid E) = \frac{P(A_i)\,P(E \mid A_i)}{P(A_1)\,P(E \mid A_1) + P(A_2)\,P(E \mid A_2) + \cdots + P(A_n)\,P(E \mid A_n)}$$

If you draw a tree diagram to represent the experiment, then this formula can also be remembered by noting that

$$P(A_i \mid E) = \frac{\text{The product of the probabilities along the limb through } A_i}{\text{The sum of the products of the probabilities along each limb terminating at } E}$$

Concept Questions page 498

1. An *a priori probability* gives the likelihood that an event *will* occur and an *a posteriori probability* gives the probability that an event *did occur* after the outcomes of an experiment have been observed. Examples will vary.

3. It represents the a posteriori probability that the component having the property described by E was produced in factory A.

Exercises page 499

1.

$P(A) = .35$ → A: $.57$ → D; $.43$ → D^c

$P(B) = .35$ → B: $.71$ → D; $.29$ → D^c

$P(C) = .30$ → C: $.33$ → D; $.67$ → D^c

3. a. $P(D^c) = \dfrac{15 + 10 + 20}{35 + 35 + 30} = .45$

b. $P(B \mid D^c) = \dfrac{10}{15 + 10 + 20} \approx .22$

5. a. $P(D) = \dfrac{25+20+15}{50+40+35} = .48$ **b.** $P(B \mid D) = \dfrac{20}{25+20+15} \approx .33$

7. a. $P(A) \cdot P(D \mid A) = (.4)(.2) = .08$ **b.** $P(B) \cdot P(D \mid B) = (.6)(.25) = .15$

c. $P(A \mid D) = \dfrac{P(A) \cdot P(D \mid A)}{P(A) \cdot P(D \mid A) + P(B) \cdot P(D \mid B)} = \dfrac{.4(.2)}{.08 + .15} \approx .348$

9. a. $P(A) \cdot P(D \mid A) = \frac{1}{3} \cdot \frac{1}{4} = \frac{1}{12}$ **b.** $P(B) \cdot P(D \mid B) = \frac{1}{2} \cdot \frac{1}{2} = \frac{1}{4}$

c. $P(C) \cdot P(D \mid C) = \dfrac{1}{6} \cdot \dfrac{1}{3} = \dfrac{1}{18}$

d. $P(A \mid D) = \dfrac{P(A) \cdot P(D \mid A)}{P(A) \cdot P(D \mid A) + P(B) \cdot P(D \mid B) + P(C) \cdot P(C \mid B)} = \dfrac{\frac{1}{12}}{\frac{1}{12} + \frac{1}{4} + \frac{1}{18}} = \dfrac{1}{12} \cdot \dfrac{36}{14} = \dfrac{3}{14}$

11. a. $P(B) = P(A) \cdot P(B \mid A) + P(A^c) \cdot P(B \mid A^c) = (.3)(.2) + (.7)(.3) = .27$

b. $P(A \mid B) = \dfrac{P(A \cap B)}{P(B)} = \dfrac{(.3)(.2)}{.27} \approx .22$

c. $P(B^c) = P(A) \cdot P(B^c \mid A) + P(A^c) \cdot P(B^c \mid A)$

$= (.3)(.8) + (.7)(.7) = .73$

d. $P(A \mid B^c) = \dfrac{P(A \cap B^c)}{P(B^c)} = \dfrac{.3(.8)}{.73} \approx .33$

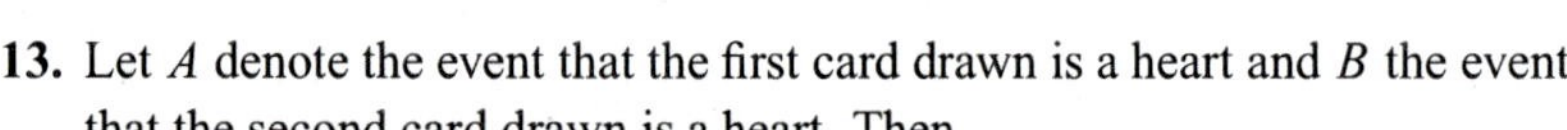

13. Let A denote the event that the first card drawn is a heart and B the event that the second card drawn is a heart. Then

$$P(A \mid B) = \frac{P(A) \cdot P(B \mid A)}{P(A) \cdot P(B \mid A) + P(A^c) \cdot P(B \mid A^c)} = \frac{\frac{1}{4} \cdot \frac{12}{51}}{\frac{1}{4} \cdot \frac{12}{51} + \frac{3}{4} \cdot \frac{13}{51}} = \frac{4}{17}.$$

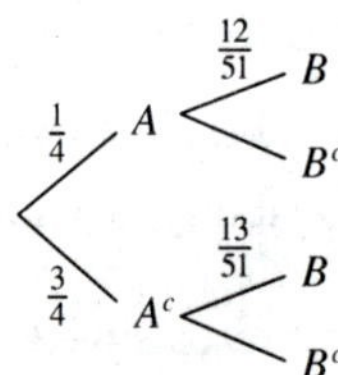

15.

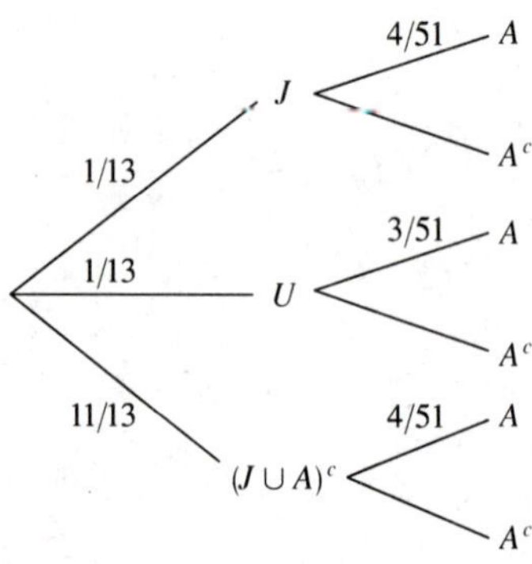

$$P(J \mid A) = \frac{\frac{1}{13} \cdot \frac{4}{51}}{\frac{1}{13} \cdot \frac{4}{51} + \frac{1}{13} \cdot \frac{3}{51} + \frac{11}{13} \cdot \frac{4}{51}} = \frac{\frac{4}{13 \cdot 51}}{\frac{51}{13 \cdot 51}}$$

$= \frac{4}{51} \approx .0784.$

17.

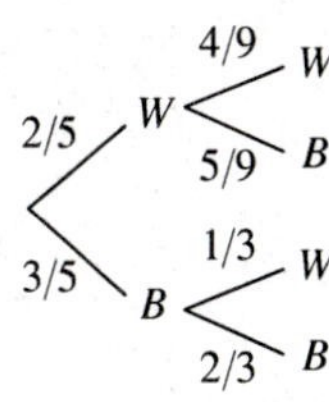

19. Referring to the tree diagram in Exercise 17, we see that the probability that the transferred ball was black given that the second ball was white is

$$P(B \mid W) = \frac{\frac{3}{5} \cdot \frac{1}{3}}{\frac{2}{5} \cdot \frac{4}{9} + \frac{3}{5} \cdot \frac{1}{3}} = \frac{9}{17}.$$

21. Let D denote the event that a senator selected at random is a Democrat, R the event that a senator selected at random is a Republican, and M the event that a senator has served in the military. We see that the probability that a senator selected at random who has served in the military is a Republican is

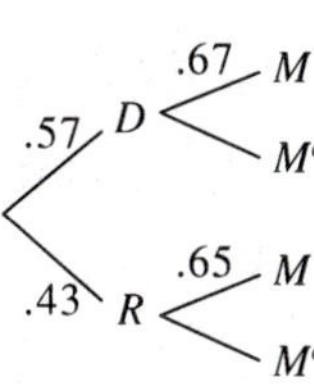

$$P(R \mid M) = \frac{P(R)\,P(M \mid R)}{P(M)} = \frac{(.43)(.65)}{(.57)(.67) + (.43)(.65)} \approx .423.$$

23. Let H_2 denote the event that the coin tossed is the two-headed coin, H_B the event that the coin is the biased coin, and H_F the event that the coin is the fair coin.

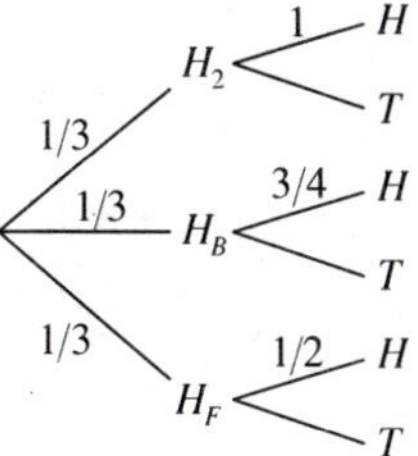

a. $P(H) = \frac{1}{3} \cdot 1 + \frac{1}{3} \cdot \frac{3}{4} + \frac{1}{3} \cdot \frac{1}{2} = \frac{1}{3} + \frac{1}{4} + \frac{1}{6} = \frac{9}{12} = \frac{3}{4}.$

b. $P(H_F \mid H) = \dfrac{\frac{1}{3} \cdot \frac{1}{2}}{\frac{3}{4}} = \frac{2}{9}$

25. Let D denote the event that the person tested has the disease and E the event that the test result is positive. Then the required probability is $P(D \mid E) = \dfrac{P(D)\,P(E \mid D)}{P(D)\,P(E \mid D) + P(D^c)\,P(E \mid D^c)} = \dfrac{(.003)(.95)}{(.003)(.95) + (.997)(.02)} \approx .125.$

27. $P(\text{III} \mid D) = \dfrac{(.30)(.02)}{(.35)(.015) + (.35)(.01) + (.30)(.02)} = \dfrac{.006}{.01475} \approx .407$

29. Let x denote the age of an adult selected at random from the population, and let R denote the event that the adult is a renter.

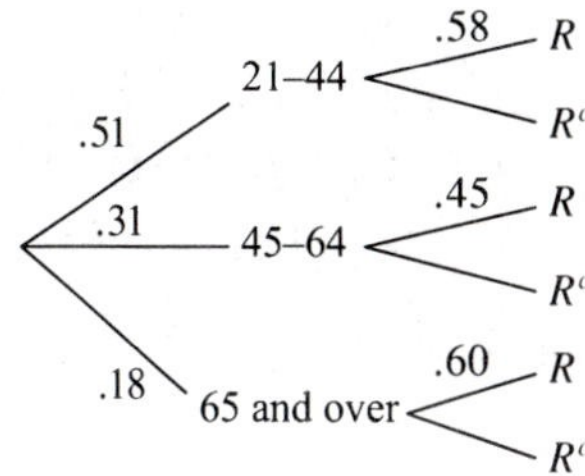

a. $P(R) = (.51)(.58) + (.31)(.45) + (.18)(.60) \approx .543.$

b. $P(21 \le x \le 44 \mid R) = \dfrac{(.51)(.58)}{.543} \approx .545.$

c. $P(E) = 1 - P(21 \le x \le 44 \mid R) = 1 - .545 = .455.$

31. Let D and R denote the events that the respondent is a Democratic and a Republican voter, respectively; and let S, O, and X denote the events that the respondent supports, opposes, and either doesn't know or refuses, respectively. Then

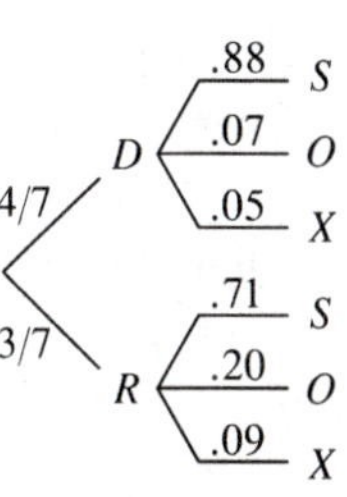

$$P(R \mid S) = \frac{\frac{3}{7}(.71)}{\frac{4}{7} \cdot (.88) + \frac{3}{7} \cdot (.71)} \approx .3770.$$

33. Let M and F denote the events that a person arrested for crime in 1988 was male and female, respectively; and let U denote the event that the person was under the age of 18.

a. $P(U) = (.89)(.30) + (.11)(.27) \approx .297$.

b. $P(F \mid U) = \dfrac{(.11)(.27)}{(.89)(.30) + (.11)(.27)} \approx .100$.

35. Let D denote the event that the person has the disease, and let Y denote the event that the test is positive. We see that the required probability is

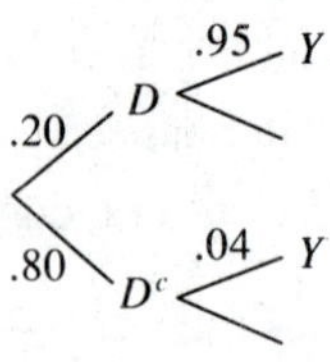

$$P(D \mid Y) = \frac{P(D) \cdot P(Y \mid D)}{P(D) \cdot P(Y \mid D) + P(D^c) \cdot P(Y \mid D^c)}$$
$$= \frac{(.2)(.95)}{(.2)(.95) + (.8)(.04)} \approx .856.$$

37. a. The probability is
$P(C)P(F) + P(C^c)P(F) = (.625)(.63) + (.375)(.47) = .57$.

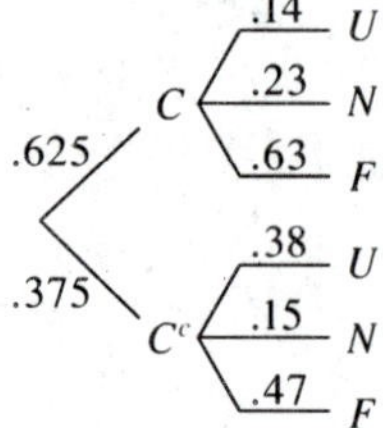

b. The probability is
$$\frac{P(C)P(F)}{P(C)P(F) + P(C^c)P(F)} = \frac{(.625)(.63)}{(.57)} \approx .691.$$

39. Let N and D denote the events that a employee was placed by Nancy and Darla, respectively; and let S denote the event that the employee placed by one of these women was satisfactory. Then

$$P(D \mid S^c) = \frac{(.55)(.3)}{(.45)(.2) + (.55)(.3)} \approx .647.$$

41. Let D, R, and I denote the events that a voter selected at random was a registered Democrat, Republican, or Independent, respectively; and let V denote the event that a voter selected at random voted for the incumbent senator. Then the probability that a randomly selected voter who voted for the incumbent was a registered Republican is

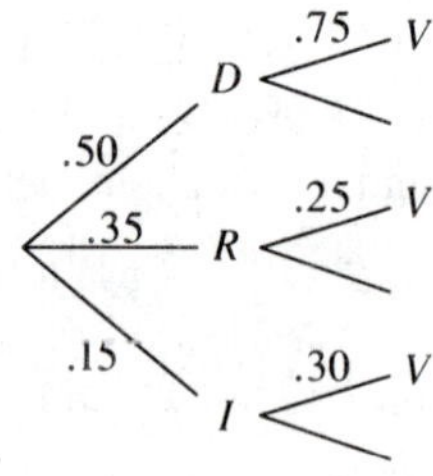

$$P(R \mid V) = \frac{(.35)(.25)}{(.5)(.75) + (.35)(.25) + (.15)(.30)} \approx .172.$$

43. Let A, B, C, and D denote the events that a guest's age is 21–34, 35–44, 45–54, and 55 and over, respectively; and let O denote the event that a man keeps his paper money in order of denomination. Then the required probability is $P(B \mid O) = \dfrac{\frac{3}{8} \cdot \frac{61}{100}}{\frac{5}{16} \cdot \frac{9}{10} + \frac{3}{8} \cdot \frac{61}{100} + \frac{3}{16} \cdot \frac{8}{10} + \frac{1}{8} \cdot \frac{8}{10}} \approx .301$.

45. a. Let A, B, C, D, and E denote the events that a respondent's annual household income is less than 15,000, 15,000–29,999, 30,000–49,999, 50,000–74,999, and 75,000 and higher, respectively; and let R, M, and P denote the probabilities that a person considers himself rich, middle class, and poor, respectively. The probability that a respondent chosen at random calls himself or herself middle class is thus $(.112)(.24) + (.186)(.60) + (.245)(.86) + (.219)(.90) + (.238)(.91) \approx .763.$

b. $$P(C \mid M) = \frac{.245(.86)}{.112(.24) + .186(.60) + .245(.86) + .219(.9) + .238(.91)} \approx .276.$$

c. Using the results of part (b), the required probability is $1 - .276 = .724$.

47. a. $$P(F) = (.24)(.38) + (.08)(.60) + (.08)(.66) + (.07)(.58) + (.09)(.52) + (.44)(.48) = .4906.$$

b. $P(M \mid B) = .62$.

c. $P(B \mid F) = \dfrac{.24(.38)}{.4906} \approx .186.$

49. Let A, B, C, and D denote the events that a person in the survey belongs to the Millennial Generation, Generation X, the Baby Boomer generation, and the Silent Generation, respectively, and let S denote the event that a person in the survey has slept with a cell phone nearby. The required probability is

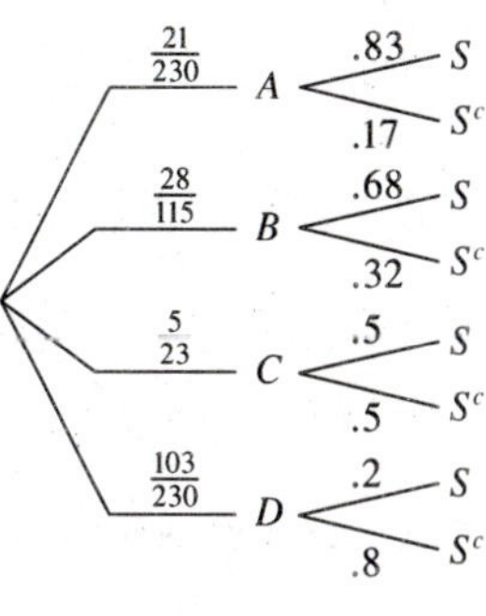

$$P(S^c \mid A) = \frac{P(A)P(S^c \mid A)}{\left[\begin{array}{l} P(A)P(S^c \mid A) + P(B)P(S^c \mid B) \\ \quad + P(C)P(S^c \mid C) + P(D)P(S^c \mid D) \end{array}\right]}$$

$$= \frac{\frac{21}{230}(.17)}{\frac{21}{230}(.17) + \frac{28}{115}(.32) + \frac{5}{23}(.5) + \frac{103}{230}(.80)} \approx .028.$$

51. Let N, G, H, C, U, and X denote the categories no diploma, GED diploma, high school graduate, some college, undergraduate level, and graduate degree, respectively. If a person selected at random from this group was a smoker, the probability that he or she has a graduate degree is given by

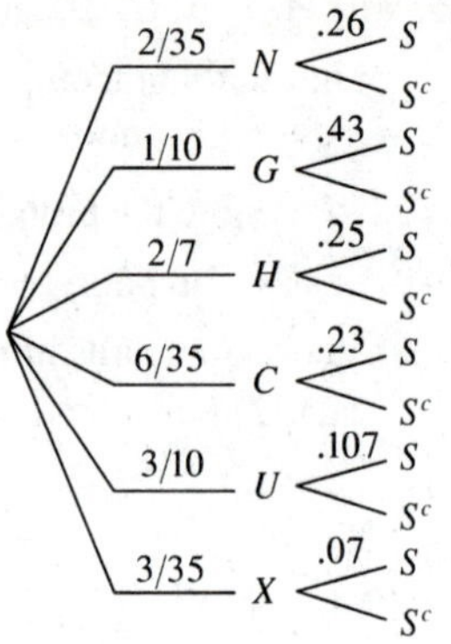

2/35 N .26 S S^c
1/10 G .43 S S^c
2/7 H .25 S S^c
6/35 C .23 S S^c
3/10 U .107 S S^c
3/35 X .07 S S^c

$$P(G \mid S) = \frac{\frac{3}{35} \cdot \frac{7}{100}}{\frac{2}{35} \cdot \frac{26}{100} + \frac{1}{10} \cdot \frac{43}{100} + \frac{2}{7} \cdot \frac{25}{100} + \frac{6}{35} \cdot \frac{23}{100} + \frac{21}{70} \cdot \frac{107}{1000} + \frac{3}{35} \cdot \frac{7}{100}}$$

$\approx .029.$

8.4 Distributions of Random Variables

Problem-Solving Tips

1. A **random variable** is a rule that assigns a number to each outcome of a chance experiment.

2. A **probability distribution of a random variable** gives the distinct values of the random variable X and the probabilities associated with these values.

3. A **histogram** is the graph of the probability distribution of a random variable.

Concept Questions page 511

1. A random variable is a rule that assigns a number to each outcome of a chance experiment. Examples vary.

3. To construct a histogram for a probability distribution, follow these steps:

a. Locate the values of the random variable on a number line.

b. Above each such number on the number line, erect a rectangle with width 1 and height equal to the probability associated with that value of the random variable.

Exercises page 511

1. a,b.

Outcome	GGG	GGR	GRG	RGG	GRR	RGR	RRG	RRR
Value	3	2	2	2	1	1	1	0

c. {GGG}

3. X may assume the values in the set $S = \{1, 2, 3, \ldots\}$.

5. The event that the sum of the dice rolls is 7 is $E = \{(1,6), (2,5), (3,4), (4,3), (5,2), (6,1)\}$ and $P(E) = \frac{6}{36} = \frac{1}{6}$.

7. X may assume the value of any positive integer. The random variable is infinite discrete.

9. $\{x \mid x \geq 0\}$. The random variable is continuous.

11. X may assume the value of any positive integer. The random variable is infinite discrete.

13. No. The probability assigned to a value of the random variable X cannot be negative. Here $P(X = 0) = -.2$, which is proscribed.

15. No. The sum of the probabilities exceeds 1.

17. We must have $.1 + .4 + a + .1 + .2 = 1$, so $a = .2$.

19. a. $P(X = -10) = .20$

b. $P(X \geq 5) = .1 + .25 + .1 + .15 = .60$

c. $P(-5 \leq X \leq 5) = .15 + .05 + .1 = .30$

d. $P(X \leq 20) = .20 + .15 + .05 + .1 + .25 + .1 + .15 = 1$

e. $P(X < 5) = P(X = -10) + P(X = -5) + P(X = 0) = .20 + .15 + .05 = .4$

f. $P(X = 3) = 0$

21.

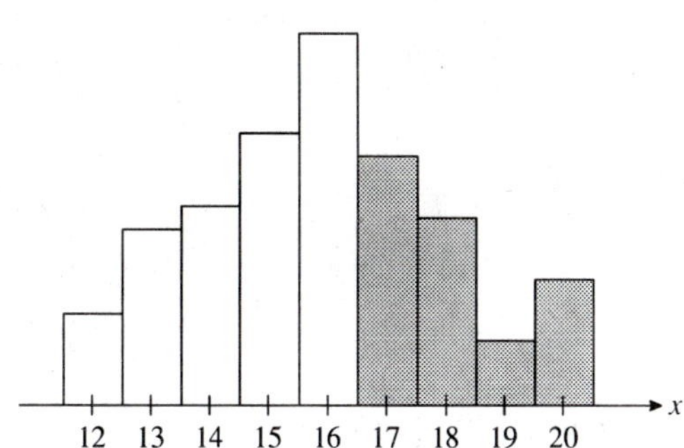

23. a.

x	1	2	3	4	5	6
$P(X = x)$	$\frac{1}{6}$	$\frac{1}{6}$	$\frac{1}{6}$	$\frac{1}{6}$	$\frac{1}{6}$	$\frac{1}{6}$

y	1	2	3	4	5	6
$P(Y = y)$	$\frac{1}{6}$	$\frac{1}{6}$	$\frac{1}{6}$	$\frac{1}{6}$	$\frac{1}{6}$	$\frac{1}{6}$

b.

$x + y$	2	3	4	5	6	7	8	9	10	11	12
$P(X + Y = x + y)$	$\frac{1}{36}$	$\frac{2}{36}$	$\frac{3}{36}$	$\frac{4}{36}$	$\frac{5}{36}$	$\frac{6}{36}$	$\frac{5}{36}$	$\frac{4}{36}$	$\frac{3}{36}$	$\frac{2}{36}$	$\frac{1}{36}$

25. a.

x	$P(X = x)$
0	.017
1	.067
2	.033
3	.117
4	.233
5	.133
6	.167
7	.1
8	.05
9	.067
10	.017

b.

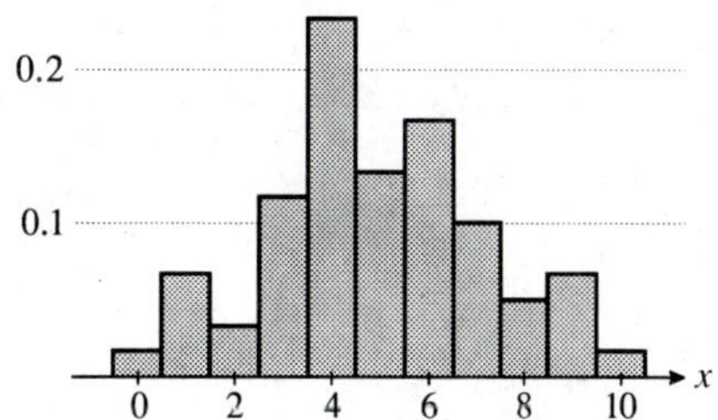

c. $P(X = 1) + P(X = 2) + P(X = 3) = .067 + .033 + .117 = .217$

27. a.

x	1	2	3	4	5	6	7	8	9	10
$P(X = x)$	.007	.029	.021	.079	.164	.15	.20	.207	.114	.029

b. $P(X = 6) + P(X = 7) + P(X = 8) + P(X = 9) + P(X = 10) = .15 + .20 + .207 + .114 + .029 = .7$

29. True. This follows from the definition.

Technology Exercises page 516

1.

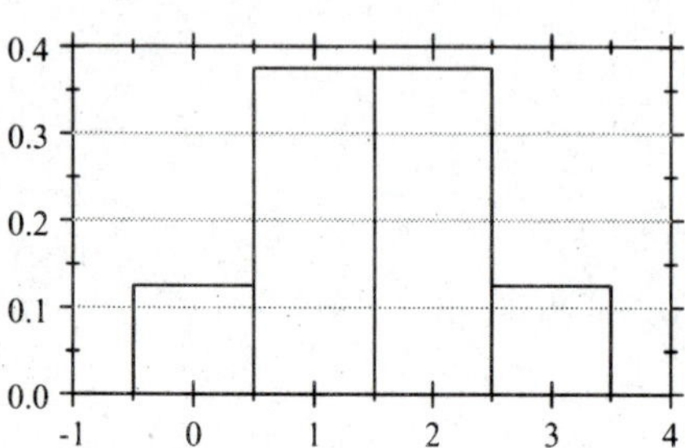

It is the same as Figure 13 on page 508 of the text.

3.

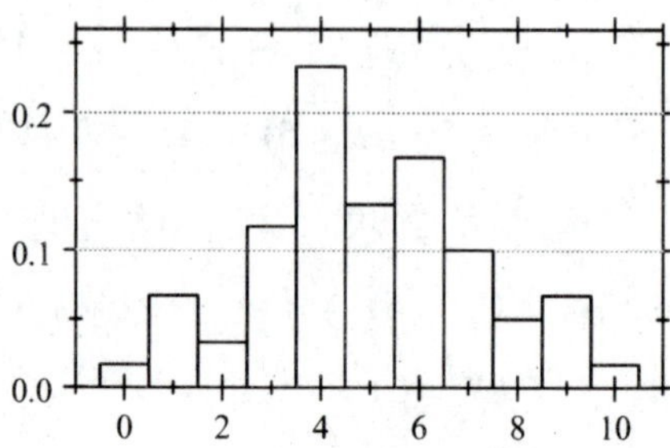

8.5 Expected Value

Problem-Solving Tips

1. The **expected value** of a random variable X is given by $E(X) = x_1p_1 + x_2p_2 + \cdots + x_np_n$ where $x_1, x_2, \ldots x_n$ are the values assumed by X and $p_1, p_2, \ldots, p_n$ are the associated probabilities.

2. If $P(E)$ is the probability of an event E occurring, then the **odds in favor** of E occurring are $\dfrac{P(E)}{P(E^c)}$ and the **odds against** E occurring are $\dfrac{P(E^c)}{P(E)}$.

3. If the odds in favor of an event E occurring are a to b, then the probability of E occurring is $P(E) = \dfrac{a}{a+b}$.

Concept Questions page 525

1. The expected value of a random variable X is given by $E(X) = x_1p_1 + x_2p_2 + \cdots + x_np_n$. Examples will vary.

3. a. The odds in favor of E occurring are $\dfrac{P(E)}{P(E^c)}$.

b. The odds in favor of E occurring are $\dfrac{a}{a+b}$.

Exercises page 525

1. a. The student's grade point average is given by $\dfrac{(2)(4)(3) + (3)(3)(3) + (4)(2)(3) + (1)(1)(3)}{(10)(3)} = 2.6$.

b.

x	0	1	2	3	4
$P(X = x)$	0	.1	.4	.3	.2

$E(X) = 1(.1) + 2(.4) + 3(.3) + 4(.2) = 2.6$.

3. $E(X) = -5(.12) + -1(.16) + 0(.26) + 1(.22) + 5(.14) + 8(.1) = .96$.

5. $E(X) = 0(.07) + 25(.12) + 50(.17) + 75(.14) + 100(.28) + 125(.18) + 150(.04) = 78.5$, or \$78.50.

7. A customer entering the store is expected to buy $E(X) = (0)(.42) + (1)(.36) + (2)(.14) + (3)(.05) + (4)(.03) = .91$, or .91 DVDs.

9. The expected number of accidents is given by $E(X) = (0)(.935) + (1)(.03) + (2)(.02) + (3)(.01) + (4)(.005) = .12$.

11. The expected number of machines that will break down on a given day is given by $E(X) = (0)(.43) + (1)(.19) + (2)(.12) + (3)(.09) + (4)(.04) + (5)(.03) + (6)(.03) + (7)(.02) + (8)(.05) = 1.73$.

13. The associated probabilities are $\frac{3}{50}, \frac{8}{50}, \ldots$, and $\frac{5}{50}$, respectively. Therefore, the expected interest rate is $(3.9)\left(\frac{3}{50}\right) + 4\left(\frac{8}{50}\right) + 4.1\left(\frac{12}{50}\right) + 4.2\left(\frac{14}{50}\right) + 4.3\left(\frac{8}{50}\right) + 4.4\left(\frac{5}{50}\right) = 4.162$, or approximately 4.16%.

15. The expected net earnings of a person who buys one ticket are $-1(.997) + 24(.002) + 99(.0006) + 499(.0002) + 1999(.0002) = -.39$, a loss of \$.39 per ticket.

17. The expected gain of the insurance company is given by $E(X) = .992(260) - (19{,}740)(.008) = 100$, or \$100.

19. His expected profit is
$E = (580{,}000 - 450{,}000)(.24) + (570{,}000 - 450{,}000)(.4) + (560{,}000 - 450{,}000)(.36) = 118{,}800$, or \$118,800.

21. City A: $E(X) = (10{,}000{,}000)(.2) - 250{,}000 = 1{,}750{,}000$, or \$1.75 million. City B: $E(X) = (7{,}000{,}000)(.3) - 200{,}000 = 1{,}900{,}000$, or \$1.9 million. We see that the company should bid for the rights in City B.

23. The expected number of houses sold per year at Company A is given by

$$E(X) = (12)(.02) + (13)(.03) + (14)(.05) + (15)(.07) + (16)(.07) + (17)(.16) + (18)(.17) + (19)(.13) + (20)(.11) + (21)(.09) + (22)(.06) + (23)(.03) + (24)(.01) = 18.09.$$

The expected number of houses sold per year at Company B is given by

$$E(X) = (6)(.01) + (7)(.04) + (8)(.07) + (9)(.06) + (10)(.11) + (11)(.12) + (12)(.19) + (13)(.17) + (14)(.13) + (15)(.04) + (16)(.03) + (17)(.02) + (18)(.01) = 11.77.$$

Thus, Sally's expected commission at Company A is $(.03)(308{,}000)(18.09) = 167{,}151.60$, or \$167,151.60. Her expected commission at Company B is $(.03)(474{,}000)(11.77) = 167{,}369.40$, or \$167,369.40. Based on these expectations, she should accept the job offer from Company B.

25. Maria expects her business to grow at the rate of $(5)(.12) + (4.5)(.24) + (3)(.4) + (0)(.2) + (-.5)(.04) = 2.86$, or 2.86%/year during the upcoming year.

27. The expected value of the winnings on a \$1 bet placed on a split is $E(X) = 17 \cdot \frac{2}{38} + (-1) \cdot \frac{36}{38} \approx -.0526$, a loss of 5.3 cents.

29. The expected value of a player's winnings are $(1)\left(\frac{18}{37}\right) + (-1)\left(\frac{19}{37}\right) = -\frac{1}{37} \approx -.027$, a loss of 2.7 cents per bet.

31. The odds in favor of E occurring are $\dfrac{P(E)}{P(E^c)} = \dfrac{.4}{.6}$, or 2 to 3. The odds against E occurring are 3 to 2.

33. The probability of E not occurring is given by $P(E) = \frac{2}{3+2} = \frac{2}{5} = .4$.

35. The probability that she will win her match is $P(E) = \frac{7}{7+5} = \frac{7}{12} \approx .583$.

37. The probability that the business deal will not go through is $P(E) = \frac{5}{5+9} = \frac{5}{14} \approx .357$.

39. Let X and Y be random variables that assume values $x_1, x_2, \ldots, x_n$ and $y_1, y_2, \ldots, y_n$ with probabilities $p_1, p_2, \ldots, p_n$, respectively.

a. $E(c) = cp_1 + cp_2 + \cdots + cp_n = c(1) = c$

b. $E(cX) = cx_1p_1 + cx_2p_2 + \cdots + cx_np_n = c(x_1p_1 + x_2p_2 + \cdots + x_np_n) = cE(X)$

c. $E(X+Y) = (x_1+y_1)p_1 + (x_2+y_2)p_2 + \cdots + (x_n+y_n)p_n$

$$= (x_1p_1 + x_2p_2 + \cdots + x_np_n) + (y_1p_1 + y_2p_2 + \cdots + y_np_n) = E(X) + E(Y)$$

d. $E(X-Y) = (x_1-y_1)p_1 + (x_2-y_2)p_2 + \cdots + (x_n-y_n)p_n$

$$= (x_1p_1 + x_2p_2 + \cdots + x_np_n) - (y_1p_1 + y_2p_2 + \cdots + y_np_n) = E(X) - E(Y)$$

41. a. The mean is $\dfrac{40 + 45 + 2(50) + 55 + 2(60) + 2(75) + 2(80) + 4(85) + 2(90) + 2(95) + 100}{20} = 74$, the mode is 85 (the value that appears most frequently), and the median is 80 (the middle value).

b. The mode is the least representative of this set of test scores.

43. We first arrange the numbers in increasing order:

$$0, 0, \underbrace{1, 1, \ldots, 1}_{9\text{ times}}, \underbrace{2, 2, \ldots, 2}_{16\text{ times}}, \underbrace{3, 3, \ldots, 3}_{12\text{ times}}, \underbrace{4, 4, \ldots, 4}_{8\text{ times}}, \underbrace{5, 5, \ldots, 5}_{6\text{ times}}, 6, 6, 6, 6, 7, 7, 8$$

There are 60 numbers, so the median is 3. This is close to the mean of 3.1 obtained in Example 1.

45. The average is $\frac{1}{10}(16.1 + 16 + 15.8 + 16 + 15.9 + 16.1 + 15.9 + 16 + 16 + 16.2) = 16$, or 16 oz. Next, we arrange the numbers in increasing order: 15.8, 15.9, 15.9, 16, 16, 16, 16, 16.1, 16.1, 16.2, so the median is $\frac{16+16}{2} = 16$, or 16 oz. The mode is also 16 oz.

47. True. This follows from the definition.

8.6 Variance and Standard Deviation

Problem-Solving Tips

1. The **variance** of a random variable X is a measure of the spread of a probability distribution about its mean. The variance of a random variable X is given by $\text{Var}(X) = p_1(x_1-\mu)^2 + p_2(x_2-\mu)^2 + \cdots + p_n(x_n-\mu)^2$, where $x_1, x_2, \ldots, x_n$ are the values assumed by X and $p_1 = P(X = x_1), p_2 = P(X = x_2), \ldots, p_n = P(X = x_n)$. The **standard deviation** of a random variable X is $\sigma = \sqrt{\text{Var}(X)}$.

2. **Chebychev's Inequality** gives the proportion of values of a random variable X lying within k standard deviations of the expected value of X. The probability that a randomly chosen outcome of the experiment lies between $\mu - k\sigma$ and $\mu + k\sigma$ is $P(\mu - k\sigma \le X \le \mu + k\sigma) \ge 1 - \dfrac{1}{k^2}$.

Concept Questions page 535

1. If a random variable has the probability distribution shown at right and expected value $E(X) = \mu$, then the variance of the random variable X is

x	x_1	x_2	x_3	$\cdots$	x_n
$P(X=x)$	p_1	p_2	p_3	$\cdots$	p_n

$\text{Var}(X) = p_1(x_1-\mu)^2 + p_2(x_2-\mu)^2 + \cdots + p_n(x_n-\mu)^2$ and the standard variation of the random variable X is given by $\sigma = \sqrt{\text{Var}(X)}$.

Exercises page 535

1. $\mu = (1)(.4) + (2)(.3) + 3(.2) + (4)(.1) = 2$,
$\text{Var}(X) = (.4)(1-2)^2 + (.3)(2-2)^2 + (.2)(3-2)^2 + (.1)(4-2)^2 = .4 + 0 + .2 + .4 = 1$, and $\sigma = \sqrt{1} = 1$.

3. $\mu = (-2)\left(\frac{1}{16}\right) + (-1)\left(\frac{4}{16}\right) + 0\left(\frac{6}{16}\right) + 1\left(\frac{4}{16}\right) + 2\left(\frac{1}{16}\right) = \frac{0}{16} = 0$,
$\text{Var}(X) = \frac{1}{16}(-2-0)^2 + \frac{4}{16}(-1-0)^2 + \frac{6}{16}(0-0)^2 + \frac{4}{16}(1-0)^2 + \frac{1}{16}(2-0)^2 = 1$, and $\sigma = \sqrt{1} = 1$.

5. $\mu = .1(430) + (.2)(480) + (.4)(520) + (.2)(565) + (.1)(580) = 518$,
$\text{Var}(X) = .1(430-518)^2 + (.2)(480-518)^2 + (.4)(520-518)^2 + (.2)(565-518)^2 + (.1)(580-518)^2 = 1891$, and $\sigma = \sqrt{1891} \approx 43.5$.

7. The mean of the histogram in Figure (b) is more concentrated about its mean than the histogram in Figure (a). Therefore, the histogram in Figure (a) has the larger variance.

9. $E(X) = 1(.1) + 2(.2) + 3(.3) + 4(.2) + 5(.2) = 3.2$, so
$\text{Var}(X) = (.1)(1-3.2)^2 + (.2)(2-3.2)^2 + (.3)(3-3.2)^2 + (.2)(4-3.2)^2 + (.2)(5-3.2)^2 = 1.56$.

11. $\mu = \frac{1+2+3+4+5+6+7+8}{8} = 4.5$, so
$$V(X) = \tfrac{1}{8}(1-4.5)^2 + \tfrac{1}{8}(2-4.5)^2 + \tfrac{1}{8}(3-4.5)^2 + \tfrac{1}{8}(4-4.5)^2 + \tfrac{1}{8}(5-4.5)^2 + \tfrac{1}{8}(6-4.5)^2 + \tfrac{1}{8}(7-4.5)^2 + \tfrac{1}{8}(8-4.5)^2 = 5.25.$$

13. **a.** Let X be the annual birth rate during the years 1997–2006.

b.

x	13.9	14.0	14.1	14.2	14.5	14.6	14.7
$P(X=x)$	.1	.2	.2	.1	.2	.1	.1

c. $E(X) = (.1)(13.9) + (.2)(14.0) + (.2)(14.1) + (.1)(14.2) + (.2)(14.5) + (.1)(14.6) + (.1)(14.7) = 14.26$,
$$\text{Var}(X) = (.1)(13.9-14.26)^2 + (.2)(14.0-14.26)^2 + (.2)(14.1-14.26)^2 + (.1)(14.2-14.26)^2 + (.2)(14.5-14.26)^2 + (.1)(14.6-14.26)^2 + (.1)(14.7-14.26)^2 = .0744,$$
and $\sigma = \sqrt{.0744} \approx .2728$.

15. **a.** For Mutual Fund A, $\mu_X = (.2)(-4) + (.5)(8) + (.3)(10) = 6.2$, or \$620, and
$\text{Var}(X) = (.2)(-4-6.2)^2 + (.5)(8-6.2)^2 + (.3)(10-6.2)^2 = 26.76$, or \$267,600.
For Mutual Fund B, $\mu_X = (.2)(-2) + (.4)(6) + (.4)(8) = 5.2$, or \$520, and
$\text{Var}(X) = (.2)(-2-5.2)^2 + (.4)(6-5.2)^2 + (.4)(8-5.2)^2 = 13.76$, or \$137,600.

b. Mutual Fund A

c. Mutual Fund B

17. $\text{Var}(X) = (.4)(1)^2 + (.3)(2)^2 + (.2)(3)^2 + (.1)(4)^2 - (2)^2 = 1$.

19. $\mu_X = \left(\frac{10}{500}\cdot 380 + \frac{20}{500}\cdot 390 + \frac{75}{500}\cdot 400 + \frac{85}{500}\cdot 410 + \frac{70}{500}\cdot 420 + \frac{90}{500}\cdot 450 + \frac{90}{500}\cdot 480 + \frac{55}{500}\cdot 500 + \frac{5}{500}\cdot 550\right)$

$= 439.6$, or \$439,600,

$$\begin{aligned}\text{Var}(X) = \Big[&\tfrac{10}{500}(380-439.6)^2 + \tfrac{20}{500}(390-439.6)^2 + \tfrac{75}{500}(400-439.6)^2 \\ &+ \tfrac{85}{500}(410-439.6)^2 + \tfrac{70}{500}(420-439.6)^2 + \tfrac{90}{500}(450-439.6)^2 \\ &+ \tfrac{90}{500}(480-439.6)^2 + \tfrac{55}{500}(500-439.6)^2 + \tfrac{5}{500}(550-439.6)^2\Big]\left[(1000)^2\right] \approx 1.44384 \times 10^9,\end{aligned}$$

and $\sigma_X = \sqrt{1.44384 \times 10^9} \approx 37{,}998$, or \$37,998.

21. Let X denote the random variable that is the average occupancy rate. The probability distribution of X is as follows:

x	94.7	95.1	95.2	95.6	96.1
Relative frequency	1	1	1	1	1
$P(X = x)$	$\frac{1}{5}$	$\frac{1}{5}$	$\frac{1}{5}$	$\frac{1}{5}$	$\frac{1}{5}$

$\mu_X = \frac{1}{5}(94.7 + 95.1 + 95.2 + 95.6 + 96.1) = 95.34$, or 95.34%,

$$\begin{aligned}\text{Var}(X) &= \tfrac{1}{5}\left[(94.7-95.34)^2 + (95.1-95.34)^2 + (95.2-95.34)^2 + (95.6-95.34)^2 + (96.1-95.34)^2\right] \\ &\approx .2264,\end{aligned}$$

and $\sigma_X = \sqrt{.2264} \approx .4758$, or approximately .5%.

23. $\mu_X = 22\left(\frac{1332}{27,127}\right) + 27\left(\frac{4219}{27,127}\right) + 32\left(\frac{6345}{27,127}\right) + 37\left(\frac{7598}{27,127}\right) + 42\left(\frac{7633}{27,127}\right) \approx 34.9456$, or 34.95,

$$\begin{aligned}\text{Var}(X) \approx &\left(\tfrac{1332}{27,127}\right)(22-34.9456)^2 + \left(\tfrac{4219}{27,127}\right)(27-34.9456)^2 + \left(\tfrac{6345}{27,127}\right)(32-34.9456)^2 \\ &+ \left(\tfrac{7598}{27,127}\right)(37-34.9456)^2 + \left(\tfrac{7633}{27,127}\right)(42-34.9456)^2 \approx 35.2622,\end{aligned}$$

and $\sigma_X \approx \sqrt{35.2622} \approx 5.938$, or approximately 5.94.

25. $\mu_X = \frac{1}{6}(-41 - 29 - 29 - 25 - 16 - 13) = -25.5$, or -25.5%,

$$\begin{aligned}\text{Var}(X) = \tfrac{1}{6}\{&[-41-(-25.5)]^2 + [-29-(-25.5)]^2 + [-29-(-25.5)]^2 \\ &+ [-25-(-25.5)]^2 + [-16-(-25.5)]^2 + [-13-(-25.5)]^2\} = 85.25,\end{aligned}$$

and $\sigma_X = \sqrt{85.25} \approx 9.23$.

27. $\mu_X = \frac{1}{7}(16.3 + 15.4 + 22.2 + 17.2 + 23.2 + 30.4 + 26.4) \approx 21.59$, or approximately 21.59%,

$$\begin{aligned}\text{Var}(X) \approx \tfrac{1}{7}\big[&(16.3-21.59)^2 + (15.4-21.59)^2 + (22.2-21.59)^2 + (17.2-21.59)^2 \\ &+ (23.2-21.59)^2 + (30.4-21.59)^2 + (26.4-21.59)^2\big] \approx 27.04,\end{aligned}$$

and $\sigma_X \approx \sqrt{27.04} = 5.2$.

29. $\mu_X = \frac{1}{7}(7.2 + 8.2 + 12.1 + 8.7 + 11.1 + 10.9 + 9.3) \approx 9.64$, or approximately 9.64%,

$$\begin{aligned}\text{Var}(X) \approx \tfrac{1}{7}\big[&(7.2-9.64)^2 + (8.2-9.64)^2 + (12.1-9.64)^2 + (8.7-9.64)^2 \\ &+ (11.1-9.64)^2 + (10.9-9.64)^2 + (9.3-9.64)^2\big] \approx 2.6853,\end{aligned}$$

and $\sigma_X \approx \sqrt{2.6853} \approx 1.64$.

31. $\mu_X = \frac{1}{9}(67 + 72 + 83 + 88 + 89 + 91 + 113 + 121 + 127) \approx 94.5556$, or approximately 94.56%,

$$\text{Var}(X) \approx \frac{1}{9}\left[(67 - 94.5556)^2 + (72 - 94.5556)^2 + (83 - 94.5556)^2 + (88 - 94.5556)^2 + (89 - 94.5556)^2 + (91 - 94.5556)^2 + (113 - 94.5556)^2 + (121 - 94.5556)^2 + (127 - 94.5556)^2\right] \approx 397.8025,$$

and $\sigma_X \approx \sqrt{397.8025} \approx 19.94$.

33.

x	1342	1428	1545	1707	1807	1815
Relative Frequency	1	1	1	1	1	1
$P(X = x)$	$\frac{1}{6}$	$\frac{1}{6}$	$\frac{1}{6}$	$\frac{1}{6}$	$\frac{1}{6}$	$\frac{1}{6}$

$\mu_x = \frac{1}{6}(1342 + 1428 + 1545 + 1707 + 1807 + 1815) \approx 1607.33$,

$$\text{Var}(X) \approx \frac{1}{6}\left[(1342 - 1607.33)^2 + (1428 - 1607.33)^2 + (1545 - 1607.33)^2 + (1707 - 1607.33)^2 + (1807 - 1607.33)^2 + (1815 - 1607.33)^2\right] \approx 33{,}228.889,$$

and $\sigma_X = \sqrt{32{,}228.889} \approx 182.2879$, so the average number of hours worked is approximately 1607 and the standard deviation is approximately 182 hours.

35. $\mu_X = \frac{1}{10}(3.76 + 3.7 + 3.66 + 3.5 + 3.46 + 3.46 + 3.43 + 3.2 + 2.88 + 2.19) = 3.324$,

$$\text{Var}(X) = \frac{1}{10}\left[(3.76 - 3.324)^2 + (3.7 - 3.324)^2 + (3.66 - 3.324)^2 + (3.5 - 3.324)^2 + 2(3.46 - 3.324)^2 + (3.43 - 3.324)^2 + (3.2 - 3.324)^2 + (2.88 - 3.324)^2 + (2.19 - 3.324)^2\right] \approx .202204,$$

and $\sigma_X \approx \sqrt{.202204} \approx .4497$.

37. a. Using Chebychev's inequality $P(\mu - k\sigma \le X \le \mu + k\sigma) \ge 1 - 1/k^2$ with $\mu - k\sigma = 42 - k(2) = 38$ and $k = 2$, we have $P(\mu - k\sigma \le X \le \mu + k\sigma) \ge 1 - \frac{1}{2^2} = 1 - \frac{1}{4} = \frac{3}{4}$, or at least .75.

b. Using Chebychev's inequality with $\mu - k\sigma = 42 - k(2) = 32$ and $k = 5$, we have $P(\mu - k\sigma \le X \le \mu + k\sigma) \ge 1 - \frac{1}{5^2} = 1 - \frac{1}{25} = \frac{24}{25}$, or at least .96.

39. Here $\mu = 50$ and $\sigma = 1.4$. We require that $c = k\sigma$, so $k = \dfrac{c}{1.4}$. Next, we solve $.96 = 1 - \left(\dfrac{1.4}{c}\right)^2$, obtaining $\dfrac{1.96}{c^2} = .04$, $c^2 = \frac{1.96}{.04} = 49$, and $c = 7$.

41. Using Chebychev's inequality with $\mu - k\sigma = 24 - k(3) = 20$ and $k = \frac{4}{3}$, we have $P(\mu - k\sigma \le X \le \mu + k\sigma) \ge 1 - \frac{1}{(4/3)^2} = 1 - \frac{9}{16} = \frac{7}{16}$, or at least .4375.

43. Using Chebychev's inequality with $\mu - k\sigma = 52{,}000 - k(500) = 50{,}000$ and $k = 4$, we have $P(\mu - k\sigma \le X \le \mu + k\sigma) \ge 1 - \frac{1}{4^2} = 1 - \frac{1}{16} = \frac{15}{16}$, or at least .9375.

45. True. This follows from the definition.

Technology Exercises page 541

1. a.

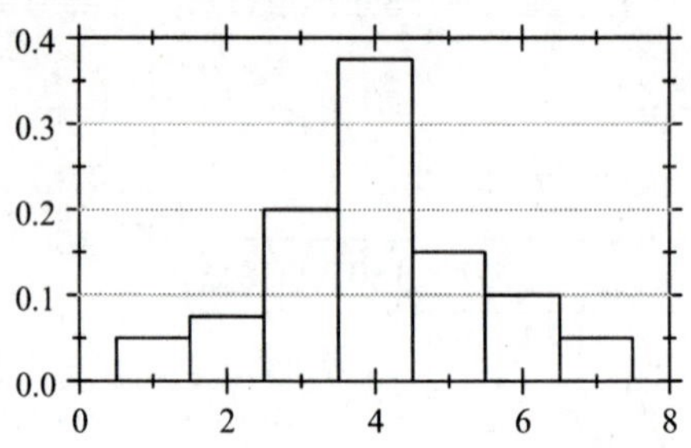

b. $\mu = 4$ and $\sigma \approx 1.40$.

3. a.

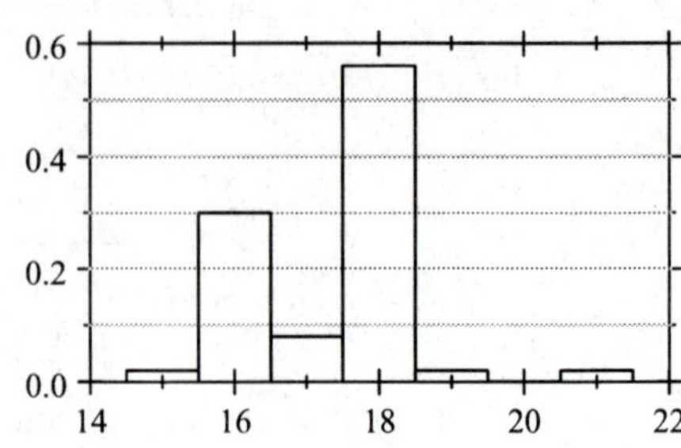

b. $\mu = 17.34$ and $\sigma \approx 1.11$.

5. a. Let X denote the random variable that gives the weight of a carton of sugar.

b.

x	4.96	4.97	4.98	4.99	5.00	5.01	5.02	5.03	5.04	5.05	5.06
$P(X = x)$	$\frac{3}{30}$	$\frac{4}{30}$	$\frac{4}{30}$	$\frac{1}{30}$	$\frac{1}{30}$	$\frac{5}{30}$	$\frac{3}{30}$	$\frac{3}{30}$	$\frac{4}{30}$	$\frac{1}{30}$	$\frac{1}{30}$

c. $\mu = 5.00467 \approx 5.00$, $\text{Var}(X) \approx .0009$, and $\sigma \approx \sqrt{.0009} = .03$.

7. a.

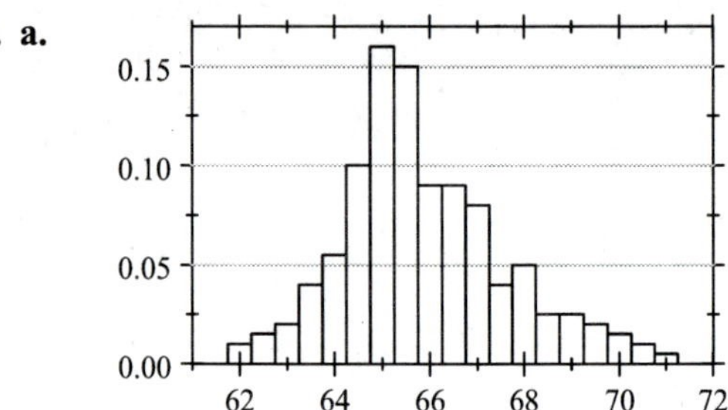

b. $\mu = 65.875$ and $\sigma = 1.73$.

CHAPTER 8 Concept Review Questions page 542

1. conditional

3. *a posteriori* probability

5. finite, infinite, continuous

7. a. $\dfrac{P(E)}{P(E^c)}$ **b.** $\dfrac{a}{a+b}$

CHAPTER 8 Review Exercises page 543

1. $P(A \cap E) = (.3)(.6) = .18$.

3. $P(C \cap E) = (.2)(.3) = .06$.

5. $P(A \mid E) = \frac{.18}{.49} \approx .37$.

7. $P(E)P(F) = (.6)(.32) = .192 \neq P(E \cap F) = 0.22$, and so E and F are not independent.

9. a. $P(A) = 1 - P(A^c) = 1 - \frac{1}{8} = \frac{7}{8}$.

b. $P(B) = 1 - P(B^c) = 1 - \frac{1}{8} = \frac{7}{8}$.

c. $P(A \cap B) = \frac{7}{8}$ and $P(A) \cdot P(B) = \frac{7}{8} \cdot \frac{7}{8} = \frac{49}{64}$. Because $P(A \cap B) \neq P(A) \cdot P(B)$, they are not independent events.

11. $P(E) = \dfrac{7 \cdot 6 \cdot 5 \cdot 4 \cdot 3}{7^5} \approx .150$.

13. Let E, F, and G denote the events that the first toss is even, the second toss is odd, and the third toss is a 1, respectively. Then $P(E) = \frac{1}{2}$, $P(F) = \frac{1}{2}$, and $P(G) = \frac{1}{6}$. Because the outcomes are independent, the required probability is $P(E \cap F \cap G) = P(E) \cdot P(F) \cdot P(G) = \frac{1}{2} \cdot \frac{1}{2} \cdot \frac{1}{6} = \frac{1}{24}$.

15. The probability that all three cards are aces is $\dfrac{C(4,3)}{C(52,3)} \approx .00018$.

17.

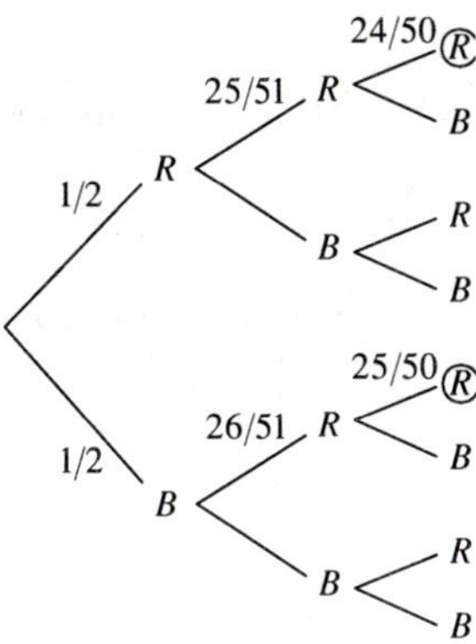

The required probability is

$\frac{1}{2} \cdot \frac{25}{51} \cdot \frac{24}{50} + \frac{1}{2} \cdot \frac{26}{51} \cdot \frac{25}{50} = \frac{1250}{5100} \approx .245$.

19.

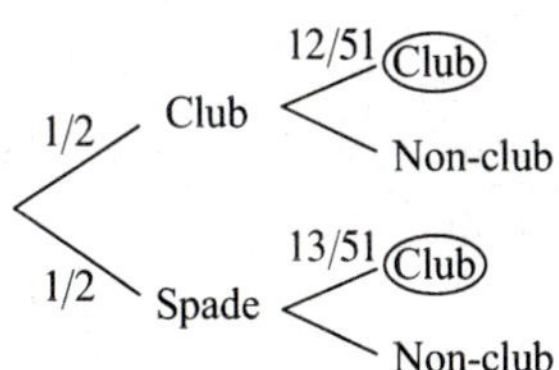

The probability that the second card is a club given that the first card was black is $\frac{1}{2} \cdot \frac{12}{51} + \frac{1}{2} \cdot \frac{13}{51} \approx .245$.

21.

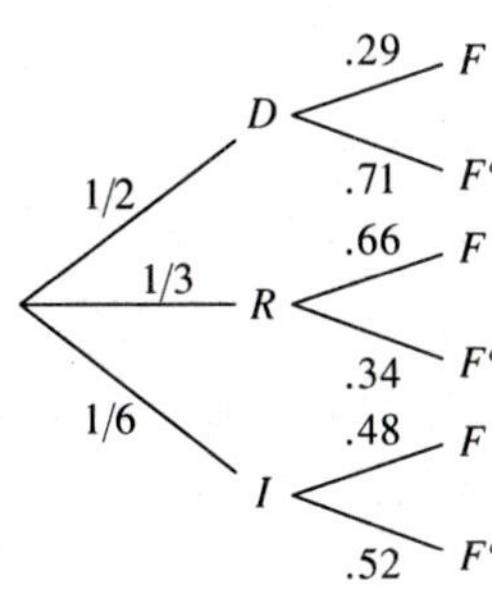

The required probability is

$$P(I \mid F) = \frac{\frac{1}{6}(.48)}{\frac{1}{2}(.29) + \frac{1}{3}(.66) + \frac{1}{6}(.48)} \approx .180,$$ or

approximately 18%.

23. a. $S = \{\text{WWW, BWW, WBW, WWB, BBW, BWB, WBB, BBB}\}$

b.

Outcome	WWW	BWW	WBW	WWB	BBW	BWB	WBB	BBB
Value	0	1	1	1	2	2	2	3

c.

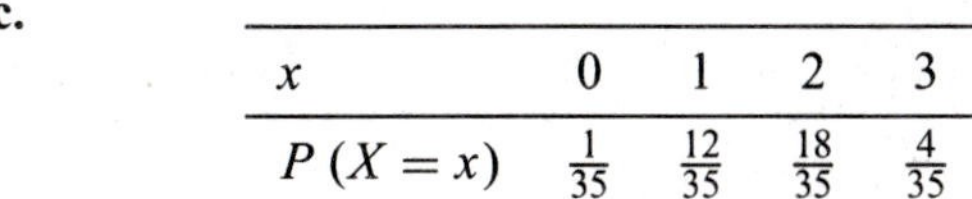

x	0	1	2	3
$P(X = x)$	$\frac{1}{35}$	$\frac{12}{35}$	$\frac{18}{35}$	$\frac{4}{35}$

d. 0.5 0.4 0.3 0.2 0.1 0 1 2 3 x

25. a. $P(1 \le X \le 4) = .1 + .2 + .3 + .2 = .8$.

b. $\mu = 0(.1) + 1(.1) + 2(.2) + 3(.3) + 4(.2) + 5(.1) = 2.7$,
$\text{Var}(X) = .1(0-2.7)^2 + .1(1-2.7)^2 + .2(2-2.7)^2 + .3(3-2.7)^2 + .2(4-2.7)^2 + .1(5-2.7)^2 \approx 2.01$, and $\sigma \approx \sqrt{2.01} \approx 1.42$.

27. The company's expected gross profit is $E(X) = (.3)(80{,}000) + (.60)(75{,}000) + (.10)(70{,}000) - 64{,}000 = 12{,}000$, or \$12,000.

29.

x	24.0	27.2	26.0	28.0	28.6
Relative frequency	1	1	1	1	1
$P(X = x)$	$\frac{1}{5}$	$\frac{1}{5}$	$\frac{1}{5}$	$\frac{1}{5}$	$\frac{1}{5}$

$\mu_X = \frac{1}{5}(24.0 + 27.2 + 26.0 + 28.0 + 28.6) = 26.76$,

$$\text{Var}(X) = \tfrac{1}{5}\left[(24.0-26.76)^2 + (27.2-26.76)^2 + (26.0-26.76)^2 + (28.0-26.76)^2 + (28.6-26.76)^2\right] \approx 2.6624,$$

and $\sigma_X \approx \sqrt{2.6624} \approx 1.6317$. Thus, the average quarterly revenue is €26.76 billion and the standard deviation is €1.6317 billion.

CHAPTER 8 Before Moving On... page 544

1. Because A and B are independent, $P(A \cap B) = P(A) \cdot P(B) = .3(.6) = .18$, so $P(A \cup B) = P(A) + P(B) - P(A \cap B) = .3 + .6 - .18 = .72$.

2. $P(A \mid D) = \dfrac{.4(.2)}{.4(.2) + .6(.3)} \approx .308$.

3.

x	−3	−2	0	1	2	3
$P(X = x)$	.05	.1	.25	.3	.2	.1

4. a. $P(X \le 0) = P(X = 0) + P(X = -1) + P(X = -3) + P(X = -4) = .28 + .32 + .14 + .06 = .8$.

b. $P(-4 \le X \le 1) = 1 - P(X = 3) = 1 - .08 = .92$.

5. $\mu_X = (-3)(.08) + (-1)(.24) + 0(.32) + 1(.16) + 3(.12) + 5(.08) = .44$, $\text{Var}(X) = .08(-3-.44)^2 + .24(-1-.44)^2 + .32(0-.44)^2 + .16(1-.44)^2 + .12(3-.44)^2 + .08(5-.44)^2 \approx 4.0064$, and $\sigma_X \approx \sqrt{4.0064} \approx 2.0016$.

9 THE DERIVATIVE

9.1 Limits

Problem-Solving Tips

In this section, an important theorem was introduced (properties of limits on page 102). After you read Theorem 1, try to express the theorem in your own words. While you will not usually be required to prove the theorem in this course, you will be asked to understand the results of the theorem. For example, Theorem 1 gives us the properties of limits that allow us to evaluate sums, differences, products, quotients, powers, and constant multiples of functions at specified values, with certain restrictions. You should be able to use limit notation to write out each of these properties. You should also be able to use these properties to evaluate the limits of functions.

Here are some tips for solving the problems in the exercises that follow:

1. **To find the limit of a function** $f(x)$ **as** $x \to a$, where a is a real number, first try substituting a for x in the rule for f and simplify the result.

2. **To evaluate the limit of a quotient** that has the indeterminate form $0/0$:

   a. Replace the given function with an appropriate one that takes on the same values as the original function everywhere except at $x = a$.

   b. Evaluate the limit of this function as x approaches a.

Concept Questions page 562

1. The values of $f(x)$ can be made as close to 3 as we please by taking x sufficiently close to $x = 2$.

3. **a.** $$\begin{aligned}\lim_{x\to 4}\sqrt{x}\left(2x^2+1\right) &= \lim_{x\to 4}\left(\sqrt{x}\right)\lim_{x\to 4}\left(2x^2+1\right) && \text{(Property 4)}\\ &= \sqrt{4}\left[2(4)^2+1\right] && \text{(Properties 1 and 3)}\\ &= 66\end{aligned}$$

   b. $$\begin{aligned}\lim_{x\to 1}\left(\frac{2x^2+x+5}{x^4+1}\right)^{3/2} &= \left(\lim_{x\to 1}\frac{2x^2+x+5}{x^4+1}\right)^{3/2} && \text{(Property 1)}\\ &= \left(\frac{2+1+5}{1+1}\right)^{3/2} && \text{(Properties 2, 3, and 5)}\\ &= 4^{3/2} = 8\end{aligned}$$

5. $\lim_{x\to\infty} f(x) = L$ means $f(x)$ can be made as close to L as we please by taking x sufficiently large.
   $\lim_{x\to-\infty} f(x) = M$ means $f(x)$ can be made as close to M as we please by taking negative x as large as we please in absolute value.

Exercises page 562

1. $\lim_{x \to -2} f(x) = 3.$

3. $\lim_{x \to 3} f(x) = 3.$

5. $\lim_{x \to -2} f(x) = 3.$

7. The limit does not exist. If we consider any value of x to the right of $x = -2$, $-2 < f(x) \leq 0$. If we consider values of x to the left of $x = -2$, $0 \leq f(x) \leq 2$. Because $f(x)$ does not approach a fixed number as x approaches $x = -2$, we conclude that the limit does not exist.

9.

x	1.9	1.99	1.999	2.001	2.01	2.1
$f(x)$	4.61	4.9601	4.9960	5.004	5.0401	5.41

$\lim_{x \to 2} (x^2 + 1) = 5.$

11.

x	−0.1	−0.01	−0.001	0.001	0.01	0.1
$f(x)$	−1	−1	−1	1	1	1

The limit does not exist.

13.

x	0.9	0.99	0.999	1.001	1.01	1.1
$f(x)$	100	10,000	1,000,000	1,000,000	10,000	100

The limit does not exist.

15.

x	0.9	0.99	0.999	1.001	1.01	1.1
$f(x)$	2.9	2.99	2.999	3.001	3.01	3.1

$$\lim_{x \to 1} \frac{x^2 + x - 2}{x - 1} = 3.$$

17.

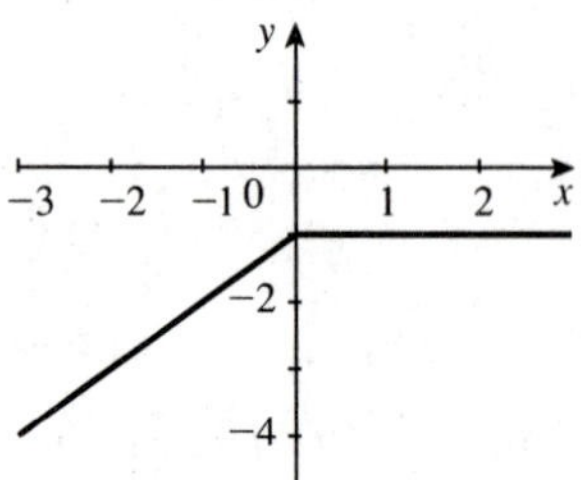

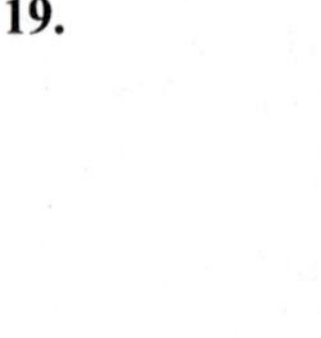

$\lim_{x \to 0} f(x) = -1.$

19.

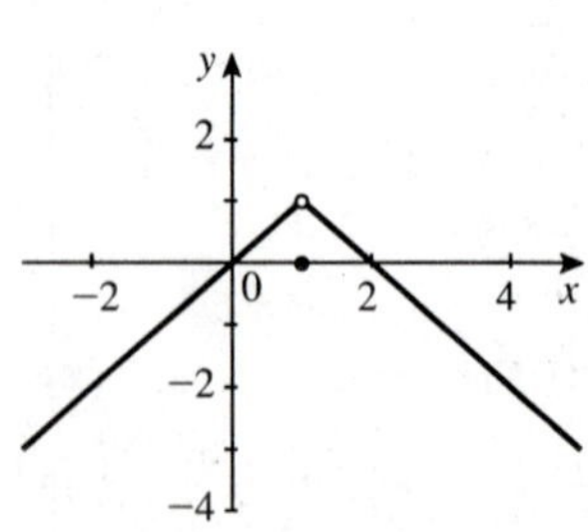

$\lim_{x \to 1} f(x) = 1.$

21.

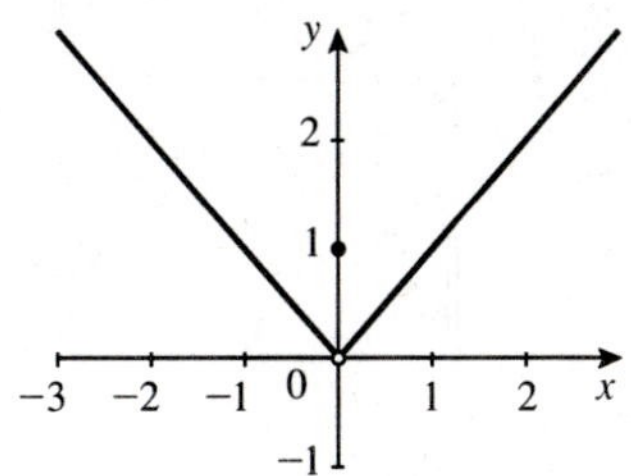

$\lim_{x\to 0} f(x) = 0.$

23. $\lim_{x\to 2} 4 = 4.$

25. $\lim_{x\to 3} x = 3.$

27. $\lim_{x\to 1} \left(1 - 3x^2\right) = 1 - 3(1)^2 = -2.$

29. $\lim_{x\to 1} \left(2x^3 - 3x^2 + x + 2\right) = 2(1)^3 - 3(1)^2 + 1 + 2$

$= 2.$

31. $\lim_{s\to 0} \left(2s^2 - 1\right)(2s + 4) = (-1)(4) = -4.$

33. $\lim_{x\to 2} \frac{2x+1}{x+4} = \frac{2(2)+1}{2+4} = \frac{5}{6}.$

35. $\lim_{x\to 2} \sqrt{x+2} = \sqrt{2+2} = 2.$

37. $\lim_{x\to -3} \sqrt{2x^4 + x^2} = \sqrt{2(-3)^4 + (-3)^2} = \sqrt{162+9}$

$= \sqrt{171} = 3\sqrt{19}.$

39. $\lim_{x\to -1} \frac{\sqrt{x^2+8}}{2x+5} = \frac{\sqrt{(-1)^2+8}}{2(-1)+5} = \frac{\sqrt{9}}{3} = 1.$

41. $\lim_{x\to a} \left[f(x) - g(x)\right] = \lim_{x\to a} f(x) - \lim_{x\to a} g(x)$

$= 3 - 4 = -1.$

43. $\lim_{x\to a} \left[4f(x) - 3g(x)\right] = \lim_{x\to a} 4f(x) - \lim_{x\to a} 3g(x)$

$= 4(3) - 3(4) = 0.$

45. $\lim_{x\to a} \sqrt{g(x)} = \lim_{x\to a} \sqrt{4} = 2.$

47. $\lim_{x\to a} \frac{2f(x) - g(x)}{f(x)g(x)} = \frac{2(3) - (4)}{(3)(4)} = \frac{2}{12} = \frac{1}{6}.$

49. $\lim_{x\to 1} \frac{x^2-1}{x-1} = \lim_{x\to 1} \frac{(x-1)(x+1)}{x-1} = \lim_{x\to 1} (x+1)$

$= 1 + 1 = 2.$

51. $\lim_{x\to 0} \frac{x^2 - x}{2x} = \lim_{x\to 0} \frac{x(x-1)}{2x} = \frac{1}{2} \lim_{x\to 0} (x-1)$

$= \frac{1}{2}(0-1) = -\frac{1}{2}.$

53. $\lim_{x\to -5} \frac{x^2 - 25}{x+5} = \lim_{x\to -5} \frac{(x+5)(x-5)}{x+5}$

$= \lim_{x\to -5} (x-5) = -10.$

55. $\lim_{x\to 1} \frac{x}{x-1}$ does not exist.

57. $\lim_{x\to -2} \frac{x^2 - x - 6}{x^2 + x - 2} = \lim_{x\to -2} \frac{(x-3)(x+2)}{(x+2)(x-1)} = \lim_{x\to -2} \frac{x-3}{x-1} = \frac{-2-3}{-2-1} = \frac{5}{3}.$

59. $\lim_{x\to 1} \frac{\sqrt{x}-1}{x-1} = \lim_{x\to 1} \frac{\sqrt{x}-1}{x-1} \cdot \frac{\sqrt{x}+1}{\sqrt{x}+1} = \lim_{x\to 1} \frac{x-1}{(x-1)\left(\sqrt{x}+1\right)} = \lim_{x\to 1} \frac{1}{\sqrt{x}+1} = \frac{1}{2}.$

61. $\lim_{x\to 1} \dfrac{2x-2}{x^3+x^2-2x} = \lim_{x\to 1} \dfrac{2(x-1)}{x(x-1)(x+2)} = \lim_{x\to 1} \dfrac{2}{x(x+2)} = \dfrac{2}{3}.$

63. $\lim_{x\to\infty} f(x) = \infty$ (does not exist) and $\lim_{x\to-\infty} f(x) = \infty$ (does not exist).

65. $\lim_{x\to\infty} f(x) = 0$ and $\lim_{x\to-\infty} f(x) = 0$.

67. $\lim_{x\to\infty} f(x) = -\infty$ (does not exist) and $\lim_{x\to-\infty} f(x) = -\infty$ (does not exist).

69. $f(x) = \dfrac{1}{x^2+1}$.

x	1	10	100	1000
$f(x)$	0.5	0.009901	0.0001	0.000001

x	-1	-10	-100	-1000
$f(x)$	0.5	0.009901	0.0001	0.000001

$\lim_{x\to\infty} f(x) = \lim_{x\to-\infty} f(x) = 0.$

71. $f(x) = 3x^3 - x^2 + 10$.

x	1	5	10	100	1000
$f(x)$	12	360	2910	2.99×10^6	2.999×10^9

x	-1	-5	-10	-100	-1000
$f(x)$	6	-390	-3090	-3.01×10^6	-3.0×10^9

$\lim_{x\to\infty} f(x) = \infty$ (does not exist) and $\lim_{x\to-\infty} f(x) = -\infty$ (does not exist).

73. $\lim_{x\to\infty} \dfrac{3x+2}{x-5} = \lim_{x\to\infty} \dfrac{3+\dfrac{2}{x}}{1-\dfrac{5}{x}} = \dfrac{3}{1} = 3.$

75. $\lim_{x\to-\infty} \dfrac{3x^3+x^2+1}{x^3+1} = \lim_{x\to-\infty} \dfrac{3+\dfrac{1}{x}+\dfrac{1}{x^3}}{1+\dfrac{1}{x^3}} = 3.$

77. $\lim_{x\to-\infty} \dfrac{x^4+1}{x^3-1} = \lim_{x\to-\infty} \dfrac{x+\dfrac{1}{x^3}}{1-\dfrac{1}{x^3}} = -\infty$; that is, the limit does not exist.

79. $\lim_{x\to\infty} \dfrac{x^5-x^3+x-1}{x^6+2x^2+1} = \lim_{x\to\infty} \dfrac{\dfrac{1}{x}-\dfrac{1}{x^3}+\dfrac{1}{x^5}-\dfrac{1}{x^6}}{1+\dfrac{2}{x^4}+\dfrac{1}{x^6}} = 0.$

81. a. The cost of removing 50% of the pollutant is $C(50) = \dfrac{0.5(50)}{100-50} = 0.5$, or \$500,000. Similarly, we find that the cost of removing 60%, 70%, 80%, 90%, and 95% of the pollutant is \$750,000, \$1,166,667, \$2,000,000, \$4,500,000, and \$9,500,000, respectively.

b. $\lim_{x\to 100} \frac{0.5x}{100-x} = \infty$, which means that the cost of removing the pollutant increases without bound if we wish to remove almost all of the pollutant.

83. $\lim_{x\to\infty} \overline{C}(x) = \lim_{x\to\infty} 2.2 + \frac{2500}{x} = 2.2$, or \$2.20 per DVD. In the long run, the average cost of producing x DVDs approaches \$2.20/disc.

85. a. $T(1) = \frac{120}{1+4} = 24$, or \$24 million. $T(2) = \frac{120(2)^2}{8} = 60$, or \$60 million. $T(3) = \frac{120(3)^2}{13} \approx 83.1$, or \$83.1 million.

b. In the long run, the movie will gross $\lim_{x\to\infty} \frac{120x^2}{x^2+4} = \lim_{x\to\infty} \frac{120}{1+\frac{4}{x^2}} = 120$, or \$120 million.

87. $\lim_{t\to\infty} \frac{1000}{1+199e^{-0.8t}} = 1000$, or 1000 children.

89. $\lim_{x\to\infty} \frac{ax}{x+b} = \lim_{x\to\infty} \frac{a}{1+\frac{b}{x}} = a$. As the amount of substrate becomes very large, the initial speed approaches the constant a moles per liter per second.

91. True. $\lim_{x\to 0} f(x)g(x) = \left[\lim_{x\to 0} f(x)\right]\left[\lim_{x\to 0} g(x)\right] = (4)(0) = 0$.

93. False. Let $f(x) = (x-3)^2$ and $g(x) = x-3$. Then $\lim_{x\to 3} f(x) = 0$ and $\lim_{x\to 3} g(x) = 0$, but $\lim_{x\to 3} \frac{f(x)}{g(x)} = \lim_{x\to 3} \frac{(x-3)^2}{x-3} = \lim_{x\to 3} (x-3) = 0$.

95. False. Neither of the limits $\lim_{x\to 1} \frac{2x}{x-1}$ and $\lim_{x\to 1} \frac{2}{x-1}$ exists.

97. Consider the functions $f(x) = \begin{cases} -1 & \text{if } x < 0 \\ 1 & \text{if } x \ge 0 \end{cases}$ and $g(x) = \begin{cases} 1 & \text{if } x < 0 \\ -1 & \text{if } x \ge 0 \end{cases}$ Then $\lim_{x\to 0} f(x)$ and $\lim_{x\to 0} g(x)$ do not exist, but $\lim_{x\to 0} [f(x)g(x)] = \lim_{x\to 0} (-1) = -1$. This example does not contradict Theorem 1 because the hypothesis of Theorem 1 is that $\lim_{x\to 0} f(x)$ and $\lim_{x\to 0} g(x)$ both exist. It does not say anything about the situation where one or both of these limits fails to exist.

Using Technology page 568

1. 5 **3.** 3 **5.** $\frac{2}{3}$ **7.** $e^2 \approx 7.38906$

9.

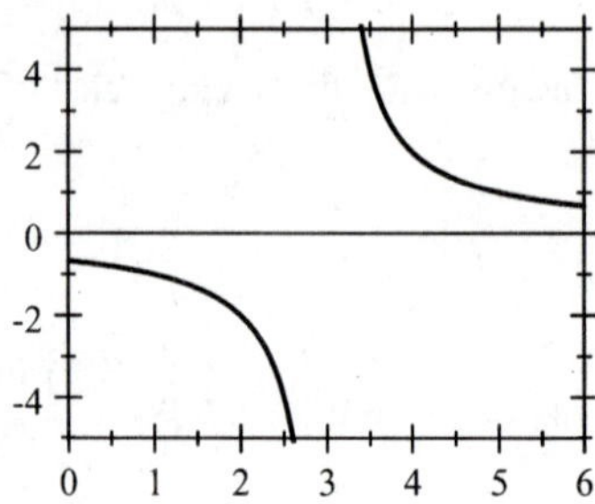

From the graph we see that $f(x)$ does not approach any finite number as x approaches 3.

11. a.

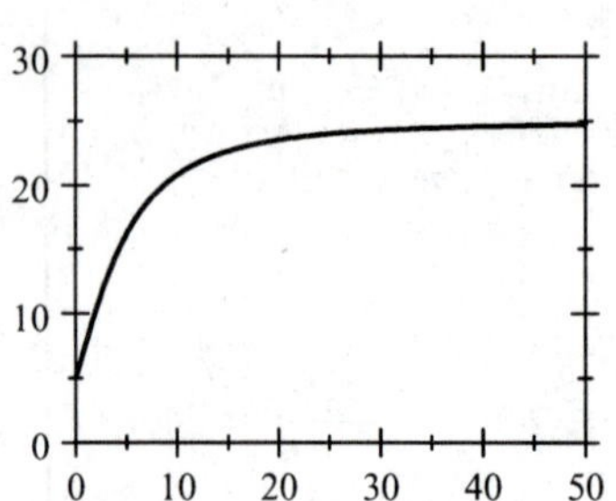

b. $\lim_{t\to\infty} \frac{25t^2 + 125t + 200}{t^2 + 5t + 40} = 25$, so in the long run the population will approach 25,000.

c.

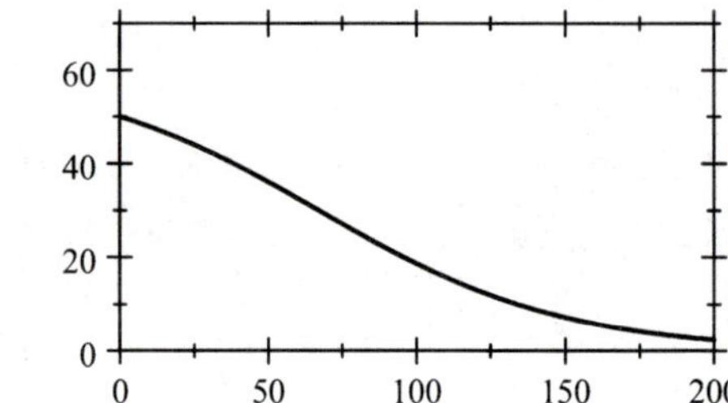

13. a. The initial population of crocodiles is $P(0) = \frac{300}{6} = 50$.

b. $\lim_{t\to\infty} P(t) = \lim_{t\to\infty} \frac{300e^{-0.024t}}{5e^{-0.024t} + 1} = \frac{0}{0+1} = 0$.

9.2 One-Sided Limits and Continuity

Problem-Solving Tips

The problem-solving skills that you learned in earlier sections are building blocks for the rest of the course. You can't skip a section or a concept and hope to understand the material in a new section. It just won't work. If you don't build a strong foundation, you won't be able to understand the later concepts. For example, in this section we discussed one-sided limits. You need to understand the definition of a limit before you can understand what is meant by a one-sided limit. That means you should be able to express the definition of a limit in your own words. If you can't grasp a new concept, it may well be that you still don't understand a previous concept. If so, you need to go back and review the earlier section before you go on.

As another example, the continuity of polynomial and rational functions is discussed on page 572 of the text. If you don't remember how to identify polynomial and rational functions, go back to Chapter 1 and review this material.

Here are some tips for solving the problems in the exercises that follow:

1. **To evaluate the limit of a piecewise-defined function** at a real number a, follow the same procedure that you used to evaluate a piecewise-defined function. First find the subdomain that a lies in, then use the rule for that subdomain to find the limit of f at a.

2. **To determine the values of x at which a function is continuous**, check to see if the function is a polynomial or rational function. A polynomial function $y = P(x)$ is continuous at every value of x and a rational function is continuous at every value of x where the denominator is nonzero.

Concept Questions page 576

1. $\lim_{x\to 3^-} f(x) = 2$ means $f(x)$ can be made as close to 2 as we please by taking x sufficiently close to but to the left of $x = 3$. $\lim_{x\to 3^+} f(x) = 4$ means $f(x)$ can be made as close to 4 as we please by taking x sufficiently close to but to the right of $x = 3$.

3. **a.** f is continuous at a if $\lim_{x\to a} f(x) = f(a)$.

   b. f is continuous on an interval I if f is continuous at each point in I.

5. Refer to page 574 in the text. Answers will vary.

Exercises page 577

1. $\lim_{x\to 2^-} f(x) = 3$ and $\lim_{x\to 2^+} f(x) = 2$, so $\lim_{x\to 2} f(x)$ does not exist.

3. $\lim_{x\to -1^-} f(x) = \infty$ and $\lim_{x\to -1^+} f(x) = 2$, so $\lim_{x\to -1} f(x)$ does not exist.

5. $\lim_{x\to 1^-} f(x) = 0$ and $\lim_{x\to 1^+} f(x) = 2$, so $\lim_{x\to 1} f(x)$ does not exist.

7. $\lim_{x\to 0^-} f(x) = -2$ and $\lim_{x\to 0^+} f(x) = 2$, so $\lim_{x\to 0} f(x)$ does not exist.

9. True. 11. True. 13. False. 15. True. 17. False. 19. True.

21. $\lim_{x\to 1^+} (2x + 5) = 7$.

23. $\lim_{x\to 2^-} \frac{x-4}{x+2} = \frac{2-4}{2+2} = -\frac{1}{2}$.

25. $\lim_{x\to 0^+} \frac{1}{x}$ does not exist because $\frac{1}{x} \to \infty$ as $x \to 0$ from the right.

27. $\lim_{x\to 0^+} \frac{x-1}{x^2+1} = \frac{-1}{1} = -1$.

29. $\lim_{x\to 0^+} \sqrt{x} = \sqrt{\lim_{x\to 0^+} x} = 0$.

31. $\lim_{x\to -2^+} \left(2x + \sqrt{2+x}\right) = \lim_{x\to -2^+} 2x + \lim_{x\to -2^+} \sqrt{2+x} = -4 + 0 = -4$.

33. $\lim_{x\to 1^-} \frac{1+x}{1-x} = \infty$; that is, the limit does not exist.

35. $\lim_{x\to 2^-} \frac{x^2-4}{x-2} = \lim_{x\to 2^-} \frac{(x+2)(x-2)}{x-2} = \lim_{x\to 2^-} (x+2) = 4$.

37. $\lim_{x\to 0^-} f(x) = \lim_{x\to 0^-} 2x = 0$ and $\lim_{x\to 0^+} f(x) = \lim_{x\to 0^+} x^2 = 0$.

39. The function is discontinuous at $x = 0$. Conditions 2 and 3 are violated.

41. The function is continuous everywhere.

43. The function is discontinuous at $x = 0$. Condition 3 is violated.

45. f is continuous for all values of x.

47. f is continuous for all values of x. Note that $x^2 + 1 \geq 1 > 0$.

49. f is discontinuous at $x = \frac{1}{2}$, where the denominator is 0. Thus, f is continuous on $\left(-\infty, \frac{1}{2}\right)$ and $\left(\frac{1}{2}, \infty\right)$.

51. Observe that $x^2 + x - 2 = (x + 2)(x - 1) = 0$ if $x = -2$ or $x = 1$, so f is discontinuous at these values of x. Thus, f is continuous on $(-\infty, -2)$, $(-2, 1)$, and $(1, \infty)$.

53. f is continuous everywhere since all three conditions are satisfied.

55. f is continuous everywhere since all three conditions are satisfied.

57. Because the denominator $x^2 - 1 = (x - 1)(x + 1) = 0$ if $x = -1$ or 1, we see that f is discontinuous at -1 and 1.

59. Because $x^2 - 3x + 2 = (x - 2)(x - 1) = 0$ if $x = 1$ or 2, we see that the denominator is zero at these points and so f is discontinuous at these numbers.

61. The function f is discontinuous at $x = 1, 2, 3, \ldots, 12$ because the limit of f does not exist at these points.

63. Having made steady progress up to $x = x_1$, Michael's progress comes to a standstill at that point. Then at $x = x_2$ a sudden breakthrough occurs and he then continues to solve the problem.

65. Conditions 2 and 3 are not satisfied at any of these points.

67.

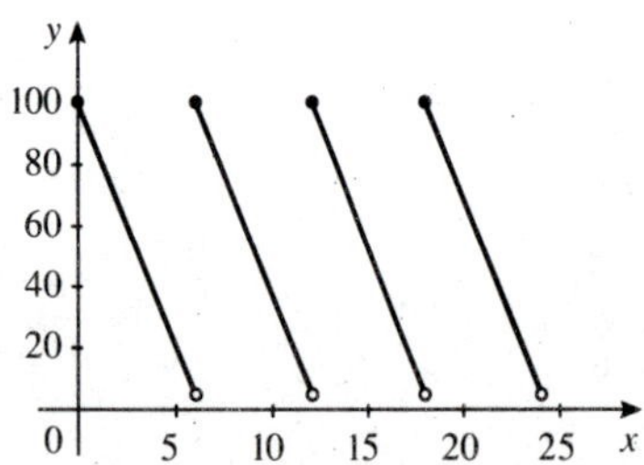

f is discontinuous at $x = 6, 12, 18$, and 24.

69.

$$f(x) = \begin{cases} 2 & \text{if } 0 < x \leq \frac{1}{2} \\ 3 & \text{if } \frac{1}{2} < x \leq 1 \\ \vdots & \vdots \\ 10 & \text{if } x > 4 \end{cases}$$

f is discontinuous at $x = \frac{1}{2}, 1, 1\frac{1}{2}, \ldots, 4$.

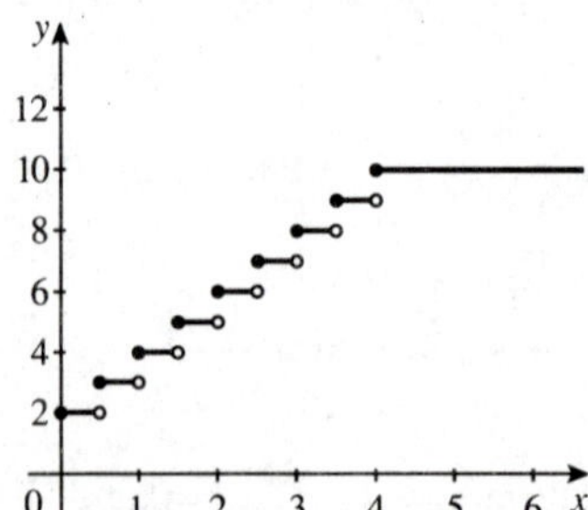

71. a. $\lim_{t \to 0^+} S(t) = \lim_{t \to 0^+} \frac{a}{t} + b = \infty$. As the time taken to excite the tissue is made shorter and shorter, the electric current gets stronger and stronger.

b. $\lim_{t\to\infty} \frac{a}{t} + b = b$. As the time taken to excite the tissue is made longer and longer, the electric current gets weaker and weaker and approaches b.

73. We require that $f(1) = 1 + 2 = 3 = \lim_{x\to 1^+} kx^2 = k$, so $k = 3$.

75. a. Yes, because if $f + g$ were continuous at a, then $g = (f + g) - f$ would be continuous (the difference of two continuous functions is continuous), and this would imply that g is continuous, a contradiction.

b. No. Consider the functions f and g defined by $f(x) = \begin{cases} -1 & \text{if } x < 0 \\ 1 & \text{if } x \geq 0 \end{cases}$ and $g(x) = \begin{cases} 1 & \text{if } x < 0 \\ -1 & \text{if } x \geq 0 \end{cases}$

Both f and g are discontinuous at $x = 0$, but $f + g$ is continuous everywhere.

77. a. f is a polynomial of degree 2 and is therefore continuous everywhere, including the interval $[1, 3]$.

b. $f(1) = 3$ and $f(3) = -1$ and so f must have at least one zero in $(1, 3)$.

79. a. f is a polynomial of degree 3 and is therefore continuous on $[-1, 1]$.

b. $f(-1) = (-1)^3 - 2(-1)^2 + 3(-1) + 2 = -1 - 2 - 3 + 2 = -4$ and $f(1) = 1 - 2 + 3 + 2 = 4$. Because $f(-1)$ and $f(1)$ have opposite signs, we see that f has at least one zero in $(-1, 1)$.

81. $f(0) = 6$, $f(3) = 3$, and f is continuous on $[0, 3]$. Thus, the Intermediate Value Theorem guarantees that there is at least one value of x for which $f(x) = 4$. Solving $f(x) = x^2 - 4x + 6 = 4$, we find $x^2 - 4x + 2 = 0$. Using the quadratic formula, we find that $x = 2 \pm \sqrt{2}$. Because $2 + \sqrt{2}$ does not lie in $[0, 3]$, we see that $x = 2 - \sqrt{2} \approx 0.59$.

83. $x^5 + 2x - 7 = 0$

Step	Interval in which a root lies
1	(1, 2)
2	(1, 1.5)
3	(1.25, 1.5)
4	(1.25, 1.375)
5	(1.3125, 1.375)
6	(1.3125, 1.34375)
7	(1.328125, 1.34375)
8	(1.3359375, 1.34375)

We see that a root is approximately 1.34.

85. a. $h(0) = 4 + 64(0) - 16(0) = 4$ and $h(2) = 4 + 64(2) - 16(4) = 68$.

b. The function h is continuous on $[0, 2]$. Furthermore, the number 32 lies between 4 and 68. Therefore, the Intermediate Value Theorem guarantees that there is at least one value of t in $(0, 2]$ such that $h(t) = 32$, that is, Joan must see the ball at least once during the time the ball is in the air.

c. We solve $h(t) = 4 + 64t - 16t^2 = 32$, obtaining $16t^2 - 64t + 28 = 0$, $4t^2 - 16t + 7 = 0$, and $(2t - 1)(2t - 7) = 0$. Thus, $t = \frac{1}{2}$ or $t = \frac{7}{2}$. Joan sees the ball on its way up half a second after it was thrown and again 3 seconds later when it is on its way down. Note that the ball hits the ground when $t \approx 4.06$, but Joan sees it approximately half a second before it hits the ground.

87. False. Take $f(x)=\begin{cases}-1 & \text{if } x<2\\ 4 & \text{if } x=2\\ 1 & \text{if } x>2\end{cases}$ Then $f(2)=4$, but $\lim_{x\to 2} f(x)$ does not exist.

89. False. Consider $f(x)=\begin{cases}0 & \text{if } x<2\\ 3 & \text{if } x\geq 2\end{cases}$ Then $\lim_{x\to 2^+} f(x)=f(2)=3$, but $\lim_{x\to 2^-} f(x)=0$.

91. False. Consider $f(x)=\begin{cases}2 & \text{if } x<5\\ 3 & \text{if } x>5\end{cases}$ Then $f(5)$ is not defined, but $\lim_{x\to 5^-} f(x)=2$.

93. False. Let $f(x)=\begin{cases}x & \text{if } x\neq 0\\ 1 & \text{if } x=0\end{cases}$ Then $\lim_{x\to 0^+} f(x)=\lim_{x\to 0^-} f(x)$, but $f(0)=1$.

95. False. Let $f(x)=\begin{cases}1/x & \text{if } x\neq 0\\ 0 & \text{if } x=0\end{cases}$ Then f is continuous for all $x\neq 0$ and $f(0)=0$, but $\lim_{x\to 0} f(x)$ does not exist.

97. The statement is false. The Intermediate Value Theorem says that there is at least one number c in $[a,b]$ such that $f(c)=M$ if M is a number between $f(a)$ and $f(b)$.

99. **a.** Both $g(x)=x$ and $h(x)=\sqrt{1-x^2}$ are continuous on $[-1,1]$ and so $f(x)=x-\sqrt{1-x^2}$ is continuous on $[-1,1]$.

b. $f(-1)=-1$ and $f(1)=1$, and so f has at least one zero in $(-1,1)$.

c. Solving $f(x)=0$, we have $x=\sqrt{1-x^2}$, $x^2=1-x^2$, and $2x^2=1$, so $x=\frac{\pm\sqrt{2}}{2}$.

101. Consider the function f defined by $f(x)=\begin{cases}-1 & \text{if } -1\leq x<0\\ 1 & \text{if } 0\leq x<1\end{cases}$ Then $f(-1)=-1$ and $f(1)=1$, but if we take the number $\frac{1}{2}$, which lies between $y=-1$ and $y=1$, there is no value of x such that $f(x)=\frac{1}{2}$.

Using Technology page 583

1.

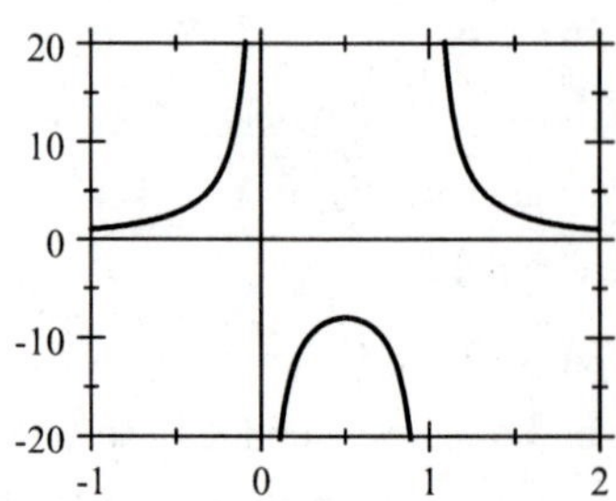

The function is discontinuous at $x=0$ and $x=1$.

3.

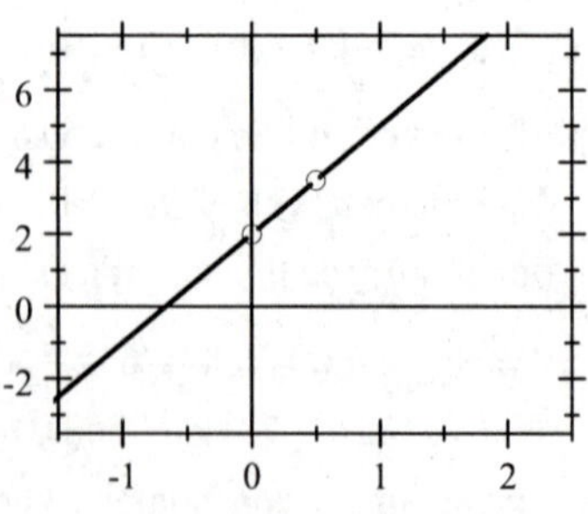

The function is discontinuous at $x=0$ and $\frac{1}{2}$.

5.

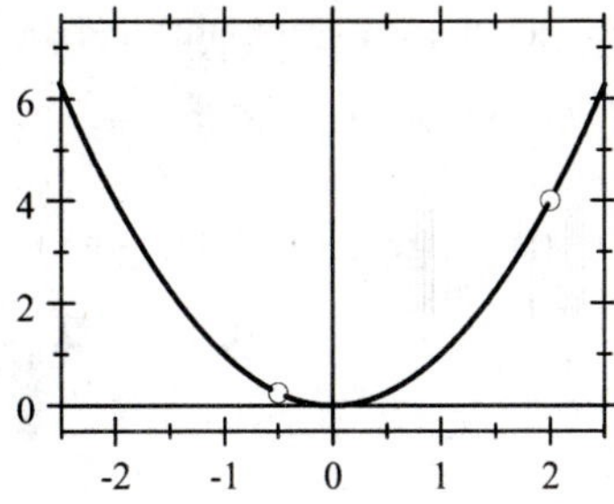

The function is discontinuous at $x = -\frac{1}{2}$ and 2.

7.

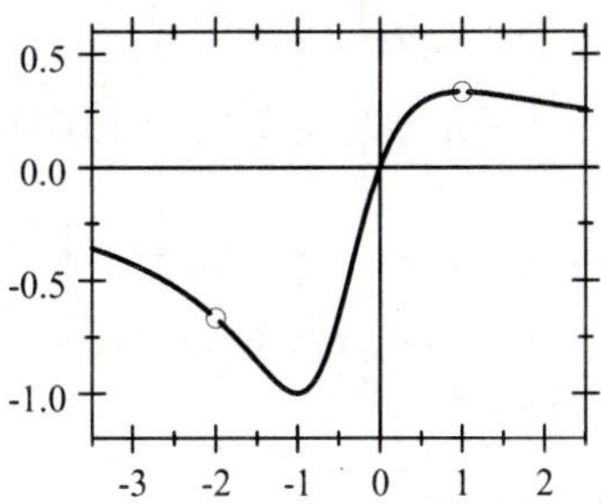

The function is discontinuous at $x = -2$ and 1.

9.

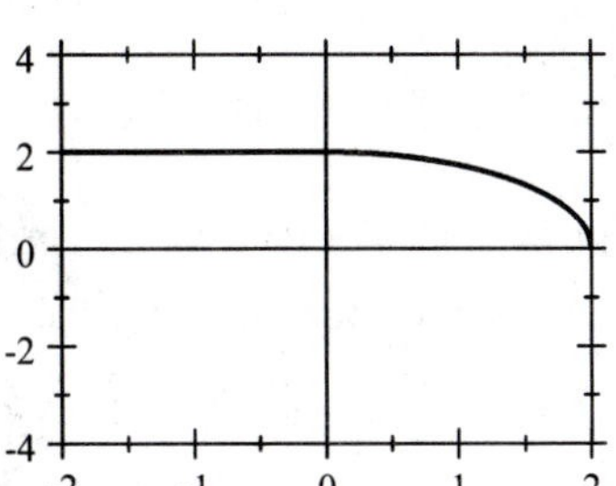

11.

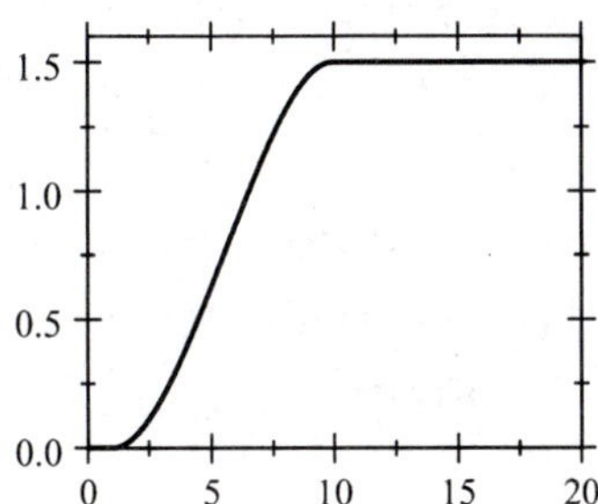

9.3 The Derivative

Problem-Solving Tips

When you solve an applied problem, it is important to understand the question in mathematical terms. For example, if you are given a function $f(t)$ describing the size of a country's population at any time t and asked to find the rate of change of that country's population at any time t, this means that you need to find the derivative of the given function; that is, find $f'(t)$. If you are then asked to find the population of the country at a specified time, say $t = 2$, you need to evaluate the function at $t = 2$; that is, find $f(2)$. On the other hand, if you are asked to find the rate of change of the population at time $t = 2$, then you need to evaluate the derivative of the function at the value $t = 2$; that is find $f'(2)$. Here again, the key is to be familiar with the terminology and notation introduced in the chapter.

Here are some tips for solving the problems in the exercises that follow:

1. **To find the slope of the tangent line to the graph of a function at an arbitrary point** on the graph of that function, find the derivative of f; that is, find $f'(x)$.

2. **To find the slope of the tangent line to the graph of a function at a given point** (x_0, y_0) on the graph of that function, find $f'(x)$ and then evaluate $f'(x_0)$.

Concept Questions page 595

1. **a.** $m = \dfrac{f(2+h) - f(2)}{h}$

   b. The slope of the tangent line is $\lim\limits_{h \to 0} \dfrac{f(2+h) - f(2)}{h}$.

3. a. The expression $\dfrac{f(x+h)-f(x)}{h}$ gives (i) the slope of the secant line passing through the points $(x, f(x))$ and $(x+h, f(x+h))$, and (ii) the average rate of change of f over the interval $[x, x+h]$.

b. The expression $\lim\limits_{h\to 0}\dfrac{f(x+h)-f(x)}{h}$ gives (i) the slope of the tangent line to the graph of f at the point $(x, f(x))$, and (ii) the instantaneous rate of change of f at x.

Exercises page 596

1. The rate of change of the average infant's weight when $t = 3$ is $\frac{7.5}{5}$, or 1.5 lb/month. The rate of change of the average infant's weight when $t = 18$ is $\frac{3.5}{6}$, or approximately 0.58 lb/month. The average rate of change over the infant's first year of life is $\frac{22.5-7.5}{12}$, or 1.25 lb/month.

3. The rate of change of the percentage of households watching television at 4 p.m. is $\frac{12.3}{4}$, or approximately 3.1 percent per hour. The rate at 11 p.m. is $\frac{-42.3}{2} = -21.15$, that is, it is dropping off at the rate of approximately 21.2% per hour.

5. a. Car A is travelling faster than Car B at t_1 because the slope of the tangent line to the graph of f is greater than the slope of the tangent line to the graph of g at t_1.

b. Their speed is the same because the slope of the tangent lines are the same at t_2.

c. Car B is travelling faster than Car A.

d. They have both covered the same distance and are once again side by side at t_3.

7. a. P_2 is decreasing faster at t_1 because the slope of the tangent line to the graph of g at t_1 is greater than the slope of the tangent line to the graph of f at t_1.

b. P_1 is decreasing faster than P_2 at t_2.

c. Bactericide B is more effective in the short run, but bactericide A is more effective in the long run.

9. $f(x) = 13$.

Step 1 $f(x+h) = 13$.

Step 2 $f(x+h) - f(x) = 13 - 13 = 0$.

Step 3 $\dfrac{f(x+h)-f(x)}{h} = \dfrac{0}{h} = 0$.

Step 4 $f'(x) = \lim\limits_{h\to 0}\dfrac{f(x+h)-f(x)}{h} = \lim\limits_{h\to 0} 0 = 0$.

11. $f(x) = 2x + 7$.

Step 1 $f(x+h) = 2(x+h) + 7$.

Step 2 $f(x+h) - f(x) = [2(x+h)+7] - (2x+7) = 2h$.

Step 3 $\dfrac{f(x+h)-f(x)}{h} = \dfrac{2h}{h} = 2$.

Step 4 $f'(x) = \lim\limits_{h\to 0}\dfrac{f(x+h)-f(x)}{h} = \lim\limits_{h\to 0} 2 = 2$.

13. $f(x) = 2x^2$.

Step 1 $f(x+h) = 2(x+h)^2 = 2x^2 + 4xh + 2h^2$.

Step 2 $f(x+h) - f(x) = \left(2x^2 + 4xh + 2h^2\right) - 2x^2 = 4xh + 2h^2 = h(4x + 2h)$.

Step 3 $\dfrac{f(x+h) - f(x)}{h} = \dfrac{h(4x+2h)}{h} = 4x + 2h$.

Step 4 $f'(x) = \lim\limits_{h\to 0} \dfrac{f(x+h) - f(x)}{h} = \lim\limits_{h\to 0}(4x + 2h) = 4x$.

15. $f(x) = -x^2 + 3x$.

Step 1 $f(x+h) = -(x+h)^2 + 3(x+h) = -x^2 - 2xh - h^2 + 3x + 3h$.

Step 2 $f(x+h) - f(x) = \left(-x^2 - 2xh - h^2 + 3x + 3h\right) - \left(-x^2 + 3x\right) = -2xh - h^2 + 3h$

$= h(-2x - h + 3)$.

Step 3 $\dfrac{f(x+h) - f(x)}{h} = \dfrac{h(-2x - h + 3)}{h} = -2x - h + 3$.

Step 4 $f'(x) = \lim\limits_{h\to 0} \dfrac{f(x+h) - f(x)}{h} = \lim\limits_{h\to 0}(-2x - h + 3) = -2x + 3$.

17. $f(x) = 2x + 7$.

Step 1 $f(x+h) = 2(x+h) + 7 = 2x + 2h + 7$.

Step 2 $f(x+h) - f(x) = 2x + 2h + 7 - 2x - 7 = 2h$.

Step 3 $\dfrac{f(x+h) - f(x)}{h} = \dfrac{2h}{h} = 2$.

Step 4 $f'(x) = \lim\limits_{h\to 0} \dfrac{f(x+h) - f(x)}{h} = \lim\limits_{h\to 0} 2 = 2$.

Therefore, $f'(x) = 2$. In particular, the slope at $x = 2$ is 2. Therefore, an equation of the tangent line is $y - 11 = 2(x - 2)$ or $y = 2x + 7$.

19. $f(x) = 3x^2$.We first compute $f'(x) = 6x$ (see Exercise 13). Because the slope of the tangent line is $f'(1) = 6$, we use the point-slope form of the equation of a line and find that an equation is $y - 3 = 6(x - 1)$, or $y = 6x - 3$.

21. $f(x) = -1/x$. We first compute $f'(x)$ using the four-step process:

Step 1 $f(x+h) = -\dfrac{1}{x+h}$.

Step 2 $f(x+h) - f(x) = -\dfrac{1}{x+h} + \dfrac{1}{x} = \dfrac{-x + (x+h)}{x(x+h)} = \dfrac{h}{x(x+h)}$.

Step 3 $\dfrac{f(x+h) - f(x)}{h} = \dfrac{\dfrac{h}{x(x+h)}}{h} = \dfrac{1}{x(x+h)}$.

Step 4 $f'(x) = \lim\limits_{h\to 0} \dfrac{f(x+h) - f(x)}{h} = \lim\limits_{h\to 0} \dfrac{1}{x(x+h)} = \dfrac{1}{x^2}$.

The slope of the tangent line is $f'(3) = \frac{1}{9}$. Therefore, an equation is $y - \left(-\frac{1}{3}\right) = \frac{1}{9}(x - 3)$, or $y = \frac{1}{9}x - \frac{2}{3}$.

23. **a.** $f(x) = 2x^2 + 1$.

Step 1 $f(x+h) = 2(x+h)^2 + 1 = 2x^2 + 4xh + 2h^2 + 1$.

Step 2 $f(x+h) - f(x) = (2x^2 + 4xh + 2h^2 + 1) - (2x^2 + 1)$

$$= 4xh + 2h^2 = h(4x + 2h).$$

Step 3 $\dfrac{f(x+h) - f(x)}{h} = \dfrac{h(4x+2h)}{h} = 4x + 2h$.

Step 4 $f'(x) = \lim\limits_{h \to 0} \dfrac{f(x+h) - f(x)}{h} = \lim\limits_{h \to 0} (4x + 2h) = 4x$.

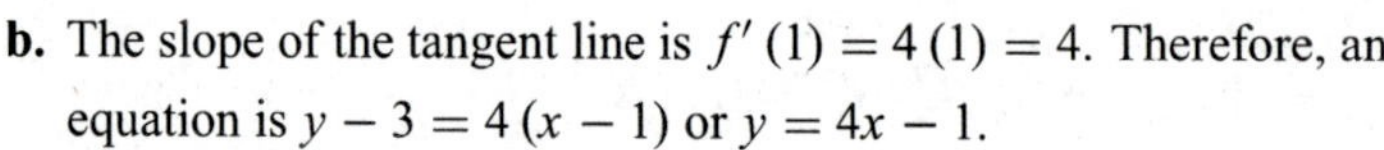

b. The slope of the tangent line is $f'(1) = 4(1) = 4$. Therefore, an equation is $y - 3 = 4(x - 1)$ or $y = 4x - 1$.

c.

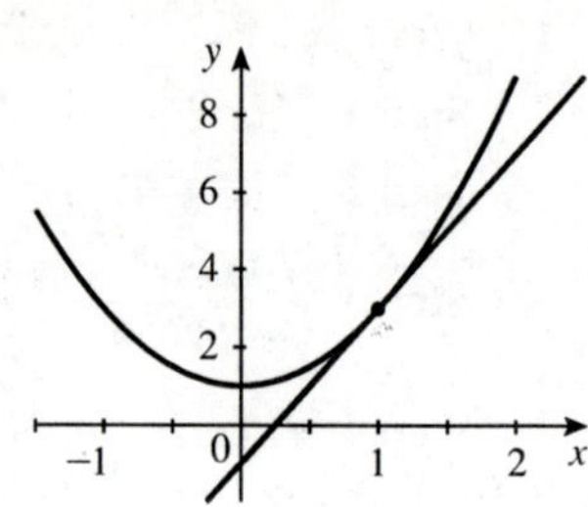

25. **a.** $f(x) = x^2 - 2x + 1$. We use the four-step process:

Step 1 $f(x+h) = (x+h)^2 - 2(x+h) + 1 = x^2 + 2xh + h^2 - 2x - 2h + 1$.

Step 2 $f(x+h) - f(x) = (x^2 + 2xh + h^2 - 2x - 2h + 1) - (x^2 - 2x + 1) = 2xh + h^2 - 2h$

$$= h(2x + h - 2).$$

Step 3 $\dfrac{f(x+h) - f(x)}{h} = \dfrac{h(2x+h-2)}{h} = 2x + h - 2$.

Step 4 $f'(x) = \lim\limits_{h \to 0} \dfrac{f(x+h) - f(x)}{h} = \lim\limits_{h \to 0} (2x + h - 2)$

$$= 2x - 2.$$

b. At a point on the graph of f where the tangent line to the curve is horizontal, $f'(x) = 0$. Then $2x - 2 = 0$, or $x = 1$. Because $f(1) = 1 - 2 + 1 = 0$, we see that the required point is $(1, 0)$.

c.

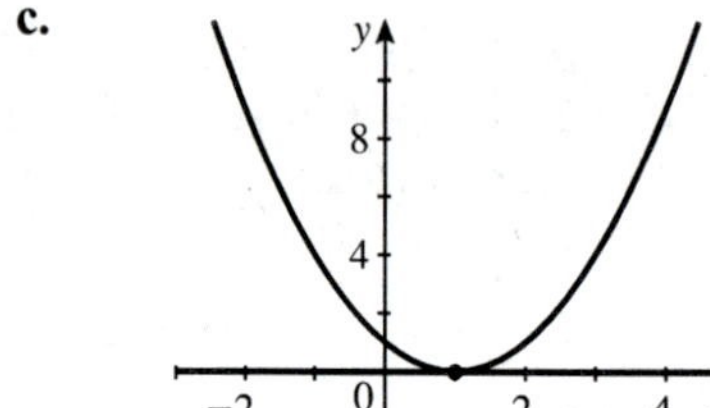

d. It is changing at the rate of 0 units per unit change in x.

27. **a.** $f(x) = x^2 + x$, so $\dfrac{f(3) - f(2)}{3 - 2} = \dfrac{(3^2 + 3) - (2^2 + 2)}{1} = 6$,

$\dfrac{f(2.5) - f(2)}{2.5 - 2} = \dfrac{(2.5^2 + 2.5) - (2^2 + 2)}{0.5} = 5.5$, and $\dfrac{f(2.1) - f(2)}{2.1 - 2} = \dfrac{(2.1^2 + 2.1) - (2^2 + 2)}{0.1} = 5.1$.

b. We first compute $f'(x)$ using the four-step process.

Step 1 $f(x+h) = (x+h)^2 + (x+h) = x^2 + 2xh + h^2 + x + h$.

Step 2 $f(x+h) - f(x) = (x^2 + 2xh + h^2 + x + h) - (x^2 + x) = 2xh + h^2 + h = h(2x + h + 1)$.

Step 3 $\dfrac{f(x+h) - f(x)}{h} = \dfrac{h(2x+h+1)}{h} = 2x + h + 1$.

Step 4 $f'(x) = \lim\limits_{h \to 0} \dfrac{f(x+h) - f(x)}{h} = \lim\limits_{h \to 0} (2x + h + 1) = 2x + 1$.

The instantaneous rate of change of y at $x = 2$ is $f'(2) = 2(2) + 1$, or 5 units per unit change in x.

c. The results of part (a) suggest that the average rates of change of f at $x = 2$ approach 5 as the interval $[2, 2 + h]$ gets smaller and smaller ($h = 1$, 0.5, and 0.1). This number is the instantaneous rate of change of f at $x = 2$ as computed in part (b).

29. a. $f(t) = 2t^2 + 48t$. The average velocity of the car over the time interval [20, 21] is
$\dfrac{f(21) - f(20)}{21 - 20} = \dfrac{\left[2(21)^2 + 48(21)\right] - \left[2(20)^2 + 48(20)\right]}{1} = 130\ \dfrac{\text{ft}}{\text{s}}$. Its average velocity over [20, 20.1] is
$\dfrac{f(20.1) - f(20)}{20.1 - 20} = \dfrac{\left[2(20.1)^2 + 48(20.1)\right] - \left[2(20)^2 + 48(20)\right]}{0.1} = 128.2\ \dfrac{\text{ft}}{\text{s}}$. Its average velocity over
[20, 20.01] is $\dfrac{f(20.01) - f(20)}{20.01 - 20} = \dfrac{\left[2(20.01)^2 + 48(20.01)\right] - \left[2(20)^2 + 48(20)\right]}{0.01} = 128.02\ \dfrac{\text{ft}}{\text{s}}$.

b. We first compute $f'(t)$ using the four-step process.

Step 1 $f(t+h) = 2(t+h)^2 + 48(t+h) = 2t^2 + 4th + 2h^2 + 48t + 48h$.

Step 2 $f(t+h) - f(t) = \left(2t^2 + 4th + 2h^2 + 48t + 48h\right) - \left(2t^2 + 48t\right) = 4th + 2h^2 + 48h$

$= h(4t + 2h + 48)$.

Step 3 $\dfrac{f(t+h) - f(t)}{h} = \dfrac{h(4t + 2h + 48)}{h} = 4t + 2h + 48$.

Step 4 $f'(t) = \lim\limits_{t\to 0} \dfrac{f(t+h) - f(t)}{h} = \lim\limits_{t\to 0} (4t + 2h + 48) = 4t + 48$.

The instantaneous velocity of the car at $t = 20$ is $f'(20) = 4(20) + 48$, or 128 ft/s.

c. Our results show that the average velocities do approach the instantaneous velocity as the intervals over which they are computed decreases.

31. a. We solve the equation $16t^2 = 400$ and find $t = 5$, which is the time it takes the screwdriver to reach the ground.

b. The average velocity over the time interval [0, 5] is $\dfrac{f(5) - f(0)}{5 - 0} = \dfrac{16(25) - 0}{5} = 80$, or 80 ft/s.

c. The velocity of the screwdriver at time t is

$$v(t) = \lim_{h\to 0} \frac{f(t+h) - f(t)}{h} = \lim_{h\to 0} \frac{16(t+h)^2 - 16t^2}{h} = \lim_{h\to 0} \frac{16t^2 + 32th + 16h^2 - 16t^2}{h}$$
$$= \lim_{h\to 0} \frac{(32t + 16h)h}{h} = 32t.$$

In particular, the velocity of the screwdriver when it hits the ground (at $t = 5$) is $v(5) = 32(5) = 160$, or 160 ft/s.

33. a. We write $V = f(p) = \dfrac{1}{p}$. The average rate of change of V is $\dfrac{f(3) - f(2)}{3 - 2} = \dfrac{\frac{1}{3} - \frac{1}{2}}{1} = -\dfrac{1}{6}$, a decrease of $\frac{1}{6}$ liter/atmosphere.

b. $V'(t) = \lim\limits_{h\to 0} \dfrac{\dfrac{f(p+h) - f(p)}{h}}{h} = \lim\limits_{h\to 0} \dfrac{\dfrac{1}{p+h} - \dfrac{1}{p}}{h} = \lim\limits_{h\to 0} \dfrac{p - (p+h)}{hp(p+h)} = \lim\limits_{h\to 0} -\dfrac{1}{p(p+h)} = -\dfrac{1}{p^2}$. In particular, the rate of change of V when $p = 2$ is $V'(2) = -\dfrac{1}{2^2}$, a decrease of $\frac{1}{4}$ liter/atmosphere.

35. a. $P(x) = -\frac{1}{3}x^2 + 7x + 30$. Using the four-step process, we find that

$$P'(x) = \lim_{h\to 0} \frac{P(x+h) - P(x)}{h} = \lim_{h\to 0} \frac{-\frac{1}{3}\left(x^2 + 2xh + h^2\right) + 7x + 7h + 30 - \left(-\frac{1}{3}x^2 + 7x + 30\right)}{h}$$
$$= \lim_{h\to 0} \frac{-\frac{2}{3}xh - \frac{1}{3}h^2 + 7h}{h} = \lim_{h\to 0} \left(-\tfrac{2}{3}x - \tfrac{1}{3}h + 7\right) = -\tfrac{2}{3}x + 7.$$

b. $P'(10) = -\frac{2}{3}(10) + 7 \approx 0.333$, or approximately \$333 per \$1000 spent on advertising.

$P'(30) = -\frac{2}{3}(30) + 7 = -13$, a decrease of \$13,000 per \$1000 spent on advertising.

37. $N(t) = t^2 + 2t + 50$. We first compute $N'(t)$ using the four-step process.

Step 1 $N(t+h) = (t+h)^2 + 2(t+h) + 50 = t^2 + 2th + h^2 + 2t + 2h + 50.$

Step 2 $N(t+h) - N(t) = \left(t^2 + 2th + h^2 + 2t + 2h + 50\right) - \left(t^2 + 2t + 50\right) = 2th + h^2 + 2h = h(2t + h + 2).$

Step 3 $\dfrac{N(t+h) - N(t)}{h} = 2t + h + 2.$

Step 4 $N'(t) = \lim_{h \to 0}(2t + h + 2) = 2t + 2.$

The rate of change of the country's GNP two years from now is $N'(2) = 2(2) + 2 = 6$, or \$6 billion/yr. The rate of change four years from now is $N'(4) = 2(4) + 2 = 10$, or \$10 billion/yr.

39. a. $f'(h)$ gives the instantaneous rate of change of the temperature with respect to height at a given height h, in °F per foot.

b. Because the temperature decreases as the altitude increases, the sign of $f'(h)$ is negative.

c. Because $f'(1000) = -0.05$, the change in the air temperature as the altitude changes from 1000 ft to 1001 ft is approximately $-0.05°$ F.

41. $\dfrac{f(a+h) - f(a)}{h}$ gives the average rate of change of the seal population over the time interval $[a, a+h]$.

$\lim_{h \to 0} \dfrac{f(a+h) - f(a)}{h}$ gives the instantaneous rate of change of the seal population at $x = a$.

43. $\dfrac{f(a+h) - f(a)}{h}$ gives the average rate of change of the country's industrial production over the time interval $[a, a+h]$. $\lim_{h \to 0} \dfrac{f(a+h) - f(a)}{h}$ gives the instantaneous rate of change of the country's industrial production at $x = a$.

45. $\dfrac{f(a+h) - f(a)}{h}$ gives the average rate of change of the atmospheric pressure over the altitudes $[a, a+h]$.

$\lim_{h \to 0} \dfrac{f(a+h) - f(a)}{h}$ gives the instantaneous rate of change of the atmospheric pressure with respect to altitude at $x = a$.

47. a. f has a limit at $x = a$.

b. f is not continuous at $x = a$ because $f(a)$ is not defined.

c. f is not differentiable at $x = a$ because it is not continuous there.

49. a. f has a limit at $x = a$.

b. f is continuous at $x = a$.

c. f is not differentiable at $x = a$ because f has a kink at the point $x = a$.

51. a. f does not have a limit at $x = a$ because it is unbounded in the neighborhood of a.

b. f is not continuous at $x = a$.

c. f is not differentiable at $x = a$ because it is not continuous there.

53. $s(t) = -0.1t^3 + 2t^2 + 24t$. Our computations yield the following results: 32.1, 30.939, 30.814, 30.8014, and 30.8001. The motorcycle's instantaneous velocity at $t = 2$ is approximately 30.8 ft/s.

55. False. Let $f(x) = |x|$. Then f is continuous at $x = 0$, but is not differentiable there.

57. Observe that the graph of f has a kink at $x = -1$. We have $\dfrac{f(-1+h) - f(-1)}{h} = 1$ if $h > 0$, and -1 if $h < 0$, so that $\displaystyle\lim_{h\to 0} \frac{f(-1+h) - f(-1)}{h}$ does not exist.

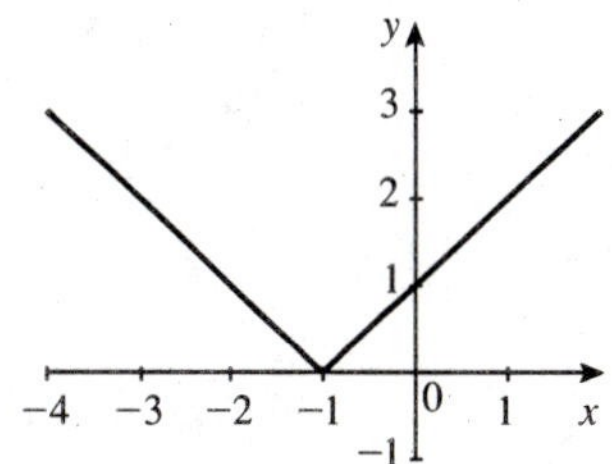

59. For continuity, we require that $f(1) = 1 = \displaystyle\lim_{x\to 1^+} (ax + b) = a + b$, or $a + b = 1$. Using the four-step process, we have $f'(x) = \begin{cases} 2x & \text{if } x < 1 \\ a & \text{if } x > 1 \end{cases}$ In order that the derivative exist at $x = 1$, we require that $\displaystyle\lim_{x\to 1^-} 2x = \lim_{x\to 1^+} a$, or $2 = a$. Therefore, $b = -1$ and so $f(x) = \begin{cases} x^2 & \text{if } x \le 1 \\ 2x - 1 & \text{if } x > 1 \end{cases}$

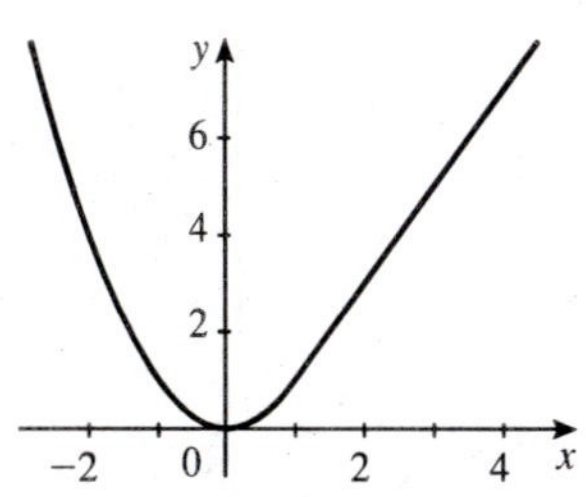

61. We have $f(x) = x$ if $x > 0$ and $f(x) = -x$ if $x < 0$. Therefore, when $x > 0$, $f'(x) = \displaystyle\lim_{h\to 0} \frac{f(x+h) - f(x)}{h} = \lim_{h\to 0} \frac{x+h-x}{h} = \lim_{h\to 0} \frac{h}{h} = 1$, and when $x < 0$, $f'(x) = \displaystyle\lim_{h\to 0} \frac{f(x+h) - f(x)}{h} = \lim_{h\to 0} \frac{-x-h-(-x)}{h} = \lim_{h\to 0} \frac{-h}{h} = -1$. Because the right-hand limit does not equal the left-hand limit, we conclude that $\displaystyle\lim_{h\to 0} f(x)$ does not exist.

Using Technology page 602

1. a. $y = 9x - 11$

b.

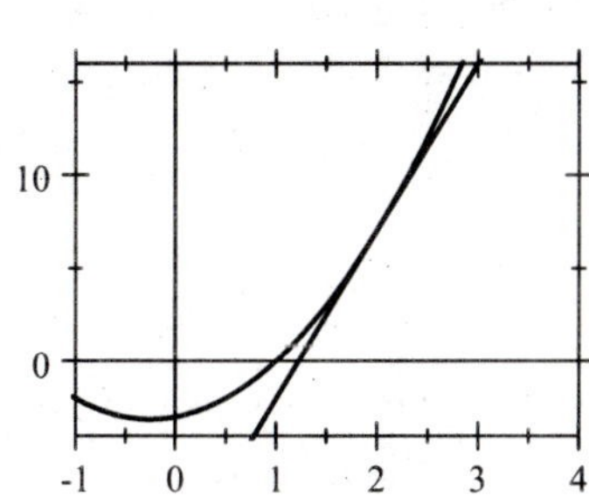

3. a. Using the numerical derivative operation, we find the derivative of f at $(8, 2)$ to be $f'(8) = \frac{1}{12}$, so an equation of the tangent line is $y - 2 = \frac{1}{12}(x - 8)$ or $y = \frac{1}{12}x + \frac{4}{3}$.

b.

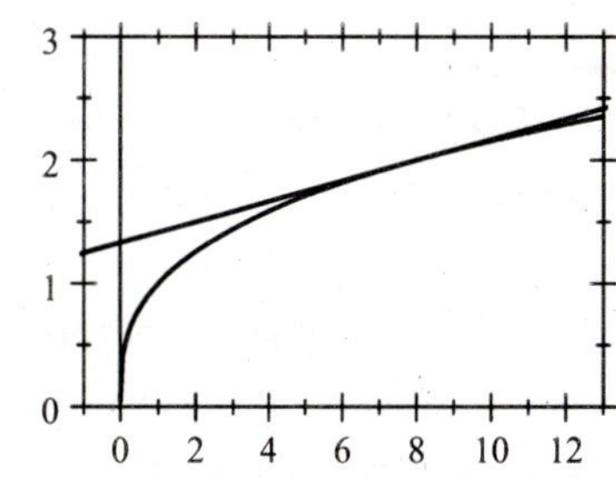

5. a. 4 **b.** $y = 4x - 1$

c.

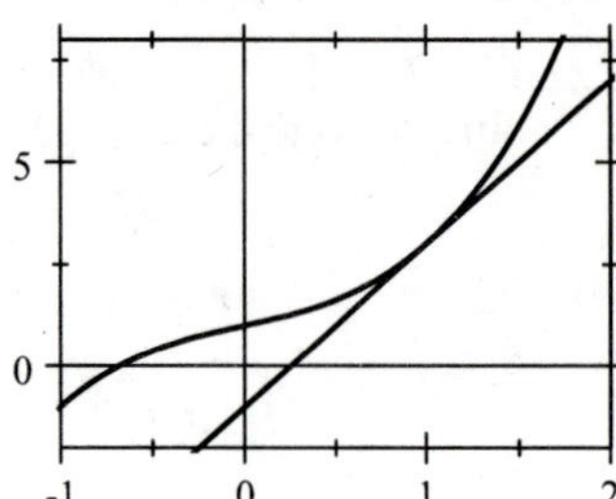

7. a. 4.02 **b.** $y = 4.02x - 3.57$

c.

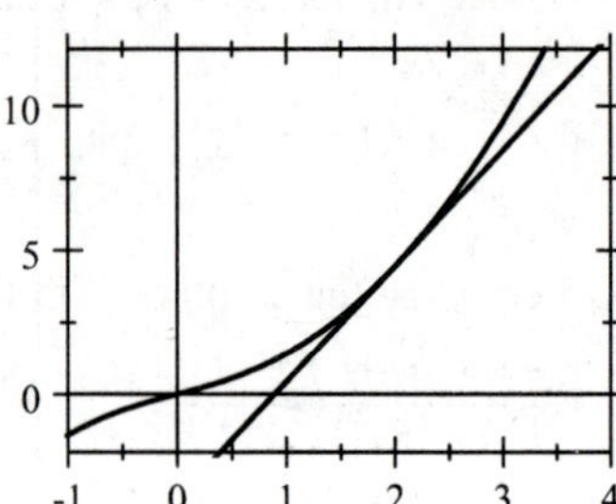

9. a.

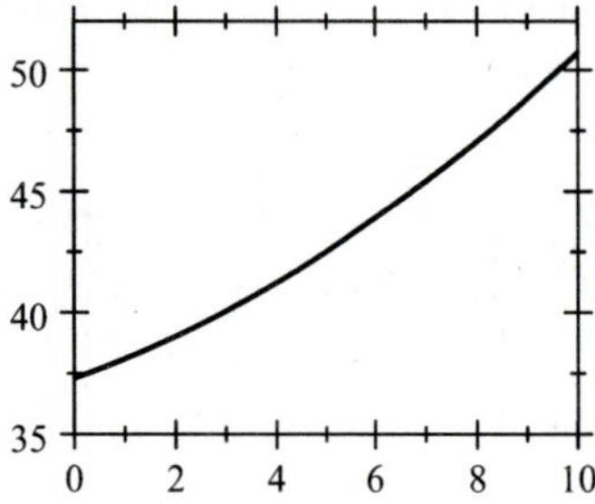

b. 41.22 cents/mile

c. 1.22 cents/mile/yr

9.4 Basic Rules of Differentiation

Problem-Solving Tips

In this section, you are given four basic rules for finding the derivative of a function. As you work through the exercises that follow, first decide which rule(s) you need to find the derivative of the given function. Then write out your solution. After doing this a few times, you should have the formulas memorized. The key here is to try not to look at the formula in the text, and to work the problem just as if you were taking a test. If you train yourself to work in this manner, writing tests will become a lot easier. Also, make sure to distinguish between the notation dy/dx and d/dx. The first notation is used for the derivative of a function y, where as the second notation tells us to find the derivative of the function that follows with respect to x.

Here are some tips for solving the problems in the exercises that follow:

1. **To find the derivative of a function involving radicals**, first rewrite the expression in exponential form. For example, if $f(x) = 2x - 5\sqrt{x}$, rewrite the function in the form $f(x) = 2x - 5x^{1/2}$.

2. **To find the point on the graph of f where the tangent line is horizontal**, set $f'(x) = 0$ and solve for x. (Here we are making use of the fact that the slope of a horizontal line is zero.) This yields the x-value of the point on the graph where the tangent line is horizontal. To find the corresponding y-value, evaluate the function f at this value of x.

Concept Questions page 610

1. **a.** The derivative of a constant is zero.

   b. The derivative of $f(x) = x^n$ is n times x raised to the $(n-1)$th power.

   c. The derivative of a constant times a function is the constant times the derivative of the function.

   d. The derivative of the sum is the sum of the derivatives.

3. **a.** $F'(x) = \frac{d}{dx}[af(x) + bg(x)] = \frac{d}{dx}[af(x)] + \frac{d}{dx}[bg(x)] = af'(x) + bg'(x)$.

   b. $F'(x) = \frac{d}{dx}\left[\frac{f(x)}{a}\right] = \frac{1}{a}\frac{d}{dx}[f(x)] = \frac{f'(x)}{a}$.

Exercises page 610

1. $f'(x) = \frac{d}{dx}(-3) = 0$.

3. $f'(x) = \frac{d}{dx}(x^5) = 5x^4$.

5. $f'(x) = \frac{d}{dx}(x^{3.1}) = 3.1x^{2.1}$.

7. $f'(x) = \frac{d}{dx}(4x^2) = 8x$.

9. $f'(r) = \frac{d}{dr}(\pi r^2) = 2\pi r$.

11. $f'(x) = \frac{d}{dx}(6x^{1/3}) = \frac{1}{3}(6)x^{(1/3-1)} = 2x^{-2/3}$.

13. $f'(x) = \frac{d}{dx}(3\sqrt{x}) = \frac{d}{dx}(3x^{1/2}) = \frac{1}{2}(3)x^{-1/2} = \frac{3}{2}x^{-1/2} = \frac{3}{2\sqrt{x}}$.

15. $f'(x) = \frac{d}{dx}(7x^{-12}) = (-12)(7)x^{-12-1} = -84x^{-13}$.

17. $f'(x) = \frac{d}{dx}(4x^2 - 2x + 7) = 8x - 2$.

19. $f'(x) = \frac{d}{dx}(-x^3 + 2x^2 - 6) = -3x^2 + 4x$.

21. $f'(x) = \frac{d}{dx}(0.03x^2 - 0.4x + 10) = 0.06x - 0.4$.

23. $f(x) = \frac{2x^3 - 4x^2 + 3}{x} = 2x^2 - 4x + \frac{3}{x}$, so $f'(x) = \frac{d}{dx}(2x^2 - 4x + 3x^{-1}) = 4x - 4 - \frac{3}{x^2}$.

25. $f'(x) = \frac{d}{dx}(4x^4 - 3x^{5/2} + 2) = 16x^3 - \frac{15}{2}x^{3/2}$.

27. $f'(x) = \frac{d}{dx}(5x^{-1} + 4x^{-2}) = -5x^{-2} - 8x^{-3} = -\frac{5}{x^2} - \frac{8}{x^3}$.

29. $f'(t) = \frac{d}{dt}(4t^{-4} - 3t^{-3} + 2t^{-1}) = -16t^{-5} + 9t^{-4} - 2t^{-2} = -\frac{16}{t^5} + \frac{9}{t^4} - \frac{2}{t^2}$.

31. $f'(x) = \frac{d}{dx}(3x - 5x^{1/2}) = 3 - \frac{5}{2}x^{-1/2} = 3 - \frac{5}{2\sqrt{x}}$.

33. $f'(x) = \frac{d}{dx}(2x^{-2} - 3x^{-1/3}) = -4x^{-3} + x^{-4/3} = -\frac{4}{x^3} + \frac{1}{x^{4/3}}$.

35. $f'(x) = \frac{d}{dx}(2x^3 - 4x) = 6x^2 - 4$.

    a. $f'(-2) = 6(-2)^2 - 4 = 20$.

    b. $f'(0) = 6(0) - 4 = -4$.

    c. $f'(2) = 6(2)^2 - 4 = 20$.

37. The given limit is $f'(1)$, where $f(x) = x^3$. Because $f'(x) = 3x^2$, we have $\lim_{h \to 0} \frac{(1+h)^3 - 1}{h} = f'(1) = 3$.

39. Let $f(x) = 3x^2 - x$. Then $\lim_{h\to 0} \frac{3(2+h)^2 - (2+h) - 10}{h} = \lim_{h\to 0} \frac{f(2+h) - f(2)}{h}$ because $f(2+h) - f(2) = 3(2+h)^2 - (2+h) - [3(4) - 2] = 3(2+h)^2 - (2+h) - 10$. But the last limit is simply $f'(2)$. Because $f'(x) = 6x - 1$, we have $f'(2) = 11$. Therefore, $\lim_{h\to 0} \frac{3(2+h)^2 - (2+h) - 10}{h} = 11$.

41. $f(x) = 2x^2 - 3x + 4$. The slope of the tangent line at any point $(x, f(x))$ on the graph of f is $f'(x) = 4x - 3$. In particular, the slope of the tangent line at the point $(2, 6)$ is $f'(2) = 4(2) - 3 = 5$. An equation of the required tangent line is $y - 6 = 5(x - 2)$ or $y = 5x - 4$.

43. $f(x) = x^4 - 3x^3 + 2x^2 - x + 1$, so $f'(x) = 4x^3 - 9x^2 + 4x - 1$. The slope is $f'(2) = 4(2)^3 - 9(2)^2 + 4(2) - 1 = 3$. An equation of the tangent line is $y - (-1) = 3(x - 2)$ or $y = 3x - 7$.

45. a. $f'(x) = 3x^2$. At a point where the tangent line is horizontal, $f'(x) = 0$, or $3x^2 = 0$, and so $x = 0$. Therefore, the point is $(0, 0)$.

b.

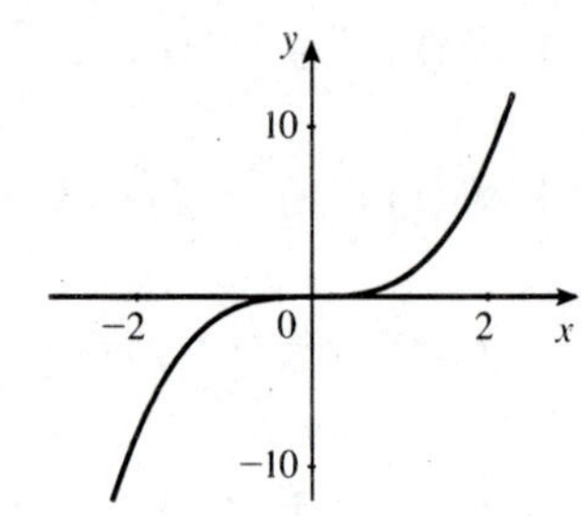

47. a. $f(x) = x^3 + 1$. The slope of the tangent line at any point $(x, f(x))$ on the graph of f is $f'(x) = 3x^2$. At the point(s) where the slope is 12, we have $3x^2 = 12$, so $x = \pm 2$. The required points are $(-2, -7)$ and $(2, 9)$.

b. The tangent line at $(-2, -7)$ has equation $y - (-7) = 12[x - (-2)]$, or $y = 12x + 17$, and the tangent line at $(2, 9)$ has equation $y - 9 = 12(x - 2)$, or $y = 12x - 15$.

c.

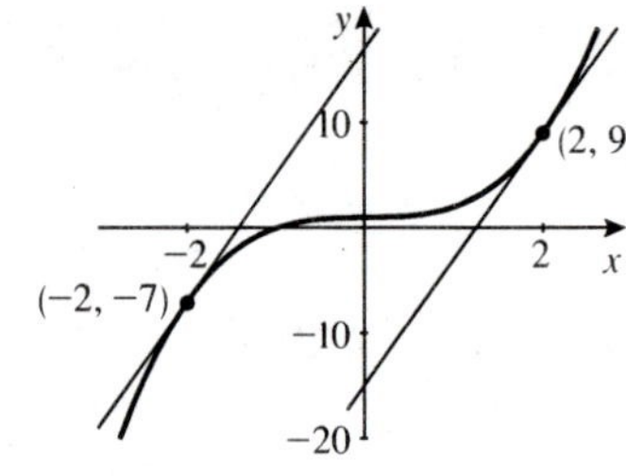

49. $f(x) = \frac{1}{4}x^4 - \frac{1}{3}x^3 - x^2$, so $f'(x) = x^3 - x^2 - 2x$.

a. $f'(x) = x^3 - x^2 - 2x = -2x$ implies $x^3 - x^2 = 0$, so $x^2(x - 1) = 0$. Thus, $x = 0$ or $x = 1$. $f(1) = \frac{1}{4}(1)^4 - \frac{1}{3}(1)^3 - (1)^2 = -\frac{13}{12}$ and $f(0) = \frac{1}{4}(0)^4 - \frac{1}{3}(0)^3 - (0)^2 = 0$. We conclude that the corresponding points on the graph are $\left(1, -\frac{13}{12}\right)$ and $(0, 0)$.

b. $f'(x) = x^3 - x^2 - 2x = 0$ implies $x\left(x^2 - x - 2\right) = 0$, $x(x - 2)(x + 1) = 0$, and so $x = 0, 2$, or -1. $f(0) = 0$, $f(2) = \frac{1}{4}(2)^4 - \frac{1}{3}(2)^3 - (2)^2 = 4 - \frac{8}{3} - 4 = -\frac{8}{3}$, and $f(-1) = \frac{1}{4}(-1)^4 - \frac{1}{3}(-1)^3 - (-1)^2 = \frac{1}{4} + \frac{1}{3} - 1 = -\frac{5}{12}$. We conclude that the corresponding points are $(0, 0)$, $\left(2, -\frac{8}{3}\right)$, and $\left(-1, -\frac{5}{12}\right)$.

c. $f'(x) = x^3 - x^2 - 2x = 10x$ implies $x^3 - x^2 - 12x = 0$, $x\left(x^2 - x - 12\right) = 0$, $x(x - 4)(x + 3) = 0$, so $x = 0, 4$, or -3. $f(0) = 0$, $f(4) = \frac{1}{4}(4)^4 - \frac{1}{3}(4)^3 - (4)^2 = 48 - \frac{64}{3} = \frac{80}{3}$, and $f(-3) = \frac{1}{4}(-3)^4 - \frac{1}{3}(-3)^3 - (-3)^2 = \frac{81}{4} + 9 - 9 = \frac{81}{4}$. We conclude that the corresponding points are $(0, 0)$, $\left(4, \frac{80}{3}\right)$, and $\left(-3, \frac{81}{4}\right)$.

51. $V(r) = \frac{4}{3}\pi r^3$, so $V'(r) = 4\pi r^2$.

a. $V'\left(\frac{2}{3}\right) = 4\pi\left(\frac{4}{9}\right) = \frac{16}{9}\pi$ cm^3/cm.

b. $V'\left(\frac{5}{4}\right) = 4\pi\left(\frac{25}{16}\right) = \frac{25}{4}\pi$ cm^3/cm.

53. a. $N(1) = 16.3\left(1^{0.8766}\right) = 16.3(1) = 16.3$, or 16.3 million cameras.

b. $N'(t) = (16.3)(0.8766)\,t^{-0.1234}$, so $N'(1) \approx 14.29$, or approximately 14.3 million cameras per year.

c. $N(5) = 16.3\left(5^{0.8766}\right) \approx 66.82$, or approximately 66.8 million cameras.

d. $N'(5) = (16.3)(0.8766)\left(5^{-0.1234}\right) \approx 11.71$, or approximately 11.7 million cameras per year.

55. a.

1970 ($t = 1$)	1980 ($t = 2$)	1990 ($t = 3$)	2000 ($t = 4$)
49.6%	41.1%	36.9%	34.1%

b. $P'(t) = (49.6)\left(-0.27t^{-1.27}\right) = -\dfrac{13.392}{t^{1.27}}$. In 1980, $P'(2) \approx -5.6$, or decreasing at 5.6%/decade. In 1990, $P'(3) \approx -3.3$, or decreasing at 3.3%/decade.

57. a. The number of viewers will be $N(8) = 52(8)^{0.531} \approx 156.87$, or approximately 157 million.

b. The projected number is changing at the rate of $N'(8) = 52(0.531)\,t^{-0.469}\big|_{t=8} = 52(0.531)\,8^{-0.469} \approx 10.41$, or approximately 10.4 million viewers/year.

59. a. $f(t) = 120t - 15t^2$, so $v = f'(t) = 120 - 30t$.

b. $v(0) = 120$ ft/sec

c. Setting $v = 0$ gives $120 - 30t = 0$, or $t = 4$. Therefore, the stopping distance is $f(4) = 120(4) - 15\left(4^2\right)$ or 240 ft.

61. a. At the beginning of 1980, $P(0) = 5\%$. At the beginning of 1990, $P(10) = -0.0105\left(10^2\right) + 0.735(10) + 5 = 11.3$, or 11.3%. At the beginning of 2000, $P(20) = -0.0105(20)^2 + 0.735(20) + 5 = 15.5$, or 15.5%.

b. $P'(t) = -0.021t + 0.735$. At the beginning of 1985, $P'(5) = -0.021(5) + 0.735 = 0.63$, or 0.63%/yr. At the beginning of 1990, $P'(10) = -0.021(10) + 0.735 = 0.525$, or 0.525%/yr.

63. a. $f(t) = 5.303t^2 - 53.977t + 253.8$. The rate of change of the groundfish population at any time t is given by $f'(t) = 10.606t - 53.977$. The rate of change at the beginning of 1994 is given by $f'(5) = 10.606(5) - 53.977 = -0.947$, so the population is decreasing at the rate of 0.9 thousand metric tons/yr. At the beginning of 1996, the rate of change is $f'(7) = 10.606(7) - 53.977 = 20.265$, so the population is increasing at the rate of approximately 20.3 thousand metric tons/yr.

b. Yes. The groundfish population was decreasing before the introduction of the restrictions, and increasing after they were introduced.

65. $I(t) = -0.2t^3 + 3t^2 + 100$, so $I'(t) = -0.6t^2 + 6t$.

a. In 2007, it is changing at a rate of $I'(5) = -0.6(25) + 6(5)$, or 15 points/yr. In 2009, it is $I'(7) = -0.6(49) + 6(7)$, or 12.6 points/yr. In 2012, it is $I'(10) = -0.6(100) + 6(10)$, or 0 points/yr.

b. The average rate of increase of the CPI over the period from 2007 to 2012 is $\frac{I(10) - I(5)}{10 - 5} = \frac{[-0.2(1000) + 3(100) + 100] - [-0.2(125) + 3(25) + 100]}{5} = \frac{200 - 150}{5} = 10$, or 10 points/yr.

67. a. $f'(x) = \frac{d}{dx}\left[0.0001x^{5/4} + 10\right] = \frac{5}{4}\left(0.0001x^{1/4}\right) = 0.000125x^{1/4}$.

b. $f'(10{,}000) = 0.000125(10{,}000)^{1/4} = 0.00125$, or \$0.00125/radio.

69. a. $f(t) = 20t - 40\sqrt{t} + 50$, so $f'(t) = 20 - 40\left(\frac{1}{2}\right)t^{-1/2} = 20\left(1 - \frac{1}{\sqrt{t}}\right)$.

b. $f(0) = 20(0) - 40\sqrt{0} + 50 = 50$, so $f(1) = 20(1) - 40\sqrt{1} + 50 = 30$ and $f(2) = 20(2) - 40\sqrt{2} + 50 \approx 33.43$. The average velocities at 6, 7, and 8 a.m. are 50, 30, and 33.43 mph, respectively.

c. $f'\left(\frac{1}{2}\right) = 20 - 20\left(\frac{1}{2}\right)^{-1/2} \approx -8.28$, $f'(1) = 20 - 20(1)^{-1/2} = 0$, and $f'(2) = 20 - 20(2)^{-1/2} \approx 5.86$. At 6:30 a.m. the average velocity is decreasing at the rate of 8.28 mph/hr, at 7 a.m. it is not changing, and at 8 a.m. it is increasing at the rate of 5.86 mph.

71. $N(t) = 2t^3 + 3t^2 - 4t + 1000$, so $N'(t) = 6t^2 + 6t - 4$. $N'(2) = 6(4) + 6(2) - 4 = 32$, or 32 turtles/yr; and $N'(8) = 6(64) + 6(8) - 4 = 428$, or 428 turtles/yr. The population ten years after implementation of the conservation measures will be $N(10) = 2\left(10^3\right) + 3\left(10^2\right) - 4(10) + 1000$, or 3260 turtles.

73. a. $P(t) = 0.0004t^3 + 0.0036t^2 + 0.8t + 12$. At the beginning of 1991, $P(0) = 12\%$. At the beginning of 2004, $P(13) = 0.0004(13)^3 + 0.0036(13)^2 + 0.8(13) + 12 \approx 23.9\%$.

b. $P'(t) = 0.0012t^2 + 0.0072t + 0.8$. At the beginning of 1991, $P'(0) = 0.8$, or 0.8%/yr. At the beginning of 2004, $P'(13) = 0.0012(13)^2 + 0.0072(13) + 0.8 \approx 1.1$, or approximately 1.1%/yr.

75. a. At any time t, the function $D = g + f$ at t, $D(t) = (g + f)(t) = g(t) + f(t)$, gives the total population aged 65 and over of the developed and the underdeveloped/emerging countries.

b. $D(t) = g(t) + f(t) = \left(0.46t^2 + 0.16t + 287.8\right) + (3.567t + 175.2) = 0.46t^2 + 3.727t + 463$, so $D'(t) = 0.92t + 3.727$. Therefore, $D'(10) = 0.92(10) + 3.727 = 12.927$, which says that the combined population is growing at the rate of approximately 13 million people per year in 2010.

77. True. $\frac{d}{dx}\left[2f(x) - 5g(x)\right] = \frac{d}{dx}\left[2f(x)\right] - \frac{d}{dx}\left[5g(x)\right] = 2f'(x) - 5g'(x)$.

79.
$$\frac{d}{dx}\left(x^3\right) = \lim_{h \to 0} \frac{(x+h)^3 - x^3}{h} = \lim_{h \to 0} \frac{x^3 + 3x^2h + 3xh^2 + h^3 - x^3}{h} = \lim_{h \to 0} \frac{h\left(3x^2 + 3xh + h^2\right)}{h}$$
$$= \lim_{h \to 0}\left(3x^2 + 3xh + h^2\right) = 3x^2.$$

Using Technology page 616

1. 1

3. 0.4226

5. 0.1613

7. a.

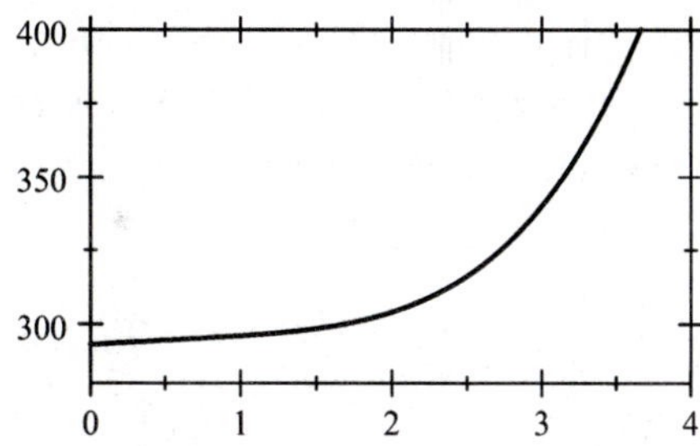

b. 3.4295 parts/million per 40 years;
105.4332 parts/million per 40 years

9. a. $f(t) = 0.611t^3 + 9.702t^2 + 32.544t + 473.5$

b.

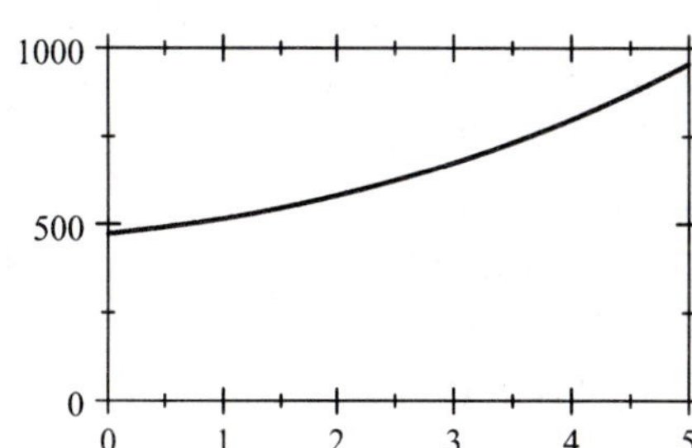

c. At the beginning of 2000, the assets of the hedge funds were increasing at the rate of \$53.781 billion/yr, and at the beginning of 2003, they were increasing at the rate of \$139.488 billion/yr.

9.5 The Product and Quotient Rules; Higher-Order Derivatives

Problem-Solving Tips

The answers at the back of the book for the exercises in this section are given in both simplified and unsimplified terms. Here, as with all of your homework, you should make it a practice to analyze your errors. If you do not get the right answer for the unsimplified form, it means that you are not applying the rules for differentiating correctly. In this case you need to review the rules, making sure that you can write out each rule. If you have the correct answer for the unsimplified form but the incorrect answer for the simplified form, it probably means that you have made an algebraic error. You may need to review the rules for simplifying algebraic expressions given on page 12 of the text and then work some of the exercises given in section 1.2 to get back into practice. In any case, you will need to simplify your answers when you work the problems on the applications of the derivative in the next chapter, so you should get in the habit of doing so now.

Here are some tips for solving the problems in the exercises that follow:

1. **To find the derivative of a function involving radicals**, first rewrite the expression in exponential form. For example, if $f(x) = 2x - 5\sqrt{x}$, rewrite the function in the form $f'^{1/2}$.

2. **To find the point on the graph of f where the tangent line is horizontal**, set $f'(x) = 0$ and solve for x. (Here we are making use of the fact that the slope of a horizontal line is zero.) This yields the x-value of the point on the graph where the tangent line is horizontal. To find the corresponding y-value, evaluate the function f at this value of x.

Concept Questions page 625

1. a. The derivative of the product of two functions is equal to the first function times the derivative of the second function plus the second function times the derivative of the first function.

   b. The derivative of the quotient of two functions is equal to the quotient whose numerator is given by the denominator times the derivative of the numerator minus the numerator times the derivative of the denominator and whose denominator is the square of the denominator of the quotient.

3. a. The second derivative of f is the derivative of f'.

   b. To find the second derivative of f, we differentiate f'.

Exercises page 625

1. $f(x) = 2x(x^2+1)$, so $f'(x) = 2x\frac{d}{dx}(x^2+1) + (x^2+1)\frac{d}{dx}(2x) = 2x(2x) + (x^2+1)(2) = 6x^2+2$.

3. $f(t) = (t-1)(2t+1)$, so
$f'(t) = (t-1)\frac{d}{dt}(2t+1) + (2t+1)\frac{d}{dt}(t-1) = (t-1)(2) + (2t+1)(1) = 4t-1$.

5. $f(x) = (3x+2)(x^2-2)$, so
$f'(x) = (3x+2)\frac{d}{dx}(x^2-2) + (x^2-2)\frac{d}{dx}(3x+2) = (3x+2)(2x) + (x^2-2)(3) = 9x^2+4x-6$.

7. $f(x) = (x^3-1)(2x+1)$, so
$f'(x) = (x^3-1)\frac{d}{dx}(2x+1) + (2x+1)\frac{d}{dx}(x^3-1) = (x^3-1)(2) + (2x+1)(3x^2) = 8x^3+3x^2-2$.

9. $f(w) = (w^3-w^2+w-1)(w^2+2)$, so

$$\begin{aligned} f'(w) &= (w^3-w^2+w-1)\frac{d}{dw}(w^2+2) + (w^2+2)\frac{d}{dw}(w^3-w^2+w-1) \\ &= (w^3-w^2+w-1)(2w) + (w^2+2)(3w^2-2w+1) \\ &= 2w^4-2w^3+2w^2-2w+3w^4-2w^3+w^2+6w^2-4w+2 = 5w^4-4w^3+9w^2-6w+2. \end{aligned}$$

11. $f(x) = (5x^2+1)(2\sqrt{x}-1)$, so

$$\begin{aligned} f'(x) &= (5x^2+1)\tfrac{d}{dx}(2x^{1/2}-1) + (2x^{1/2}-1)\tfrac{d}{dx}(5x^2+1) = (5x^2+1)(x^{-1/2}) + (2x^{1/2}-1)(10x) \\ &= 5x^{3/2}+x^{-1/2}+20x^{3/2}-10x = \frac{25x^2-10x\sqrt{x}+1}{\sqrt{x}}. \end{aligned}$$

13. $f(x) = (x^2-5x+2)\left(x-\frac{2}{x}\right)$, so

$$\begin{aligned} f'(x) &= (x^2-5x+2)\frac{d}{dx}\left(x-\frac{2}{x}\right) + \left(x-\frac{2}{x}\right)\frac{d}{dx}(x^2-5x+2) \\ &= \frac{(x^2-5x+2)(x^2+2)}{x^2} + \frac{(x^2-2)(2x-5)}{x} = \frac{(x^2-5x+2)(x^2+2)+x(x^2-2)(2x-5)}{x^2} \\ &= \frac{x^4+2x^2-5x^3-10x+2x^2+4+2x^4-5x^3-4x^2+10x}{x^2} = \frac{3x^4-10x^3+4}{x^2}. \end{aligned}$$

15. $f(x) = \dfrac{1}{x-2}$, so $f'(x) = \dfrac{(x-2)\frac{d}{dx}(1)-(1)\frac{d}{dx}(x-2)}{(x-2)^2} = \dfrac{0-1(1)}{(x-2)^2} = -\dfrac{1}{(x-2)^2}$.

17. $f(x) = \dfrac{2x-1}{2x+1}$, so

$$f'(x) = \frac{(2x+1)\frac{d}{dx}(2x-1) - (2x-1)\frac{d}{dx}(2x+1)}{(2x+1)^2} = \frac{(2x+1)(2) - (2x-1)(2)}{(2x+1)^2} = \frac{4}{(2x+1)^2}.$$

19. $f(x) = \dfrac{1}{x^2+1}$, so $f'(x) = \dfrac{(x^2+1)\frac{d}{dx}(1) - (1)\frac{d}{dx}(x^2+1)}{(x^2+1)^2} = \dfrac{(x^2+1)(0) - 1(2x)}{(x^2+1)^2} = -\dfrac{2x}{(x^2+1)^2}$.

21. $f(s) = \dfrac{s^2-4}{s+1}$, so

$$f'(s) = \frac{(s+1)\dfrac{d}{ds}(s^2-4) - (s^2-4)\dfrac{d}{ds}(s+1)}{(s+1)^2} = \frac{(s+1)(2s) - (s^2-4)(1)}{(s+1)^2} = \frac{s^2+2s+4}{(s+1)^2}.$$

23. $f(x) = \dfrac{\sqrt{x}+1}{x^2+1}$, so

$$f'(x) = \frac{(x^2+1)\frac{d}{dx}(x^{1/2}) - (x^{1/2}+1)\frac{d}{dx}(x^2+1)}{(x^2+1)^2} = \frac{(x^2+1)\left(\frac{1}{2}x^{-1/2}\right) - (x^{1/2}+1)(2x)}{(x^2+1)^2}$$

$$= \frac{\left(\frac{1}{2}x^{-1/2}\right)\left[(x^2+1) - (x^{1/2}+1)4x^{3/2}\right]}{(x^2+1)^2} = \frac{1-3x^2-4x^{3/2}}{2\sqrt{x}(x^2+1)^2}.$$

25. $f(x) = \dfrac{x^2+2}{x^2+x+1}$, so

$$f'(x) = \frac{(x^2+x+1)\frac{d}{dx}(x^2+2) - (x^2+2)\frac{d}{dx}(x^2+x+1)}{(x^2+x+1)^2}$$

$$= \frac{(x^2+x+1)(2x) - (x^2+2)(2x+1)}{(x^2+x+1)^2} = \frac{2x^3+2x^2+2x-2x^3-x^2-4x-2}{(x^2+x+1)^2} = \frac{x^2-2x-2}{(x^2+x+1)^2}.$$

27. $f(x) = \dfrac{(x+1)(x^2+1)}{x-2} = \dfrac{(x^3+x^2+x+1)}{x-2}$, so

$$f'(x) = \frac{(x-2)\frac{d}{dx}(x^3+x^2+x+1) - (x^3+x^2+x+1)\frac{d}{dx}(x-2)}{(x-2)^2}$$

$$= \frac{(x-2)(3x^2+2x+1) - (x^3+x^2+x+1)}{(x-2)^2}$$

$$= \frac{3x^3+2x^2+x-6x^2-4x-2-x^3-x^2-x-1}{(x-2)^2} = \frac{2x^3-5x^2-4x-3}{(x-2)^2}.$$

29. $f(x) = \dfrac{x}{x^2-4} - \dfrac{x-1}{x^2+4} = \dfrac{x(x^2+4)-(x-1)(x^2-4)}{(x^2-4)(x^2+4)} = \dfrac{x^2+8x-4}{(x^2-4)(x^2+4)}$, so

$$f'(x) = \frac{(x^2-4)(x^2+4)\frac{d}{dx}(x^2+8x-4)-(x^2+8x-4)\frac{d}{dx}(x^4-16)}{(x^2-4)^2(x^2+4)^2}$$

$$= \frac{(x^2-4)(x^2+4)(2x+8)-(x^2+8x-4)(4x^3)}{(x^2-4)^2(x^2+4)^2}$$

$$= \frac{2x^5+8x^4-32x-128-4x^5-32x^4+16x^3}{(x^2-4)^2(x^2+4)^2} = \frac{-2x^5-24x^4+16x^3-32x-128}{(x^2-4)^2(x^2+4)^2}.$$

31. $h'(x) = f(x)g'(x) + f'(x)g(x)$ by the Product Rule. Therefore, $h'(1) = f(1)g'(1) + f'(1)g(1) = (2)(3) + (-1)(-2) = 8$.

33. Using the Quotient Rule followed by the Product Rule, we have

$$h'(x) = \frac{[x+g(x)]\frac{d}{dx}[xf(x)] - xf(x)\frac{d}{dx}[x+g(x)]}{[x+g(x)]^2} = \frac{[x+g(x)][xf'(x)+f(x)] - xf(x)[1+g'(x)]}{[x+g(x)]^2}.$$

Therefore,

$$h'(1) = \frac{[1+g(1)][f'(1)+f(1)] - f(1)[1+g'(1)]}{[1+g(1)]^2} = \frac{(1-2)(-1+2)-2(1+3)}{(1-2)^2} = \frac{-1-8}{1} = -9.$$

35. $f(x) = (2x-1)(x^2+3)$, so

$$f'(x) = (2x-1)\tfrac{d}{dx}(x^2+3) + (x^2+3)\tfrac{d}{dx}(2x-1) = (2x-1)(2x) + (x^2+3)(2)$$

$$= 6x^2-2x+6 = 2(3x^2-x+3).$$

At $x=1$, $f'(1) = 2[3(1)^2-(1)+3] = 2(5) = 10$.

37. $f(x) = \dfrac{x}{x^4-2x^2-1}$, so

$$f'(x) = \frac{(x^4-2x^2-1)\frac{d}{dx}(x) - x\frac{d}{dx}(x^4-2x^2-1)}{(x^4-2x^2-1)^2} = \frac{(x^4-2x^2-1)(1)-x(4x^3-4x)}{(x^4-2x^2-1)^2}$$

$$= \frac{-3x^4+2x^2-1}{(x^4-2x^2-1)^2}.$$

Therefore, $f'(-1) = \dfrac{-3+2-1}{(1-2-1)^2} = -\dfrac{2}{4} = -\dfrac{1}{2}$.

39. $f(x) = (x^3+1)(x^2-2)$, so

$f'(x) = (x^3+1)\frac{d}{dx}(x^2-2) + (x^2-2)\frac{d}{dx}(x^3+1) = (x^3+1)(2x) + (x^2-2)(3x^2)$. The slope of the tangent line at $(2, 18)$ is $f'(2) = (8+1)(4) + (4-2)(12) = 60$. An equation of the tangent line is $y - 18 = 60(x-2)$, or $y = 60x - 102$.

41. $f(x) = \dfrac{x+1}{x^2+1}$, so

$$f'(x) = \frac{(x^2+1)\frac{d}{dx}(x+1) - (x+1)\frac{d}{dx}(x^2+1)}{(x^2+1)^2} = \frac{(x^2+1)(1) - (x+1)(2x)}{(x^2+1)^2} = \frac{-x^2-2x+1}{(x^2+1)^2}.$$ At $x = 1$, $f'(1) = \dfrac{-1-2+1}{4} = -\dfrac{1}{2}$. Therefore, the slope of the tangent line at $x = 1$ is $-\frac{1}{2}$ and an equation is $y - 1 = -\frac{1}{2}(x-1)$ or $y = -\frac{1}{2}x + \frac{3}{2}$.

43. $f(x) = 2x^2 - 2x + 1$, so $f'(x) = 4x - 2$ and $f''(x) = 4$.

45. $f(x) = 2x^3 - 3x^2 + 1$, so $f'(x) = 6x^2 - 6x$ and $f''(x) = 12x - 6 = 6(2x-1)$.

47. $h(t) = t^4 - 2t^3 + 6t^2 - 3t + 10$, so $h'(t) = 4t^3 - 6t^2 + 12t - 3$ and $h''(t) = 12t^2 - 12t + 12 = 12(t^2 - t + 1)$.

49. $f(x) = 3x^4 - 2x^3$, so $f'(x) = 12x^3 - 6x^2$, $f''(x) = 36x^2 - 12x$, and $f'''(x) = 72x - 12 = 12(6x-1)$.

51. $f(x) = \dfrac{1}{x}$, so $f'(x) = \dfrac{d}{dx}(x^{-1}) = -x^{-2} = -\dfrac{1}{x^2}$, $f''(x) = 2x^{-3} = \dfrac{2}{x^3}$, and $f'''(x) = -6x^{-4} = -\dfrac{6}{x^4}$.

53. Using the Product Rule, we find
$g'(x) = \frac{d}{dx}\left[x^2 f(x)\right] = x^2\frac{d}{dx}\left[f(x)\right] + f(x)\frac{d}{dx}(x^2) = x^2 f'(x) + 2xf(x)$. Therefore,
$g'(2) = 2^2 f'(2) + 2(2)f(2) = (4)(-1) + 4(3) = 8$.

55. $f(x) = (x^3+1)(3x^2-4x+2)$, so

$$\begin{aligned} f'(x) &= (x^3+1)\tfrac{d}{dx}(3x^2-4x+2) + (3x^2-4x+2)\tfrac{d}{dx}(x^3+1) \\ &= (x^3+1)(6x-4) + (3x^2-4x+2)(3x^2) \\ &= 6x^4 + 6x - 4x^3 - 4 + 9x^4 - 12x^3 + 6x^2 = 15x^4 - 16x^3 + 6x^2 + 6x - 4. \end{aligned}$$

At $x = 1$, $f'(1) = 15(1)^4 - 16(1)^3 + 6(1)^2 + 6(1) - 4 = 7$. Thus, the slope of the tangent line at the point $x = 1$ is 7 and an equation is $y - 2 = 7(x-1)$, or $y = 7x - 5$.

57. $f(x) = (x^2+1)(2-x)$, so
$f'(x) = (x^2+1)\frac{d}{dx}(2-x) + (2-x)\frac{d}{dx}(x^2+1) = (x^2+1)(-1) + (2-x)(2x) = -3x^2 + 4x - 1$. At a point where the tangent line is horizontal, we have $f'(x) = -3x^2 + 4x - 1 = 0$ or $3x^2 - 4x + 1 = (3x-1)(x-1) = 0$, giving $x = \frac{1}{3}$ or $x = 1$. Because $f\left(\frac{1}{3}\right) = \left(\frac{1}{9}+1\right)\left(2-\frac{1}{3}\right) = \frac{50}{27}$ and $f(1) = 2(2-1) = 2$, we see that the required points are $\left(\frac{1}{3}, \frac{50}{27}\right)$ and $(1, 2)$.

59. $f(x) = (x^2+6)(x-5)$, so

$$\begin{aligned} f'(x) &= (x^2+6)\tfrac{d}{dx}(x-5) + (x-5)\tfrac{d}{dx}(x^2+6) = (x^2+6)(1) + (x-5)(2x) \\ &= x^2 + 6 + 2x^2 - 10x = 3x^2 - 10x + 6. \end{aligned}$$

At a point where the slope of the tangent line is -2, we have $f'(x) = 3x^2 - 10x + 6 = -2$. This gives $3x^2 - 10x + 8 = (3x-4)(x-2) = 0$, so $x = \frac{4}{3}$ or $x = 2$. Because $f\left(\frac{4}{3}\right) = \left(\frac{16}{9}+6\right)\left(\frac{4}{3}-5\right) = -\frac{770}{27}$ and $f(2) = (4+6)(2-5) = -30$, the required points are $\left(\frac{4}{3}, -\frac{770}{27}\right)$ and $(2, -30)$.

61. $y = \dfrac{1}{1+x^2}$, so $y' = \dfrac{(1+x^2)\frac{d}{dx}(1) - (1)\frac{d}{dx}(1+x^2)}{(1+x^2)^2} = \dfrac{-2x}{(1+x^2)^2}$. Thus, the slope of the tangent line at $\left(1, \frac{1}{2}\right)$ is $y'|_{x=1} = \left.\dfrac{-2x}{(1+x^2)^2}\right|_{x=1} = \dfrac{-2}{4} = -\dfrac{1}{2}$ and an equation of the tangent line is $y - \frac{1}{2} = -\frac{1}{2}(x-1)$, or $y = -\frac{1}{2}x + 1$. Next, the slope of the required normal line is 2 and its equation is $y - \frac{1}{2} = 2(x-1)$, or $y = 2x - \frac{3}{2}$.

63. $C(x) = \dfrac{0.5x}{100-x}$, so $C'(x) = \dfrac{(100-x)(0.5) - 0.5x(-1)}{(100-x)^2} = \dfrac{50}{(100-x)^2}$. $C'(80) = \dfrac{50}{20^2} = 0.125$, $C'(90) = \dfrac{50}{10^2} = 0.5$, $C'(95) = \dfrac{50}{5^2} = 2$, and $C'(99) = \dfrac{50}{1} = 50$. The rates of change of the cost of removing 80%, 90%, 95%, and 99% of the toxic waste are 0.125, 0.5, 2, and 50 million dollars per 1% increase in waste removed. It is too costly to remove all of the pollutant.

65. $N(t) = \dfrac{10{,}000}{1+t^2} + 2000$, so $N'(t) = \dfrac{d}{dt}\left[10{,}000\left(1+t^2\right)^{-1} + 2000\right] = -\dfrac{10{,}000}{\left(1+t^2\right)^2}(2t) = -\dfrac{20{,}000t}{\left(1+t^2\right)^2}$. The rates of change after 1 minute and 2 minutes are $N'(1) = -\dfrac{20{,}000}{\left(1+1^2\right)^2} = -5000$ and $N'(2) = -\dfrac{20{,}000(2)}{\left(1+2^2\right)^2} = -1600$. The population of bacteria after one minute is $N(1) = \dfrac{10{,}000}{1+1} + 2000 = 7000$, and the population after two minutes is $N(2) = \dfrac{10{,}000}{1+4} + 2000 = 4000$.

67. a. $R(x) = xd(x) = \dfrac{50x}{0.01x^2+1}$.

b. $R'(x) = \dfrac{d}{dx}\left(\dfrac{50x}{0.01x^2+1}\right) = 50\dfrac{d}{dx}\left(\dfrac{x}{0.01x^2+1}\right) = 50 \cdot \dfrac{(0.01x^2+1)(1) - x(0.02x)}{(0.01x^2+1)^2} = \dfrac{50(1-0.01x^2)}{(0.01x^2+1)^2}$.

c. $R'(8) \approx 6.69$, $R'(10) = 0$, and $R'(12) \approx -3.70$, so the revenue is increasing at the rate of approximately \$6700 per thousand watches at a sales level of 8000 watches per week, the revenue is stable at a sales level of 10,000 watches per week, and the revenue is decreasing by approximately \$3700 per thousand watches at a sales level of 12,000 watches per week.

69. a. $N(t) = \dfrac{60t+180}{t+6}$, so

$$N'(t) = \frac{(t+6)\dfrac{d}{dt}(60t+180) - (60t+180)\dfrac{d}{dt}(t+6)}{(t+6)^2} = \frac{(t+6)(60) - (60t+180)(1)}{(t+6)^2} = \frac{180}{(t+6)^2}.$$

b. $N'(1) = \dfrac{180}{(1+6)^2} \approx 3.7$, $N'(3) = \dfrac{180}{(3+6)^2} \approx 2.2$, $N'(4) = \dfrac{180}{(4+6)^2} = 1.8$, and $N'(7) = \dfrac{180}{(7+6)^2} \approx 1.1$. We conclude that the rates at which the average student is increasing his or her speed one week, three weeks, four weeks, and seven weeks into the course are approximately 3.7, 2.2, 1.8, and 1.1 words per minute, respectively.

c. Yes.

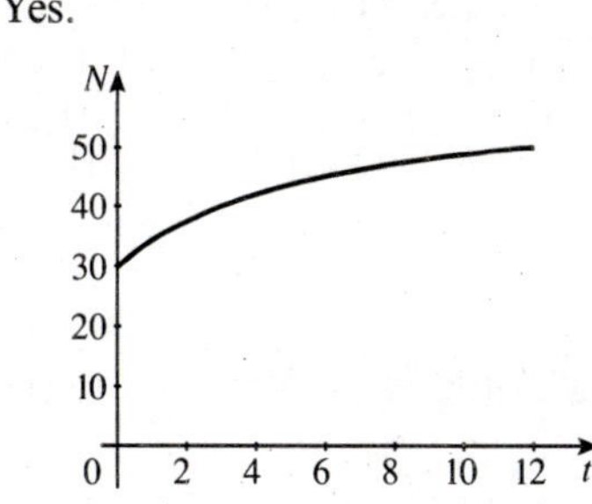

d. $N(12) = \dfrac{60(12)+180}{12+6} = 50$, or 50 words/minute.

71. $f(t) = \dfrac{0.055t + 0.26}{t+2}$, so $f'(t) = \dfrac{(t+2)(0.055) - (0.055t+0.26)(1)}{(t+2)^2} = -\dfrac{0.15}{(t+2)^2}$. At the beginning, the formaldehyde level is changing at the rate of $f'(0) = -\dfrac{0.15}{4} = -0.0375$; that is, it is decreasing at the rate of 0.0375 parts per million per year. Next, $f'(3) = -\dfrac{0.15}{5^2} = -0.006$, and so the level is decreasing at the rate of 0.006 parts per million per year at the beginning of the fourth year (when $t = 3$).

73. Its velocity at any time t is $v(t) = \frac{d}{dt}(16t^2) = 32t$. The hammer strikes the ground when $16t^2 = 256$ or $t = 4$ (we reject the negative root). Therefore, its velocity at the instant it strikes the ground is $v(4) = 32(4) = 128$ ft/sec. Its acceleration at time t is $a(t) = \frac{d}{dt}(32t) = 32$. In particular, its acceleration at $t = 4$ is 32 ft/sec^2.

75. $N(t) = -0.1t^3 + 1.5t^2 + 100$.

a. $N'(t) = -0.3t^2 + 3t = 0.3t(10-t)$. Because $N'(t) > 0$ for $t = 0, 1, 2, \ldots, 7$, it is evident that $N(t)$ (and therefore the crime rate) was increasing from 2005 through 2012.

b. $N''(t) = -0.6t + 3 = 0.6(5-t)$. Now $N''(4) = 0.6 > 0$, $N''(5) = 0$, $N''(6) = -0.6 < 0$, and $N''(7) = -1.2 < 0$. This shows that the rate of the rate of change was decreasing beyond $t = 5$ (in the year 2010). This shows that the program was working.

77. $N(t) = 0.00037t^3 - 0.0242t^2 + 0.52t + 5.3$ for $0 \le t \le 10$, so $N'(t) = 0.00111t^2 - 0.0484t + 0.52$ and $N''(t) = 0.00222t - 0.0484$. Therefore, $N(8) = 0.00037(8)^3 - 0.0242(8)^2 + 0.52(8) + 5.3 \approx 8.1$, $N'(8) = 0.00111(8)^2 - 0.0484(8) + 0.52 \approx 0.204$, and $N''(8) = 0.00222(8) - 0.0484 \approx -0.031$. We conclude that at the beginning of 1998, there were 8.1 million persons receiving disability benefits, the number was increasing at the rate of 0.2 million/yr, and the rate of the rate of change of the number of persons was decreasing at the rate of 0.03 million persons/yr^2.

79. False. Take $f(x) = x$ and $g(x) = x$. Then $f(x)g(x) = x^2$, so $\frac{d}{dx}[f(x)g(x)] = \frac{d}{dx}(x^2) = 2x \ne f'(x)g'(x) = 1$.

81. False. Let $f(x) = x^3$. Then $\dfrac{d}{dx}\left[\dfrac{f(x)}{x^2}\right] = \dfrac{d}{dx}\left(\dfrac{x^3}{x^2}\right) = \dfrac{d}{dx}(x) = 1 \ne \dfrac{f'(x)}{2x} = \dfrac{3x^2}{2x} = \frac{3}{2}x$.

83. Let $f(x) = u(x)v(x)$ and $g(x) = w(x)$. Then $h(x) = f(x)g(x)$. Therefore, $h'(x) = f'(x)g(x) + f(x)g'(x)$. But $f'(x) = u(x)v'(x) + u'(x)v(x)$, so

$$h'(x) = [u(x)v'(x) + u'(x)v(x)]g(x) + u(x)v(x)w'(x)$$
$$= u(x)v(x)w'(x) + u(x)v'(x)w(x) + u'(x)v(x)w(x).$$

Using Technology page 631

1. 0.8750 **3.** 0.0774 **5.** -0.5000 **7.** 31,312 per year

9. -18 **11.** 15.2762 **13.** -0.6255 **15.** 0.1973

17. $f''(6) = -68.46214$. This tells us that at the beginning of 1988, the rate of the rate at which banks were failing was decreasing at 68 banks per year per year.

9.6 The Chain Rule

Problem-Solving Tips

1. It is often easier to find the derivative of a quotient when the numerator is a constant by using the general power rule instead of the quotient rule. For example, to find the derivative of $f(x) = -\frac{1}{\sqrt{2x^2-1}}$ in Self-Check Exercise 1 of this section, we first rewrite the function in the form $f(x) = -(2x^2-1)^{-1/2}$ and then use the general power rule to find the derivative.

2. **To simplify a function involving the powers of an expression**, factor out the lowest power of the expression. For example, to simplify $5(x+1)^{1/2} - 3(x+1)^{-1/2}$, factor out $(x+1)^{-1/2}$, which is the lowest power of $x+1$ in the expression.

Concept Questions page 639

1. The derivative of $h(x) = g(f(x))$ is equal to the derivative of g evaluated at $f(x)$ times the derivative of f.

3. $(g \circ f)'(t) = [(g \circ f)(t)]' = g'(f(t))f'(t)$ describes the rate of change of the revenue as a function of time.

Exercises page 639

1. $f(x) = (2x-1)^3$, so $f'(x) = 3(2x-1)^2 \frac{d}{dx}(2x-1) = 3(2x-1)^2(2) = 6(2x-1)^2$.

3. $f(x) = (x^2+2)^5$, so $f'(x) = 5(x^2+2)^4(2x) = 10x(x^2+2)^4$.

5. $f(x) = (2x-x^2)^4$, so $f'(x) = 4(2x-x^2)^3 \frac{d}{dx}(2x-x^2) = 4(2x-x^2)^3(2-2x) = 8x^3(1-x)(2-x)^3$.

7. $f(x) = (2x+1)^{-2}$, so $f'(x) = -2(2x+1)^{-3} \frac{d}{dx}(2x+1) = -2(2x+1)^{-3}(2) = -4(2x+1)^{-3}$.

9. $f(x) = (x^2-4)^{5/2}$, so $f'(x) = \frac{5}{2}(x^2-4)^{3/2} \frac{d}{dx}(x^2-4) = \frac{5}{2}(x^2-4)^{3/2}(2x) = 5x(x^2-4)^{3/2}$.

11. $f(x) = \sqrt{3x-2} = (3x-2)^{1/2}$, so $f'(x) = \frac{1}{2}(3x-2)^{-1/2}(3) = \frac{3}{2}(3x-2)^{-1/2} = \frac{3}{2\sqrt{3x-2}}$.

13. $f(x) = \sqrt[3]{1-x^2}$, so

$$f'(x) = \tfrac{d}{dx}\left(1-x^2\right)^{1/3} = \tfrac{1}{3}\left(1-x^2\right)^{-2/3}\tfrac{d}{dx}\left(1-x^2\right) = \tfrac{1}{3}\left(1-x^2\right)^{-2/3}(-2x) = -\tfrac{2}{3}x\left(1-x^2\right)^{-2/3}$$
$$= \frac{-2x}{3\left(1-x^2\right)^{2/3}}.$$

15. $f(x) = \dfrac{1}{(2x+3)^3} = (2x+3)^{-3}$, so $f'(x) = -3(2x+3)^{-4}(2) = -6(2x+3)^{-4} = -\dfrac{6}{(2x+3)^4}$.

17. $f(t) = \dfrac{1}{\sqrt{2t-4}}$, so $f'(t) = \frac{d}{dt}(2t-4)^{-1/2} = -\frac{1}{2}(2t-4)^{-3/2}(2) = -(2t-4)^{-3/2} = -\dfrac{1}{(2t-4)^{3/2}}$.

19. $y = \dfrac{1}{\left(4x^4+x\right)^{3/2}}$, so $\frac{dy}{dx} = \frac{d}{dx}\left(4x^4+x\right)^{-3/2} = -\frac{3}{2}\left(4x^4+x\right)^{-5/2}\left(16x^3+1\right) = -\frac{3}{2}\left(16x^3+1\right)\left(4x^4+x\right)^{-5/2}$.

21. $f(x) = \left(3x^2+2x+1\right)^{-2}$, so

$$f'(x) = -2\left(3x^2+2x+1\right)^{-3}\tfrac{d}{dx}\left(3x^2+2x+1\right) = -2\left(3x^2+2x+1\right)^{-3}(6x+2)$$
$$= -4(3x+1)\left(3x^2+2x+1\right)^{-3}.$$

23. $f(x) = \left(x^2+1\right)^3 - \left(x^3+1\right)^2$, so

$$f'(x) = 3\left(x^2+1\right)^2\tfrac{d}{dx}\left(x^2+1\right) - 2\left(x^3+1\right)\tfrac{d}{dx}\left(x^3+1\right) = 3\left(x^2+1\right)^2(2x) - 2\left(x^3+1\right)\left(3x^2\right)$$
$$= 6x\left[\left(x^2+1\right)^2 - x\left(x^3+1\right)\right] = 6x\left(2x^2-x+1\right).$$

25. $f(t) = \left(t^{-1}-t^{-2}\right)^3$, so $f'(t) = 3\left(t^{-1}-t^{-2}\right)^2\frac{d}{dt}\left(t^{-1}-t^{-2}\right) = 3\left(t^{-1}-t^{-2}\right)^2\left(-t^{-2}+2t^{-3}\right)$.

27. $f(x) = \sqrt{x+1} + \sqrt{x-1} = (x+1)^{1/2} + (x-1)^{1/2}$, so
$f'(x) = \frac{1}{2}(x+1)^{-1/2}(1) + \frac{1}{2}(x-1)^{-1/2}(1) = \frac{1}{2}\left[(x+1)^{-1/2} + (x-1)^{-1/2}\right]$.

29. $f(x) = 2x^2(3-4x)^4$, so

$$f'(x) = 2x^2(4)(3-4x)^3(-4) + (3-4x)^4(4x) = 4x(3-4x)^3(-8x+3-4x)$$
$$= 4x(3-4x)^3(-12x+3) = (-12x)(4x-1)(3-4x)^3.$$

31. $f(x) = (x-1)^2(2x+1)^4$, so

$$f'(x) = (x-1)^2\tfrac{d}{dx}(2x+1)^4 + (2x+1)^4\tfrac{d}{dx}(x-1)^2 \quad \text{(by the Product Rule)}$$
$$= (x-1)^2(4)(2x+1)^3\tfrac{d}{dx}(2x+1) + (2x+1)^4(2)(x-1)\tfrac{d}{dx}(x-1)$$
$$= 8(x-1)^2(2x+1)^3 + 2(x-1)(2x+1)^4 = 2(x-1)(2x+1)^3(4x-4+2x+1)$$
$$= 6(x-1)(2x-1)(2x+1)^3.$$

33. $f(x) = \left(\dfrac{x+3}{x-2}\right)^3$, so

$$f'(x) = 3\left(\frac{x+3}{x-2}\right)^2 \frac{d}{dx}\left(\frac{x-3}{x-2}\right) = 3\left(\frac{x+3}{x-2}\right)^2 \left[\frac{(x-2)(1)-(x+3)(1)}{(x-2)^2}\right]$$
$$= 3\left(\frac{x+3}{x-2}\right)^2 \left[-\frac{5}{(x-2)^2}\right] = -\frac{15(x+3)^2}{(x-2)^4}.$$

35. $s(t) = \left(\dfrac{t}{2t+1}\right)^{3/2}$, so

$$s'(t) = \frac{3}{2}\left(\frac{t}{2t+1}\right)^{1/2} \frac{d}{dt}\left(\frac{t}{2t+1}\right) = \frac{3}{2}\left(\frac{t}{2t+1}\right)^{1/2} \left[\frac{(2t+1)(1)-t(2)}{(2t+1)^2}\right]$$
$$= \frac{3}{2}\left(\frac{t}{2t+1}\right)^{1/2} \left[\frac{1}{(2t+1)^2}\right] = \frac{3t^{1/2}}{2(2t+1)^{5/2}}.$$

37. $g(u) = \left(\dfrac{u+1}{3u+2}\right)^{1/2}$, so

$$g'(u) = \frac{1}{2}\left(\frac{u+1}{3u+2}\right)^{-1/2} \frac{d}{du}\left(\frac{u+1}{3u+2}\right) = \frac{1}{2}\left(\frac{u+1}{3u+2}\right)^{-1/2} \left[\frac{(3u+2)(1)-(u+1)(3)}{(3u+2)^2}\right]$$
$$= -\frac{1}{2\sqrt{u+1}\,(3u+2)^{3/2}}.$$

39. $f(x) = \dfrac{x^2}{(x^2-1)^4}$, so

$$f'(x) = \frac{(x^2-1)^4 \dfrac{d}{dx}(x^2) - (x^2)\dfrac{d}{dx}(x^2-1)^4}{\left[(x^2-1)^4\right]^2} = \frac{(x^2-1)^4(2x) - x^2(4)(x^2-1)^3(2x)}{(x^2-1)^8}$$
$$= \frac{(x^2-1)^3(2x)(x^2-1-4x^2)}{(x^2-1)^8} = \frac{(-2x)(3x^2+1)}{(x^2-1)^5}.$$

41. $h(x) = \dfrac{(3x^2+1)^3}{(x^2-1)^4}$, so

$$h'(x) = \frac{(x^2-1)^4(3)(3x^2+1)^2(6x) - (3x^2+1)^3(4)(x^2-1)^3(2x)}{(x^2-1)^8}$$
$$= \frac{2x(x^2-1)^3(3x^2+1)^2\left[9(x^2-1)-4(3x^2+1)\right]}{(x^2-1)^8} = -\frac{2x(3x^2+13)(3x^2+1)^2}{(x^2-1)^5}.$$

43. $f(x) = \dfrac{\sqrt{2x+1}}{x^2-1}$, so

$$f'(x) = \frac{(x^2-1)\left(\frac{1}{2}\right)(2x+1)^{-1/2}(2) - (2x+1)^{1/2}(2x)}{(x^2-1)^2} = \frac{(2x+1)^{-1/2}\left[(x^2-1)-(2x+1)(2x)\right]}{(x^2-1)^2}$$
$$= -\frac{3x^2+2x+1}{\sqrt{2x+1}\,(x^2-1)^2}.$$

45. $g(t) = \dfrac{(t+1)^{1/2}}{(t^2+1)^{1/2}}$, so

$$g'(t) = \frac{(t^2+1)^{1/2}\dfrac{d}{dt}(t+1)^{1/2} - (t+1)^{1/2}\dfrac{d}{dt}(t^2+1)^{1/2}}{t^2+1}$$
$$= \frac{(t^2+1)^{1/2}\left(\frac{1}{2}\right)(t+1)^{-1/2}(1) - (t+1)^{1/2}\left(\frac{1}{2}\right)(t^2+1)^{-1/2}(2t)}{t^2+1}$$
$$= \frac{\frac{1}{2}(t+1)^{-1/2}(t^2+1)^{-1/2}\left[(t^2+1) - 2t(t+1)\right]}{t^2+1} = -\frac{t^2+2t-1}{2\sqrt{t+1}\,(t^2+1)^{3/2}}.$$

47. $f(x) = (x^2+2)^5$, so $f'(x) = 5(x^2+2)^4(2x) = 10x(x^2+2)^4$ and
$f''(x) = 10(x^2+2)^4 + 10x(4)(x^2+2)^3(2x) = 10(x^2+2)^3\left[(x^2+2)+8x^2\right] = 10(9x^2+2)(x^2+2)^3$.

49. $y = g(u) = u^{4/3}$, so $\frac{dy}{du} = \frac{4}{3}u^{1/3}$, and $u = f(x) = 3x^2-1$, so $\dfrac{du}{dx} = 6x$. Thus,
$\dfrac{dy}{dx} = \dfrac{dy}{du}\cdot\dfrac{du}{dx} = \frac{4}{3}u^{1/3}(6x) = \frac{4}{3}(3x^2-1)^{1/3}6x = 8x(3x^2-1)^{1/3}$.

51. $y = u^{-2/3}$ and $u = 2x^3 - x + 1$, so $\dfrac{dy}{du} = -\frac{2}{3}u^{-5/3} = -\dfrac{2}{3u^{5/3}}$ and $\dfrac{du}{dx} = 6x^2 - 1$, so
$\dfrac{dy}{dx} = \dfrac{dy}{du}\cdot\dfrac{du}{dx} = -\dfrac{2(6x^2-1)}{3u^{5/3}} = -\dfrac{2(6x^2-1)}{3(2x^3-x+1)^{5/3}}$.

53. $y = \sqrt{u} + \dfrac{1}{\sqrt{u}}$ and $u = x^3 - x$, so $\dfrac{dy}{du} = \frac{1}{2}u^{-1/2} - \frac{1}{2}u^{-3/2}$ and $\dfrac{du}{dx} = 3x^2 - 1$, so
$\dfrac{dy}{dx} = \dfrac{dy}{du}\cdot\dfrac{du}{dx} = \left[\dfrac{1}{2\sqrt{x^3-x}} - \dfrac{1}{2(x^3-x)^{3/2}}\right](3x^2-1) = \dfrac{(3x^2-1)(x^3-x-1)}{2(x^3-x)^{3/2}}$.

55. $g(x) = f(2x+1)$. Let $u = 2x+1$, so $\dfrac{du}{dx} = 2$. Using the Chain Rule, we have
$g'(x) = f'(u)\dfrac{du}{dx} = f'(2x+1)\cdot 2 = 2f'(2x+1)$.

57. $F(x) = g(f(x))$, so $F'(x) = g'(f(x))f'(x)$. Thus, $F'(2) = g'(3)(-3) = (4)(-3) = -12$.

59. Let $g(x) = x^2 + 1$. Then $F(x) = f(g(x))$. Next, $F'(x) = f'(g(x))g'(x)$ and
$F'(1) = f'(2)(2x) = (3)(2) = 6$.

61. No. Suppose $h = g(f(x))$. Let $f(x) = x$ and $g(x) = x^2$. Then $h = g(f(x)) = g(x) = x^2$ and
$h'(x) = 2x \neq g'(f'(x)) = g'(1) = 2(1) = 2$.

63. $f(x) = (1-x)(x^2-1)^2$, so
$f'(x) = (1-x)2(x^2-1)(2x) + (-1)(x^2-1)^2 = (x^2-1)(4x-4x^2-x^2+1) = (x^2-1)(-5x^2+4x+1)$.
Therefore, the slope of the tangent line at $(2,-9)$ is $f'(2) = \left[(2)^2-1\right]\left[-5(2)^2+4(2)+1\right] = -33$. Then an equation of the line is $y + 9 = -33(x-2)$, or $y = -33x + 57$.

65. $f(x) = x\sqrt{2x^2+7}$, so $f'(x) = \sqrt{2x^2+7} + x\left(\frac{1}{2}\right)\left(2x^2+7\right)^{-1/2}(4x)$. The slope of the tangent line at $x = 3$ is $f'(3) = \sqrt{25} + \left(\frac{3}{2}\right)(25)^{-1/2}(12) = \frac{43}{5}$, so an equation is $y - 15 = \frac{43}{5}(x-3)$, or $y = \frac{43}{5}x - \frac{54}{5}$.

67. $N(t) = (60+2t)^{2/3}$, so $N'(t) = \frac{2}{3}(60+2t)^{-1/3}\frac{d}{dt}(60+2t) = \frac{4}{3}(60+2t)^{-1/3}$. The rate of increase at the end of the second week is $N'(2) = \frac{4}{3}(64)^{-1/3} = \frac{1}{3}$, or approximately 0.333 million/week. At the end of the 12th week, $N'(12) = \frac{4}{3}(84)^{-1/3} \approx 0.3$, or 0.3 million/week. The number of viewers in the 2nd week is $N(2) = (60+4)^{2/3} = 16$, or 16 million, and the number of viewers in the 24th week is $N(24) = (60+48)^{2/3} \approx 22.7$, or approximately 22.7 million.

69. $P(t) = 33.55(t+5)^{0.205}$. $P'(t) = 33.55(0.205)(t+5)^{-0.795}(1) = 6.87775(t+5)^{-0.795}$. The rate of change at the beginning of 2000 is $P'(20) = 6.87775(25)^{-0.795} \approx 0.5322$, or 0.53%/yr. The percent of these mothers was $P(20) = 33.55(25)^{0.205} \approx 64.90$, or 64.9%.

71. a. $P(0) = 100$ and $P(1) \approx 30.0$. The probability of survival at the moment of diagnosis is 100%. The probability of survival 1 year after diagnosis is approximately 30%.

b. $P'(t) = \dfrac{d}{dt}\left[100(1+0.14t)^{-9.2}\right] = 100(-9.2)(1+0.14t)^{-10.2}(0.14) = -\dfrac{920 \cdot 0.14}{(1+0.14t)^{10.2}} = -\dfrac{128.8}{(1+014t)^{10.2}}$.

Thus, $P'(0) = -128.8$ and $P'(1) \approx -33.84$. At the moment of diagnosis, the probability of survival is falling at the rate of approximately 129% per year. After 1 year, the probability of survival is dropping at the rate of approximately 34% per year.

73. $C(t) = 0.01\left(0.2t^2+4t+64\right)^{2/3}$.

a.
$$\begin{aligned} C'(t) &= 0.01\left(\tfrac{2}{3}\right)\left(0.2t^2+4t+64\right)^{-1/3}\frac{d}{dt}\left(0.2t^2+4t+64\right) \\ &= (0.01)\left(\tfrac{2}{3}\right)(0.4t+4)\left(0.2t^2+4t+64\right)^{-1/3} \approx 0.027(0.1t+1)\left(0.2t^2+4t+64\right)^{-1/3}. \end{aligned}$$

b. $C'(5) = 0.027(0.5+1)[0.2(25)+4(5)+64]^{-1/3} \approx 0.009$, or 0.009 parts per million per year.

75. a. $A(t) = 0.03t^3(t-7)^4 + 60.2$, so

$A'(t) = 0.03\left[3t^2(t-7)^4 + t^3(4)(t-7)^3\right] = 0.03t^2(t-7)^3[3(t-7)+4t] = 0.21t^2(t-3)(t-7)^3$.

b. $A'(1) = 0.21(-2)(-6)^3 = 90.72$, $A'(3) = 0$, and $A'(4) = 0.21(16)(1)(-3)^3 = -90.72$. The amount of pollutant is increasing at the rate of 90.72 units/hr at 8 a.m. The rate of change is 0 units/hr at 10 a.m. and -90.72 units/hr at 11 a.m.

77. $P(t) = \dfrac{300\sqrt{\frac{1}{2}t^2+2t+25}}{t+25} = \dfrac{300\left(\frac{1}{2}t^2+2t+25\right)^{1/2}}{t+25}$, so

$$P'(t) = 300\left[\frac{(t+25)\frac{1}{2}\left(\frac{1}{2}t^2+2t+25\right)^{-1/2}(t+2)-\left(\frac{1}{2}t^2+2t+25\right)^{1/2}(1)}{(t+25)^2}\right]$$

$$= 300\left[\frac{\left(\frac{1}{2}t^2+2t+25\right)^{-1/2}\left[(t+25)(t+2)-2\left(\frac{1}{2}t^2+2t+25\right)\right]}{2(t+25)^2}\right] = \frac{3450t}{(t+25)^2\sqrt{\frac{1}{2}t^2+2t+25}}.$$

Ten seconds into the run, the athlete's pulse rate is increasing at $P'(10) = \dfrac{3450(10)}{(35)^2\sqrt{50+20+25}} \approx 2.9$, or approximately 2.9 beats per minute per second. Sixty seconds into the run, it is increasing at $P'(60) = \dfrac{3450(60)}{(85)^2\sqrt{1800+120+25}} \approx 0.65$, or approximately 0.7 beats per minute per second. Two minutes into the run, it is increasing at $P'(120) = \dfrac{3450(120)}{(145)^2\sqrt{7200+240+25}} \approx 0.23$, or approximately 0.2 beats per minute per second. The pulse rate two minutes into the run is given by $P(120) = \dfrac{300\sqrt{7200+240+25}}{120+25} \approx 178.8$, or approximately 179 beats per minute.

79. The area is given by $A = \pi r^2$. The rate at which the area is increasing is given by dA/dt, that is, $\dfrac{dA}{dt} = \dfrac{d}{dt}\left(\pi r^2\right) = \dfrac{d}{dt}\left(\pi r^2\right)\dfrac{dr}{dt} = 2\pi r\dfrac{dr}{dt}$. If $r = 40$ and $dr/dt = 2$, then $\dfrac{dA}{dt} = 2\pi(40)(2) = 160\pi$, that is, it is increasing at the rate of 160π, or approximately 503 ft^2/sec.

81. $f(t) = 6.25t^2 + 19.75t + 74.75$ and $g(x) = -0.00075x^2 + 67.5$, so $\dfrac{dS}{dt} = g'(x)f'(t) = (-0.0015x)(12.5t + 19.75)$. When $t = 4$, we have $x = f(4) = 6.25(16) + 19.75(4) + 74.75 = 253.75$ and $\left.\dfrac{dS}{dt}\right|_{t=4} = (-0.0015)(253.75)[12.5(4) + 19.75] \approx -26.55$; that is, the average speed will be dropping at the rate of approximately 27 mph per decade. The average speed of traffic flow at that time will be $S = g(f(4)) = -0.00075\left(253.75^2\right) + 67.5 \approx 19.2$, or approximately 19 mph.

83. $N(x) = 1.42x$ and $x(t) = \dfrac{7t^2+140t+700}{3t^2+80t+550}$. The number of construction jobs as a function of time is $n(t) = N(x(t))$. Using the Chain Rule,

$$n'(t) = \frac{dN}{dx}\cdot\frac{dx}{dt} = 1.42\frac{dx}{dt} = (1.42)\left[\frac{\left(3t^2+80t+550\right)(14t+140)-\left(7t^2+140t+700\right)(6t+80)}{\left(3t^2+80t+550\right)^2}\right]$$

$$= \frac{1.42\left(140t^2+3500t+21000\right)}{\left(3t^2+80t+550\right)^2}.$$

$n'(12) = \dfrac{1.42\left[140(12)^2+3500(12)+21000\right]}{\left[3(12)^2+80(12)+550\right]^2} \approx 0.0313115$, or approximately 31,312 jobs/year/month.

85. $x = f(p) = 10\sqrt{\dfrac{50-p}{p}}$, so

$$\frac{dx}{dp} = \frac{d}{dp}\left[10\left(\frac{50-p}{p}\right)^{1/2}\right] = (10)\left(\tfrac{1}{2}\right)\left(\frac{50-p}{p}\right)^{-1/2}\frac{d}{dp}\left(\frac{50-p}{p}\right)$$

$$= 5\left(\frac{50-p}{p}\right)^{-1/2}\cdot\frac{d}{dp}\left(\frac{50}{p}-1\right) = 5\left(\frac{50-p}{p}\right)^{-1/2}\left(-\frac{50}{p^2}\right) = -\frac{250}{p^2\left(\dfrac{50-p}{p}\right)^{1/2}}$$

and $\left.\dfrac{dx}{dp}\right|_{p=25} = -\left.\dfrac{250}{p^2\left(\dfrac{50-p}{p}\right)^{1/2}}\right|_{p=25} = -\dfrac{250}{(625)\left(\dfrac{25}{25}\right)^{1/2}} = -0.4$. Thus, the quantity demanded is falling at the rate of 0.4 (1000) or 400 sports watches per dollar increase in price.

87. True. This is just the statement of the Chain Rule.

89. True. $\frac{d}{dx}\sqrt{f(x)} = \frac{d}{dx}[f(x)]^{1/2} = \frac{1}{2}[f(x)]^{-1/2}f'(x) = \dfrac{f'(x)}{2\sqrt{f(x)}}$.

91. Let $f(x) = x^{1/n}$ so that $[f(x)]^n = x$. Differentiating both sides with respect to x, we get $n[f(x)]^{n-1}f'(x) = 1$, so $f'(x) = \dfrac{1}{n[f(x)]^{n-1}} = \dfrac{1}{n[x^{1/n}]^{n-1}} = \dfrac{1}{nx^{1-(1/n)}} = \dfrac{1}{n}x^{(1/n)-1}$, as was to be shown.

Using Technology page 644

1. 0.5774 **3.** 0.9390 **5.** −4.9498

7. a. Using the numerical derivative operation, we find that $N'(0) = 5.41450$, so the rate of change of the number of people watching TV on mobile phones at the beginning of 2007 is approximately 5.415 million/year.

b. $N'(4) \approx 2.5136$, so the corresponding rate of change at the beginning of 2011 is expected to be approximately 2.5136 million/year.

9.7 Differentiation of Exponential and Logarithmic Functions

Problem-Solving Tips

1. The derivative of e^x is equal to e^x. By the Chain Rule, $\frac{d}{dx}(e^{3x}) = 3e^{3x}$ and $\frac{d}{dx}\left(e^{2x^2-1}\right) = 4xe^{2x^2-1}$. Note that the exponents in the original function and the derivative are the same.

2. Don't confuse functions of the type e^x with functions of the type x^r. The latter is a *power function* and its exponent is a *constant*; whereas the exponent in an *exponential function* such as e^x is a *variable*. A different rule is used to differentiate the two types of function. Thus, $\frac{d}{dx}(x^2e^x) = x^2\frac{d}{dx}(e^x) + e^x\frac{d}{dx}(x^2) = x^2e^x + e^x\cdot 2x = xe^x(x+2)$.

Concept Questions

page 652

1. a. $f'(x) = e^x$

b. $g'(x) = e^{f(x)} \cdot f'(x)$

3. a. $f'(x) = \dfrac{1}{x}$

b. $g'(x) = \dfrac{f'(x)}{f(x)}$

Exercises

page 652

1. $f(x) = e^{3x}$, so $f'(x) = 3e^{3x}$

3. $g(t) = e^{-t}$, so $g'(t) = -e^{-t}$

5. $f(x) = e^x + x^2$, so $f'(x) = e^x + 2x$.

7. $f(x) = x^3e^x$, so $f'(x) = x^3e^x + e^x\left(3x^2\right) = x^2e^x(x+3)$.

9. $f(x) = \dfrac{e^x}{x}$, so $f'(x) = \dfrac{x(e^x) - e^x(1)}{x^2} = \dfrac{e^x(x-1)}{x^2}$.

11. $f(x) = 3\left(e^x + e^{-x}\right)$, so $f'(x) = 3\left(e^x - e^{-x}\right)$.

13. $f(w) = \dfrac{e^w + 2}{e^w} = 1 + \dfrac{2}{e^w} = 1 + 2e^{-w}$, so $f'(w) = -2e^{-w} = -\dfrac{2}{e^w}$.

15. $f(x) = 2e^{3x-1}$, so $f'(x) = 2e^{3x-1}(3) = 6e^{3x-1}$.

17. $h(x) = e^{-x^2}$, so $h'(x) = e^{-x^2}(-2x) = -2xe^{-x^2}$.

19. $f(x) = 3e^{1/x}$, so $f'(x) = 3e^{1/x} \cdot \dfrac{d}{dx}\left(\dfrac{1}{x}\right) = 3e^{1/x}\left(-\dfrac{1}{x^2}\right) = -\dfrac{3e^{1/x}}{x^2}$.

21. $f(x) = (e^x + 1)^{25}$, so $f'(x) = 25(e^x+1)^{24}e^x = 25e^x(e^x+1)^{24}$.

23. $f(x) = e^{\sqrt{x}}$, so $f'(x) = e^{\sqrt{x}}\dfrac{d}{dx}\left(x^{1/2}\right) = e^{\sqrt{x}}\frac{1}{2}x^{-1/2} = \dfrac{e^{\sqrt{x}}}{2\sqrt{x}}$.

25. $f(x) = (x-1)e^{3x+2}$, so $f'(x) = (x-1)(3)e^{3x+2} + e^{3x+2} = e^{3x+2}(3x-3+1) = e^{3x+2}(3x-2)$.

27. $f(x) = \dfrac{e^x - 1}{e^x + 1}$, so $f'(x) = \dfrac{(e^x+1)(e^x) - (e^x-1)(e^x)}{(e^x+1)^2} = \dfrac{e^x(e^x+1-e^x+1)}{(e^x+1)^2} = \dfrac{2e^x}{(e^x+1)^2}$.

29. $f(x) = e^{-4x} + e^{3x}$, so $f'(x) = -4e^{-4x} + 3e^{3x}$ and $f''(x) = 16e^{-4x} + 9e^{3x}$.

31. $f(x) = 2xe^{3x}$, so $f'(x) = 2e^{3x} + 2xe^{3x}(3) = 2(3x+1)e^{3x}$ and $f''(x) = 6e^{3x} + 2(3x+1)e^{3x}(3) = 6(3x+2)e^{3x}$.

33. $y = f(x) = e^{2x-3}$, so $f'(x) = 2e^{2x-3}$. To find the slope of the tangent line to the graph of f at $x = \frac{3}{2}$, we compute $f'\left(\frac{3}{2}\right) = 2e^{3-3} = 2$. Next, using the point-slope form of the equation of a line, we find that $y - 1 = 2\left(x - \frac{3}{2}\right) = 2x - 3$, or $y = 2x - 2$.

35. $f(x) = 5\ln x$, so $f'(x) = 5\left(\dfrac{1}{x}\right) = \dfrac{5}{x}$.

37. $f(x) = \ln(x+1)$, so $f'(x) = \dfrac{1}{x+1}$.

39. $f(x) = \ln x^8$, so $f'(x) = \dfrac{8x^7}{x^8} = \dfrac{8}{x}$.

41. $f(x) = \ln x^{1/2}$, so $f'(x) = \dfrac{\frac{1}{2}x^{-1/2}}{x^{1/2}} = \dfrac{1}{2x}$.

43. $f(x) = \ln\left(\dfrac{1}{x^2}\right) = \ln x^{-2}$, so $f'(x) = -\dfrac{2x^{-3}}{x^{-2}} = -\dfrac{2}{x}$.

45. $f(x) = \ln\left(4x^2 - 5x + 3\right)$, so $f'(x) = \dfrac{8x - 5}{4x^2 - 5x + 3}$.

47. $f(x) = \ln\left(\dfrac{2x}{x+1}\right) = \ln 2x - \ln(x+1)$, so

$$f'(x) = \frac{2}{2x} - \frac{1}{x+1} = \frac{1}{x} - \frac{1}{x+1} = \frac{(x+1) - x}{x(x+1)} = \frac{x+1-x}{x(x+1)} = \frac{1}{x(x+1)}.$$

49. $f(x) = x^2 \ln x$, so $f'(x) = x^2\left(\frac{1}{x}\right) + (\ln x)(2x) = x + 2x\ln x = x(1 + 2\ln x)$.

51. $f(x) = \dfrac{2\ln x}{x}$, so $f'(x) = \dfrac{x\left(\frac{2}{x}\right) - 2\ln x}{x^2} = \dfrac{2(1 - \ln x)}{x^2}$.

53. $f(u) = \ln(u-2)^3$, so $f'(u) = \dfrac{3(u-2)^2}{(u-2)^3} = \dfrac{3}{u-2}$.

55. $f(x) = (\ln x)^{1/2}$, so $f'(x) = \frac{1}{2}(\ln x)^{-1/2}\left(\frac{1}{x}\right) = \dfrac{1}{2x\sqrt{\ln x}}$.

57. $f(x) = (\ln x)^2$, so $f'(x) = 2(\ln x)\left(\dfrac{1}{x}\right) = \dfrac{2\ln x}{x}$.

59. $f(x) = \ln\left(x^3 + 1\right)$, so $f'(x) = \dfrac{3x^2}{x^3 + 1}$.

61. $f(x) = e^x \ln x$, so $f'(x) = e^x \ln x + e^x\left(\frac{1}{x}\right) = \dfrac{e^x(x\ln x + 1)}{x}$.

63. $f(x) = \ln 2 + \ln x$, so $f'(x) = \dfrac{1}{x}$ and $f''(x) = -\dfrac{1}{x^2}$.

65. $f(x) = \ln\left(x^2 + 2\right)$, so $f'(x) = \dfrac{2x}{\left(x^2 + 2\right)}$ and $f''(x) = \dfrac{\left(x^2+2\right)(2) - 2x(2x)}{\left(x^2+2\right)^2} = \dfrac{2\left(2 - x^2\right)}{\left(x^2+2\right)^2}$.

67. $y = x\ln x$. The slope of the tangent line at any point is $y' = \ln x + x\left(\frac{1}{x}\right) = \ln x + 1$. In particular, the slope of the tangent line at $(1, 0)$ is $m = \ln 1 + 1 = 1$. Thus, an equation of the tangent line is $y - 0 = 1(x - 1)$, or $y = x - 1$.

69. $P(t) = 20.6e^{-0.009t}$, so $P'(t) = 20.6(-0.009)e^{-0.009t} = -0.1854e^{-0.009t}$. Thus, $P'(10) = -0.1694$, $P'(20) = -0.1549$ and $P'(30) = -0.1415$. This tells us that the percentage of the total population relocating each year was decreasing at the rate of 0.17% in 1970, 0.15% in 1980, and 0.14% in 1990.

71. a. $P(t) = 0.07e^{0.54t}$. The population at the beginning of 2000 was $P(0) = 0.07$, or 70,000. The population at the beginning of 2030 will be $P(3) \approx 0.3537$, or approximately 353,700.

b. $P'(t) = 0.0378e^{0.54t}$. The population was changing at the rate of $P'(0) = 0.0378$, or 37,800/decade, at the beginning of 2000. At the beginning of 2030, it is projected to be changing at the rate of $P'(3) \approx 0.191$, or increasing by approximately 191,000/decade.

73. a. $N(t) = 130.7e^{-0.11t^2} + 50$. The number of deaths in 1950 was $N(0) = 130.7 + 50 = 180.7$, or approximately 181 per 100,000 people.

b. $N'(t) = (130.7)(-0.1155)(2t)e^{-0.1155t^2}$

$= -30.1917te^{-0.1155t^2}$.

The rates of change of the number of deaths per 100,000 people per decade are given in the table.

Year	1950	1960	1970	1980
Rate	0	−27	−38	−32

c. The number is given by $N(6) \approx 52.04$, or approximately 52.

75. $f'(t) = \dfrac{d}{dt}\left(93.1e^{-0.1626t}\right) = 93.1\left(-0.1626e^{-0.1626t}\right) = -15.13806e^{-0.1626t}$, so $f'(1) \approx -12.9$, $f'(2) \approx -10.9$, $f'(3) \approx -9.3$, and $f'(4) \approx -7.9$. Thus, the required rates are approximately 12.9%, 10.9%, 9.3%, and 7.9% per year.

77. a. The temperature inside the house is given by $T(0) = 30 + 40e^0 = 70$, or 70°F.

b. The reading is changing at the rate of $T'(1) = 40(-0.98)e^{-0.98t}\big|_{t=1} \approx -14.7$. Thus, it is dropping at the rate of approximately 14.7°F/min.

c. The temperature outdoors is given by $\lim_{t\to\infty} T(t) = \lim_{t\to\infty}\left(30 + 40e^{-0.98t}\right) = 30 + 0 = 30$, or 30°F.

79. The demand equation is $p(x) = 100e^{-0.0002x} + 150$. Next, $p'(x) = 100(-0.0002)e^{-0.0002x} = -0.02e^{-0.0002x}$.

a. To find the rate of change of the price per bottle when $x = 1000$, we compute $p'(1000) = -0.02e^{-0.0002(1000)} = -0.02e^{-0.2} \approx -0.0164$, or -1.64 cents per bottle. To find the rate of change of the price per bottle when $x = 2000$, we compute $p'(2000) = -0.02e^{-0.0002(2000)} = -0.02e^{-0.4} \approx -0.0134$, or -1.34 cents per bottle.

b. The price per bottle when $x = 1000$ is given by $p(1000) = 100e^{-0.0002(1000)} + 150 \approx 231.87$, or \$231.87/bottle. The price per bottle when $x = 2000$ is given by $p(2000) = 100e^{-0.0002(2000)} + 150 \approx 217.03$, or \$217.03/bottle.

81. $f(t) = 1.5 + 1.8te^{-1.2t}$, so $f'(t) = 1.8\frac{d}{dt}\left(te^{-1.2t}\right) = 1.8\left[e^{-1.2t} + te^{-1.2t}(-1.2)\right] = 1.8e^{-1.2t}(1 - 1.2t)$. $f'(0) \approx 1.8$, $f'(1) \approx -0.11$, $f'(2) \approx -0.23$, and $f'(3) \approx -0.13$. Thus, measured in barrels per \$1000 of output per decade, the amount of oil used is increasing by 1.8 in 1965, decreasing by 0.11 in 1966, and so on.

83. $f(x) = 7.2956\ln\left(0.0645012x^{0.95} + 1\right)$, so

$$f'(x) = 7.2956 \cdot \frac{\frac{d}{dx}\left(0.0645012x^{0.95} + 1\right)}{0.0645012x^{0.95} + 1} = \frac{7.2956(0.0645012)\left(0.95x^{-0.05}\right)}{0.0645012x^{0.95} + 1} = \frac{0.447046207}{x^{0.05}\left(0.0645012x^{0.95} + 1\right)}.$$

Thus, $f'(100) \approx 0.05799$, or approximately 0.0580%/kg, and $f'(500) \approx 0.01330$, or approximately 0.0133%/kg.

85. a. $P(t) = 28.5 + 14.42\ln t$. The proportion in 2001 is $P(1) = 28.5$. $P'(t) = \dfrac{14.42}{t}$, and so the proportion in 2001 is changing at the rate of $P'(1) = 14.42\%/\text{yr}$.

b. The proportion in 2006 is expected to be $P(6) \approx 54.34$. Its rate of change is expected to be $P'(6) = \frac{14.42}{6} \approx 2.4\%/\text{yr.}$

87. False. $f(x) = e^{\pi}$ is a constant function and so $f'(x) = 0$.

89. If $x \leq 0$, then $|x| = -x$. Therefore, $\ln|x| = \ln(-x)$. Writing $f(x) = \ln|x|$, we have $|x| = -x = e^{f(x)}$. Differentiating both sides with respect to x and using the Chain Rule,we have $-1 = e^{f(x)} \cdot f'(x)$, so $f'(x) = -\dfrac{1}{e^{f(x)}} = -\dfrac{1}{-x} = \dfrac{1}{x}$.

Using Technology page 656

1. 5.4366

3. 12.3929

5. 0.1861

7. a.

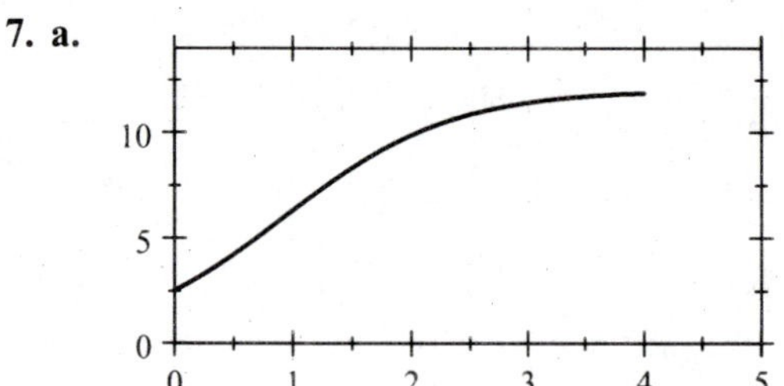

b. We estimate $P'(1) \approx 4.272$ billion people per half century.

9. a. Using the function evaluation capabilities of a graphing utility, we find $f(11) = 153.024$ and $g(11) \approx 235.181$. This tells us that the number of violent-crime arrests will be 153,024 at the beginning of the year 2000, but if trends like inner-city drug use and wider availability of guns continue, then the number of arrests will be 235,181.

b. Using the differentiation capability of a graphing utility, we find $f'(11) = -0.634$ and $g'(11) \approx 18.401$. This tells us that the number of violent-crime arrests will be decreasing at the rate of 634 per year at the beginning of the year 2000, but if the trends like inner-city drug use and wider availability of guns continue, then the number of arrests will be increasing at the rate of 18,401 per year at the beginning of the year 2000.

11. a. $P(10) = \dfrac{74}{1 + 2.6e^{-0.166(10)+0.04536(10)^2-0.0066(10)^3}} \approx 69.63\%$.

b. $P'(10) \approx 5.09361$, or approximately 5.09%/decade.

9.8 Marginal Functions in Economics

Problem-Solving Tips

1. **The marginal cost function is the derivative of the cost function.** Similarly, the marginal profit function and the marginal revenue function are the derivatives of the profit function and the revenue function, respectively. The key word here is "marginal'': it indicates that we are dealing with the derivative of a function.

2. **The average cost function** is given by $\overline{C}(x) = \frac{C(x)}{x}$ and the **marginal average cost function** is given by $\overline{C}'(x)$.

3. Remember that the revenue is increasing on an interval where the demand is inelastic, decreasing on an interval where the demand is elastic, and stationary at the point where the demand is unitary.

Concept Questions

page 663

1. **a.** The marginal cost function is the derivative of the cost function.

   b. The average cost function is equal to the total cost function divided by the total number of the commodity produced.

   c. The marginal average cost function is the derivative of the average cost function.

   d. The marginal revenue function is the derivative of the revenue function.

   e. The marginal profit function is the derivative of the profit function.

Exercises

page 663

1. **a.** $C(x)$ is always increasing because as x, the number of units produced, increases, the amount of money that must be spent on production also increases.

   b. This occurs at $x = 4$, a production level of 4000. You can see this by looking at the slopes of the tangent lines for x less than, equal to, and a little larger then $x = 4$.

3. **a.** The actual cost incurred in the production of the 1001st disc is given by

$$C(1001) - C(1000) = \left[2000 + 2(1001) - 0.0001(1001)^2\right] - \left[2000 + 2(1000) - 0.0001(1000)^2\right]$$
$$= 3901.7999 - 3900 = 1.7999, \text{ or approximately } \$1.80.$$

The actual cost incurred in the production of the 2001st disc is given by

$$C(2001) - C(2000) = \left[2000 + 2(2001) - 0.0001(2001)^2\right] - \left[2000 + 2(2000) - 0.0001(2000)^2\right]$$
$$= 5601.5999 - 5600 = 1.5999, \text{ or approximately } \$1.60.$$

   b. The marginal cost is $C'(x) = 2 - 0.0002x$. In particular, $C'(1000) = 2 - 0.0002(1000) = 1.80$, or \$1.80, and $C'(2000) = 2 - 0.0002(2000) = 1.60$, or \$1.60.

5. **a.** $\overline{C}(x) = \dfrac{C(x)}{x} = \dfrac{100x + 200{,}000}{x} = 100 + \dfrac{200{,}000}{x}$.

   b. $\overline{C}'(x) = \dfrac{d}{dx}(100) + \dfrac{d}{dx}\left(200{,}000x^{-1}\right) = -200{,}000x^{-2} = -\dfrac{200{,}000}{x^2}$.

   c. $\lim\limits_{x\to\infty} \overline{C}(x) = \lim\limits_{x\to\infty}\left(100 + \dfrac{200{,}000}{x}\right) = 100$. This says that the average cost approaches \$100 per unit if the production level is very high.

7. $\overline{C}(x) = \dfrac{C(x)}{x} = \dfrac{2000 + 2x - 0.0001x^2}{x} = \dfrac{2000}{x} + 2 - 0.0001x$, so

$$\overline{C}'(x) = -\frac{2000}{x^2} + 0 - 0.0001 = -\frac{2000}{x^2} - 0.0001.$$

9. **a.** $R'(x) = \dfrac{d}{dx}\left(8000x - 100x^2\right) = 8000 - 200x$.

   b. $R'(39) = 8000 - 200(39) = 200$, $R'(40) = 8000 - 200(40) = 0$, and $R'(41) = 8000 - 200(41) = -200$.

   c. This suggests the total revenue is maximized if the price charged per passenger is \$40.

11. **a.** $P(x) = R(x) - C(x) = \left(-0.04x^2 + 800x\right) - (200x + 300{,}000) = -0.04x^2 + 600x - 300{,}000$.

   b. $P'(x) = -0.08x + 600$.

   c. $P'(5000) = -0.08(5000) + 600 = 200$ and $P'(8000) = -0.08(8000) + 600 = -40$.

d. 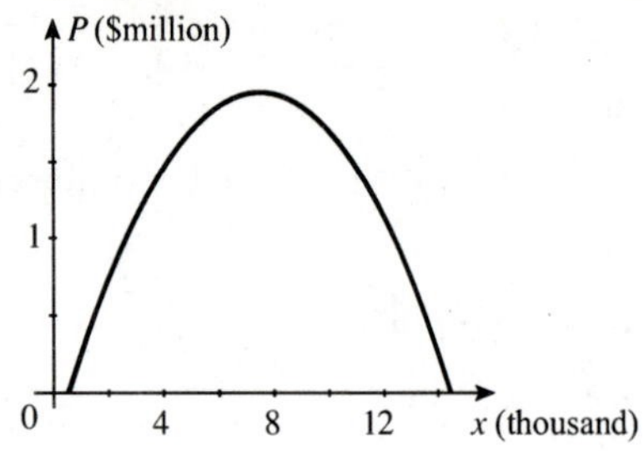

The profit realized by the company increases as production increases, peaking at a production level of 7500 units. Beyond this level of production, the profit begins to fall.

13. a. The revenue function is $R(x) = px = (600 - 0.05x)\,x = 600x - 0.05x^2$ and the profit function is

$$P(x) = R(x) - C(x) = \left(600x - 0.05x^2\right) - \left(0.000002x^3 - 0.03x^2 + 400x + 80{,}000\right)$$
$$= -0.000002x^3 - 0.02x^2 + 200x - 80{,}000.$$

b. $C'(x) = \dfrac{d}{dx}\left(0.000002x^3 - 0.03x^2 + 400x + 80{,}000\right) = 0.000006x^2 - 0.06x + 400,$

$R'(x) = \dfrac{d}{dx}\left(600x - 0.05x^2\right) = 600 - 0.1x$, and

$P'(x) = \dfrac{d}{dx}\left(-0.000002x^3 - 0.02x^2 + 200x - 80{,}000\right) = -0.000006x^2 - 0.04x + 200.$

c. $C'(2000) = 0.000006\,(2000)^2 - 0.06\,(2000) + 400 = 304$, and this says that at a production level of 2000 units, the cost for producing the 2001st unit is \$304. $R'(2000) = 600 - 0.1\,(2000) = 400$, and this says that the revenue realized in selling the 2001st unit is \$400. $P'(2000) = R'(2000) - C'(2000) = 400 - 304 = 96$, and this says that the revenue realized in selling the 2001st unit is \$96.

d. 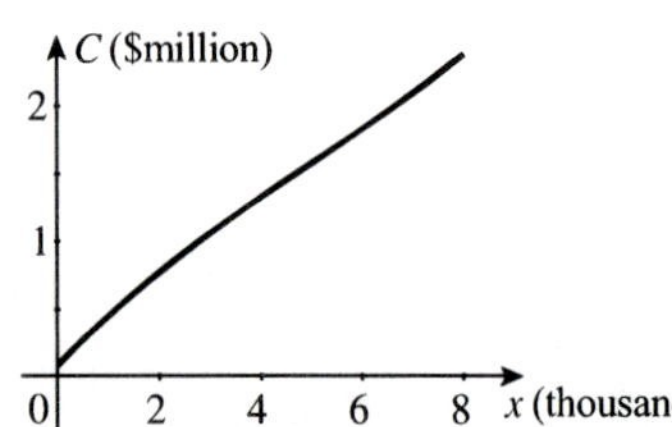

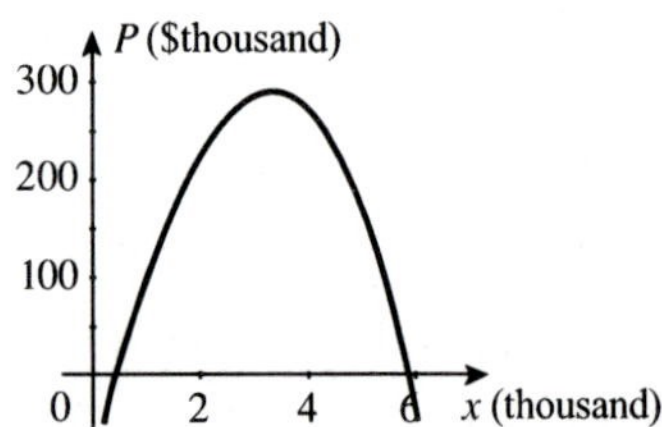

15. a. $\overline{C}(x) = \dfrac{C(x)}{x} = \dfrac{0.000002x^3 - 0.03x^2 + 400x + 80{,}000}{x} = 0.000002x^2 - 0.03x + 400 + \dfrac{80{,}000}{x}.$

b. $\overline{C}'(x) = 0.000004x - 0.03 - \dfrac{80{,}000}{x^2}.$

c. $\overline{C}'(5000) = 0.000004\,(5000) - 0.03 - \dfrac{80{,}000}{5000^2} = -0.0132$, and this says that at a production level of 5000 units, the average cost of production is dropping at the rate of approximately a penny per unit.

$\overline{C}'(10{,}000) = 0.000004\,(10000) - 0.03 - \dfrac{80{,}000}{10{,}000^2} = 0.0092,$

and this says that, at a production level of 10,000 units, the average cost of production is increasing at the rate of approximately a penny per unit.

d. 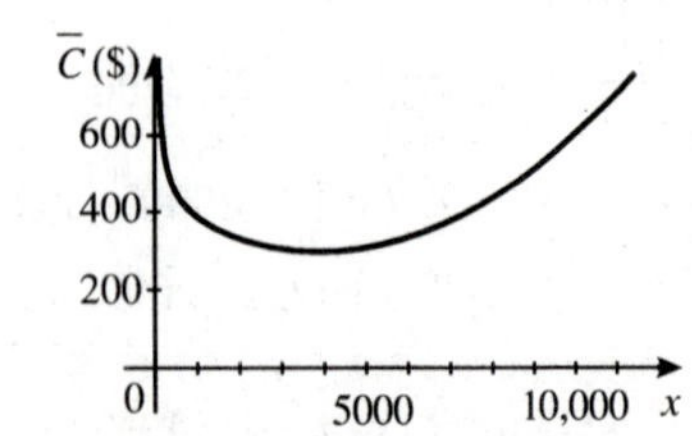

17. **a.** $R(x) = px = \dfrac{50x}{0.01x^2 + 1}$.

b. $R'(x) = \dfrac{(0.01x^2 + 1)\,50 - 50x\,(0.02x)}{(0.01x^2 + 1)^2} = \dfrac{50 - 0.5x^2}{(0.01x^2 + 1)^2}$.

c. $R'(2) = \dfrac{50 - 0.5\,(4)}{[0.01\,(4) + 1]^2} \approx 44.379$. This result says that at a sales level of 2000 units, the revenue increases at the rate of approximately $44,380 per 1000 units.

19. $C(x) = 0.873x^{1.1} + 20.34$, so $C'(x) = 0.873\,(1.1)\,x^{0.1} = 0.9603x^{0.1}$. $C'(10) = 0.9603\,(10)^{0.1} \approx 1.21$.

21. The consumption function is given by $C(x) = 0.712x + 95.05$. The marginal propensity to consume is given by $\frac{dC}{dx} = 0.712$. The marginal propensity to save is given by $\frac{dS}{dx} = 1 - \frac{dC}{dx} = 1 - 0.712 = 0.288$.

23. **a.** $R(x) = px = 100xe^{-0.0001x}$.

b. $R'(x) = 100e^{-0.0001x} + 100xe^{-0.0001x}\,(-0.0001) = 100\,(1 - 0.0001x)\,e^{-0.0001x}$.

c. $R'(10{,}000) = 100\,[1 - 0.0001\,(10{,}000)]\,e^{-1} = 0$, or \$0/pair.

25. False. In fact, it makes good sense to *increase* the level of production since, in this instance, the profit increases by $f'(a)$ units per unit increase in x.

CHAPTER 9 Concept Review Questions page 667

1. $f(x)$, L, a

3. **a.** L, x **b.** M, negative, absolute

5. **a.** continuous **b.** discontinuous **c.** every

7. **a.** $[a, b]$, $f(c) = M$ **b.** $f(x) = 0$, (a, b)

9. **a.** $\dfrac{f(a+h) - f(a)}{h}$ **b.** $\lim\limits_{h \to 0} \dfrac{f(a+h) - f(a)}{h}$

11. **a.** $f(x)\,g'(x) + g(x)\,f'(x)$ **b.** $\dfrac{g(x)\,f'(x) - f(x)\,g'(x)}{[g(x)]^2}$

13. Marginal cost, marginal revenue, marginal profit, marginal average cost

CHAPTER 9 Review Exercises page 668

1. $\lim\limits_{x \to 0} (5x - 3) = 5\,(0) - 3 = -3$.

3. $\lim\limits_{x \to -1} (3x^2 + 4)\,(2x - 1) = \left[3\,(-1)^2 + 4\right][2\,(-1) - 1] = -21$.

5. $\lim_{x\to 2} \frac{x+3}{x^2-9} = \frac{2+3}{4-9} = -1.$

7. $\lim_{x\to 3} \sqrt{2x^3-5} = \sqrt{2(27)-5} = 7.$

9. $\lim_{x\to 1^+} \frac{x-1}{x(x-1)} = \lim_{x\to 1^+} \frac{1}{x} = 1.$

11. $\lim_{x\to\infty} \frac{x^2}{x^2-1} = \lim_{x\to\infty} \frac{1}{1-\frac{1}{x^2}} = 1.$

13. $\lim_{x\to\infty} \frac{3x^2+2x+4}{2x^2-3x+1} = \lim_{x\to\infty} \frac{3+\frac{2}{x}+\frac{4}{x^2}}{2-\frac{3}{x}+\frac{1}{x^2}} = \frac{3}{2}.$

15. $\lim_{x\to 2^+} f(x) = \lim_{x\to 2^+} (-x+3) = -2+3 = 1$ and
$\lim_{x\to 2^-} f(x) = \lim_{x\to 2^-} (2x-3) = 2(2)-3 = 4-3 = 1.$
Therefore, $\lim_{x\to 2} f(x) = 1.$

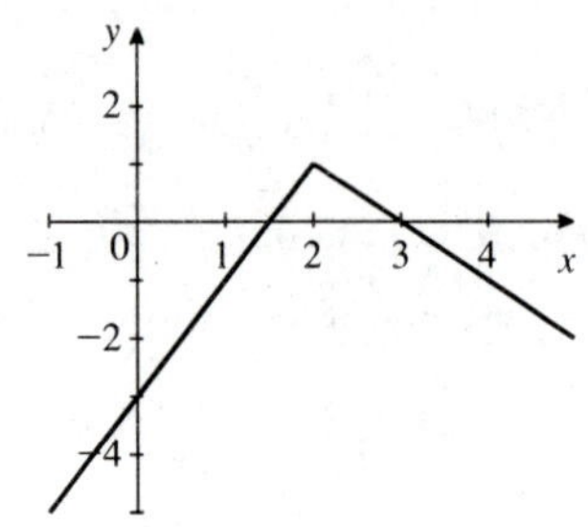

17. The function is discontinuous at $x = 2$.

19. Because $\lim_{x\to -1} f(x) = \lim_{x\to -1} \frac{1}{(x+1)^2} = \infty$ (does not exist), we see that f is discontinuous at $x = -1$.

21. a. Let $f(x) = x^2+2$. Then the average rate of change of y over $[1, 2]$ is $\frac{f(2)-f(1)}{2-1} = \frac{(4+2)-(1+2)}{1} = 3.$

Over $[1, 1.5]$, it is $\frac{f(1.5)-f(1)}{1.5-1} = \frac{(2.25+2)-(1+2)}{0.5} = 2.5$. Over $[1, 1.1]$, it is $\frac{f(1.1)-f(1)}{1.1-1} = \frac{(1.21+2)-(1+2)}{0.1} = 2.1.$

b. Computing $f'(x)$ using the four-step process., we obtain
$f'(x) = \lim_{h\to 0} \frac{f(x+h)-f(x)}{h} = \lim_{h\to 0} \frac{h(2x+h)}{h} = \lim_{h\to 0} (2x+h) = 2$. Therefore, the instantaneous rate of change of f at $x = 1$ is $f'(1) = 2$, or 2 units per unit change in x.

23. $f(x) = -\frac{1}{x}$. We use the four-step process:

Step 1 $f(x+h) = -\frac{1}{x+h}.$

Step 2 $f(x+h) - f(x) = -\frac{1}{x+h} - \left(-\frac{1}{x}\right) = -\frac{1}{x+h} + \frac{1}{x} = \frac{h}{x(x+h)}.$

Step 3 $\frac{f(x+h)-f(x)}{h} = \frac{1}{x(x+h)}.$

Step 4 $f'(x) = \lim_{h\to 0} \frac{f(x+h)-f(x)}{h} = \lim_{h\to 0} \frac{1}{x(x+h)} = \frac{1}{x^2}.$

25. $f(x) = -x^2$. We use the four-step process:

Step 1 $f(x+h) = -(x+h)^2 = -x^2 - 2xh - h^2$.

Step 2 $f(x+h) - f(x) = (-x^2 - 2xh - h^2) - (-x^2) = -2xh - h^2 = h(-2x - h)$.

Step 3 $\dfrac{f(x+h) - f(x)}{h} = -2x - h$.

Step 4 $f'(x) = \lim_{h\to 0} \dfrac{f(x+h) - f(x)}{h} = \lim_{h\to 0} (-2x - h) = -2x$.

The slope of the tangent line is $f'(2) = -2(2) = -4$. An equation of the tangent line is $y - (-4) = -4(x-2)$, or $y = -4x + 4$.

27. $f'(x) = \dfrac{d}{dx}(3x^5 - 2x^4 + 3x^2 - 2x + 1) = 15x^4 - 8x^3 + 6x - 2$.

29. $g'(x) = \dfrac{d}{dx}(-2x^{-3} + 3x^{-1} + 2) = 6x^{-4} - 3x^{-2}$.

31. $g'(t) = \dfrac{d}{dt}(2t^{-1/2} + 4t^{-3/2} + 2) = -t^{-3/2} - 6t^{-5/2}$.

33. $f'(t) = \dfrac{d}{dt}(t + 2t^{-1} + 3t^{-2}) = 1 - 2t^{-2} - 6t^{-3} = 1 - \dfrac{2}{t^2} - \dfrac{6}{t^3}$.

35. $h'(x) = \dfrac{d}{dx}(x^2 - 2x^{-3/2}) = 2x + 3x^{-5/2} = 2x + \dfrac{3}{x^{5/2}}$.

37. $g(t) = \dfrac{t^2}{2t^2+1}$, so $g'(t) = \dfrac{(2t^2+1)\dfrac{d}{dt}(t^2) - t^2\dfrac{d}{dt}(2t^2+1)}{(2t^2+1)^2} = \dfrac{(2t^2+1)(2t) - t^2(4t)}{(2t^2+1)^2} = \dfrac{2t}{(2t^2+1)^2}$.

39. $f(x) = \dfrac{\sqrt{x}-1}{\sqrt{x}+1} = \dfrac{x^{1/2}-1}{x^{1/2}+1}$, so

$$f'(x) = \frac{(x^{1/2}+1)\left(\frac{1}{2}x^{-1/2}\right) - (x^{1/2}-1)\left(\frac{1}{2}x^{-1/2}\right)}{(x^{1/2}+1)^2} = \frac{\frac{1}{2} + \frac{1}{2}x^{-1/2} - \frac{1}{2} + \frac{1}{2}x^{-1/2}}{(x^{1/2}+1)^2} = \frac{x^{-1/2}}{(x^{1/2}+1)^2}$$

$$= \frac{1}{\sqrt{x}\left(\sqrt{x}+1\right)^2}.$$

41. $f(x) = \dfrac{x^2(x^2+1)}{x^2-1} = \dfrac{x^4+x^2}{x^2-1}$, so

$$f'(x) = \frac{(x^2-1)\frac{d}{dx}(x^4+x^2) - (x^4+x^2)\frac{d}{dx}(x^2-1)}{(x^2-1)^2} = \frac{(x^2-1)(4x^3+2x) - (x^4+x^2)(2x)}{(x^2-1)^2}$$

$$= \frac{4x^5 + 2x^3 - 4x^3 - 2x - 2x^5 - 2x^3}{(x^2-1)^2} = \frac{2x^5 - 4x^3 - 2x}{(x^2-1)^2} = \frac{2x(x^4 - 2x^2 - 1)}{(x^2-1)^2}.$$

43. $f(x) = (3x^3 - 2)^8$, so $f'(x) = 8(3x^3-2)^7(9x^2) = 72x^2(3x^3-2)^7$.

45. $f'(t) = \dfrac{d}{dt}(2t^2+1)^{1/2} = \frac{1}{2}(2t^2+1)^{-1/2}\dfrac{d}{dt}(2t^2+1) = \frac{1}{2}(2t^2+1)^{-1/2}(4t) = \dfrac{2t}{\sqrt{2t^2+1}}$.

47. $s(t) = (3t^2 - 2t + 5)^{-2}$, so

$$s'(t) = -2(3t^2 - 2t + 5)^{-3}(6t - 2) = -4(3t^2 - 2t + 5)^{-3}(3t - 1) = -\frac{4(3t - 1)}{(3t^2 - 2t + 5)^3}.$$

49. $f(x) = xe^{2x}$, so $f'(x) = e^{2x} + xe^{2x}(2) = (1 + 2x)e^{2x}$.

51. $g(t) = \sqrt{t}e^{-2t}$, so $g'(t) = \frac{1}{2}t^{-1/2}e^{-2t} + \sqrt{t}e^{-2t}(-2) = \dfrac{1 - 4t}{2\sqrt{t}e^{2t}}$.

53. $y = \dfrac{e^{2x}}{1 + e^{-2x}}$, so $y' = \dfrac{(1 + e^{-2x})e^{2x}(2) - e^{2x} \cdot e^{-2x}(-2)}{(1 + e^{-2x})^2} = \dfrac{2(e^{2x} + 2)}{(1 + e^{-2x})^2}$.

55. $f(x) = xe^{-x^2}$, so $f'(x) = e^{-x^2} + xe^{-x^2}(-2x) = (1 - 2x^2)e^{-x^2}$.

57. $f(x) = x^2e^x + e^x$, so $f'(x) = 2xe^x + x^2e^x + e^x = (x^2 + 2x + 1)e^x = (x + 1)^2e^x$.

59. $f(x) = \ln\left(e^{x^2} + 1\right)$, so $f'(x) = \dfrac{e^{x^2}(2x)}{e^{x^2} + 1} = \dfrac{2xe^{x^2}}{e^{x^2} + 1}$.

61. $f(x) = \dfrac{\ln x}{x + 1}$, so $f'(x) = \dfrac{(x + 1)\left(\frac{1}{x}\right) - \ln x}{(x + 1)^2} = \dfrac{1 + \frac{1}{x} - \ln x}{(x + 1)^2} = \dfrac{x - x\ln x + 1}{x(x + 1)^2}$.

63. $y = \ln(e^{4x} + 3)$, so $y' = \dfrac{e^{4x}(4)}{e^{4x} + 3} = \dfrac{4e^{4x}}{e^{4x} + 3}$.

65. $f(x) = \dfrac{\ln x}{1 + e^x}$, so

$$f'(x) = \frac{(1 + e^x)\frac{d}{dx}\ln x - \ln x\frac{d}{dx}(1 + e^x)}{(1 + e^x)^2} = \frac{(1 + e^x)\left(\frac{1}{x}\right) - (\ln x)e^x}{(1 + e^x)^2} = \frac{1 + e^x - xe^x\ln x}{x(1 + e^x)^2}$$
$$= \frac{1 + e^x(1 - x\ln x)}{x(1 + e^x)^2}.$$

67. $h(x) = \left(x + \dfrac{1}{x}\right)^2 = (x + x^{-1})^2$, so

$$h'(x) = 2(x + x^{-1})(1 - x^{-2}) = 2\left(x + \frac{1}{x}\right)\left(1 - \frac{1}{x^2}\right) = 2\left(\frac{x^2 + 1}{x}\right)\left(\frac{x^2 - 1}{x^2}\right) = \frac{2(x^2 + 1)(x^2 - 1)}{x^3}.$$

69. $h(t) = (t^2 + t)^4(2t^2)$, so

$$h'(t) = (t^2 + t)^4\frac{d}{dt}(2t^2) + 2t^2\frac{d}{dt}(t^2 + t)^4 = (t^2 + t)^4(4t) + 2t^2 \cdot 4(t^2 + t)^3(2t + 1)$$
$$= 4t(t^2 + t)^3\left[(t^2 + t) + 4t^2 + 2t\right] = 4t^2(5t + 3)(t^2 + t)^3.$$

71. $g(x) = x^{1/2}(x^2-1)^3$, so

$$g'(x) = \frac{d}{dx}\left[x^{1/2}(x^2-1)^3\right] = x^{1/2}\cdot 3(x^2-1)^2(2x) + (x^2-1)^3\cdot \tfrac{1}{2}x^{-1/2}$$
$$= \tfrac{1}{2}x^{-1/2}(x^2-1)^2\left[12x^2 + (x^2-1)\right] = \frac{(13x^2-1)(x^2-1)^2}{2\sqrt{x}}.$$

73. $h(x) = \dfrac{(3x+2)^{1/2}}{4x-3}$, so

$$h'(x) = \frac{(4x-3)\frac{1}{2}(3x+2)^{-1/2}(3) - (3x+2)^{1/2}(4)}{(4x-3)^2} = \frac{\frac{1}{2}(3x+2)^{-1/2}\left[3(4x-3) - 8(3x+2)\right]}{(4x-3)^2}$$
$$= -\frac{12x+25}{2\sqrt{3x+2}\,(4x-3)^2}.$$

75. $f(x) = 2x^4 - 3x^3 + 2x^2 + x + 4$, so $f'(x) = \dfrac{d}{dx}\left(2x^4 - 3x^3 + 2x^2 + x + 4\right) = 8x^3 - 9x^2 + 4x + 1$ and

$$f''(x) = \frac{d}{dx}\left(8x^3 - 9x^2 + 4x + 1\right) = 24x^2 - 18x + 4 = 2\left(12x^2 - 9x + 2\right).$$

77. $h(t) = \dfrac{t}{t^2+4}$, so $h'(t) = \dfrac{(t^2+4)(1) - t(2t)}{(t^2+4)^2} = \dfrac{4-t^2}{(t^2+4)^2}$ and

$$h''(t) = \frac{(t^2+4)^2(-2t) - (4-t^2)\,2(t^2+4)(2t)}{(t^2+4)^4} = \frac{-2t(t^2+4)\left[(t^2+4) + 2(4-t^2)\right]}{(t^2+4)^4} = \frac{2t(t^2-12)}{(t^2+4)^3}.$$

79. $h(x) = \dfrac{e^x}{1+e^x}$, so $h'(x) = \dfrac{(1+e^x)e^x - e^x(e^x)}{(1+e^x)^2} = \dfrac{e^x(1+e^x-e^x)}{(1+e^x)^2} = \dfrac{e^x}{(1+e^x)^2}$ and

$$h''(x) = \frac{(1+e^x)^2 e^x - e^x(2)(1+e^x)e^x}{(1+e^x)^4} = \frac{(1+e^x)e^x(1+e^x-2e^x)}{(1+e^x)^2} = \frac{e^x(1-e^x)}{(1+e^x)^3}.$$

81. $y = \ln(3x+1)$, so $y' = \dfrac{3}{3x+1}$ and $y'' = 3\dfrac{d}{dx}(3x+1)^{-1} = -3(3x+1)^{-2}(3) = -\dfrac{9}{(3x+1)^2}$.

83. $f'(x) = \frac{d}{dx}\left(2x^2+1\right)^{1/2} = \frac{1}{2}\left(2x^2+1\right)^{-1/2}(4x) = 2x\left(2x^2+1\right)^{-1/2}$, so

$$f''(x) = 2\left(2x^2+1\right)^{-1/2} + 2x\cdot\left(-\tfrac{1}{2}\right)\left(2x^2+1\right)^{-3/2}(4x) = 2\left(2x^2+1\right)^{-3/2}\left[(2x^2+1) - 2x^2\right] = \frac{2}{\left(2x^2+1\right)^{3/2}}.$$

85. $h'(x) = g'(f(x))\,f'(x)$. But $g'(x) = 1 - \dfrac{1}{x^2}$ and $f'(x) = e^x$, so $f(0) = e^0 = 1$, $f'(0) = e^0 = 1$, and $g'(1) = 1 - \frac{1}{1} = 0$. Therefore, $h'(0) = g'(f(0))\,f'(0) = g'(1)\,f'(0) = 0\cdot 1 = 0$.

87. $f(x) = 2x^3 - 3x^2 - 16x + 3$ and $f'(x) = 6x^2 - 6x - 16$.

a. To find the point(s) on the graph of f where the slope of the tangent line is equal to -4, we solve $6x^2 - 6x - 16 = -4$, obtaining $6x^2 - 6x - 12 = 0$, $6\left(x^2 - x - 2\right) = 0$, and $6(x-2)(x+1) = 0$. Thus, $x = 2$ or $x = -1$. Now $f(2) = 2(2)^3 - 3(2)^2 - 16(2) + 3 = -25$ and $f(-1) = 2(-1)^3 - 3(-1)^2 - 16(-1) + 3 = 14$, so the points are $(2, -25)$ and $(-1, 14)$.

b. Using the point-slope form of the equation of a line, we find that an equation of the tangent line at $(2, -25)$ is $y - (-25) = -4(x - 2)$, $y + 25 = -4x + 8$, or $y = -4x - 17$, and an equation of the tangent line at $(-1, 14)$ is $y - 14 = -4(x + 1)$, or $y = -4x + 10$.

89. $y = \left(4 - x^2\right)^{1/2}$, so $y' = \frac{1}{2}\left(4 - x^2\right)^{-1/2}(-2x) = -\dfrac{x}{\sqrt{4 - x^2}}$. The slope of the tangent line is obtained by letting $x = 1$, giving $m = -\frac{1}{\sqrt{3}} = -\frac{\sqrt{3}}{3}$. Therefore, an equation of the tangent line at $x = 1$ is $y - \sqrt{3} = -\frac{\sqrt{3}}{3}(x - 1)$, or $y = -\frac{\sqrt{3}}{3}x + \frac{4\sqrt{3}}{3}$.

91. $y = e^{-2x}$, so $y' = -2e^{-2x}$. This gives the slope of the tangent line to the graph of $y = e^{-2x}$ at any point (x, y). In particular, the slope of the tangent line at $\left(1, e^{-2}\right)$ is $y'(1) = -2e^{-2}$. An equation of the tangent line is $y - e^{-2} = -2e^{-2}(x - 1)$, or $y = \dfrac{1}{e^2}(-2x + 3)$.

93. $f(x) = (2x - 1)^{-1}$, so $f'(x) = -2(2x - 1)^{-2}$, $f''(x) = 8(2x - 1)^{-3} = \dfrac{8}{(2x - 1)^3}$, and $f'''(x) = -48(2x - 1)^4 = -\dfrac{48}{(2x - 1)^4}$. Because $(2x - 1)^4 = 0$ when $x = \frac{1}{2}$, we see that the domain of f''' is $\left(-\infty, \frac{1}{2}\right) \cup \left(\frac{1}{2}, \infty\right)$.

95. a. $C'(x)$ gives the instantaneous rate of change of the total manufacturing cost c in dollars when x units of a certain product are produced.

b. Positive

c. Approximately \$20.

97. a. $P(t) = 0.01484t^2 + 0.446t + 15$, so $P(0) = 15$, or 15%, and $P(22) = 0.01484(22)^2 + 0.446(22) + 15 \approx 31.99$, or approximately 31.99%.

b. $P'(t) = 2(0.01484)t + 0.446 = 0.02968t + 0.446$, so $P'(2) = 0.02968(2) + 0.446 = 0.50536$, or approximately 0.51%/yr, and $P'(20) = 0.02968(20) + 0.446 = 1.0396$, or approximately 1.04%/yr.

99. He can expect to live $f(100) = 46.9[1 + 1.09(100)]^{0.1} \approx 75.0433$, or approximately 75.04 years. $f'(t) = 46.9(0.1)(1 + 1.09t)^{-0.9}(1.09) = 5.1121(1 + 1.09t)^{-0.9}$, so the required rate of change is $f'(100) = 5.1121[1 + 1.09(100)]^{-0.9} \approx 0.074$, or approximately 0.07 yr/yr.

101. $C(x) = 2500 + 2.2x$.

a. The marginal cost is $C'(x) = 2.2$. The marginal cost when $x = 1000$ is $C'(1000) = 2.2$. The marginal cost when $x = 2000$ is $C'(2000) = 2.2$.

b. $\overline{C}(x) = \dfrac{C(x)}{x} = \dfrac{2500 + 2.2x}{x} = 2.2 + \dfrac{2500}{x}$, so $\overline{C}'(x) = -\dfrac{2500}{x^2}$.

103. a. $D(1) = 4000 - 3000e^{-0.06} \approx 1175$, $D(12) = 4000 - 3000e^{-0.72} \approx 2540$, and $D(24) = 4000 - 3000e^{-1.44} \approx 3289$.

b. $\lim_{t \to \infty} D(t) = \lim_{t \to \infty}\left(4000 - 3000e^{-0.06t}\right) = 4000$.

CHAPTER 9 Before Moving On... page 671

1. $\lim\limits_{x\to-1}\dfrac{x^2+4x+3}{x^2+3x+2}=\lim\limits_{x\to-1}\dfrac{(x+3)(x+1)}{(x+2)(x+1)}=2.$

2. a. $\lim\limits_{x\to1^-}f(x)=\lim\limits_{x\to1^-}\left(x^2-1\right)=0.$

b. $\lim\limits_{x\to1^+}f(x)=\lim\limits_{x\to1^+}x^3=1.$

Because $\lim\limits_{x\to1^-}f(x)\neq\lim\limits_{x\to1^+}f(x)$, f is not continuous at 1.

3. The slope of the tangent line at any point is

$$\begin{aligned}\lim_{h\to0}\frac{f(x+h)-f(x)}{h}&=\lim_{h\to0}\frac{\left[(x+h)^2-3(x+h)+1\right]-\left(x^2-3x+1\right)}{h}\\&=\lim_{h\to0}\frac{x^2+2xh+h^2-3x-3h+1-x^2+3x-1}{h}\\&=\lim_{h\to0}\frac{h(2x+h-3)}{h}=\lim_{h\to0}(2x+h-3)=2x-3.\end{aligned}$$

Therefore, the slope at 1 is $2(1)-3=-1$. An equation of the tangent line is $y-(-1)=-1(x-1)$, or $y+1=-x+1$, or $y=-x$.

4. $f(x)=2x^3-3x^{1/3}+5x^{-2/3}$, so $f'(x)=2\left(3x^2\right)-3\left(\frac{1}{3}x^{-2/3}\right)+5\left(-\frac{2}{3}x^{-5/3}\right)=6x^2-x^{-2/3}-\frac{10}{3}x^{-5/3}$.

5. $$\begin{aligned}g'(x)&=\frac{d}{dx}\left[x\left(2x^2-1\right)^{1/2}\right]=\left(2x^2-1\right)^{1/2}+x\left(\tfrac{1}{2}\right)\left(2x^2-1\right)^{-1/2}\frac{d}{dx}\left(2x^2-1\right)\\&=\left(2x^2-1\right)^{1/2}+\tfrac{1}{2}x\left(2x^2-1\right)^{-1/2}(4x)=\left(2x^2-1\right)^{-1/2}\left[\left(2x^2-1\right)+2x^2\right]=\frac{4x^2-1}{\sqrt{2x^2-1}}.\end{aligned}$$

6. $$\frac{dy}{dx}=\frac{\left(x^2+x+1\right)(2)-(2x+1)(2x+1)}{\left(x^2+x+1\right)^2}=\frac{2x^2+2x+2-\left(4x^2+4x+1\right)}{\left(x^2+x+1\right)^2}=-\frac{2x^2+2x-1}{\left(x^2+x+1\right)^2}.$$

7. $f(x)=\dfrac{1}{\sqrt{x+1}}$, so $f'(x)=\dfrac{d}{dx}(x+1)^{-1/2}=-\frac{1}{2}(x+1)^{-3/2}=-\dfrac{1}{2(x+1)^{3/2}}$,

so $f''(x)=-\frac{1}{2}\left(-\frac{3}{2}\right)(x+1)^{-5/2}=\frac{3}{4}(x+1)^{-5/2}=\dfrac{3}{4(x+1)^{5/2}}$ and

$f'''(x)=\frac{3}{4}\left(-\frac{5}{2}\right)(x+1)^{-7/2}=-\frac{15}{8}(x+1)^{-7/2}=-\dfrac{15}{8(x+1)^{7/2}}$.

8. $f(x)=e^{\sqrt{x}}$, so $f'(x)=\dfrac{d}{dx}e^{x^{1/2}}=e^{x^{1/2}}\dfrac{d}{dx}\left(x^{1/2}\right)=e^{x^{1/2}}\left(\frac{1}{2}x^{-1/2}\right)=\dfrac{e^{\sqrt{x}}}{2\sqrt{x}}$.

9. $y=x\ln\left(x^2+1\right)$, so $\dfrac{dy}{dx}=x\dfrac{d}{dx}\ln\left(x^2+1\right)+\ln\left(x^2+1\right)\dfrac{d}{dx}(x)=x\cdot\dfrac{2x}{x^2+1}+\ln\left(x^2+1\right)=\dfrac{2x^2}{x^2+1}+\ln\left(x^2+1\right)$.

Thus, $\left.\dfrac{dy}{dx}\right|_{x=1}=\dfrac{2}{1+1}+\ln2=1+\ln2$.

10 APPLICATIONS OF THE DERIVATIVE

10.1 Applications of the First Derivative

Problem-Solving Tips

1. The critical number of a function f is any number x in the domain of f such that $f'(x) = 0$ or $f'(x)$ does not exist. Note that the definition requires that x be in the domain of f. For example, consider the function $f(x) = x + \frac{1}{x}$ in Example 8 on page 683 of the text. Even though f' is discontinuous at $x = 0$, this value does not qualify as a critical number because it does not lie in the domain of f.

2. Note that when you use test values to find the sign of a derivative over an interval, you don't need to evaluate the derivative at a test value. You need only find the **sign** of the derivative at that test value. For example, to find the sign of $f'(x) = \dfrac{(x+1)(x-1)}{x^2}$ in the interval $(0, 1)$ using the test value $x = \frac{1}{2}$, we simply note that the numerator is the product of a positive number and a negative number, so it is negative. Because x^2 is always positive, the denominator is always positive. The quotient of a negative number and a positive number is negative, so f' is negative over the interval $(0, 1)$.

Concept Questions page 685

1. **a.** f is increasing on I if whenever x_1 and x_2 are in I with $x_1 < x_2$, then $f(x_1) < f(x_2)$.

   b. f is decreasing on I if whenever x_1 and x_2 are in I with $x_1 < x_2$, then $f(x_1) > f(x_2)$.

3. **a.** f has a relative maximum at $x = a$ if there is an open interval I containing a such that $f(x) \le f(a)$ for all x in I.

   b. f has a relative minimum at $x = a$ if there is an open interval I containing a such that $f(x) \ge f(a)$ for all x in I.

5. See page 682 of the text.

Exercises page 686

1. f is decreasing on $(-\infty, 0)$ and increasing on $(0, \infty)$.

3. f is increasing on $(-\infty, -1)$ and $(1, \infty)$, and decreasing on $(-1, 1)$.

5. f is increasing on $(0, 2)$ and decreasing on $(-\infty, 0)$ and $(2, \infty)$.

7. f is decreasing on $(-\infty, -1)$ and $(1, \infty)$ and increasing on $(-1, 1)$.

9. Increasing on $(20.2, 20.6)$ and $(21.7, 21.8)$, constant on $(19.6, 20.2)$ and $(20.6, 21.1)$, and decreasing on $(21.1, 21.7)$ and $(21.8, 22.7)$.

11. **a.** Positive **b.** Positive **c.** Zero **d.** Zero

e. Negative **f.** Negative **g.** Positive

13. $f(x) = 3x + 5$, so $f'(x) = 3 > 0$ for all x. Thus, f is increasing on $(-\infty, \infty)$.

15. $f(x) = x^2 - 3x$, so $f'(x) = 2x - 3$. f' is continuous everywhere and is equal to zero when $x = \frac{3}{2}$. From the sign diagram, we see that f is decreasing on $\left(-\infty, \frac{3}{2}\right)$ and increasing on $\left(\frac{3}{2}, \infty\right)$.

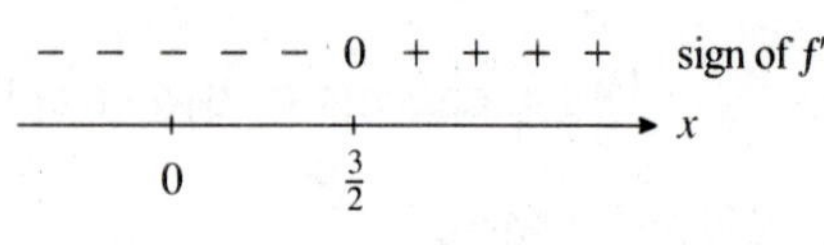

17. $g(x) = x - x^3$, so $g'(x) = 1 - 3x^2$ is continuous everywhere and is equal to zero when $1 - 3x^2 = 0$, or $x = \pm\frac{\sqrt{3}}{3}$. From the sign diagram, we see that f is decreasing on $\left(-\infty, -\frac{\sqrt{3}}{3}\right)$ and $\left(\frac{\sqrt{3}}{3}, \infty\right)$ and increasing on $\left(-\frac{\sqrt{3}}{3}, \frac{\sqrt{3}}{3}\right)$.

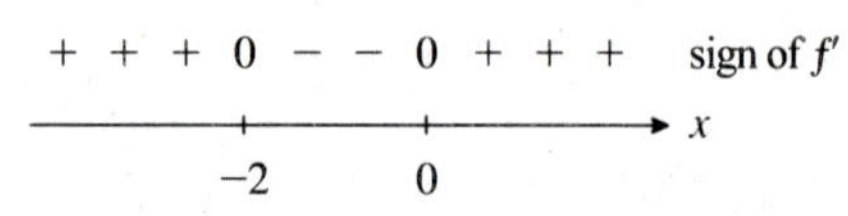

19. $g(x) = x^3 + 3x^2 + 1$, so $g'(x) = 3x^2 + 6x = 3x(x + 2)$. From the sign diagram, we see that g is increasing on $(-\infty, -2)$ and $(0, \infty)$ and decreasing on $(-2, 0)$.

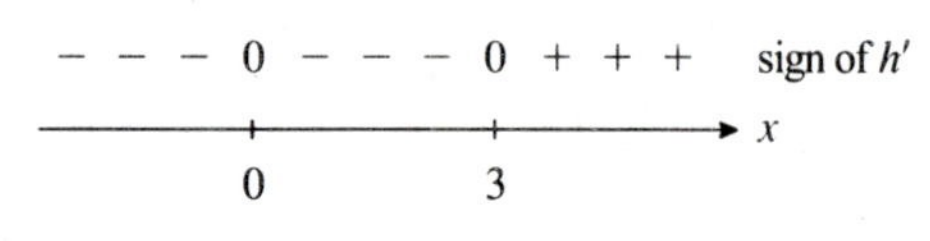

21. $f(x) = \frac{1}{3}x^3 - 3x^2 + 9x + 20$, so $f'(x) = x^2 - 6x + 9 = (x - 3)^2 > 0$ for all x except $x = 3$, at which point $f'(3) = 0$. Therefore, f is increasing on $(-\infty, \infty)$.

23. $h(x) = x^4 - 4x^3 + 10$, so $h'(x) = 4x^3 - 12x^2 = 4x^2(x - 3) = 0$ if $x = 0$ or 3. From the sign diagram of h', we see that h is increasing on $(3, \infty)$ and decreasing on $(-\infty, 3)$.

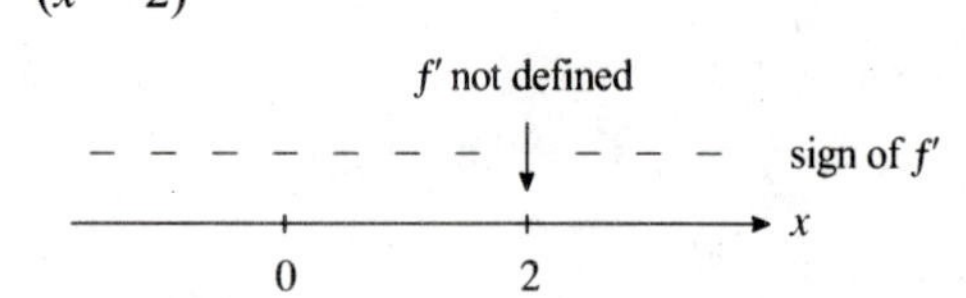

25. $f(x) = \dfrac{1}{x - 2} = (x - 2)^{-1}$, so $f'(x) = -1(x - 2)^{-2}(1) = -\dfrac{1}{(x - 2)^2}$ is discontinuous at $x = 2$ and is continuous and nonzero everywhere else. From the sign diagram, we see that f is decreasing on $(-\infty, 2)$ and $(2, \infty)$.

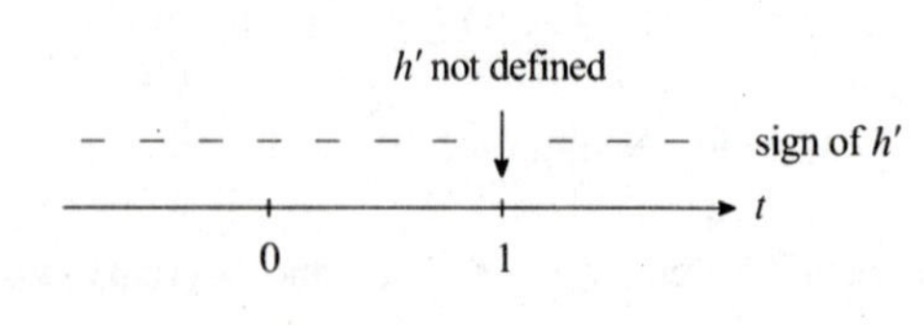

27. $h(t) = \dfrac{t}{t - 1}$, so $h'(t) = \dfrac{(t - 1)(1) - t(1)}{(t - 1)^2} = -\dfrac{1}{(t - 1)^2}$. From the sign diagram, we see that $h'(t) < 0$ whenever h' is defined. We conclude that h is decreasing on $(-\infty, 1)$ and $(1, \infty)$.

h′ not defined

– – – – – – – – ↓ – – – sign of h′

t

0 1

29. $f(x) = -x^{3/5}$, so $f'(x) = -\frac{3}{5}x^{-2/5} = -\dfrac{3}{5x^{2/5}}$. Observe that $f'(x)$ is not defined at $x = 0$, but is negative everywhere else. Therefore, f is decreasing on $(-\infty, \infty)$.

31. $f(x) = \sqrt{x+1}$, so $f'(x) = \dfrac{d}{dx}(x+1)^{1/2} = \frac{1}{2}(x+1)^{-1/2} = \dfrac{1}{2\sqrt{x+1}}$, and we see that $f'(x) > 0$ if $x > -1$. Therefore, f is increasing on $(-1, \infty)$.

33. $f(x) = \sqrt{16-x^2} = \left(16-x^2\right)^{1/2}$, so $f'(x) = \frac{1}{2}\left(16-x^2\right)^{-1/2}(-2x) = -\dfrac{x}{\sqrt{16-x^2}}$. Because the domain of f is $[-4, 4]$, we consider the sign diagram for f' on this interval. We see that f is increasing on $(-4, 0)$ and decreasing on $(0, 4)$.

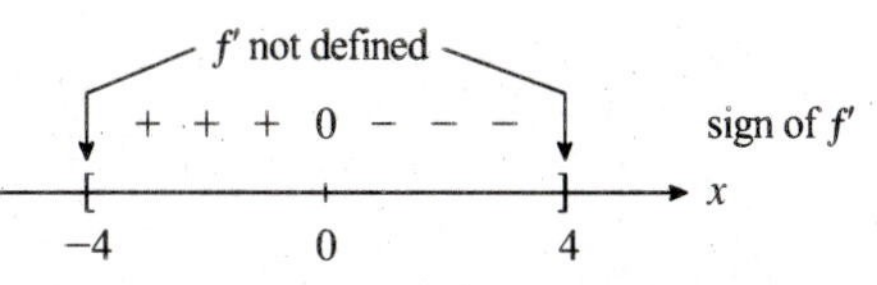

35. $f(x) = x^2e^{-x}$, so $f'(x) = 2xe^{-x} + x^2e^{-x}(-1) = x(2-x)e^{-x}$. Observe that $f'(x) = 0$ if $x = 0$ or 2. The sign diagram of f' shows that f is increasing on $(0, 2)$ and decreasing on $(-\infty, 0)$ and $(2, \infty)$.

− − − 0 + + + 0 − − − sign of f'

x: 0, 2

37. $f(x) = \dfrac{\ln x}{x}$, so $f'(x) = \dfrac{x\frac{1}{x} - \ln x}{x^2} = \dfrac{1-\ln x}{x^2}$. Observe that $f'(x) = 0$ if $1 - \ln x = 0$, or $x = e$. The sign diagram of f' on $(0, \infty)$ shows that f is increasing on $(0, e)$ and decreasing on (e, ∞).

+ + + + 0 − − − − sign of f'

x: 0, e

39. $f'(x) = \dfrac{d}{dx}\left(x^{-1} - x\right) = -\dfrac{1}{x^2} - 1 = -\dfrac{1+x^2}{x^2} < 0$ for all $x \neq 0$. Therefore, f is decreasing on $(-\infty, 0)$ and $(0, \infty)$.

41. f has a relative maximum of $f(0) = 1$ and relative minima of $f(-1) = 0$ and $f(1) = 0$.

43. f has a relative maximum of $f(-1) = 2$ and a relative minimum of $f(1) = -2$.

45. f has a relative maximum of $f(1) = 3$ and a relative minimum of $f(2) = 2$.

47. f has a relative minimum at $(0, 2)$.

49. a

51. d

53. $f(x) = x^2 - 4x$, so $f'(x) = 2x - 4 = 2(x-2)$ has a critical point at $x = 2$. From the sign diagram, we see that $f(2) = -4$ is a relative minimum by the First Derivative Test.

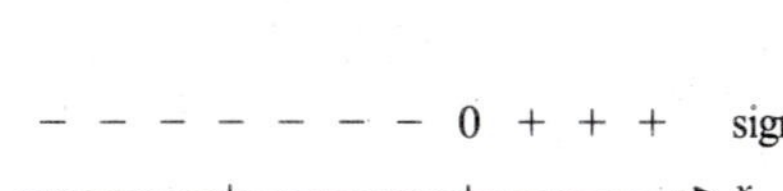

55. $h(t) = -t^2 + 6t + 6$, so $h'(t) = -2t + 6 = -2(t-3) = 0$ if $t = 3$, a critical number. The sign diagram and the First Derivative Test imply that h has a relative maximum at 3 with value $h(3) = -9 + 18 + 6 = 15$.

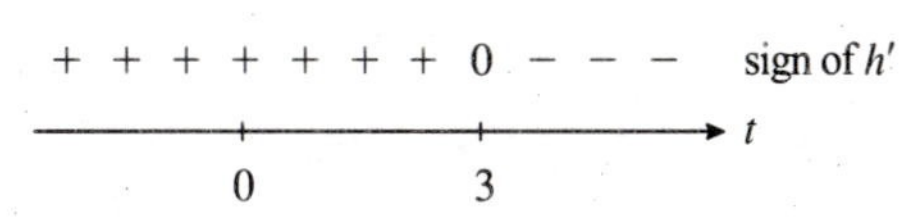

57. $f(x) = x^{5/3}$, so $f'(x) = \frac{5}{3}x^{2/3}$, $x = 0$ as the critical number of f. From the sign diagram, we see that f' does not change sign as we move across $x = 0$, and conclude that f has no relative extremum.

+ + + + + 0 + + + + + sign of f'

x

0

59. $g(x) = x^3 - 3x^2 + 5$, so $g'(x) = 3x^2 - 6x = 3x(x-2) = 0$ if $x = 0$ or 2. From the sign diagram, we see that the critical number $x = 0$ gives a relative maximum, whereas $x = 2$ gives a relative minimum. The values are $g(0) = 5$ and $g(2) = 8 - 12 + 5 = 1$.

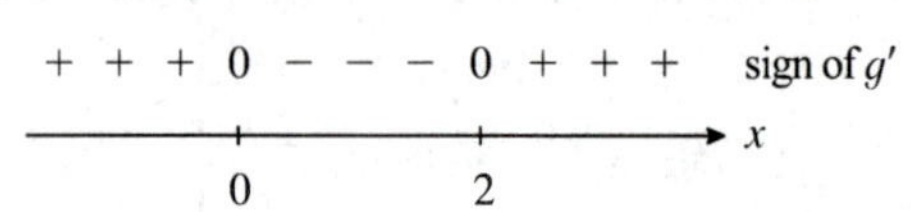

61. $f(x) = \frac{1}{2}x^4 - x^2$, so $f'(x) = 2x^3 - 2x = 2x\left(x^2 - 1\right) = 2x(x+1)(x-1)$ is continuous everywhere and has zeros at $x = -1$, 0, and 1, the critical numbers of f. Using the First Derivative Test and the sign diagram of f', we see that $f(-1) = -\frac{1}{2}$ and $f(1) = -\frac{1}{2}$ are relative minima of f and $f(0) = 0$ is a relative maximum of f.

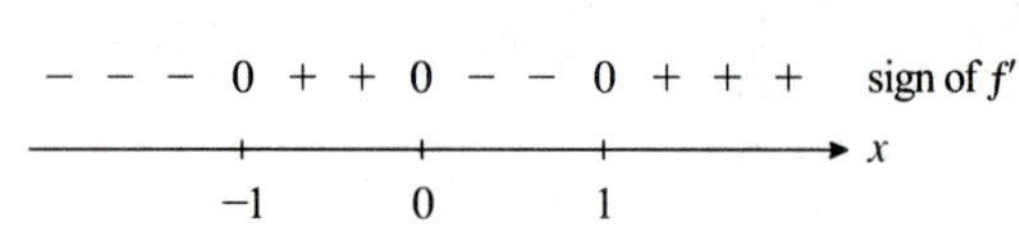

63. $F(x) = \frac{1}{3}x^3 - x^2 - 3x + 4$, so $F'(x) = x^2 - 2x - 3 = (x-3)(x+1) = 0$ gives $x = -1$ and $x = 3$ as critical numbers. From the sign diagram, we see that $x = -1$ gives a relative maximum and $x = 3$ gives a relative minimum. The values are $F(-1) = -\frac{1}{3} - 1 + 3 + 4 = \frac{17}{3}$ and $F(3) = 9 - 9 - 9 + 4 = -5$.

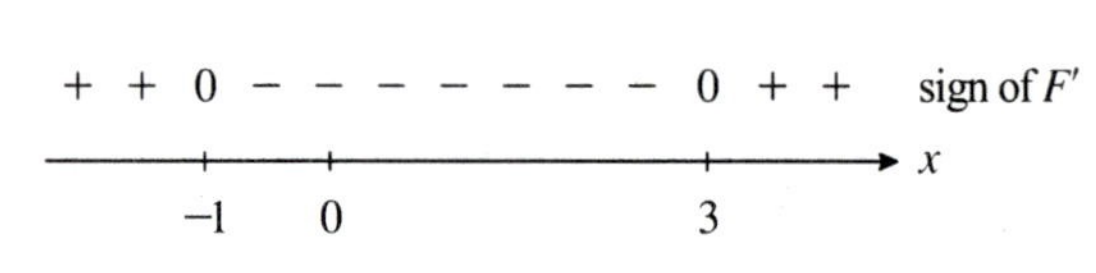

65. $g(x) = x^4 - 4x^3 + 20$. Setting $g'(x) = 4x^3 - 12x^2 = 4x^2(x-3) = 0$ gives $x = 0$ and $x = 3$ as critical numbers. From the sign diagram, we see that $x = 3$ gives a relative minimum. Its value is $g(3) = 3^4 - 4(3)^3 + 20 = -7$.

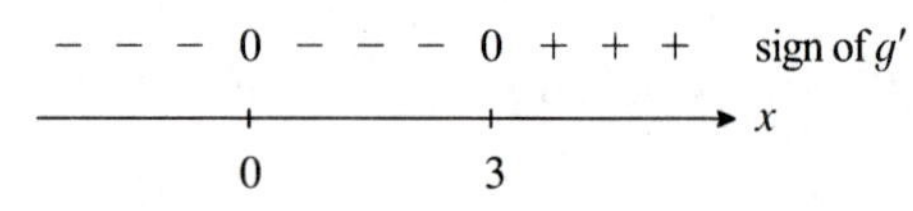

67. $g'(x) = \dfrac{d}{dx}\left(1 + \dfrac{1}{x}\right) = -\dfrac{1}{x^2}$. Observe that g' is nonzero for all values of x. Furthermore, g' is not defined at $x = 0$, but $x = 0$ is not in the domain of g. Therefore, g has no critical number and hence no relative extremum.

69. $f(x) = x + \dfrac{9}{x} + 2$, so $f'(x) = 1 - \dfrac{9}{x^2} = \dfrac{x^2 - 9}{x^2} = \dfrac{(x+3)(x-3)}{x^2} = 0$ gives $x = -3$ and $x = 3$ as critical numbers. From the sign diagram, we see that $(-3, -4)$ is a relative maximum and $(3, 8)$ is a relative minimum.

f' not defined

+ + + 0 − − ↓ − − 0 + + + sign of f'

x

−3 0 3

71. $f(x) = \frac{x}{1+x^2}$, so $f'(x) = \frac{(1+x^2)(1) - x(2x)}{(1+x^2)^2} = \frac{1-x^2}{(1+x^2)^2} = \frac{(1-x)(1+x)}{(1+x^2)^2} = 0$ if $x = \pm 1$, and these are critical numbers of f. From the sign diagram of f', we see that f has a relative minimum at $\left(-1, -\frac{1}{2}\right)$ and a relative maximum at $\left(1, \frac{1}{2}\right)$.

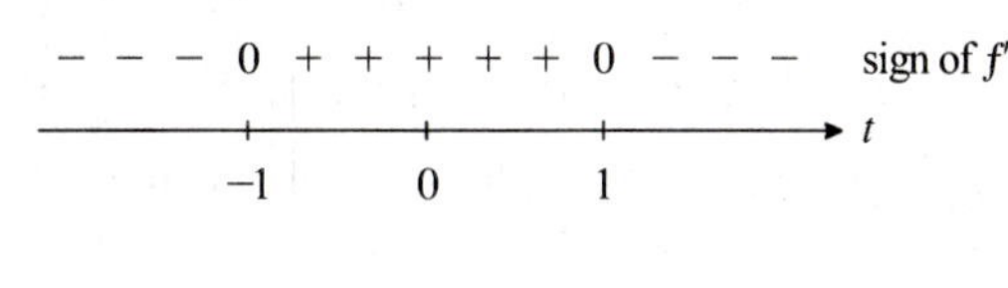

73. $f(x) = 2xe^{-x}$, so $f'(x) = 2e^{-x} - 2xe^{-x} = 2(1-x)e^{-x}$. Setting $f'(x) = 0$ gives $1 - x = 0$, so the only critical number is 1. (Note that e^{-x} is never 0). From the sign diagram of f', we see that f has a relative maximum at $\left(1, 2e^{-1}\right)$.

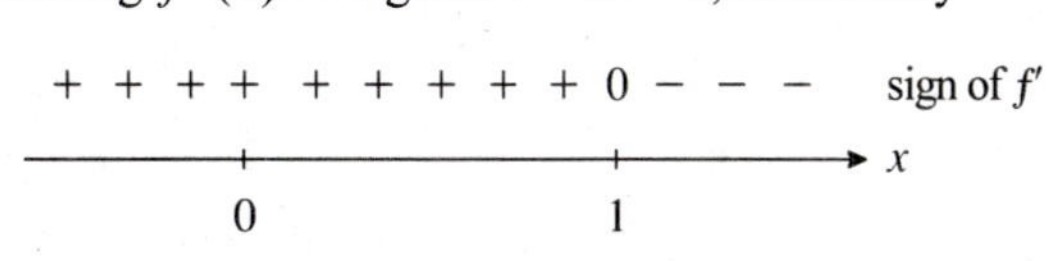

75. $f(x) = x - \ln x$, so $f'(x) = 1 - \frac{1}{x} = \frac{x-1}{x} = 0$ if $x = 1$, a critical point of f. Note that f is defined only for $x > 0$. From the sign diagram of f', we see that f has a relative minimum at $(1, 1)$.

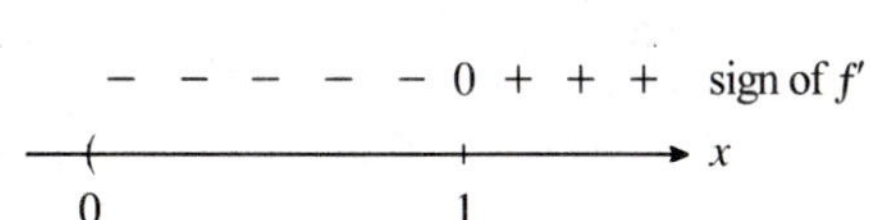

77. $h(t) = -16t^2 + 64t + 80$, so $h'(t) = -32t + 64 = -32(t - 2)$. The sign diagram shows us that the stone is rising in the time interval $(0, 2)$ and falling when $t > 2$. It hits the ground when $h(t) = -16t^2 + 64t + 80 = 0$, that is, $t^2 - 4t - 5 = (t-5)(t+1) = 0$ or $t = 5$. (We reject the root $t = -1$.)

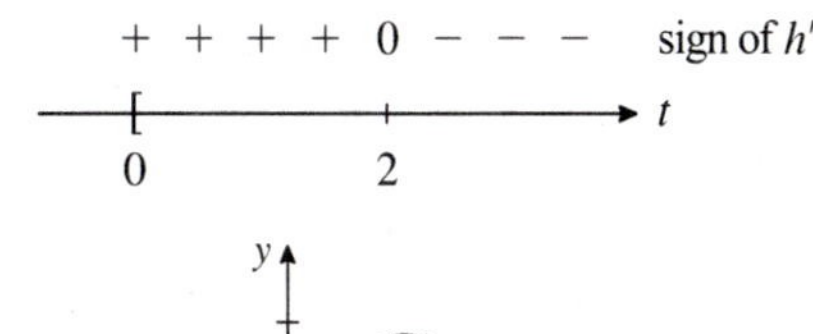

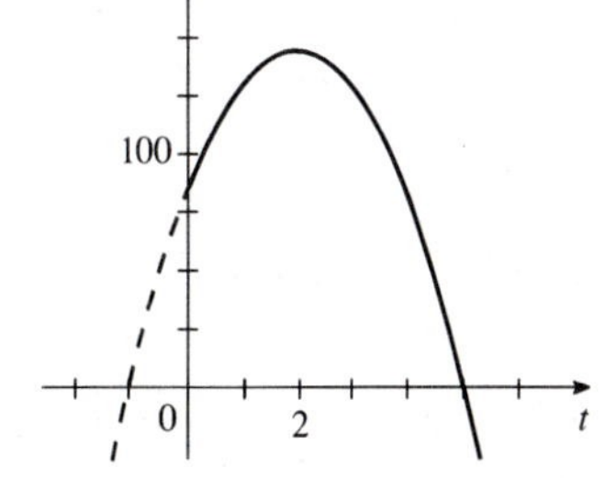

79. $N'(t) = \frac{d}{dt}\left(0.63t^2 + 1.02t + 2.7\right) = 1.26t + 1.02$. Since $N'(t) > 1.02 > 0$ for t in $(0, 20)$, we conclude that N is increasing on $(0, 20)$. Thus, the number of cell phone subscribers is increasing throughout the period under consideration.

81. $h(t) = -\frac{1}{3}t^3 + 16t^2 + 33t + 10$, so $h'(t) = -t^2 + 32t + 33 = -(t+1)(t-33)$. The sign diagram for h' shows that the rocket is ascending on the time interval $(0, 33)$ and descending on $(33, T)$ for some positive number T. The parachute is deployed 33 seconds after liftoff.

+ + + + 0 – – – sign of h'

t: 0, 33

83. $f(t) = 20t - 40\sqrt{t} + 50 = 20t - 40t^{1/2} + 50$, so $f'(t) = 20 - 40\left(\frac{1}{2}t^{-1/2}\right) = 20\left(1 - \frac{1}{\sqrt{t}}\right) = \frac{20(\sqrt{t} - 1)}{\sqrt{t}}$.

Thus, f' is continuous on $(0, 4)$ and is equal to zero at $t = 1$. From the sign diagram, we see that f is decreasing on $(0, 1)$ and increasing on $(1, 4)$. We conclude that the average speed decreases from 6 a.m. to 7 a.m. and then picks up from 7 a.m. to 10 a.m.

− 0 + + + + + + sign of f'

0 1 4 t

85. a. $f'(t) = \frac{d}{dt}\left(-0.05t^3 + 0.56t^2 + 5.47t + 7.5\right) = -0.15t^2 + 1.12t + 5.47$. Setting $f'(t) = 0$ gives $-0.15t^2 + 1.12t + 5.47 = 0$. Using the quadratic formula, we find $t = \frac{-1.12 \pm \sqrt{(1.12)^2 - 4(-0.15)(5.47)}}{-0.3}$; that is, $t \approx -3.37$ or 10.83. Because f' is continuous, the only critical numbers of f are $t \approx -3.4$ and $t \approx 10.8$, both of which lie outside the interval of interest. Nevertheless, this result can be used to tell us that f' does not change sign in the interval $(-3.4, 10.8)$. Using $t = 0$ as the test number, we see that $f'(0) = 5.47 > 0$ and so we see that f is increasing on $(-3.4, 10.8)$ and, in particular, in the interval $(0, 10)$. Thus, we conclude that f is increasing on $(0, 10)$.

b. The result of part (a) tells us that sales in the web hosting industry will be increasing from 1999 through 2009.

87. $A(t) = -96.6t^4 + 403.6t^3 + 660.9t^2 + 250$, so
$A'(t) = -386.4t^3 + 1210.8t^2 + 1321.8t = t\left(-386.4t^2 + 1210.8t + 1321.8\right)$. Solving $A'(t) = 0$, we find $t = 0$ and $t = \frac{-1210.8 \pm \sqrt{(1210.8)^2 - 4(-386.4)(1321.8)}}{-2(386.4)} = \frac{-1210.8 \pm 1873.2}{-2(386.4)} \approx 4$. Because t lies in the interval $[0, 5]$, we see that the continuous function A' has zeros at $t = 0$ and $t = 4$.

From the sign diagram, we see that f is increasing on $(0, 4)$ and decreasing on $(4, 5)$. We conclude that the cash in the Central Provident Trust Funds will be increasing from 2005 to 2045 and decreasing from 2045 to 2055.

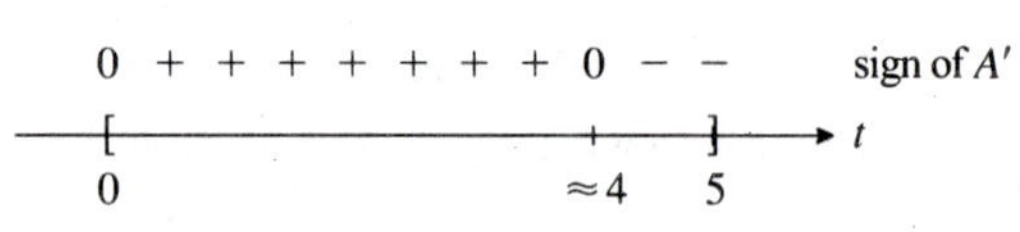

89. a. In 2005, the percentage was $f(0) = \frac{5.3\sqrt{0} - 300}{\sqrt{0} - 10} = 30$ (%). In 2015, it will be $f(10) = \frac{5.3\sqrt{10} - 300}{\sqrt{10} - 10} \approx 41.4$ (%).

b. $f'(t) = \frac{\left(t^{1/2} - 10\right)(5.3)\left(\frac{1}{2}t^{-1/2}\right) - \left(5.3t^{1/2} - 300\right)\left(\frac{1}{2}t^{-1/2}\right)}{\left(t^{1/2} - 10\right)^2} = \frac{247}{2\sqrt{t}\left(\sqrt{t} - 10\right)^2} > 0$. Thus, f is increasing on $(0, 10)$, indicating that the percentage of small and lower-midsize vehicles is increasing from 2005 through 2015.

91. $A(t) = \frac{136}{1 + 0.25(t - 4.5)^2} + 28$, so

$$A'(t) = \frac{d}{dt}\left[\frac{136}{1 + 0.25(t - 4.5)^2} + 28\right] = -\frac{136}{\left[1 + 0.25(t - 4.5)^2\right]^2} \cdot 2(0.25)(t - 4.5) = -\frac{68(t - 4.5)}{\left[1 + 0.25(t - 4.5)^2\right]^2}.$$

Observe that $A'(t) > 0$ if $t < 4.5$ and $A'(t) < 0$ if $t > 4.5$, so the pollution is increasing from 7 a.m. to 11:30 a.m. and decreasing from 11:30 a.m. to 6 p.m.

93. $N(t) = 5.3e^{0.095t^2 - 0.85t}$.

a. $N'(t) = 5.3e^{0.095t^2 - 0.85t}(0.19t - 0.85)$. Because $N'(t)$ is negative for $0 < t < 4$, we see that $N(t)$ is decreasing over that interval.

b. To find the rate at which the number of polio cases was decreasing at the beginning of 1959, we compute $N'(0) = 5.3e^{0.095(0^2) - 0.85(0)}(0.85) \approx 5.3(-0.85) = -4.505$, or 4505 cases per year per year (t is measured in thousands). To find the rate at which the number of polio cases was decreasing at the beginning of 1962, we compute $N'(3) = 5.3e^{0.095(9) - 0.85(3)}(0.57 - 0.85) \approx -0.272$, or 272 cases per year per year.

95. a. $G(t) = (D - S)(t) = D(t) - S(t) = \left(0.0007t^2 + 0.0265t + 2\right) - \left(-0.0014t^2 + 0.0326t + 1.9\right)$

$= 0.0021t^2 - 0.0061t + 0.1$.

b. $G'(t) = 0.0042t - 0.0061 = 0$ implies $t \approx 1.5$. From the sign diagram of G', we see that G is decreasing on $(0, 1.5)$ and increasing on $(1.5, 15)$. This shows that the gap between the demand and supply of nurses was increasing from 2000 through the middle of 2001 but starts widening from the middle of 2001 through 2015.

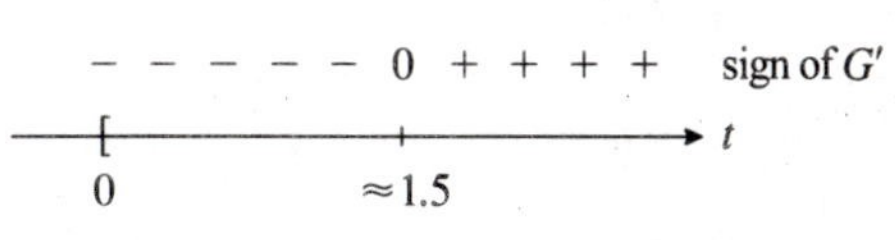

c. The relative minimum of G occurs at $t = 1.5$ and is $f(1.5) \approx 0.0956$. This says that the smallest shortage is approximately 96,000.

97. True. Let $a < x_1 < x_2 < b$. Then $f(x_2) > f(x_1)$ and $g(x_2) > g(x_1)$. Therefore, $(f + g)(x_2) = f(x_2) + g(x_2) > f(x_1) + g(x_1) = (f + g)(x_1)$, and so $f + g$ is increasing on (a, b).

99. True. Let $a < x_1 < x_2 < b$. Then $f(x_1) < f(x_2)$ and $g(x_1) < g(x_2)$. We find

$$\begin{aligned}(fg)(x_2) - (fg)(x_1) &= f(x_2)g(x_2) - f(x_1)g(x_1)\\ &= f(x_2)g(x_2) - f(x_2)g(x_1) + f(x_2)g(x_1) - f(x_1)g(x_1)\\ &= f(x_2)\left[g(x_2) - g(x_1)\right] + g(x_1)\left[f(x_2) - f(x_1)\right] > 0,\end{aligned}$$

so $(fg)(x_2) > (fg)(x_1)$ and fg is increasing on (a, b).

101. False. Let $f(x) = |x|$. Then f has a relative minimum at $x = 0$, but $f'(0)$ does not exist.

103. $f'(x) = 3x^2 + 1$ is continuous on $(-\infty, \infty)$ and is always greater than or equal to 1, so f has no critical number in $(-\infty, \infty)$. Therefore, f has no relative extremum on $(-\infty, \infty)$.

105. We require that $f'(-1) = 0$; that is, $f'(-1) = \left(3ax^2 + 12x + b\right)\big|_{x=-1} = 3a - 12 + b = 0$, and $f'(2) = 0$, or $f'(2) = \left(3ax^2 + 12x + b\right)\big|_{x=2} = 12a + 24 + b = 0$. Solving the system $\begin{cases} 3a + b = 12 \\ 12a + b = -24 \end{cases}$ simultaneously gives $a = -4$ and $b = 24$.

107. a. $f'(x) = -2x$ if $x \neq 0$, $f'(-1) = 2$, and $f'(1) = -2$, so $f'(x)$ changes sign from positive to negative as we move across $x = 0$.

b. f does not have a relative maximum at $x = 0$ because $f(0) = 2$ but a neighborhood of $x = 0$, for example $\left(-\frac{1}{2}, \frac{1}{2}\right)$, contains numbers with values larger than 2. This does not contradict the First Derivative Test because f is not continuous at $x = 0$.

109. $f(x) = ax^2 + bx + c$. Setting $f'(x) = 2ax + b = 2a\left(x + \frac{b}{2a}\right) = 0$ gives $x = -\frac{b}{2a}$ as the only critical number of f. If $a < 0$, the sign diagram shows that $x = -\frac{b}{2a}$ gives a relative maximum. Similarly, it can be shown that if $a > 0$, then $x = -\frac{b}{2a}$ gives a relative minimum.

$+ + + + + 0 - - -$ sign of f'

$0 \quad -\frac{b}{2a} \quad x$

111. a. $f'(x) = 3x^2 + 1$, and so $f'(x) > 1$ on the interval $(0, 1)$. Therefore, f is increasing on $(0, 1)$.

b. $f(0) = -1$ and $f(1) = 1 + 1 - 1 = 1$. Thus, the Intermediate Value Theorem guarantees that there is at least one root of $f(x) = 0$ in $(0, 1)$. Because f is increasing on $(0, 1)$, the graph of f can cross the x-axis at only one point in $(0, 1)$, and so $f(x) = 0$ has exactly one root.

Using Technology page 693

1. a. f is decreasing on $(-\infty, -0.2934)$ and increasing on $(-0.2934, \infty)$.

b. Relative minimum $f(-0.2934) = -2.5435$.

3. a. f is increasing on $(-\infty, -1.6144)$ and $(0.2390, \infty)$ and decreasing on $(-1.6144, 0.2390)$.

b. Relative maximum $f(-1.6144) = 26.7991$, relative minimum $f(0.2390) = 1.6733$.

5. a. f is decreasing on $(-\infty, -1)$ and $(0.33, \infty)$ and increasing on $(-1, 0.33)$.

b. Relative maximum $f(0.33) = 1.11$, relative minimum $f(-1) = -0.63$.

7. a. f is decreasing on $(-\infty, 0.40)$ and increasing on $(0.40, \infty)$.

b. Relative minimum $(0.40, 0.79)$.

9. a.

b. Increasing on $(0, 3.6676)$, decreasing on $(3.6676, 6)$.

11. The PSI is increasing on the interval $(0, 4.5)$ and decreasing on $(4.5, 11)$. It is highest when $t = 4.5$ (at 11:30 a.m.) and has value 164.

10.2 Applications of the Second Derivative

Problem-Solving Tips

1. If $f''(x) > 0$, then the graph of f "holds water" and we say the graph of f is concave upward.

If $f''(x) < 0$, then the graph of f "loses water" and we say the graph of f is concave downward.

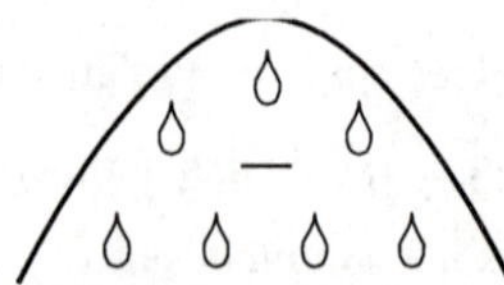

2. **To find the inflection points of a function** f, determine the number(s) in the domain of f for which $f''(x) = 0$ or $f''(x)$ does not exist. Note that each of these numbers c provides us with a candidate $(c, f(c))$ for an inflection point of f.

3. Note that the second derivative test is not valid when $f''(c) = 0$ or $f''(c)$ does not exist. In these cases you need to use the first derivative test to determine the relative extrema.

Concept Questions page 703

1. **a.** f is concave upward on (a, b) if f' is increasing on (a, b). f is concave downward on (a, b) if f' is decreasing on (a, b).

   b. For the procedure for determining where f is concave upward and where f is concave downward, see page 695 of the text.

3. The Second Derivative Test is stated on page 702 of the text. In general, if f'' is easy to compute, then use the Second Derivative Test. However, keep in mind that (1) in order to use this test f'' must exist, (2) the test is inconclusive if $f''(c) = 0$, and (3) the test is inconvenient to use if f'' is difficult to compute.

Exercises page 704

1. f is concave downward on $(-\infty, 0)$ and concave upward on $(0, \infty)$. f has an inflection point at $(0, 0)$.

3. f is concave downward on $(-\infty, 0)$ and $(0, \infty)$.

5. f is concave upward on $(-\infty, 0)$ and $(1, \infty)$ and concave downward on $(0, 1)$. $(0, 0)$ and $(1, -1)$ are inflection points of f.

7. f is concave downward on $(-\infty, -2)$ and $(-2, 2)$ and $(2, \infty)$.

9. **a.** f is concave upward on $(0, 2)$, $(4, 6)$, $(7, 9)$, and $(9, 12)$ and concave downward on $(2, 4)$ and $(6, 7)$.

   b. f has inflection points at $\left(2, \frac{5}{2}\right)$, $(4, 2)$, $(6, 2)$, and $(7, 3)$.

11. (a)

13. (b)

15. **a.** $D_1'(t) > 0$, $D_2'(t) > 0$, $D_1''(t) > 0$, and $D_2''(t) < 0$ on $(0, 12)$.

    b. With or without the proposed promotional campaign, the deposits will increase, but with the promotion, the deposits will increase at an increasing rate whereas without the promotion, the deposits will increase at a decreasing rate.

17. (c)

19. (d)

21. **a.** Between 8 a.m. and 10 a.m. the rate of change of the rate of smartphone assembly is increasing; between 10 a.m. and 12 noon, that rate is decreasing.

    b. If you look at the tangent lines to the graph of N, you will see that the tangent line at P has the greatest slope. This means that the rate at which the average worker is assembling smartphones is greatest—that is, the worker is most efficient—when $t = 2$, at 10 a.m.

23. The significance of the inflection point Q is that the restoration process is working at its peak at the time t_0 corresponding to the t-coordinate of Q.

25. $f(x) = 4x^2 - 12x + 7$, so $f'(x) = 8x - 12$ and $f''(x) = 8$. Thus, $f''(x) > 0$ everywhere, and so f is concave upward everywhere.

27. $f(x) = \dfrac{1}{x^4} = x^{-4}$, so $f'(x) = -\dfrac{4}{x^5}$ and $f''(x) = \dfrac{20}{x^6} > 0$ for all values of x in $(-\infty, 0)$ and $(0, \infty)$, and so f is concave upward on its domain.

29. $f(x) = 2x^2 - 3x + 4$, so $f'(x) = 4x - 3$ and $f''(x) = 4 > 0$ for all values of x. Thus, f is concave upward on $(-\infty, \infty)$.

31. $f(x) = 1 - x^3$, so $f'(x) = -3x^2$ and $f''(x) = -6x$. From the sign diagram of f'', we see that f is concave upward on $(-\infty, 0)$ and concave downward on $(0, \infty)$.

+ + + + + 0 − − − − − sign of f''
x
0

33. $f(x) = x^4 - 6x^3 + 2x + 8$, so $f'(x) = 4x^3 - 18x^2 + 2$ and $f''(x) = 12x^2 - 36x = 12x(x - 3)$. The sign diagram of f'' shows that f is concave upward on $(-\infty, 0)$ and $(3, \infty)$ and concave downward on $(0, 3)$.

+ + + 0 − − − 0 + + + sign of f''
x
0 3

35. $f(x) = x^{4/7}$, so $f'(x) = \frac{4}{7}x^{-3/7}$ and $f''(x) = -\frac{12}{49}x^{-10/7} = -\dfrac{12}{49x^{10/7}}$. Observe that $f''(x) < 0$ for all $x \neq 0$, so f is concave downward on $(-\infty, 0)$ and $(0, \infty)$.

37. $f(x) = (4 - x)^{1/2}$, so $f'(x) = \frac{1}{2}(4 - x)^{-1/2}(-1) = -\frac{1}{2}(4 - x)^{-1/2}$ and $f''(x) = \frac{1}{4}(4 - x)^{-3/2}(-1) = -\dfrac{1}{4(4 - x)^{3/2}} < 0$ whenever it is defined, so f is concave downward on $(-\infty, 4)$.

39. $f'(x) = \dfrac{d}{dx}(x - 3)^{-1} = -(x - 3)^{-2}$ and $f''(x) = 2(x - 3)^{-3} = \dfrac{2}{(x - 3)^3}$. The sign diagram of f'' shows that f is concave downward on $(-\infty, 3)$ and concave upward on $(3, \infty)$.

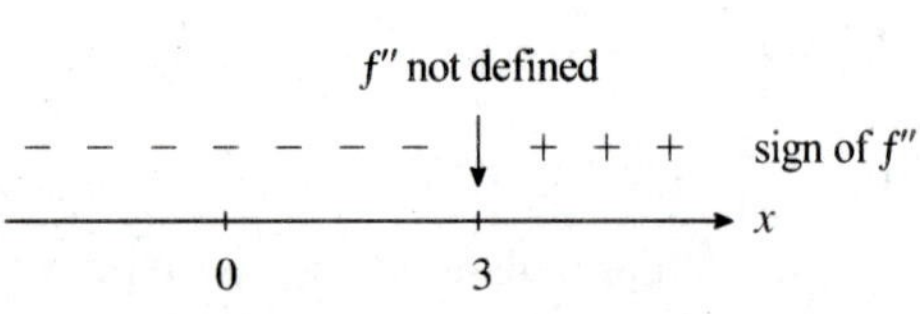

41. $f'(x) = \dfrac{d}{dx}(2 + x^2)^{-1} = -(2 + x^2)^{-2}(2x) = -2x(2 + x^2)^{-2}$ and

$f''(x) = -2(2 + x^2)^{-2} - 2x(-2)(2 + x^2)^{-3}(2x) = 2(2 + x^2)^{-3}\left[-(2 + x^2) + 4x^2\right] = \dfrac{2(3x^2 - 2)}{(2 + x^2)^3} = 0$ if $x = \pm\sqrt{\frac{2}{3}} = \frac{\sqrt{6}}{3}$. From the sign diagram of f'', we see that f is concave upward on $\left(-\infty, -\frac{\sqrt{6}}{3}\right)$ and $\left(\frac{\sqrt{6}}{3}, \infty\right)$ and concave downward on $\left(-\frac{\sqrt{6}}{3}, \frac{\sqrt{6}}{3}\right)$.

+ + + 0 − − − − − 0 + + + sign of f''
x
$-\frac{\sqrt{6}}{3}$ 0 $\frac{\sqrt{6}}{3}$

43. $h(t) = \dfrac{t^2}{t-1}$, so $h'(t) = \dfrac{(t-1)(2t) - t^2(1)}{(t-1)^2} = \dfrac{t^2 - 2t}{(t-1)^2}$ and

$$h''(t) = \frac{(t-1)^2(2t-2) - (t^2 - 2t)\,2(t-1)}{(t-1)^4} = \frac{(t-1)(2t^2 - 4t + 2 - 2t^2 + 4t)}{(t-1)^4} = \frac{2}{(t-1)^3}.$$

The sign diagram of h'' shows that h is concave downward on $(-\infty, 1)$ and concave upward on $(1, \infty)$.

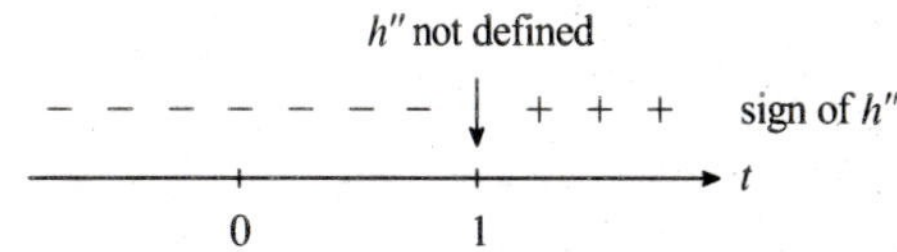

45. $g(x) = x + \dfrac{1}{x^2}$, so $g'(x) = 1 - 2x^{-3}$ and $g''(x) = 6x^{-4} = \dfrac{6}{x^4} > 0$ whenever $x \neq 0$. Therefore, g is concave upward on $(-\infty, 0)$ and $(0, \infty)$.

47. $g(t) = (2t-5)^{1/3}$, so $g'(t) = \frac{1}{3}(2t-5)^{-2/3}(2) = \frac{2}{3}(2t-5)^{-2/3}$ and $g''(t) = -\frac{4}{9}(2t-5)^{-5/3} \cdot 2 = -\dfrac{8}{9(2t-5)^{5/3}}$. The sign diagram of g'' shows that g is concave upward on $\left(-\infty, \frac{5}{2}\right)$ and concave downward on $\left(\frac{5}{2}, \infty\right)$.

g″ not defined
\+ + + + + + + − − − sign of g″
t: 0, $\frac{5}{2}$

49. $f(x) = \frac{1}{2}e^x - \frac{1}{2}e^{-x}$, so $f'(x) = \frac{1}{2}(e^x + e^{-x})$ and $f''(x) = \frac{1}{2}(e^x - e^{-x})$. Setting $f''(x) = 0$ gives $e^x = e^{-x}$ or $e^{2x} = 1$, and so $x = 0$. From the sign diagram for f'',we conclude that f is concave upward on $(0, \infty)$ and concave downward on $(-\infty, 0)$.

− − − − 0 + + + + sign of f″
x: 0

51. $f(x) = x^2 + \ln x^2$, so $f'(x) = 2x + \dfrac{2x}{x^2} = 2x + \dfrac{2}{x}$ and $f''(x) = 2 - \dfrac{2}{x^2}$. To find the intervals of concavity for f, we first set $f''(x) = 0$, giving $2 - \dfrac{2}{x^2} = 0$, $2 = \dfrac{2}{x^2}$, $2x^2 = 2$, $x^2 = 1$, and so $x = \pm 1$.

From the sign diagram for f'', we see that f is concave upward on $(-\infty, -1)$ and $(1, \infty)$ and concave downward on $(-1, 0)$ and $(0, 1)$.

f″ not defined
\+ + + 0 − − − − 0 + + + sign of f″
x: −1, 0, 1

53. $f(x) = x^3 - 2$, so $f'(x) = 3x^2$ and $f''(x) = 6x$. $f''(x)$ is continuous everywhere and has a zero at $x = 0$. From the sign diagram of f'', we conclude that $(0, -2)$ is an inflection point of f.

− − − − − 0 + + + + + sign of f″
x: 0

55. $f(x) = 6x^3 - 18x^2 + 12x - 20$, so $f'(x) = 18x^2 - 36x + 12$ and $f''(x) = 36x - 36 = 36(x-1) = 0$ if $x = 1$. The sign diagram of f'' shows that f has an inflection point at $(1, -20)$.

− − − − − − − − 0 + + + sign of f″
x: 0, 1

57. $f(x) = 3x^4 - 4x^3 + 1$, so $f'(x) = 12x^3 - 12x^2$ and $f''(x) = 36x^2 - 24x = 12x(3x - 2) = 0$ if $x = 0$ or $\frac{2}{3}$. These are candidates for inflection points. The sign diagram of f'' shows that $(0, 1)$ and $\left(\frac{2}{3}, \frac{11}{27}\right)$ are inflection points of f.

+ + + 0 − − − 0 + + + sign of f''; x: 0, $\frac{2}{3}$

59. $g(t) = t^{1/3}$, so $g'(t) = \frac{1}{3}t^{-2/3}$ and $g''(t) = -\frac{2}{9}t^{-5/3} = -\frac{2}{9t^{5/3}}$. Observe that $t = 0$ is in the domain of g. Next, since $g''(t) > 0$ if $t < 0$ and $g''(t) < 0$ if $t > 0$, we see that $(0, 0)$ is an inflection point of g.

61. $f(x) = (x - 1)^3 + 2$, so $f'(x) = 3(x - 1)^2$ and $f''(x) = 6(x - 1)$. Observe that $f''(x) < 0$ if $x < 1$ and $f''(x) > 0$ if $x > 1$ and so $(1, 2)$ is an inflection point of f.

63. $f(x) = 2e^{-x^2} = 2\left(e^{-x^2}\right)$, so $f'(x) = 2(-2x)e^{-x^2} = -4xe^{-x^2}$ and $f''(x) = -4x(-2x)e^{-x^2} - 4e^{-x^2} = -4e^{-x^2}(-2x^2 + 1) = 4e^{-x^2}(2x^2 - 1)$. Setting $f''(x) = 0$ gives $2x^2 = 1$, $x^2 = \frac{1}{2}$, and so $x = \pm\frac{\sqrt{2}}{2}$. The sign diagram for f'' shows that $\left(-\frac{\sqrt{2}}{2}, 2e^{-1/2}\right)$ and $\left(\frac{\sqrt{2}}{2}, 2e^{-1/2}\right)$ are inflection points.

+ + + 0 − − − − − 0 + + + sign of f''; x: $-\frac{\sqrt{2}}{2}$, 0, $\frac{\sqrt{2}}{2}$

65. $f(x) = x^2 \ln x$, so $f'(x) = 2x \ln x + x^2\left(\frac{1}{x}\right) = 2x \ln x + x$ and $f''(x) = 2 \ln x + 2x\left(\frac{1}{x}\right) + 1 = 2 \ln x + 3 = 0$ implies that $\ln x = -\frac{3}{2}$, so $x = e^{-3/2}$. From the sign diagram of f'', we see that $\left(e^{-3/2}, -\frac{3}{2}e^{-3}\right)$ is an inflection point of f.

− − − − 0 + + + + sign of f''; x: 0, $e^{-3/2}$

67. $f(x) = -x^2 + 2x + 4$, so $f'(x) = -2x + 2$. The critical number of f is $x = 1$. Because $f''(x) = -2$ and $f''(1) = -2 < 0$, we conclude that $f(1) = 5$ is a relative maximum of f.

69. $f(x) = 2x^3 + 1$, so $f'(x) = 6x^2 = 0$ if $x = 0$ and this is a critical number of f. Next, $f''(x) = 12x$, and so $f''(0) = 0$. Thus, the Second Derivative Test fails. But the First Derivative Test shows that $(0, 0)$ is not a relative extremum.

71. $f(x) = \frac{1}{3}x^3 - 2x^2 - 5x - 5$, so $f'(x) = x^2 - 4x - 5 = (x - 5)(x + 1)$ and this gives $x = -1$ and $x = 5$ as critical numbers of f. Next, $f''(x) = 2x - 4$. Because $f''(-1) = -6 < 0$, we see that $\left(-1, -\frac{7}{3}\right)$ is a relative maximum of f. Next, $f''(5) = 6 > 0$ and this shows that $\left(5, -\frac{115}{3}\right)$ is a relative minimum of f.

73. $g(t) = t + \frac{9}{t}$, so $g'(t) = 1 - \frac{9}{t^2} = \frac{t^2 - 9}{t^2} = \frac{(t + 3)(t - 3)}{t^2}$, showing that $t = \pm 3$ are critical numbers of g. Now, $g''(t) = 18t^{-3} = \frac{18}{t^3}$. Because $g''(-3) = -\frac{18}{27} < 0$, the Second Derivative Test implies that g has a relative maximum at $(-3, -6)$. Also, $g''(3) = \frac{18}{27} > 0$ and so g has a relative minimum at $(3, 6)$.

75. $f(x) = \frac{x}{1 - x}$, so $f'(x) = \frac{(1 - x)(1) - x(-1)}{(1 - x)^2} = \frac{1}{(1 - x)^2}$ is never zero. Thus, there is no critical number and f has no relative extremum.

77. $f(t) = t^2 - \frac{16}{t}$, so $f'(t) = 2t + \frac{16}{t^2} = \frac{2t^3 + 16}{t^2} = \frac{2(t^3+8)}{t^2}$. Setting $f'(t) = 0$ gives $t = -2$ as a critical number. Next, we compute $f''(t) = \frac{d}{dt}(2t + 16t^{-2}) = 2 - 32t^{-3} = 2 - \frac{32}{t^3}$. Because $f''(-2) = 2 - \frac{32}{(-8)} = 6 > 0$, we see that $(-2, 12)$ is a relative minimum.

79. $g(s) = \frac{s}{1+s^2}$, so $g'(s) = \frac{(1+s^2)(1) - s(2s)}{(1+s^2)^2} = \frac{1-s^2}{(1+s^2)^2} = 0$ gives $s = -1$ and $s = 1$ as critical numbers of g. Next, we compute $g''(s) = \frac{(1+s^2)^2(-2s) - (1-s^2)\,2(1+s^2)(2s)}{(1+s^2)^4} = \frac{2s(1+s^2)(-1-s^2-2+2s^2)}{(1+s^2)^4} = \frac{2s(s^2-3)}{(1+s^2)^3}$. Now $g''(-1) = \frac{1}{2} > 0$, and so $g(-1) = -\frac{1}{2}$ is a relative minimum of g. Next, $g''(1) = -\frac{1}{2} < 0$ and so $g(1) = \frac{1}{2}$ is a relative maximum of g.

81. $g(t) = e^{t^2-2t}$, so $g'(t) = e^{t^2-2t}(2t-2) = 2(t-1)e^{t^2-2t}$ and $g''(t) = 2\left[(t-1)(2t-2)e^{t^2-2t} + e^{t^2-2t}(1)\right] = 2(2t^2 - 4t + 3)e^{t^2-2t}$. Setting $g'(t) = 0$ gives $2(t-1) = 0$, so $t = 1$ is the only critical number. $g''(1) = 2(2-4+3)e^{1-2} = 2e^{-1} > 0$, so $\left(1, \frac{1}{e}\right)$ is a relative minimum.

83. $f(x) = \ln(x^2+1)$, so $f'(x) = \frac{2x}{x^2+1}$ and $f''(x) = \frac{(x^2+1)(2) - 2x(2x)}{(x^2+1)^2} = \frac{2(1-x^2)}{(x^2+1)^2}$. Setting $f'(x) = 0$ gives $x = 0$, and $f''(0) = \frac{2(1)}{(0^2+1)^2} = 2 > 0$, so we see that $(0, 0)$ is a relative minimum.

85.

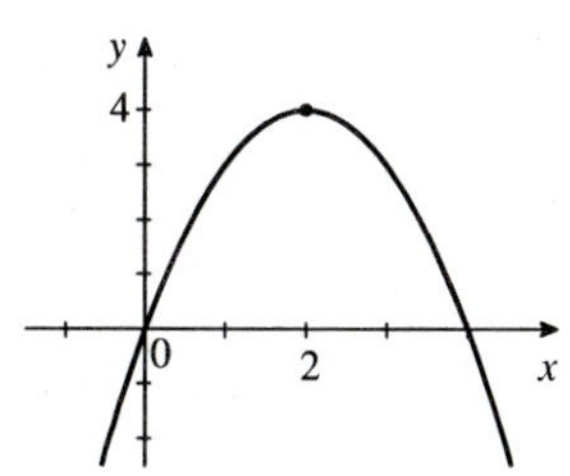

87.

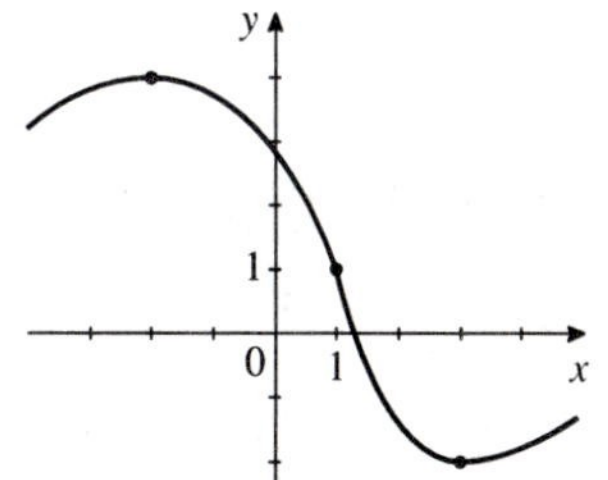

89.

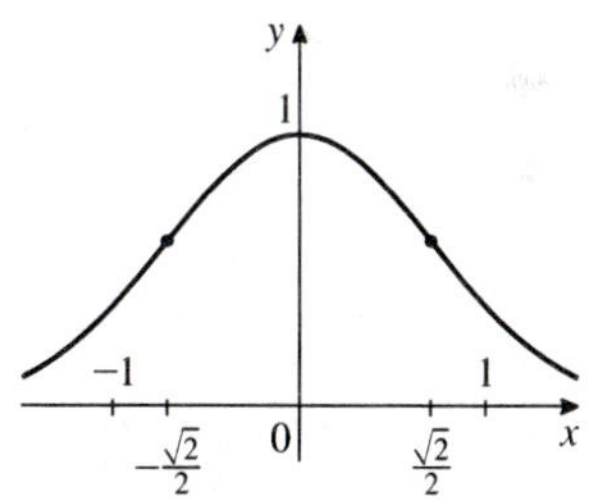

91. **a.** $N'(t)$ is positive because N is increasing on $(0, 12)$.

b. $N''(t) < 0$ on $(0, 6)$ and $N''(t) > 0$ on $(6, 12)$.

c. The rate of growth of the number of help-wanted advertisements was decreasing over the first six months of the year and increasing over the last six months.

93. $f(t)$ increases at an increasing rate until the water level reaches the middle of the vase (this corresponds to the inflection point of f). At this point, $f(t)$ is increasing at the fastest rate. Though $f(t)$ still increases until the vase is filled, it does so at a decreasing rate.

95. **a.** $f'(t) = \frac{d}{dt}\left(0.43t^{0.43}\right) = \left(0.43^2\right)t^{-0.57} = \frac{0.1849}{t^{0.57}}$ is positive if $t \geq 1$. This shows that f is increasing for $t > 1$, and this implies that the average state cigarette tax was increasing during the period in question.

b. $f''(t) = \frac{d}{dt}\left(0.1849t^{-0.57}\right) = (0.1849)(-0.57)t^{-1.57} = -\frac{0.105393}{t^{1.57}}$ is negative if $t \geq 1$. Thus, the rate of the increase of the cigarette tax is decreasing over the period in question.

97. a. $S'(t) = \frac{d}{dt}\left(0.195t^2 + 0.32t + 23.7\right) = 0.39t + 0.32 > 0$ on $[0, 7]$, so sales were increasing through the years in question.

b. $S''(t) = 0.39 > 0$ on $[0, 7]$, so sales continued to accelerate through the years.

99. a. $f(t) = 0.71e^{0.7t}$. The frequency for a 70 year old is given by $f(1) \approx 1.43$ (%), and for an 80 year old it is $f(3) \approx 5.80$ (%).

b. $f'(t) = \frac{d}{dt}\left(0.71e^{0.7t}\right) = 0.7\left(0.71e^{0.7t}\right) = 0.497e^{0.7t}$, which is positive for $0 < t < 5$, and so f is increasing on $(0, 5)$. This says that the frequency of Alzheimer's disease increases with age in the age range under consideration.

c. $f''(t) = \frac{d}{dt}\left(0.497e^{0.7t}\right) = 0.3479e^{0.7t}$, which is positive for $0 < t < 5$, and so f is concave upward on $(0, 5)$. This says that the frequency of Alzheimer's disease is increasing at an increasing rate in the age range under consideration.

101. a. $A'(t) = \frac{d}{dt}\left[0.92(t+1)^{0.61}\right] = 0.92(0.61)(t+1)^{-0.39} = \frac{0.5612}{(t+1)^{0.39}} > 0$ on $(0, 4)$, so A is increasing on $(0, 4)$. This tells us that the spending is increasing over the years in question.

b. $A''(t) = (0.5612)(-0.39)(t+1)^{-1.39} = -\frac{0.218868}{(t+1)^{1.39}} < 0$ on $(0, 4)$, so A'' is concave downward on $(0, 4)$. This tells us that the spending is increasing but at a decreasing rate.

103. a. $R(t) = -0.2t^3 + 1.64t^2 + 1.31t + 3.2$, so $R'(t) = -0.6t^2 + 3.28t + 1.31$ and $R''(t) = -1.2t + 3.28$.

b. Setting $R'(t) = 0$ and solving for t gives $t = \frac{-3.28 \pm \sqrt{(3.28)^2 - 4(-0.6)(1.31)}}{2(-0.6)} \approx -0.374$ or 5.840. Both roots lie outside the interval $(0, 4)$. Because $R'(0) = 1.31 > 0$, we conclude that $R'(t) > 0$ for all t in $(0, 4)$.

c. Setting $R''(t) = 0$ gives $-1.2t + 3.28 = 0$, so $t \approx 2.73$. From the sign diagram, we see that R has an inflection point at approximately $(2.73, 14.93)$. This says that between 2004 and 2008, Google's revenue was increasing fastest in late September 2006.

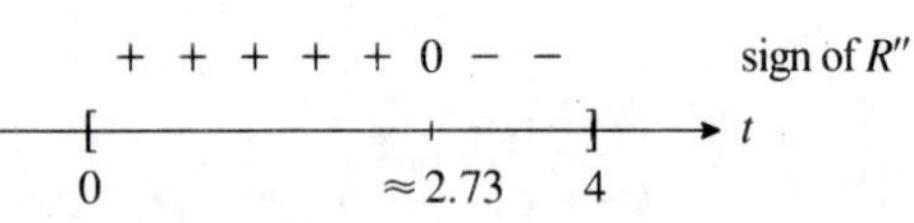

105. a. $N(t) = -2.42t^3 - 24.5t^2 - 123.3t + 506$. The number of measles deaths in 1999 is given by $N(0) = 506$, or 506,000. The number of measles deaths in 2005 is given by $N(6) = -2.42(6^3) + 24.5(6^2) - 123.3(6) + 506 = 125.48$, or approximately 125,480.

b. $N'(t) = \frac{d}{dt}\left(-2.42t^3 + 24.5t^2 - 123.3t + 506\right) = -7.26t^2 + 49t - 123.3$. Because the discriminant $(49)^2 - 4(-7.26)(-123.3) = -1179.6 < 0$, we see that $N'(t)$ has no zero. Because $N'(0) = -123.3 < 0$, we conclude that $N'(t) < 0$ on $(0, 6)$. This shows that N is decreasing on $(0, 6)$, so the number of measles deaths was dropping from 1999 through 2005.

c. $N''(t) = -14.52t + 49 = 0$ implies that $t \approx 3.37$, so the number of measles deaths was decreasing most rapidly in April 2002. The rate is given by $N'(3.37) = -7.26(3.37)^2 + 49(3.37) - 123.3 \approx -40.62$, or approximately -41 deaths/yr/yr.

107. a. $R'(t) = \frac{d}{dt}\left(0.00731t^4 - 0.174t^3 + 1.528t^2 + 0.48t + 19.3\right) = 0.02924t^3 - 0.522t^2 + 3.056t + 0.48$ and $R''(t) = 0.08772t^2 - 1.044t + 3.056$. Solving the equation $R''(t) = 0$, we obtain $t = \dfrac{1.044 \pm \sqrt{(-1.044)^2 - 4(0.08772)(3.056)}}{2(0.08772)} \approx 5.19$ or 6.71. (Note that 6.71 lies outside the interval of interest.)

From the sign diagram of R'', we see that the inflection point is approximately $(5.19, 43.93)$. We see that the dependency ratio will be increasing at the greatest pace around $t = 5.19$, that is, around 2052.

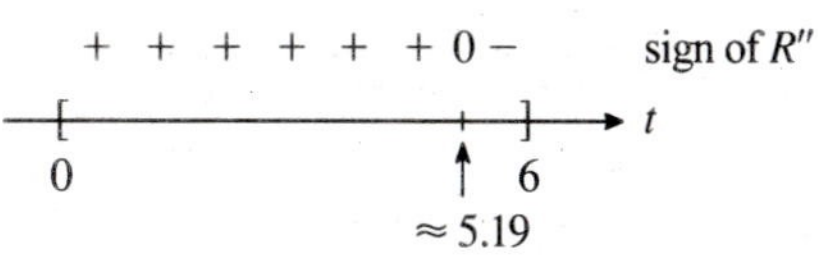

b. The dependency ratio will be $R(5.19) \approx 43.93$, or approximately 44.

109. False. Let $f(x) = x + \frac{1}{x}$ (see Example 2). Then f is concave downward on $(-\infty, 0)$ and concave upward on $(0, \infty)$, but f does not have an inflection point at 0.

111. True. Suppose the degree of P is $n \geq 3$. Thus $P''(x) = 0$ can have at most $n - 2$ zeros.

10.3 Curve Sketching

Problem-Solving Tips

1. **To find the horizontal asymptotes of a function** f, find the limit of f as $x \to \infty$ and as $x \to -\infty$. If the limit is equal to a real number b, then $y = b$ is a horizontal asymptote of f.

2. **To find the vertical asymptotes of a rational function** $f(x) = P(x)/Q(x)$, determine the values a for which $Q(a) = 0$. If $Q(a) = 0$ but $P(a) \neq 0$, then the line $x = a$ is a vertical asymptote of f.

3. If a line $x = a$ is a vertical asymptote of the graph of a rational function f, then the denominator of $f(x)$ is equal to zero at $x = a$. However, if both numerator and denominator of $f(x)$ are equal to zero, then $x = a$ is not necessarily a vertical asymptote.

Concept Questions page 718

1. a. See the definition on page 712 of the text. **b.** See the definition on page 714 of the text.

3. See the procedure given on page 713 of the text.

Exercises page 719

1. $y = 0$ is a horizontal asymptote.

3. $y = 0$ is a horizontal asymptote and $x = 0$ is a vertical asymptote.

5. $y = 0$ is a horizontal asymptote and $x = -1$ and $x = 1$ are vertical asymptotes.

7. $y = 3$ is a horizontal asymptote and $x = 0$ is a vertical asymptote.

9. $y = 0$ is a horizontal asymptote.

11. $\lim_{x\to\infty} \frac{1}{x} = 0$, and so $y = 0$ is a horizontal asymptote. Next, since the numerator of the rational expression is not equal to zero and the denominator is zero at $x = 0$, we see that $x = 0$ is a vertical asymptote.

13. $f(x) = -\frac{2}{x^2}$, so $\lim_{x\to\infty} f(x) = \lim_{x\to\infty}\left(-\frac{2}{x^2}\right) = 0$. Thus, $y = 0$ is a horizontal asymptote. Next, the denominator of $f(x)$ is equal to zero at $x = 0$. Because the numerator of $f(x)$ is not equal to zero at $x = 0$, we see that $x = 0$ is a vertical asymptote.

15. $\lim_{x\to\infty} \frac{x-2}{x+2} = \lim_{x\to\infty} \frac{1-\frac{2}{x}}{1+\frac{2}{x}} = 1$, and so $y = 1$ is a horizontal asymptote. Next, the denominator is equal to zero at $x = -2$ and the numerator is not equal to zero at this number, so $x = -2$ is a vertical asymptote.

17. $h(x) = x^3 - 3x^2 + x + 1$. $h(x)$ is a polynomial function, and therefore it does not have any horizontal or vertical asymptotes.

19. $\lim_{t\to\infty} \frac{t^2}{t^2-16} = \lim_{t\to\infty} \frac{1}{1-\frac{16}{t^2}} = 1$, and so $y = 1$ is a horizontal asymptote. Next, observe that the denominator of the rational expression $t^2 - 16 = (t+4)(t-4) = 0$ if $t = -4$ or $t = 4$. But the numerator is not equal to zero at these numbers, so $t = -4$ and $t = 4$ are vertical asymptotes.

21. $\lim_{x\to\infty} \frac{3x}{x^2-x-6} = \lim_{x\to\infty} \frac{\frac{3}{x}}{1-\frac{1}{x}-\frac{6}{x^2}} = 0$ and so $y = 0$ is a horizontal asymptote. Next, observe that the denominator $x^2 - x - 6 = (x-3)(x+2) = 0$ if $x = -2$ or $x = 3$. But the numerator $3x$ is not equal to zero at these numbers, so $x = -2$ and $x = 3$ are vertical asymptotes.

23. $\lim_{t\to\infty}\left[2 + \frac{5}{(t-2)^2}\right] = 2$ and $\lim_{t\to-\infty}\left[2 + \frac{5}{(t-2)^2}\right] = 2$, and so $y = 2$ is a horizontal asymptote. Next observe that $\lim_{t\to2^+} g(t) = \lim_{t\to2^-}\left[2 + \frac{5}{(t-2)^2}\right] = \infty$, and so $t = 2$ is a vertical asymptote.

25. $\lim_{x\to\infty} \frac{x^2-2}{x^2-4} = \lim_{x\to\infty} \frac{1-\frac{2}{x^2}}{1-\frac{4}{x^2}} = 1$, and so $y = 1$ is a horizontal asymptote. Next, observe that the denominator $x^2 - 4 = (x+2)(x-2) = 0$ if $x = -2$ or 2. Because the numerator $x^2 - 2$ is not equal to zero at these numbers, the lines $x = -2$ and $x = 2$ are vertical asymptotes.

27. $g(x) = \frac{x^3-x}{x(x+1)}$. Rewrite $g(x)$ as $\frac{x^2-1}{x+1}$ for $x \neq 0$, and note that $\lim_{x\to-\infty} g(x) = \lim_{x\to-\infty} \frac{x-\frac{1}{x}}{1+\frac{1}{x}} = -\infty$ and $\lim_{x\to\infty} g(x) = \infty$. Therefore, there is no horizontal asymptote. Next, note that the denominator of $g(x)$ is equal to zero at $x = 0$ and $x = -1$. However, since the numerator of $g(x)$ is also equal to zero when $x = 0$, we see that $x = 0$ is not a vertical asymptote. Also, the numerator of $g(x)$ is equal to zero when $x = -1$, so $x = -1$ is not a vertical asymptote.

29. f is the derivative function of the function g. Observe that at a relative maximum or minimum of g, $f(x) = 0$.

31.

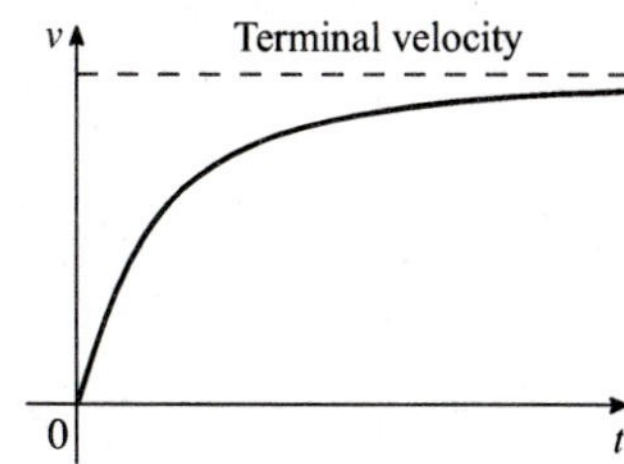

33.

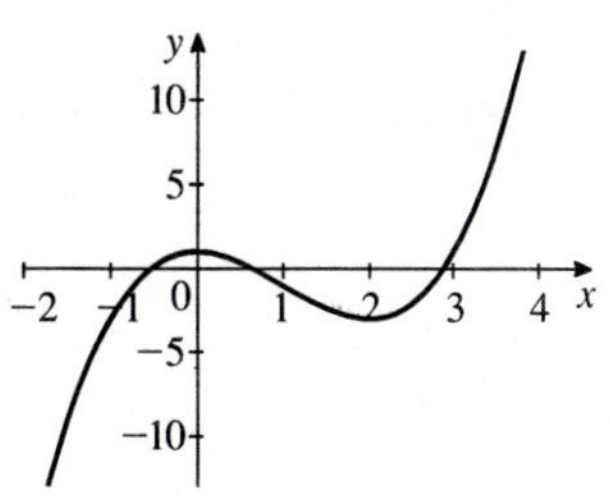

35.

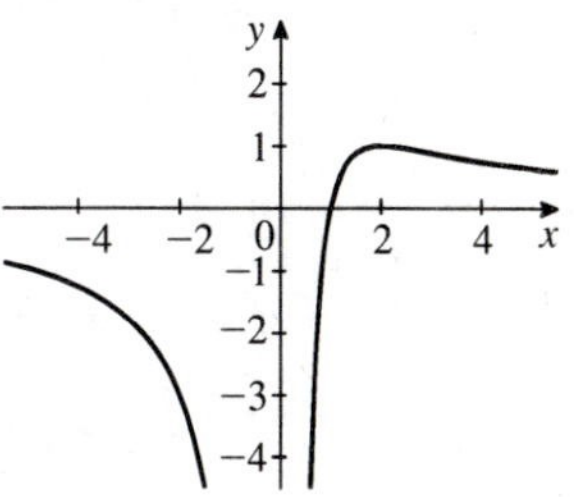

37. $g(x) = 4 - 3x - 2x^3$. We first gather the following information on f.

1. The domain of f is $(-\infty, \infty)$.
2. Setting $x = 0$ gives $y = 4$ as the y-intercept. Setting $y = g(x) = 0$ gives a cubic equation which is not easily solved, and we will not attempt to find the x-intercepts.
3. $\lim_{x\to-\infty} g(x) = \infty$ and $\lim_{x\to\infty} g(x) = -\infty$.
4. The graph of g has no asymptote.
5. $g'(x) = -3 - 6x^2 = -3(2x^2 + 1) < 0$ for all values of x and so g is decreasing on $(-\infty, \infty)$.
6. The results of step 5 show that g has no critical number and hence no relative extremum.
7. $g''(x) = -12x = 0$ if $x = 0$. Because $g''(x) > 0$ for $x < 0$ and $g''(x) < 0$ for $x > 0$, we see that g is concave upward on $(-\infty, 0)$ and concave downward on $(0, \infty)$.
8. From the results of step 7, we see that $(0, 4)$ is an inflection point of g.

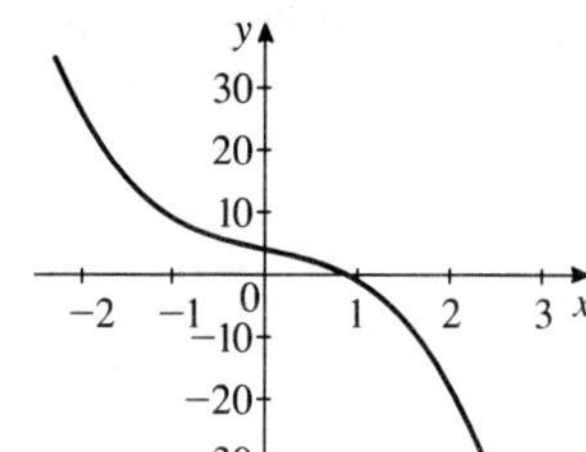

39. $h(x) = x^3 - 3x + 1$. We first gather the following information on h.

1. The domain of h is $(-\infty, \infty)$.
2. Setting $x = 0$ gives 1 as the y-intercept. We will not find the x-intercept.
3. $\lim_{x\to-\infty} (x^3 - 3x + 1) = -\infty$ and $\lim_{x\to\infty} (x^3 - 3x + 1) = \infty$.
4. There is no asymptote because $h(x)$ is a polynomial.
5. $h'(x) = 3x^2 - 3 = 3(x + 1)(x - 1)$, and we see that $x = -1$ and $x = 1$ are critical numbers. From the sign diagram, we see that h is increasing on $(-\infty, -1)$ and $(1, \infty)$ and decreasing on $(-1, 1)$.

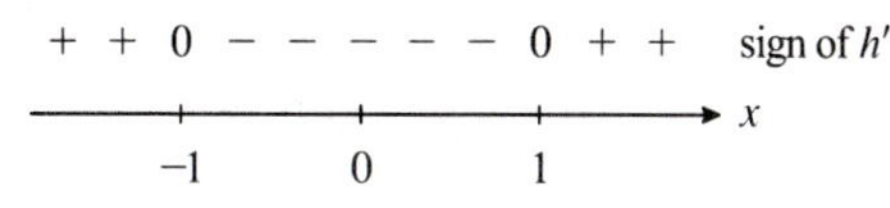

6. The results of step 5 show that $(-1, 3)$ is a relative maximum and $(1, -1)$ is a relative minimum.
7. $h''(x) = 6x = 0$ if $x = 0$, $h''(x) < 0$ if $x < 0$, and $h''(x) > 0$ if $x > 0$. Thus, the graph of h is concave downward on $(-\infty, 0)$ and concave upward on $(0, \infty)$.
8. The results of step 7 show that $(0, 1)$ is an inflection point of h.

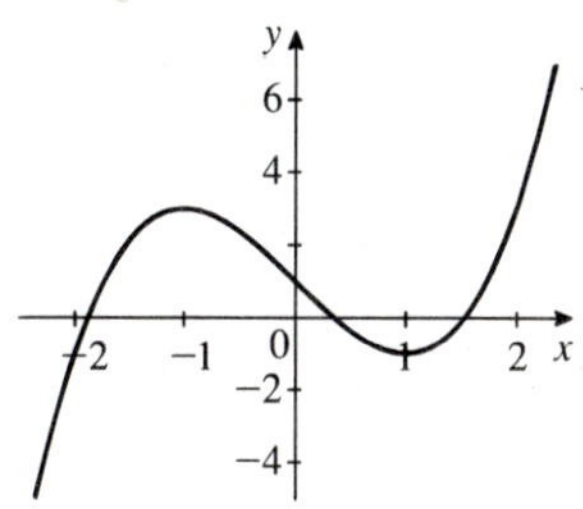

41. $f(x) = -2x^3 + 3x^2 + 12x + 2$. We first gather the following information on f.

1. The domain of f is $(-\infty, \infty)$.
2. Setting $x = 0$ gives 2 as the y-intercept.
3. $\lim_{x\to-\infty}\left(-2x^3 + 3x^2 + 12x + 2\right) = \infty$ and $\lim_{x\to\infty}\left(-2x^3 + 3x^2 + 12x + 2\right) = -\infty$
4. There is no asymptote because $f(x)$ is a polynomial function.
5. $f'(x) = -6x^2 + 6x + 12 = -6\left(x^2 - x - 2\right) = -6(x-2)(x+1) = 0$ if $x = -1$ or $x = 2$, the critical numbers of f. From the sign diagram, we see that f is decreasing on $(-\infty, -1)$ and $(2, \infty)$ and increasing on $(-1, 2)$.

   – – 0 + + + + + 0 – – sign of f'
   x: −1, 0, 2

6. The results of step 5 show that $(-1, -5)$ is a relative minimum and $(2, 22)$ is a relative maximum.
7. $f''(x) = -12x + 6 = 0$ if $x = \frac{1}{2}$. The sign diagram of f''shows that the graph of f is concave upward on $\left(-\infty, \frac{1}{2}\right)$ and concave downward on $\left(\frac{1}{2}, \infty\right)$.

   + + + + + + 0 – – – – sign of f''
   x: 0, $\frac{1}{2}$

8. The results of step 7 show that $\left(\frac{1}{2}, \frac{17}{2}\right)$ is an inflection point.

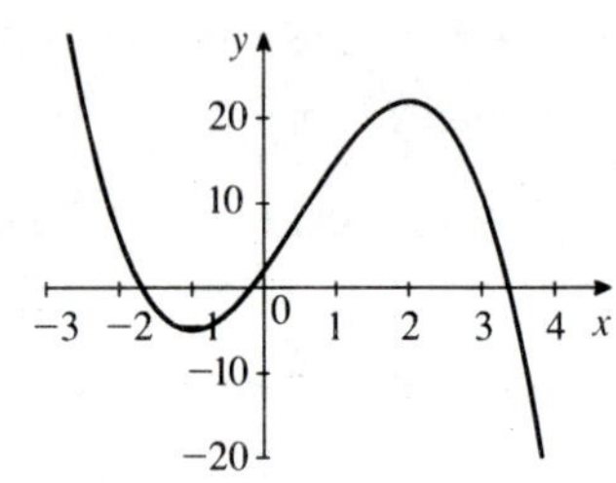

43. $h(x) = \frac{3}{2}x^4 - 2x^3 - 6x^2 + 8$. We first gather the following information on h.

1. The domain of h is $(-\infty, \infty)$.
2. Setting $x = 0$ gives 8 as the y-intercept.
3. $\lim_{x\to-\infty} h(x) = \lim_{x\to\infty} h(x) = \infty$
4. There is no asymptote.
5. $h'(x) = 6x^3 - 6x^2 - 12x = 6x\left(x^2 - x - 2\right) = 6x(x-2)(x+1) = 0$ if $x = -1$, 0, or 2, and these are the critical numbers of h. The sign diagram of h' shows that h is increasing on $(-1, 0)$ and $(2, \infty)$ and decreasing on $(-\infty, -1)$ and $(0, 2)$.

   – – 0 + 0 – – – 0 + + sign of h'
   x: −1, 0, 2

6. The results of step 5 show that $\left(-1, \frac{11}{2}\right)$ and $(2, -8)$ are relative minima of h and $(0, 8)$ is a relative maximum of h.
7. $h''(x) = 18x^2 - 12x - 12 = 6\left(3x^2 - 2x - 2\right)$. The zeros of h'' are $x = \frac{2\pm\sqrt{4+24}}{6} \approx -0.5$ or 1.2. The sign diagram of h'' shows that the graph of h is concave upward on $(-\infty, -0.5)$ and $(1.2, \infty)$ and concave downward on $(-0.5, 1.2)$.

   + + 0 – – – – – – – 0 + + sign of h''
   x: ≈−0.5, 0, ≈1.2

8. The results of step 7 also show that $(-0.5, 6.8)$ and $(1.2, -1)$ are inflection points.

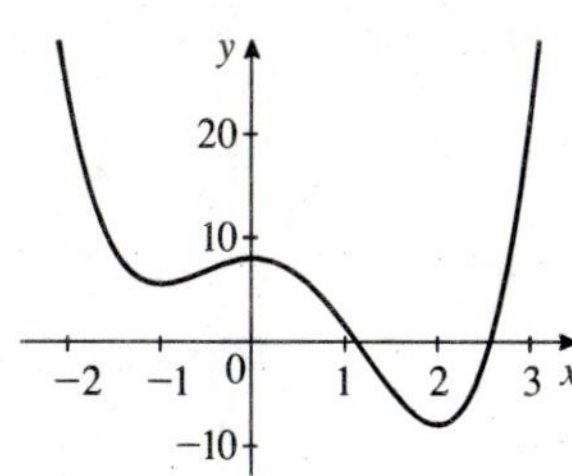

45. $f(t) = \sqrt{t^2 - 4}$. We first gather the following information on f.

1. The domain of f is found by solving $t^2 - 4 \geq 0$ to obtain $(-\infty, -2] \cup [2, \infty)$.

2. Because $t \neq 0$, there is no y-intercept. Next, setting $y = f(t) = 0$ gives the t-intercepts as -2 and 2.

3. $\lim_{t \to -\infty} f(t) = \lim_{t \to \infty} f(t) = \infty$.

4. There is no asymptote.

5. $f'(t) = \frac{1}{2}\left(t^2 - 4\right)^{-1/2}(2t) = t\left(t^2 - 4\right)^{-1/2} = \dfrac{t}{\sqrt{t^2 - 4}}$. Setting $f'(t) = 0$ gives $t = 0$. But $t = 0$ is not in the domain of f and so there is no critical number. From the sign diagram for f', we see that f is increasing on $(2, \infty)$ and decreasing on $(-\infty, -2)$.

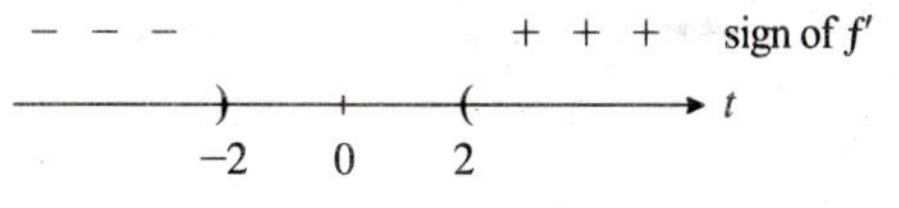

6. From the results of step 5 we see that there is no relative extremum.

7. $f''(t) = \left(t^2 - 4\right)^{-1/2} + t\left(-\frac{1}{2}\right)\left(t^2 - 4\right)^{-3/2}(2t)$

$$= \left(t^2 - 4\right)^{-3/2}\left(t^2 - 4 - t^2\right) = -\frac{4}{\left(t^2 - 4\right)^{3/2}}.$$

8. Because $f''(t) < 0$ for all t in the domain of f, we see that f is concave downward everywhere. From the results of step 7, we see that there is no inflection point.

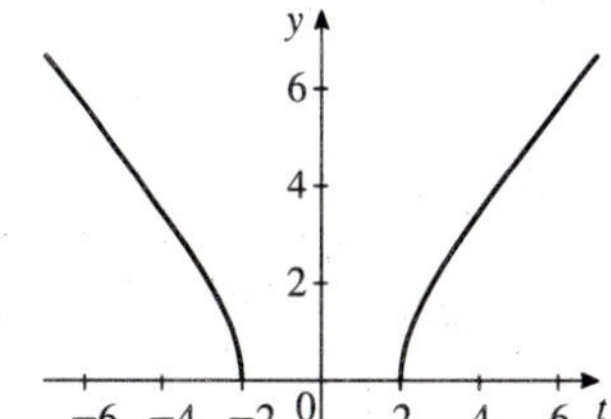

47. $g(x) = \frac{1}{2}x - \sqrt{x}$. We first gather the following information on g.

1. The domain of g is $[0, \infty)$.

2. The y-intercept is 0. To find the x-intercept(s), set $y = 0$, giving $\frac{1}{2}x - \sqrt{x} = 0$, $x = 2\sqrt{x}$, $x^2 = 4x$, $x(x - 4) = 0$, and so $x = 0$ or $x = 4$.

3. $\lim_{x \to \infty}\left(\frac{1}{2}x - \sqrt{x}\right) = \lim_{x \to \infty} \frac{1}{2}x\left(1 - \frac{2}{\sqrt{x}}\right) = \infty$.

4. There is no asymptote.

5. $g'(x) = \frac{1}{2} - \frac{1}{2}x^{-1/2} = \frac{1}{2}x^{-1/2}\left(x^{1/2} - 1\right) = \dfrac{\sqrt{x} - 1}{2\sqrt{x}}$, which is zero when $x = 1$. From the sign diagram for g', we see that g is decreasing on $(0, 1)$ and increasing on $(1, \infty)$.

− − − 0 + + + sign of g'

0 1 x

6. From the results of part 5, we see that $g(1) = -\frac{1}{2}$ is a relative minimum.

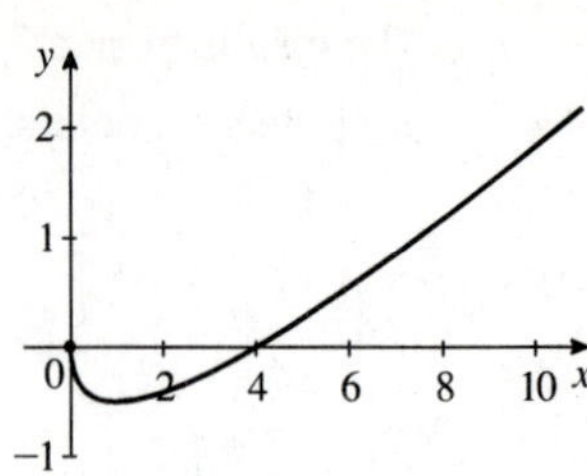

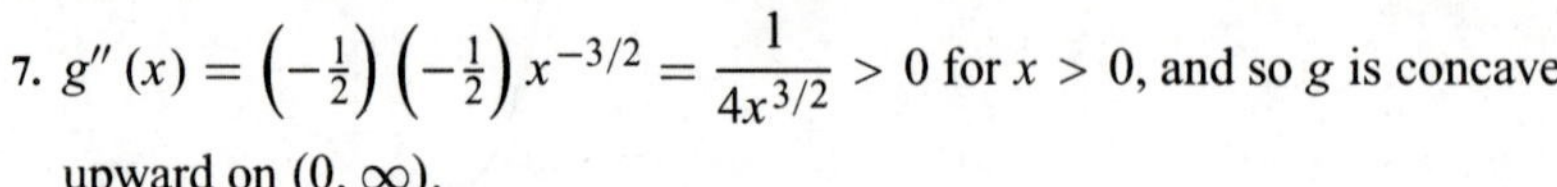

7. $g''(x) = \left(-\frac{1}{2}\right)\left(-\frac{1}{2}\right)x^{-3/2} = \dfrac{1}{4x^{3/2}} > 0$ for $x > 0$, and so g is concave upward on $(0, \infty)$.

8. There is no inflection point.

49. $g(x) = \dfrac{2}{x-1}$. We first gather the following information on g.

1. The domain of g is $(-\infty, 1) \cup (1, \infty)$.

2. Setting $x = 0$ gives -2 as the y-intercept. There is no x-intercept because $\dfrac{2}{x-1} \neq 0$ for all x.

3. $\lim_{x\to-\infty} \dfrac{2}{x-1} = 0$ and $\lim_{x\to\infty} \dfrac{2}{x-1} = 0$.

4. The results of step 3 show that $y = 0$ is a horizontal asymptote. Furthermore, the denominator of $g(x)$ is equal to zero at $x = 1$ but the numerator is not equal to zero there. Therefore, $x = 1$ is a vertical asymptote.

5. $g'(x) = -2(x-1)^{-2} = -\dfrac{2}{(x-1)^2} < 0$ for all $x \neq 1$ and so g is decreasing on $(-\infty, 1)$ and $(1, \infty)$.

6. Because g has no critical number, there is no relative extremum.

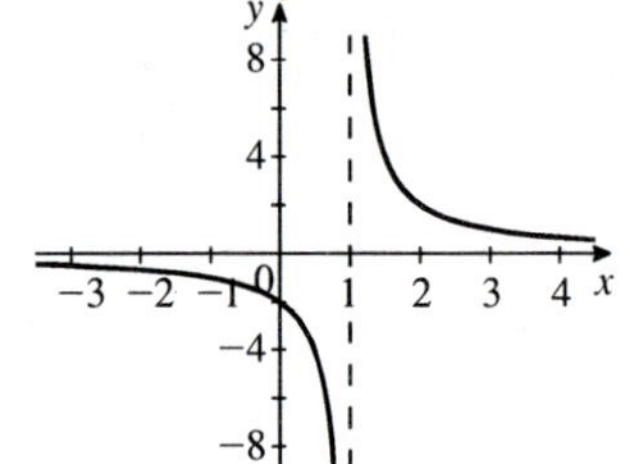

7. $g''(x) = \dfrac{4}{(x-1)^3}$ and so $g''(x) < 0$ if $x < 1$ and $g''(x) > 0$ if $x > 1$.
Therefore, the graph of g is concave downward on $(-\infty, 1)$ and concave upward on $(1, \infty)$.

8. Because g is undefined at $x = 1$, we see that g has no inflection point.

51. $h(x) = \dfrac{x+2}{x-2}$. We first gather the following information on h.

1. The domain of h is $(-\infty, 2) \cup (2, \infty)$.

2. Setting $x = 0$ gives $y = -1$ as the y-intercept. Next, setting $y = 0$ gives $x = -2$ as the x-intercept.

3. $\lim_{x\to\infty} h(x) = \lim_{x\to-\infty} \dfrac{1+\frac{2}{x}}{1-\frac{2}{x}} = \lim_{x\to-\infty} h(x) = 1.$

4. Setting $x - 2 = 0$ gives $x = 2$. Furthermore, $\lim_{x\to2^+} \dfrac{x+2}{x-2} = \infty$ and $\lim_{x\to2^-} \dfrac{x+2}{x-2} = -\infty$, so $x = 2$ is a vertical asymptote of h. Also, from the results of step 3, we see that $y = 1$ is a horizontal asymptote of h.

5. $h'(x) = \dfrac{(x-2)(1)-(x+2)(1)}{(x-2)^2} = -\dfrac{4}{(x-2)^2}$. We see that h has no critical number. (Note that $x = 2$ is not in the domain of h.) The sign diagram of h' shows that h is decreasing on $(-\infty, 2)$ and $(2, \infty)$.

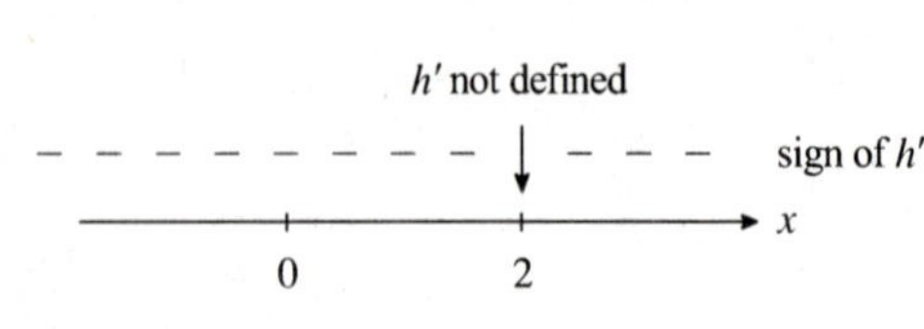

6. From the results of step 5, we see that there is no relative extremum.

7. $h''(x) = \dfrac{8}{(x-2)^3}$. Note that $x = 2$ is not a candidate for an inflection point because $h(2)$ is not defined. Because $h''(x) < 0$ for $x < 2$ and $h''(x) > 0$ for $x > 2$, we see that h is concave downward on $(-\infty, 2)$ and concave upward on $(2, \infty)$.

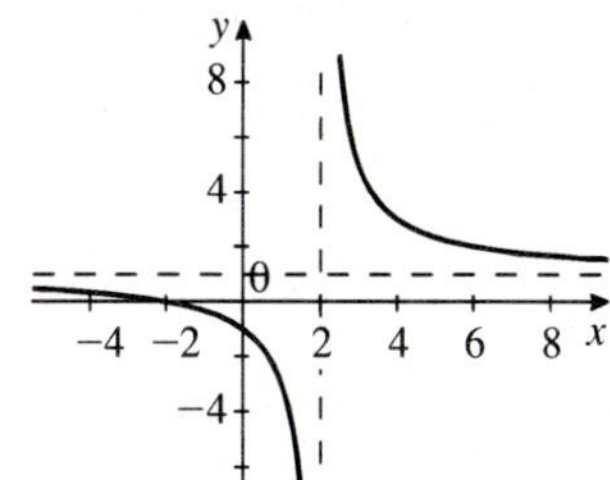

8. From the results of step 7, we see that there is no inflection point.

53. $f(t) = \dfrac{t^2}{1+t^2}$. We first gather the following information on f.

1. The domain of f is $(-\infty, \infty)$.
2. Setting $t = 0$ gives the y-intercept as 0. Similarly, setting $y = 0$ gives the t-intercept as 0.
3. $\lim_{t\to-\infty} \dfrac{t^2}{1+t^2} = \lim_{t\to\infty} \dfrac{t^2}{1+t^2} = 1$.
4. The results of step 3 show that $y = 1$ is a horizontal asymptote. There is no vertical asymptote since the denominator is never zero.
5. $f'(t) = \dfrac{(1+t^2)(2t) - t^2(2t)}{(1+t^2)^2} = \dfrac{2t}{(1+t^2)^2} = 0$, if $t = 0$, the only critical number of f. Because $f'(t) < 0$ if $t < 0$ and $f'(t) > 0$ if $t > 0$, we see that f is decreasing on $(-\infty, 0)$ and increasing on $(0, \infty)$.
6. The results of step 5 show that $(0, 0)$ is a relative minimum.
7. $f''(t) = \dfrac{(1+t^2)^2(2) - 2t(2)(1+t^2)(2t)}{(1+t^2)^4} = \dfrac{2(1+t^2)\left[(1+t^2) - 4t^2\right]}{(1+t^2)^4} = \dfrac{2(1-3t^2)}{(1+t^2)^3} = 0$ if $t = \pm\frac{\sqrt{3}}{3}$.

   The sign diagram of f'' shows that f is concave downward on $\left(-\infty, -\frac{\sqrt{3}}{3}\right)$ and $\left(\frac{\sqrt{3}}{3}, \infty\right)$ and concave upward on $\left(-\frac{\sqrt{3}}{3}, \frac{\sqrt{3}}{3}\right)$.

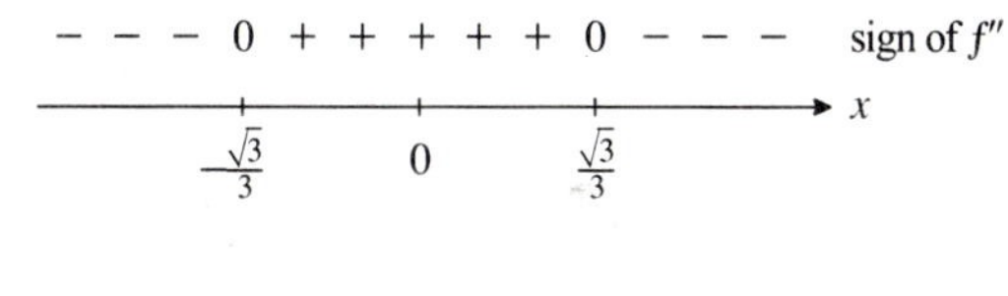

8. The results of step 7 show that $\left(-\frac{\sqrt{3}}{3}, \frac{1}{4}\right)$ and $\left(\frac{\sqrt{3}}{3}, \frac{1}{4}\right)$ are inflection points.

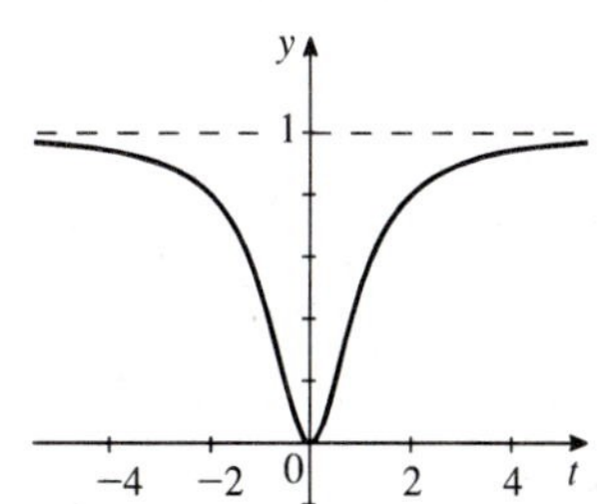

55. $f(t) = e^t - t$. We first gather the following information on f.

1. The domain of f is $(-\infty, \infty)$.
2. Setting $t = 0$ gives 1 as the y-intercept.
3. $\lim_{t\to-\infty} (e^t - t) = \infty$ and $\lim_{t\to\infty} (e^t - t) = \infty$.
4. There is no asymptote.
5. $f'(t) = e^t - 1$ if $t = 0$, a critical point of f. From the sign diagram for f', we see that f is decreasing on $(-\infty, 0)$ and increasing on $(0, \infty)$.

− − − − 0 + + + + sign of f'

0 (t axis)

6. From the results of part 5, we see that $(0, 1)$ is a relative minimum of f.

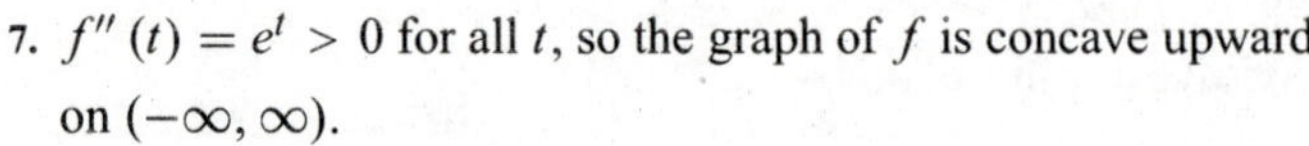

7. $f''(t) = e^t > 0$ for all t, so the graph of f is concave upward on $(-\infty, \infty)$.

8. There is no inflection point.

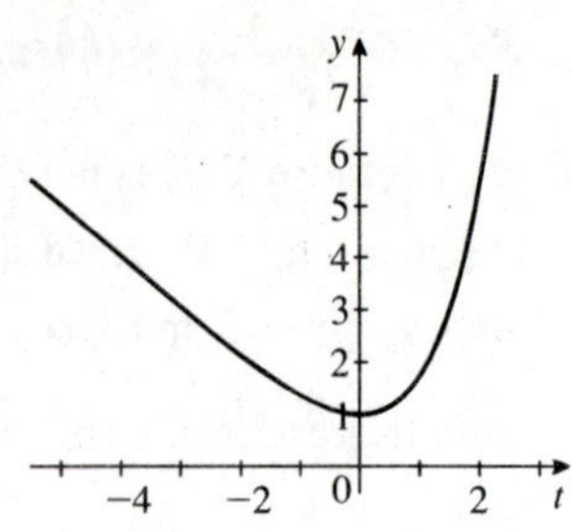

57. $f(x) = 2 - e^{-x}$. We first gather the following information on f.

1. The domain of f is $(-\infty, \infty)$.

2. Setting $x = 0$ gives 1 as the y-intercept.

3. $\lim_{x\to-\infty} \left(2 - e^{-x}\right) = -\infty$ and $\lim_{x\to\infty} \left(2 - e^{-x}\right) = 2$.

4. From the results of part 3, we see that $y = 2$ is a horizontal asymptote of f.

5. $f'(x) = e^{-x} > 0$ for all x in $(-\infty, \infty)$, so f is increasing on $(-\infty, \infty)$.

6. Because there is no critical point, f has no relative extremum.

7. $f''(x) = -e^{-x} < 0$ for all x in $(-\infty, \infty)$ and so the graph of f is concave downward on $(-\infty, \infty)$.

8. There is no inflection point.

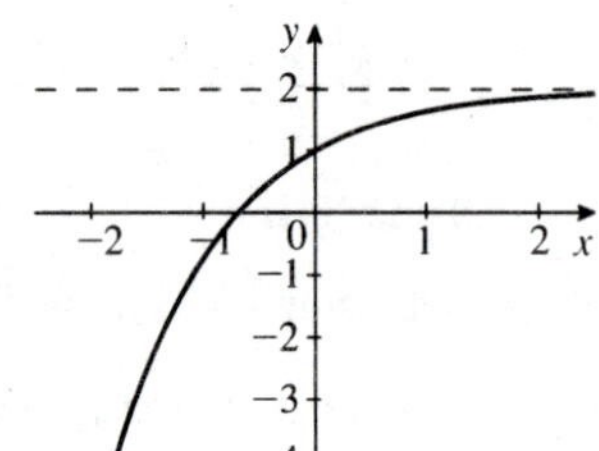

59. $f(x) = \ln(x - 1)$. We first gather the following information on f.

1. The domain of f is obtained by requiring that $x - 1 > 0$. We find the domain to be $(1, \infty)$.

2. Because $x \neq 0$, there is no y-intercept. Next, setting $y = 0$ gives $x - 1 = 1$, so the x-intercept is 2.

3. $\lim_{x\to1^+} \ln(x - 1) = -\infty$.

4. There is no horizontal asymptote. Because $\lim_{x\to1^+} \ln(x - 1) = -\infty$, $x = 1$ is a vertical asymptote.

5. $f'(x) = \dfrac{1}{x - 1} > 0$ for $x > 1$, so f has no critical number.

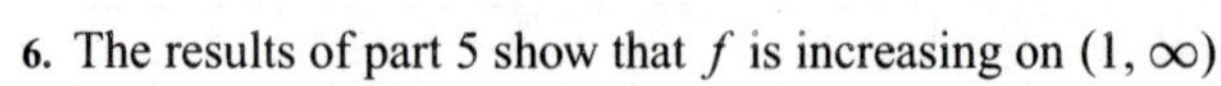

6. The results of part 5 show that f is increasing on $(1, \infty)$.

7. $f''(x) = -\dfrac{1}{(x - 1)^2}$. Because $f''(x) < 0$ for $x > 1$, we see that f is concave downward on $(1, \infty)$.

8. From the results of part 7, we see that f has no inflection point.

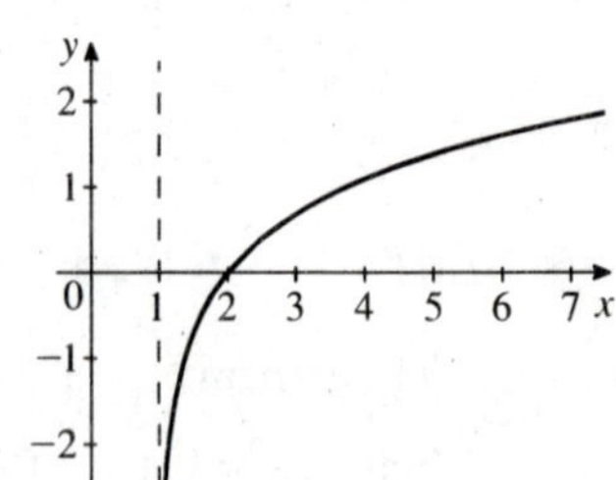

61. a. The denominator of $C(x)$ is equal to zero if $x = 100$. Also, $\lim_{x\to100^-} \dfrac{0.5x}{100 - x} = \infty$ and $\lim_{x\to100^+} \dfrac{0.5x}{100 - x} = -\infty$. Therefore, $x = 100$ is a vertical asymptote of C.

b. No, because the denominator is equal to zero in that case.

63. a. Because $\lim_{t\to\infty} C(t) = \lim_{t\to\infty} \frac{0.2t}{t^2+1} = \lim_{t\to\infty} \left(\frac{0.2}{t+\frac{1}{t^2}}\right) = 0$, $y = 0$ is a horizontal asymptote.

b. Our results reveal that as time passes, the concentration of the drug decreases and approaches zero.

65. $G(t) = -0.2t^3 + 2.4t^2 + 60$. We first gather the following information on G.

1. The domain of G is restricted to $[0, 8]$.
2. Setting $t = 0$ gives 60 as the y-intercept.

Step 3 is unnecessary in this case because of the restricted domain.

4. There is no asymptote because G is a polynomial function.
5. $G'(t) = -0.6t^2 + 4.8t = -0.6t(t-8) = 0$ if $t = 0$ or $t = 8$, critical numbers of G. But $G'(t) > 0$ on $(0, 8)$, so G is increasing on its domain.
6. The results of step 5 tell us that there is no relative extremum.
7. $G''(t) = -1.2t + 4.8 = -1.2(t-4) = 0$ if $t = 4$. The sign diagram of G'' shows that G is concave upward on $(0, 4)$ and concave downward on $(4, 8)$.

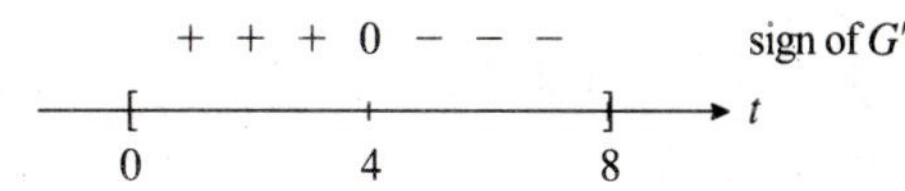

8. The results of step 7 show that $(4, 85.6)$ is an inflection point.

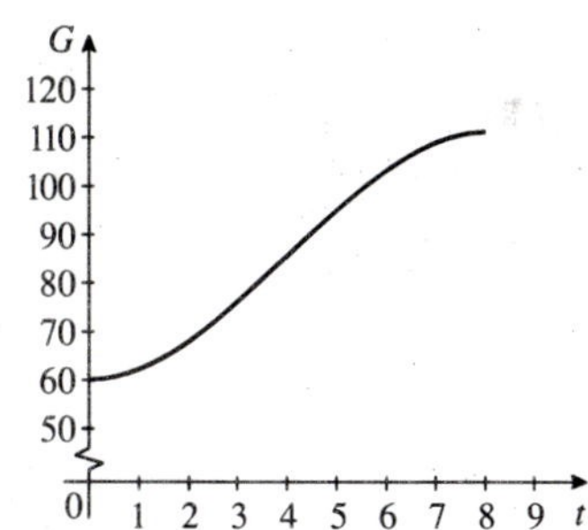

67. $N(t) = -\frac{1}{2}t^3 + 3t^2 + 10t$, $0 \le t \le 4$. We first gather the following information on N.

1. The domain of N is restricted to $[0, 4]$.
2. The x- and y-intercepts are 0.

Step 3 does not apply because the domain of $N(t)$ is $[0, 4]$.

4. There is no asymptote.
5. $N'(t) = -\frac{3}{2}t^2 + 6t + 10 = -\frac{1}{2}(3t^2 - 12t - 20) > 0$ on $(0, 4)$. Therefore, N is increasing on $(0, 4)$.
6. There is no relative extremum in $(0, 4)$.
7. $N''(t) = -3t + 6 = -3(t-2) = 0$ at $t = 2$. From the sign diagram of N'', we see that N is concave upward on $(0, 2)$ and concave downward on $(2, 4)$.

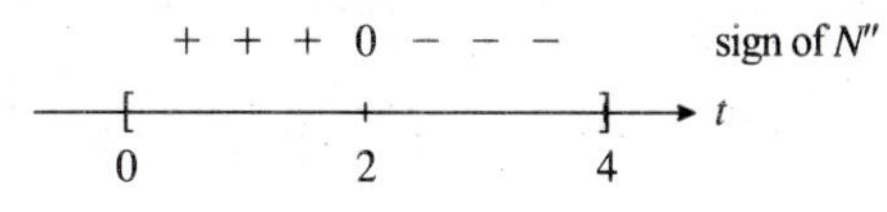

8. The point $(2, 28)$ is an inflection point.

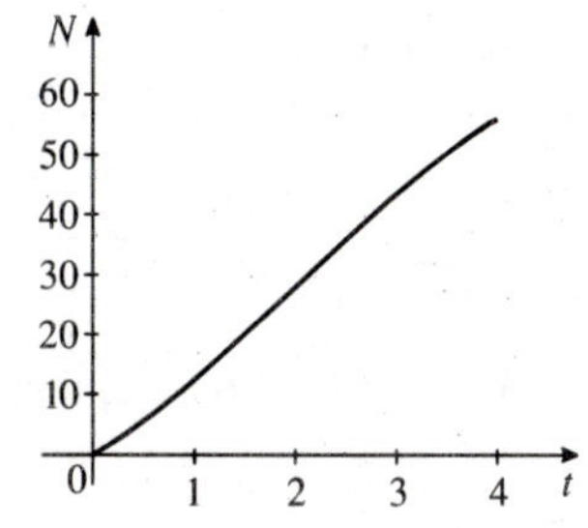

69. $T(x) = \dfrac{120x^2}{x^2+4}$. We first gather the following information on T.

1. The domain of T is $[0, \infty)$.

2. Setting $x = 0$ gives 0 as the y-intercept. If $y = 0$, then $x = 0$, so the x-intercept is also 0.

3. $\lim_{x\to\infty} \dfrac{120x^2}{x^2+4} = 120$.

4. The result of step 3 shows that $y = 120$ is a horizontal asymptote.

5. $T'(x) = 120\left[\dfrac{(x^2+4)\,2x - x^2(2x)}{(x^2+4)^2}\right] = \dfrac{960x}{(x^2+4)^2}$. Because $T'(x) > 0$ if $x > 0$, we see that T is increasing on $(0, \infty)$.

6. There is no relative extremum.

7. $T''(x) = 960\left[\dfrac{(x^2+4)^2 - x(2)(x^2+4)(2x)}{(x^2+4)^4}\right] = \dfrac{960(x^2+4)\left[(x^2+4) - 4x^2\right]}{(x^2+4)^4} = \dfrac{960(4-3x^2)}{(x^2+4)^3}$.

   The sign diagram for T'' shows that T is concave downward on $\left(\frac{2\sqrt{3}}{3}, \infty\right)$ and concave upward on $\left(0, \frac{2\sqrt{3}}{3}\right)$.

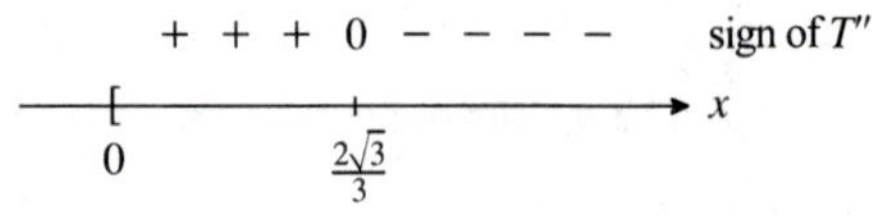

8. We see from the results of step 7 that $\left(\frac{2\sqrt{3}}{3}, 30\right)$ is an inflection point.

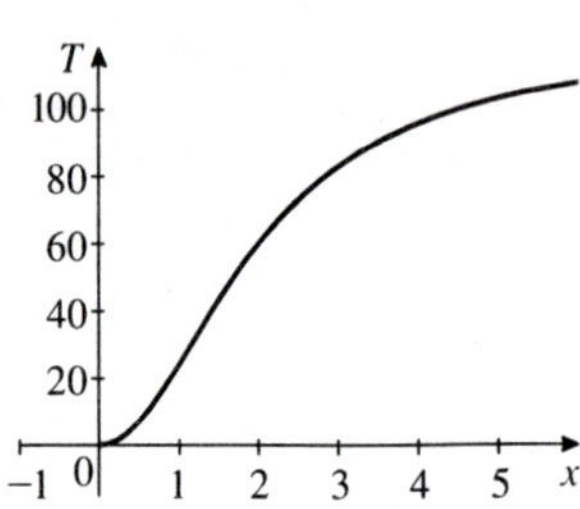

71. $C(x) = \dfrac{0.5x}{100-x}$. We first gather the following information on C.

1. The domain of C is $[0, 100)$.

2. Setting $x = 0$ gives the y-intercept as 0. Similarly, setting $y = 0$ gives 0 as the x-intercept.

3. $\lim_{x\to 100^-} \dfrac{0.5x}{100-x} = \infty$.

4. From the result of step 3, we see that $x = 100$ is a vertical asymptote.

5. $C'(x) = 0.5\left[\dfrac{(100-x)(1) - x(-1)}{(100-x)^2}\right] = \dfrac{50}{(100-x)^2} > 0$ for all x in the domain of C. Therefore C is increasing on $(0, 100)$.

6. There is no relative extremum.

7. $C''(x) = -\dfrac{100}{(100-x)^3}$, so $C''(x) > 0$ if $x < 100$ and the graph of C is concave upward on $(0, 100)$.

8. There is no inflection point.

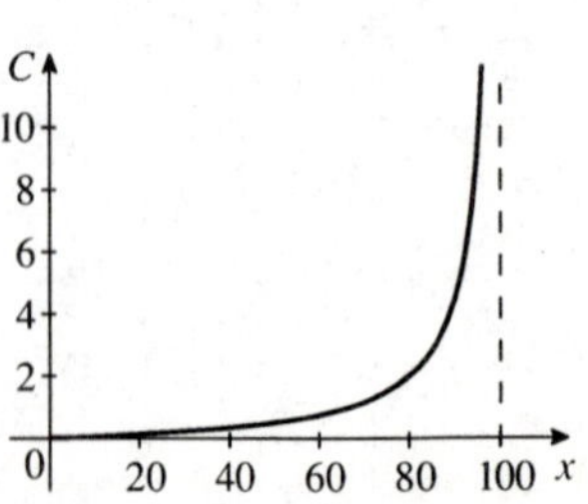

73. **a.** $N(0) = \dfrac{3000}{1+99} = 30.$

b. $N'(x) = 3000\dfrac{d}{dx}\left(1+99e^{-x}\right)^{-1} = -3000\left(1+99e^{-x}\right)^{-2}\left(-99e^{-x}\right) = \dfrac{297{,}000e^{-x}}{\left(1+99e^{-x}\right)^{2}}$. Because $N'(x) > 0$ for all x in $(0, \infty)$, we see that N is increasing on $(0, \infty)$.

c. From the graph of N, we see that the total number of students who contracted influenza during that particular epidemic is approximately $\lim\limits_{x\to\infty} \dfrac{3000}{1+99e^{-x}} = 3000.$

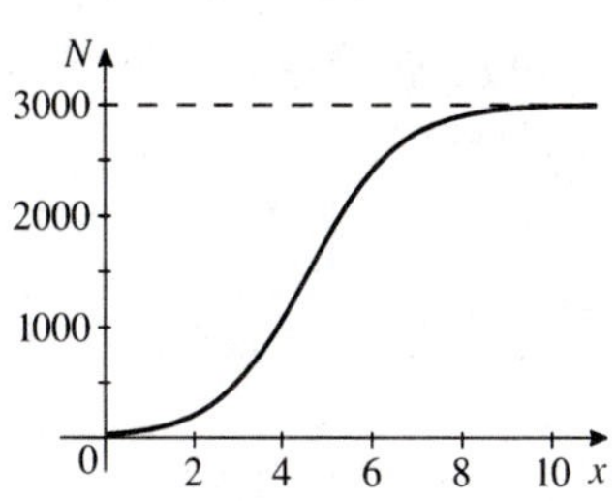

Using Technology page 725

1.

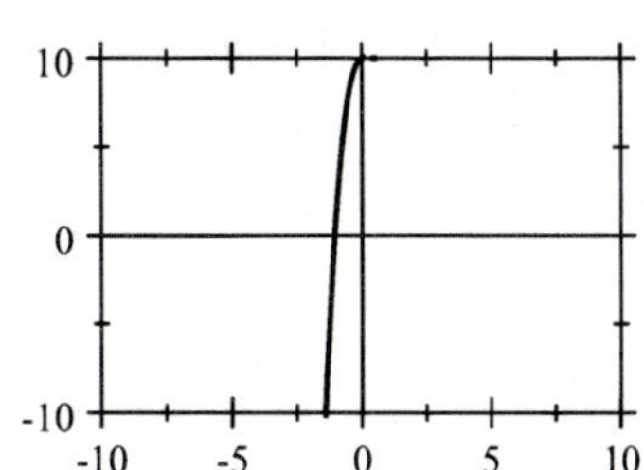

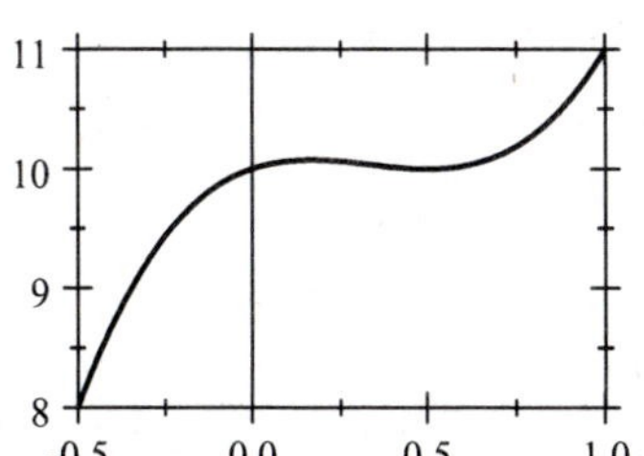

$f(x) = 4x^3 - 4x^2 + x + 10$, so $f'(x) = 12x^2 - 8x + 1 = (6x-1)(2x-1) = 0$ if $x = \frac{1}{6}$ or $x = \frac{1}{2}$. The second graph shows that f has a maximum at $x = \frac{1}{6}$ and a minimum at $x = \frac{1}{2}$.

3.

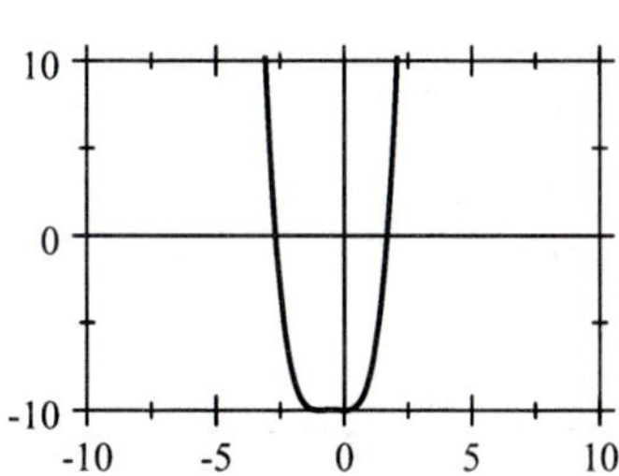

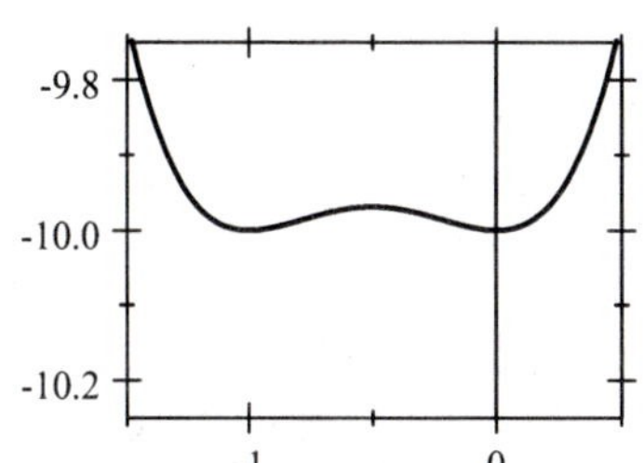

$f(x) = \frac{1}{2}x^4 + x^3 + \frac{1}{2}x^2 - 10$, so $f'(x) = 2x^3 + 3x^2 + x = x(x+1)(2x+1) = 0$ if $x = -1, -\frac{1}{2}$, or 0. The second graph shows that x has minima at $x = -1$ and $x = 0$ and a maximum at $x = -\frac{1}{2}$.

5. -0.9733, 2.3165, and 4.6569.

7. 1.5142

9. -0.7680, 1.6783

11.

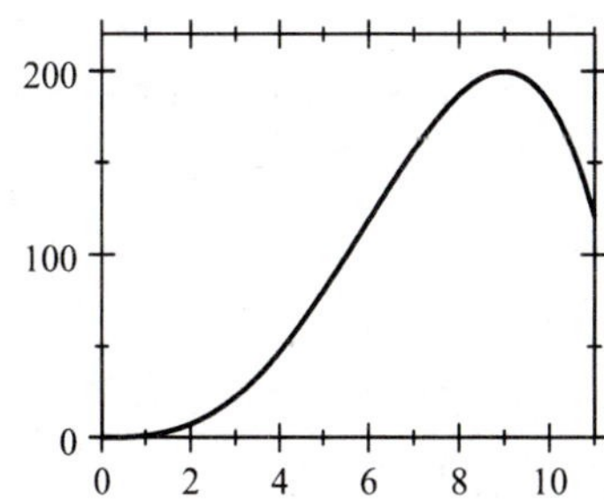

10.4 Optimization I

Problem-Solving Tips

1. **To determine the absolute maximum and absolute minimum** of a continuous function f on a closed interval $[a, b]$, find the critical numbers of f that lie in (a, b). Then compute the value of f at each critical number of f and compute $f(a)$ and $f(b)$. The largest and smallest of these values are the absolute maximum and absolute minimum values of f, respectively.

2. Note that the procedure in Tip 1 holds only for a *continuous* function f over a *closed* interval $[a, b]$.

Concept Questions page 733

1. **a.** A function f has an absolute maximum at a if $f(x) \le f(a)$ for all x in the domain of f.

   b. A function f has an absolute minimum at a if $f(x) \ge f(a)$ for all x in the domain of f.

Exercises page 733

1. f has no absolute extremum.

3. f has an absolute minimum at $(0, 0)$ and no absolute maximum.

5. f has an absolute minimum at $(0, -2)$ and an absolute maximum at $(1, 3)$.

7. f has an absolute minimum at $\left(\frac{3}{2}, -\frac{27}{16}\right)$ and an absolute maximum at $(-1, 3)$.

9. The graph of $f(x) = 2x^2 + 3x - 4$ is a parabola that opens upward. Therefore, the vertex of the parabola is the absolute minimum of f. To find the vertex, we solve the equation $f'(x) = 4x + 3 = 0$, finding $x = -\frac{3}{4}$. We conclude that the absolute minimum value is $f\left(-\frac{3}{4}\right) = -\frac{41}{8}$.

11. Because $\lim_{x\to-\infty} x^{1/3} = -\infty$ and $\lim_{x\to\infty} x^{1/3} = \infty$, we see that h is unbounded. Therefore, it has no absolute extremum.

13. $f(x) = \dfrac{1}{1+x^2}$. Using the techniques of graphing, we sketch the graph of f. The absolute maximum of f is $f(0) = 1$. Alternatively, observe that $1 + x^2 \ge 1$ for all real values of x. Therefore, $f(x) \le 1$ for all x, and we see that the absolute maximum is attained when $x = 0$.

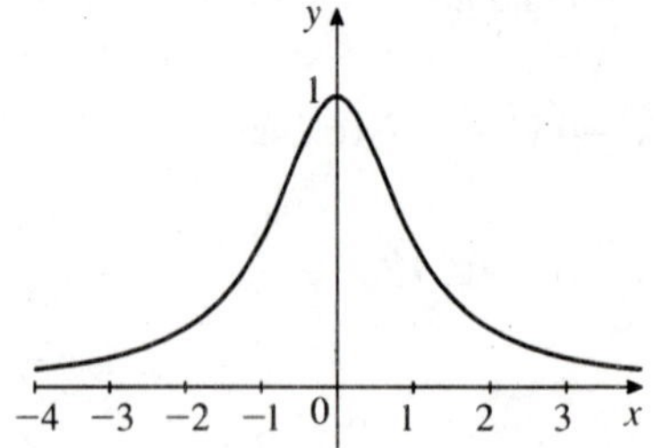

15. $f(x) = x^2 - 2x - 3$ and $f'(x) = 2x - 2 = 0$, so $x = 1$ is a critical number. From the table, we conclude that the absolute maximum value is $f(-2) = 5$ and the absolute minimum value is $f(1) = -4$.

x	-2	1	3
$f(x)$	5	-4	0

17. $f(x) = -x^2 + 4x + 10$; The function f is continuous and defined on the closed interval $[0, 5]$. $f'(x) = -2x + 4$, and so $x = 2$ is a critical number. From the table, we conclude that $f(2) = 14$ is the absolute maximum value and $f(5) = 5$ is the absolute minimum value.

x	0	2	5
$f(x)$	10	14	5

19. The function $f(x) = x^3 + 3x^2 - 1$ is continuous and defined on the closed interval $[-3, 2]$ and differentiable on $(-3, 2)$. The critical numbers of f are found by solving $f'(x) = 3x^2 + 6x = 3x(x+2) = 0$, giving $x = -2$ and $x = 0$. From the table, we see that the absolute maximum value of f is $f(2) = 19$ and the absolute minimum value is $f(-3) = f(0) = -1$.

x	-3	-2	0	2
$f(x)$	-1	3	-1	19

21. The function $g(x) = 3x^4 + 4x^3$ is continuous on the closed interval $[-2, 1]$ and differentiable on $(-2, 1)$. The critical numbers of g are found by solving $g'(x) = 12x^3 + 12x^2 = 12x^2(x+1) = 0$, giving $x = 0$ and $x = -1$. From the table, we see that $g(-2) = 16$ is the absolute maximum value of g and $g(-1) = -1$ is the absolute minimum value of g.

x	-2	-1	0	1
$g(x)$	16	-1	0	7

23. $f(x) = \dfrac{x+1}{x-1}$ on $[2, 4]$. Next, we compute $f'(x) = \dfrac{(x-1)(1) - (x+1)(1)}{(x-1)^2} = -\dfrac{2}{(x-1)^2}$. Because there is no critical number ($x = 1$ is not in the domain of f), we need only test the endpoints. We conclude that $f(4) = \frac{5}{3}$ is the absolute minimum value and $f(2) = 3$ is the absolute maximum value.

25. $f(x) = 4x + \dfrac{1}{x}$ is continuous on $[1, 2]$ and differentiable on $(1, 2)$. To find the critical numbers of f, we solve $f'(x) = 4 - \dfrac{1}{x^2} = 0$, obtaining $x = \pm\frac{1}{2}$. Because these critical numbers lie outside the interval $[1, 2]$, they are not candidates for the absolute extrema of f. Evaluating f at the endpoints of the interval $[1, 2]$, we find that the absolute maximum value of f is $f(2) = \frac{17}{2}$, and the absolute minimum value of f is $f(1) = 5$.

27. $f(x) = \frac{1}{2}x^2 - 2\sqrt{x} = \frac{1}{2}x^2 - 2x^{1/2}$. To find the critical numbers of f, we solve $f'(x) = x - x^{-1/2} = 0$, or $x^{3/2} - 1 = 0$, obtaining $x = 1$. From the table, we conclude that $f(3) \approx 1.04$ is the absolute maximum value and $f(1) = -\frac{3}{2}$ is the absolute minimum value.

x	0	1	3
$f(x)$	0	$-\frac{3}{2}$	$\frac{9}{2} - 2\sqrt{3} \approx 1.04$

29.

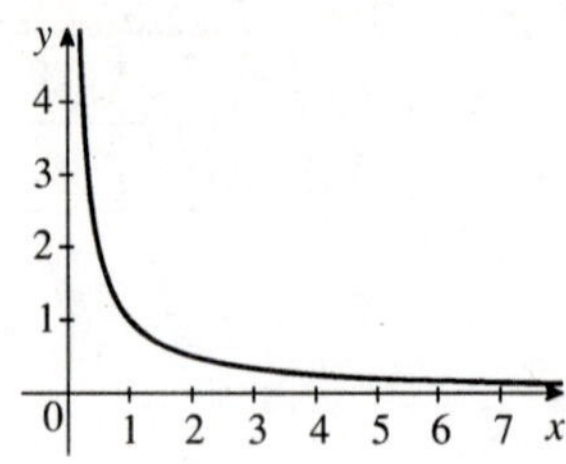

From the graph of $f(x) = \dfrac{1}{x}$ for $x > 0$, we conclude that f has no absolute extremum.

31. $f(x) = 3x^{2/3} - 2x$. The function f is continuous on $[0, 3]$ and differentiable on $(0, 3)$. To find the critical numbers of f, we solve $f'(x) = 2x^{-1/3} - 2 = 0$, obtaining $x = 1$ as the critical number. From the table, we conclude that the absolute maximum value is $f(1) = 1$ and the absolute minimum value is $f(0) = 0$.

x	0	1	3
$f(x)$	0	1	$3^{5/3} - 6 \approx 0.24$

33. $f(x) = x^{2/3}(x^2 - 4)$, so $f'(x) = x^{2/3}(2x) + \frac{2}{3}x^{-1/3}(x^2 - 4) = \frac{2}{3}x^{-1/3}\left[3x^2 + (x^2 - 4)\right] = \dfrac{8(x^2 - 1)}{3x^{1/3}} = 0$. Observe that f' is not defined at $x = 0$. Furthermore, $f'(x) = 0$ at $x \pm 1$. So the critical numbers of f are -1 and 0, and 1. From the table, we see that f has absolute minima at $(-1, -3)$ and $(1, -3)$ and absolute maxima at $(0, 0)$ and $(2, 0)$.

x	-1	0	1	2
$f(x)$	-3	0	-3	0

35. $f(x) = 2e^{-x^2}$, so $f'(x) = -4xe^{-x^2} = 0$ if $x = 0$, the only critical point of f. From the table, we see that f has an absolute minimum value of $2e^{-1}$ attained at $x = -1$ and $x = 1$. It has an absolute maximum at $(0, 2)$.

x	-1	0	1
$f(x)$	$2e^{-1}$	2	$2e^{-1}$

37. $g(x) = (2x - 1)e^{-x}$, so

$g'(x) = 2e^{-x} + (2x - 1)e^{-x}(-1) = (3 - 2x)e^{-x} = 0$ if $x = \frac{3}{2}$.

The graph of g shows that $\left(\frac{3}{2}, 2e^{-3/2}\right)$ is an absolute maximum, and $(0, -1)$ is an absolute minimum.

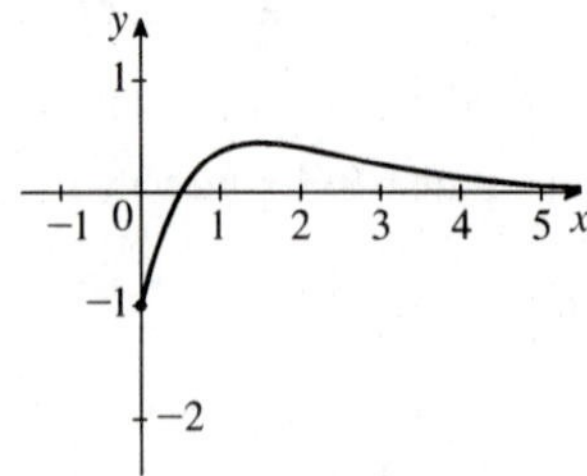

39. $f(x) = x - \ln x$, so $f'(x) = 1 - \dfrac{1}{x} = \dfrac{x-1}{x} = 0$ if $x = 1$, a critical point of f. From the table, we see that f has an absolute minimum at $(1, 1)$ and an absolute maximum at $(3, 3 - \ln 3)$.

x	$\frac{1}{2}$	1	3
$f(x)$	$\frac{1}{2} + \ln 2$	1	$3 - \ln 3$

41. $h(t) = -16t^2 + 64t + 80$. To find the maximum value of h, we solve $h'(t) = -32t + 64 = -32(t-2) = 0$, giving $t = 2$ as the critical number of h. Furthermore, this value of t gives rise to the absolute maximum value of h since the graph of h is a parabola that opens downward. The maximum height is given by $h(2) = -16(4) + 64(2) + 80 = 144$, or 144 feet.

43. $P(t) = 0.0135t^2 - 1.126t + 41.2$, so $P'(t) = 0.027t - 1.126 = 0$ implies $t \approx 41.7$, a critical number of P. $P''(t) = 0.027$ and $P''(41.7) = 0.027 > 0$, so $t = 41.7$ gives a minimum of P. This occurred around September of 1991. $P(41.7) \approx 17.72$, or 17.72%.

45. $N(t) = 0.81t - 1.14\sqrt{t} + 1.53$, so $N'(t) = 0.81 - 1.14\left(\frac{1}{2}t^{-1/2}\right) = 0.81 - \dfrac{0.57}{t^{1/2}}$. Setting $N'(t) = 0$ gives $t^{1/2} = \dfrac{0.57}{0.81}$, so $t \approx 0.4952$ is a critical number of N. Evaluating $N(t)$ at the endpoints $t = 0$ and $t = 6$ as well as at the critical number, we see that the absolute maximum of N occurs at $t = 6$ and the absolute minimum occurs at $t \approx 0.5$. Our results tell us that the number of nonfarm full-time self-employed women over the time interval from 1963 to 1993 reached its maximum of approximately 3.6 million in 1993.

t	0	0.4952	6
$N(t)$	1.53	1.13	3.60

47. $P(x) = -0.000002x^3 + 6x - 400$, so $P'(x) = -0.000006x^2 + 6 = 0$ if $x = \pm 1000$. We reject the negative root. Next, we compute $P''(x) = -0.000012x$. Because $P''(1000) = -0.012 < 0$, the Second Derivative Test shows that $x = 1000$ gives a relative maximum of f. From physical considerations, or from a sketch of the graph of f, we see that the maximum profit is realized if 1000 cases are produced per day. That profit is $P(1000) = -0.000002(1000)^3 + 6(1000) - 400$, or \$3600/day.

49. The revenue is $R(x) = px = -0.0004x^2 + 10x$ and the profit is $P(x) = R(x) - C(x) = -0.0004x^2 + 10x - \left(400 + 4x + 0.0001x^2\right) = -0.0005x^2 + 6x - 400$. $P'(x) = -0.001x + 6 = 0$ if $x = 6000$, a critical number. Because $P''(x) = -0.001 < 0$ for all x, we see that the graph of P is a parabola that opens downward. Therefore, a level of production of 6000 rackets/day will yield a maximum profit.

51. The cost function is $C(x) = V(x) + 20{,}000 = 0.000001x^3 - 0.01x^2 + 50x + 20{,}000$, so the profit function is

$$P(x) = R(x) - C(x) = -0.02x^2 + 150x - 0.000001x^3 + 0.01x^2 - 50x - 20{,}000$$
$$= -0.000001x^3 - 0.01x^2 + 100x - 20{,}000.$$

We want to maximize P on $[0, 7500]$. $P'(x) = -0.000003x^2 - 0.02x + 100$. Setting $P'(x) = 0$ gives $3x^2 + 20{,}000x - 100{,}000{,}000 = 0$, so or $x = \dfrac{-20{,}000 \pm \sqrt{20{,}000^2 + 1{,}200{,}000{,}000}}{6} = -10{,}000$ or 3,333.33.

Thus, $x = 3333.33$ is a critical number in the interval $[0, 7500]$. From the table, we see that a level of production of 3,333 pagers per week will yield a maximum profit of \$165,185.20 per week.

x	0	3333.33	7500
$P(x)$	−20,000	165,185.2	−254,375

53. a. $\overline{C}(x) = \dfrac{C(x)}{x} = 0.0025x + 80 + \dfrac{10{,}000}{x}$.

b. $\overline{C}'(x) = 0.0025 - \dfrac{10{,}000}{x^2} = 0$ if $0.0025x^2 = 10{,}000$, or $x = 2000$. Because $\overline{C}''(x) = \dfrac{20{,}000}{x^3}$, we see that $\overline{C}''(x) > 0$ for $x > 0$ and so $\overline{C}$ is concave upward on $(0, \infty)$. Therefore, $x = 2000$ yields a minimum.

c. We solve $\overline{C}(x) = C'(x)$: $0.0025x + 80 + \dfrac{10{,}000}{x} = 0.005x + 80$, so $0.0025x^2 = 10{,}000$ and $x = 2000$.

d. It appears that we can solve the problem in two ways.

55. $\overline{C}(x) = \dfrac{C(x)}{x}$, so $\overline{C}'(x) = \dfrac{xC'(x) - C(x)}{x^2} = 0$. This implies that $xC'(x) - C(x) = x^2$, so $\dfrac{C(x)}{x} = C'(x)$. This shows that at a level of production where the average cost is minimized, the average cost $\dfrac{C(x)}{x}$ is equal to the marginal cost $C'(x)$.

57. The demand equation is $p = \sqrt{800 - x} = (800 - x)^{1/2}$, so the revenue function is $R(x) = xp = x(800 - x)^{1/2}$. To find the maximum of R, we compute

$$R'(x) = \tfrac{1}{2}(800 - x)^{-1/2}(-1)(x) + (800 - x)^{1/2} = \tfrac{1}{2}(800 - x)^{-1/2}[-x + 2(800 - x)]$$
$$= \tfrac{1}{2}(800 - x)^{-1/2}(1600 - 3x).$$

Next, $R'(x) = 0$ implies $x = 800$ or $x = \frac{1600}{3}$, the critical numbers of R.

From the table, we conclude that $R\left(\frac{1600}{3}\right) \approx 8709$ is the absolute maximum value. Therefore, the revenue is maximized by producing $\frac{1600}{3} \approx 533$ dresses.

x	0	$\frac{1600}{3}$	800
$R(x)$	0	8709	0

59. $P(t) = 80{,}000e^{\sqrt{t}/2 - 0.09t} = 80{,}000e^{(1/2)t^{1/2} - 0.09t}$, so $P'(t) = 80{,}000\left(\frac{1}{4}t^{-1/2} - 0.09\right)e^{(1/2)t^{1/2} - 0.09t}$. Setting $P'(t) = 0$, we have $\frac{1}{4}t^{-1/2} = 0.09$, so $t^{-1/2} = 0.36$, $\dfrac{1}{\sqrt{t}} = 0.36$, and $t = \left(\frac{1}{0.36}\right)^2 \approx 7.72$. Evaluating $P(t)$ at each of its endpoints and at the point $t = 7.72$, we find $P(0) = 80{,}000$, $P(7.72) \approx 160{,}207.69$, and $P(8) \approx 160{,}170.71$. We conclude that P is optimized at $t = 7.72$. The optimal price is approximately $160,208.

61. $f(t) = 100\left(\dfrac{t^2 - 4t + 4}{t^2 + 4}\right)$.

a. $f'(t) = 100\left[\dfrac{(t^2 + 4)(2t - 4) - (t^2 - 4t + 4)(2t)}{(t^2 + 4)^2}\right] = \dfrac{400(t^2 - 4)}{(t^2 + 4)^2} = \dfrac{400(t - 2)(t + 2)}{(t^2 + 4)^2}$.

From the sign diagram for f', we see that $t = 2$ gives a relative minimum, and we conclude that the oxygen content is the lowest 2 days after the organic waste has been dumped into the pond.

- - - 0 + + + sign of f'
0 2 t

b. $f''(t) = 400\left[\dfrac{(t^2+4)^2(2t) - (t^2-4)\,2(t^2+4)(2t)}{(t+4)^4}\right] = 400\left[\dfrac{(2t)(t^2+4)(t^2+4-2t^2+8)}{(t^2+4)^4}\right]$

$= -\dfrac{800t(t^2-12)}{(t^2+4)^3}.$

$f''(t) = 0$ when $t = 0$ and $t = \pm 2\sqrt{3}$. We reject $t = 0$ and $t = -2\sqrt{3}$. From the sign diagram for f'', we see that $t = 2\sqrt{3}$ gives an inflection point of f and we conclude that this is an absolute maximum. Therefore, the rate of oxygen regeneration is greatest 3.5 days after the organic waste has been dumped into the pond.

0 + + + 0 − − − sign of f''

t: 0, $2\sqrt{3}$

63. We compute $\overline{R}'(x) = \dfrac{xR'(x) - R(x)}{x^2}$. Setting $\overline{R}'(x) = 0$ gives $xR'(x) - R(x) = 0$, or $R'(x) = \dfrac{R(x)}{x} = \overline{R}(x)$, so a critical number of $\overline{R}$ occurs when $\overline{R}(x) = R'(x)$. Next, we compute $\overline{R}''(x) = \dfrac{x^2\left[R'(x) + xR''(x) - R'(x)\right] - \left[xR'(x) - R(x)\right](2x)}{x^4} = \dfrac{R''(x)}{x} < 0$. Thus, by the Second Derivative Test, the critical number does give the maximum revenue.

65. $G(t) = -0.2t^3 + 2.4t^2 + 60$, so the growth rate is $G'(t) = -0.6t^2 + 4.8t$. To find the maximum growth rate, we compute $G''(t) = -1.2t + 4.8$. Setting $G''(t) = 0$ gives $t = 4$ as a critical number. From the table, we see that G' is maximal at $t = 4$; that is, the growth rate is greatest in 2008.

t	0	4	8
$G'(t)$	0	9.6	0

67. $P(t) = 0.04363t^3 - 0.267t^2 - 1.59t + 14.7$, so $P'(t) = 0.13089t^2 - 0.534t - 1.59 = 0$ gives

$t = \dfrac{0.534 \pm \sqrt{(0.534)^2 - 4(0.13089)(-1.59)}}{2(0.13089)} \approx -2$ or 6.08. We reject the negative root.

From the table, we see that P has an absolute minimum when $t \approx 6.08$, and this number corresponds to approximately early 1970.

t	0	6.08	9
$P(t)$	14.7	4.97	10.57

69. $A'(t) = \dfrac{d}{dt}\left(-0.00005t^3 - 0.000826t^2 + 0.0153t + 4.55\right) = -0.00015t^2 - 0.001652t + 0.0153$. Using the quadratic formula to solve $f'(t) = 0$ with $a = -0.00015$, $b = -0.001652$, and $c = 0.0153$, we have

$t = \dfrac{-(-0.001652) \pm \sqrt{(-0.001652)^2 - 4(-0.00015)(0.0153)}}{2(-0.00015)} \approx -17.01$ or 5.997.

We disregard the negative root. From the table, we see that A has an absolute maximum when $t \approx 6$, so the cortex of children of average intelligence reaches a maximum thickness around the time the children are 6 years old.

t	5	6	19
$A(t)$	4.5996	4.601	4.200

71. a. On $[0, 3]$, $C(t) = 0.6t^2 + 2.4t + 7.6$, so $C'(t) = 1.2t + 2.4 = 0$ implies $t = -2$ which lies outside the interval $[0, 3]$. (We evaluate C at each relevant point below.)

On $[3, 5]$, $C(t) = 3t^2 + 18.8t - 63.2$, so $C'(t) = 6t + 18.8 = 0$ implies $t \approx -3.13$ which lies outside the interval $[3, 5]$.

On $[5, 8]$, $C(t) = -3.3167t^3 + 80.1t^2 - 642.583t + 1730.8025$, so $C'(t) = -9.9501t^2 + 160.2t - 642.583 = 0$ implies $t = \dfrac{-160.2 \pm \sqrt{160.2^2 - 4(-9.9501)(-642.583)}}{2(-9.9501)} \approx 7.58$ or 8.52. Only the critical number $t = 7.58$ lies

inside the interval $[5, 8]$.

From the table, we see that the investment peaked when $t = 5$, that is, in the year 2000. The amount invested was \$105.8 billion.

t	0	3	5	7.58	8
$C(t)$	7.6	20.2	105.8	17.8	18.4

b. Investment was lowest (at \$7.6 billion) when $t = 0$, that is, in 1995.

73. $R = D^2\left(\dfrac{k}{2} - \dfrac{D}{3}\right) = \dfrac{kD^2}{2} - \dfrac{D^3}{3}$, so $\dfrac{dR}{dD} = \dfrac{2kD}{2} - \dfrac{3D^2}{3} = kD - D^2 = D(k - D)$. Setting $\dfrac{dR}{dD} = 0$, we have $D = 0$ or $k = D$. We consider only $k = D$ because $D > 0$. If $k > D$, $\dfrac{dR}{dD} > 0$ and if $k < D$, $\dfrac{dR}{dD} < 0$. Therefore $k = D$ gives a relative maximum. The nature of the problem suggests that $k = D$ gives the absolute maximum of R. We can also verify this by graphing R.

75. Setting $P' = 0$ gives $P' = \dfrac{d}{dR}\left[\dfrac{E^2R}{(R+r)^2}\right] = E^2\left[\dfrac{(R+r)^2 - R(2)(R+r)}{(R+r)^4}\right] = \dfrac{E^2(r-R)}{(R+r)^3} = 0$. Therefore, $R = r$ is a critical number of P. Because $P'' = E^2\dfrac{(R+r)^3(-1) - (r-R)(3)(R+r)^2}{(R+r)^6} = \dfrac{2E^2(R-2r)}{(R+r)^4}$ and $P''(r) = \dfrac{-2E^2r}{(2r)^4} = -\dfrac{E^2}{8r^3} < 0$, the Second Derivative Test and physical considerations both imply that $R = r$ gives a relative maximum value of P. The maximum power is $P = \dfrac{E^2r}{(2r)^2} = \dfrac{E^2}{4r}$ watts.

77. $R'(x) = \dfrac{d}{dx}[kx(Q-x)] = k\dfrac{d}{dx}(Qx - x^2) = k(Q - 2x)$ is continuous everywhere and has a zero at $\frac{1}{2}Q$; this is the only critical number of R in $(0, Q)$. $R(0) = 0$, $R\left(\frac{1}{2}Q\right) = \frac{1}{4}kQ^2$, and $R(Q) = 0$, so the absolute maximum value of R is $R\left(\frac{1}{2}Q\right) = \frac{1}{4}kQ^2$, showing that the rate of chemical reaction is greatest when exactly half of the original substrate has been transformed.

79. a. $A(t) = \begin{cases} 100e^{-1.4t} & \text{if } 0 \le t < 1 \\ 100\left(1 + e^{1.4}\right)e^{-1.4t} & \text{if } t \ge 1 \end{cases}$ so $A'(t) = \begin{cases} -140e^{-1.4t} & \text{if } 0 < t < 1 \\ -140\left(1 + e^{1.4}\right)e^{-1.4t} & \text{if } t > 1 \end{cases}$ Thus, after 12 hours the amount of drug is changing at the rate of $A'\left(\frac{1}{2}\right) = -140e^{-0.7} \approx -69.52$, or decreasing at the rate of 70 mg/day. After 2 days, it is changing at the rate of $A'(2) = -140\left(1 + e^{1.4}\right)e^{-2.8} \approx -43.04$, or decreasing at the rate of 43 mg/day.

b. From the graph of A, we see that the maximum occurs at $t = 1$, that is, at the time when she takes the second dose.

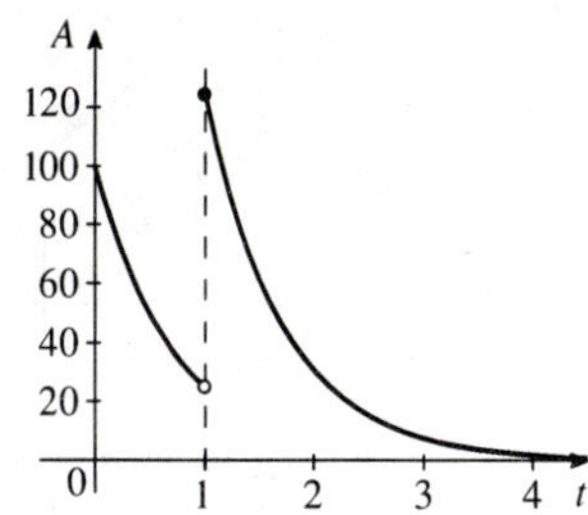

c. The maximum amount is
$A(1) = 100\left(1 + e^{1.4}\right)e^{-1.4} \approx 124.66$, or 125 mg.

81. False. Let $f(x) = \begin{cases} -|x| & \text{if } x \neq 0 \\ -1 & \text{if } x = 0 \end{cases}$ on $[-1, 1]$.

83. False. Let $f(x) = \begin{cases} -x & \text{if } -1 \leq x < 0 \\ \frac{1}{2} & \text{if } 0 \leq x \leq 1 \end{cases}$ Then f is discontinuous at $x = 0$, but f has an absolute maximum value of 1, attained at $x = -1$.

85. Because $f(x) = c$ for all x, the function f satisfies $f(x) \leq c$ for all x and so f has absolute maxima at all values of x. Similarly, f has absolute minima at all values of x.

87. **a.** f is not continuous at $x = 0$ because $\lim_{x \to 0} f(x)$ does not exist.

b. $\lim_{x \to 0^-} f(x) = \lim_{x \to 0^-} \frac{1}{x} = -\infty$ and $\lim_{x \to 0^+} f(x) = \lim_{x \to 0^+} \frac{1}{x} = \infty$.

c.

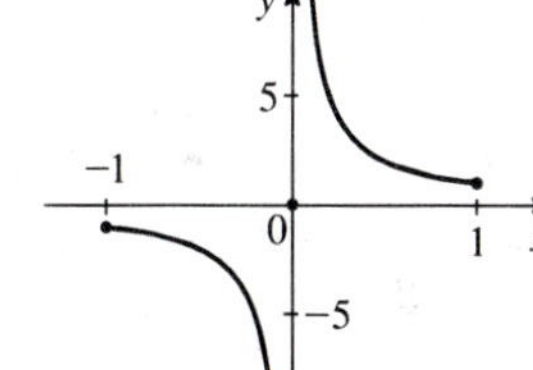

Using Technology page 739

1. Absolute maximum value 145.9, absolute minimum value -4.3834.

3. Absolute maximum value 16, absolute minimum value -0.1257.

5. Absolute maximum value 11.8922, absolute minimum value 0.

7. Absolute maximum value 2.8889, absolute minimum value 0.

9. a.

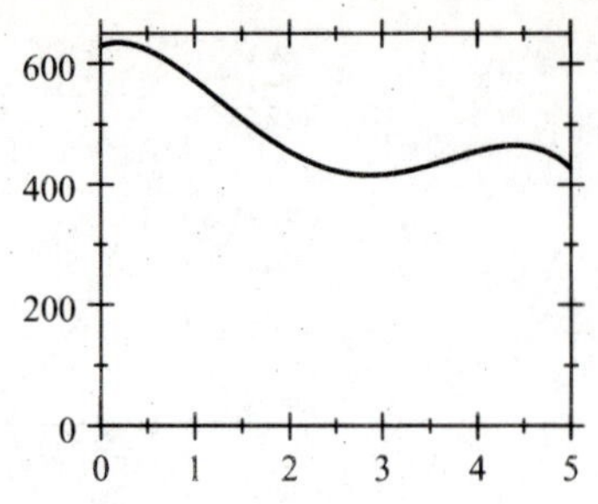

b. Using the function for finding the absolute minimum of f on $[0, 5]$, we see that the absolute minimum value of f is approximately 415.56, occurring when $t \approx 2.87$. This proves the assertion.

11. a. $N(t) = 1.2576t^4 - 26.357t^3 + 127.98t^2 + 82.3t + 43$, so $N'(t) = 5.0304t^3 - 79.071t^2 + 255.96t + 82.3$.

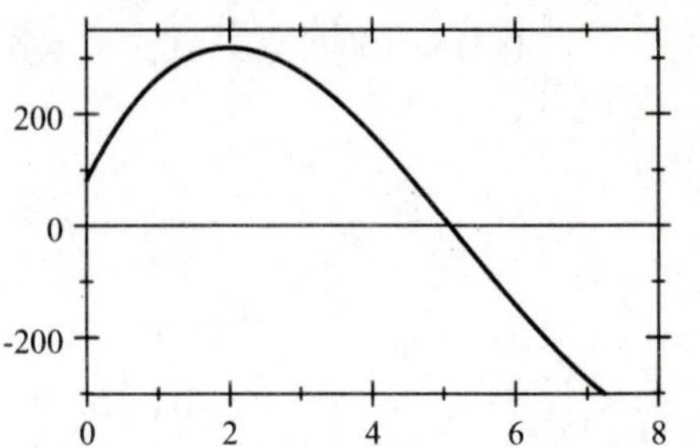

From the graph, we see that $N'(t)$ has a maximum when $t \approx 2$, on February 8.

b. From the graph in part (a), the maximum number of sickouts occurred when $N'(t) = 0$, that is, when $t \approx 5$. We calculate $N(5) \approx 1145$ canceled flights.

10.5 Optimization II

Problem-Solving Tips

Follow the guidelines given on page 740 of the text to solve the optimization problems that follow. Remember, Theorem 3 in Section 10.4 provides us with a method of computing the absolute extrema of a continuous function over a closed interval $[a, b]$. If the problem involves a function that is to be optimized over an interval that is not closed, then use the graphical method to find the optimal values of f. You might review Example 4 on page 743 of the text to make sure you understand how to use the graphical method.

Concept Questions page 746

1. We could solve the problem by sketching the graph of f and checking to see if there is an absolute extremum.

Exercises page 746

1. Let x and y denote the lengths of two adjacent sides of the rectangle. We want to maximize $A = xy$. But the perimeter is $2x + 2y$ and this is equal to 100, so $2x + 2y = 100$, and therefore $y = 50 - x$. Thus, $A = f(x) = x(50 - x) = -x^2 + 50x$, $0 \leq x \leq 50$. We allow the "degenerate"cases $x = 0$ and $x = 50$. $A' = -2x + 50 = 0$ implies that $x = 25$ is a critical number of f. $A(0) = 0$, $A(25) = 625$, and $A(50) = 0$, so we see that A is maximized for $x = 25$. The required dimensions are 25 ft by 25 ft.

3. We have $2x + y = 3000$ and we want to maximize the function $A = f(x) = xy = x(3000 - 2x) = 3000x - 2x^2$ on the interval $[0, 1500]$. The critical number of A is obtained by solving $f'(x) = 3000 - 4x = 0$, giving $x = 750$. From the table of values, we conclude that $x = 750$ yields the absolute maximum value of A. Thus, the required dimensions are 750×1500 yards. The maximum area is 1,125,000 yd^2.

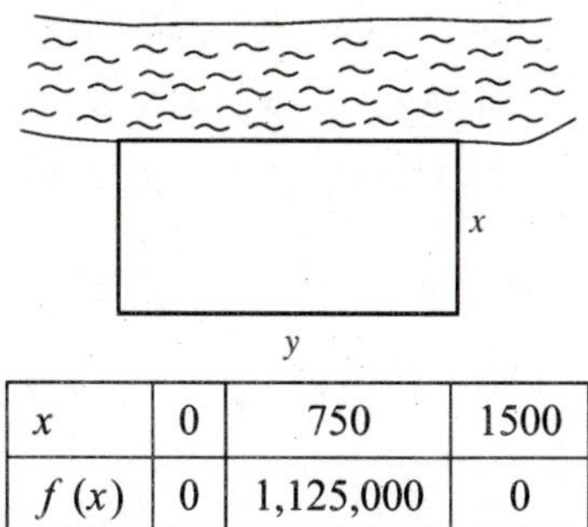

x	0	750	1500
$f(x)$	0	1,125,000	0

5. Let x denote the length of the side made of wood and y the length of the side made of steel. The cost of construction is $C = 6(2x) + 3y$, but $xy = 800$, so $y = \dfrac{800}{x}$. Therefore, $C = f(x) = 12x + 3\left(\dfrac{800}{x}\right) = 12x + \dfrac{2400}{x}$. To minimize C, we compute $f'(x) = 12 - \dfrac{2400}{x^2} = \dfrac{12x^2 - 2400}{x^2} = \dfrac{12(x^2 - 200)}{x^2}$. Setting $f'(x) = 0$ gives $x = \pm\sqrt{200}$ as critical numbers of f. The sign diagram of f' shows that $x = \pm\sqrt{200}$ gives a relative minimum of f.

$- - - \; 0 \; + + +$ sign of f'

$0 \qquad \sqrt{200} \qquad x$

$f''(x) = \dfrac{4800}{x^3} > 0$ if $x > 0$, and so f is concave upward for $x > 0$. Therefore, $x = \sqrt{200} = 10\sqrt{2}$ yields the absolute minimum. Thus, the dimensions of the enclosure should be $10\sqrt{2}$ ft $\times$ $40\sqrt{2}$ ft, or 14.1 ft $\times$ 56.6 ft.

7. Let the dimensions of each square that is cut out be $x'' \times x''$. Then the dimensions of the box are $(8 - 2x)''$ by $(8 - 2x)''$ by x'', and its volume is be $V = f(x) = x(8 - 2x)^2$. We want to maximize f on $[0, 4]$.

$$\begin{aligned} f'(x) &= (8 - 2x)^2 + x(2)(8 - 2x)(-2) \quad \text{(by the Product Rule)} \\ &= (8 - 2x)[(8 - 2x) - 4x] \\ &= (8 - 2x)(8 - 6x) \\ &= 0 \text{ if } x = 4 \text{ or } \tfrac{4}{3}. \end{aligned}$$

The latter is a critical number in $(0, 4)$. From the table, we see that $x = \frac{4}{3}$ yields an absolute maximum for f, so the dimensions of the box should be $\frac{16}{3}'' \times \frac{16}{3}'' \times \frac{4}{3}''$.

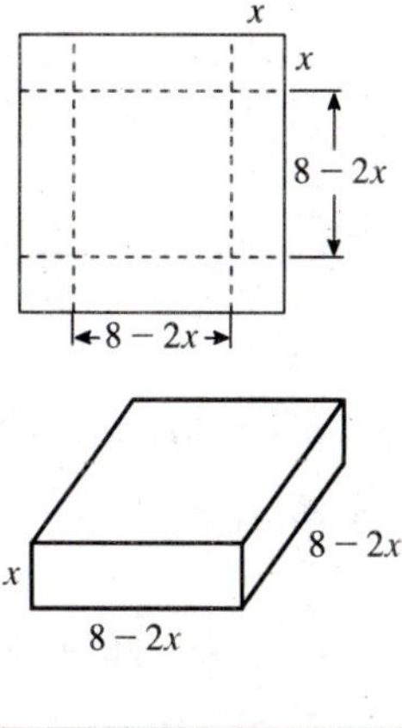

x	0	$\frac{4}{3}$	4
$f(x)$	0	$\frac{1024}{27}$	0

9. Let x denote the length of the sides of the box and y denote its height. Referring to the figure, we see that the volume of the box is given by $x^2y = 128$. The amount of material used is given by $S = f(x) = 2x^2 + 4xy = 2x^2 + 4x\left(\frac{128}{x^2}\right) = 2x^2 + \frac{512}{x}$. We want to minimize f subject to the condition that $x > 0$.

Now $f'(x) = 4x - \frac{512}{x^2} = \frac{4x^3 - 512}{x^2} = \frac{4(x^3 - 128)}{x^2}$. Setting $f'(x) = 0$ yields $x \approx 5.04$, a critical number of f. Next, $f''(x) = 4 + \frac{1024}{x^3} > 0$ for all $x > 0$. Thus, the graph of f is concave upward, and so $x = 5.04$ yields an absolute minimum of f. The required dimensions are $5.04'' \times 5.04'' \times 5.04''$.

11. The length plus the girth of the box is $4x + h = 108$ and $h = 108 - 4x$. Then $V = x^2h = x^2(108 - 4x) = 108x^2 - 4x^3$ and $V' = 216x - 12x^2$. We want to maximize V on the interval $[0, 27]$. Setting $V'(x) = 0$ and solving for x, we obtain $x = 18$ and $x = 0$. We calculate $V(0) = 0$, $V(18) = 11{,}664$, and $V(27) = 0$. Thus, the dimensions of the box are $18'' \times 18'' \times 36''$ and its maximum volume is approximately 11,664 in^3.

13. We take $2\pi r + \ell = 108$. We want to maximize $V = \pi r^2\ell = \pi r^2(-2\pi r + 108) = -2\pi^2r^3 + 108\pi r^2$ subject to the condition that $0 \le r \le \frac{54}{\pi}$. Now $V'(r) = -6\pi^2r^2 + 216\pi r = -6\pi r(\pi r - 36) = 0$ implies that $r = 0$ and $r = \frac{36}{\pi}$ are critical numbers of V. From the table, we conclude that the maximum volume occurs when $r = \frac{36}{\pi} \approx 11.5$ inches and $\ell = 108 - 2\pi\left(\frac{36}{\pi}\right) = 36$ inches and the volume of the parcel is $46{,}656/\pi$ in^3.

r	0	$\frac{36}{\pi}$	$\frac{54}{\pi}$
V	0	$\frac{46{,}656}{\pi}$	0

15. Let y denote the height and x the width of the cabinet. Then $y = \frac{3}{2}x$. Because the volume is to be 2.4 ft^3, we have $xyd = 2.4$, where d *is* the depth of the cabinet. Thus, $x\left(\frac{3}{2}x\right)d = 2.4$, so $d = \frac{2.4(2)}{3x^2} = \frac{1.6}{x^2}$. The cost for constructing the cabinet is

$$C = 40(2xd + 2yd) + 20(2xy) = 80\left[\frac{1.6}{x} + \left(\frac{3x}{2}\right)\left(\frac{1.6}{x^2}\right)\right] + 40x\left(\frac{3x}{2}\right) = \frac{320}{x} + 60x^2,$$ so

$C'(x) = -\frac{320}{x^2} + 120x = \frac{120x^3 - 320}{x^2} = 0$ if $x = \sqrt[3]{\frac{8}{3}} = \frac{2}{\sqrt[3]{3}} = \frac{2}{3}\sqrt[3]{9}$. Therefore, $x = \frac{2}{3}\sqrt[3]{9}$ is a critical number of C. The sign diagram shows that $x = \frac{2}{3}\sqrt[3]{9}$ gives a relative minimum. Next, $C''(x) = \frac{640}{x^3} + 120 > 0$ for all $x > 0$, telling us that the graph of C is concave upward, so $x = \frac{2}{3}\sqrt[3]{9}$ yields an absolute minimum. The required dimensions are $\frac{2}{3}\sqrt[3]{9}' \times \sqrt[3]{9}' \times \frac{2}{5}\sqrt[3]{9}'$.

$- - - \; 0 \; + + + +$ sign of C'

$0 \qquad \frac{2\sqrt[3]{9}}{3} \qquad x$

17. We want to maximize the function $R(x) = (200 + x)(300 - x) = -x^2 + 100x + 60{,}000$. Now $R'(x) = -2x + 100 = 0$ gives $x = 50$, and this is a critical number of R. Because $R''(x) = -2 < 0$, we see that $x = 50$ gives an absolute maximum of R. Therefore, the number of passengers should be 250. The fare will then be \$250/passenger and the revenue will be \$62,500.

19. Let x denote the number of people beyond 20 who sign up for the cruise. Then the revenue is $R(x) = (20 + x)(600 - 4x) = -4x^2 + 520x + 12{,}000$. We want to maximize R on the closed bounded interval $[0, 70]$. $R'(x) = -8x + 520 = 0$ implies $x = 65$, a critical number of R. Evaluating R at this critical number and the endpoints, we see that R is maximized if $x = 65$. Therefore, 85 passengers will result in a maximum revenue of \$28,900. The fare in this case is \$340/passenger.

x	0	65	70
$R(x)$	12,000	28,900	28,800

21. The fuel consumption is $x/600$ gallons/mile, the fuel cost is \$3/gallon, and the labor cost is $18/x$ dollars per mile. Therefore, the total cost is $C(x) = \dfrac{18}{x} + \dfrac{x}{200}$. We calculate $C'(x) = -\dfrac{18}{x^2} + \dfrac{1}{200} = 0$, giving $-\dfrac{18}{x^2} = -\dfrac{1}{200}$, $x^2 = 18(200)$, $x^2 = 3600$, and so $x = 60$. Next, $C''(x) = \dfrac{48}{x^3} > 0$ for all $x > 0$ so C is concave upward. Therefore, $x = 60$ gives the absolute minimum. The most economical speed is 60 mph.

23. We want to maximize $S = kh^2w$. But $h^2 + w^2 = 24^2$, or $h^2 = 576 - w^2$, so $S = f(w) = kw(576 - w^2) = k(576w - w^3)$ for $0 \le w \le 24$. Now, setting $f'(w) = k(576 - 3w^2) = 0$ gives $w = \pm\sqrt{192} \approx \pm 13.86$. Only the positive root is a critical number of interest. Next, we find $f''(w) = -6kw$, and in particular, $f''\left(\sqrt{192}\right) = -6\sqrt{192}k < 0$, so that $w \approx 13.86$ gives a relative maximum of f. Because $f''(w) < 0$ for $w > 0$, we see that the graph of f is concave downward on $(0, 24)$, and so $w = \sqrt{192}$ gives an absolute maximum of f. We find $h^2 = 576 - 192 = 384$ and so $h \approx 19.60$, so the width and height of the log should be approximately 13.86 inches and 19.60 inches, respectively.

25. We want to minimize $C(x) = 1.50(10{,}000 - x) + 2.50\sqrt{3000^2 + x^2}$ subject to $0 \le x \le 10{,}000$. Now $C'(x) = -1.50 + 2.5\left(\frac{1}{2}\right)(9{,}000{,}000 + x^2)^{-1/2}(2x) = -1.50 + \dfrac{2.50x}{\sqrt{9{,}000{,}000 + x^2}} = 0$ if $2.5x = 1.50\sqrt{9{,}000{,}000 + x^2}$, or $6.25x^2 = 2.25(9{,}000{,}000 + x^2)$, or $4x^2 = 20{,}250{,}000$, giving $x = 2250$. From the table, we see that $x = 2250$, or 2250 ft, gives the absolute minimum.

x	0	2250	10,000
$C(x)$	22,500	21,000	26,101

27. The time of flight is $T = f(x) = \dfrac{12 - x}{6} + \dfrac{\sqrt{x^2 + 9}}{4}$, so $f'(x) = -\dfrac{1}{6} + \dfrac{1}{4}\left(\dfrac{1}{2}\right)(x^2 + 9)^{-1/2}(2x) = -\dfrac{1}{6} + \dfrac{x}{4\sqrt{x^2 + 9}} = \dfrac{3x - 2\sqrt{x^2 + 9}}{12\sqrt{x^2 + 9}}$. Setting $f'(x) = 0$ gives $3x = 2\sqrt{x^2 + 9}$, $9x^2 = 4(x^2 + 9)$, and $5x^2 = 36$. Therefore, $x = \pm\frac{6}{\sqrt{5}} = \pm\frac{6\sqrt{5}}{5}$. Only the critical number $x = \frac{6\sqrt{5}}{5}$ is of interest. The nature of the problem suggests $x \approx 2.68$ gives an absolute minimum for T.

29. The area enclosed by the rectangular region of the racetrack is $A = (\ell)(2r) = 2r\ell$. The length of the racetrack is $2\pi r + 2\ell$, and is equal to 1760. That is, $2(\pi r + \ell) = 1760$, and $\pi r + \ell = 880$. Therefore, we want to maximize $A = f(r) = 2r(880 - \pi r) = 1760r - 2\pi r^2$. The restriction on r is $0 \le r \le \frac{880}{\pi}$. To maximize A, we compute $f'(r) = 1760 - 4\pi r$. Setting $f'(r) = 0$ gives $r = \frac{1760}{4\pi} = \frac{440}{\pi} \approx 140$. Because $f(0) = f\left(\frac{880}{\pi}\right) = 0$, we see that the maximum rectangular area is enclosed if we take $r = \frac{440}{\pi}$ and $\ell = 880 - \pi\left(\frac{440}{\pi}\right) = 440$. So $r \approx 140$ and $\ell = 440$. The total area enclosed is $2r\ell + \pi r^2 = 2\left(\frac{440}{\pi}\right)(440) + \pi\left(\frac{440}{\pi}\right)^2 = \frac{2(440)^2}{\pi} + \frac{440^2}{\pi} = \frac{580{,}800}{\pi} \approx 184{,}874 \text{ ft}^2$.

31. Let x denote the number of bottles in each order. We want to minimize $C(x) = 200\left(\frac{2{,}000{,}000}{x}\right) + \frac{x}{2}(0.40) = \frac{400{,}000{,}000}{x} + 0.2x$. We compute $C'(x) = -\frac{400{,}000{,}000}{x^2} + 0.2$. Setting $C'(x) = 0$ gives $x^2 = \frac{400{,}000{,}000}{0.2} = 2{,}000{,}000{,}000$, or $x \approx 44{,}721$, a critical number of C. $C''(x) = \frac{800{,}000{,}000}{x^3} > 0$ for all $x > 0$, and we see that the graph of C is concave upward and so $x = 44{,}721$ gives an absolute minimum of C. Therefore, there should be $2{,}000{,}000/x \approx 45$ orders per year (since we can not have fractions of an order.) Each order should be for $2{,}000{,}000/45 \approx 44{,}445$ bottles.

33. **a.** Because the sales are assumed to be steady and D units are expected to be sold per year, the number of orders per year is D/x. Because is costs \$$K$ per order, the ordering cost is KD/x. The purchasing cost is PD (cost per item times number purchased). Finally, the holding cost is $\frac{1}{2}xh$ (the average number on hand times holding cost per item). Therefore, $C(x) = \frac{KD}{x} + pD + \frac{hx}{2}$.

b. $C'(x) = -\frac{KD}{x^2} + \frac{h}{2} = 0$ implies $\frac{KD}{x^2} = \frac{h}{2}$, so $x^2 = \frac{2KD}{h}$ and $x = \pm\sqrt{\frac{2KD}{h}}$. We reject the negative root. So $x = \sqrt{\frac{2KD}{h}}$ is the only critical number. Next, $C''(x) = \frac{2KD}{x^3} > 0$ for $x > 0$, so $C''\left(\sqrt{\frac{2KD}{h}}\right) > 0$ and the Second Derivative Test shows that $x = \sqrt{\frac{2KD}{h}}$ does give a relative minimum. Because C is concave upward, this is also the absolute minimum.

CHAPTER 10 Concept Review Questions page 751

1. **a.** $f(x_1) < f(x_2)$ **b.** $f(x_1) > f(x_2)$

3. **a.** $f(x) \le f(c)$ **b.** $f(x) \ge f(c)$

5. **a.** $f'(x)$ **b.** > 0 **c.** concavity **d.** relative maximum; relative extremum

7. 0, 0

9. **a.** $f(x) \le f(c)$, absolute maximum value **b.** $f(x) \ge f(c)$, open interval

CHAPTER 10 Review Exercises page 752

1. a. $f(x) = \frac{1}{3}x^3 - x^2 + x - 6$, so $f'(x) = x^2 - 2x + 1 = (x-1)^2$. $f'(x) = 0$ gives $x = 1$, the critical number of f. Now $f'(x) > 0$ for all $x \neq 1$. Thus, f is increasing on $(-\infty, \infty)$.

b. Because $f'(x)$ does not change sign as we move across the critical number $x = 1$, the First Derivative Test implies that $x = 1$ does not give a relative extremum of f.

c. $f''(x) = 2(x-1)$. Setting $f''(x) = 0$ gives $x = 1$ as a candidate for an inflection point of f. Because $f''(x) < 0$ for $x < 1$, and $f''(x) > 0$ for $x > 1$, we see that f is concave downward on $(-\infty, 1)$ and concave upward on $(1, \infty)$.

d. The results of part (c) imply that $\left(1, -\frac{17}{3}\right)$ is an inflection point.

3. a. $f(x) = x^4 - 2x^2$, so $f'(x) = 4x^3 - 4x = 4x\left(x^2 - 1\right) = 4x(x+1)(x-1)$. The sign diagram of f' shows that f is decreasing on $(-\infty, -1)$ and $(0, 1)$ and increasing on $(-1, 0)$ and $(1, \infty)$.

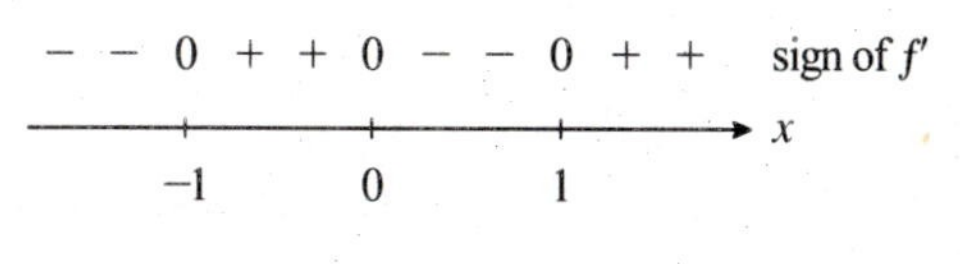

b. The results of part (a) and the First Derivative Test show that $(-1, -1)$ and $(1, -1)$ are relative minima and $(0, 0)$ is a relative maximum.

c. $f''(x) = 12x^2 - 4 = 4\left(3x^2 - 1\right) = 0$ if $x = \pm\frac{\sqrt{3}}{3}$. The sign diagram shows that f is concave upward on $\left(-\infty, -\frac{\sqrt{3}}{3}\right)$) and $\left(\frac{\sqrt{3}}{3}, \infty\right)$ and concave downward on $\left(-\frac{\sqrt{3}}{3}, \frac{\sqrt{3}}{3}\right)$.

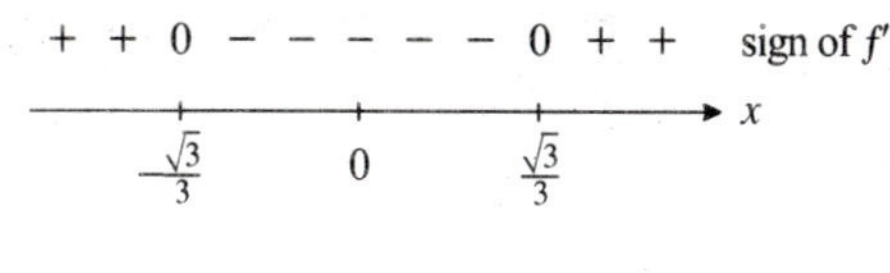

d. The results of part (c) show that $\left(-\frac{\sqrt{3}}{3}, -\frac{5}{9}\right)$ and $\left(\frac{\sqrt{3}}{3}, -\frac{5}{9}\right)$ are inflection points.

5. a. $f(x) = \dfrac{x^2}{x-1}$, so $f'(x) = \dfrac{(x-1)(2x) - x^2(1)}{(x-1)^2} = \dfrac{x^2 - 2x}{(x-1)^2} = \dfrac{x(x-2)}{(x-1)^2}$.

The sign diagram of f' shows that f is increasing on $(-\infty, 0)$ and $(2, \infty)$ and decreasing on $(0, 1)$ and $(1, 2)$.

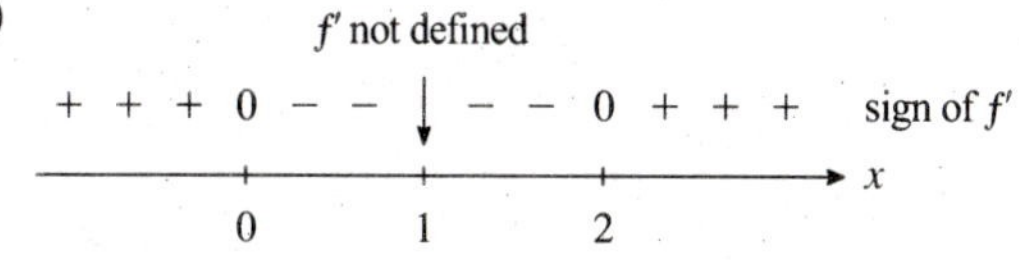

b. The results of part (a) show that $(0, 0)$ is a relative maximum and $(2, 4)$ is a relative minimum.

c. $f''(x) = \dfrac{(x-1)^2(2x-2) - x(x-2)2(x-1)}{(x-1)^4} = \dfrac{2(x-1)\left[(x-1)^2 - x(x-2)\right]}{(x-1)^4} = \dfrac{2}{(x-1)^3}$. Because $f''(x) < 0$ if $x < 1$ and $f''(x) > 0$ if $x > 1$, we see that f is concave downward on $(-\infty, 1)$ and concave upward on $(1, \infty)$.

d. Because $x = 1$ is not in the domain of f, there is no inflection point.

7. a. $f(x) = (1-x)^{1/3}$, so $f'(x) = -\frac{1}{3}(1-x)^{-2/3} = -\dfrac{1}{3(1-x)^{2/3}}$. The sign diagram for f' shows that f is decreasing on $(-\infty, \infty)$.

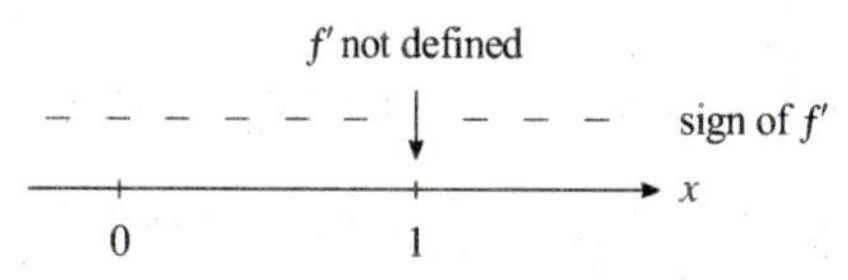

b. There is no relative extremum.

c. $f''(x) = -\frac{2}{9}(1-x)^{-5/3} = -\dfrac{2}{9(1-x)^{5/3}}$. The sign diagram for f'' shows that f is concave downward on $(-\infty, 1)$ and concave upward on $(1, \infty)$.

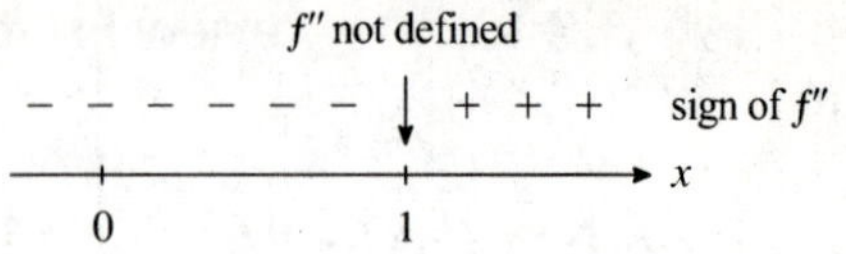

d. $x = 1$ is a candidate for an inflection point of f. Referring to the sign diagram for f'', we see that $(1, 0)$ is an inflection point.

9. a. $f(x) = \dfrac{2x}{x+1}$, so $f'(x) = \dfrac{(x+1)(2) - 2x(1)}{(x+1)^2} = \dfrac{2}{(x+1)^2} > 0$ if $x \neq -1$. Therefore f is increasing on $(-\infty, -1)$ and $(-1, \infty)$.

b. Because there is no critical number, f has no relative extremum.

c. $f''(x) = -4(x+1)^{-3} = -\dfrac{4}{(x+1)^3}$. Because $f''(x) > 0$ if $x < -1$ and $f''(x) < 0$ if $x > -1$, we see that f is concave upward on $(-\infty, -1)$ and concave downward on $(-1, \infty)$.

d. There is no inflection point because $f''(x) \neq 0$ for all x in the domain of f.

11. a. $f(x) = (4-x)e^x$, so $f'(x) = (4-x)e^x + e^x(-1) = e^x(3-x)$. Setting $f'(x) = 0$ gives $x = 3$ as the only critical number of f. The sign diagram for f' shows that f is increasing on $(-\infty, 3)$ and decreasing on $(3, \infty)$.

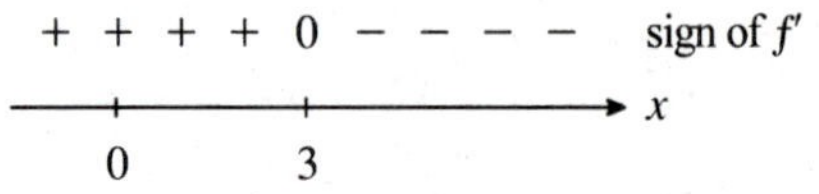

b. f has a relative maximum of $f(3) = e^3$.

c. $f''(x) = e^x(-1) + (3-x)e^x = e^x(2-x)$. Setting $f''(x) = e^x(2-x) = 0$ gives $x = 2$. The sign diagram for f'' shows that f is concave upward on $(-\infty, 2)$ and concave downward on $(2, \infty)$.

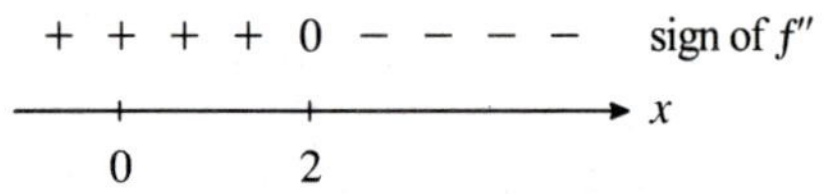

d. f has an inflection point at $(2, 2e^2)$.

13. $f(x) = x^2 - 5x + 5$.

1. The domain of f is $(-\infty, \infty)$.
2. Setting $x = 0$ gives 5 as the y-intercept.
3. $\lim_{x\to-\infty}(x^2 - 5x + 5) = \lim_{x\to\infty}(x^2 - 5x + 5) = \infty$.
4. There is no asymptote because f is a quadratic function.
5. $f'(x) = 2x - 5 = 0$ if $x = \frac{5}{2}$. The sign diagram shows that f is increasing on $\left(\frac{5}{2}, \infty\right)$ and decreasing on $\left(-\infty, \frac{5}{2}\right)$.

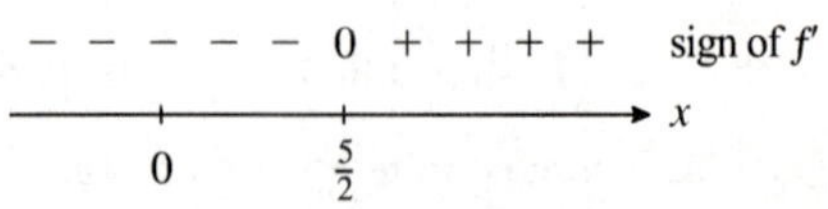

6. The First Derivative Test implies that $\left(\frac{5}{2}, -\frac{5}{4}\right)$ is a relative minimum.

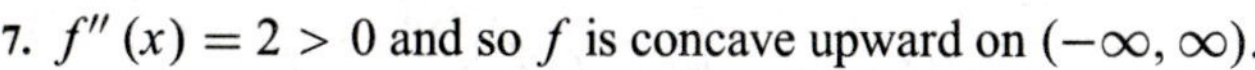

7. $f''(x) = 2 > 0$ and so f is concave upward on $(-\infty, \infty)$.

8. There is no inflection point.

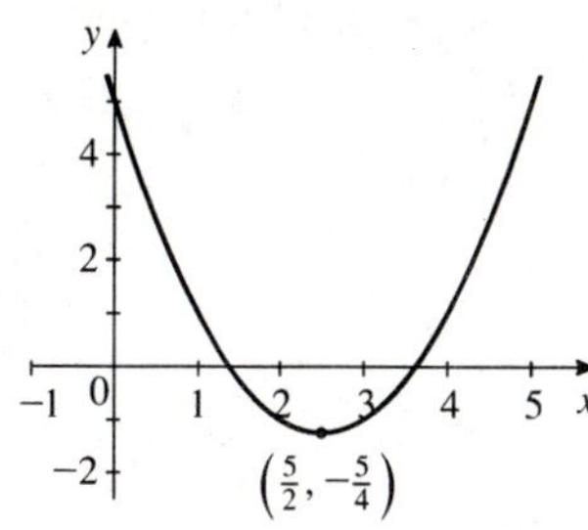

15. $g(x) = 2x^3 - 6x^2 + 6x + 1$.

1. The domain of g is $(-\infty, \infty)$.

2. Setting $x = 0$ gives 1 as the y-intercept.

3. $\lim_{x\to-\infty} g(x) = -\infty$ and $\lim_{x\to\infty} g(x) = \infty$.

4. There is no vertical or horizontal asymptote.

5. $g'(x) = 6x^2 - 12x + 6 = 6\left(x^2 - 2x + 1\right) = 6(x-1)^2$. Because $g'(x) > 0$ for all $x \neq 1$, we see that g is increasing on $(-\infty, 1)$ and $(1, \infty)$.

6. $g'(x)$ does not change sign as we move across the critical number $x = 1$, so there is no extremum.

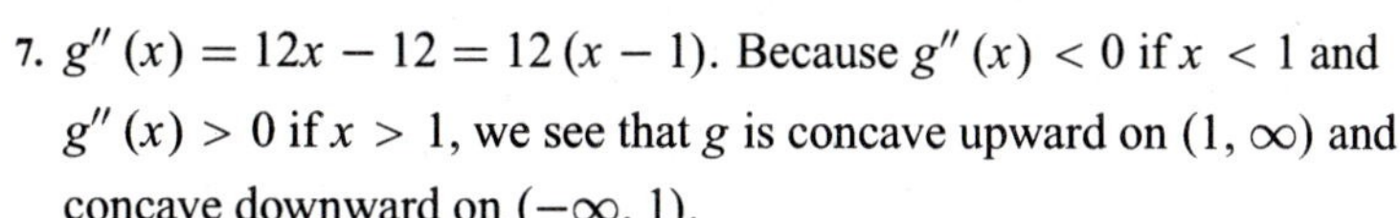

7. $g''(x) = 12x - 12 = 12(x-1)$. Because $g''(x) < 0$ if $x < 1$ and $g''(x) > 0$ if $x > 1$, we see that g is concave upward on $(1, \infty)$ and concave downward on $(-\infty, 1)$.

8. The point $x = 1$ gives rise to the inflection point $(1, 3)$.

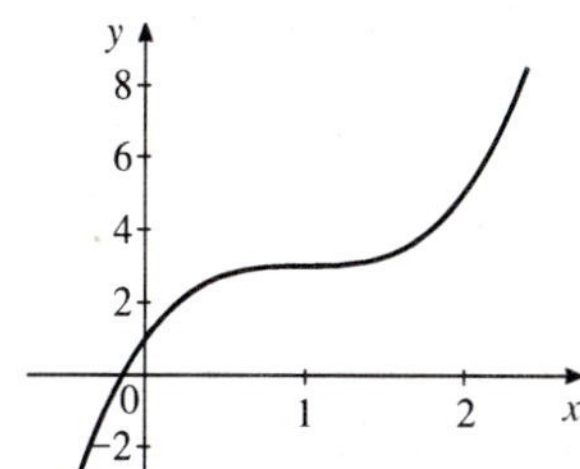

17. $h(x) = x\sqrt{x-2}$.

1. The domain of h is $[2, \infty)$.

2. There is no y-intercept. Setting $y = 0$ gives 2 as the x-intercept.

3. $\lim_{x\to\infty} x\sqrt{x-2} = \infty$.

4. There is no asymptote.

5. $h'(x) = (x-2)^{1/2} + x\left(\frac{1}{2}\right)(x-2)^{-1/2} = \frac{1}{2}(x-2)^{-1/2}[2(x-2)+x] = \dfrac{3x-4}{2\sqrt{x-2}} > 0$ on $(2, \infty)$, and so h is increasing on $(2, \infty)$.

6. Because h has no critical number in $(2, \infty)$, there is no relative extremum.

7. $h''(x) = \dfrac{1}{2}\left[\dfrac{(x-2)^{1/2}(3) - (3x-4)\frac{1}{2}(x-2)^{-1/2}}{x-2}\right] = \dfrac{(x-2)^{-1/2}[6(x-2)-(3x-4)]}{4(x-2)} = \dfrac{3x-8}{4(x-2)^{3/2}}$.

   The sign diagram for h'' shows that h is concave downward on $\left(2, \frac{8}{3}\right)$ and concave upward on $\left(\frac{8}{3}, \infty\right)$.

   – – – – 0 + + + + sign of h''
   x: 2, $\frac{8}{3}$

8. The results of step 7 tell us that $\left(\frac{8}{3}, \frac{8\sqrt{6}}{9}\right)$ is an inflection point.

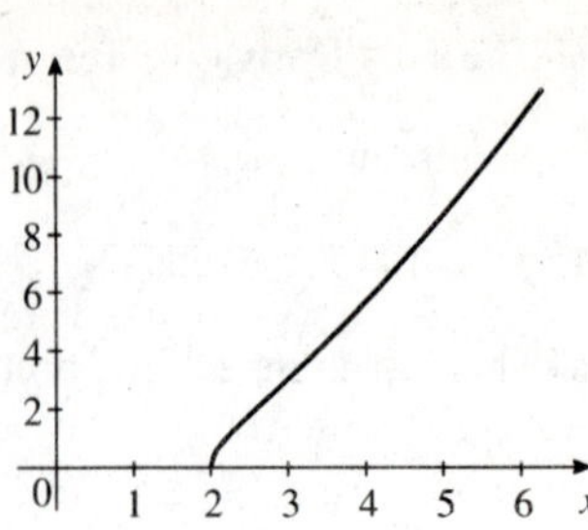

19. $f(x) = \dfrac{x-2}{x+2}$.

1. The domain of f is $(-\infty, -2) \cup (-2, \infty)$.

2. Setting $x = 0$ gives -1 as the y-intercept. Setting $y = 0$ gives 2 as the x-intercept.

3. $\displaystyle\lim_{x\to-\infty} \frac{x-2}{x+2} = \lim_{x\to\infty} \frac{x-2}{x+2} = 1$.

4. The results of step 3 tell us that $y = 1$ is a horizontal asymptote. Next, observe that the denominator of $f(x)$ is equal to zero at $x = -2$, but its numerator is not equal to zero there. Therefore, $x = -2$ is a vertical asymptote.

5. $f'(x) = \dfrac{(x+2)(1) - (x-2)(1)}{(x+2)^2} = \dfrac{4}{(x+2)^2}$. The sign diagram of f' tells us that f is increasing on $(-\infty, -2)$ and $(-2, \infty)$.

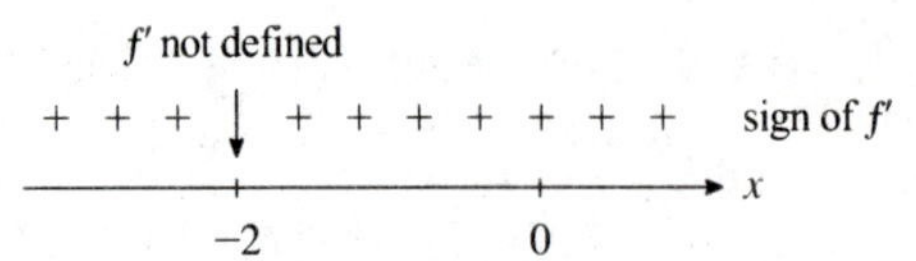

6. The results of step 5 tell us that there is no relative extremum.

7. $f''(x) = -\dfrac{8}{(x+2)^3}$. The sign diagram of f'' shows that f is concave upward on $(-\infty, -2)$ and concave downward on $(-2, \infty)$.

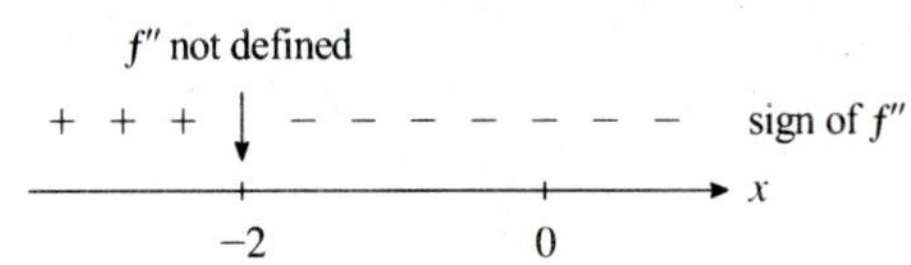

8. There is no inflection point.

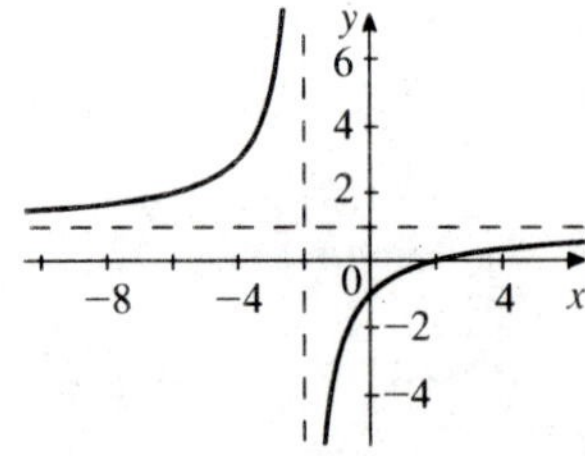

21. $f(x) = xe^{-2x}$. We first gather the following information on f.

1. The domain of f is $(-\infty, \infty)$.

2. Setting $x = 0$ gives 0 as the y-intercept.

3. $\displaystyle\lim_{x\to-\infty} xe^{-2x} = -\infty$ and $\displaystyle\lim_{x\to\infty} xe^{-2x} = 0$.

4. The results of part 3 show that $y = 0$ is a horizontal asymptote.

5. $f'(x) = e^{-2x} + xe^{-2x}(-2) = (1-2x)e^{-2x}$. Observe that $f'(x) = 0$ at $x = \frac{1}{2}$, a critical number of f.

The sign diagram of f' shows that f is increasing on $\left(-\infty, \frac{1}{2}\right)$ and decreasing on $\left(\frac{1}{2}, \infty\right)$.

+ + + + + 0 − − − sign of f'

x: 0, $\frac{1}{2}$

6. The results of part 5 show that $\left(\frac{1}{2}, \frac{1}{2}e^{-1}\right)$ is a relative maximum.

7. $f''(x) = -2e^{-2x} + (1-2x)e^{-2x}(-2) = 4(x-1)e^{-2x} = 0$ if $x = 1$. The sign diagram of f'' shows that the graph of f is concave downward on $(-\infty, 1)$ and concave upward on $(1, \infty)$.

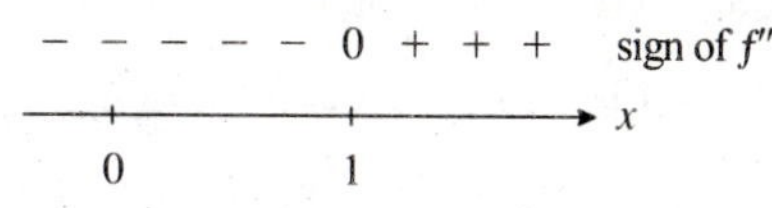

8. f has an inflection point at $(1, 1/e^2)$.

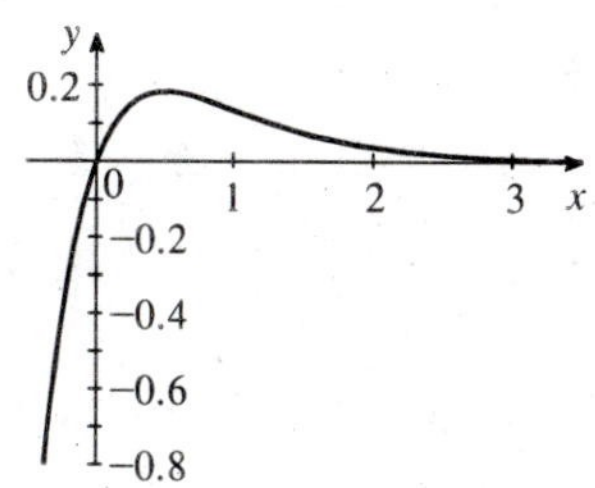

23. $\lim_{x\to-\infty} \frac{1}{2x+3} = \lim_{x\to\infty} \frac{1}{2x+3} = 0$ and so $y = 0$ is a horizontal asymptote. Because the denominator is equal to zero at $x = -\frac{3}{2}$ but the numerator is not equal to zero there, we see that $x = -\frac{3}{2}$ is a vertical asymptote.

25. $\lim_{x\to-\infty} \frac{5x}{x^2-2x-8} = \lim_{x\to\infty} \frac{5x}{x^2-2x-8} = 0$, so $y = 0$ is a horizontal asymptote. Next, note that the denominator is zero if $x^2 - 2x - 8 = (x-4)(x+2) = 0$, that is, if $x = -2$ or $x = 4$. Because the numerator is not equal to zero at these points, we see that $x = -2$ and $x = 4$ are vertical asymptotes.

27. $f(x) = 2x^2 + 3x - 2$, so $f'(x) = 4x + 3$. Setting $f'(x) = 0$ gives $x = -\frac{3}{4}$ as a critical number of f. Next, $f''(x) = 4 > 0$ for all x, so f is concave upward on $(-\infty, \infty)$. Therefore, $f\left(-\frac{3}{4}\right) = -\frac{25}{8}$ is an absolute minimum of f. There is no absolute maximum.

29. $g(t) = \sqrt{25-t^2} = (25-t^2)^{1/2}$. Differentiating $g(t)$, we have $g'(t) = \frac{1}{2}(25-t^2)^{-1/2}(-2t) = -\frac{t}{\sqrt{25-t^2}}$. Setting $g'(t) = 0$ gives $t = 0$ as a critical number of g. The domain of g is given by solving the inequality $25 - t^2 \geq 0$ or $(5-t)(5+t) \geq 0$ which implies that $t \in [-5, 5]$. From the table, we conclude that $g(0) = 5$ is the absolute maximum of g and $g(-5) = 0$ and $g(5) = 0$ is the absolute minimum value of g.

t	-5	0	5
$g(t)$	0	5	0

31. $h(t) = t^3 - 6t^2$, so $h'(t) = 3t^2 - 12t = 3t(t-4) = 0$ if $t = 0$ or $t = 4$. But only $t = 4$ lies in $(2, 5)$, so $t = 4$ is a critical number of h. From the table, we see that h has an absolute minimum at $(4, -32)$ and an absolute maximum at $(2, -16)$.

t	2	4	5
$h(t)$	-16	-32	-25

33. $f(x) = x - \frac{1}{x}$ on $[1, 3]$, so $f'(x) = 1 + \frac{1}{x^2}$. Because $f'(x)$ is never zero, f has no critical number. Calculating $f(x)$ at the endpoints, we see that $f(1) = 0$ is the absolute minimum value and $f(3) = \frac{8}{3}$ is the absolute maximum value.

35. $f(t) = 3te^{-t}$ on $[-2, 2]$, so $f'(t) = 3e^{-t} + 3t(-e^{-t}) = 3e^{-t}(1-t)$. Setting $f'(t) = 0$ gives $t = 1$ as the only critical number of f. From the sign diagram of f' we see that $f(1) = 3e^{-1} = 3/e$ is a local maximum value of f. Also, $f(-2) = -6e^2$ and $f(2) = 6e^{-2}$, so f has an absolute maximum at $(1, 3/e)$ and an absolute minimum value at $(-2, -6e^2)$.

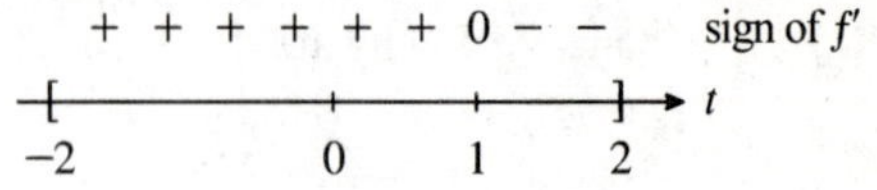

+ + + + + + 0 − − sign of f'

−2 0 1 2 t

37. $f(s) = s\sqrt{1-s^2}$ on $[-1, 1]$. The function f is continuous on $[-1, 1]$ and differentiable on $(-1, 1)$. Next, $f'(s) = (1-s^2)^{1/2} + s\left(\frac{1}{2}\right)(1-s^2)^{-1/2}(-2s) = \dfrac{1-2s^2}{\sqrt{1-s^2}}$. Setting $f'(s) = 0$, we find that $s = \pm\frac{\sqrt{2}}{2}$ are critical numbers of f. From the table, we see that $f\left(-\frac{\sqrt{2}}{2}\right) = -\frac{1}{2}$ is the absolute minimum value and $f\left(\frac{\sqrt{2}}{2}\right) = \frac{1}{2}$ is the absolute maximum value of f.

x	-1	$-\frac{\sqrt{2}}{2}$	$\frac{\sqrt{2}}{2}$	1
$f(x)$	0	$-\frac{1}{2}$	$\frac{1}{2}$	0

39. a. The sign of R_1' is negative and the sign of R_2' is positive on $(0, T)$. The sign of R_1'' is negative and the sign of R_2'' is positive on $(0, T)$.

b. The revenue of the neighborhood bookstore is decreasing at an increasing rate, while the revenue of the new bookstore is increasing at an increasing rate.

41. We want to maximize $P(x) = -x^2 + 8x + 20$. Now, $P'(x) = -2x + 8 = 0$ if $x = 4$, a critical number of P. Because $P''(x) = -2 < 0$, the graph of P is concave downward. Therefore, the critical number $x = 4$ yields an absolute maximum. So, to maximize profit, the company should spend \$4000 per month on advertising.

43. $S(x) = -0.002x^3 + 0.6x^2 + x + 500$, so $S'(x) = -0.006x^2 + 1.2x + 1$ and $S''(x) = -0.012x + 1.2$. $x = 100$ is a candidate for an inflection point of S. The sign diagram for S'' shows that $(100, 4600)$ is an inflection point of S.

+ + + 0 − − − sign of S''

0 100 200 x

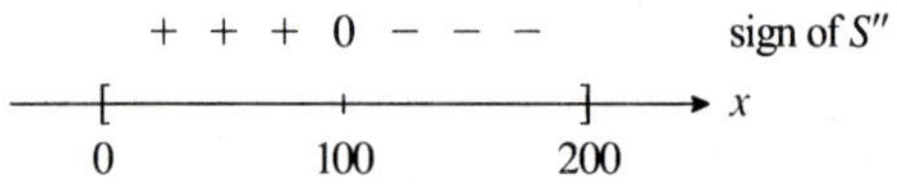

45. $C(x) = 0.0001x^3 - 0.08x^2 + 40x + 5000$, so $C'(x) = 0.0003x^2 - 0.16x + 40$ and $C''(x) = 0.0006x - 0.16$. Thus, $x = 266.67$ is a candidate for an inflection point of C. The sign diagram for C'' shows that C has an inflection point at $(266.67, 11874.08)$. Thus, 267 calculators will be produced at a cost of approximately \$11,874.

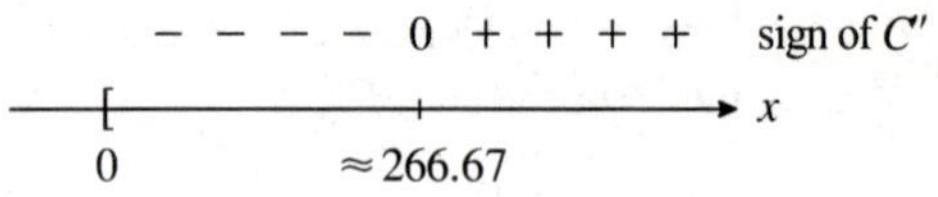

− − − − 0 + + + + sign of C''

0 ≈266.67 x

47. a. $f(t) = \dfrac{150\sqrt{t} + 766}{59 - \sqrt{t}}$. $f(0) \approx 12.98$, so the proportion in 2005 was approximately 13.0%. The projected proportion in 2015 is given by $f(10) \approx 22.21$, or approximately 22.2%.

b. $f'(t) = \dfrac{(59 - t^{1/2})(150)\left(\frac{1}{2}t^{-1/2}\right) - (150t^{1/2} + 766)\left(-\frac{1}{2}t^{-1/2}\right)}{(59 - t^{1/2})^2} = \dfrac{4.808}{\sqrt{t}(59 - t^{1/2})^2} > 0$ for $0 < t < 10$, so f is increasing on $(0, 10)$. This says that the percentage of small and lower-midsize vehicles will be growing over the period from 2005 to 2015.

49. The revenue is $R(x) = px = x\left(-0.0005x^2 + 60\right) = -0.0005x^3 + 60x$. Therefore, the total profit is $P(x) = R(x) - C(x) = -0.0005x^3 + 0.001x^2 + 42x - 4000$. $P'(x) = -0.0015x^2 + 0.002x + 42$, and setting $P'(x) = 0$ gives $3x^2 - 4x - 84{,}000 = 0$. Solving for x, we find $x = \dfrac{4 \pm \sqrt{16 - 4(3)(-84{,}000)}}{2(3)} = \dfrac{4 \pm 1004}{6} = 168$ or -167. We reject the negative root. Next, $P''(x) = -0.003x + 0.002$ and $P''(168) = -0.003(168) + 0.002 = -0.502 < 0$. By the Second Derivative Test, $x = 168$ gives a relative maximum. Therefore, the required level of production is 168 DVDs.

51. a. $C(x) = 0.001x^2 + 100x + 4000$, so $\overline{C}(x) = \dfrac{C(x)}{x} = \dfrac{0.001x^2 + 100x + 4000}{x} = 0.001x + 100 + \dfrac{4000}{x}$.

b. $\overline{C}'(x) = 0.001 - \dfrac{4000}{x^2} = \dfrac{0.001x^2 - 4000}{x^2} = \dfrac{0.001\left(x^2 - 4{,}000{,}000\right)}{x^2}$. Setting $\overline{C}'(x) = 0$ gives $x = \pm 2000$. We reject the negative root. The sign diagram of $\overline{C}'$shows that $x = 2000$ gives rise to a relative minimum of $\overline{C}$. Because $\overline{C}''(x) = \dfrac{8000}{x^3} > 0$ if $x > 0$, we see that $\overline{C}$ is concave upward on $(0, \infty)$, and so $x = 2000$ yields an absolute minimum. The required production level is 2000 units.

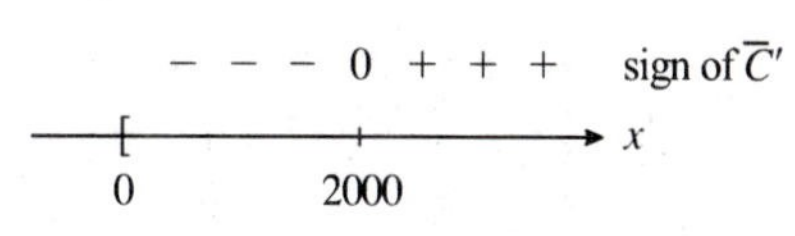

53. a. $P(t) = -0.0002t^3 + 0.018t^2 - 0.36t + 10$, so $P'(t) = -0.0006t^2 + 0.036t - 0.36$. Setting $P'(t) = 0$ gives $-0.0006t^2 + 0.036t - 0.36 = 0$, or $t^2 - 60t + 600 = 0$. Thus, $t = \dfrac{60 \pm \sqrt{60^2 - 4(1)(600)}}{2} \approx 12.7$ or 47.3. We reject the root 47.3 because it lies outside $[0, 30]$. The sign diagram for P' shows that P is decreasing on $(0, 12.7)$ and increasing on $(12.7, 30)$.

– – – 0 + + + + + sign of P'

0 ≈12.68 30 t

b. From the table, we see that the absolute minimum of P occurs at $t = 12.7$, and $P(12.7) \approx 7.9$.

t	0	12.7	10
$P(t)$	10	7.9	8

c. The percentage of women 65 and older in the workforce was decreasing from 1970 to September 1982 and increasing from September 1982 to 2000. It reached a minimum value of 7.9% in September 1982.

55. The volume is $V = f(x) = x(10 - 2x)^2$ in.^{3} for $0 \le x \le 5$. To maximize V, we compute

$$f'(x) = 12x^2 - 80x + 100 = 4\left(3x^2 - 20x + 25\right)$$
$$= 4(3x - 5)(x - 5).$$

Setting $f'(x) = 0$ gives $x = \frac{5}{3}$ and 5 as critical numbers of f. From the table, we see that the box has a maximum volume of 74.07 in.^{3}.

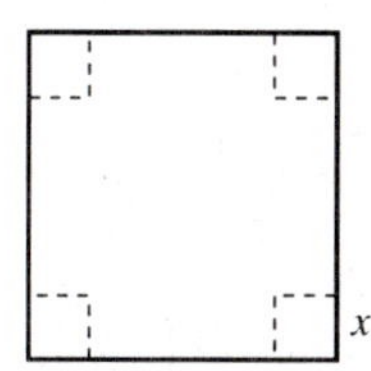

x	0	$\frac{5}{3}$	5
$f(x)$	0	$\frac{2000}{27} \approx 74.07$	0

57. $C(x) = 30(2)(2x)(x) + 20(2)(2xh + xh) = 120x^2 + 120xh$. But $x(2x)h = 4$, so $h = \dfrac{2}{x^2}$. Therefore,

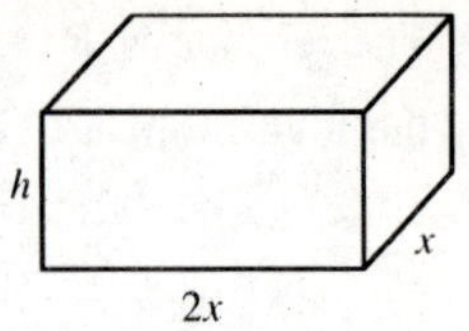

$C(x) = 120x^2 + 120x\left(\dfrac{2}{x^2}\right) = 120x^2 + \dfrac{240}{x}$, so $C'(x) = 240x - \dfrac{240}{x^2}$.

Setting $C'(x) = 0$ gives $240x - \dfrac{240}{x^2} = 0$, or $x^3 = 1$. Therefore, $x = 1$. $C''(x) = 240 + \dfrac{480}{x^3}$, and in particular, $C''(1) > 0$. Therefore, the cost is minimized by taking $x = 1$. The required dimensions are 1 ft × 2 ft × 2 ft.

59. $f(x) = x^2 + ax + b$, so $f'(x) = 2x + a$. We require that $f'(2) = 0$, so $(2)(2) + a = 0$, and $a = -4$. Next, $f(2) = 7$ implies that $f(2) = 2^2 + (-4)(2) + b = 7$, so $b = 11$. Thus, $f(x) = x^2 - 4x + 11$. Because the graph of f is a parabola that opens upward, $(2, 7)$ is a relative minimum.

61. Because $(a, f(a))$ is an inflection point of f, $f''(a) = 0$ or $f''(a)$ is not defined. This shows that a is a critical number of f'. Next, f changes concavity at $(a, f(a))$. If the concavity changes from concave downward to concave upward [that is, $f''(x) < 0$ for $x < a$ and $f''(x) > 0$ for $x > a$], then f' has a relative minimum at a. On the other hand, if the concavity changes from concave upward to concave downward, [$f''(x) > 0$ for $x < a$ and $f''(x) < 0$ for $x > a$], then f' has a relative maximum at a. In either case, f' has a relative extremum at a.

CHAPTER 10 Before Moving On... page 755

1. $f(x) = \dfrac{x^2}{1-x}$, so $f'(x) = \dfrac{(1-x)(2x) - x^2(-1)}{(1-x)^2} = \dfrac{2x - 2x^2 + x^2}{(1-x)^2} = \dfrac{x(2-x)}{(1-x)^2}$; f' is not defined at 1 and has zeros at 0 and 2.

The sign diagram of f' shows that f is decreasing on $(-\infty, 0)$ and $(2, \infty)$ and increasing on $(0, 1)$ and $(1, 2)$.

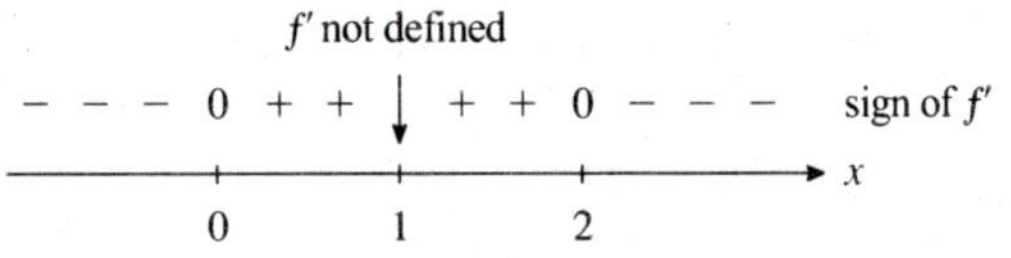

2. $f(x) = 4xe^{-x}$, so $f'(x) = 4x\frac{d}{dx}(e^{-x}) + e^{-x}\frac{d}{dx}(4x) = -4e^{-x} + 4e^{-x} = 4e^{-x}(1-x)$ and $f''(x) = 4e^{-x}(-1) + (1-x)(-4e^{-x}) = 4e^{-x}(-1-1+x) = 4e^{-x}(x-2)$. Because e^{-x} is never 0, the only critical number of f is 1. Also, because $f''(1) = 4e^{-1}(-1) < 0$, we see that f has a relative maximum at $(1, 4/e)$. From the sign diagram for f'', we see that $(2, 8e^{-2})$ is an inflection point of f.

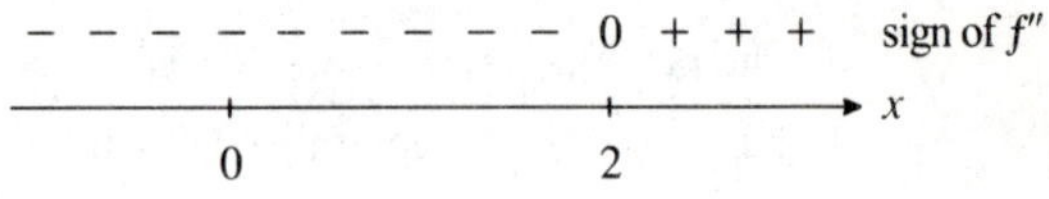

3. $f(x) = \frac{1}{3}x^3 - \frac{1}{4}x^2 - \frac{1}{2}x + 1$, so $f'(x) = x^2 - \frac{1}{2}x - \frac{1}{2}$ and $f''(x) = 2x - \frac{1}{2} = 0$ gives $x = \frac{1}{4}$. The sign diagram of f'' shows that f is concave downward on $\left(-\infty, \frac{1}{4}\right)$ and concave upward on $\left(\frac{1}{4}, \infty\right)$.

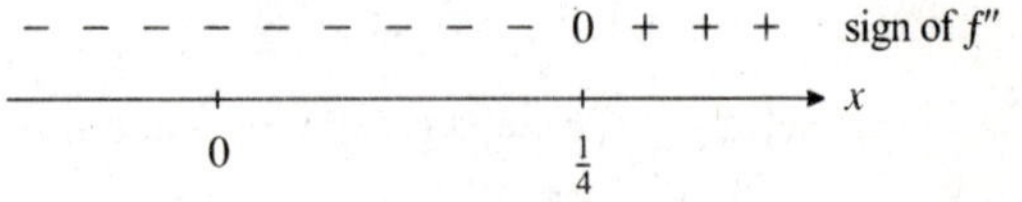

Because $f\left(\frac{1}{4}\right) = \frac{1}{3}\left(\frac{1}{4}\right)^3 - \frac{1}{4}\left(\frac{1}{4}\right)^2 - \frac{1}{2}\left(\frac{1}{4}\right) + 1 = \frac{83}{96}$, the inflection point is $\left(\frac{1}{4}, \frac{83}{96}\right)$.

4. $f(x) = 2x^3 - 9x^2 + 12x - 1$.

1. The domain of f is $(-\infty, \infty)$.
2. Setting $y = f(x) = 0$ gives -1 as the y-intercept of f.
3. $\lim_{x\to-\infty} f(x) = -\infty$ and $\lim_{x\to\infty} f(x) = \infty$.
4. There is no asymptote.
5. $f'(x) = 6x^2 - 18x + 12 = 6(x^2 - 3x + 2) = 6(x-2)(x-1)$.

   The sign diagram of f' shows that f is increasing on $(-\infty, 1)$ and $(2, \infty)$ and decreasing on $(1, 2)$.

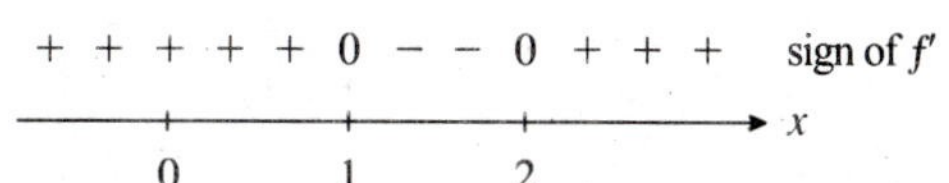

6. We see that $(1, 4)$ is a relative maximum and $(2, 3)$ is a relative minimum.
7. $f''(x) = 12x - 18 = 6(2x - 3)$. The sign diagram of f'' shows that f is concave downward on $\left(-\infty, \frac{3}{2}\right)$ and concave upward on $\left(\frac{3}{2}, \infty\right)$.

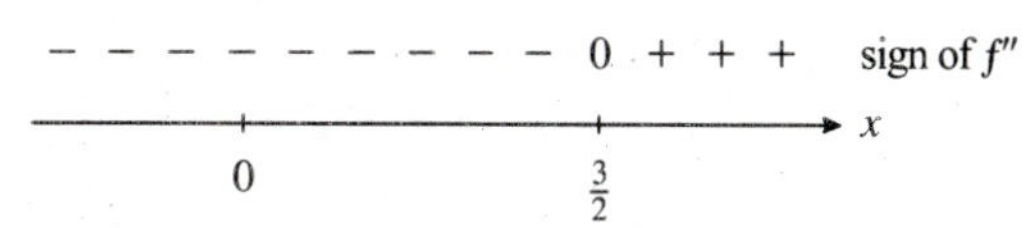

8. $f\left(\frac{3}{2}\right) = 2\left(\frac{3}{2}\right)^3 - 9\left(\frac{3}{2}\right)^2 + 12\left(\frac{3}{2}\right) - 1 = \frac{7}{2}$, so $\left(\frac{3}{2}, \frac{7}{2}\right)$ is an inflection point of f.

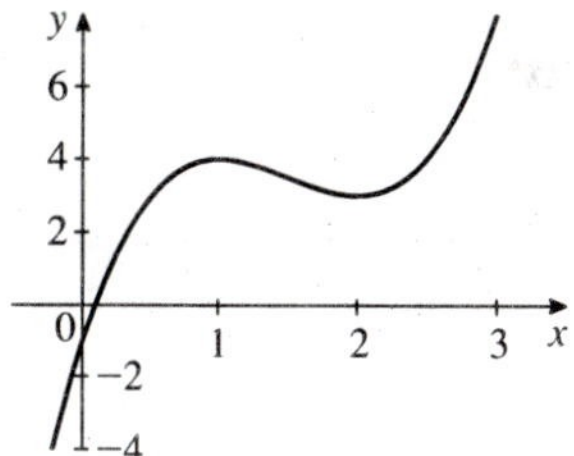

5. $f(x) = 2x^3 + 3x^2 - 1$ is continuous on the closed interval $[-2, 3]$. $f'(x) = 6x^2 + 6x = 6x(x+1)$, so the critical numbers of f are -1 and 0. From the table, we see that the absolute maximum value of f is 80 and the absolute minimum value is -5.

x	-2	-1	0	3
y	-5	0	-1	80

6. The amount of material used (the surface area) is $A = \pi r^2 + 2\pi rh$. But $V = \pi r^2 h = 1$, and so $h = \frac{1}{\pi r^2}$. Therefore, $A = \pi r^2 + 2\pi r\left(\frac{1}{\pi r^2}\right) = \pi r^2 + \frac{2}{r}$, so $A' = 2\pi r - \frac{2}{r^2} = 0$ implies $2\pi r = \frac{2}{r^2}$, $r^3 = \frac{1}{\pi}$, and $r = \frac{1}{\sqrt[3]{\pi}}$. Because $A'' = 2\pi + \frac{4}{r^3} > 0$ for $r > 0$, we see that $r = \frac{1}{\sqrt[3]{\pi}}$ does give an absolute minimum. Also, $h = \frac{1}{\pi r^2} = \frac{1}{\pi} \cdot \pi^{2/3} = \frac{1}{\pi^{1/3}} = \frac{1}{\sqrt[3]{\pi}}$. Therefore, the radius and height should each be $\frac{1}{\sqrt[3]{\pi}}$ ft.

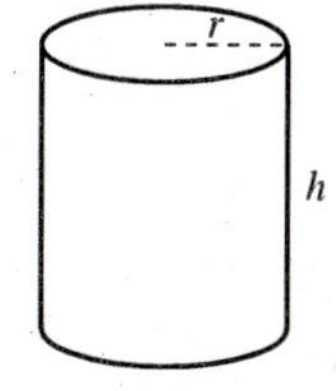

11 INTEGRATION

11.1 Antiderivatives and the Rules of Integration

Problem-Solving Tips

1. Get into the habit of using the correct notation for integration. The indefinite integral of $f(x)$ with respect to x is written $\int f(x)\,dx$. It is incorrect to write $\int f(x)$ without indicating that you are integrating with respect to x. You will appreciate how important the correct notation is if you use CAS or graphic calculator with the capability to do symbolic integration. If you don't enter this information (the variable with respect to which you are performing the integration) into your calculator or computer, the integration will not be performed.

2. If you are finding an indefinite integral, be sure to include a constant of integration in your answer. Remember that $\int f(x)\,dx$ is the family of functions given by $F(x)+C$, where $F'(x)=f(x)$.

3. It's very easy to check your answer if you are finding an indefinite integral. Just take the derivative of your answer and you should get the integrand. [You are verifying that $F'(x)=f(x)$.] If not, you know immediately that you have made an error.

Concept Questions page 766

1. An antiderivative of a continuous function f on an interval I is a function F such that $F'(x)=f(x)$ for every x in I. For example, an antiderivative of $f(x)=x^2$ on $(-\infty,\infty)$ is the function $F(x)=\frac{1}{3}x^3$ on $(-\infty,\infty)$.

3. The indefinite integral of f is the family of functions $F(x)+C$, where F is an antiderivative of f and C is an arbitrary constant.

Exercises page 766

1. $F(x)=\frac{1}{3}x^3+2x^2-x+2$, so $F'(x)=x^2+4x-1=f(x)$.

3. $F(x)=\left(2x^2-1\right)^{1/2}$, so $F'(x)=\frac{1}{2}\left(2x^2-1\right)^{-1/2}(4x)=\dfrac{2x}{\sqrt{2x^2-1}}=f(x)$.

5. a. $G'(x) = \frac{d}{dx}(2x) = 2 = f(x)$, and so G is an antiderivative of f.

b. $F(x) = G(x) + C = 2x + C$, where C is an arbitrary constant.

c. 

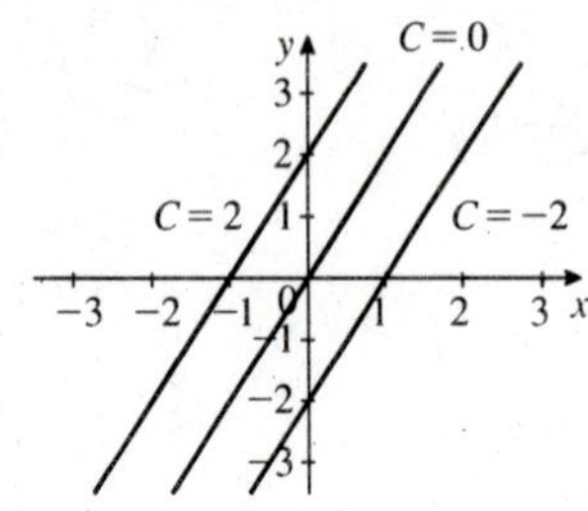

7. a. $G'(x) = \frac{d}{dx}\left(\frac{1}{3}x^3\right) = x^2 = f(x)$

b. $F(x) = G(x) + C = \frac{1}{3}x^3 + C$

c.

C=0 C=3 C=−3

9. $\int 6\,dx = 6x + C.$

11. $\int x^3\,dx = \frac{1}{4}x^4 + C.$

13. $\int x^{-4}\,dx = -\frac{1}{3}x^{-3} + C.$

15. $\int x^{2/3}\,dx = \frac{3}{5}x^{5/3} + C.$

17. $\int x^{-5/4}\,dx = -4x^{-1/4} + C.$

19. $\int \frac{2}{x^3}\,dx = 2\int x^{-3}\,dx = \frac{2}{2}\left(-x^{-2}\right) + C = -\frac{1}{x^2} + C.$

21. $\int \pi\sqrt{t}\,dt = \pi\int t^{1/2}\,dt = \pi\left(\frac{2}{3}t^{3/2}\right) + C$
$= \frac{2\pi}{3}t^{3/2} + C.$

23. $\int (3 - 4x)\,dx = \int 3\,dx - 4\int x\,dx = 3x - 2x^2 + C.$

25. $\int \left(x^2 + x + x^{-3}\right)dx = \int x^2\,dx + \int x\,dx + \int x^{-3}\,dx = \frac{1}{3}x^3 + \frac{1}{2}x^2 - \frac{1}{2}x^{-2} + C.$

27. $\int 5e^x\,dx = 5e^x + C.$

29. $\int (1 + x + e^x)\,dx = x + \frac{1}{2}x^2 + e^x + C.$

31. $\int \left(4x^3 - \frac{2}{x^2} - 1\right)dx = \int \left(4x^3 - 2x^{-2} - 1\right)dx = x^4 + 2x^{-1} - x + C = x^4 + \frac{2}{x} - x + C.$

33. $\int \left(x^{5/2} + 2x^{3/2} - x\right)dx = \frac{2}{7}x^{7/2} + \frac{4}{5}x^{5/2} - \frac{1}{2}x^2 + C.$

35. $\int \left(x^{1/2} + 2x^{-1/2}\right)dx = \frac{2}{3}x^{3/2} + 4x^{1/2} + C.$

37. $\int \left(\frac{u^3 + 2u^2 - u}{3u}\right)du = \frac{1}{3}\int \left(u^2 + 2u - 1\right)du = \frac{1}{9}u^3 + \frac{1}{3}u^2 - \frac{1}{3}u + C.$

39. $\int (2t + 1)(t - 2)\,dt = \int \left(2t^2 - 3t - 2\right)dt = \frac{2}{3}t^3 - \frac{3}{2}t^2 - 2t + C.$

41. $\int \frac{1}{x^2}\left(x^4 - 2x^2 + 1\right)dx = \int \left(x^2 - 2 + x^{-2}\right)dx = \frac{1}{3}x^3 - 2x - x^{-1} + C = \frac{1}{3}x^3 - 2x - \frac{1}{x} + C.$

43. $\int \frac{ds}{(s+1)^{-2}} = \int (s+1)^2\,ds = \int \left(s^2 + 2s + 1\right)ds = \frac{1}{3}s^3 + s^2 + s + C.$

45. $\int \left(e^t + t^e\right)dt = e^t + \frac{1}{e+1}t^{e+1} + C.$

47. $\int \frac{x^3 + x^2 - x + 1}{x^2}\,dx = \int \left(x + 1 - \frac{1}{x} + \frac{1}{x^2}\right)dx = \frac{1}{2}x^2 + x - \ln|x| - \frac{1}{x} + C$

49. $\displaystyle\int \frac{\left(x^{1/2}-1\right)^2}{x^2}\,dx = \int \frac{x-2x^{1/2}+1}{x^2}\,dx = \int \left(x^{-1}-2x^{-3/2}+x^{-2}\right)dx$

$\displaystyle= \ln|x| + 4x^{-1/2} - x^{-1} + C = \ln|x| + \frac{4}{\sqrt{x}} - \frac{1}{x} + C.$

51. $\int f'(x)\,dx = \int (3x+1)\,dx = \frac{3}{2}x^2 + x + C$. The condition $f(1) = 3$ gives $f(1) = \frac{3}{2} + 1 + C = 3$, so $C = \frac{1}{2}$. Therefore, $f(x) = \frac{3}{2}x^2 + x + \frac{1}{2}$.

53. $f'(x) = 3x^2 + 4x - 1$, so $f(x) = x^3 + 2x^2 - x + C$. Using the given initial condition, we have $f(2) = 8 + 2(4) - 2 + C = 9$, so $16 - 2 + C = 9$, or $C = -5$. Therefore, $f(x) = x^3 + 2x^2 - x - 5$.

55. $\displaystyle f(x) = \int f'(x)\,dx = \int \left(1 + \frac{1}{x^2}\right)dx = \int \left(1 + x^{-2}\right)dx = x - \frac{1}{x} + C$. Using the given initial condition, we have $f(1) = 1 - 1 + C = 3$, so $C = 3$. Therefore, $f(x) = x - \dfrac{1}{x} + 3$.

57. $\displaystyle f(x) = \int \frac{x+1}{x}\,dx = \int \left(1 + \frac{1}{x}\right)dx = x + \ln|x| + C$. Using the initial condition, we have $f(1) = 1 + \ln 1 + C = 1 + C = 1$, so $C = 0$. Thus, $f(x) = x + \ln|x|$.

59. $f(x) = \int f'(x)\,dx = \int \frac{1}{2}x^{-1/2}\,dx = \frac{1}{2}\left(2x^{1/2}\right) + C = x^{1/2} + C$, and $f(2) = \sqrt{2} + C = \sqrt{2}$ implies $C = 0$. Thus, $f(x) = \sqrt{x}$.

61. $f'(x) = e^x + x$, so $f(x) = e^x + \frac{1}{2}x^2 + C$ and $f(0) = e^0 + \frac{1}{2}(0) + C = 1 + C$. Thus, $3 = 1 + C$, and so $2 = C$. Therefore, $f(x) = e^x + \frac{1}{2}x^2 + 2$.

63. Because $f(t) \geq g(t)$ for all t in $[0, 180]$, we see that the amount on deposit at branch A is always growing at a rate faster than that of the amount on deposit at branch B. We conclude that branch A will have a larger amount on deposit that branch B at the end of 180 business days.

65. The position of the car is $s(t) = \int f(t)\,dt = \int 2\sqrt{t}\,dt = \int 2t^{1/2}\,dt = 2\left(\frac{2}{3}t^{3/2}\right) + C = \frac{4}{3}t^{3/2} + C$. $s(0) = 0$ implies that $s(0) = C = 0$, so $s(t) = \frac{4}{3}t^{3/2}$.

67. $C(x) = \int C'(x)\,dx = \int \left(0.000009x^2 - 0.009x + 8\right)dx = 0.000003x^3 - 0.0045x^2 + 8x + k$. $C(0) = k = 120$, and so $C(x) = 0.000003x^3 - 0.0045x^2 + 8x + 120$. Thus, $C(500) = 0.000003(500)^3 - 0.0045(500)^2 + 8(500) + 120 = \$3370$.

69. $P'(x) = -0.004x + 20$, so $P(x) = -0.002x^2 + 20x + C$. Because $C = -16{,}000$, we find that $P(x) = -0.002x^2 + 20x - 16{,}000$. The company realizes a maximum profit when $P'(x) = 0$, that is, when $x = 5000$ units. Next, $P(5000) = -0.002(5000)^2 + 20(5000) - 16{,}000 = 34{,}000$. Thus, the maximum profit of \$34,000 is realized at a production level of 5000 units.

71. **a.** $f(t) = \int r(t)\,dt = \int (0.0058t + 0.159)\,dt = 0.0029t^2 + 0.159t + C$. $f(0) = 1.6$, and so $0 + 0 + C = 1.6$, or $C = 1.6$. Therefore, $f(t) = 0.0029t^2 + 0.159t + 1.6$.

b. The national health expenditure in 2015 will be $f(13) = 0.0029\left(13^2\right) + 0.159(13) + 1.6 = 4.1571$, or approximately \$4.16 trillion.

73. a. The number of subscribers in year t is given by
$N(t) = \int r(t)\,dt = \int \left(-0.375t^2 + 2.1t + 2.45\right) dt = -0.125t^3 + 1.05t^2 + 2.45t + C$. To find C, note that $N(0) = 1.5$. This gives $N(0) = C = 1.5$. Therefore, $N(t) = -0.125t^3 + 1.05t^2 + 2.45t + 1.5$.

b. $N(5) = -0.125\left(5^3\right) + 1.05\left(5^2\right) + 2.45(5) + 1.5 = 24.375$, or 24.375 million subscribers.

75. a. The approximate average credit card debt per U.S. household in year t is
$A(t) = \int D(t)\,dt = \int \left(-4.479t^2 + 69.8t + 279.5\right) dt = -1.493t^3 + 34.9t^2 + 279.5t + C$. Using the condition $A(0) = 2917$, we find $A(0) = C = 2917$. Therefore, $A(t) = -1.493t^3 + 34.9t^2 + 279.5t + 2917$.

b. The average credit card debt per U.S. household in 2003 was
$A(13) = -1.493\left(13^3\right) + 34.9\left(13^2\right) + 279.5(13) + 2917 \approx 9168.479$, or approximately \$9168.

77. a. The number of gastric bypass surgeries performed in year t is
$N(t) = \int R(t)\,dt = \int \left(9.399t^2 - 13.4t + 14.07\right) dt = 3.133t^3 - 6.7t^2 + 14.07t + C$. Using the condition $N(0) = 36.7$, we find $N(0) = C = 36.7$. Therefore, $N(t) = 3.133t^3 - 6.7t^2 + 14.07t + 36.7$.

b. The number of bypass surgeries performed in 2003 was
$N(3) = 3.133\left(3^3\right) - 6.7\left(3^2\right) + 14.07(3) + 36.7 = 103.201$, or approximately 103,201.

79. a. We have the initial-value problem $C'(t) = 12.288t^2 - 150.5594t + 695.23$ with $C(0) = 3142$. Integrating, we find $C(t) = \int C'(t)\,dt = \int \left(12.288t^2 - 150.5594t + 695.23\right) dt = 4.096t^3 - 75.2797t^2 + 695.23t + k$. Using the initial condition, we find $C(0) = 0 + k = 3142$, and so $k = 3142$. Therefore, $C(t) = 4.096t^3 - 75.2797t^2 + 695.23t + 3142$.

b. The average out-of-pocket costs for beneficiaries is 2005 are
$C(1) = 4.096\left(1^3\right) - 75.2797\left(1^2\right) + 695.23(1) + 3142 = 3766.0463$, or approximately \$3766.05.

81. The number of new subscribers at any time is $N(t) = \int \left(100 + 210t^{3/4}\right) dt = 100t + 120t^{7/4} + C$. The given condition implies that $N(0) = 5000$. Using this condition, we find $C = 5000$. Therefore, $N(t) = 100t + 120t^{7/4} + 5000$. The number of subscribers 16 months from now is $N(16) = 100(16) + 120(16)^{7/4} + 5000$, or 21,960.

83. $h(t) = \int h'(t)\,dt = \int \left(-3t^2 + 192t\right) dt = -t^3 + 96t^2 + C = -t^3 + 96t^2 + C$. $h(0) = C = 0$ implies $h(t) = -t^3 + 96t^2$. The altitude 30 seconds after liftoff is $h(30) = -30^3 + 96(30)^2 = 59{,}400$ ft.

85. a.
$$S(t) = \int R(t)\,dt = \int \left(3t^3 - 17.9445t^2 + 28.7222t + 26.632\right) dt$$
$$= 0.75t^4 - 5.9815t^3 + 14.3611t^2 + 26.632t + C.$$
$S(0) = 108$, so $0 + C + 108$ and $C = 108$. Thus, $S(t) = 0.75t^4 - 5.9815t^3 + 14.3611t^2 + 26.632t + 108$.

b. The total sales of organic milk in 2004 were
$S(5) = 0.75(5)^4 - 5.9815(5)^3 + 14.3611(5)^2 + 26.632(5) + 108 = 321.25$, or \$321.25 million.

87. a. The number of health-care agencies in year t is $N(t) = \int \left(-0.0000372t^2 + 0.00372t - 0.186\right) dt$. Using the condition $N(0) = 9.3$, we obtain $C = 9.3$, so $N(t) = -0.0000124t^3 + 0.00186t^2 - 0.186t + 9.3$.

b. The number of health-care agencies in 2002 is
$N(14) = -0.0000124(14)^3 + 0.00186(14)^2 - 0.186(14) + 9.3 \approx 7.03$, or 7030.

c. The number of health-care agencies in 2005 is
$N(17) = -0.0000124(17)^3 + 0.00186(17)^2 - 0.186(17) + 9.3 \approx 6.61$, or 6610.

89. $v(r) = \int v'(r)\,dr = \int -kr\,dr = -\frac{1}{2}kr^2 + C$. But $v(R) = -\frac{1}{2}kR^2 + C = 0$, so $C = \frac{1}{2}kR^2$. Therefore, $v(R) = -\frac{1}{2}kr^2 + \frac{1}{2}kR^2 = \frac{1}{2}k\left(R^2 - r^2\right)$.

91. Denote the constant deceleration by k. Then $f''(t) = -k$, so $f'(t) = v(t) = -kt + C_1$. Next, the given condition implies that $v(0) = 88$. This gives $C_1 = 88$, so $f'(t) = -kt + 88$. Now $s = f(t) = \int f'(t)\,dt = \int(-kt + 88)\,dt = -\frac{1}{2}kt^2 + 88t + C_2$, and $f(0) = 0$ gives $s = f(t) = -\frac{1}{2}kt^2 + 88t$. Because the car is brought to rest in 9 seconds, we have $v(9) = -9k + 88 = 0$, or $k = \frac{88}{9} \approx 9.78$, so the deceleration is 9.78 ft/sec^2. The distance covered is $s = f(9) = -\frac{1}{2}\left(\frac{88}{9}\right)(81) + 88(9) = 396$, so the stopping distance is 396 ft.

93. The time taken by runner A to cross the finish line is $t = \frac{200}{22} = \frac{100}{11}$ sec. Let a be the constant acceleration of runner B as he begins to spurt. Then $\dfrac{dv}{dt} = a$, so the velocity of runner B as he runs towards the finish line is $v = \int a\,dt = at + c$. At $t = 0$, $v = 20$ and so $v = at = 0$ and $c = 20$. Now $\frac{ds}{dt} = v = at + 20$, so $s = \int(at + 20)\,dt = \frac{1}{2}at^2 + 20t + k$, where k is the constant of integration. Next, $s(0) = 0$ gives $s = \frac{1}{2}at^2 + 20t = \left(\frac{1}{2}at + 20\right)t$. In order for runner B to cover 220 ft in $\frac{100}{11}$ sec, we must have $\left[\frac{1}{2}a\left(\frac{100}{11}\right) + 20\right]\frac{100}{11} = 220$, so $\frac{50}{11}a + 20 = \frac{220 \cdot 11}{100} = \frac{121}{5}$, $\frac{50}{11}a = \frac{121}{5} - 20 = \frac{21}{5}$, and $a = \frac{21}{5} \cdot \frac{11}{50} = 0.924$ ft/sec^2. Therefore, runner B must have an acceleration of at least 0.924 ft/sec^2.

95. a. We have the initial-value problem $R'(t) = 8\sqrt{2}t^{1/2} - 32t^3$ with $R(0) = 0$. Integrating, we find $R(t) = \int\left(8\sqrt{2}t^{1/2} - 32t^3\right)dt = \frac{16\sqrt{2}}{3}t^{3/2} - 8t^4 + C$. $R(0) = 0$ implies that $C = 0$, so $R(t) = \frac{16\sqrt{2}}{3}t^{3/2} - 8t^4$.

b. $R\left(\frac{1}{2}\right) = \frac{16\sqrt{2}}{3}\left(\frac{1}{2}\right)^{3/2} - 8\left(\frac{1}{2}\right)^4 \approx 2.166$, so after $\frac{1}{2}$ hr, approximately 2.2 inches of rain had fallen.

97. True. See the proof in Section 11.1 of the text.

99. True. Use the Sum Rule followed by the Constant Multiple Rule.

11.2 Integration by Substitution

Problem-Solving Tips

1. Here are some tips for using the method of substitution.

a. The idea is to replace the given integral by a simpler integral, so look for a substitution $u = g(x)$ that simplifies the integral.

b. Check to see that $du = g'(x)\,dx$ appears in the integral.

2. Look through Problems 1–50 to familiarize yourself with the types of functions that can be integrated using the method of substitution. Even if you don't complete every problem, check to see if you can set up the given integral so that the method of substitution can be used to complete the integration.

Concept Questions page 778

1. To find $I = \int f(g(x))g'(x)\,dx$ by the Method of Substitution, let $u = g(x)$, so that $du = g'(x)\,dx$. Making the substitution, we obtain $I = \int f(u)\,du$, which can be integrated with respect to u. Finally, replace u by $u = g(x)$ to evaluate the integral.

Exercises page 778

1. Let $u = 4x + 3$, so $du = 4\,dx$ and $dx = \frac{1}{4}\,du$. Then $\int 4(4x+3)^4\,dx = \int u^4\,du = \frac{1}{5}u^5 + C = \frac{1}{5}(4x+3)^5 + C$.

3. Let $u = x^3 - 2x$, so $du = (3x^2 - 2)\,dx$. Then $\int (x^3 - 2x)^2 (3x^2 - 2)\,dx = \int u^2\,du = \frac{1}{3}u^3 + C = \frac{1}{3}(x^3 - 2x)^3 + C$.

5. Let $u = 2x^2 + 3$, so $du = 4x\,dx$. Then $\displaystyle\int \frac{4x}{(2x^2+3)^3}\,dx = \int \frac{1}{u^3}\,du = \int u^{-3}\,du = -\frac{1}{2}u^{-2} + C = -\frac{1}{2(2x^2+3)^2} + C$.

7. Let $u = t^3 + 2$, so $du = 3t^2\,dt$ and $t^2\,dt = \frac{1}{3}\,du$. Then $\int 3t^2\sqrt{t^3+2}\,dt = \int u^{1/2}\,du = \frac{2}{3}u^{3/2} + C = \frac{2}{3}(t^3+2)^{3/2} + C$.

9. Let $u = x^2 - 1$, so $du = 2x\,dx$ and $x\,dx = \frac{1}{2}\,du$. Then $\int 2(x^2-1)^9 x\,dx = \int u^9\,du = \frac{1}{10}u^{10} + C = \frac{1}{10}(x^2-1)^{10} + C$.

11. Let $u = 1 - x^5$, so $du = -5x^4\,dx$ and $x^4\,dx = -\frac{1}{5}\,du$. Then $\displaystyle\int \frac{x^4}{1-x^5}\,dx = -\frac{1}{5}\int \frac{du}{u} = -\frac{1}{5}\ln|u| + C = -\frac{1}{5}\ln\left|1 - x^5\right| + C$.

13. Let $u = x - 2$, so $du = dx$. Then $\displaystyle\int \frac{2}{x-2}\,dx = 2\int \frac{du}{u} = 2\ln|u| + C = \ln u^2 + C = \ln(x-2)^2 + C$.

15. Let $u = 0.3x^2 - 0.4x + 2$. Then $du = (0.6x - 0.4)\,dx = 2(0.3x - 0.2)\,dx$. Thus, $\displaystyle\int \frac{0.3x - 0.2}{0.3x^2 - 0.4x + 2}\,dx = \int \frac{1}{2u}\,du = \frac{1}{2}\ln|u| + C = \frac{1}{2}\ln(0.3x^2 - 0.4x + 2) + C$.

17. Let $u = 3x^2 - 1$, so $du = 6x\,dx$ and $x\,dx = \frac{1}{6}\,du$. Then $\displaystyle\int \frac{2x}{3x^2-1}\,dx = \frac{1}{3}\int \frac{du}{u} = \frac{1}{3}\ln|u| + C = \frac{1}{3}\ln\left|3x^2 - 1\right| + C$.

19. Let $u = -2x$, so $du = -2\,dx$ and $dx = -\frac{1}{2}\,du$. Then $\int e^{-2x}\,dx = -\frac{1}{2}\int e^u\,du = -\frac{1}{2}e^u + C = -\frac{1}{2}e^{-2x} + C$.

21. Let $u = 2 - x$, so $du = -dx$ and $dx = -du$. Then $\int e^{2-x}\,dx = -\int e^u\,du = -e^u + C = -e^{2-x} + C$.

23. Let $u = -x^2$, so $du = -2x\,dx$ and $x\,dx = -\frac{1}{2}\,du$. Then $\int xe^{-x^2}\,dx = \int -\frac{1}{2}e^u\,du = -\frac{1}{2}e^u + C = -\frac{1}{2}e^{-x^2} + C$.

25. $\int (e^x - e^{-x})\,dx = \int e^x\,dx - \int e^{-x}\,dx = e^x - \int e^{-x}\,dx$. To evaluate the second integral on the right, let $u = -x$ so $du = -dx$ and $dx = -du$. Then $\int (e^x - e^{-x})\,dx = e^x + \int e^u\,du = e^x + e^u + C = e^x + e^{-x} + C$.

27. Let $u = 1 + e^x$, so $du = e^x\,dx$. Then $\displaystyle\int \frac{2e^x}{1+e^x}\,dx = 2\int \frac{du}{u} = 2\ln|u| + C = 2\ln(1+e^x) + C$.

29. Let $u = \sqrt{x} = x^{1/2}$. Then $du = \frac{1}{2}x^{-1/2}\,dx$ and $2\,du = x^{-1/2}\,dx$, so
$\displaystyle\int \frac{e^{\sqrt{x}}}{\sqrt{x}}\,dx = \int 2e^u\,du = 2e^u + C = 2e^{\sqrt{x}} + C.$

31. Let $u = e^{3x} + x^3$, so $du = \left(3e^{3x} + 3x^2\right)dx = 3\left(e^{3x} + x^2\right)dx$ and $\left(e^{3x} + x^2\right)dx = \frac{1}{3}du$. Then
$\displaystyle\int \frac{e^{3x} + x^2}{\left(e^{3x} + x^3\right)^3}\,dx = \frac{1}{3}\int \frac{du}{u^3} = \frac{1}{3}\int u^{-3}\,du = -\frac{u^{-2}}{6} + C = -\frac{1}{6\left(e^{3x} + x^3\right)^2} + C.$

33. Let $u = e^{2x} + 1$, so $du = 2e^{2x}\,dx$ and $\frac{1}{2}\,du = e^{2x}\,dx$. Then
$\int e^{2x}\left(e^{2x} + 1\right)^3 dx = \int \frac{1}{2}u^3\,du = \frac{1}{8}u^4 + C = \frac{1}{8}\left(e^{2x} + 1\right)^4 + C.$

35. Let $u = \ln 5x$, so $du = \frac{1}{x}\,dx$. Then $\displaystyle\int \frac{\ln 5x}{x}\,dx = \int u\,du = \frac{1}{2}u^2 + C = \frac{1}{2}(\ln 5x)^2 + C.$

37. Let $u = \ln x$, so $du = \frac{1}{x}\,dx$. Then $\displaystyle\int \frac{2}{x\ln x}\,dx = 2\int \frac{du}{u} = 2\ln|u| + C = 2\ln|\ln x| + C.$

39. Let $u = \ln x$, so $du = \frac{1}{x}\,dx$. Then $\displaystyle\int \frac{\sqrt{\ln x}}{x}\,dx = \int \sqrt{u}\,du = \frac{2}{3}u^{3/2} + C = \frac{2}{3}(\ln x)^{3/2} + C.$

41. $\displaystyle\int \left(xe^{x^2} - \frac{x}{x^2+2}\right)dx = \int xe^{x^2}\,dx - \int \frac{x}{x^2+2}\,dx$. To evaluate the first integral, let $u = x^2$, so $du = 2x\,dx$ and $x\,dx = \frac{1}{2}\,du$. Then $\int xe^{x^2}\,dx = \frac{1}{2}\int e^u\,du + C_1 = \frac{1}{2}e^u + C_1 = \frac{1}{2}e^{x^2} + C_1$. To evaluate the second integral, let $u = x^2 + 2$, so $du = 2x\,dx$ and $x\,dx = \frac{1}{2}\,du$. Then $\displaystyle\int \frac{x}{x^2+2}\,dx = \frac{1}{2}\int \frac{du}{u} = \frac{1}{2}\ln|u| + C_2 = \frac{1}{2}\ln\left(x^2+2\right) + C_2$.
Therefore, $\displaystyle\int \left(xe^{x^2} - \frac{x}{x^2+2}\right)dx = \frac{1}{2}e^{x^2} - \frac{1}{2}\ln\left(x^2+2\right) + C.$

43. Let $u = \sqrt{x} - 1$, so $du = \frac{1}{2}x^{-1/2}\,dx = \frac{1}{2\sqrt{x}}\,dx$ and $dx = 2\sqrt{x}\,du$. Also, we have $\sqrt{x} = u + 1$, so $x = (u+1)^2 = u^2 + 2u + 1$ and $dx = 2(u+1)\,du$. Thus,

$$\begin{aligned}\int \frac{x+1}{\sqrt{x}-1}\,dx &= \int \frac{u^2+2u+2}{u}\cdot 2(u+1)\,du = 2\int \frac{\left(u^3+3u^2+4u+2\right)}{u}\,du\\ &= 2\int\left(u^2+3u+4+\tfrac{2}{u}\right)du = 2\left(\tfrac{1}{3}u^3+\tfrac{3}{2}u^2+4u+2\ln|u|\right)+C\\ &= 2\left[\tfrac{1}{3}\left(\sqrt{x}-1\right)^3+\tfrac{3}{2}\left(\sqrt{x}-1\right)^2+4\left(\sqrt{x}-1\right)+2\ln\left|\sqrt{x}-1\right|\right]+C.\end{aligned}$$

45. Let $u = x - 1$, so $du = dx$. Also, $x = u + 1$, and so

$$\begin{aligned}\int x(x-1)^5\,dx &= \int (u+1)u^5\,du = \int \left(u^6+u^5\right)du = \tfrac{1}{7}u^7 + \tfrac{1}{6}u^6 + C = \tfrac{1}{7}(x-1)^7 + \tfrac{1}{6}(x-1)^6 + C\\ &= \frac{(6x+1)(x-1)^6}{42} + C.\end{aligned}$$

47. Let $u = 1 + \sqrt{x}$, so $du = \frac{1}{2}x^{-1/2}\,dx$ and $dx = 2\sqrt{x}\,du = 2(u-1)\,du$. Then

$$\int \frac{1-\sqrt{x}}{1+\sqrt{x}}dx = \int \left(\frac{1-(u-1)}{u}\right) \cdot 2(u-1)\,du = 2\int \frac{(2-u)(u-1)}{u}\,du = 2\int \frac{-u^2+3u-2}{u}\,du$$

$$= 2\int \left(-u+3-\frac{2}{u}\right) du = -u^2 + 6u - 4\ln|u| + C_1$$

$$= -\left(1+\sqrt{x}\right)^2 + 6\left(1+\sqrt{x}\right) - 4\ln\left(1+\sqrt{x}\right) + C_1$$

$$= -1 - 2\sqrt{x} - x + 6 + 6\sqrt{x} - 4\ln\left(1+\sqrt{x}\right) + C_1 = -x + 4\sqrt{x} - 4\ln\left(1+\sqrt{x}\right) + C.$$

49. $I = \int v^2(1-v)^6\,dv$. Let $u = 1 - v$, so $du = -dv$. Also, $1 - u = v$, and so $(1-u)^2 = v^2$. Therefore,

$$I = \int -\left(1-2u+u^2\right)u^6 du = \int -\left(u^6 - 2u^7 + u^8\right) du = -\left(\tfrac{1}{7}u^7 - \tfrac{1}{4}u^8 + \tfrac{1}{9}u^9\right) + C$$

$$= -u^7\left(\tfrac{1}{7} - \tfrac{1}{4}u + \tfrac{1}{9}u^2\right) + C = -\tfrac{1}{252}(1-v)^7\left[36 - 63(1-v) + 28\left(1-2v+v^2\right)\right]$$

$$= -\tfrac{1}{252}(1-v)^7\left(36 - 63 + 63v + 28 - 56v + 28v^2\right) + C = -\tfrac{1}{252}(1-v)^7\left(28v^2 + 7v + 1\right) + C.$$

51. $f(x) = \int f'(x)\,dx = 5\int(2x-1)^4\,dx$. Let $u = 2x - 1$, so $du = 2\,dx$ and $dx = \frac{1}{2}\,du$. Then $f(x) = \frac{5}{2}\int u^4\,du = \frac{1}{2}u^5 + C = \frac{1}{2}(2x-1)^5 + C$. Next, $f(1) = 3$ implies $\frac{1}{2} + C = 3$, so $C = \frac{5}{2}$. Therefore, $f(x) = \frac{1}{2}(2x-1)^5 + \frac{5}{2}$.

53. $f(x) = \int -2xe^{-x^2+1}\,dx$. Let $u = -x^2 + 1$, so $du = -2x\,dx$. Then $f(x) = \int e^u du = e^u + C = e^{-x^2+1} + C$. The condition $f(1) = 0$ implies $f(1) = 1 + C = 0$, so $C = -1$. Therefore, $f(x) = e^{-x^2+1} - 1$.

55. The number of subscribers at time t is $N(t) = \int R(t)\,dt = \int 3.36(t+1)^{0.05}\,dt$. Let $u = t + 1$, so $du = dt$. Then $N = 3.36\int u^{0.05}\,du = 3.2u^{1.05} + C = 3.2(t+1)^{1.05} + C$. To find C, use the condition $N(0) = 3.2$ to calculate $N(0) = 3.2 + C = 3.2$, so $C = 0$. Therefore, $N(t) = 3.2(t+1)^{1.05}$. If the projection holds true, then the number of subscribers at the beginning of 2008 is $N(4) = 3.2(4+1)^{1.05} \approx 17.341$, or 17.341 million.

57. $N(t) = \int 2000(1+0.2t)^{-3/2}\,dt$. Let $u = 1 + 0.2t$, so $du = 0.2\,dt$ and $5\,du = dt$. Then $N(t) = (5)(2000)\int u^{-3/2}\,du = -20{,}000u^{-1/2} + C = -20{,}000(1+0.2t)^{-1/2} + C$. Next, $N(0) = -20{,}000(1)^{-1/2} + C = 1000$. Therefore, $C = 21{,}000$ and $N(t) = -\dfrac{20{,}000}{\sqrt{1+0.2t}} + 21{,}000$. In particular, $N(5) = -\frac{20{,}000}{\sqrt{2}} + 21{,}000 \approx 6858$, or approximately 6858 students.

59. $p(x) = -\displaystyle\int \frac{250x}{\left(16+x^2\right)^{3/2}}\,dx = -250\int \frac{x}{\left(16+x^2\right)^{3/2}}\,dx$. Let $u = 16 + x^2$, so $du = 2x\,dx$ and $x\,dx = \frac{1}{2}\,du$.

Then $p(x) = -\frac{250}{2}\int u^{-3/2}\,du = (-125)(-2)u^{-1/2} + C = \dfrac{250}{\sqrt{16+x^2}} + C$. $p(3) = \dfrac{250}{\sqrt{16+9}} + C = 50$ implies $C = 0$, and so $p(x) = \dfrac{250}{\sqrt{16+x^2}}$.

61. Let $u = 2t + 4$, so $du = 2\,dt$. Then $r(t) = \displaystyle\int \frac{30}{\sqrt{2t+4}}\,dt = 30\int \tfrac{1}{2}u^{-1/2}\,du = 30u^{1/2} + C = 30\sqrt{2t+4} + C$. $r(0) = 60 + C = 0$, so $C = -60$. Therefore, $r(t) = 30\left(\sqrt{2t+4} - 2\right)$. Then $r(16) = 30\left(\sqrt{36} - 2\right) = 120$ ft, so the polluted area is $\pi r^2 = \pi(120)^2 = 14{,}400\pi$, or $14{,}400\pi$ ft^2.

63. Let $u = 1 + 2.449e^{-0.3277t}$, so $du = -0.8025373e^{-0.3277t}\,dt$ and $e^{-0.3277t}\,dt = -1.246048003\,du$. Then $h(t) = \int \frac{52.8706e^{-0.3277t}}{\left(1 + 2.449e^{-0.3277t}\right)^2}\,dt = 52.8706\,(-1.246048003) \int \frac{du}{u^2} = 65.8793u^{-1} + C = \frac{65.8793}{1 + 2.449e^{-0.3277t}} + C$. $h(0) = \frac{65.8793}{1 + 2.449} + C = 19.4$, so $C \approx 0.3$ and $h(t) = \frac{65.8793}{1 + 2.449e^{-0.3277t}} + 0.3$ and hence $h(8) = \frac{65.8793}{1 + 2.449e^{-0.3277(8)}} + 0.3 \approx 56.22$, or 56.22 inches.

65. $A(t) = \int A'(t)\,dt = r \int e^{-at}\,dt$. Let $u = -at$, so $du = -a\,dt$ and $dt = -\frac{1}{a}\,du$. Then $A(t) = r\left(-\frac{1}{a}\right) \int e^u\,du = -\frac{r}{a}e^u + C = -\frac{r}{a}e^{-at} + C$. $A(0) = 0$ implies $-\frac{r}{a} + C = 0$, so $C = \frac{r}{a}$. Therefore, $A(t) = -\frac{r}{a}e^{-at} + \frac{r}{a} = \frac{r}{a}\left(1 - e^{-at}\right)$.

67. True. Let $I = \int x f\left(x^2\right) dx$ and put $u = x^2$. Then $du = 2x\,dx$ and $x\,dx = \frac{1}{2}\,du$, so $I = \frac{1}{2} \int f(u)\,du = \frac{1}{2} \int f(x)\,dx$.

11.3 Area and the Definite Integral

Problem-Solving Tips

1. In Sections 11.1 and 11.2, we found the indefinite integral of a function, that is, $\int f(x)\,dx = F(x) + C$. Note that our answer is a *family of functions* $F(x) + C$ for which $F'(x) = f(x)$. In this section, we found the definite integral of a function, $\int_a^b f(x)\,dx = \lim_{n\to\infty} \left[f(x_1)\,\Delta x + f(x_2)\,\Delta x + \cdots + f(x_n)\,\Delta x\right]$. Note that the answer here is a *number*.

2. The geometric interpretation of a definite integral is as follows: If f is continuous on $[a, b]$, then $\int_a^b f(x)\,dx$ is equal to the area of the region above the x-axis between the x-axis and the graph of f over $[a, b]$ minus the area of the corresponding region below the x-axis.

Concept Questions page 788

1. See page 785 in the text.

Exercises page 788

1. $\frac{1}{3}(1.9 + 1.5 + 1.8 + 2.4 + 2.7 + 2.5) = \frac{12.8}{3} \approx 4.27$.

3. a.

$A = \frac{1}{2}(2)(6) = 6$.

b. $\Delta x = \frac{2}{4} = \frac{1}{2}$, so $x_1 = 0$, $x_2 = \frac{1}{2}$, $x_3 = 1$, $x_4 = \frac{3}{2}$. Thus,

$$A \approx \frac{1}{2}\left[3(0) + 3\left(\frac{1}{2}\right) + 3(1) + 3\left(\frac{3}{2}\right)\right] = \frac{9}{2} = 4.5.$$

c. $\Delta x = \frac{2}{8} = \frac{1}{4}$, so $x_1 = 0, \ldots, x_8 = \frac{7}{4}$. Thus,

$$A \approx \frac{1}{4}\left[3(0) + 3\left(\frac{1}{4}\right) + 3\left(\frac{1}{2}\right) + 3\left(\frac{3}{4}\right) + 3(1) + 3\left(\frac{5}{4}\right) + 3\left(\frac{3}{2}\right) + 3\left(\frac{7}{4}\right)\right] = \frac{21}{4} = 5.25.$$

d. Yes.

5. a.

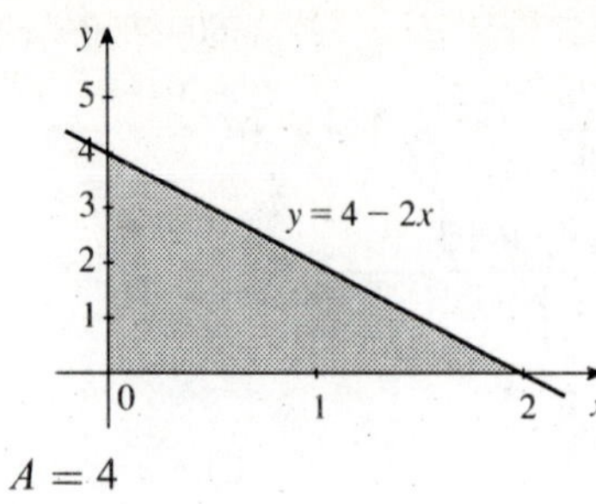

$A = 4$

b. $\Delta x = \frac{2}{5} = 0.4$, so $x_1 = 0, x_2 = 0.4, x_3 = 0.8, x_4 = 1.2, x_5 = 1.6$.
Thus, $A \approx 0.4\{[4-2(0)]+[4-2(0.4)]+[4-2(0.8)] + [4-2(1.2)]+[4-2(1.6)]\} = 4.8$.

c. $\Delta x = \frac{2}{10} = 0.2$, so $x_1 = 0, x_2 = 0.2, x_3 = 0.4, \ldots, x_{10} = 1.8$.
Thus,
$$A \approx 0.2\{[4-2(0)]+[4-2(0.2)]+[4-2(0.4)] + [4-2(0.6)]+[4-2(0.8)]+[4-2(1.0)]+[4-2(1.2)] + [4-2(1.4)]+[4-2(1.6)]+[4-2(1.8)]\} = 4.4.$$

d. Yes.

7. a. $\Delta x = \frac{4-2}{2} = 1$, so $x_1 = 2.5, x_2 = 3.5$. The Riemann sum is $1\left(2.5^2+3.5^2\right) = 18.5$.

b. $\Delta x = \frac{4-2}{5} = 0.4$, so $x_1 = 2.2, x_2 = 2.6, x_3 = 3.0, x_4 = 3.4, x_5 = 3.8$. The Riemann sum is $0.4\left(2.2^2+2.6^2+3.0^2+3.4^2+3.8^2\right) = 18.64$.

c. $\Delta x = \frac{4-2}{10} = 0.2$, so $x_1 = 2.1, x_2 = 2.3, x_2 = 2.5, \ldots, x_{10} = 3.9$. The Riemann sum is $0.2\left(2.1^2+2.3^2+2.5^2+2.7^2+2.9^2+3.1^2+3.3^2+3.5^2+3.7^2+3.9^2\right) = 18.66$.

d. The area appears to be $18\frac{2}{3}$.

9. a. $\Delta x = \frac{4-2}{2} = 1$, so $x_1 = 3, x_2 = 4$. The Riemann sum is $(1)\left(3^2+4^2\right) = 25$.

b. $\Delta x = \frac{4-2}{5} = 0.4$, so $x_1 = 2.4, x_2 = 2.8, x_3 = 3.2, x_4 = 3.6, x_5 = 4$. The Riemann sum is $0.4\left(2.4^2+2.8^2+\cdots+4^2\right) = 21.12$.

c. $\Delta x = \frac{4-2}{10} = 0.2$, so $x_1 = 2.2, x_2 = 2.4, x_3 = 2.6, \ldots, x_{10} = 4$. The Riemann sum is $0.2\left(2.2^2+2.4^2+2.6^2+2.8^2+3.0^2+3.2^2+3.4^2+3.6^2+3.8^2+4^2\right) = 19.88$.

d. The area appears to be approximately 19.9.

11. a. $\Delta x = \frac{1}{2}$, so $x_1 = 0, x_2 = \frac{1}{2}$. The Riemann sum is $f(x_1)\Delta x + f(x_2)\Delta x = \left[(0)^3 + \left(\frac{1}{2}\right)^3\right]\frac{1}{2} = \frac{1}{16} = 0.0625$.

b. $\Delta x = \frac{1}{5}$, so $x_1 = 0, x_2 = \frac{1}{5}, x_3 = \frac{2}{5}, x_4 = \frac{3}{5}, x_5 = \frac{4}{5}$. The Riemann sum is
$$f(x_1)\Delta x + f(x_2)\Delta x + \cdots + f(x_5)\Delta x = \left[0^3 + \left(\tfrac{1}{5}\right)^3 + \left(\tfrac{2}{5}\right)^3 + \left(\tfrac{3}{5}\right)^3 + \left(\tfrac{4}{5}\right)^3\right]\frac{1}{5} = \frac{100}{625} = 0.16.$$

c. $\Delta x = \frac{1}{10}$, so $x_1 = 0, x_2 = \frac{1}{10}, x_3 = \frac{2}{10}, \ldots, x_{10} = \frac{9}{10}$. The Riemann sum is
$$f(x_1)\Delta x + f(x_2)\Delta x + \cdots + f(x_{10})\Delta x$$
$$= \left[0^3 + \left(\tfrac{1}{10}\right)^3 + \left(\tfrac{2}{10}\right)^3 + \left(\tfrac{3}{10}\right)^3 + \left(\tfrac{4}{10}\right)^3 + \left(\tfrac{5}{10}\right)^3 + \left(\tfrac{6}{10}\right)^3 + \left(\tfrac{7}{10}\right)^3 + \left(\tfrac{8}{10}\right)^3 + \left(\tfrac{9}{10}\right)^3\right]\frac{1}{10}$$
$$= \frac{2025}{10{,}000} = 0.2025 \approx 0.2.$$

d. The Riemann sums seem to approach 0.2.

13. $\Delta x = \frac{2-0}{5} = \frac{2}{5}$, so $x_1 = \frac{1}{5}, x_2 = \frac{3}{5}, x_3 = \frac{5}{5}, x_4 = \frac{7}{5}, x_5 = \frac{9}{5}$. Thus,
$$A \approx \left\{\left[\left(\tfrac{1}{5}\right)^2+1\right]+\left[\left(\tfrac{3}{5}\right)^2+1\right]+\left[\left(\tfrac{5}{5}\right)^2+1\right]+\left[\left(\tfrac{7}{5}\right)^2+1\right]+\left[\left(\tfrac{9}{5}\right)^2+1\right]\right\}\left(\tfrac{2}{5}\right) = \frac{580}{125} = 4.64.$$

15. $\Delta x = \frac{3-1}{4} = \frac{1}{2}$, so $x_1 = \frac{3}{2}, x_2 = \frac{4}{2} = 2, x_3 = \frac{5}{2}, x_4 = 3$. Thus, $A \approx \left(\frac{1}{3/2} + \frac{1}{2} + \frac{1}{5/2} + \frac{1}{3}\right)\frac{1}{2} \approx 0.95$.

17. $A \approx 20\left[f(10)+f(30)+f(50)+f(70)+f(90)\right]=20(80+100+110+100+80)=9400\text{ ft}^2$.

11.4 The Fundamental Theorem of Calculus

Concept Questions page 798

1. See the Fundamental Theorem of Calculus on page 790 of the text.

Exercises page 799

1. $A=\int_1^4 2\,dx=2x|_1^4=2(4-1)=6$. The region is a rectangle with area $3\cdot 2=6$.

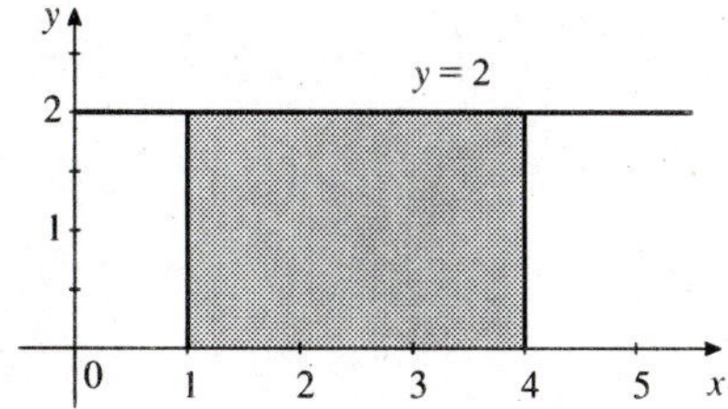

3. $A=\int_1^3 2x\,dx=x^2\Big|_1^3=9-1=8$. The region is a parallelogram with area $\frac{1}{2}(3-1)(2+6)=8$.

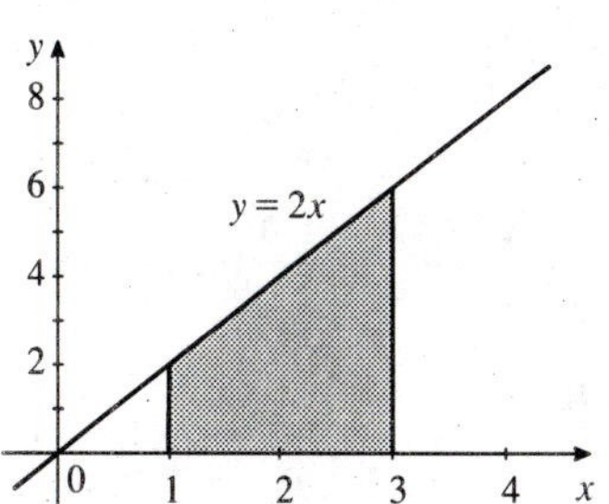

5. $A=\int_{-1}^2(2x+3)\,dx=\left(x^2+3x\right)\Big|_{-1}^2=(4+6)-(1-3)=12$.

7. $A=\int_{-1}^2\left(-x^2+4\right)dx=\left(-\frac{1}{3}x^3+4x\right)\Big|_{-1}^2=\left(-\frac{8}{3}+8\right)-\left(\frac{1}{3}-4\right)=9$.

9. $A=\displaystyle\int_1^2\frac{1}{x}\,dx=\ln x|_1^2=\ln 2-\ln 1=\ln 2$.

11. $A=\int_1^9\sqrt{x}\,dx=\frac{2}{3}x^{3/2}\Big|_1^9=\frac{2}{3}(27-1)=\frac{52}{3}$.

13. $A=\int_{-8}^{-1}\left(1-x^{1/3}\right)dx=\left(x-\frac{3}{4}x^{4/3}\right)\Big|_{-8}^{-1}=\left(-1-\frac{3}{4}\right)-(-8-12)=\frac{73}{4}$.

15. $A=\int_0^2 e^x\,dx=e^x|_0^2=e^2-1\approx 6.39$.

17. $\int_2^4 3\,dx=3x|_2^4=3(4-2)=6$.

19. $\int_1^2(2x+3)\,dx=\left(x^2+3x\right)\Big|_1^2=(4+6)-(1+3)=6$.

21. $\int_{-1}^3 2x^2\,dx=\frac{2}{3}x^3\Big|_{-1}^3=\frac{2}{3}(27)-\frac{2}{3}(-1)=\frac{56}{3}$.

23. $\int_{-2}^2\left(x^2-1\right)dx=\left(\frac{1}{3}x^3-x\right)\Big|_{-2}^2=\left(\frac{8}{3}-2\right)-\left(-\frac{8}{3}+2\right)=\frac{4}{3}$.

25. $\int_1^8 2x^{1/3}\,dx=2\cdot\frac{3}{4}x^{4/3}\Big|_1^8=\frac{3}{2}(16-1)=\frac{45}{2}$.

27. $\int_0^1 (x^3 - 2x^2 + 1)\,dx = \left(\frac{1}{4}x^4 - \frac{2}{3}x^3 + x\right)\Big|_0^1 = \frac{1}{4} - \frac{2}{3} + 1 = \frac{7}{12}.$

29. $\displaystyle\int_1^4 \frac{1}{x}\,dx = \ln x|_1^4 = \ln 4 - \ln 1 = \ln 4.$

31. $\int_0^4 x\,(x^2 - 1)\,dx = \int_0^4 (x^3 - x)\,dx = \left(\frac{1}{4}x^4 - \frac{1}{2}x^2\right)\Big|_0^4 = 64 - 8 = 56.$

33. $\int_1^3 (t^2 - t)^2\,dt = \int_1^3 (t^4 - 2t^3 + t^2)\,dt = \left(\frac{1}{5}t^5 - \frac{1}{2}t^4 + \frac{1}{3}t^3\right)\Big|_1^3 = \left(\frac{243}{5} - \frac{81}{2} + \frac{27}{3}\right) - \left(\frac{1}{5} - \frac{1}{2} + \frac{1}{3}\right) = \frac{512}{30} = \frac{256}{15}.$

35. $\int_{-3}^{-1} x^{-2}dx = -\frac{1}{x}\Big|_{-3}^{-1} = 1 - \frac{1}{3} = \frac{2}{3}.$

37. $\displaystyle\int_1^4 \left(\sqrt{x} - \frac{1}{\sqrt{x}}\right)dx = \int_1^4 (x^{1/2} - x^{-1/2})\,dx = \left(\frac{2}{3}x^{3/2} - 2x^{1/2}\right)\Big|_1^4 = \left(\frac{16}{3} - 4\right) - \left(\frac{2}{3} - 2\right) = \frac{8}{3}.$

39. $\displaystyle\int_1^4 \frac{3x^3 - 2x^2 + 4}{x^2}\,dx = \int_1^4 (3x - 2 + 4x^{-2})\,dx = \left(\frac{3}{2}x^2 - 2x - \frac{4}{x}\right)\Big|_1^4$
$= (24 - 8 - 1) - \left(\frac{3}{2} - 2 - 4\right) = \frac{39}{2}.$

41. a. $C(300) - C(0) = \int_0^{300} (0.0003x^2 - 0.12x + 20)\,dx = (0.0001x^3 - 0.06x^2 + 20x)\big|_0^{300}$
$= 0.0001\,(300)^3 - 0.06\,(300)^2 + 20\,(300) = 3300.$

Therefore, $C(300) = 3300 + C(0) = 3300 + 800 = \$4100.$

b. $\int_{200}^{300} C'(x)\,dx = (0.0001x^3 - 0.06x^2 + 20x)\big|_{200}^{300}$
$= [0.0001\,(300)^3 - 0.06\,(300)^2 + 20\,(300)] - [0.0001\,(200)^3 - 0.06\,(200)^2 + 20\,(200)] = \$900.$

43. a. The profit is

$\int_0^{200} (-0.0003x^2 + 0.02x + 20)\,dx + P(0) = (-0.0001x^3 + 0.01x^2 + 20x)\big|_0^{200} + P(0)$
$= 3600 + P(0) = 3600 - 800$, or \$2800.

b. $\int_{200}^{220} P'(x)\,dx = P(220) - P(200) = (-0.0001x^3 + 0.01x^2 + 20x)\big|_{200}^{220} = 219.20$, or \$219.20.

45. a. $f(t) = \int R(t)\,dt = \int 0.8256t^{-0.04}\,dt = \frac{0.8256}{0.96}t^{0.96} + C = 0.86t^{0.96} + C$. $f(1) = 0.9$, and so $0.86 + C = 0.9$ and $C = 0.04$. Thus, $f(t) = 0.86t^{0.96} + 0.04$.

b. In 2012, mobile phone ad spending is projected to be $f(6) = 0.86\,(6^{0.96}) + 0.04 \approx 4.84$, or approximately \$4.84 billion.

47. The distance is $\int_0^{20} v(t)\,dt = \int_0^{20} (-t^2 + 20t + 440)\,dt = \left(-\frac{1}{3}t^3 + 10t^2 + 440t\right)\Big|_0^{20} \approx 10{,}133.3$ ft.

49. a. The percentage of these households in decade t is
$P(t) = \int R(t)\,dt = \int (0.8499t^2 - 3.872t + 5)\,dt = 0.2833t^3 - 1.936t^2 + 5t + C$. The condition $P(0) = 5.6$ gives $P(0) = C = 5.6$, so $P(t) = 0.2833t^3 - 1.936t^2 + 5t + 5.6$.

b. The percentage of these households in 2010 is $P(4) = 0.2833\,(4^3) - 1.936\,(4^2) + 5\,(4) + 5.6 = 12.7552$, or approximately 12.8%.

c. The percentage of these households in 2000 was $P(3) = 0.2833(3^3) - 1.936(3^2) + 5(3) + 5.6 = 10.8251$. Therefore, the net increase in the percentage of these households from 1970 ($t = 0$) to 2000 ($t = 3$) is $P(3) - P(0) = 10.8251 - 5.6 = 5.2251$, or approximately 5.2%.

51. The number of solar panels produced during the second year is approximately $N = \int_{12}^{24} \left(\frac{4t}{1+t^2} + 3t^{1/2} \right) dt$. To find $I = \int \frac{4t}{1+t^2}\,dt$, let $u = 1 + t^2$, so $du = 2t\,dt$. Then $I = 2\int \frac{du}{u} = \ln|u| + C = \ln\left(1+t^2\right)^2 + C$. Using this result, we find $N = \left[2\ln\left(1+t^2\right) + 2t^{3/2}\right]_{12}^{24} \approx 154.77$, or 15,477 panels.

53. The increase in the senior population over the period in question is $\int_0^3 f(t)\,dt = \int_0^3 \frac{85}{1 + 1.859e^{-0.66t}}\,dt$.

Multiplying the numerator and denominator of the integrand by $e^{0.66t}$ gives $\int_0^3 f(t)\,dt = 85\int_0^3 \frac{e^{0.66t}}{e^{0.66t} + 1.859}\,dt$.

Now let $u = e^{0.66t} + 1.859$, so $du = 0.66e^{0.66t}\,dt$ and $e^{0.66t}\,dt = \frac{du}{0.66}$. We first evaluate the indefinite integral $I = \int \frac{85}{1 + 1.859e^{-0.66t}}\,dt = \frac{85}{0.66}\int \frac{du}{u} = \frac{85}{0.66}\ln|u| + C = \frac{85}{0.66}\ln\left(e^{0.66t} + 1.859\right) + C$. Thus,

$\int_0^3 f(t)\,dt = \frac{85}{0.66}\left[\ln\left(e^{1.98} + 1.859\right) - \ln\left(e^0 + 1.859\right)\right] \approx 149.14$, or approximately 149.14 million people.

55. $f(x) = x^4 - 2x^2 + 2$, so $f'(x) = 4x^3 - 4x = 4x\left(x^2 - 1\right) = 4x(x+1)(x-1)$. Setting $f'(x) = 0$ gives $x = -1$, 0, and 1 as critical numbers. Now calculate $f''(x) = 12x^2 - 4 = 4\left(3x^2 - 1\right)$ and use the second derivative test: $f''(-1) = 8 > 0$, so $(-1, 1)$ is a relative minimum; $f''(0) = -4 < 0$, so $(0, 2)$ is a relative maximum; and $f''(1) = 8 > 0$, so $(1, 1)$ is a relative minimum. The graph of f is symmetric with respect to the y-axis because $f(-x) = (-x)^4 - 2(-x)^2 + 2 = x^4 - 2x^2 + 2 = f(x)$. Thus, the required area is the area under the graph of f between $x = 0$ and $x = 1$, that is, $A = \int_0^1 \left(x^4 - 2x^2 + 2\right) dx = \left(\frac{1}{5}x^5 - \frac{2}{3}x^3 + 2x\right)\Big|_0^1 = \frac{1}{5} - \frac{2}{3} + 2 = \frac{23}{15}$.

57. False. The integrand $f(x) = 1/x^3$ is discontinuous at $x = 0$.

59. False. $f(x)$ is not nonnegative on $[0, 2]$.

Using Technology page 802

1. 6.1787 **3.** 0.7873 **5.** −0.5888 **7.** 2.7044

9. 3.9973 **11.** 46%, 24% **13.** 333,209 **15.** 3,761,490

11.5 Evaluating Definite Integrals

Concept Questions page 809

1. *Approach I:* We first find the indefinite integral. Let $u = x^3 + 1$, so that $du = 3x^2dx$ and or $x^2dx = \frac{1}{3}du$. Then $\int x^2 \left(x^3+1\right)^2 dx = \frac{1}{3}\int u^2\,du = \frac{1}{9}u^3 + C = \frac{1}{9}\left(x^3+1\right)^3 + C$. Therefore, $\int_0^1 x^2 \left(x^3+1\right)^2 dx = \frac{1}{9}\left(x^3+1\right)^3\Big|_0^1 = \frac{1}{9}(8-1) = \frac{7}{9}$.

   Approach II: Transform the definite integral in x into an integral in u: Let $u = x^3+1$, so that $du = 3x^2\,dx$ and $x^2\,dx = \frac{1}{3}\,du$. Next, find the limits of integration with respect to u. If $x = 0$, then $u = 0^3 + 1 = 1$ and if $x = 1$, then $u = 1^3 + 1 = 2$. Therefore, $\int_0^1 x^2\left(x^3+1\right)^2 dx = \frac{1}{3}\int_1^2 u^2\,du = \frac{1}{9}u^3\Big|_1^2 = \frac{1}{9}(8-1) = \frac{7}{9}$.

Exercises page 809

1. Let $u = x^2 - 1$, so $du = 2x\,dx$ and $x\,dx = \frac{1}{2}\,du$. If $x = 0$, then $u = -1$ and if $x = 2$, then $u = 3$, so $\int_0^2 x\left(x^2-1\right)^3 dx = \frac{1}{2}\int_{-1}^3 u^3\,du = \frac{1}{8}u^4\Big|_{-1}^3 = \frac{1}{8}(81) - \frac{1}{8}(1) = 10$.

3. Let $u = 5x^2 + 4$, so $du = 10x\,dx$ and $x\,dx = \frac{1}{10}\,du$. If $x = 0$, then $u = 4$ and if $x = 1$, then $u = 9$, so $\int_0^1 x\sqrt{5x^2+4}\,dx = \frac{1}{10}\int_4^9 u^{1/2}\,du = \frac{1}{15}u^{3/2}\Big|_4^9 = \frac{1}{15}(27) - \frac{1}{15}(8) = \frac{19}{15}$.

5. Let $u = x^3 + 1$, so $du = 3x^2\,dx$ and $x^2\,dx = \frac{1}{3}\,du$. If $x = 0$, then $u = 1$ and if $x = 2$, then $u = 9$, so $\int_0^2 x^2\left(x^3+1\right)^{3/2} dx = \frac{1}{3}\int_1^9 u^{3/2}\,du = \frac{2}{15}u^{5/2}\Big|_1^9 = \frac{2}{15}(243) - \frac{2}{15}(1) = \frac{484}{15}$.

7. Let $u = 2x + 1$, so $du = 2\,dx$ and $dx = \frac{1}{2}\,du$. If $x = 0$, then $u = 1$ and if $x = 1$ then $u = 3$, so
$$\int_0^1 \frac{1}{\sqrt{2x+1}}\,dx = \frac{1}{2}\int_1^3 \frac{1}{\sqrt{u}}\,du = \frac{1}{2}\int_1^3 u^{-1/2}\,du = u^{1/2}\big|_1^3 = \sqrt{3} - 1.$$

9. Let $u = 3x - 1$, so $du = 3\,dx$ and $dx = \frac{1}{3}\,du$. If $x = 1$, then $u = 2$ and if $x = 2$, then $u = 5$, so $\int_1^2 (3x-1)^4\,dx = \frac{1}{3}\int_2^5 u^4\,du = \frac{1}{15}u^5\Big|_2^5 = \frac{1}{15}(3125 - 32) = \frac{1031}{5}$.

11. Let $u = x^3 + 1$, so $du = 3x^2\,dx$ and $x^2\,dx = \frac{1}{3}\,du$. If $x = -1$, then $u = 0$ and if $x = 1$, then $u = 2$, so $\int_{-1}^1 x^2\left(x^3+1\right)^4 dx = \frac{1}{3}\int_0^2 u^4\,du = \frac{1}{15}u^5\Big|_0^2 = \frac{32}{15}$.

13. Let $u = x - 1$, so $x = u + 1$ and $du = dx$. If $x = 1$, then $u = 0$ and if $x = 5$, then $u = 4$, so $\int_1^5 x\sqrt{x-1}\,dx = \int_0^4 (u+1)\,u^{1/2}\,du = \int_0^4 \left(u^{3/2} + u^{1/2}\right) du = \left(\frac{2}{5}u^{5/2} + \frac{2}{3}u^{3/2}\right)\Big|_0^4 = \frac{2}{5}(32) + \frac{2}{3}(8) = \frac{272}{15}$.

15. Let $u = x^2$, so $du = 2x\,dx$ and $x\,dx = \frac{1}{2}\,du$. If $x = 0$, then $u = 0$ and if $x = 2$, then $u = 4$, so $\int_0^2 2xe^{x^2}\,dx = \int_0^4 e^u\,du = e^u\big|_0^4 = e^4 - 1$.

17. $\int_0^1 \left(e^{2x} + x^2 + 1\right) dx = \left(\frac{1}{2}e^{2x} + \frac{1}{3}x^3 + x\right)\Big|_0^1 = \left(\frac{1}{2}e^2 + \frac{1}{3} + 1\right) - \frac{1}{2} = \frac{1}{2}e^2 + \frac{5}{6}$.

19. Let $u = x^2 + 1$, so $du = 2x\,dx$ and $x\,dx = \frac{1}{2}\,du$. If $x = -1$, when $u = 2$ and if $x = 1$, then $u = 2$. Thus $\int_{-1}^{1} xe^{x^2+1}\,dx = \frac{1}{2}\int_{2}^{2} e^u\,du = \frac{1}{2}e^u\Big|_2^2 = 0$ because the upper and lower limits are equal.

21. Let $u = x - 2$, so $du = dx$. If $x = 3$, then $u = 1$ and if $x = 6$, then $u = 4$, so $\displaystyle\int_3^6 \frac{1}{x-2}\,dx = \int_1^4 \frac{du}{u} = \ln|u|\Big|_1^4 = \ln 4$.

23. Let $u = x^3 + 3x^2 - 1$, so $du = (3x^2 + 6x)\,dx = 3\left(x^2 + 2x\right)dx$. If $x = 1$, then $u = 3$, and if $x = 2$, then $u = 19$, so $\displaystyle\int_1^2 \frac{x^2 + 2x}{x^3 + 3x^2 - 1}\,dx = \frac{1}{3}\int_3^{19} \frac{du}{u} = \frac{1}{3}\ln u\Big|_3^{19} = \frac{1}{3}\left(\ln 19 - \ln 3\right)$.

25. $\displaystyle\int_1^2 \left(4e^{2u} - \frac{1}{u}\right)du = 2e^{2u} - \ln u\Big|_1^2 = \left(2e^4 - \ln 2\right) - \left(2e^2 - 0\right) = 2e^4 - 2e^2 - \ln 2$.

27. $\int_1^2 \left(2e^{-4x} - x^{-2}\right)dx = \left(-\frac{1}{2}e^{-4x} + \frac{1}{x}\right)\Big|_1^2 = \left(-\frac{1}{2}e^{-8} + \frac{1}{2}\right) - \left(-\frac{1}{2}e^{-4} + 1\right) = -\frac{1}{2}e^{-8} + \frac{1}{2}e^{-4} - \frac{1}{2}$
$= \frac{1}{2}\left(e^{-4} - e^{-8} - 1\right)$.

29. $A = \int_{-1}^{2} \left(x^2 - 2x + 2\right)dx = \left(\frac{1}{3}x^3 - x^2 + 2x\right)\Big|_{-1}^{2} = \left(\frac{8}{3} - 4 + 4\right) - \left(-\frac{1}{3} - 1 - 2\right) = 6$.

31. $\displaystyle A = \int_1^2 \frac{dx}{x^2} = \int_1^2 x^{-2}\,dx = -\frac{1}{x}\Big|_1^2 = -\frac{1}{2} - (-1) = \frac{1}{2}$.

33. $A = \int_{-1}^{2} e^{-x/2}\,dx = -2e^{-x/2}\Big|_{-1}^{2} = -2\left(e^{-1} - e^{1/2}\right) = 2\left(\sqrt{e} - 1/e\right)$.

35. The average value is $\frac{1}{2}\int_0^2 (2x + 3)\,dx = \frac{1}{2}\left(x^2 + 3x\right)\Big|_0^2 = \frac{1}{2}(10) = 5$.

37. The average value is $\frac{1}{2}\int_1^3 \left(2x^2 - 3\right)dx = \frac{1}{2}\left(\frac{2}{3}x^3 - 3x\right)\Big|_1^3 = \frac{1}{2}\left(9 + \frac{7}{3}\right) = \frac{17}{3}$.

39. The average value is

$\frac{1}{3}\int_{-1}^{2} \left(x^2 + 2x - 3\right)dx = \frac{1}{3}\left(\frac{1}{3}x^3 + x^2 - 3x\right)\Big|_{-1}^{2} = \frac{1}{3}\left[\left(\frac{8}{3} + 4 - 6\right) - \left(-\frac{1}{3} + 1 + 3\right)\right]$
$= \frac{1}{3}\left(\frac{8}{3} - 2 + \frac{1}{3} - 4\right) = -1$.

41. The average value is $\frac{1}{4}\int_0^4 (2x + 1)^{1/2}\,dx = \left(\frac{1}{4}\right)\left(\frac{1}{2}\right)\left(\frac{2}{3}\right)(2x + 1)^{3/2}\Big|_0^4 = \frac{1}{12}(27 - 1) = \frac{13}{6}$.

43. The average value is $\frac{1}{2}\int_0^2 xe^{x^2}\,dx = \dfrac{1}{4}e^{x^2}\Big|_0^2 = \frac{1}{4}\left(e^4 - 1\right)$.

45. The amount of oil produced worldwide between 1980 and the end of the twentieth century is given by $\int_0^{20} 3.5e^{0.05t}\,dt$. Let $u = 0.05t$, so $du = 0.05\,dt$. If $t = 0$, then $u = 0$ and if $t = 20$, then $u = 1$. Thus, the amount produced was $\int_0^{20} 3.5e^{0.05t}\,dt = \frac{3.5}{0.05}e^u\Big|_0^1 = 70\,(e - 1) \approx 120.3$, or 120.3 billion metric tons.

47. The amount is $\int_1^2 t\left(\frac{1}{2}t^2+1\right)^{1/2} dt$. Let $u=\frac{1}{2}t^2+1$, so $du = t\,dt$. If $t=1$, then $u=\frac{3}{2}$ and if $t=2$, then $u=3$. Thus, $\int_1^2 t\left(\frac{1}{2}t^2+1\right)^{1/2} dt = \int_{3/2}^3 u^{1/2}\,du = \frac{2}{3}u^{3/2}\Big|_{3/2}^3 = \frac{2}{3}\left[(3)^{3/2}-\left(\frac{3}{2}\right)^{3/2}\right] \approx \$2.24$ million.

49. The tractor depreciates by

$$V=\int_0^5 R(t)\,dt=\int_0^5 13{,}388.61e^{-0.22314t}\,dt=\frac{13{,}388.61}{-0.22314}e^{-0.22314t}\Big|_0^5\approx -60{,}000.94e^{-0.22314t}\Big|_0^5$$

$$\approx -60{,}000.94\,(-0.672314)\approx 40{,}339.47, \text{ or } \$40{,}339.$$

51. The average spending per year between 2005 and 2011 is

$$A=\frac{1}{7-1}\int_1^7 0.86t^{0.96}\,dt=\frac{0.86}{6}\cdot\frac{1}{1.96}t^{1.96}\Big|_1^7=\frac{0.86}{6\,(1.96)}\left(7^{1.96}-1\right)\approx 3.24, \text{ or } \$3.24 \text{ billion per year.}$$

53. **a.** The gasoline consumption in 2017 is given by $A(10)=0.014\left(10^2\right)+1.93\,(10)+140=160.7$, or 160.7 billion gallons per year.

b. The average consumption per year between 2007 and 2017 is given by

$A=\frac{1}{10-0}\int_0^{10}\left(0.014t^2+1.93t+140\right)dt=\frac{1}{10}\left(\frac{0.014}{3}t^3+\frac{1.93}{2}t^2+140t\right)\Big|_0^{10}\approx 150.12$, or approximately 150.1 billion gallons per year per year.

55. The average amount of money per year spent by lobbyists from 1998 through 2008 was approximately

$\frac{1}{10-0}\int_0^{10}\left(1.84t^2+12.51t+174\right)dt=\frac{1}{10}\left(\frac{1.84}{3}t^3+\frac{12.51}{2}t^2+174t\right)\Big|_0^{10}\approx 297.9$, or \$297.9 million per year.

57. The average yearly sales of the company over its first 5 years of operation is given by

$$\frac{1}{5-0}\int_0^5 t\left(0.2t^2+4\right)^{1/2}dt=\frac{1}{5}\left[\left(\frac{5}{2}\right)\left(\frac{2}{3}\right)\left(0.2t^2+4\right)^{3/2}\right]_0^5 \quad (\text{let } u=0.2t^2+4)$$

$$=\frac{1}{5}\left[\frac{5}{3}(5+4)^{3/2}-\frac{5}{3}(4)^{3/2}\right]=\frac{1}{3}(27-8)=\frac{19}{3}, \text{ or about } \$6.33 \text{ million.}$$

59. The average concentration of the drug is

$$\frac{1}{4}\int_0^4\frac{0.2t}{t^2+1}\,dt=\frac{0.2}{4}\int_0^4\frac{t}{t^2+1}\,dt=\frac{0.2}{(4)(2)}\ln\left(t^2+1\right)\Big|_0^4=0.025\ln 17\approx 0.071,$$

or 0.071 milligrams per cm^3.

61. $\frac{1}{5-0}\int_0^5 p\,dt=\frac{1}{5}\int_0^5\left(18-3e^{-2t}-6e^{-t/3}\right)dt=\frac{1}{5}\left(18t+\frac{3}{2}e^{-2t}+18e^{-t/3}\right)\Big|_0^5$

$$=\frac{1}{5}\left[18\,(5)+\frac{3}{2}e^{-10}+18e^{-5/3}-\frac{3}{2}-18\right]\approx 14.78, \text{ or } \$14.78.$$

63. The average content of oxygen in the pond over the first 10 days is

$$A=\frac{1}{10-0}\int_0^{10}100\left(\frac{t^2+10t+100}{t^2+20t+100}\right)dt=\frac{100}{10}\int_0^{10}\left[1-\frac{10}{t+10}+\frac{100}{(t+10)^2}\right]dt$$

$$=10\int_0^{10}\left[1-10\,(t+10)^{-1}+100\,(t+10)^{-2}\right]dt.$$

Using the substitution $u=t+10$ for the third integral, we have

$$A=10\left[t-10\ln(t+10)-\frac{100}{t+10}\right]\Big|_0^{10}=10\left\{\left[10-10\ln 20-\frac{100}{20}\right]-\left[-10\ln 10-10\right]\right\}$$

$$=10\,(10-10\ln 20-5+10\ln 10+10)\approx 80.6853, \text{ or approximately } 80.7\%.$$

65. $\int_a^a f(x)\,dx = F(x)|_a^a = F(a) - F(a) = 0$, where $F'(x) = f(x)$.

67. $\int_1^3 x^2\,dx = \frac{1}{3}x^3\Big|_1^3 = 9 - \frac{1}{3} = \frac{26}{3} = -\int_3^1 x^2\,dx = -\frac{1}{3}x^3\Big|_3^1 = -\frac{1}{3} + 9 = \frac{26}{3}$.

69. $\int_1^9 2\sqrt{x}\,dx = \frac{4}{3}x^{3/2}\Big|_1^9 = \frac{4}{3}(27-1) = \frac{104}{3}$ and $2\int_1^9 \sqrt{x}\,dx = 2\left(\frac{2}{3}x^{3/2}\right)\Big|_1^9 = \frac{104}{3}$.

71. $\int_0^3 \left(1+x^3\right)dx = x + \frac{1}{4}x^4\Big|_0^3 = 3 + \frac{81}{4} = \frac{93}{4}$ and

$\int_0^1 \left(1+x^3\right)dx + \int_1^3 \left(1+x^3\right)dx = \left(x + \frac{1}{4}x^4\right)\Big|_0^1 + \left(x + \frac{1}{4}x^4\right)\Big|_1^3 = \left(1 + \frac{1}{4}\right) + \left(3 + \frac{81}{4}\right) - \left(1 + \frac{1}{4}\right) = \frac{93}{4}$,

demonstrating Property 5.

73. $\int_3^3 \left(1+\sqrt{x}\right)e^{-x}\,dx = 0$ by Property 1 of the definite integral.

75. a. $\int_{-1}^2 \left[2f(x) + g(x)\right]dx = 2\int_{-1}^2 f(x)\,dx + \int_{-1}^2 g(x)\,dx = 2(-2) + 3 = -1$.

b. $\int_{-1}^2 \left[g(x) - f(x)\right]dx = \int_{-1}^2 g(x)\,dx - \int_{-1}^2 f(x)\,dx = 3 - (-2) = 5$.

c. $\int_{-1}^2 \left[2f(x) - 3g(x)\right]dx = 2\int_{-1}^2 f(x)\,dx - 3\int_{-1}^2 g(x)\,dx = 2(-2) - 3(3) = -13$.

77. True. This follows from Property 1 of the definite integral.

79. False. Only a constant can be "moved out" of the integral.

81. True. This follows from Properties 3 and 4 of the definite integral.

Using Technology page 813

1. 7.716667 **3.** 17.564865 **5.** 10,140 **7.** 60.45 mg/day

11.6 Area between Two Curves

Problem-Solving Tips

Note that the formula for the area between the graphs of two continuous functions f and g, where $f(x) \geq g(x)$ on $[a, b]$, is given by $\int_a^b \left[f(x) - g(x)\right]dx$. The condition $f(x) \geq g(x)$ on $[a, b]$ tells us that we cannot interchange $f(x)$ and $g(x)$ in this formula, as that would yield a negative answer, and area cannot be negative.

Concept Questions page 820

1. $\int_a^b \left[f(x) - g(x)\right]dx$

Exercises page 820

1. $-\int_0^6 \left(x^3 - 6x^2\right)dx = \left(-\frac{1}{4}x^4 + 2x^3\right)\Big|_0^6 = -\frac{1}{4}(6)^4 + 2(6)^3 = 108$.

3. $A = -\int_{-1}^0 x\sqrt{1-x^2}\,dx + \int_0^1 x\sqrt{1-x^2}dx = 2\int_0^1 x\left(1-x^2\right)^{1/2}dx$ by symmetry. Let $u = 1 - x^2$, so $du = -2x\,dx$ and $x\,dx = -\frac{1}{2}\,du$. If $x = 0$, then $u = 1$ and if $x = 1$, then $u = 0$, so

$A = (2)\left(-\frac{1}{2}\right)\int_0^1 u^{1/2}\,du = -\frac{2}{3}u^{3/2}\Big|_1^0 = \frac{2}{3}$.

5. $A = -\int_0^4 (x - 2\sqrt{x})\,dx = \int_0^4 (-x + 2x^{1/2})\,dx = \left(-\frac{1}{2}x^2 + \frac{4}{3}x^{3/2}\right)\Big|_0^4 = -8 + \frac{32}{3} = \frac{8}{3}$.

7. The required area is given by

$$\int_{-1}^0 (x^2 - x^{1/3})\,dx + \int_0^1 (x^{1/3} - x^2)\,dx = \left(\frac{1}{3}x^3 - \frac{3}{4}x^{4/3}\right)\Big|_{-1}^0 + \left(\frac{3}{4}x^{4/3} - \frac{1}{3}x^3\right)\Big|_0^1 = -\left(-\frac{1}{3} - \frac{3}{4}\right) + \left(\frac{3}{4} - \frac{1}{3}\right) = \frac{3}{2}.$$

9. The required area is given by $-\int_{-1}^2 -x^2 dx = \frac{1}{3}x^3\Big|_{-1}^2 = \frac{8}{3} + \frac{1}{3} = 3$.

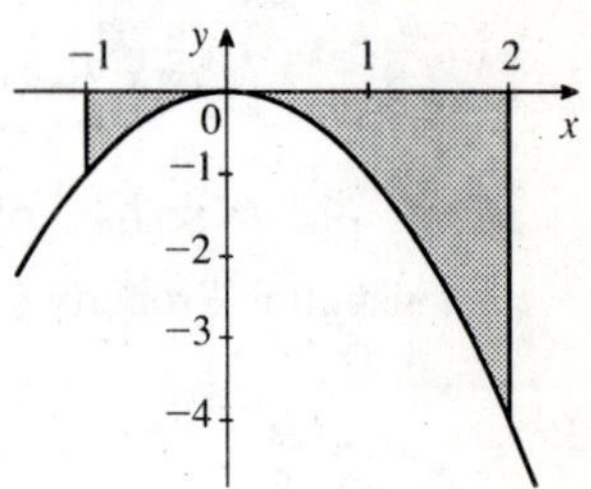

11. $y = x^2 - 5x + 4 = (x - 4)(x - 1) = 0$ if $x = 1$ or 4, the x-intercepts of the graph of f. Thus,

$$A = -\int_1^3 (x^2 - 5x + 4)\,dx = \left(-\frac{1}{3}x^3 + \frac{5}{2}x^2 - 4x\right)\Big|_1^3$$
$$= \left(-9 + \frac{45}{2} - 12\right) - \left(-\frac{1}{3} + \frac{5}{2} - 4\right) = \frac{10}{3}.$$

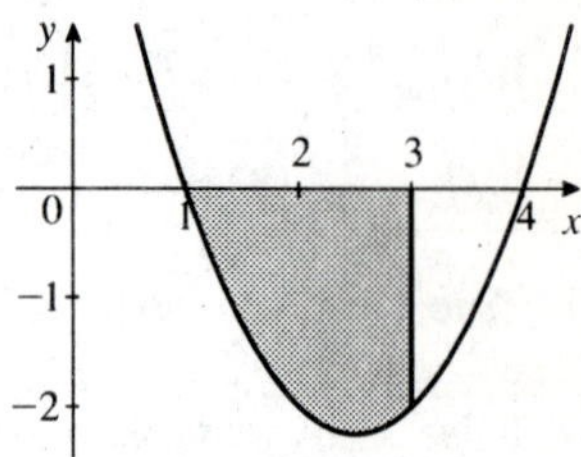

13. The required area is given by

$$-\int_0^9 -(1 + \sqrt{x})\,dx = \left(x + \frac{2}{3}x^{3/2}\right)\Big|_0^9 = 9 + 18 = 27.$$

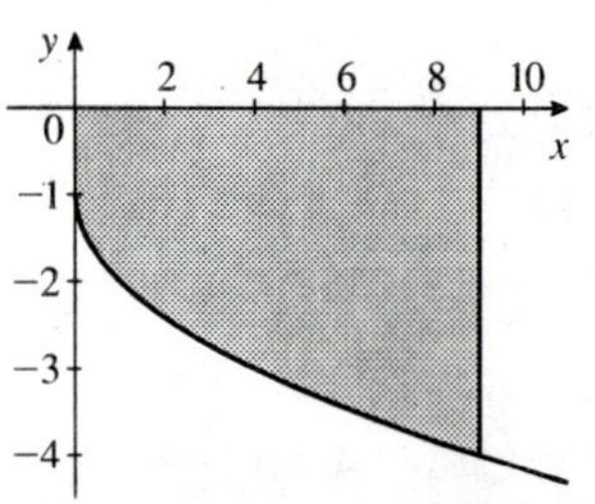

15. $-\int_{-2}^4 (-e^{x/2})\,dx = 2e^{x/2}\big|_{-2}^4 = 2(e^2 - e^{-1})$.

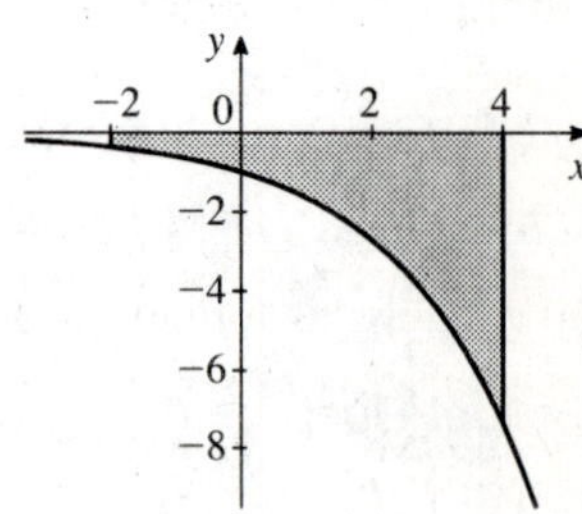

17. $A = \int_1^3 [(x^2 + 3) - 1]\,dx = \int_1^3 (x^2 + 2)\,dx = \left(\frac{1}{3}x^3 + 2x\right)\Big|_1^3$

$$= (9 + 6) - \left(\frac{1}{3} + 2\right) = \frac{38}{3}.$$

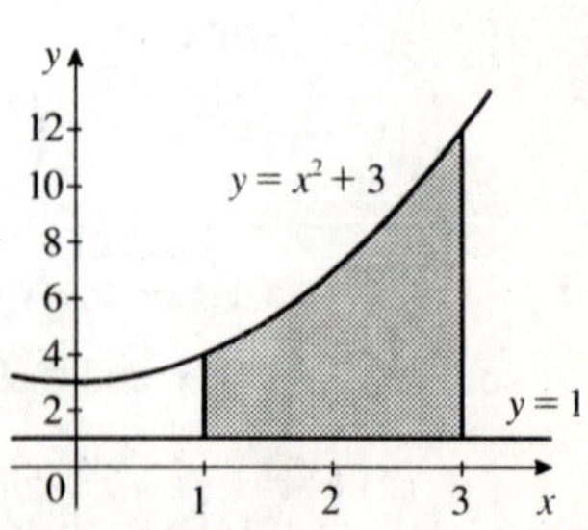

19. $A = \int_0^2 \left(-x^2 + 2x + 3 + x - 3\right) dx = \int_0^2 \left(-x^2 + 3x\right) dx$
$= \left(-\frac{1}{3}x^3 + \frac{3}{2}x^2\right)\Big|_0^2 = -\frac{1}{3}(8) + \frac{3}{2}(4) = 6 - \frac{8}{3} = \frac{10}{3}.$

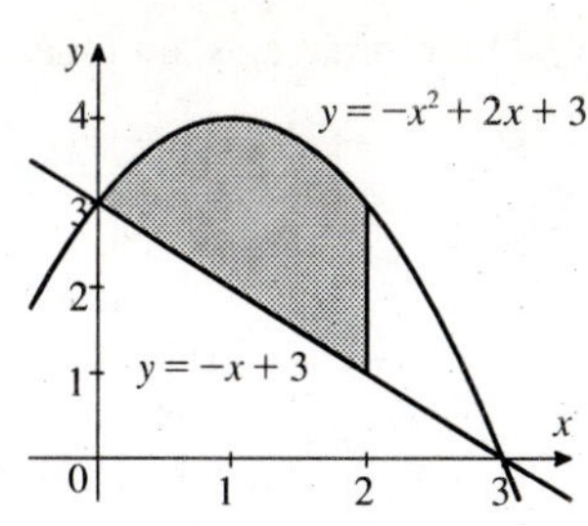

21. $A = \int_{-1}^2 \left[\left(x^2 + 1\right) - \frac{1}{3}x^3\right] dx = \int_{-1}^2 \left(-\frac{1}{3}x^3 + x^2 + 1\right) dx$
$= \left(-\frac{1}{12}x^4 + \frac{1}{3}x^3 + x\right)\Big|_{-1}^2$
$= \left(-\frac{4}{3} + \frac{8}{3} + 2\right) - \left(-\frac{1}{12} - \frac{1}{3} - 1\right) = \frac{19}{4}.$

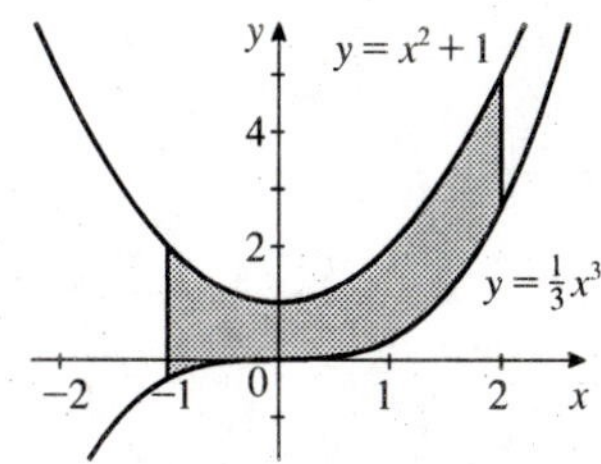

23. $A = \int_1^4 \left[(2x - 1) - \frac{1}{x}\right] dx = \int_1^4 \left(2x - 1 - \frac{1}{x}\right) dx$
$= \left(x^2 - x - \ln x\right)\Big|_1^4 = (16 - 4 - \ln 4) - (1 - 1 - \ln 1)$
$= 12 - \ln 4 \approx 10.6.$

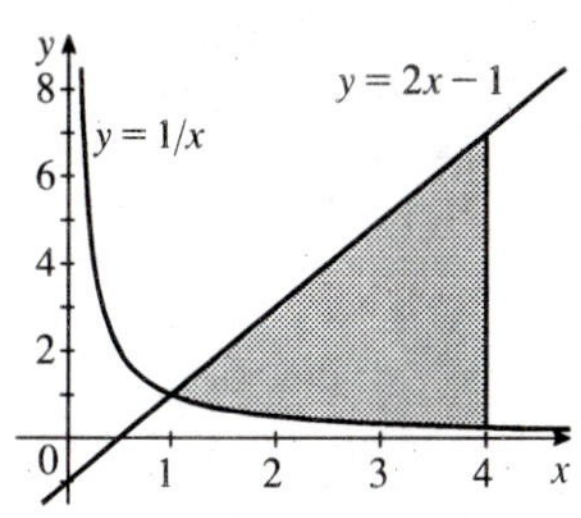

25. $A = \int_1^2 \left(e^x - \frac{1}{x}\right) dx = \left(e^x - \ln x\right)\Big|_1^2 = \left(e^2 - \ln 2\right) - e = e^2 - e - \ln 2.$

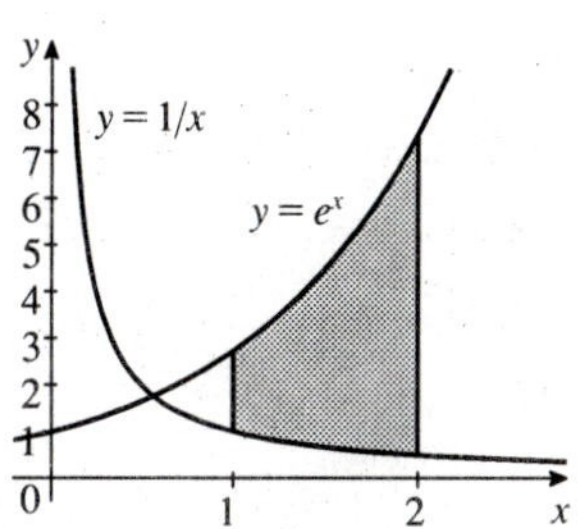

27. $A = -\int_{-1}^0 x\,dx + \int_0^2 x\,dx = -\frac{1}{2}x^2\Big|_{-1}^0 + \frac{1}{2}x^2\Big|_0^2 = \frac{1}{2} + 2 = \frac{5}{2}.$

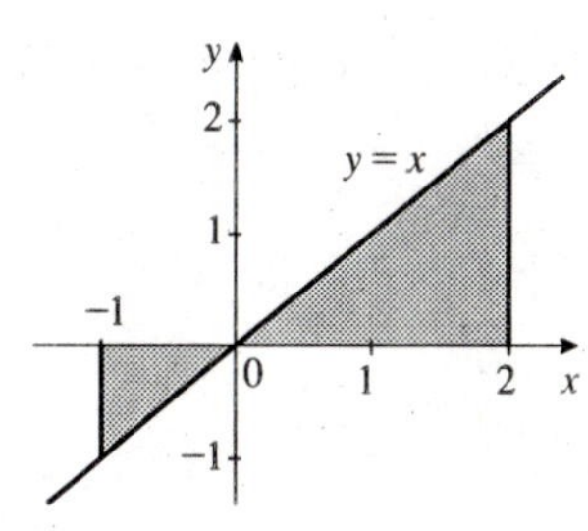

29. The x-intercepts are found by solving

$x^2 - 4x + 3 = (x-3)(x-1) = 0$, giving $x = 1$ or 3. Thus,

$$A = -\int_{-1}^{1}(-x^2+4x-3)\,dx + \int_{1}^{2}(-x^2+4x-3)\,dx$$
$$= \left(\tfrac{1}{3}x^3 - 2x^2 + 3x\right)\Big|_{-1}^{1} + \left(-\tfrac{1}{3}x^3 + 2x^2 - 3x\right)\Big|_{1}^{2}$$
$$= \left(\tfrac{1}{3} - 2 + 3\right) - \left(-\tfrac{1}{3} - 2 - 3\right) + \left(-\tfrac{8}{3} + 8 - 6\right) - \left(-\tfrac{1}{3} + 2 - 3\right)$$
$$= \tfrac{22}{3}.$$

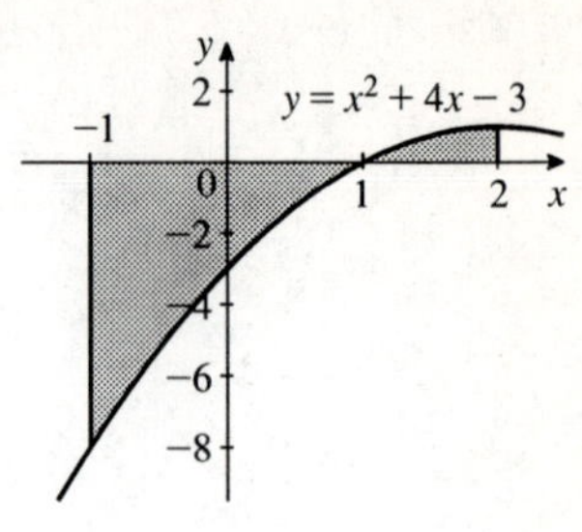

31. $$A = \int_{0}^{1}(x^3 - 4x^2 + 3x)\,dx - \int_{1}^{2}(x^3 - 4x^2 + 3x)\,dx$$
$$= \left(\tfrac{1}{4}x^4 - \tfrac{4}{3}x^3 + \tfrac{3}{2}x^2\right)\Big|_{0}^{1} - \left(\tfrac{1}{4}x^4 - \tfrac{4}{3}x^3 + \tfrac{3}{2}x^2\right)\Big|_{1}^{2}$$
$$= \left(\tfrac{1}{4} - \tfrac{4}{3} + \tfrac{3}{2}\right) - \left(4 - \tfrac{32}{3} + 6\right) + \left(\tfrac{1}{4} - \tfrac{4}{3} + \tfrac{3}{2}\right) = \tfrac{3}{2}.$$

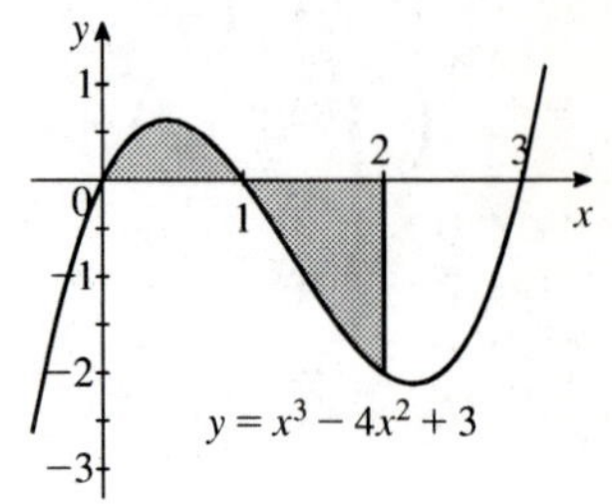

33. $$A = -\int_{-1}^{0}(e^x - 1)\,dx + \int_{0}^{3}(e^x - 1)dx$$
$$= (-e^x + x)\big|_{-1}^{0} + (e^x - x)\big|_{0}^{3}$$
$$= -1 - (-e^{-1} - 1) + (e^3 - 3) - 1 = e^3 - 4 + \tfrac{1}{e} \approx 16.5.$$

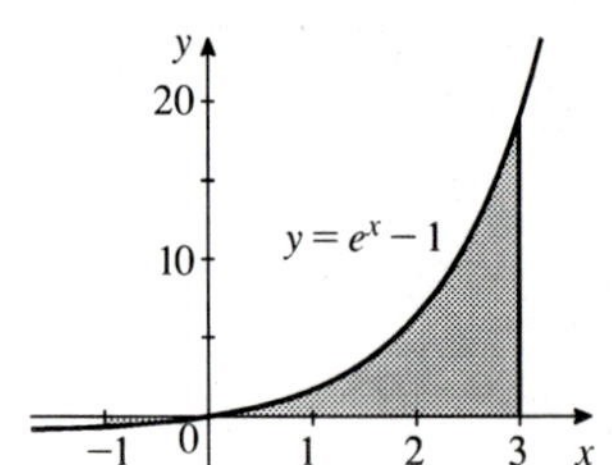

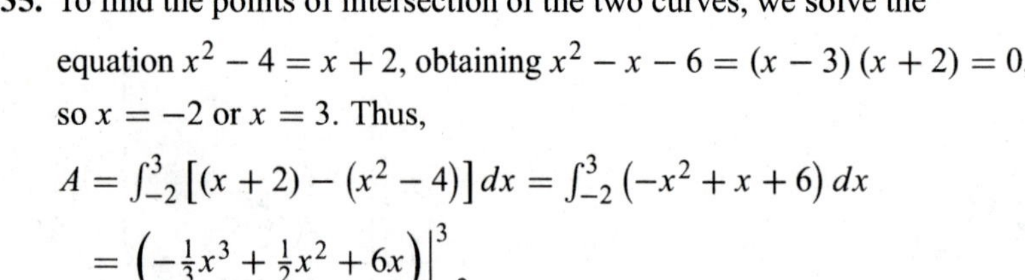

35. To find the points of intersection of the two curves, we solve the equation $x^2 - 4 = x + 2$, obtaining $x^2 - x - 6 = (x-3)(x+2) = 0$, so $x = -2$ or $x = 3$. Thus,

$$A = \int_{-2}^{3}\left[(x+2) - (x^2 - 4)\right]dx = \int_{-2}^{3}(-x^2 + x + 6)\,dx$$
$$= \left(-\tfrac{1}{3}x^3 + \tfrac{1}{2}x^2 + 6x\right)\Big|_{-2}^{3}$$
$$= \left(-9 + \tfrac{9}{2} + 18\right) - \left(\tfrac{8}{3} + 2 - 12\right) = \tfrac{125}{6}.$$

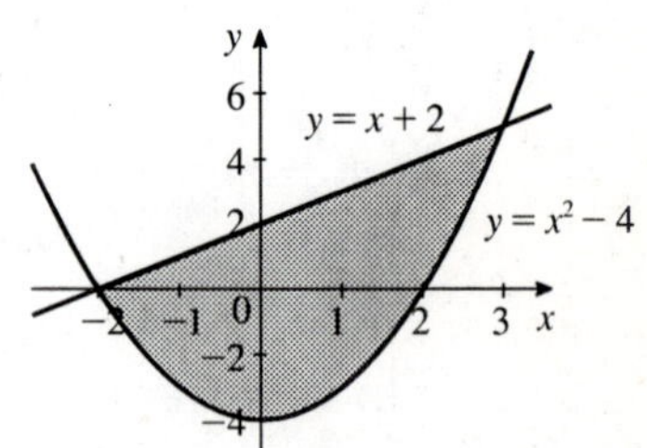

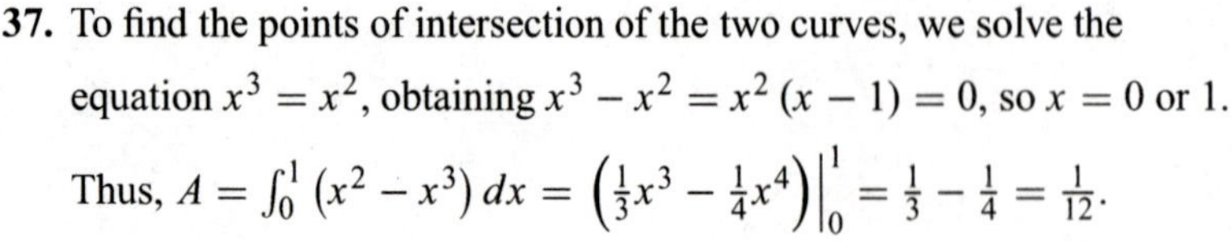

37. To find the points of intersection of the two curves, we solve the equation $x^3 = x^2$, obtaining $x^3 - x^2 = x^2(x-1) = 0$, so $x = 0$ or 1.

Thus, $A = \int_{0}^{1}(x^2 - x^3)\,dx = \left(\tfrac{1}{3}x^3 - \tfrac{1}{4}x^4\right)\Big|_{0}^{1} = \tfrac{1}{3} - \tfrac{1}{4} = \tfrac{1}{12}$.

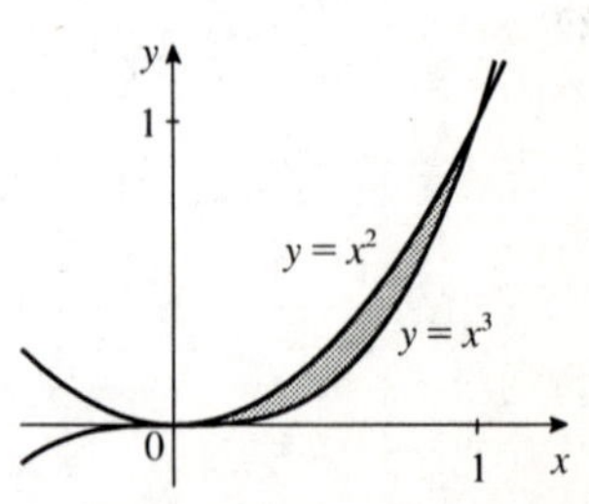

39. To find the points of intersection of the two curves, we solve the equation $x^3 - 6x^2 + 9x = x^2 - 3x$, obtaining $x^3 - 7x^2 + 12x = x(x-4)(x-3) = 0$, so $x = 0, 3$, or 4. Thus,

$$A = \int_0^3 \left[\left(x^3 - 6x^2 + 9x\right) - \left(x^2 - 3x\right)\right] dx + \int_3^4 \left[\left(x^2 - 3x\right) - \left(x^3 - 6x^2 + 9x\right)\right] dx$$
$$= \int_0^3 \left(x^3 - 7x^2 + 12x\right) dx - \int_3^4 \left(x^3 - 7x^2 + 12x\right) dx$$
$$= \left(\tfrac{1}{4}x^4 - \tfrac{7}{3}x^3 + 6x^2\right)\Big|_0^3 - \left(\tfrac{1}{4}x^4 - \tfrac{7}{3}x^3 + 6x^2\right)\Big|_3^4$$
$$= \left(\tfrac{81}{4} - 63 + 54\right) - \left(64 - \tfrac{448}{3} + 96\right) + \left(\tfrac{81}{4} - 63 + 54\right) = \tfrac{71}{6}.$$

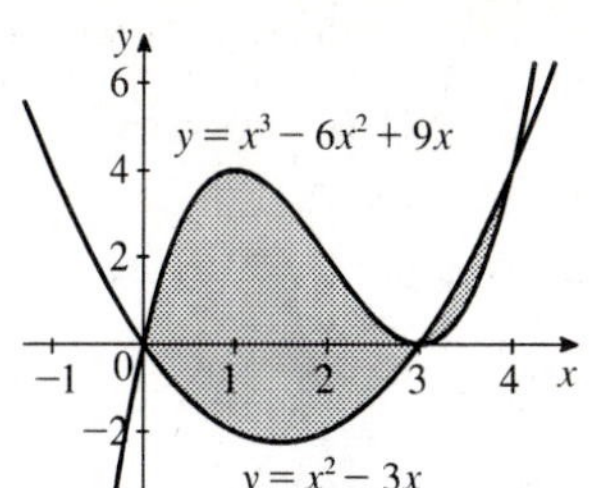

41. To find the points at which the graph intersects the x-axis, we solve the equation $x\sqrt{9 - x^2} = 0$, obtaining $x = -3$, 0, and 3. By symmetry, $A = 2\int_0^3 x\left(9 - x^2\right)^{1/2} dx$. We integrate using the substitution $u = 9 - x^2$, so $du = -2x\,dx$. If $x = 0$, then $u = 9$ and if $x = 3$, then $u = 0$, so

$$A = 2\int_9^0 -\tfrac{1}{2}u^{1/2}\,du = -\int_9^0 u^{1/2}\,du = -\left.\tfrac{2}{3}u^{3/2}\right|_9^0 = \tfrac{2}{3}(9)^{3/2} = 18.$$

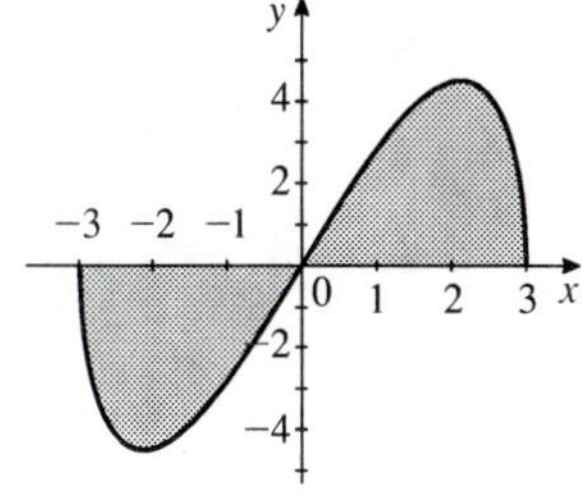

43. $S = \int_0^b \left[g(x) - f(x)\right] dx$ gives the additional revenue that the company would realize if it used a different advertising agency.

45. The shortfall is $\int_{2010}^{2050} \left[f(t) - g(t)\right] dt$.

47. a. $\int_{T_1}^{T} \left[g(t) - f(t)\right] dt - \int_0^{T_1} \left[f(t) - g(t)\right] dt = A_2 - A_1$.

b. The number $A_2 - A_1$ gives the distance car 2 is ahead of car 1 after T seconds.

49. Mexican oil profits from hedging in 2009 are given by $P = 70 \cdot 8 - \int_0^8 f(t)\,dt - \left[\int_8^{12} f(t)\,dt - 70 \cdot 4\right] = 840 - \int_0^{12} f(t)\,dt$ (dollars).

51. The additional amount of coal that will be produced is

$$\int_0^{20} \left(3.5e^{0.05t} - 3.5e^{0.01t}\right) dt = 3.5\int_0^{20} \left(e^{0.05t} - e^{0.01t}\right) dt = 3.5\left(20e^{0.05t} - 100e^{0.01t}\right)\Big|_0^{20}$$
$$= 3.5\left[\left(20e - 100e^{0.2}\right) - (20 - 100)\right] \approx 42.8 \text{ billion metric tons.}$$

53. If the campaign is mounted, there will be

$\int_0^5 \left(60e^{0.02t} + t^2 - 60\right) dt = \left(3000e^{0.02t} + \tfrac{1}{3}t^3 - 60t\right)\Big|_0^5 = 3315.5 + \tfrac{125}{3} - 300 - 3000 \approx 57.179$, or 57,179 fewer people.

55. True. If $f(x) \geq g(x)$ on $[a, b]$, then the area of the region is $\int_a^b \left[f(x) - g(x)\right] dx = \int_a^b |f(x) - g(x)|\,dx$. If $f(x) \leq g(x)$ on $[a, b]$, then the area of the region is $\int_a^b \left[g(x) - f(x)\right] dx = \int_a^b \{-\left[f(x) - g(x)\right]\}\,dx = \int_a^b |f(x) - g(x)|\,dx$.

57. False. Take $f(x) = x$ and $g(x) = 0$ on $[0, 1]$. Then the area bounded by the graphs of f and g on $[0, 1]$ is $A = \int_0^1 (x - 0)\, dx = \frac{1}{2}x^2\Big|_0^1 = \frac{1}{2}$ and so $A^2 = \frac{1}{4}$. However, $\int_0^1 [f(x) - g(x)]^2\, dx = \int_0^1 x^2\, dx = \frac{1}{3}$.

59. The area of R' is
$A = \int_a^b \{[f(x) + C] - [g(x) + C]\}\, dx = \int_a^b [f(x) + C - g(x) - C]\, dx = \int_a^b [f(x) - g(x)]\, dx$.

Using Technology

page 825

1. a.

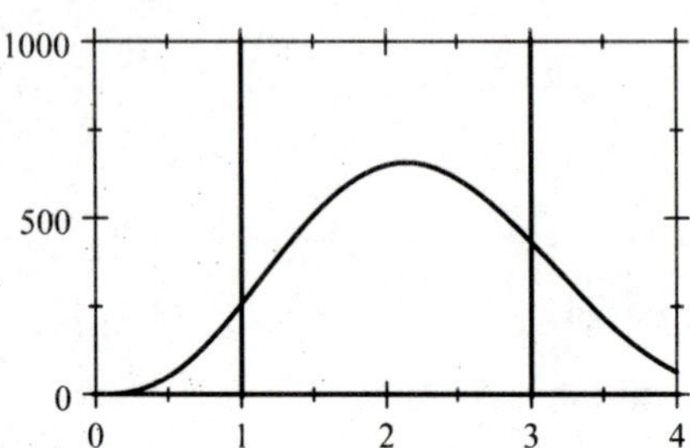

b. $A \approx 1074.2857$.

3. a. 

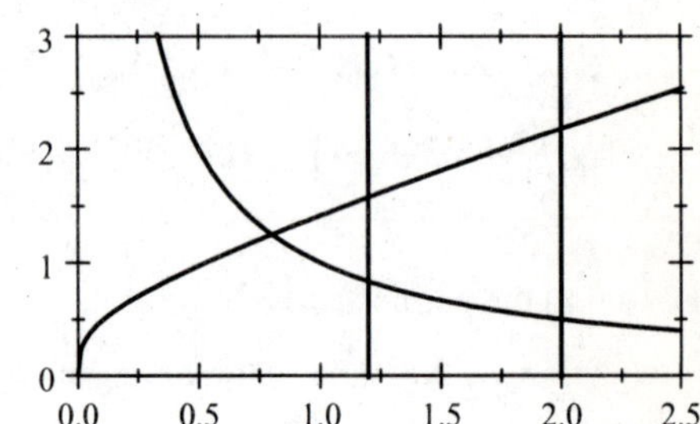

b. $A \approx 0.9961$.

5. a.

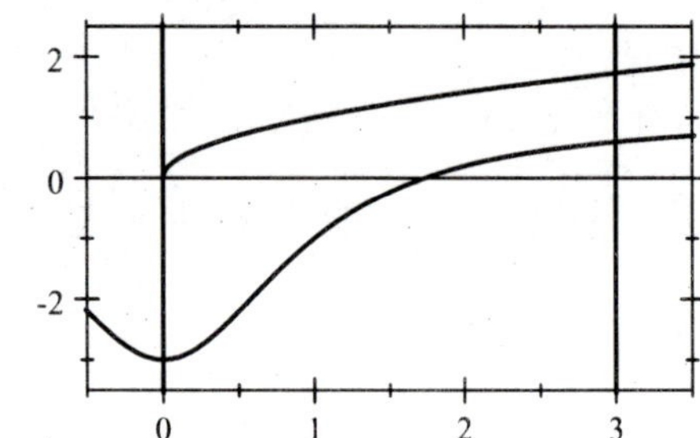

b. $A \approx 5.4603$.

7. a.

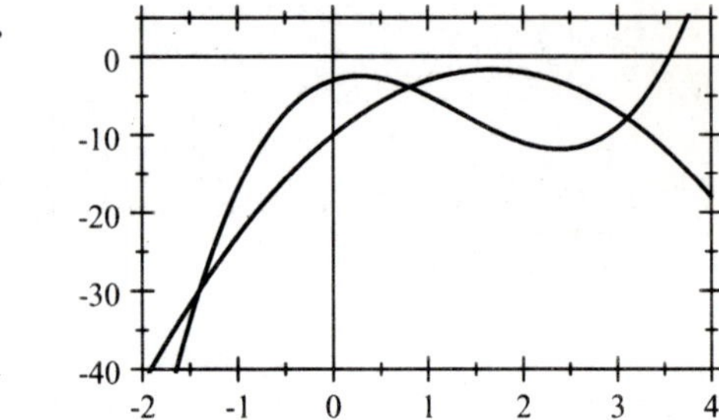

b. $A \approx 25.8549$.

9. a.

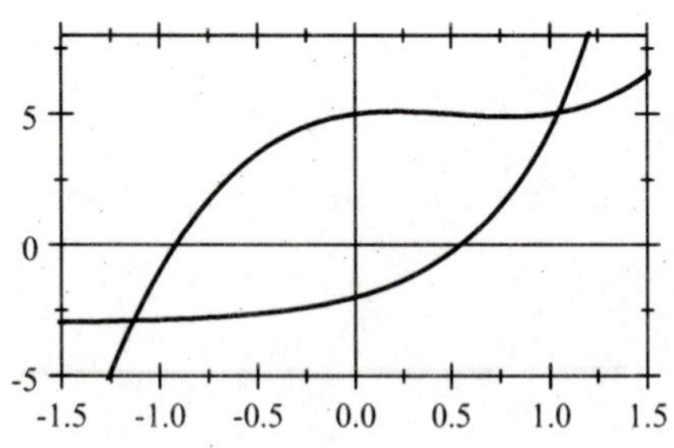

b. $A \approx 10.5144$.

11. a.

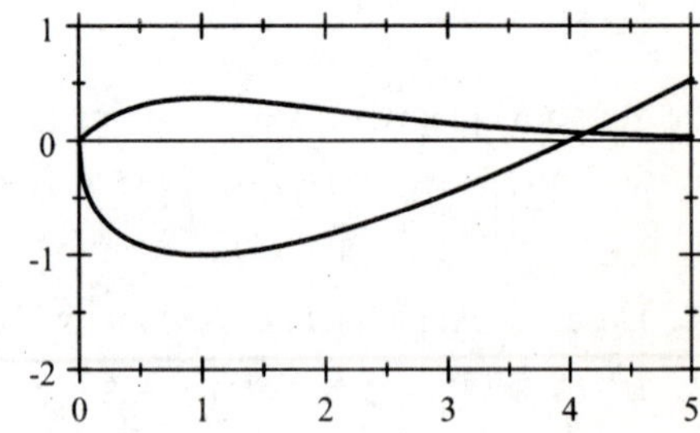

b. $A \approx 3.5799$.

13. The area of the larger region is $\int_{0.65}^{5.89} \left[\left(3x^2 + 10x - 11\right) - \left(2x^3 - 8x^2 + 4x - 3\right)\right] dx \approx 207.43$.

11.7 Applications of the Definite Integral to Business and Economics

Concept Questions page 835

1. a. See the definition on page 827 of the text. **b.** See the definition on page 827 of the text.

3. See the definition on page 832 of the text.

Exercises page 835

1. When $p = 4$, $-0.01x^2 - 0.1x + 6 = 4$, so $x^2 + 10x - 200 = 0$, and therefore $(x-10)(x+20) = 0$, giving $x = 10$ or -20. We reject the root $x = -20$ and find that the equilibrium price occurs at $x = 10$. The consumers' surplus is thus $CS = \int_0^{10} (-0.01x^2 - 0.1x + 6)\,dx - (4)(10) = \left(-\frac{0.01}{3}x^3 - 0.05x^2 + 6x\right)\Big|_0^{10} - 40 \approx 11.667$, or \$11,667.

3. Setting $p = 10$, we have $\sqrt{225 - 5x} = 10$, $225 - 5x = 100$, and so $x = 25$. Then $CS = \int_0^{25} \sqrt{225 - 5x}\,dx - (10)(25) = \int_0^{25} (225 - 5x)^{1/2}\,dx - 250$. To evaluate the integral, let $u = 225 - 5x$, so $du = -5\,dx$ and $dx = -\frac{1}{5}\,du$. If $x = 0$, then $u = 225$ and if $x = 25$, then $u = 100$, so $CS = -\frac{1}{5}\int_{225}^{100} u^{1/2}\,du - 250 = -\frac{2}{15}u^{3/2}\Big|_{225}^{100} - 250 = -\frac{2}{15}(1000 - 3375) - 250 = 66.667$, or \$6,667.

5. To find the equilibrium point, we solve $0.01x^2 + 0.1x + 3 = -0.01x^2 - 0.2x + 8$, finding $0.02x^2 + 0.3x - 5 = 0$, $2x^2 + 30x - 500 = (2x - 20)(x + 25) = 0$, and so $x = -25$ or 10. Thus, the equilibrium point is $(10, 5)$. Then $PS = (5)(10) - \int_0^{10} (0.01x^2 + 0.1x + 3)\,dx = 50 - \left(\frac{0.01}{3}x^3 + 0.05x^2 + 3x\right)\Big|_0^{10} = 50 - \frac{10}{3} - 5 - 30 = \frac{35}{3}$, or approximately \$11,667.

7. a. Setting $p = 250$, we have $100 + 80e^{0.05x} = 250$, $e^{0.05x} = \frac{150}{80} = \frac{15}{8}$, $\ln e^{0.05x} = \ln\frac{15}{8}$, $0.05x = \ln\frac{15}{8}$, and $x \approx 12.572$. The number of matresses the supplier will make available in the market is approximately 1257.

b. Taking $\overline{p} = 250$ and $\overline{x} = 12.572$ and using Formula (17), we find $PS \approx 12.572 \cdot 250 - \int_0^{12.572} (100 + 80e^{0.05x})\,dx = 3143 - \left(100x + \frac{80}{0.05}e^{0.05x}\right)\Big|_0^{12.572} \approx 485.826$, and so the producers' surplus is approximately \$48,583.

9. If $p = 160$, then we have $100\left(0.5x + \frac{0.4}{1+x}\right) = 160$, so $50x + \frac{40}{1+x} = 160$, $50x^2 + 50x + 40 = 160 + 160x$, $50x^2 - 110x - 120 = 0$, $5x^2 - 11x - 12 = 0$, and $(5x + 4)(x - 3) = 0$. Thus, $x = -\frac{4}{5}$ or $x = 3$. We reject the negative root, and using Formula (17) with $\overline{p} = 160$ and $\overline{x} = 3$, we have

$$PS = 3 \cdot 160 - \int_0^3 100\left(0.5x + \frac{0.4}{1+x}\right)dx = 480 - 100\left[0.25x^2 + 0.4\ln(1+x)\right]_0^3 = 480 - 100(2.25 + 0.4\ln 4) \approx 199.548.$$

Therefore, the producers' surplus is approximately \$199,548.

11. To find the market equilibrium, we solve $-0.2x^2 + 80 = 0.1x^2 + x + 40$, obtaining $0.3x^2 + x - 40 = 0$, $3x^2 + 10x - 400 = 0$, $(3x + 40)(x - 10) = 0$, and so $x = -\frac{40}{3}$ or $x = 10$. We reject the negative root. The corresponding equilibrium price is \$60, the consumers' surplus is $CS = \int_0^{10} (-0.2x^2 + 80)\,dx - (60)(10) = \left(-\frac{0.2}{3}x^3 + 80x\right)\Big|_0^{10} - 600 \approx 133.33$, or \$13,333, and the producers' surplus is $PS = 600 - \int_0^{10} (0.1x^2 + x + 40)\,dx = 600 - \left(\frac{0.1}{3}x^3 + \frac{1}{2}x^2 + 40x\right)\Big|_0^{10} \approx 116.67$, or \$11,667.

13. Here $P = 200{,}000$, $r = 0.08$, and $T = 5$, so $PV = \int_0^5 200{,}000e^{-0.08t}\,dt = -\frac{200{,}000}{0.08}e^{-0.08t}\Big|_0^5 = -2{,}500{,}000\left(e^{-0.4} - 1\right) \approx 824{,}199.88$, or approximately \$824,200.

15. Here $P = 250$, $m = 12$, $T = 20$, and $r = 0.08$, so $A = \dfrac{mP}{r}\left(e^{rT} - 1\right) = \dfrac{12(250)}{0.08}\left(e^{1.6} - 1\right) \approx 148{,}238.72$, or approximately \$148,239.

17. Here $P = 150$, $m = 12$, $T = 15$, and $r = 0.06$, so $A = \dfrac{12(150)}{0.06}\left(e^{0.9} - 1\right) \approx 43{,}788.09$, or approximately \$43,788.

19. Here $P = 2000$, $m = 1$, $T = 15.75$, and $r = 0.05$, so $A = \dfrac{1(2000)}{0.05}\left(e^{0.7875} - 1\right) \approx 47{,}915.79$, or approximately \$47,916.

21. Here $P = 1200$, $m = 12$, $T = 15$, and $r = 0.06$, so $PV = \dfrac{12(1200)}{0.06}\left(1 - e^{-0.9}\right) \approx 142{,}423.28$, or approximately \$142,423.

23. We want the present value of an annuity with $P = 300$, $m = 12$, $T = 10$, and $r = 0.08$, so $PV = \dfrac{12(300)}{0.08}\left(1 - e^{-0.8}\right) \approx 24{,}780.20$, or approximately \$24,780.

25. a.

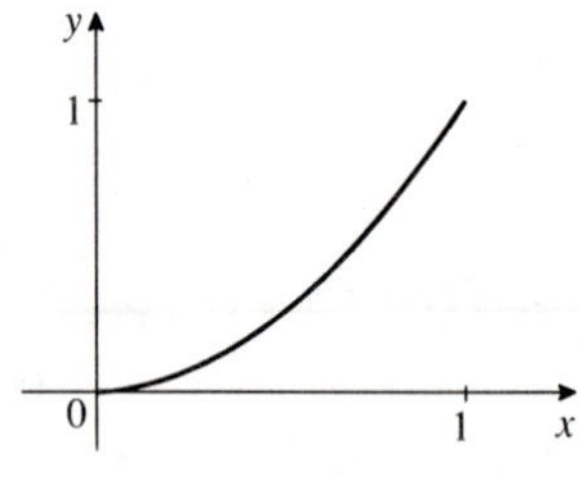

b. $f(0.4) = \frac{15}{16}(0.4)^2 + \frac{1}{16}(0.4) = 0.175$ and $f(0.9) = \frac{15}{16}(0.9)^2 + \frac{1}{16}(0.9) \approx 0.816$. Thus, the lowest 40% of earners receive 17.5% of the total income and the lowest 90% of earners receive 81.6%.

27. a. $L_1 = 2\int_0^1 [x - f(x)]\,dx = 2\int_0^1 \left(x - \frac{13}{14}x^2 - \frac{1}{14}x\right)dx = 2\int_0^1 \left(\frac{13}{14}x - \frac{13}{14}x^2\right)dx = \frac{13}{7}\int_0^1 (x - x^2)\,dx$

$= \frac{13}{7}\left(\frac{1}{2}x^2 - \frac{1}{3}x^3\right)\Big|_0^1 = \frac{13}{7}\left(\frac{1}{2} - \frac{1}{3}\right) = \frac{13}{7} \cdot \frac{1}{6} = \frac{13}{42} \approx 0.3095$ and

$L_2 = 2\int_0^1 \left(x - \frac{9}{11}x^4 - \frac{2}{11}x\right)dx = 2\int_0^1 \left(\frac{9}{11}x - \frac{9}{11}x^4\right)dx = 2\left(\frac{9}{11}\right)\int_0^1 (x - x^4)\,dx$

$= \frac{18}{11}\left(\frac{1}{2}x^2 - \frac{1}{5}x^5\right)\Big|_0^1 = \frac{18}{11}\left(\frac{1}{2} - \frac{1}{5}\right) \approx 0.4909.$

b. College teachers have a more equitable income distribution.

Using Technology page 838

1. The consumer's surplus is \$18,000,000 and the producer's surplus is \$11,700,000.

3. The consumer's surplus is \$33,120 and the producer's surplus is \$2880.

5. Investment A will generate a higher net income.

CHAPTER 11 Concept Review Questions page 840

1. a. $F'(x) = f(x)$ **b.** $F(x) + C$

3. a. unknown **b.** function

5. a. $\int_a^b f(x)\,dx$ **b.** minus

7. a. $\frac{1}{b-a}\int_a^b f(x)\,dx$ **b.** area, area

9. a. $\int_0^{\overline{x}} D(x)\,dx - \overline{p}\,\overline{x}$ **b.** $\overline{p}\,\overline{x} - \int_0^{\overline{x}} S(x)\,dx$

11. $\frac{mP}{r}\left(e^{rT} - 1\right)$

CHAPTER 11 Review Exercises page 841

1. $\int (x^3 + 2x^2 - x)\,dx = \frac{1}{4}x^4 + \frac{2}{3}x^3 - \frac{1}{2}x^2 + C.$

3. $\int \left(x^4 - 2x^3 + \frac{1}{x^2}\right)dx = \frac{x^5}{5} - \frac{x^4}{2} - \frac{1}{x} + C.$

5. $\int x\left(2x^2 + x^{1/2}\right)dx = \int \left(2x^3 + x^{3/2}\right)dx = \frac{1}{2}x^4 + \frac{2}{5}x^{5/2} + C.$

7. $\int \left(x^2 - x + \frac{2}{x} + 5\right)dx = \int x^2\,dx - \int x\,dx + 2\int \frac{dx}{x} + 5\int dx = \frac{1}{3}x^3 - \frac{1}{2}x^2 + 2\ln|x| + 5x + C.$

9. Let $u = 3x^2 - 2x + 1$, so $du = (6x - 2)\,dx = 2(3x - 1)\,dx$ or $(3x - 1)\,dx = \frac{1}{2}du$. So $\int (3x - 1)\left(3x^2 - 2x + 1\right)^{1/3} dx = \frac{1}{2}\int u^{1/3} du = \frac{3}{8}u^{4/3} + C = \frac{3}{8}\left(3x^2 - 2x + 1\right)^{4/3} + C.$

11. Let $u = x^2 - 2x + 5$, so $du = 2(x - 1)\,dx$ and $(x - 1)\,dx = \frac{1}{2}\,du$. Then $\int \frac{x-1}{x^2 - 2x + 5}\,dx = \frac{1}{2}\int \frac{du}{u} = \frac{1}{2}\ln|u| + C = \frac{1}{2}\ln\left(x^2 - 2x + 5\right) + C.$

13. Put $u = x^2 + x + 1$, so $du = (2x + 1)\,dx = 2\left(x + \frac{1}{2}\right)dx$ and $\left(x + \frac{1}{2}\right)dx = \frac{1}{2}du$. Then $\int \left(x + \frac{1}{2}\right)e^{x^2+x+1}\,dx = \frac{1}{2}\int e^u\,du = \frac{1}{2}e^u + C = \frac{1}{2}e^{x^2+x+1} + C.$

15. Let $u = \ln x$, so $du = \frac{1}{x}\,dx$. Then $\int \frac{(\ln x)^5}{x}\,dx = \int u^5\,du = \frac{1}{6}u^6 + C = \frac{1}{6}(\ln x)^6 + C$.

17. Let $u = x^2 + 1$, so $x^2 = u - 1$, $du = 2x\,dx$, $x\,dx = \frac{1}{2}\,du$. Then

$$\int x^3 (x^2+1)^{10}\,dx = \int x \cdot x^2 (x^2+1)^{10}\,dx = \tfrac{1}{2}\int (u-1)\,u^{10}\,du = \tfrac{1}{2}\int \left(u^{11} - u^{10}\right) du$$
$$= \tfrac{1}{2}\left(\tfrac{1}{12}u^{12} - \tfrac{1}{11}u^{11}\right) + C = \tfrac{1}{264}u^{11}(11u - 12) + C = \tfrac{1}{264}(x^2+1)^{11}(11x^2 - 1) + C.$$

19. Let $u = x - 2$, so $du = dx$. Then $x = u + 2$ and

$$\int \frac{x}{\sqrt{x-2}}\,dx = \int \frac{u+2}{\sqrt{u}}\,du = \int \left(u^{1/2} + 2u^{-1/2}\right) du = \int u^{1/2}\,du + 2\int u^{-1/2}\,du = \tfrac{2}{3}u^{3/2} + 4u^{1/2} + C$$
$$= \tfrac{2}{3}u^{1/2}(u+6) + C = \tfrac{2}{3}\sqrt{x-2}\,(x - 2 + 6) + C = \tfrac{2}{3}(x+4)\sqrt{x-2} + C.$$

21. $\int_0^1 (2x^3 - 3x^2 + 1)\,dx = \left(\frac{1}{2}x^4 - x^3 + x\right)\Big|_0^1 = \frac{1}{2} - 1 + 1 = \frac{1}{2}$.

23. $\int_1^4 \left(x^{1/2} + x^{-3/2}\right) dx = \left(\frac{2}{3}x^{3/2} - 2x^{-1/2}\right)\Big|_1^4 = \left(\frac{2}{3}x^{3/2} - \frac{2}{\sqrt{x}}\right)\Big|_1^4 = \left(\frac{16}{3} - 1\right) - \left(\frac{2}{3} - 2\right) = \frac{17}{3}$.

25. Let $u = x^3 - 3x^2 + 1$, so $du = (3x^2 - 6x)\,dx = 3(x^2 - 2x)\,dx$ and $(x^2 - 2x)\,dx = \frac{1}{3}\,du$. If $x = -1$, $u = -3$ and if $x = 0$, $u = 1$, so $\int_{-1}^0 12(x^2 - 2x)(x^3 - 3x^2 + 1)^3\,dx = (12)\left(\frac{1}{3}\right)\int_{-3}^1 u^3\,du = 4\left(\frac{1}{4}\right)u^4\Big|_{-3}^1 = 1 - 81 = -80$.

27. Let $u = x^2 + 1$, so $du = 2x\,dx$ and $x\,dx = \frac{1}{2}du$. If $x = 0$, then $u = 1$, and if $x = 2$, then $u = 5$, so

$$\int_0^2 \frac{x}{x^2+1}\,dx = \frac{1}{2}\int_1^5 \frac{du}{u} = \tfrac{1}{2}\ln u\Big|_1^5 = \tfrac{1}{2}\ln 5.$$

29. Let $u = 1 + 2x^2$, so $du = 4x\,dx$ and $x\,dx = \frac{1}{4}\,du$. If $x = 0$, then $u = 1$ and if $x = 2$, then $u = 9$, so

$$\int_0^2 \frac{4x}{\sqrt{1+2x^2}}\,dx = \int_1^9 \frac{du}{u^{1/2}} = 2u^{1/2}\Big|_1^9 = 2(3-1) = 4.$$

31. Let $u = 1 + e^{-x}$, so $du = -e^{-x}\,dx$ and $e^{-x}\,dx = -du$. If $x = -1$, then $u = 1 + e$ and if $x = 0$, then $u = 2$, so

$$\int_{-1}^0 \frac{e^{-x}}{(1+e^{-x})^2}\,dx = -\int_{1+e}^2 \frac{du}{u^2} = \frac{1}{u}\Big|_{1+e}^2 = \frac{1}{2} - \frac{1}{1+e} = \frac{e-1}{2(1+e)}.$$

33. $f(x) = \int f'(x)\,dx = \int (3x^2 - 4x + 1)\,dx = 3\int x^2\,dx - 4\int x\,dx + \int dx = x^3 - 2x^2 + x + C$. The given condition implies that $f(1) = 1$, so $1 - 2 + 1 + C = 1$, and thus $C = 1$. Therefore, the required function is $f(x) = x^3 - 2x^2 + x + 1$.

35. $f(x) = \int f'(x)\,dx = \int (1 - e^{-x})\,dx = x + e^{-x} + C$. Now $f(0) = 2$ implies $0 + 1 + C = 2$, so $C = 1$ and the required function is $f(x) = x + e^{-x} + 1$.

37. **a.** The integral $\int_0^T [f(t) - g(t)]\,dt$ represents the distance in feet between Car A and Car B at time T. If Car A is ahead, the integral is positive and if Car B is ahead, it is negative.

b. The distance is greatest at $t = 10$, at which point Car B's velocity exceeds that of Car A and Car B starts catching up. At that instant, the distance between the cars is $\int_0^{10} [f(t) - g(t)]\,dt$.

39. $\Delta x = \frac{2-1}{5} = \frac{1}{5}$, so $x_1 = \frac{6}{5}$, $x_2 = \frac{7}{5}$, $x_3 = \frac{8}{5}$, $x_4 = \frac{9}{5}$, $x_5 = \frac{10}{5}$. The Riemann sum is

$$f(x_1)\Delta x + \cdots + f(x_5)\Delta x = \left\{\left[-2\left(\tfrac{6}{5}\right)^2 + 1\right] + \left[-2\left(\tfrac{7}{5}\right)^2 + 1\right] + \cdots + \left[-2\left(\tfrac{10}{5}\right)^2 + 1\right]\right\}\left(\tfrac{1}{5}\right)$$
$$= \tfrac{1}{5}(-1.88 - 2.92 - 4.12 - 5.48 - 7) = -4.28.$$

41. **a.** $R(x) = \int R'(x)\,dx = \int(-0.03x + 60)\,dx = -0.015x^2 + 60x + C$. $R(0) = 0$ implies that $C = 0$, so $R(x) = -0.015x^2 + 60x$.

b. From $R(x) = px$, we have $-0.015x^2 + 60x = px$, and so $p = -0.015x + 60$.

43. **a.** We have the initial-value problem $T'(t) = 0.15t^2 - 3.6t + 14.4$ with $T(0) = 24$. Integrating, we find $T(t) = \int T'(t)\,dt = \int\left(0.15t^2 - 3.6t + 14.4\right)dt = 0.05t^3 - 1.8t^2 + 14.4t + C$. Using the initial condition, we find $T(0) = 24 = 0 + C$, so $C = 24$. Therefore, $T(t) = 0.05t^3 - 1.8t^2 + 14.4t + 24$.

b. The temperature at 10 a.m. was $T(4) = 0.05(4)^3 - 1.8(4)^2 + 14.4(4) + 24 = 56$, or 56°F.

45. $C(t) = \int C'(t)\,dt = \int\left(0.003t^2 + 0.06t + 0.1\right)dt = 0.001t^3 + 0.03t^2 + 0.1t + k$. But $C(0) = 2$, so $C(0) = k = 2$. Therefore, $C(t) = 0.001t^3 + 0.03t^2 + 0.1t + 2$. The pollution five years from now will be $C(5) = 0.001(5)^3 + 0.03(5)^2 + 0.1(5) + 2 = 3.375$, or 3.375 parts per million.

47. Using the substitution $u = 1 + 0.4t$, we find that $N(t) = \int 3000(1 + 0.4t)^{-1/2}\,dt = \frac{3000}{0.4}\cdot 2(1 + 0.4t)^{1/2} + C = 15{,}000\sqrt{1 + 0.4t} + C$. $N(0) = 100{,}000$ implies $15{,}000 + C = 100{,}000$, so $C = 85{,}000$. Therefore, $N(t) = 15{,}000\sqrt{1 + 0.4t} + 85{,}000$. The number using the subway six months from now will be $N(6) = 15{,}000\sqrt{1 + 2.4} + 85{,}000 \approx 112{,}659$.

49. Let $u = 5 - x$, so $du = -dx$. Then $p(x) = \int \frac{240}{(5-x)^2}\,dx = 240\int(5-x)^{-2}\,dx = 240\int\left(-u^{-2}\right)du = 240u^{-1} + C = \frac{240}{5-x} + C$. Next, the condition $p(2) = 50$ gives $\frac{240}{3} + C = 80 + C = 50$, so $C = -30$. Therefore, $p(x) = \frac{240}{5-x} - 30$.

51. The number will be

$$\int_0^{10}\left(0.00933t^3 + 0.019t^2 - 0.10833t + 1.3467\right)dt$$
$$= \left(0.0023325t^4 + 0.0063333t^3 - 0.054165t^2 + 1.3467t\right)\Big|_0^{10}$$
$$= 0.0023325(10)^4 + 0.0063333(10)^3 - 0.054165(10)^2 + 1.3467(10)$$

≈ 37.7, or approximately 37.7 million Americans.

53. $A = \int_{-1}^{2}\left(3x^2 + 2x + 1\right)dx = \left(x^3 + x^2 + x\right)\Big|_{-1}^{2} = \left(2^3 + 2^2 + 2\right) - \left[(-1)^3 + 1 - 1\right] = 14 - (-1) = 15$.

55. $A = \displaystyle\int_1^3 \frac{1}{x^2}\,dx = \int_1^3 x^{-2}\,dx = -\frac{1}{x}\Big|_1^3 = -\frac{1}{3} + 1 = \frac{2}{3}$.

57. $A = \int_a^b [f(x) - g(x)]\,dx = \int_0^2 (e^x - x)\,dx = \left(e^x - \frac{1}{2}x^2\right)\Big|_0^2$
$= (e^2 - 2) - (1 - 0) = e^2 - 3.$

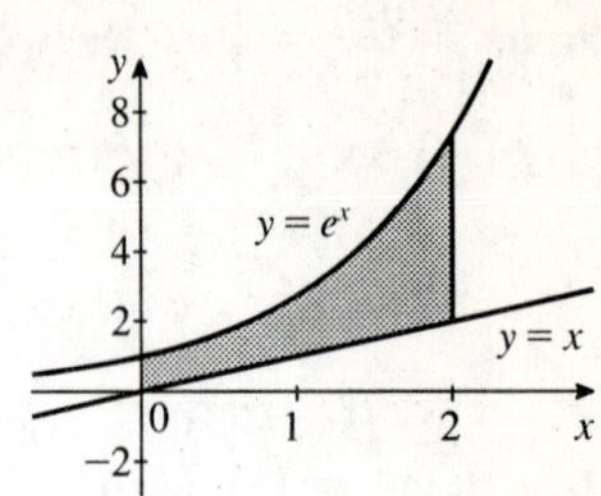

59. $A = \int_0^1 (x^3 - 3x^2 + 2x)\,dx - \int_1^2 (x^3 - 3x^2 + 2x)\,dx$
$= \left(\frac{1}{4}x^4 - x^3 + x^2\right)\Big|_0^1 - \left(\frac{1}{4}x^4 - x^3 + x^2\right)\Big|_1^2$
$= \frac{1}{4} - 1 + 1 - \left[(4 - 8 + 4) - \left(\frac{1}{4} - 1 + 1\right)\right]$
$= \frac{1}{4} + \frac{1}{4} = \frac{1}{2}.$

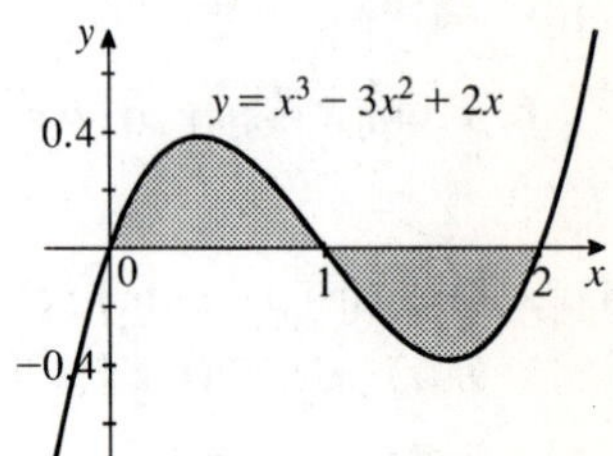

61. $A = \dfrac{1}{3}\displaystyle\int_0^3 \frac{x}{\sqrt{x^2 + 16}}\,dx = \frac{1}{3} \cdot \frac{1}{2} \cdot 2\left(x^2 + 16\right)^{1/2}\Big|_0^3 = \frac{1}{3}\left(x^2 + 16\right)^{1/2}\Big|_0^3 = \frac{1}{3}(5 - 4) = \frac{1}{3}.$

63. $\overline{v} = \frac{1}{5}\int_0^5 \left(\frac{1}{12}t^2 + 2t + 44\right)dt = \frac{1}{5}\left(\frac{1}{36}t^3 + t^2 + 44t\right)\Big|_0^5 = \frac{1}{5}\left(\frac{125}{36} + 25 + 220\right) = \dfrac{125 + 900 + 7920}{5(36)}$
≈ 49.69, or 49.7 ft/sec.

65. Because $p = 8$, we have $-0.01x^2 - 0.2x + 23 = 8$, $-0.01x^2 - 0.2x + 15 = 0$, and so $x^2 + 20x - 1500 = (x - 30)(x + 50) = 0$, giving $x = -50$ or 30. Thus,
$CS = \int_0^{30} (-0.01x^2 - 0.2x + 23)\,dx - 8(30) = \left(-\frac{0.01}{3}x^3 - 0.1x^2 + 23x\right)\Big|_0^{30} - 240$
$= -\frac{0.01}{3}(30)^3 - 0.1(900) + 23(30) - 240 = 270$, or \$270,000.

67. Use Equation (17) with $P = 4000$, $r = 0.08$, $T = 20$, and $m = 1$ to get $A = \dfrac{1 \cdot 4000}{0.08}\left(e^{1.6} - 1\right) \approx 197{,}651.62$. That is, Chi-Tai will have approximately \$197,652 in his account after 20 years.

69. Here $P = 80{,}000$, $m = 1$, $T = 10$, and $r = 0.1$, so $PV = \dfrac{1 \cdot 80{,}000}{0.1}\left(1 - e^{-1}\right) \approx 505{,}696$, or approximately \$505,696.

71. The average population will be $\frac{1}{5}\int 80{,}000e^{0.05t}\,dt = \frac{80{,}000}{5}\left(\frac{1}{0.05}\right)e^{0.05t}\Big|_0^5 = 320{,}000\left(e^{0.25} - 1\right) \approx 90{,}888.$

CHAPTER 11 Before Moving On... page 844

1. $\int\left(2x^3+\sqrt{x}+\frac{2}{x}-\frac{2}{\sqrt{x}}\right)dx=2\int x^3\,dx+\int x^{1/2}\,dx+2\int\frac{1}{x}\,dx-2\int x^{-1/2}\,dx$

$=\frac{1}{2}x^4+\frac{2}{3}x^{3/2}+2\ln|x|-4x^{1/2}+C.$

2. $f(x)=\int f'(x)\,dx=\int(e^x+x)\,dx=e^x+\frac{1}{2}x^2+C$. $f(0)=2$ implies $f(0)=e^0+0+C=2$, so $C=1$. Therefore, $f(x)=e^x+\frac{1}{2}x^2+1$.

3. Let $u=x^2+1$, so $du=2x\,dx$ or $x\,dx=\frac{1}{2}du$. Then

$\int\frac{x}{\sqrt{x^2+1}}\,dx=\frac{1}{2}\int\frac{du}{\sqrt{u}}=\frac{1}{2}\int u^{-1/2}\,du=\frac{1}{2}\left(2u^{1/2}\right)+C=\sqrt{u}+C=\sqrt{x^2+1}+C.$

4. Let $u=2-x^2$, so $du=-2x\,dx$ and $x\,dx=-\frac{1}{2}\,du$. If $x=0$, then $u=2$ and if $x=1$, then $u=1$. Therefore,

$\int_0^1 x\sqrt{2-x^2}\,dx=-\frac{1}{2}\int_2^1 u^{1/2}\,du=-\frac{1}{2}\left(\frac{2}{3}u^{3/2}\right)\Big|_2^1=-\frac{1}{3}u^{3/2}\Big|_2^1=-\frac{1}{3}\left(1-2^{3/2}\right)=\frac{1}{3}\left(2\sqrt{2}-1\right).$

5. To find the points of intersection, we solve $x^2-1=1-x$, obtaining $x^2+x-2=0$, $(x+2)(x-1)=0$, and so $x=-2$ or $x=1$. The points of intersection are $(-2,3)$ and $(1,0)$. Thus, the required area is

$A=\int_{-2}^1\left[(1-x)-\left(x^2-1\right)\right]dx=\int_{-2}^1\left(2-x-x^2\right)dx=\left(2x-\frac{1}{2}x^2-\frac{1}{3}x^3\right)\Big|_{-2}^1$

$=\left(2-\frac{1}{2}-\frac{1}{3}\right)-\left(-4-2+\frac{8}{3}\right)=\frac{9}{2}.$

12 CALCULUS OF SEVERAL VARIABLES

12.1 Functions of Several Variables

Concept Questions page 852

1. A function of two variables is a rule that assigns to each point (x, y) in a subset of the plane a unique number $f(x, y)$. For example, $f(x, y) = x^2 + 2y^2$ has the whole xy-plane as its domain.

3. a. The graph of $f(x, y)$ is the set $S = \{(x, y, z) \mid z = f(x, y), (x, y) \in D\}$, where D is the domain of f.

b. The level curve of f is the projection onto the xy-plane of the trace of $f(x, y)$ in the plane $z = k$, where k is a constant in the range of f.

Exercises page 853

1. $f(x, y) = 2x + 3y - 4$, so $f(0, 0) = 2(0) + 3(0) - 4 = -4$, $f(1, 0) = 2(1) + 3(0) - 4 = -2$, $f(0, 1) = 2(0) + 3(1) - 4 = -1$, $f(1, 2) = 2(1) + 3(2) - 4 = 4$, and $f(2, -1) = 2(2) + 3(-1) - 4 = -3$.

3. $f(x, y) = x^2 + 2xy - x + 3$, so $f(1, 2) = 1^2 + 2(1)(2) - 1 + 3 = 7$, $f(2, 1) = 2^2 + 2(2)(1) - 2 + 3 = 9$, $f(-1, 2) = (-1)^2 + 2(-1)(2) - (-1) + 3 = 1$, and $f(2, -1) = 2^2 + 2(2)(-1) - 2 + 3 = 1$.

5. $g(s, t) = 3s\sqrt{t} + t\sqrt{s} + 2$, so $g(1, 4) = 3(1)\sqrt{4} + 4\sqrt{1} + 2 = 6 + 4 + 2 = 12$, $g(4, 1) = 3(4)\sqrt{1} + \sqrt{4} + 2 = 12 + 2 + 2 = 16$, $g(0, 4) = 0 + 0 + 2 = 2$, and $g(4, 9) = 3(4)\sqrt{9} + 9\sqrt{4} + 2 = 56$.

7. $h(s, t) = s \ln t - t \ln s$, so $h(1, e) = \ln e - e \ln 1 = \ln e = 1$, $h(e, 1) = e \ln 1 - \ln e = -1$, and $h(e, e) = e \ln e - e \ln e = 0$.

9. $g(r, s, t) = re^{s/t}$, so $g(1, 1, 1) = e$, $g(1, 0, 1) = 1$, and $g(-1, -1, -1) = -e^{-1/(-1)} = -e$.

11. $f(x, y) = 2x + 3y$. The domain of f is the set of all ordered pairs (x, y), where x and y are real numbers.

13. $h(u, v) = \dfrac{uv}{u + v}$. The domain is all real values of u and v except those satisfying the equation $u = -v$.

15. $g(r, s) = \sqrt{rs}$. The domain of g is the set of all ordered pairs (r, s) satisfying $rs \geq 0$, that is the set of all ordered pairs whose members have the same sign (allowing zeros).

17. $h(x, y) = \ln(x + y + 5)$. The domain of h is the set of all ordered pairs (x, y) such that $x + y > -5$.

19. The graph shows level curves of $z = f(x, y) = 2x + 3y$ for $z = -2, -1, 0, 1$, and 2.

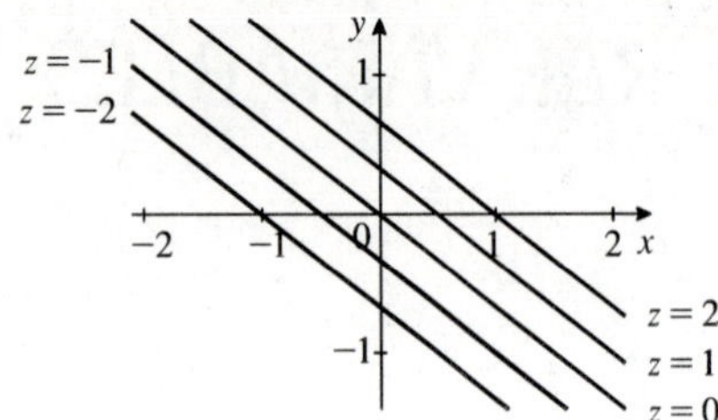

21. The graph shows level curves of $z = f(x, y) = 2x^2 + y$ for $z = -2, -1, 0, 1$, and 2.

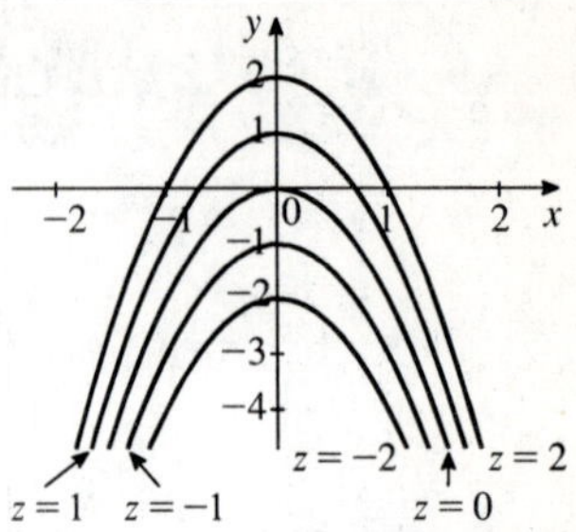

23. The graph shows level curves of $z = f(x, y) = \sqrt{16 - x^2 - y^2}$ for $z = 0, 1, 2, 3$, and 4.

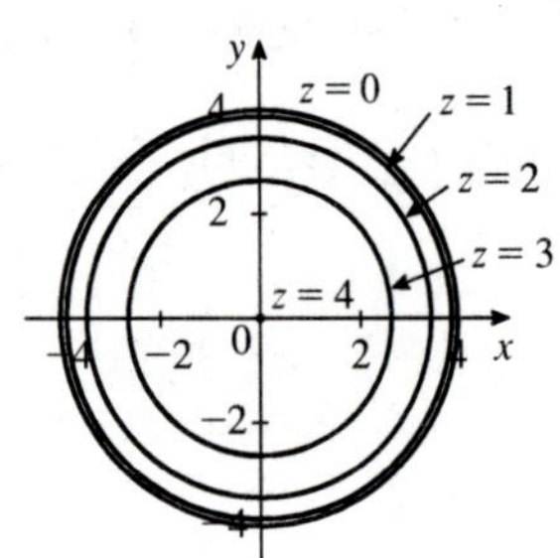

25. The level curves of f have equations $f(x, y) = \sqrt{x^2 + y^2} = C$. An equation of the curve containing the point (3, 4) satisfies $\sqrt{3^2 + 4^2} = C$, so $C = \sqrt{9 + 16} = 5$. Thus, an equation is $\sqrt{x^2 + y^2} = 5$.

27. (b)

29. No. Suppose the level curves $f(x, y) = c_1$ and $f(x, y) = c_2$ intersect at a point (x_0, y_0) and $c_1 \neq c_2$. Then $f(x_0, y_0) = c_1$ and $f(x_0, y_0) = c_2$ where $c_1 \neq c_2$. Thus, f takes on two distinct values at (x_0, y_0), contradicting the definition of a function.

31. $V = f(1.5, 4) = \pi (1.5)^2 (4) = 9\pi$, or 9π ft^3.

33. a. $M = \dfrac{80}{(1.8)^2} = 24.69$.

b. We must have $\dfrac{w}{(1.8)^2} < 25$; that is, $w < 25(1.8)^2 = 81$. Thus, the maximum weight is 81 kg.

35. a. $R(x, y) = xp + yq = x\left(200 - \frac{1}{5}x - \frac{1}{10}y\right) + y\left(160 - \frac{1}{10}x - \frac{1}{4}y\right) = -\frac{1}{5}x^2 - \frac{1}{4}y^2 - \frac{1}{5}xy + 200x + 160y$.

b. The domain of R is the set of all points (x, y) satisfying $200 - \frac{1}{5}x - \frac{1}{10}y \geq 0$, $160 - \frac{1}{10}x - \frac{1}{4}y \geq 0$, $x \geq 0$, and $y \geq 0$.

37. a.
$$\begin{aligned} R(x, y) &= xp + yq = 20x - 0.005x^2 - 0.001xy + 15y - 0.001xy - 0.003y^2 \\ &= -0.005x^2 - 0.003y^2 - 0.002xy + 20x + 15y. \end{aligned}$$

b. Because p and q must both be nonnegative, the domain of R is the set of all ordered pairs (x, y) for which $20 - 0.005x - 0.001y \geq 0$, $15 - 0.001x - 0.003y \geq 0$, $x \geq 0$, and $y \geq 0$.

39. **a.** The domain of V is the set of all ordered pairs (P, T) where P and T are positive real numbers.

b. $V = \dfrac{30.9\,(273)}{760} \approx 11.10$ liters.

41. The output is $f\,(32, 243) = 100\,(32)^{3/5}\,(243)^{2/5} = 100\,(8)\,(9) = 7200$, or \$7200 billion.

43. The number of suspicious fires is $N\,(100, 20) = \dfrac{100\left[1000 + 0.03\left(100^2\right)(20)\right]^{1/2}}{\left[5 + 0.2\,(20)\right]^2} = 103.29$, or about 103.

45. **a.** If $r = 6\%$, then $P = f\,(300000, 0.06, 30) = \dfrac{300{,}000\,(0.06)}{12\left[1 - \left(1 + \frac{0.06}{12}\right)^{-360}\right]} \approx 1798.65$, or \$1798.65. If $r = 8\%$,

then $P = f\,(300000, 0.08, 30) = \dfrac{300{,}000\,(0.08)}{12\left[1 - \left(1 + \frac{0.08}{12}\right)^{-360}\right]} \approx 2201.29$, or \$2201.29.

b. $P = f\,(300000, 0.08, 20) = \dfrac{300{,}000\,(0.08)}{12\left[1 - \left(1 + \frac{0.08}{12}\right)^{-240}\right]} \approx 2509.32$, or \$2509.32.

47. $f\,(M, 600, 10) = \dfrac{\pi^2\,(360{,}000)\,M\,(10)}{900} \approx 39{,}478.42M$, or $\dfrac{39{,}478.42}{980} \approx 40.28$ times gravity.

49. The level curves of V have equation $\dfrac{kT}{P} = C$, where C is a positive constant. The level curves are the family of straight lines $T = \dfrac{C}{k}P$ lying in the first quadrant, because k, T, and P are positive. Every point on the level curve $V = C$ gives the same volume C.

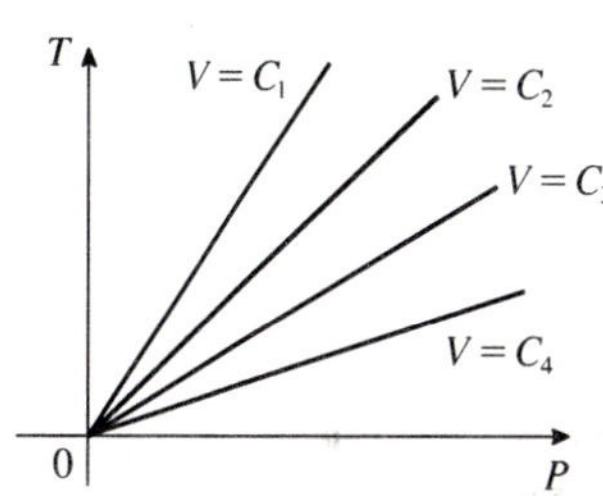

51. False. Let $h\,(x, y) = xy$. Then there is no pair of functions f and g such that $h\,(x, y) = f\,(x) + g\,(y)$.

53. False. Because $x^2 - y^2 = (x + y)\,(x - y)$, we see that $x^2 - y^2 = 0$ if $y = \pm x$. Therefore, the domain of f is $\{(x, y) \mid y \neq \pm x\}$.

55. False. Take $f\,(x, y) = \sqrt{x^2 + y^2}$, $P_1\,(-1, 1)$, and $P_2\,(1, 1)$. Then $f\,(x_1, y_1) = f\,(-1, 1) = \sqrt{(-1)^2 + 1^2} = \sqrt{2}$ and $f\,(x_2, y_2) = f\,(1, 1) = \sqrt{1^2 + 1^2} = \sqrt{2}$. So $f\,(x_1, y_1) = f\,(x_2, y_2)$, but $P\,(x_1, y_1) \neq P\,(x_2, y_2)$.

12.2 Partial Derivatives

Problem-Solving Tips

1. The expressions f_{xy} and f_{yx} denote the second partial derivatives of the function $f\,(x, y)$. Note that when this notation is used, the differentiation is carried out in the order in which x and y appear (left to right).

2. The notation $\dfrac{\partial^2 f}{\partial y\,\partial x}$ and $\dfrac{\partial^2 f}{\partial x\,\partial y}$ is also used to denote the second partial derivatives of the function $f\,(x, y)$, but in this case the differentiation is carried out in reverse order (right to left).

Concept Questions page 866

1. a. $\frac{\partial f}{\partial x}(a,b) = \frac{\partial f}{\partial x}(x,y)\Big|_{(a,b)} = \left[\lim_{h\to 0}\frac{f(x+h,y)-f(x,y)}{h}\right]_{(a,b)}$.

   b. See pages 856–858 of the text.

3. f_{xx}, f_{yy}, f_{xy}, and f_{yx}.

Exercises page 866

1. a. $f(x,y) = x^2 + 2y^2$, so $f_x(x,y) = 2x$, $f_y(x,y) = 4y$, $f_x(2,1) = 4$, and $f_y(2,1) = 4$.

   b. $f_x(2,1) = 4$ says that the slope of the tangent line to the curve of intersection of the surface $z = x^2 + 2y^2$ and the plane $y = 1$ at the point $(2,1,6)$ is 4. $f_y(2,1) = 4$ says that the slope of the tangent line to the curve of intersection of the surface $z = x^2 + 2y^2$ and the plane $x = 2$ at the point $(2,1,6)$ is 4.

   c. $f_x(2,1) = 4$ says that the rate of change of $f(x,y)$ with respect to x with y held fixed with a value of 1 is 4 units per unit change in x. $f_y(2,1) = 4$ says that the rate of change of $f(x,y)$ with respect to y with x held fixed with a value of 2 is 4 units per unit change in y.

3. $f(x,y) = 2x + 3y + 5$, so $f_x = 2$ and $f_y = 3$.

5. $g(x,y) = 3x^2 + 2y + 1$, so $g_x = 6x$ and $g_y = 2$.

7. $f(x,y) = \frac{2y}{x^2}$, so $f_x = -\frac{4y}{x^3}$ and $f_y = \frac{2}{x^2}$.

9. $g(u,v) = \frac{u-v}{u+v}$, so $\frac{\partial g}{\partial u} = \frac{(u+v)(1)-(u-v)(1)}{(u+v)^2} = \frac{2v}{(u+v)^2}$ and $\frac{\partial g}{\partial v} = \frac{(u+v)(-1)-(u-v)(1)}{(u+v)^2} = -\frac{2u}{(u+v)^2}$.

11. $f(s,t) = (s^2 - st + t^2)^3$, so $f_s = 3(s^2 - st + t^2)^2(2s - t)$ and $f_t = 3(s^2 - st + t^2)^2(2t - s)$.

13. $f(x,y) = (2x^2 + y^2)^{2/3}$, so $f_x = \frac{2}{3}(2x^2 + y^2)^{-1/3}(4x) = \frac{8}{3}x(2x^2 + y^2)^{-1/3}$ and $f_y = \frac{4}{3}y(2x^2 + y^2)^{-1/3}$.

15. $f(x,y) = e^{xy+1}$, so $f_x = ye^{xy+1}$ and $f_y = xe^{xy+1}$.

17. $f(x,y) = x\ln y + y\ln x$, so $f_x = \ln y + \frac{y}{x}$ and $f_y = \frac{x}{y} + \ln x$.

19. $g(u,v) = e^u \ln v$, so $g_u = e^u \ln v$ and $g_v = \frac{e^u}{v}$.

21. $f(x,y,z) = xyz + xy^2 + yz^2 + zx^2$, so $f_x = yz + y^2 + 2xz$, $f_y = xz + 2xy + z^2$, and $f_z = xy + 2yz + x^2$.

23. $h(r,s,t) = e^{rst}$, so $h_r = ste^{rst}$, $h_s = rte^{rst}$, and $h_t = rse^{rst}$.

25. $f(x,y) = x^2y + xy^2$, so $f_x(2,1) = (2xy + y^2)\big|_{(2,1)} = 5$ and $f_y(2,1) = (x^2 + 2xy)\big|_{(2,1)} = 8$.

27. $f(x,y) = x\sqrt{y} + y^2 = xy^{1/2} + y^2$, so $f_x(2,1) = \sqrt{y}\big|_{(2,1)} = 1$ and $f_y(2,1) = \left(\frac{x}{2\sqrt{y}} + 2y\right)\Big|_{(2,1)} = 3$.

29. $f(x,y) = \frac{x}{y}$, so $f_x(2,1) = \left.\frac{1}{y}\right|_{(2,1)} = 1$ and $f_y(2,1) = \left.-\frac{x}{y^2}\right|_{(2,1)} = -2$.

31. $f(x,y) = e^{xy}$, so $f_x(1,1) = ye^{xy}|_{(1,1)} = e$ and $f_y(1,1) = xe^{xy}|_{(1,1)} = e$.

33. $f(x,y,z) = x^2yz^3$, so $f_x(1,0,2) = 2xyz^3|_{(1,0,2)} = 0$, $f_y(1,0,2) = x^2z^3|_{(1,0,2)} = 8$, and $f_z(1,0,2) = 3x^2yz^2|_{(1,0,2)} = 0$.

35. $f(x,y) = x^2y + xy^3$, so $f_x = 2xy + y^3$ and $f_y = x^2 + 3xy^2$. Therefore, $f_{xx} = 2y$, $f_{xy} = 2x + 3y^2 = f_{yx}$, and $f_{yy} = 6xy$.

37. $f(x,y) = x^2 - 2xy + 2y^2 + x - 2y$, so $f_x = 2x - 2y + 1$ and $f_y = -2x + 4y - 2$. Therefore, $f_{xx} = 2$, $f_{xy} = -2 = f_{yx}$, and $f_{yy} = 4$.

39. $f(x,y) = \left(x^2+y^2\right)^{1/2}$, so $f_x = \frac{1}{2}\left(x^2+y^2\right)^{-1/2}(2x) = x\left(x^2+y^2\right)^{-1/2}$ and $f_y = y\left(x^2+y^2\right)^{-1/2}$. Therefore,

$$f_{xx} = \left(x^2+y^2\right)^{-1/2} + x\left(-\tfrac{1}{2}\right)\left(x^2+y^2\right)^{-3/2}(2x) = \left(x^2+y^2\right)^{-1/2} - x^2\left(x^2+y^2\right)^{-3/2}$$
$$= \left(x^2+y^2\right)^{-3/2}\left(x^2+y^2-x^2\right) = \frac{y^2}{\left(x^2+y^2\right)^{3/2}},$$
$$f_{xy} = x\left(-\tfrac{1}{2}\right)\left(x^2+y^2\right)^{-3/2}(2y) = -\frac{xy}{\left(x^2+y^2\right)^{3/2}} = f_{yx}, \text{ and}$$
$$f_{yy} = \left(x^2+y^2\right)^{-1/2} + y\left(-\tfrac{1}{2}\right)\left(x^2+y^2\right)^{-3/2}(2y) = \left(x^2+y^2\right)^{-1/2} - y^2\left(x^2+y^2\right)^{-3/2}$$
$$= \left(x^2+y^2\right)^{-3/2}\left(x^2+y^2-y^2\right) = \frac{x^2}{\left(x^2+y^2\right)^{3/2}}.$$

41. $f(x,y) = e^{-x/y}$, so $f_x = -\frac{1}{y}e^{-x/y}$ and $f_y = \frac{x}{y^2}e^{-x/y}$. Therefore, $f_{xx} = \frac{1}{y^2}e^{-x/y}$,

$$f_{xy} = -\frac{x}{y^3}e^{-x/y} + \frac{1}{y^2}e^{-x/y} = \left(\frac{-x+y}{y^3}\right)e^{-x/y} = f_{yx}, \text{ and } f_{yy} = -\frac{2x}{y^3}e^{-x/y} + \frac{x^2}{y^4}e^{-x/y} = \frac{x}{y^3}\left(\frac{x}{y} - 2\right)e^{-x/y}.$$

43. **a.** $f(x,y) = 20x^{3/4}y^{1/4}$, so $f_x(256,16) = 15\left.\left(\frac{y}{x}\right)^{1/4}\right|_{(256,16)} = 15\left(\frac{16}{256}\right)^{1/4} = 15\left(\frac{2}{4}\right) = 7.5$ and

$f_y(256,16) = 5\left.\left(\frac{x}{y}\right)^{3/4}\right|_{(256,16)} = 5\left(\frac{256}{16}\right)^{3/4} = 5(8) = 40$.

b. Yes.

45. $p(x,y) = 200 - 10\left(x - \frac{1}{2}\right)^2 - 15(y-1)^2$, so $\frac{\partial p}{\partial x}(0,1) = -20\left.\left(x - \frac{1}{2}\right)\right|_{(0,1)} = 10$. At the location $(0,1)$ in the figure, the price of land is increasing by \$10 per square foot per mile to the east. $\frac{\partial p}{\partial y}(0,1) = -30(y-1)|_{(0,1)} = 0$, so at the point $(0,1)$ in the figure, the price of land is unchanging with respect to north-south change.

47. $f(p,q) = 10{,}000 - 10p - e^{0.5q}$ and $g(p,q) = 50{,}000 - 4000q - 10p$. Thus, $\frac{\partial f}{\partial q} = -0.5e^{0.5q} < 0$ and $\frac{\partial g}{\partial p} = -10 < 0$, and so the two commodities are complementary commodities.

49. $R(x,y) = -0.2x^2 - 0.25y^2 - 0.2xy + 200x + 160y$, so $\frac{\partial R}{\partial x}(300,250) = (-0.4x - 0.2y + 200)|_{(300,250)} = -0.4(300) - 0.2(250) + 200 = 30$. This says that at sales levels of 300 finished and 250 unfinished units, revenue is increasing by \$30 per week per unit increase in finished pieces. $\frac{\partial R}{\partial y}(300,250) = -0.5y - 0.2x + 160|_{(300,250)} = -0.5(250) - 0.2(300) + 160 = -25$, and this says that at the same sales levels, revenue is decreasing by \$25 per week per unit increase in unfinished pieces.

51. a. $T = f(t,s) = 35.74 + 0.6215t - 35.75s^{0.16} + 0.4275ts^{0.16}$, so $f(32,20) = 35.74 + 0.6215(32) - 35.75(20^{0.16}) + 0.4275(32)(20^{0.16}) \approx 19.99$, or approximately 20°F.

b. $\frac{\partial T}{\partial s} = -35.75(0.16s^{-0.84}) + 0.4275t(0.16s^{-0.84}) = 0.16(-35.75 + 0.4275t)s^{-0.84}$, so $\left.\frac{\partial T}{\partial s}\right|_{(32,20)} = 0.16[-35.75 + 0.4275(32)]20^{-0.84} \approx -0.285$; that is, the wind chill will drop by 0.3° for each 1 mph increase in wind speed.

53. $N(x,y) = \frac{120\sqrt{1000+0.03x^2y}}{(5+0.2y)^2}$, so

$$\frac{\partial N}{\partial x} = \frac{\partial}{\partial x}\frac{120(1000+0.03x^2y)^{1/2}}{(5+0.2y)^2} = \frac{120\left(\frac{1}{2}\right)(1000+0.03x^2y)^{-1/2}(0.06xy)}{(5+0.2y)^2}.$$ Thus,

$\frac{\partial N}{\partial x}(100,20) = \left.\frac{3.6xy}{(5+0.2y)^2\sqrt{1000+0.03x^2y}}\right|_{(100,20)} \approx 1.06$. This means that with the level of reinvestment held constant at 20 cents per dollar deposited, the number of suspicious fires will grow at the rate of approximately 1 fire per increase of 1 person per census tract when the number of people per census tract is 100. Next,

$$\frac{\partial N}{\partial y}(100,20) = 120\frac{\partial}{\partial y}\left.\left[(1000+0.03x^2y)^{1/2}(5+0.2y)^{-2}\right]\right|_{(100,20)}$$

$$= 120\left.\left[\tfrac{1}{2}(1000+0.03x^2y)^{-1/2}(0.03x^2)(5+0.2y)^{-2} + (1000+0.03x^2y)^{1/2}(-2)(5+0.2y)^{-3}(0.2)\right]\right|_{(100,20)}$$

$$= \left.\frac{9x^2 - 1.08x^2y - 48{,}000}{(5+0.2y)^3\sqrt{1000+0.03x^2y}}\right|_{(100,20)} \approx -2.85$$

which tells us that if the number of people per census tract is constant at 100 per tract, the number of suspicious fires decreases at a rate of approximately 2.9 per increase of 1 cent per dollar deposited for reinvestment when the level of reinvestment is 20 cents per dollar deposited.

55. $V = \frac{30.9T}{P}$, so $\frac{\partial V}{\partial T} = \frac{30.9}{P}$ and $\frac{\partial V}{\partial P} = -\frac{30.9T}{P^2}$. Therefore, $\left.\frac{\partial V}{\partial T}\right|_{T=300,P=800} = \frac{30.9}{800} \approx 0.039$, or approximately 0.039 liters per degree. $\left.\frac{\partial V}{\partial P}\right|_{T=300,P=800} = -\frac{(30.9)(300)}{800^2} \approx -0.014$, or approximately -0.014 liters per millimeter of mercury.

57. $V = \frac{kT}{P}$, so $\frac{\partial V}{\partial T} = \frac{k}{P}$; $T = \frac{VP}{k}$, so $\frac{\partial T}{\partial P} = \frac{V}{k} = \frac{T}{P}$; and $P = \frac{kT}{V}$, so $\frac{\partial P}{\partial V} = -\frac{kT}{V^2} = -kT\frac{P^2}{(kT)^2} = -\frac{P^2}{kT}$.

Therefore $\frac{\partial V}{\partial T}\cdot\frac{\partial T}{\partial P}\cdot\frac{\partial P}{\partial V} = \frac{k}{P}\left(\frac{T}{P}\right)\left(-\frac{P^2}{kT}\right) = -1$.

59. $\dfrac{\partial P}{\partial x} = \dfrac{\partial}{\partial x} kx^{\alpha}y^{1-\alpha} = k\alpha x^{\alpha-1}y^{1-\alpha} = k\alpha \left(\dfrac{y}{x}\right)^{1-\alpha}$ and $\dfrac{\partial P}{\partial y} = k(1-\alpha)x^{\alpha}y^{-\alpha} = k(1-\alpha)\left(\dfrac{x}{y}\right)^{\alpha}$.

Therefore, $x\dfrac{\partial P}{\partial x} + y\dfrac{\partial P}{\partial y} = \dfrac{k\alpha x y^{1-\alpha}}{x^{1-\alpha}} + \dfrac{k(1-\alpha)yx^{\alpha}}{y^{\alpha}} = k\alpha x^{\alpha}y^{1-\alpha} + k(1-\alpha)x^{\alpha}y^{1-\alpha} = kx^{\alpha}y^{1-\alpha} = P$, as was to be shown.

61. True. This is a consequence of the definition of $f_x(a, b)$ as the rate of change of f in the x-direction at (a, b) with y held fixed.

63. False. Let $f(x, y) = xy^{5/3}$. Then $f_{xy} = \frac{5}{3}y^{2/3} = f_{yx}$, so both f_{xy} and f_{yx} exist at $(0, 0)$. However, $f_{yy} = \dfrac{10x}{9y^{1/3}}$ is not defined at $(0, 0)$.

Using Technology page 870

1. 1.3124, 0.4038. **3.** −1.8889, 0.7778. **5.** −0.3863, −0.8497.

12.3 Maxima and Minima of Functions of Several Variables

Problem-Solving Tips

1. **To find the relative extrema of a function of several variables**, first find the critical points of $f(x, y)$ by solving the simultaneous equations $f_x = 0$ and $f_y = 0$, then use the second derivative test to classify those points.

2. **To use the second derivative test**, first evaluate the function $D(x, y) = f_{xx}f_{yy} - f_{xy}^2$ for each critical point found in Tip 1.
   - If $D(a, b) > 0$ and $f_{xx}(a, b) < 0$, then $f(x, y)$ has a relative maximum at the point (a, b).
   - If $D(a, b) > 0$ and $f_{xx}(a, b) > 0$, then $f(x, y)$ has a relative minimum at the point (a, b).
   - If $D(a, b) < 0$ then $f(x, y)$ has neither a relative maximum nor a relative minimum at the point (a, b).
   - If $D(a, b) = 0$, then the test is inconclusive.

Concept Questions page 877

1. a. A function $f(x, y)$ has a relative maximum at (a, b) if $f(a, b)$ is the largest value of $f(x, y)$ for all (x, y) near (a, b).

b. $f(a, y)$ has an absolute maximum at (a, b) if $f(a, b)$ is the largest value of $f(x, y)$ for all (x, y) in the domain of f.

3. See the procedure on page 873 of the text.

Exercises page 877

1. $f(x, y) = 1 - 2x^2 - 3y^2$. To find the critical points of f, we solve the system $\begin{cases} f_x = -4x = 0 \\ f_y = -6y = 0 \end{cases}$ obtaining $(0, 0)$ as the only critical point of f. Next, $f_{xx} = -4$, $f_{xy} = 0$, and $f_{yy} = -6$. In particular, $f_{xx}(0, 0) = -4$, $f_{xy}(0, 0) = 0$, and $f_{yy}(0, 0) = -6$, giving $D(0, 0) = (-4)(-6) - 0^2 = 24 > 0$. Because $f_{xx}(0, 0) < 0$, the Second Derivative Test implies that $(0, 0)$ gives rise to a relative maximum of f. Finally, the relative maximum value of f is $f(0, 0) = 1$.

3. $f(x, y) = x^2 - y^2 - 2x + 4y + 1$. To find the critical points of f, we solve the system $\begin{cases} f_x = 2x - 2 = 0 \\ f_y = -2y + 4 = 0 \end{cases}$ obtaining $x = 1$ and $y = 2$, so $(1, 2)$ is the only critical point of f. $f_{xx} = 2$, $f_{xy} = 0$, and $f_{yy} = -2$, so $D(x, y) = f_{xx} f_{yy} - f_{xy}^2 = -4$. In particular, $D(1, 2) = -4 < 0$, so $(1, 2)$ gives a saddle point of f. Because $f(1, 2) = 1 - 4 - 2 + 8 + 1 = 4$, the saddle point is $(1, 2, 4)$.

5. $f(x, y) = x^2 + 2xy + 2y^2 - 4x + 8y - 1$. To find the critical points of f, we solve the system $\begin{cases} f_x = 2x + 2y - 4 = 0 \\ f_y = 2x + 4y + 8 = 0 \end{cases}$ obtaining $(8, -6)$ as the critical point of f. Next, $f_{xx} = 2$, $f_{xy} = 2$, and $f_{yy} = 4$. In particular, $f_{xx}(8, -6) = 2$, $f_{xy}(8, -6) = 2$, and $f_{yy}(8, -6) = 4$, giving $D = 2(4) - 4 = 4 > 0$. Because $f_{xx}(8, -6) > 0$, $(8, -6)$ gives rise to a relative minimum of f. The relative minimum value of f is $f(8, -6) = 8^2 + 2(8)(-6) + 2(-6)^2 - 4(8) + 8(-6) - 1 = -41$.

7. $f(x, y) = 2x^3 + y^2 - 9x^2 - 4y + 12x - 2$. To find the critical points of f, we solve the system $\begin{cases} f_x = 6x^2 - 18x + 12 = 0 \\ f_y = 2y - 4 = 0 \end{cases}$ The first equation is equivalent to $x^2 - 3x + 2 = 0$, or $(x - 2)(x - 1) = 0$, giving $x = 1$ or 2. The second equation of the system gives $y = 2$. Therefore, there are two critical points, $(1, 2)$ and $(2, 2)$. Next, we compute $f_{xx} = 12x - 18 = 6(2x - 3)$, $f_{xy} = 0$, and $f_{yy} = 2$.
At the point $(1, 2)$, $f_{xx}(1, 2) = 6(2 - 3) = -6$, $f_{xy}(1, 2) = 0$, and $f_{yy}(1, 2) = 2$, so $D(1, 2) = (-6)(2) - 0 = -12 < 0$ and we conclude that $(1, 2)$ gives a saddle point of f. Because $f(1, 2) = 2(1) + 4 - 9(1) - 4(2) + 12(1) - 2 = -1$, the saddle point is $(1, 2, -1)$.
At the point $(2, 2)$, $f_{xx}(2, 2) = 6(4 - 3) = 6$, $f_{xy}(2, 2) = 0$, and $f_{yy}(2, 2) = 2$, so $D(2, 2) = (6)(2) - 0 = 12 > 0$. Because $f_{xx}(2, 2) > 0$, we see that $(2, 2)$ gives a relative minimum with value $f(2, 2) = 2(2^3) + 4 - 9(4) - 4(2) + 12(2) - 2 = -2$.

9. $f(x, y) = x^3 + y^2 - 2xy + 7x - 8y + 4$. To find the critical points of f, we solve the system $\begin{cases} f_x = 3x^2 - 2y + 7 = 0 \\ f_y = 2y - 2x - 8 = 0 \end{cases}$ Adding the two equations gives $3x^2 - 2x - 1 = (3x + 1)(x - 1) = 0$. Therefore, $x = -\frac{1}{3}$ or 1. Substituting each of these values of x into the second equation gives $y = \frac{11}{3}$ and $y = 5$, respectively. Therefore, $\left(-\frac{1}{3}, \frac{11}{3}\right)$ and $(1, 5)$ are critical points of f. Next, $f_{xx} = 6x$, $f_{xy} = -2$, and $f_{yy} = 2$, so $D(x, y) = 12x - 4 = 4(3x - 1)$. Then $D\left(-\frac{1}{3}, \frac{11}{3}\right) = 4(-1 - 1) = -8 < 0$, and so $\left(-\frac{1}{3}, \frac{11}{3}\right)$ gives a saddle point. Because $f\left(-\frac{1}{3}, \frac{11}{3}\right) = -\frac{319}{27}$, the saddle point is $\left(-\frac{1}{3}, \frac{11}{3}, -\frac{319}{27}\right)$. Next, $D(1, 5) = 4(3 - 1) = 8 > 0$, and since $f_{xx}(1, 5) = 6 > 0$, we see that $(1, 5)$ gives a relative minimum with value $f(1, 5) = -13$.

11. $f(x,y)=x^3-3xy+y^3-2$. To find the critical points of f, we solve the system $\begin{cases} f_x = 3x^2-3y=0 \\ f_y=-3x+3y^2=0 \end{cases}$

The first equation gives $y=x^2$, and substituting this into the second equation gives $-3x+3x^4=3x\left(x^3-1\right)=0$. Therefore, $x=0$ or 1. Substituting these values of x into the first equation gives $y=0$ and $y=1$, respectively. Therefore, $(0,0)$ and $(1,1)$ are critical points of f. Next, we find $f_{xx}=6x$, $f_{xy}=-3$, and $f_{yy}=6y$, so $D=f_{xx}f_{yy}-f_{xy}^2=36xy-9$. Because $D(0,0)=-9<0$, we see that $(0,0)$ gives a saddle point of f. Because $f(0,0)=-2$, the saddle point is $(0,0,-2)$. Next, $D(1,1)=36-9=27>0$, and since $f_{xx}(1,1)=6>0$, we see that $f(1,1)=-3$ is a relative minimum value of f.

13. $f(x,y)=xy+\dfrac{4}{x}+\dfrac{2}{y}$. Solving the system of equations $\begin{cases} f_x=y-\dfrac{4}{x^2}=0 \\ f_y=x-\dfrac{2}{y^2}=0 \end{cases}$ we obtain $y=\dfrac{4}{x^2}$.

Therefore, $x-2\left(\dfrac{x^4}{16}\right)=0$ and $8x-x^4=x\left(8-x^3\right)=0$, so $x=0$ or $x=2$. Because $x=0$ is not in the domain of f, $(2,1)$ is the only critical point of f. Next, $f_{xx}=\dfrac{8}{x^3}$, $f_{xy}=1$, and $f_{yy}=\dfrac{4}{y^3}$. Therefore, $D(2,1)=\left.\left(\dfrac{32}{x^3y^3}-1\right)\right|_{(2,1)}=4-1=3>0$ and $f_{xx}(2,1)=1>0$, so the relative minimum value of f is $f(2,1)=2+\frac{4}{2}+\frac{2}{1}=6$.

15. $f(x,y)=x^2-e^{y^2}$. Solving the system of equations $\begin{cases} f_x=2x=0 \\ f_y=-2ye^{y^2}=0 \end{cases}$ we obtain $x=0$ and $y=0$.

Therefore, $(0,0)$ is the only critical point of f. Next, $f_{xx}=2$, $f_{xy}=0$, and $f_{yy}=-2e^{y^2}-4y^2e^{y^2}$, so $D(0,0)=\left[-4e^{y^2}\left(1+2y^2\right)\right]_{(0,0)}=-4(1)<0$, and we conclude that $(0,0)$ gives a saddle point. Because $f(0,0)=-1$, the saddle point is $(0,0,-1)$.

17. $f(x,y)=e^{x^2+y^2}$. Solving the system $\begin{cases} f_x=2xe^{x^2+y^2}=0 \\ f_y=2ye^{x^2+y^2}=0 \end{cases}$ we see that $x=0$ and $y=0$

(recall that $e^{x^2+y^2}\neq 0$). Therefore, $(0,0)$ is the only critical point of f. Next, we compute $f_{xx}=2e^{x^2+y^2}+2x(2x)e^{x^2+y^2}=2\left(1+2x^2\right)e^{x^2+y^2}$, $f_{xy}=2x(2y)e^{x^2+y^2}=4xye^{x^2+y^2}$, and $f_{yy}=2\left(1+2y^2\right)e^{x^2+y^2}$. In particular, at the point $(0,0)$, $f_{xx}(0,0)=2$, $f_{xy}(0,0)=0$, and $f_{yy}(0,0)=2$. Therefore, $D(0,0)=(2)(2)-0=4>0$. Because $f_{xx}(0,0)>0$, we conclude that $(0,0)$ gives rise to a relative minimum of f. The relative minimum value is $f(0,0)=1$.

19. $f(x,y) = \ln(1+x^2+y^2)$. We solve the system of equations $\begin{cases} f_x = \dfrac{2x}{1+x^2+y^2} = 0 \\ f_y = \dfrac{2y}{1+x^2+y^2} = 0 \end{cases}$ obtaining $x = 0$ and $y = 0$. Therefore, $(0,0)$ is the only critical point of f. Next,
$f_{xx} = \dfrac{(1+x^2+y^2)\,2-(2x)(2x)}{(1+x^2+y^2)^2} = \dfrac{2+2y^2-2x^2}{(1+x^2+y^2)^2}$, $f_{yy} = \dfrac{(1+x^2+y^2)\,2-(2y)(2y)}{(1+x^2+y^2)^2} = \dfrac{2+2x^2-2y^2}{(1+x^2+y^2)^2}$,
and $f_{xy} = -2x(1+x^2+y^2)^{-2}(2y) = -\dfrac{4xy}{(1+x^2+y^2)^2}$. Therefore,
$D(x,y) = \dfrac{(2+2y^2-2x^2)(2+2x^2-2y^2)}{(1+x^2+y^2)^4} - \dfrac{16x^2y^2}{(1+x^2+y^2)^4}$. Because $D(0,0) = 4 > 0$ and $f_{xx}(0,0) = 2 > 0$, $f(0,0) = 0$ is a relative minimum value.

21. $P(x) = -0.2x^2 - 0.25y^2 - 0.2xy + 200x + 160y - 100x - 70y - 4000$
$= -0.2x^2 - 0.25y^2 - 0.2xy + 100x + 90y - 4000.$

Thus, $\begin{cases} P_x = -0.4x - 0.2y + 100 = 0 \\ P_y = -0.5y - 0.2x + 90 = 0 \end{cases}$ implies that $\begin{cases} 4x + 2y = 1000 \\ 2x + 5y = 900 \end{cases}$ Solving, we find $x = 200$ and $y = 100$. Next, $P_{xx} = -0.4$, $P_{yy} = -0.5$, $P_{xy} = -0.2$, and $D(200,100) = (-0.4)(-0.5) - (-0.2)^2 > 0$. Because $P_{xx}(200,100) < 0$, we conclude that $(200,100)$ is a relative maximum of P. Thus, the company should manufacture 200 finished and 100 unfinished units per week. The maximum profit is $P(200,100) = -0.2(200)^2 - 0.25(100)^2 - 0.2(100)(200) + 100(200) + 90(100) - 4000 = 10{,}500$, or \$10,500.

23. $p(x,y) = 200 - 10\left(x - \frac{1}{2}\right)^2 - 15(y-1)^2$. Solving the system of equations $\begin{cases} p_x = -20\left(x - \frac{1}{2}\right) = 0 \\ p_y = -30(y-1) = 0 \end{cases}$ we obtain $x = \frac{1}{2}$ and $y = 1$. We conclude that the only critical point of f is $\left(\frac{1}{2}, 1\right)$. Next, $p_{xx} = -20$, $p_{xy} = 0$, and $p_{yy} = -30$, so $D\left(\frac{1}{2}, 1\right) = (-20)(-30) = 600 > 0$. Because $p_{xx} = -20 < 0$, we conclude that $f\left(\frac{1}{2}, 1\right)$ gives a relative maximum. We conclude that the price of land is highest at $\left(\frac{1}{2}, 1\right)$.

25. We want to minimize $f(x,y) = D^2 = (x-5)^2 + (y-2)^2 + (x+4)^2 + (y-4)^2 + (x+1)^2 + (y+3)^2$. We calculate $\begin{cases} f_x = 2(x-5) + 2(x+4) + 2(x+1) = 6x = 0, \\ f_y = 2(y-2) + 2(y-4) + 2(y+3) = 6y - 6 = 0 \end{cases}$ and conclude that $x = 0$ and $y = 1$. Also, $f_{xx} = 6$, $f_{xy} = 0$, $f_{yy} = 6$, and $D(x,y) = (6)(6) = 36 > 0$. Because $f_{xx} > 0$, we conclude that the function is minimized at $(0,1)$, the desired location.

27. We want to maximize $V = \pi r^2 \ell$. But we have $2\pi r + \ell = 130 \Rightarrow \ell = 130 - 2\pi r$, so we need to maximize $V = f(r) = \pi r^2(130 - 2\pi r) = -2\pi^2 r^3 + 130\pi r^2$. Now $V' = f'(r) = -6\pi^2 r^2 + 260\pi r = -2\pi r(3\pi r - 130) = 0 \Rightarrow r = 0$ or $r = \frac{130}{3\pi}$. Since $f''\left(\frac{130}{3\pi}\right) = \left(-12\pi^2 r + 260\pi\right)\big|_{130/(3\pi)} = -12\pi^2\left(\frac{130}{3\pi}\right) + 260\pi = -260\pi < 0$, we conclude that $r = \frac{130}{3\pi}$ does yield the absolute maximum for f. We find $\ell = 130 - 2\pi\left(\frac{130}{3\pi}\right) = \frac{130}{3}$ or $43\frac{1}{3}''$. Therefore, $V = \pi r^2 \ell = \pi\left(\frac{130}{3\pi}\right)^2\left(\frac{130}{3}\right) = \frac{2{,}197{,}000}{27\pi}$ in^3.

29. Refer to the figure in the text. $xy + 2xz + 2yz = 300$, so $z(2x + 2y) = 300 - xy$, The volume is given by $V = xyz = xy\dfrac{300 - xy}{2(x+y)} = \dfrac{300xy - x^2y^2}{2(x+y)}$. We find

$$\frac{\partial V}{\partial x} = \frac{1}{2}\frac{(x+y)(300y - 2xy^2) - (300xy - x^2y^2)}{(x+y)^2} = \frac{300xy - 2x^2y^2 + 300y^2 - 2xy^3 - 300xy + x^2y^2}{2(x+y)^2}$$

$$= \frac{300y^2 - 2xy^3 - x^2y^2}{2(x+y)^2} = \frac{y^2(300 - 2xy - x^2)}{2(x+y)^2}$$

and similarly $\dfrac{\partial V}{\partial y} = \dfrac{x^2(300 - 2xy - y^2)}{2(x+y)^2}$. Setting both $\dfrac{\partial V}{\partial x}$ and $\dfrac{\partial V}{\partial y}$ equal to 0 and observing that both $x > 0$ and $y > 0$, we have the system $\begin{cases} 2yx + x^2 = 300 \\ 2yx + y^2 = 300 \end{cases}$ Subtracting, we find $y^2 - x^2 = 0$, so $(y - x)(y + x) = 0$. Thus, $y = x$ or $y = -x$. The latter is not possible since x and y are both positive. Therefore, $y = x$. Substituting this value into the first equation in the system gives $2x^2 + x^2 = 300$, so $x^2 = 100$ and $x = y = 10$. Substituting these values into the expression for z gives $z = \dfrac{300 - 10^2}{2(10 + 10)} = 5$, so the dimensions are $10'' \times 10'' \times 5''$ and the volume is 500 in^3.

31. The heating cost is $C = 2xy + 8xz + 6yz$. But $xyz = 12{,}000$, so $z = \dfrac{12{,}000}{xy}$. Therefore, $C = f(x, y) = 2xy + 8x\left(\dfrac{12{,}000}{xy}\right) + 6y\left(\dfrac{12{,}000}{xy}\right) = 2xy + \dfrac{96{,}000}{y} + \dfrac{72{,}000}{x}$. To find the minimum of f, we find the critical point of f by solving the system $\begin{cases} f_x = 2y - \dfrac{72{,}000}{x^2} = 0 \\ f_y = 2x - \dfrac{96{,}000}{y^2} = 0 \end{cases}$ The first equation gives $y = \dfrac{36{,}000}{x^2}$, which when substituted into the second equation yields $2x - 96{,}000\left(\dfrac{x^2}{36{,}000}\right)^2 = 0$, so $(36{,}000)^2 x - 48{,}000x^4 = 0$ and $x(27{,}000 - x^3) = 0$. Solving this equation, we have $x = 0$ or $x = 30$. We reject the first root because $x = 0$ lies outside the domain of f. With $x = 30$, we find $y = 40$ and $z = 10$. Next, $f_{xx} = \dfrac{144{,}000}{x^3}$, $f_{yy} = \dfrac{192{,}000}{y^3}$, and $f_{xy} = 2$. In particular, $f_{xx}(30, 40) \approx 5.33$, $f_{xy} = (30, 40) = 2$, and $f_{yy}(30, 40) = 3$.
Thus, $D(30, 40) \approx (5.33)(3) - 4 = 11.99 > 0$, and since $f_{xx}(30, 40) > 0$, we see that $(30, 40)$ gives a relative minimum. Physical considerations tell us that this is an absolute minimum. The minimal annual heating cost is $f(30, 40) = 2(30)(40) + \dfrac{96{,}000}{40} + \dfrac{72{,}000}{30} = 7200$, or \$7,200.

33. False. Let $f(x, y) = xy$. Then $f_x(0, 0) = 0$ and $f_y(0, 0) = 0$, but $(0, 0)$ does not give a relative extremum of $(0, 0)$. In fact, $f_{xx} = 0$, $f_{yy} = 0$, and $f_{xy} = 1$, so $D(x, y) = f_{xx}f_{yy} - f_{xy}^2 = -1$ and $D(0, 0) = -1$, showing that $(0, 0, 0)$ is a saddle point.

35. False. $f_x(a, b)$ and/or $f_y(a, b)$ may be undefined.

37. True. Here $f_{xx}(a, b) = -f_{yy}(a, b)$, so
$D(a, b) = f_{xx}(a, b) f_{yy}(a, b) - f_{xy}^2(a, b) = f_{xx}(a, b)\left[-f_{xx}(a, b)\right] - 0 = -f_{xx}^2(a, b) < 0$, and so f cannot have a relative extremum at (a, b).

CHAPTER 12 Concept Review Questions page 880

1. xy, ordered pair, real number, $f(x, y)$

3. $z = f(x, y)$, f, surface

5. constant, x

7. $\leq$, (a, b), $\leq$, domain

CHAPTER 12 Review Exercises page 881

1. $f(x, y) = \dfrac{xy}{x^2 + y^2}$, so $f(0, 1) = 0$, $f(1, 0) = 0$, $f(1, 1) = \dfrac{1}{1+1} = \dfrac{1}{2}$, and $f(0, 0)$ does not exist because the point $(0, 0)$ does not lie in the domain of f.

3. $h(x, y, z) = xye^z + \dfrac{x}{y}$, so $h(1, 1, 0) = 1 + 1 = 2$, $h(-1, 1, 1) = -e - 1 = -(e + 1)$, and $h(1, -1, 1) = -e - 1 = -(e + 1)$.

5. $f(x, y) = \dfrac{x - y}{x + y}$, so $D = \{(x, y) \mid y \neq -x\}$.

7. $f(x, y, z) = \dfrac{xy\sqrt{z}}{(1 - x)(1 - y)(1 - z)}$. The domain of f is the set of all ordered triples (x, y, z) of real numbers such that $z \geq 0$, $x \neq 1$, $y \neq 1$, and $z \neq 1$.

9. $z = y - x^2$

11. $z = e^{xy}$

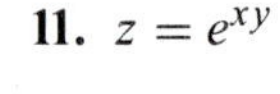

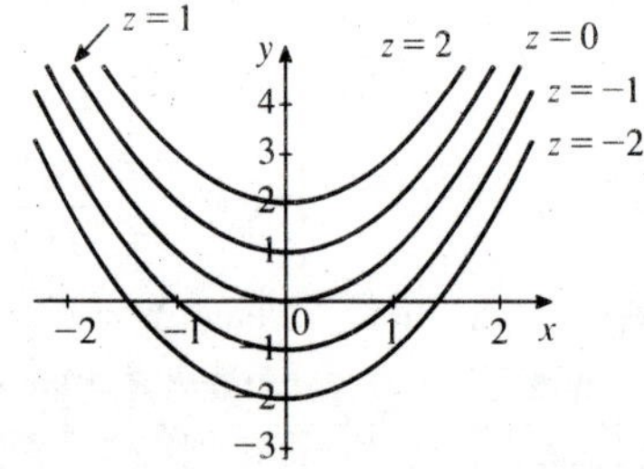

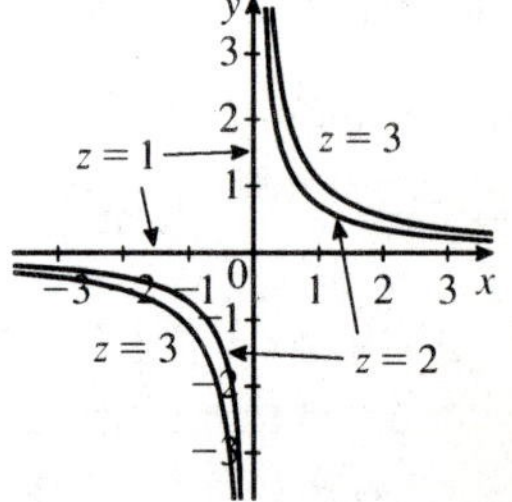

13. $f(x, y) = x\sqrt{y} + y\sqrt{x}$, so $f_x = \sqrt{y} + \dfrac{y}{2\sqrt{x}}$ and $f_y = \dfrac{x}{2\sqrt{y}} + \sqrt{x}$.

15. $f(x, y) = \dfrac{x - y}{y + 2x}$, so $f_x = \dfrac{(y + 2x) - (x - y)(2)}{(y + 2x)^2} = \dfrac{3y}{(y + 2x)^2}$ and $f_y = \dfrac{(y + 2x)(-1) - (x - y)}{(y + 2x)^2} = \dfrac{-3x}{(y + 2x)^2}$.

17. $h(x, y) = (2xy + 3y^2)^5$, so $h_x = 10y(2xy + 3y^2)^4$ and $h_y = 10(x + 3y)(2xy + 3y^2)^4$.

19. $f(x, y) = (x^2 + y^2)e^{x^2+y^2}$, so $f_x = 2xe^{x^2+y^2} + (x^2 + y^2)(2x)e^{x^2+y^2} = 2x(x^2 + y^2 + 1)e^{x^2+y^2}$ and $f_y = 2ye^{x^2+y^2} + (x^2 + y^2)(2y)e^{x^2+y^2} = 2y(x^2 + y^2 + 1)e^{x^2+y^2}$.

21. $f(x, y) = \ln\left(1 + \dfrac{x^2}{y^2}\right)$, so $f_x = \dfrac{2x/y^2}{1 + (x^2/y^2)} = \dfrac{2x}{x^2 + y^2}$ and $f_y = \dfrac{-2x^2/y^3}{1 + (x^2/y^2)} = -\dfrac{2x^2}{y(x^2 + y^2)}$.

23. $f(x,y) = x^4 + 2x^2y^2 - y^4$, so $f_x = 4x^3 + 4xy^2$ and $f_y = 4x^2y - 4y^3$. Therefore, $f_{xx} = 12x^2 + 4y^2$, $f_{xy} = 8xy = f_{yx}$, and $f_{yy} = 4x^2 - 12y^2$.

25. $g(x,y) = \dfrac{x}{x+y^2}$, so $g_x = \dfrac{(x+y^2)-x}{(x+y^2)^2} = \dfrac{y^2}{(x+y^2)^2}$ and

$g_y = \dfrac{-2xy}{(x+y^2)^2}$. Therefore, $g_{xx} = -2y^2(x+y^2)^{-3} = -\dfrac{2y^2}{(x+y^2)^3}$,

$$g_{xy} = \frac{(x+y^2)\,2y - y^2(2)(x+y^2)\,2y}{(x+y^2)^4} = \frac{2(x+y^2)(xy+y^3-2y^3)}{(x+y^2)^4} = \frac{2y(x-y^2)}{(x+y^2)^3} = g_{yx}, \text{ and}$$

$$g_{yy} = \frac{(x+y^2)^2(-2x) + 2xy(2)(x+y^2)\,2y}{(x+y^2)^4} = \frac{2x(x^2+y^2)(-x-y^2+4y^2)}{(x+y^2)^4} = \frac{2x(3y^2-x)}{(x+y^2)^3}$$

27. $h(s,t) = \ln\left(\dfrac{s}{t}\right)$. Write $h(s,t) = \ln s - \ln t$. Then $h_s = \dfrac{1}{s}$ and $h_t = -\dfrac{1}{t}$, so $h_{ss} = -\dfrac{1}{s^2}$, $h_{st} = h_{ts} = 0$, and $h_{tt} = \dfrac{1}{t^2}$.

29. $f(x,y) = 2x^2 + y^2 - 8x - 6y + 4$. To find the critical points of f, we solve the system $\begin{cases} f_x = 4x - 8 = 0 \\ f_y = 2y - 6 = 0 \end{cases}$ obtaining $x = 2$ and $y = 3$. Therefore, the sole critical point of f is $(2,3)$. Next, $f_{xx} = 4$, $f_{xy} = 0$, and $f_{yy} = 2$, so $D(2,3) = f_{xx}(2,3)\,f_{yy}(2,3) - f_{xy}(2,3)^2 = 8 > 0$. Because $f_{xx}(2,3) > 0$, we see that $f(2,3) = -13$ is a relative minimum value.

31. $f(x,y) = x^3 - 3xy + y^2$. We solve the system of equations $\begin{cases} f_x = 3x^2 - 3y = 0 \\ f_y = -3x + 2y = 0 \end{cases}$ obtaining $x^2 - y = 0$, and so $y = x^2$. Then $-3x + 2x^2 = 0$, $x(2x-3) = 0$, and so $x = 0$ or $x = \frac{3}{2}$. The corresponding values of y are $y = 0$ and $y = \frac{9}{4}$, so the critical points are $(0,0)$ and $\left(\frac{3}{2}, \frac{9}{4}\right)$. Next, $f_{xx} = 6x$, $f_{xy} = -3$, and $f_{yy} = 2$, and so $D(x,y) = 12x - 9 = 3(4x-3)$. Therefore, $D(0,0) = -9$, and so $(0,0,0)$ is a saddle point. $D\left(\frac{3}{2}, \frac{9}{4}\right) = 3(6-3) = 9 > 0$ and $f_{xx}\left(\frac{3}{2}, \frac{9}{4}\right) > 0$, and so $f\left(\frac{3}{2}, \frac{9}{4}\right) = \frac{27}{8} - \frac{81}{8} + \frac{81}{16} = -\frac{27}{16}$ is a relative minimum value.

33. $f(x,y) - f(x,y) - e^{2x^2+y^2}$. To find the critical points of f, we solve the system $\begin{cases} f_x = 4xe^{2x^2+y^2} = 0 \\ f_y = 2ye^{2x^2+y^2} = 0 \end{cases}$ giving $(0,0)$ as the only critical point of f. Next, $f_{xx} = 4\left(e^{2x^2+y^2} + 4x^2e^{2x^2+y^2}\right) = 4(1+4x^2)\,e^{2x^2+y^2}$, $f_{xy} = 8xye^{2x^2+y^2} = f_{yx}$, and $f_{yy} = 2(1+2y^2)\,e^{2x^2+y^2}$, so $D = f_{xx}(0,0)\,f_{yy}(0,0) - f_{xy}^2(0,0) = (4)(2) - 0 = 8 > 0$. Because $f_{xx}(0,0) > 0$, we see that $(0,0)$ gives a relative minimum of f. The minimum value of f is $f(0,0) = e^0 = 1$.

35. $k = \dfrac{100m}{c}$. The level curves are straight lines.

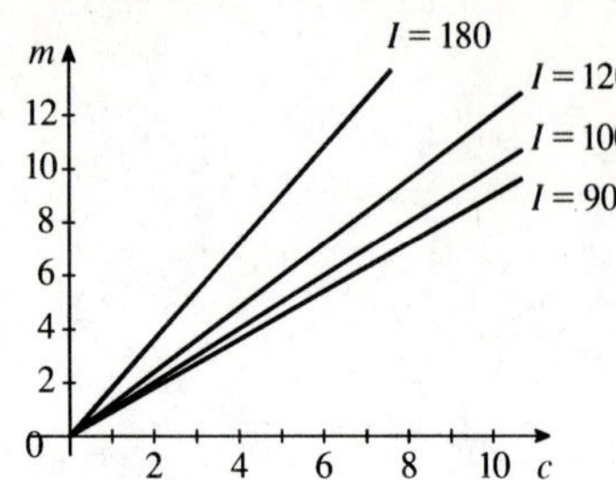

37. $f(p,q) = 900 - 9p - e^{0.4q}$ and $g(p,q) = 20{,}000 - 3000q - 4p$. We compute $\frac{\partial f}{\partial q} = -0.4e^{0.4q}$ and $\frac{\partial g}{\partial p} = -4$. Because $\frac{\partial f}{\partial q} < 0$ and $\frac{\partial g}{\partial p} < 0$ for all $p > 0$ and $q > 0$, we conclude that compact disc players and audio discs are complementary commodities.

39. We want to minimize $C(x,y) = 3(2x) + 2(x) + 3y = 8x + 3y$. The area is xy, so $xy = 303{,}750$ and $y = \dfrac{303{,}750}{x}$. Therefore, $C(x) = 8x + 3\left(\dfrac{303{,}750}{x}\right) = 8x + \dfrac{911{,}250}{x}$. We want to minimize C on the interval $(0,\infty)$. Now $C'(x) = 8 - \dfrac{911{,}250}{x^2} = \dfrac{8x^2 - 911{,}250}{x^2}$, so $C'(x) = 0$ if $8x^2 = 911{,}250$; that is, $x = \pm 337.5$. We reject the negative root. Because $C''(x) = \dfrac{1{,}822{,}500}{x^3} > 0$ for all $x > 0$, the graph of C is concave upward, and $x = 337.5$ gives an absolute minimum.Thus, $x = 337.5$ and $y = \dfrac{303{,}750}{337.5} = 900$. The dimensions of the pasture are 337.5 yd by 900 yd.

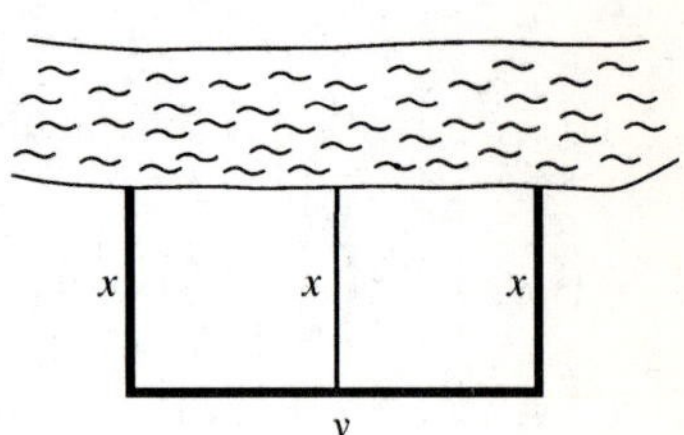

CHAPTER 12 Before Moving On... page 882

1. In order for $f(x,y) = \dfrac{\sqrt{x} + \sqrt{y}}{(1-x)(2-y)}$ to be defined, we must have $x \geq 0$, $y \geq 0$, $x \neq 1$ and $y \neq 2$. Therefore, the domain of f is $D = \{(x,y) \mid x \geq 0, y \geq 0, x \neq 1, y \neq 2\}$.

2. $f(x,y) = x + 2y^2$.

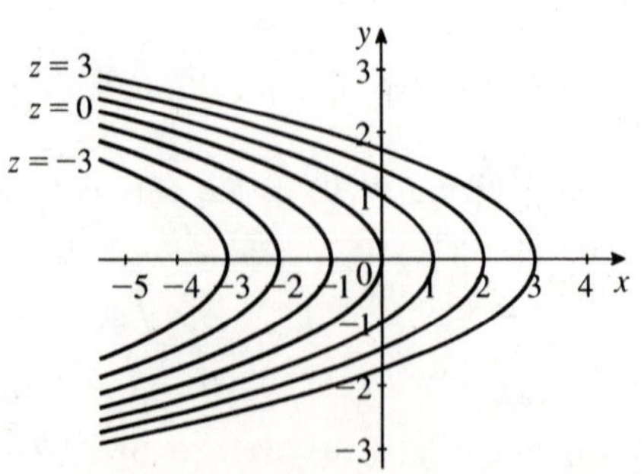

3. $f(x,y) = x^3y - 2x^2y^2 + 3xy^3$, so $f_x = 3x^2y - 4xy^2 + 3y^3$, and $f_y = x^3 - 4x^2y + 9xy^2$. $f_x(1,2) = 3 \cdot 1 \cdot 2 - 4 \cdot 1 \cdot 2^2 + 3 \cdot 2^3 = 14$ and $f_y(1,2) = 1^3 - 4 \cdot 1 \cdot 2 + 9 \cdot 1 \cdot 2^2 = 29$. At the point $(1,2)$, $f(x,y)$ increases at the rate of 14 units per unit increase in x with y held constant at 2, and $f(x,y)$ increases at the rate of 29 units per unit increase in y with x held constant at 1.

4. $f(x,y) = x^2y + e^{xy}$, so $f_x = 2xy + ye^{xy}$, $f_y = x^2 + xe^{xy}$, $f_{xx} = 2y + y^2e^{xy}$, $f_{xy} = 2x + (1+xy)e^{xy} = f_{yx}$, and $f_{yy} = x \cdot xe^{xy} = x^2e^{xy}$.

5. $f(x,y) = 2x^3 + 2y^3 - 6xy - 5$. Solving $f_x = 6x^2 - 6y = 6(x^2 - y) = 0$ and $f_y = 6y^2 - 6x = 6(y^2 - x) = 0$ simultaneously gives $y = x^2$ and $x = y^2$. Therefore, $x = x^4$, $x^4 - x = x(x^3 - 1) = 0$, and so $x = 0$ or 1. The critical points of f are $(0,0)$ and $(1,1)$. $f_{xx} = 12x$, $f_{xy} = -6$, and $f_{yy} = 12y$, so $D(x,y) = 144xy - 36$. In particular, $D(0,0) = -36 < 0$, and so $(0,0)$ does not give a relative extremum; and $D(1,1) = 108 > 0$ and $f_{xx}(1,1) = 12 > 0$, and so $f(1,1)$ gives a relative minimum value of $f(1,1) = 2(1)^3 + 2(1)^3 - 6(1)(1) - 5 = -7$.